Shock Tubes

Proceedings of the
Seventh International Shock Tube Symposium

Shock Tubes

Proceedings of the
Seventh International Shock Tube Symposium
held at University of Toronto, Toronto, Canada
23-25 June 1969

Edited by I. I. Glass
Institute for Aerospace Studies
University of Toronto

University of Toronto Press

Printed in Canada by
University of Toronto Press
Toronto and Buffalo

ISBN 0-8020-1729-0

SEVENTH INTERNATIONAL
SHOCK TUBE SYMPOSIUM

Sponsored by:
Institute for Aerospace Studies, University of Toronto (UTIAS)
United States Air Force Office of Scientific Research (AFOSR)

Endorsed by:
American Physical Society (Topical Conference, Division of Fluid Dynamics) (APS)
Canadian Association of Physicists (CAP)
German Physical Society (DPG)
Canadian Aeronautics and Space Institute (CASI)
American Institute of Aeronautics and Astronautics (AIAA)

Symposium Chairman: I. I. Glass (UTIAS)

Advisory Committee

W. Bleakney (USA)
P. Carriere (France)
R. Courant (USA)
R. J. Emrich (USA)
R. G. Fowler (USA)
W. Fiszdon (Poland)
F. N. Frenkiel (USA)
A. G. Gaydon (England)
J. G. Hall (USA)
A. Hertzberg (USA)
I. Hiraki (Japan)
D. F. Hornig (USA)
A. Kantrowitz (USA)
G. B. Kistiakowsky (USA)
O. Laporte (USA)
H. W. Liepmann (USA)
M. J. Lighthill (England)
R. K. Lobb (USA)
N. Manson (France)
A. K. Oppenheim (USA)
G. N. Patterson (Canada)
H. Reichenbach (Germany)
M. Rogers (USA)
L. I. Sedov (USSR)
Z. I. Slawsky (USA)
R. I. Soloukhin (USSR)
J. L. Stollery (England)
A. H. Taub (USA)
G. I. Taylor (England)

PREFACE

A summary of the earlier Symposia is given in the Foreword to the Fifth Symposium, which was held at the Naval Ordnance Laboratory (NOL), White Oaks (Washington) in 1965. The roles played by Mr. R. R. Birukoff and Dr. T. H. Schiffman in organizing the First Symposium at MIT, in February 1957, and the Second and Third Symposia, are appreciated. The Fourth Symposium was organized by W. J. Taylor, the Fifth by Z. I. Slawsky and J. F. Moulton Jr., and the Sixth by Dr. H. Reichenbach and their Committees.

The present Symposium was held in the Concert Hall of the Edward Johnson Building on the Main Campus of the University of Toronto, on 23 to 25 June 1969. It was attended by 288 scientists and engineers and 35 wives from many countries: 75 from Canada, 152 from the USA, 9 from England, 1 from Belgium, 1 from Denmark, 1 from Finland, 11 from France, 16 from Germany, 1 from Holland, 1 from Italy, 1 from Poland, 1 from Rumania, 7 from Russia, 1 from Switzerland, 3 from Israel, 4 from Australia, 3 from Japan. Judging from many appreciative comments, the Symposium was very successful both technically and socially, largely as a result of the great amount of work done by the Advisory Committee, the Papers Committee, the Local Program Committee, the Ladies Program Committee, the Speakers, the Session Chairmen, as well as the lively discussions.

The Seventh Symposium consisted of six sessions. Each session began with an invited paper, which set the tone and scope of the session, followed by six contributed papers reporting on recent relevant research and development in the particular area. The first session was an exception. It contained the Paul Vieille Commemorative lecture presented by Professor A. Hertzberg, and consequently had two fewer contributed papers. Professor T. V. Bazhenova's paper (Session I) was presented by Academician R. I. Soloukhin, in her absence; Professor L. M. Biberman's paper (Session V) was replaced by Dr. R. W. Getzinger's paper; an additional paper was also presented by Professor P. C. deBoer in Session VI. Unfortunately, the complete papers of Professor A. Hertzberg and Dr. A. C. Kolb are not available for publication. A few additional papers, which could not be included in the sessions themselves, were accepted for the Proceedings; these follow the relevant session papers.

Session I dealt with basic shock-tube flows, in the form of a review of the past and present research problems and the possibilities for the shock tube in the future. The questions of test time and uniformity of flow in the hot region induced by incident and reflected shock waves are still far from being completely understood. It was refreshing to have a paper on shock-wave structure in solids; one in liquids would have been a desirable complementary paper.

Session II considered driving techniques, and an excellent review of this topic was given by Dr. W. R. Warren. Several techniques to increase performance in electrical and piston-driven shock tubes, as well as the novel "Slingshot" approach, were discussed.

Session III was a special session dealing with explosive drivers, which have recently come into prominence and offer some promise of obtaining uniform, high-temperature test gases for aerophysical research and testing. The use of novel, explosive-driven cylindrical and spherical implosions to generate such flows and the sophisticated numerical methods used to predict the results were quite impressive.

Session IV addressed itself mainly to theoretical and experimental research in electromagnetic shock tubes. It would seem that there are still numerous problems to be solved in this very large area. Similar remarks apply to Session V, which dealt with chemical kinetics and spectroscopy. There is little doubt that the shock tube will continue to be a major research facility in these areas for many years to come.

Session VI concerned itself with a review of shock-tube diagnostics, instrumentation, and fundamental data as well as recent measurements of physical quantities. Academician R. I. Soloukhin ably brought the subject up to date, and he and the other contributors showed that many novel, very-rapid-response techniques for measuring physical quantities will be developed in the future to deal with hypervelocity flows of short duration.

Unfortunately, the program did not deal with current uses of shock tubes for sonic-boom problems and other social and industrial applications, nor were its uses as an aerodynamic test facility considered, owing to the terms of reference of this meeting. These important aspects will hopefully be dealt with at the next, the Eighth Symposium in London in 1971.

In retrospect, it can be stated that there are many worthwhile problems still to be solved in many areas by using shock-tube facilities. However, to be viable today this type of research also may have to be applied to new fields directed by social and economic forces. These ideas were made amply clear in the banquet address given by Dr. G. N. Patterson, Director, Institute for Aerospace Studies, which is included in the Proceedings, for the benefit of our readers.

The Symposium was made possible through the efforts of many people. Space limitations do not permit me to express my appreciation to all of those who provided invaluable assistance. However, it is a particular pleasure to acknowledge the help I received from Mr. Milton Rogers (AFOSR), Dr. G. N. Patterson (UTIAS), Dr. R. J. Emrich (Lehigh), Dr. O. Laporte (Michigan), Dr. J. G. Hall (Buffalo), Dr. A. J. Howsmon (UTIAS), and Dr. P. A. Sullivan (UTIAS). Mrs. Jean V. Dublack (UTIAS), Mrs. Willa F. Ryan (UTIAS), and especially Miss L. Ourom (Associate Editor, University of Toronto Press) provided me with valuable editorial assistance. I wish also to express my appreciation to the many authors who submitted papers to the Symposium in the areas concerned and in others, and my regrets that all could not be included in the Proceedings.

I. I. GLASS
February 1970

CONTENTS

SESSION IV: Electromagnetic Shock Tubes

Chairman: P. SAVIC, National Research Council, Ottawa

SESSION IV: Electromagnetic Shock Tubes

Chairman: P. SAVIC, National Research Council, Ottawa

BANQUET

Chairman: I. I. GLASS, Institute for Aerospace Studies,
University of Toronto, Toronto, Canada

Shock Tubes

Proceedings of the
Seventh International Shock Tube Symposium

SHOCK TUBE RESEARCH, PAST, PRESENT AND FUTURE: A CRITICAL SURVEY

A. HERTZBERG
University of Washington, Seattle

SUMMARY

The first description of the shock tube was given by Vieille in 1889. In his paper, he adequately described the properties of the shock tube in order to explain his experimental results. The early nature of this work is not surprising in view of mankind's long interest in explosive weapons. However, it is well to note that Vieille's original work was stimulated by the compelling problems of mine safety; problems which, astonishingly, still remain with us. The early nature of this work is not surprising, but the limited interest in expanding and developing this work is. Not until the work, in the late 1940's, of Dr. Bleakney's group at Princeton, was the foundation of modern shock tube practice developed. Bleakney conclusively demonstrated that the shock tube was an excellent tool for studying the nature of shock wave propagation in gases and developed much of the early experimental techniques. The publications of this group stimulated many other research efforts and the growth of interest was rapid.

The technology developed by these studies led to exploitation of the shock tube in many application areas which had not been initially anticipated. In particular, it was recognized almost simultaneously by Kantrowitz and Laporte that the shock tube was a superb tool for the study of high temperature phenomena in gases. The gas temperature could be rapidly raised by a shock and maintained for a brief period. Gas conditions hitherto not possible in the laboratory were now available. In addition, the temperature throughout the gas was uniform at a known enthalpy level. This feature of shock heated gases permitted many unique studies. The early high-temperature research programs laid a foundation for a surprisingly wide series of studies into high temperature gas phenomena utilizing the shock tube. Many special forms of the shock tube were devised to facilitate this work. In particular, the rapid development of instrumentation techniques made it possible to devise shock tubes which could not only quickly heat the gas, but quench the ensuing chemical reaction so that the debris could be studied in detail. Many of the programs established in the early 1950's are still active and the shock tube has developed into a powerful tool for the study of high temperature gas phenomena and gas phase chemical kinetics.

In his early work, Bleakney demonstrated that a quasi-steady flow region existed behind the shock wave so that the shock tube could be used as a wind tunnel with a brief, but useful period of steady flow. This work, coupled with the development of the shock tube as a device for generating high enthalpy gases, formed the foundation of the development of the present technology of the shock tube wind tunnel. It was demonstrated that by terminating the shock tube with a nozzle, the flow Mach number limitations behind the incident shock wave could be circumvented and the flow expanded to hypersonic Mach numbers. The development of the intercontinental ballistic missile and the requirement for manned re-entry helped spur this development and the shock tube wind tunnel rapidly became a source of valuable data on hypersonic flow problems. The success of the early shock tube wind tunnel programs in producing needed data further spurred their development. At the present time, the shock tube wind tunnel has become a standard research device in many universities and research establishments.

It is beyond the scope of this summary to do more than make note of many other diverse applications of shock tube technology. For example, the use of shock tubes to simulate a blast wave phenomena has been widely accepted and many novel approaches have been made to improve its usefulness for this particular type of test work. Variations of the shock tube wind tunnel have been used for the creation of high energy monoenergetic molecular beams. Shock tubes have been used to study the phenomena relating to the internal combusion engine and aid in the development of a better understanding of explosive and detonation phenomena. Special variations of the shock tube have been devised which enable the study of the flow structure of interacting multiple rocket configurations, as well as the high altitude structure of the rocket plume. These comments are only a sampling of current applications of shock tube technology.

The question could now be asked: what is the future of the shock tube as a research tool? While it is very easy to make prognostications, the author is acutely aware that history has a nasty habit of ignoring sober judgement. Looking back, I have rarely been proud about the outcome of my calculated guesses, and I have tended to underestimate the growth of technology. None the less I will venture a few comments. Much of the work that we are carrying out will certainly be obsolete. The changing demands of technology often dictate the abandonment of important work and often with little courtesy. However, new applications and new problems will still make demands on the fundamental capabilities of this very interesting research tool. As an example, the development of the shock tube into a useful short duration hypersonic wind tunnel necessitated the development of complex and effective instrumentation. Also our understanding of the wave processes in shock tubes has grown enormously over the years. These developments have made it possible for a variation of the shock tube called the Ludweig tube to be considered for high Reynolds number aerodynamic research. Thus our measurement technology, largely developed for hypersonic flows, makes it possible to consider the shock tube as an effective high Reynolds number supersonic tunnel of moderate cost. In other areas the special knowledge of rate processes in high temperature gas physics which the shock tube has given us has suggested new and novel applications. No doubt we will find in this special knowledge further prods to develop and extend our technology. I also expect the shock tube to be widely used in studies of phenomena relating to air pollution and perhaps produce the basis of new thermodynamic engines to help eliminate this problem.

We are approaching a highly mature shock tube technology and often the work may seem to become dull and played out however, new exciting applications or techniques always seem to come along and rescue us. I personally feel sanguine about the future.

ON THE DISCONTINUITIES PRODUCED BY THE SUDDEN RELEASE OF COMPRESSED GASES*

M.P. VIEILLE
presented by M. Sarrau

Studies of the propagation of plane waves in media at rest were reported previously, and it was shown that the phenomenon of discontinuities predicted theoretically could be observed either by measuring the propagation speed or by studying the deformation of the wave front (wave head).

The propagation speeds were measured by recording simultaneously on the same rotating cylinder the effect on light pistons, located at the ends of a closed tube, produced by the exciting phenomenon which was generated in the neighbourhood of one of the tube ends.

This method relies on the assumption that the initial wave is symmetrical (with respect to the source of the disturbance): this wave may be obtained, without the use of explosives, by utilizing compressed gases enclosed in spherical glass bulbs having a diameter equal to that of the tube; the rupture of these bulbs produced by increasing the pressure results in a very strong disturbance.

However, as far as the intensity of the disturbance is concerned, one is limited by the relatively small volume of the sphere which must fit inside the cylindrical tube. Consequently the attenuation of the wave can be considerable and I was able to observe speeds of only 430 m/sec, that is approximately 100 m/sec greater than the speed of sound for the case of a tube 22 mm in diameter, burst pressures of 50 atmospheres and path lengths (propagation distance) of 1 m in air.

*This is the first known published paper on shock-tube flows. The original text from Comptes Rendus de L'Académie des Sciences, 129, pp. 1228-30, 1899, follows. It is being reproduced for its scientific and historical interest and as background material for The Paul Vieille Lecture. I wish to express my thanks to one of the attendees for his spontaneous suggestion to include the original Paul Vieille paper and its translation in the Proceedings. It is gratifying to know that the Paul Vieille Lecture, initiated by the undersigned, is being continued at the Eighth Symposium. Editor.

On the other hand, by dividing a tube into two sections with the use of a diaphragm which is ruptured by slowly increasing the pressure of the gas on one side, one can choose at will the driving gas mass. As a result, however, the phenomenon ceases to be symmetrical and the propagation speeds must be recorded laterally on the same side of the diaphragm. The measurement is made by mounting the light pistons normally to the axis of the tube and flush with its inside surface. These pistons are supported by springs whose deflections are recorded on a rotating cylinder parallel to the tube axis.

Strong light diaphragms can be made of colloidion sheets. The burst pressures for these diaphragms are much more repeatable than those for diaphragms made of either tempered or untempered glass or steel which were tried initially. The rupture of the diaphragms occurs along radial regularly spaced lines and the segments (petals) thus formed are either torn off or folded against the tube wall thus promoting very rapid gas flow. The burst pressure for 22 mm diameter diaphragms is 27 atmospheres for a thickness of 0.29 mm and 14 atmospheres for a thickness of 0.11 mm.

I have also used paper diaphragms which burst at 2 atmospheres.

I have first studied the variation of propagation speed in air with burst pressure and what the attenuation law might be.

In these experiments the diaphragm was 271 mm from one end of the tube, thus creating a small volume chamber (driver) approximately 100 cc. The tube was more than 6 m long and it was instrumented at various points with identical sensors. Each measurement included five or six runs in which the sensors were alternated in order to eliminate individual response time effects.

The following table summarizes the observed results:

Diaphragm Type	Absolute Burst Pressure	Average propagation speeds measured between the following points from the diaphragm		
		m m	m m	m m
mm	atm	0.052 to 0.458	0.458 to 1.888	0.458 to 4.015
		m/sec	m/sec	m/sec
Colloidion 0.29	27	625.4	606.4	570.5
Colloidion 0.11	16	"	540.2	"
Paper	2	"	"	390.2

From these results one concludes that a small mass of air compressed to 27 atmospheres is sufficient to produce propagation speeds, in air at one atmosphere in the cylindrical tube, in excess of 600 m/sec and that these speeds remain within the same order of magnitude for several meters, the attenuation being approximately 20 m/sec/m within the limits observed.

Much weaker burst conditions still produce propagation speeds much larger than the speed of sound. Consequently explosives do not play any essential role in phenomena of propagation at great speeds; this I mentioned previously.

Furthermore, I have examined how the capacity of the reservoir (driver) influences the propagation speed.

For this experiment, the diaphragm was placed 2.508 m from one end of the tube in order to increase the length and volume of the chamber (driver) by approximately one order of magnitude.

The tests were made by using the strongest diaphragms, 0.29 mm thick, which have a burst pressure of 27 atmospheres. A speed of 608.9 m/sec measured between points 0.458 m and 1.888 m from the diaphragm was determined in a series of six runs in which the sensors were interchanged. This differs little from the speed of 606.4 m/sec which was obtained under identical conditions with the small 100 cc reservoir (driver).

This leads one to conclude that the discontinuity which determines the propagation speed remains the same in these various cases and that the reservoir capacity influences only the way in which this discontinuity is maintained.

Direct study of the propagated wave shape justified this conclusion and shows that according to theory the magnitude of the discontinuity which causes propagation speeds of 600 m/sec is well below the pressure of 27 atmospheres necessary for rupturing the diaphragms.

This research will be the subject of a future report.

MÉCANIQUE APPLIQUÉE. — *Sur les discontinuités produites par la détente brusque de gaz comprimés.* Note de M. **Paul Vieille**, présentée par M. Sarrau.

« J'ai étudié, dans de précédentes Communications, la propagation par ondes planes, dans un milieu en repos, de perturbations brusques produites par la combustion d'explosifs, et montré que le phénomène de discontinuité prévu par la théorie pouvait être mis en évidence, soit par la mesure des vitesses de propagation, soit par l'étude de la déformation du front de l'onde.

» La méthode d'enregistrement des vitesses de propagation était fondée sur l'inscription simultanée, sur un même cylindre tournant, de pistons légers placés aux extrémités d'un tube fermé, le phénomène excitateur étant produit au voisinage de l'une des extrémités.

» La méthode suppose l'onde initiale symétrique : cette onde peut être obtenue, sans l'emploi des explosifs, à l'aide de gaz comprimés dans des ampoules en verre sphériques du diamètre du tube et dont la rupture par pulvérisation sous pression croissante présente une brusquerie très grande.

» Mais on est limité dans cette voie, au point de vue de l'intensité du phénomène, par la capacité relativement faible de la sphère inscrite dans le tube cylindrique. Le coefficient d'affaiblissement de l'onde devient, dans ce cas, considérable, et je n'ai pu observer, dans des tubes de 22^{mm}, sur des parcours de 1^{m} dans l'air et pour des pressions de rupture de 50 atmosphères, que des vitesses de 430^{m} supérieures de 100^{m} environ à la vitesse normale du son.

» On peut, au contraire, en partageant un tube en deux parties par un diaphragme dont on provoque la rupture par compression lente du gaz dans l'un des segments, donner à la masse excitatrice une valeur quelconque. Mais, en même temps, le phénomène cesse d'être symétrique et les mesures de vitesse de propagation exigent l'enregistrement latéral d'un même côté du diaphragme. L'enregistrement a été obtenu par des pistons légers, normaux à l'axe du tube, et affleurant par leur base sa paroi interne. Ces pistons sont contrebutés par des ressorts dont la déformation s'inscrit parallèlement à l'axe du tube, sur un même cylindre tournant.

» On obtient des diaphragmes de faible masse et de grande résistance en utilisant des lames de collodion. Les pressions de rupture de ces lames présentent une régularité bien supérieure à celle des lames de verre trempé ou non trempé, ou de clinquant d'acier, que j'avais expérimentées tout d'abord. La rupture des diaphragmes s'opère suivant des lignes rayonnant du centre et régulièrement espacées, et les segments ainsi formés sont arrachés ou rabattus sur la paroi du tube suivant un mode très favorable à l'écoulement rapide du gaz. Cette rupture s'opère sous le diamètre de 22^{mm} à 27^{atm} absolues pour l'épaisseur de $0^{mm},29$, et à 14^{atm} pour l'épaisseur de $0^{mm},11$.

» J'ai également utilisé des diaphragmes en papier, rompant sous la pression de 2^{atm} absolues.

» J'ai étudié tout d'abord comment les vitesses de propagation dans l'air variaient avec la pression de rupture et quelle était la loi d'amortissement des vitesses.

» Dans ces expériences, le diaphragme était voisin de l'un des fonds du tube (271^{mm}) et formait une chambre de petite capacité, 100^{cc} environ. Le tube avait plus de 6^{m} de longueur et recevait en divers points des enregistreurs identiques. Chaque mesure de vitesse comportait cinq à six déterminations effectuées en permutant alternativement les enregistreurs en vue d'éliminer leurs retards propres.

» Le Tableau suivant résume les résultats observés :

Nature du diaphragme.	Pressions absolues de rupture.	Vitesses moyennes de propagation mesurées entre des points distants du diaphragme de		
		$0^{m},052$ et $0^{m},458$.	$0^{m},458$ et $1^{m},888$.	$0^{m},458$ et $4^{m},015$.
mm	atm	m	m	m
Collodion 0,29...	27	625,4	606,4	570,5
Collodion 0,11...	16	»	540,2	»
Papier..........	2	»	»	390,2

» Il résulte de ces nombres que la détente d'une faible masse d'air comprimé à 27 atmosphères suffit à assurer dans l'air sous la pression atmosphérique des vitesses de propagation en tube cylindrique supérieures à 600^{m} et que ces vitesses se soutiennent sur plusieurs mètres avec le même ordre de grandeur, la décroissance de la vitesse étant de 20^{m} environ par mètre de parcours dans les limites observées.

» Des détentes brusques beaucoup plus faibles assurent encore des vitesses de propagation très supérieures à celles du son. Les explosifs ne jouent donc aucun rôle essentiel dans les phénomènes de propagation à grande vitesse que j'ai antérieurement signalés.

» En second lieu, j'ai cherché comment la capacité du réservoir influait sur la vitesse de propagation.

» A cet effet, le diaphragme était disposé à $2^{m},508$ de l'un des fonds de façon décupler environ la longueur et le volume de la chambre à gaz.

» Les essais ont porté sur les diaphragmes les plus résistants de $0^{mm},29$ avec rupture à 27 atmosphères. La vitesse mesurée entre les points distants du diaphragme de $0^{m},458$ et $1^{m},888$ a été trouvée dans six expériences comportant la permutation alternée des enregistreurs de $608^{m},9$. Elle diffère donc à peine de la vitesse de $606^{m},4$ obtenue dans des conditions identiques avec le petit réservoir de 100^{cc}.

» Il y a donc lieu de penser que la discontinuité qui assure la vitesse de propagation reste la même dans ces divers cas et que l'influence de la capacité du réservoir s'exerce seulement sur le mode d'alimentation de cette discontinuité.

» L'étude directe de la forme de l'onde propagée justifie cette prévision et permet de montrer que la grandeur de la discontinuité qui assure des vitesses de propagation de 600^{m} est, conformément aux données théoriques, bien inférieure à la pression de 27 atmosphères qui produit la rupture des diaphragmes.

» Cette étude fera l'objet d'une prochaine Communication. »

SHOCK-STRUCTURE IN SOLIDS

H. G. HOPKINS
Department of Mathematics
University of Manchester Institute of Science and Technology

SUMMARY

The propagation of large-amplitude deformations in solids is characterised by rapid changes in physical quantities. From a macroscopic point of view, these changes may appear to be discontinuous, and the disturbance is then called a shock wave. In reality, the changes are continuous, albeit rapid, across a transition region of finite thickness. The term "structure" refers to the values of quantities within this transition region. Details of shock-structure may have considerable physical significance, since they relate to fundamental dissipative processes. Theoretical studies of shock-structure have been made on the basis of continuum mechanics, or of lattice dynamics, or on these two bases combined but biased towards the former. These predominantly one-dimensional studies are limited in extent, and also few experimental data are yet reported. The aim of this paper is to discuss the present status of knowledge of shock-structure in crystalline solids, and to comment on the future development of the subject.

INTRODUCTION

The propagation of large-amplitude disturbances, usually mechanical and thermodynamical in nature, in physical media, whether gaseous, fluid, or solid, occurs at finite velocities characteristic of the properties of the medium. The nature and manner of the propagation of such disturbances is being studied extensively, both experimentally and theoretically. Across the leading part of the disturbance, various properties of the medium appear, from a macroscopic point of view, to change discontinuously. Whenever the particle velocity is discontinuous, the discontinuity is called a shock wave, across which there is a sudden jump in mechanical variables such as density and stress (as well as particle velocity) and in thermodynamical variables such as temperature, entropy, and internal energy. The simplest situation is that of a normal compression shock. In an ordinary shock wave the equation of state is unaltered: it is the same ahead of, as it is behind, the discontinuity. Other types of shock waves do not have such simple character, e.g., as with deflagration or detonation waves (involving chemical reactions) or with waves associated with phase transitions in solids (involving atomic lattice re-arrangements), and then there are different equations of state on each side of the discontinuity. In gaseous mixtures, such as air containing water vapour, condensation shocks, in which a particular component separates out, are possible. Also important in very strong shocks are physical effects such as dissociation, ionization, and relaxation. However, the discontinuities are not discontinuities in the strict sense: there is always some finite (i.e., non-zero) thickness across which the physical properties change continuously (albeit rapidly) provided that the medium may still be viewed as a continuum. In many (practical engineering) problems, the physical relations may be obtained by analysis which treats the discontinuity as exact. The term "structure" refers to the values of the physical properties within the small but finite thickness of the transition region. Now, the disturbance normally upsets - if only quite temporarily - the thermodynamic equilibrium of the medium, whatever may be its particular nature, and a certain characteristic time must elapse before thermodynamic equilibrium is effectively re-established. The product of this time times the velocity of propagation of the discontinuity defines a characteristic length of order the atomic or molecular spacing or greater. If the physical and chemical changes taking place within the transition region are sufficiently slow, so that its thickness is rather greater than this characteristic length, then thermodynamic quasi-equilibrium may be taken to apply. On the other hand, if the physical and chemical changes are occurring too rapidly, the essential absence of thermodynamic equilibrium must be taken into account; and then the continuum concept may have to be abandoned and instead the viewpoint of physical theory taken, viz., kinetic theory in the case of gases and lattice dynamics theory in the case of crystalline solids.

A unified account of discontinuities and transitions in diffuse and condensed media might appear at first sight to be a practical and worthwhile objective in research. It is true that whenever continuum theory is applicable (and this can include shock transitions) then this theory does itself provide a formal unification. However, the physical properties of gases, liquids, and solids are so different that practically significant results are largely controlled by quite different physical processes, and then little is to be gained from a formally unified theory. On the other hand, it is true, for example, that over certain ranges of high stress an elastic-plastic solid may behave rather like a compressible fluid, so that the relevant continuum mechanics theory is that of compressible fluid dynamics, a well-developed area in research. Thus, it should be noted that shock conditions may act to modify appreciably the more normal physical properties of materials. However, generally, in regard to problems of shock discontinuities in gases, liquids, and solids, facts of basic physical differences between these states imply the need to follow differing experimental and theoretical approaches, although so far as the latter are concerned, these may be related at a continuum mechanics level although not at a physical one.

For the diffuse gaseous state of matter there is an extensive knowledge of shock-structure, both at continuum and at physical levels. In contrast, for the condensed, liquid or solid, states of matter very much less is known, for a variety of reasons. In the first place, there is the fact that condensed media are highly incompressible and access to shock conditions in controlled laboratory experiment is rather recent. Secondly, the physical processes of momentum and energy transfer between constituent atoms or molecules are different for gaseous and condensed media: for the former, such transfer is due to collisions during random motions, governed by kinetic theory, leading in particular to the Boltzmann equation; but for the latter, specifically for crystalline solids, such processes are governed by interatomic forces, which in fact of course permit the existence of regular atomic lattice structures. Thus, the basis for theoretical investigations of shock-structure in gases lies primarily with a statistical transport equation but in solids lies with the specification of interatomic forces. Liquids constitute a different case again, perhaps closer to solids than to gases, but even when they show a regular structure (as with liquid crystals) their cohesive strength is relatively low. The situations therefore greatly differ, and significantly the knowledge of shock-structure in solids is very much less extensive than it is for gases, although the subject is now attracting increasing experimental and theoretical investigation. The purpose here is to discuss the present status of knowledge of shock-structure in crystalline solids (amorphous and porous solids will not be considered) and to comment on the future development of research in the subject. Necessarily, in the space presently available, emphasis is placed upon overall viewpoints.

A review of the present knowledge of shock-structure in solids is believed to be timely, in order to bring together the various approaches that have so far been made to the problems involved and to help to foster interest in future developments. The general arrangement of this paper is outlined as follows.

In investigations of shock-structure in solids, a clear distinction is necessary between those made at a continuum mechanics level and those made at an atomic lattice dynamics level. Nevertheless, such apparently disparate investigations are reciprocal. This is because physical knowledge is needed for a general understanding of shock waves at a continuum mechanics level, and experiments involving shock waves may be used for investigations of physical phenomena (not otherwise easily accessible) and continuum theory may be needed for the interpretation of the data obtained. Now, in a gas, treating it as an inviscid and a non heat-conducting but a compressible medium, shock discontinuities are possible and jumps in physical quantities are governed by the Rankine-Hugoniot relations for conservation of mass, momentum, and energy. Any real gas is viscous and heat-conducting to an extent and then any steep spatial gradients in particle velocity are limited by viscous forces with an accompanying dissipation of mechanical energy into thermal energy. The irreversible effects called into play always set a finite lower limit to the thickness of a wave front. In the case of gases, the situation is accessible to detailed study since kinetic theory not only permits macroscopic continuum concepts such as density, velocity, and stress, and properties such as viscosity and thermal conduction, to be related to fundamental physical structure and processes but also permits shock-structure to be investigated. The role played by kinetic theory for gases is played by lattice dynamics theory for solids, but now the same kind of program cannot presently be completely carried through. However, if one ascribes to a compressible solid the properties of viscosity and thermal conductivity, then it becomes possible to investigate problems of shock-structure in solid continua. On the other hand, problems of shock-structure may be approached quite differently through classical mechanics studies of the propagation of disturbances in a crystal lattice, and such studies require specification of the interatomic forces. The theory is at its simplest in the case of one-dimensional lattices with only nearest-neighbour interactions taken into account. However, significant extensions to situations of multi-dimensional lattices or to take account of distant-neighbour interactions have not as yet been made. Another approach is one in which the above two approaches of continuum mechanics and of lattice dynamics are combined while still retaining the formulation of problems in terms of nonlinear differential (rather than discrete) equations describing dispersive wave propagation.

Shock waves are realizable in any compressible medium - the only real circumstance of difference lies with their degree of accessibility to direct physical study - and hence problems of shock-structure arise for gases, liquids, and solids. However, whereas shocks in fluids (especially gases) have been known and studied for about 100 years, shocks in solids have only been known and studied for about 20 years. It is therefore to be expected that far more is known about shocks in gases than in solids, as indeed is clear from accounts given in the literature. For example, the very comprehensive text by Zel'dovich and Raizer[1] exemplifies just this situation: much information is reported there on shocks and shock-structure in gases and also on shocks in solids; but no information is reported there on shock-structure in solids, although this does not now represent the true position. As stated earlier, the subject of shock waves in solids is of comparatively recent study. In the U.K. the subject was introduced by D. C. Pack, W. M. Evans, and H. J. James in 1948, while work done in the U.S.A. dating from a similar time was published by R. W. Goranson and co-workers and by J. M. Walsh and R. H. Christian in 1955. Similar work done in the U.S.S.R. was published by L. V. Al'tshuler and co-workers in 1958. The first important review article was published by M. H. Rice, R. G. McQueen, and J. M. Walsh in 1958. In the course of the last 10 years there has been a steady spate of publications, both original papers and review articles, on shock waves in solids. In this paper, the bibliography given mainly comprises selected references relevant to the particular aspect of shock waves in solids being discussed. General accounts of work on shock waves in solids are given, for example, by Duvall and Fowles[2], Skidmore[3], and Al'tshuler[4].

Most of the work that has been done on shock waves and shock-structure in solids refers to one-dimensional geometries, i.e. plane, cylindrical, or spherical geometries but particularly the first type. In experimental work, the apparatus usually depends upon the exploitation of techniques involving either mechanical impact (as with light gas guns) or explosively-generated impact in order to produce the extremely high transient pressures over the plane surface of a target whose material properties are the subject of study. Most experimental work on shock waves in solids lies within the pressure range of 100 kb to 1 Mb, although pressures up to 10 Mb have been reported. Naturally, there are wide ranges of other physical quantities as well, such as density, shock and particle velocities, internal energy, entropy, and temperature.

Thus, typical values cannot be quoted, since values vary greatly with both the material and shock pressure. However, it may be noted that shock velocities (exceeding, of course, particle velocities) range up to and beyond elastic dilatational wave velocities and that densities range up to 3 times normal. Other changes (e.g., in temperature) are perhaps less dramatic, except at upper parts of pressure ranges, but may still be important. The increment in stress associated with the passage of the shock front is applied in much less than 10^{-8} sec for many materials; the duration of the shock is also small, usually being between 10^{-7} and 10^{-5} sec (see Duvall and Fowles[2]). Hence, physical changes with shock waves in solids show much difference from those in gases. The references already cited give numerous physical data.

Now let us come to the nature of detailed theoretical procedures. First, it is straightforward to treat the case of an artificial (i.e., mathematical) discontinuity due to a steadily propagating shock wave in a solid. Secondly, there is the manner of treatment of shock-structure, and in this there are a number of different procedures. In the simplest, as has been done in the case of gases, one may include in the basic governing equations terms which correspond to heat conduction and viscosity (both, of course, corresponding to irreversible thermodynamic effects) - for solids, viscosity effects are sometimes taken to be more important than thermal conduction effects - and these terms, depending upon spatial gradients of quantities, will be important within the transition region but quite negligible outside it. These terms now included must necessarily have physical constants associated with them, and in this regard one may adopt either a purely phenomenological viewpoint (so that nothing that is done is done outside continuum mechanics) or a joint phenomenological and physical viewpoint, the essential question being the basis of the choice of the constitutive equations describing viscosity and thermal conduction effects, viz., recourse being made to crystal physics theories of dislocations and so on or not. Both viewpoints have been followed in published work and are acceptable provided that one can work within the framework of continuum theory, i.e., if transition thicknesses come out to be a reasonably large number of atomic lattice spacings. On the other hand, if the continuum hypothesis fails, due to the prediction of physically-inadmissibly low transition-thicknesses, or simply if a more basic physical viewpoint is required, then there is the alternative procedure of adopting directly physical theory and, in particular, working with crystal lattice dynamics theory. Naturally, the application of the theory should be to highly nonlinear states of deformation so that it differs significantly from classical linear harmonic lattice theory. Interesting work along these lines is now beginning to appear, and it is discussed later. At present, nonlinear lattice theory is not developed in much detail or generality, but already, however, studies show much mathematical and physical interest and this presages the expected strong interplay between theory and experiment in regard to shock transitions in solids.

SHOCK DISCONTINUITIES

For the sake of completeness, we shall consider first the ideal (mathematically but not physically realizable) circumstance of a sharp discontinuity under steady plane conditions, the same equation of state being valid for the material on each side of the discontinuity surface. It may be noted that the transition region is not strictly required to be of zero thickness, but the results of the analysis given are still valid for any finite thickness, provided only that we restrict attention to steady conditions of thermodynamic equilibrium outside the transition region and do not give attention to variations in physical quantities through the transition region where the present analysis is incompletely determinate. The analysis is well-known, but nonetheless important, and it follows from the application in finite form of the classical conservation laws of mass, momentum, and energy, leading to the Rankine-Hugoniot equations.

In detail, let a plane shock travel at velocity U (with respect to fixed laboratory coordinates) into stationary material at pressure p_o, density ρ_o, and specific internal energy E_o. Following the passage of the shock, the material is accelerated to a particle velocity u and compressed to a higher density ρ, and the pressure (identified later with axial compression stress) is changed to p and the internal energy to E. There are of course other changes in thermodynamic quantities, such as temperature and entropy, but their consideration is not needed here. Then, the conditions of conservation require that

$$\rho_o U = \rho(U - u), \qquad (1)$$

$$p - p_o = \rho_o U u, \qquad (2)$$

$$pu = \rho_o U(E - E_o + \tfrac{1}{2}u^2), \qquad (3)$$

in regard to mass, momentum, and energy, respectively. These are the Rankine-Hugoniot equations, and they state 3 conditions on the 5 shock quantities, $U,u,p,\rho,E,$ given the initial state, $u_o = o$, p_o, ρ_o, E_o. Thus, the system of Eqns. (1) - (3) has two degrees of freedom, i.e., 2 of the 5 shock quantities above must be specified to make the situation a completely determinate one. If, now, the material is specified with a known equation of state, then there is a given thermodynamic relation between the thermodynamic quantities, p,ρ,E (which relation is also satisfied by the initial state if the same equation of state applies on each side of the transition region, i.e., excluding phase transitions). Only one degree of freedom now remains for the system of Eqns. (1) - (3) when coupled with $E = E(p,v)$, where $v = 1/\rho$ is the specific volume. Hence, given the initial state for a specified material, all final shock quantities are uniquely determined when just 1 of the 5 quantities $U,u,p,\rho(= 1/v),E$ is fixed, and in particular when either the pressure p or the shock velocity U is chosen to be the fixed quantity. The Rankine-Hugoniot equations are not independent of the reference system, i.e.,they involve the purely kinematical quantities U,u. However, these quantities may be eliminated to obtain the Rankine-Hugoniot relation which is solely thermodynamic in nature, viz.,

$$E - E_o = \tfrac{1}{2}(v_o - v)(p + p_o). \tag{4}$$

If, as we are assuming, there is thermodynamic equilibrium of material with the same equation of state on either side of the shock (discontinuity) then

$$E = E(p,v), \quad E_o = E(p_o,v_o), \tag{5}$$

and now Eqns. (4) and (5) may be formally combined and written as

$$H(p,v;p_o,v_o) = 0, \tag{6}$$

which corresponds to the definite Hugoniot curve $p = p(v)$ centred on a point p_o,v_o drawn in a p,v-plane. According to the second law of thermodynamics, the entropy change across a shock cannot be negative, and this condition is well-known to lead to restrictions on the shape of the Hugoniot in order for the existence of stable single shocks, and otherwise shocks may not be possible or, if possible, then not in single-discontinuity but in multiple-discontinuity form. For the transition to be classed as a shock it is necessary that, with respect to an observer travelling with shock velocity, the particle velocity towards the transition region must be supersonic, and leaving the transition region, it must be subsonic; this condition excludes the Joule-Thomson effect where the particle velocity is entirely subsonic. Although shock stability is well understood for gases (see, for example, Hayes[5]) it is more complex and less extensively investigated for solids (see Duvall[6]). The Hugoniot curve, Eqn. (6), represents the locus of all equilibrium thermodynamic states that are accessible by a single shock transition from a given initial state p_o,v_o, phase transitions excluded (see Fig.1). Entropy changes in fact occur across the shock and also increase along the Hugoniot (i.e., away from its centre p_o,v_o). Such changes must correspond to irreversible non-equilibrium thermodynamic effects involving some dissipation of mechanical energy into heat energy. The remarkable fact is that the Rankine-Hugoniot equations, which relate jumps in quantities across the steadily propagating shock, are altogether independent of the details of these dissipative processes. The solution found for the discontinuous shock transition represents in fact the only admissible mathematical continuum solution of the problem when formulated in the absence of all dissipative effects such as viscosity and thermal conduction. It is easy to see that the governing equations involve no basic linear dimension in terms of which the extent of a finite transition region could be expressed and hence the transition region necessarily has to be of zero thickness.

Now only 2 of the 5 shock quantities are easily or, indeed, directly accessible to experimental measurement, and one possible procedure for experiments with plates of material is as follows: first, the shock velocity itself is quite simple to measure; and secondly, the particle velocity is determined indirectly by the measurement of the free surface velocity u_{fs} when the shock is reflected at a free surface and then either by the use of the acoustic approximation $u = \frac{1}{2}u_{fs}$ (with neglect of entropy effects) or by refinement of it (free surface velocity is due to the incident shock wave and to the reflected rarefaction wave whose joint effect is to maintain the zero pressure condition at the free surface). Once U,u are known, the Rankine-Hugoniot equations determine the remaining 3 quantities p,ρ,E and from the single experiment one point on the Hugoniot curve is established. With experimental data available from a series of experiments, a series of points on the Hugoniot is determined and thence by continuity a segment of the Hugoniot $H(p,v;p_o,v_o)=0$ is traced in the p,v-plane. In fact, in practice, matters are often simpler, because for many materials (solids and liquids) there applies with considerable accuracy over ranges excluding phase transitions an empirical linear relation between shock and particle velocity,

$$U = U_o + Su, \tag{7}$$

where U_o,S are constants. The physical explanation of the relation (7) is not well understood (see Ruoff[7]) but once the relation is established in any particular case it does permit the simpler representation of p,E in terms of $\eta = 1 - v/v_o$, U_o, S, viz.,

$$p - p_o = \rho_o U_o^2 \eta/(1 - S\eta)^2, \tag{8}$$

$$E - E_o = \tfrac{1}{2}\{U_o\eta/(1 - S\eta)\}^2. \tag{9}$$

Shock wave measurements are used to derive a more complete thermodynamic description of high pressure states by assuming the general form of an equation of state (or some equivalent) and adjusting its parameters by combined use of the Hugoniot Eqn. (4), thermodynamic procedures, and some theory of the atomic lattice. The equation of state thus conforms to the experimental Hugoniot curve, and it may now be used to predict any other thermodynamic data or curves, such as isotherms or adiabats. In order to determine temperature, specific heat data must be available and, in particular, temperatures along the Hugoniot are found from its intersections with adiabats.

In comparison with fluids, solids show an important difference in that they possess a finite yield strength. Thus, the quantity p above is not an isotropic pressure but is the normal compression stress parallel to the direction of shock propagation (the stress state is non-isotropic and conforms to plane strain conditions), and also the "elastic" Hugoniot differs from the "plastic" Hugoniot. The Hugoniot curve in fact shows a cusp corresponding to the phase change of shear yield at the so-called Hugoniot elastic limit. The plastic Hugoniot commences with a diminished slope (in comparison with the elastic Hugoniot's slope) but rises (concavely upwards) due to increasing incompressibility (the initial rise may be affected by work-hardening). The geometry of the elastic and plastic Hugoniots is shown in Fig.2. This circumstance of a phase transition has important consequences for (stable) wave propagation. Thus, a shock state beyond the point of intersection of the elastic Hugoniot with the plastic Hugoniot may be reached from an elastic state (or, in particular, a stress-free state) by a single shock transition, Eqns. (1) - (3) holding between the two states; but for a shock state below this point of intersection, a double-shock transition is necessary, which comprises an elastic shock transition carrying the elastic material up to the yield point and a plastic shock transition carrying the plastic material up to the final state, and the two shock fronts move at different velocities and diverge in time. Hence, depending upon the driving pressure, it is possible to have either a single plastic shock front (so that the elastic precursor is over-ridden) or an elastic shock front (the precursor) followed by a plastic shock front. Evidently, in the latter case, the normal single-wave structure breaks down into a double-wave one, and the situation is that shown in Fig.3.

Somewhat analogous results apply when more normal phase transitions, due to atomic rearrangements in regular crystalline structure, take place.

The above analysis involves only the use of the equations of conservation of mass, momentum, and energy in finite form, and this is due to the fact that steady state conditions are envisaged. In practice, applied pressures are not generated instantaneously but finite rise-times obtain for pressures produced either by mechanical impact or by explosions (and in either case rarefaction waves may degrade pressures at later times). The analysis of the unsteady problem depends now upon use of the conservation equations in differential as well as in finite forms. The differential equations are of the mathematical type known as (nonlinear) hyperbolic for which solutions that remain continuous indefinitely in space and time do not exist, even when the initial data are analytic. In fact, after a certain time the nonlinear equations show a transition from a continuous to a discontinuous solution, but then the equations themselves become invalid since they cease to represent correctly the conservation laws. Thereafter, the problem has to be formulated differently, on the basis of differential equations wherever the flow is continuous and of finite equations wherever the flow is discontinuous. When this is done properly it is found a shock front is built up steadily from behind and ultimately the steady state solution already obtained is achieved. The problem has been discussed in some detail by Morland[8].

The foregoing treats elastic and plastic shock discontinuities as sharp discontinuities and the existence of shock-structure does not arise.

SHOCK-STRUCTURE: CONTINUUM THEORY

The shock transition necessarily involves irreversible dissipative processes: if the precise details of these processes are altogether neglected, as earlier, then the shock transition simply appears merely as an abrupt mathematical discontinuity. The physics, in reality, will not tolerate such a discontinuity, and if we wish to eliminate it from the theoretical analysis then terms must be included in the basic governing equations which represent irreversible dissipative mechanisms. For gases, just how this is done and what are the results is well known (see, for example, Hayes[5]). Apparently, it is essential to include viscosity in addition to thermal conductivity in order to derive a determinate and realistic description of shock-structure. This has the effect of limiting the maximum steepness of the front, and it turns out that an exact solution of the now more general equations is possible for the steady propagation of an unchanging profile. This profile was named by Rayleigh (1910) the "permanent regime profile", and it was discovered independently by G. I. Taylor at about the same time. Of course, the validity of the predicted shock-structure depends upon the validity of the gasdynamic equations of motion when viscosity and heat conduction effects are included; and therefore, in particular, the shock-transition thickness should span a sufficient number of mean free paths: if not, the predictions are physically unacceptable, and then either the particular system of continuum equations used must be modified so as to lead to a revised and physically acceptable prediction or continuum theory itself must be abandoned and resort made to a physical theory. For gases, such questions are considered in detail by Zel'dovich and Raizer[1] and Hayes[5].

Somewhat in parallel with the work on shock-structure in gases, but far less extensively as yet and some 50 years after the Rayleigh and Taylor work, shock-structure in solids has been studied on the basis of continuum theory and also on the basis of lattice dynamics theory.

The first theoretical studies of shock-structure in solids were made by Band[9] (see also Band and Duvall[10]) and by Holtzman and Cowan[11] (see also Cowan[12]). In solids, thermal conduction is not thought to play much part, that is, in the short time available, while not a great deal is known about the bulk viscosity of solids. However, the primary purpose of shock wave studies is to help elucidate the physical properties of materials at high pressures. The main objection to the hydrodynamic type model for shock wave propagation in solids is that shear yielding (taking place beyond the elastic limit) normally takes a time that is long in comparison with the time of passage of a shock front. A detailed discussion of the shear-yielding process and all other important irreversible effects should therefore lead eventually to the establishment of a definite shock profile, characteristic of the material and not necessarily steady, depending essentially on the relaxation times for the various processes involved. As a first approach towards the determination of a permanent regime profile, Band[9] introduced a single-parameter visco-elastic model of a shear-yielding solid, which permits stresses lying off the Hugoniot curve to be achieved under non-equilibrium conditions. In rather more detail, the deviation in stress between the Rayleigh line (the straight line connecting initial and final states in the p,v-plane) and the Hugoniot is proportional to the particle velocity through the shock-transition region; the coefficient of proportionality (small or zero for the elastic Hugoniot but finite for the plastic Hugoniot) is an undefined coefficient of "shear-yielding" viscosity, which introduces a length scale. This hypothesis supplies the needed constitutive equation (specified by a single physical parameter) in terms of which shock-structure becomes determinate. The analytical procedure is as follows. The assumption of the existence of a permanent regime profile means that the description of events is steady (unchanging) when viewed by an observer moving at the same velocity as the shock velocity. Then, analytically, we write all dependent variables (particle velocity, pressure, etc.) not as functions $f(x,t)$ of two independent variables x,t but as functions $F(\xi)$ of the single independent variable $\xi = x - Ut$, and then $F(\xi) = \text{const.}$ for all tracks obeying $dx/dt = U$. This procedure reduces the differential equations of the problem from partial to ordinary type, their solution to satisfy requisite conditions at $\xi = \pm \infty$ corresponding to the shock transition conditions of the Rankine-Hugoniot Eqns. (1) - (3). Band[9] shows that the determination of the permanent regime profile is reduced to depend upon simple quadratures, and he finally expresses results in non-dimensional form and applies these to cases of particular materials. He also discusses other anelastic effects in addition to shear-yielding viscosity, and permanent regime profiles are determined also for heat-conducting anelastic solids. It is important of course that such permanent regime profiles should not be confused with experimentally determined non-steady profiles obtaining after finite travel times. However, physically, it is to be expected that non-steady profiles should tend to steady profiles, i.e., provided sufficient time elapses. Also, it should be noted that any sufficiently weak elastic-plastic shock is unstable and will therefore split up into an elastic transition followed by a plastic transition, and the structure of the former will be normally finer than that of the latter.

Holtzman and Cowan's[11] work on shock-structure follows similar analytical procedures but is based upon a different constitutive law for stress relaxation through the shock transition due to viscous relaxation. Their view is that the difference between the axial (compression) stress p_x and hydrostatic pressure p through the shock transition is a measure of the viscous shearing stress and that relaxation occurs according to the equation

$$p_x - p = \frac{4}{3}\tau_0 + \frac{4}{3}\eta\,\dot{\varepsilon}_x^p \ , \qquad (10)$$

where η is a coefficient of shear viscosity, in the usual notation. The analysis closely parallels G. I. Taylor's (1910) work on permanent regime profiles for gases, and it leads to definite shock profiles whose properties (e.g., thickness) depend upon numerical values of physical constants. A wide range of values of shock thickness is predicted, decreasing with increases in shock strength and increasing with increases in viscosity, the lower limit being of the order of atomic lattice spacing. A permanent regime profile is indicated schematically in Fig.4.

In later independent work, Bland[13] has discussed permanent regime profiles in a solid whose longitudinal constitutive equation exhibits second-order convective and dissipative effects, viz.,

$$p_x = A\varepsilon_x + B\varepsilon_x^2 + \eta\,\dot{\varepsilon}_x \ , \qquad (11)$$

where A, B are constants and η is a coefficient of viscosity, which corresponds to a nonlinear Voigt element. The analysis for the determination of permanent regime profiles follows along lines similar to those earlier described in connection with Band's[9] work. Bland is also able to show that an unsteady profile ultimately adopts the steady profile form. It should be noted that plastic deformation is not involved in his work (see also Bland[14]).

For a number of reasons, it is scarcely possible to achieve experimentally the steady profile forms of the permanent regimes treated in theoretical analysis as above. These reasons include the finite rise-times at initiation of a pulse and attenuation during propagation due to various physical effects and to rarefaction waves. Thus, at best, measurements relating to unsteady profile forms are the only practical ones. The quantity most easily measured is the free-surface velocity (rather than stress) and such data can be obtained now to within very fine time-resolution indeed. Thus, the basic experimental data consist of free-surface velocity vs. time traces. Such traces show various features of the elastic precursor wave and the main plastic wave. However, the conversion of these data into physical data is not easy for this depends upon the correct identification of anelastic effects present under loading conditions, and which affect the nature of the elastic-plastic profiles, and also upon assumptions of unloading behaviour, which affect the nature of the reflection from the free surface and therefore affect the relation between free-surface velocity and particle velocity. It is of course necessary that whenever particular anelastic effects are envisaged there must in fact be sufficient time available for these to be operative. The procedure is not therefore at all easy and success in the analysis of the basic experimental data consists not so much in good correlation (often simply empirical fit) between experiment and theory but in the unequivocal establishment that certain definite physical mechanisms are indeed operative. The experimental situations and associated theoretical procedures so far considered in the literature show much variety, and this fact does not permit firm general conclusions to be made. Here, it must suffice to refer rather briefly to certain particular studies, as follows.

Duvall[15] studied the propagation of plane shock waves in a stress-relaxing medium. This concerns primarily the decay of an elastic precursor wave preceding a shock, relating this to material relaxation from an elastic, non-equilibrium state towards one of equilibrium. The physical basis for relaxation is the connection between plastic strain-rate and dislocation density and velocity and Burgers vector but in the final form the governing continuum mechanics equations are essentially those introduced earlier (in a different connection) by L. E. Malvern (1951).

Johnson and Band[16] have investigated precursor decay in a shocked, elastic-plastic-relaxing solid. The constitutive equation is based upon a dislocation model for plastic flow, and the artificial viscosity method is used in numerical analysis for the determination of unsteady shock profiles in iron. Further development of this theory depends upon further experimental work to establish mechanisms of dislocation multiplication under shock conditions.

Taylor[17] reports time-resolved free-surface velocity traces for various metals, and the presence of elastic precursors is quite clear (see Fig.5). The theoretical analysis of the experimental data is based upon a uniaxial stress-strain relation appropriate to a standard linear solid (of linear visco-elasticity theory) with parameters being determined from dislocation-based arguments.

Butcher and Munson[18] report similar data. Their analysis involves elastic relaxation with plastic strain due to dislocation movements (these are stress activated, and multiplication and barriers are mechanisms that can be included). The continuum equations for the behaviour of the precursors are again those of L. E. Malvern. A reasonable fit is achieved between measured and predicted free-surface motion.

Jones et al.[19] have discussed the prediction of elastic-plastic wave profiles in aluminum under uni-axial strain loading. They treat the elastic perfectly-plastic case with strain-rate effects. The elastic precursor delay law follows that proposed by Duvall[15]. Refinements are made by use of more complex constitutive equations coming from or suggested by dislocation dynamics.

Karnes[20] also reports free-surface velocity measurements. His analysis, however, is non strain-rate dependent, and a stress-strain relation is determined that reproduces the data, this relation apparently being constructed from a series of weak shocks.

SHOCK-STRUCTURE: LATTICE DYNAMICS THEORY

Lattice dynamics theory represents of course a complete break with continuum dynamics theory: in lattice dynamics the physical system is no longer represented by a continuum but is now represented by a more-or-less orderly array of mass-points subject to interaction forces. In the simplest possible case we should have a one-dimensional lattice consisting of a series of equal and equally-spaced discrete particles connected by identical mass-less linear springs (see, for example, Band and Bhatti[21]). As is rather well known, Newton (1686) took this as a model for his theoretical study of the propagation of sound in air; his mistake, however, was to take the isothermal rather than adiabatic modulus, and Laplace (1822) made the necessary correction. In principle, lattice theories form the basis of the complete resolution of physical problems of deformation (i.e., stress and strain) in crystalline or polycrystalline materials - that is, they would, for example, provide a physical theory (as opposed to a phenomenological theory) of elasticity and so on, and also would continue to apply whenever continuum theory fails. The reality, however, is different, since the mathematics of all but the simplest possible lattice models (even when there is a precise specification of interatomic forces) is just too difficult to handle. Nevertheless, even if the models are simple ones, results may be obtained that are qualitatively if not quantitatively correct and useful predictions or conceptual ideas may be evolved. Lattice theories start to become important (that is, essential differences from continuum theories appear) at short wavelengths when dispersion effects occur. In linear lattice theory, the mathematical analysis grows in complexity with (a) increase in lattice dimensionality (i.e., from linear, to square, to cubic lattices), (b) progressive departure away from nearest-neighbour interactions to distant-neighbour interactions, (c) the presence of free surfaces, as for example with semi-infinite lattices (sometimes the asymmetry introduced at free surfaces is avoided by recourse to "cyclic boundary conditions"), and (d) the number of different species of mass-points, or isolated departures from the normal regularity of crystal structure as with occasional vacancies or interstitials. For linear elastic lattices of whatever complexity (dimensionality, free-surfaces, non nearest-neighbour interactions, and so forth), the procedure is to write down the Newtonian equations of motion for the mass-points together with appropriate boundary conditions. This forms an assemblage of linear spatially-discrete equations in the rectangular cartesian coordinates specifying atom positions (for a cubic lattice of $N \times N \times N$ atoms there will be $3N^3$ equations). For free vibrations of the system the equations are homogeneous, and the determination of natural frequencies and associated modes is a problem in linear algebra for the determination of eigenvalues and eigenvectors of matrices. Under conditions of regularity of crystal structure and of not-too-great departures away from simply nearest-neighbour interactions the matrices are strongly banded in type. In this case, when the matrices are of block circulant form, their eigenvalues are easily found in terms of simpler sub-matrices, but otherwise the complete matrices may need to be treated, perhaps by approximate methods. Such investigations are not new but they have been much accelerated by advances in matrix theory and by the use of computers (the recent work by Dean[22] shows both these trends). In so far as there is any simplicity in lattice dynamics theory above, then this is due to its linear harmonic nature. In reality, linear harmonic vibrations have a weak interaction between themselves, depending upon and increasing with amplitudes, leading to an anharmonic situation. Anharmonic studies represent an important further stage in the development of lattice theory.

Now if we take a shock wave in a material to correspond to an extremely sharp-fronted pulse with thickness of order perhaps a few lattice spacings, then nonlinear finite strain of the lattice certainly occurs. The uncertainties with continuum models of shocks, and their almost certain breakdown at intense shocks, clearly makes a case for studies of shock propagation in lattices. Work of this kind is quite recent, dating from about 1964. In view of the evident difficulties even with linear harmonic lattice theory, it is clear that preliminary studies must be confined to quite simple situations, and in fact to date mainly one-dimensional models have only been treated and these not exhaustively.

In principle, the situation to be considered may be likened to that of a fairly sharp disturbance set up in a mechanical system of beads, with mass and inter-connected by mass-less springs and dashpots (either linear or non-linear in response) moving on a straight rigid wire (with or without friction): even quite simple models of this kind are highly instructive to study. Typically, the motion is now described by nonlinear spatially-discrete equations (and either non-steady or steady solutions may be sought) and these reflect the particle (as opposed to continuum) representation of the physical situation. These discrete equations may be treated as such or they may be approximated by differential equations whose spatial order is a measure of the adequacy of representation of the true graininess (discreteness) of the situations. Work along both these lines has been reported and this is now discussed.

Anderson[23] studied the case of a 1D (monatomic) lattice when the nearest-neighbour interaction consists of a bi-linear spring (the stiffness characteristics increasing discontinuously, concave upwards) and a linear dashpot. This model therefore involves elastic and dissipative mechanisms (see Fig.6). The discrete equations of motion are written down and it is shown that these admit a steady state solution corresponding to a permanent regime profile. Data appropriate to aluminum are used and illustrative profiles are determined. As expected, the shock-transition thickness decreases with increasing shock strength and decreasing viscosity. It should be noted that in the absence of viscosity there could be no steady state solution.

Manvi, Duvall, and Lowell[24] have also studied finite amplitude stress wave propagation in 1D lattices. In contrast to Anderson's[23] work, dissipation is not considered and attention is focused on non-steady phenomena for linear and non-linear (quadratic) nearest-neighbour interactions. For a semi-infinite (monatomic) lattice, the disturbance is initiated by a finite velocity step applied at the first mass-point and this is taken to generate a shock-like disturbance in the lattice (a piston-type problem). The discrete equations of motion for the lattice particles admit an exact solution in the linear case, and this solution is simply expressed in terms of Bessel functions of the first kind (which follows a solution due originally to Schrödinger (1914)). For the equivalent continuum, the velocity profile would be rectangular, but for the lattice there are significant differences from this situation. Thus, for any particular lattice particle, all is quiescent up to a certain time, the disturbance then rises sharply (but not abruptly), then falls, and thereafter oscillates indefinitely with ever decreasing amplitude (see Fig.7). The nonlinear lattice can be treated numerically and a comparison is made between numerical solutions and approximate analytical solutions, showing fair agreement between them. There are differences of course between the wave forms for the linear and nonlinear cases but these differences are of a quantitative rather than qualitative nature. The interesting point is that the oscillations behind the steadily advancing shock front decay: this decay is due solely to mechanical causes, since the system is not supposed damped by viscosity.

Duvall, Manvi, and Lowell[25] have investigated the above problem but with allowance now for viscous damping (linear dashpots). The system must now of course possess a steady state (constant profile) solution to which unsteady solutions should tend in time. However, they give particular attention to the approximation of the exact discrete equations by continuum equations of various spatial order. This provides an interesting tie-up with the earlier continuum study by Bland[13]. A higher approximation for the undamped lattice leads to a permanent regime profile which oscillates steadily after shock arrival. In general, higher-order continuum approximations are represented by differential equations of a family which includes, as progenitors, Burgers equation (nonlinear convection and linear viscosity effects) and the Korteweg - de Vries equation (nonlinear convection and nonlinear elasticity, but no viscosity, effects). The occurrence of the Korteweg - de Vries equation in nonlinear lattice dynamics has also been discussed by Zabusky[26]. The approximation of spatially-discrete equations by continuum differential equations is very interesting, this being the exact converse of the situation encountered in numerical analysis studies of solutions of partial differential equations. If viscosity is included, there is little reason to doubt that higher-order "Korteweg - de Vries like" equations should admit steady state solutions which may perhaps be found without too much difficulty. If viscosity is not included then such steady state solutions may not exist, but nevertheless mechanical damping may occur behind the shock front.

Finally, it may be noted that Tsai and Beckett[27] have considered shock wave propagation in a 2D lattice.

CONCLUSION

The study of shock-structure in solids is comparatively recent and the extent of experimental and theoretical knowledge is therefore limited. More experimental data are of course needed but the correct interpretation and reduction of these data is necessarily done within the framework of mathematical theory soundly based upon the adoption of admissible physical mechanisms of plastic deformation. This requires further research into problems of nonlinear stress wave propagation in anelastic solids with due attention to thermodynamic aspects especially under intense shock conditions (see, for example, Hopkins[28,29]). As yet, only quite simple, although not necessarily easy, problems of shock propagation in lattices have been considered and further research here will be of much interest in helping to elucidate the nature of physical mechanisms and processes under rather intense high-pressure conditions when continuum theory must cease to be valid.

REFERENCES

1. Zel'dovich, Ya. B., and Raizer, Yu. P. Physics of Shock Waves and High-Temperature Hydrodynamic Phenomena (edited by Hayes, W. D., and Probstein, R. F.), Vols. I, II. Academic Press, New York and London, 1966, 1967.
2. Duvall, G. E., and Fowles, G. R. Shock Waves. Article in: High Pressure Physics and Chemistry (edited by Bradley, R. S.), Vol. II, p.209. Academic Press, New York and London, 1963.
3. Skidmore, I. C. An Introduction to Shock Waves in Solids. Appl. Mat. Res. 4, 131 (1965).
4. Al'tshuler, L. V. Use of Shock Waves in High-Pressure Physics. Soviet Phys.-Usp. (English Trans.) 8, 52 (1965).
5. Hayes, W. D. The Basic Theory of Gasdynamic Discontinuities. Article in: Fundamentals of Gas Dynamics (edited by Emmons, H. W.) (High Speed Aerodynamics and Jet Propulsion, Vol. III), p.416. Princeton University Press, 1958. (Reprinted: Gasdynamic Discontinuities, Princeton University Press, 1960.)
6. Duvall, G. E., Shock Wave Stability in Solids. Article in: Les Ondes de Détonation (edited by Ribaud, G.), p.337. Editions du C.N.R.S., Paris, 1962.
7. Ruoff, A. L. Linear Shock-Velocity-Particle-Velocity Relationship. J. Appl. Phys. 38, 4976 (1967).
8. Morland, L. W. The Propagation of Plane Irrotational Waves through an Elastoplastic Medium. Phil. Trans. Roy. Soc. London 251, 341 (1959).
9. Band, W. Studies in the Theory of Shock Propagation in Solids. J. Geophys. Res. 65, 695 (1960).

10. Band, W., and Duvall, G. E. Physical Nature of Shock Propagation. Amer. J. Phys. 29, 780 (1961).
11. Holtzman, A. H., and Cowan, G. R. The Strengthening of Austenitic Manganese Steel by Plane Shock Waves. Article in: Response of Metals to High Velocity Deformation (edited by Shewmon, P. G., and Zackay, V. F.) (Metallurgical Society Conferences, Vol.9), p.447. Interscience Publishers, New York and London, 1961.
12. Cowan, G. R. Shock Deformation and the Limiting Shear Strength of Metals. Trans. Metall. Soc. A.I.M.E. 233, 1120 (1965).
13. Bland, D. R. On Shock Structure in a Solid. J. Inst. Maths Applics 1, 56 (1965).
14. Bland, D. R. Finite Elastodynamics, J. Inst. Maths Applics 2, 327 (1966).
15. Duvall, G. E. Propagation of Plane Shock Waves in a Stress-Relaxing Medium. Article in: Stress Waves in Anelastic Solids (edited by Kolsky, H., and Prager, W.) (Proceedings of I.U.T.A.M. Symposium), p.20. Springer-Verlag, Berlin, 1964.
16. Johnson, J.N., and Band, W. Investigation of Precursor Delay in Iron by the Artificial Viscosity Method. J. Appl. Phys. 38, 1578 (1967).
17. Taylor, J. W. Stress Wave Profiles in Several Metals. Article in: Dislocation Dynamics (edited by Rosenfield, A. R., et al.) (Battelle Institute Materials Science Colloquia), p.573. McGraw-Hill Book Co., New York, 1968.
18. Butcher, B. M., and Munson, D. E. The Application of Dislocation Dynamics to Impact-induced Deformation under Uniaxial Strain. Article in: loc. cit. Ref. 17, p.591.
19. Jones, A. H., Maiden, C. J., Green, S. J., and Chin, H. Prediction of Elastic-plastic Wave Profiles in Aluminum 1060-0 under Uniaxial Strain Loading. Article in: Mechanical Behavior of Materials under Dynamic Loads (edited by Lindholm, U. S.) (A.R.O. and S.-W.R.I. Symposium), p.254. Springer-Verlag, Berlin, 1968.
20. Karnes, C. H. The Plate Impact Configuration for Determining Mechanical Properties of Materials at High Strain Rates. Article in: loc. cit. Ref. 19, p.270.
21. Band, W., and Bhatti, A. D. Energy Propagation in a Finite Lattice. Amer. J. Phys. 33, 930 (1965).
22. Dean, P. Atomic Vibrations in Solids. J. Inst. Maths Applics 3, 98 (1967).
23. Anderson, G. D. Shock Wave Propagation on a One-dimensional Lattice. Poulter Research Laboratories Tech. Rep. 001-64, Stanford Research Institute, 1964.
24. Manvi, R., Duvall, G. E., and Lowell, S. C. Finite Amplitude Longitudinal Waves in Lattices. Int. J. Mech. Sci. 11, 1(1969).
25. Duvall, G. E., Manvi, R., and Lowell, S. C. Steady Shock Profile in a One-dimensional Lattice. J. Appl. Phys. (In press.)
26. Zabusky, N. J. A Synergetic Approach to Problems of Nonlinear Dispersive Wave Propagation and Interaction. Article in: Nonlinear Partial Differential Equations (edited by Ames, W. F.), p.223. Academic Press, New York and London, 1967.

27. Tsai, D. H., and Beckett, C. W. Shock Wave Propagation in a Two-dimensional Crystalline Lattice. Article in: Behaviour of Dense Media under High Dynamic Pressures (I.U.T.A.M. Symposium) (edited by Berger, J.), p.99. Gordon and Breach, New York, 1968.
28. Hopkins, H. G. Dynamic Nonelastic Deformations of Metals. Article in Applied Mechanics Surveys (edited by Abramson, H. N., Liebowitz, H., Crowley, J. M., and Juhasz, S.), p.847. Spartan Books, Washington, D.C., 1966.
29. Hopkins, H. G. The Method of Characteristics and its Application to the Theory of Stress Waves in Solids. Article in: Engineering Plasticity (edited by Heyman, J., and Leckie, F. A.), p.277. Cambridge University Press, 1968.

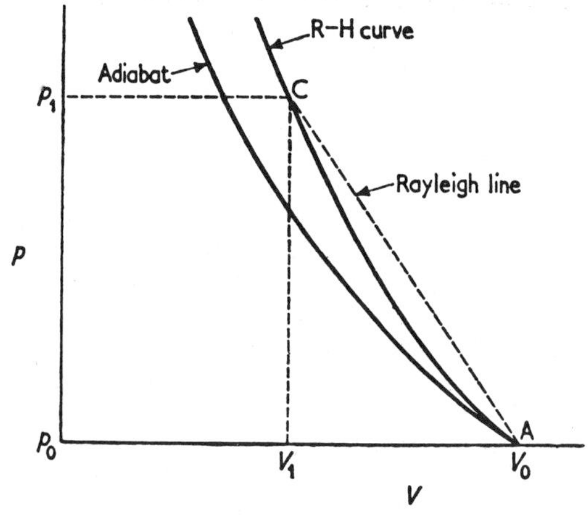

Figure 1
Rankine-Hugoniot curve.
(From Ref.2.)

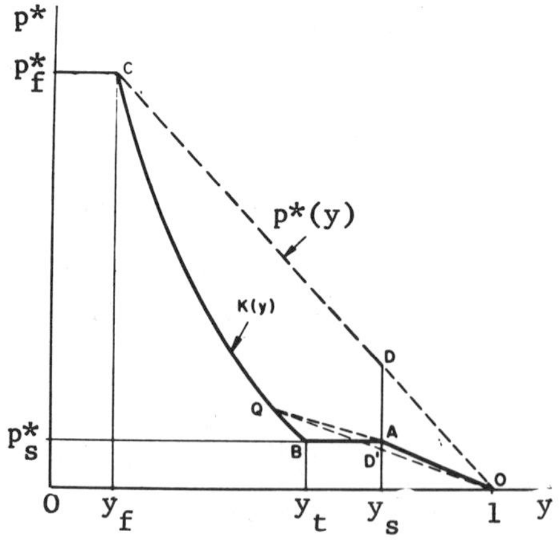

Figure 2
Elastic and plastic Hugoniots.
(From Ref.9.)

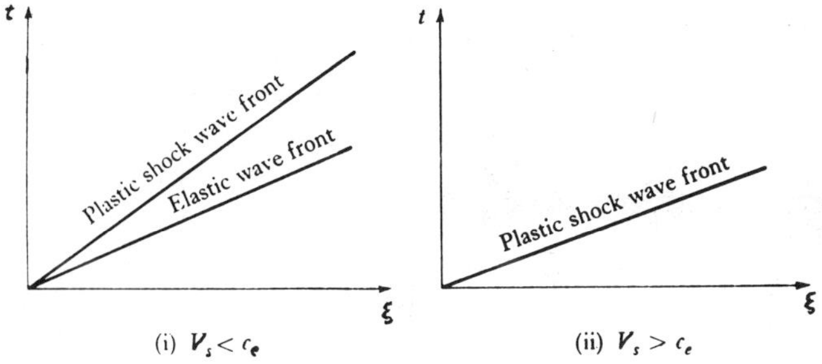

Figure 3
Elastic and plastic shock wave fronts.
(From Ref.29.)

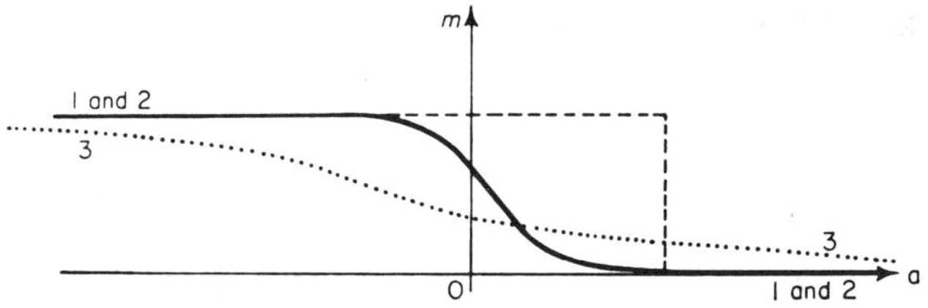

Figure 4
Permanent regime profile.
(From Ref.13.)

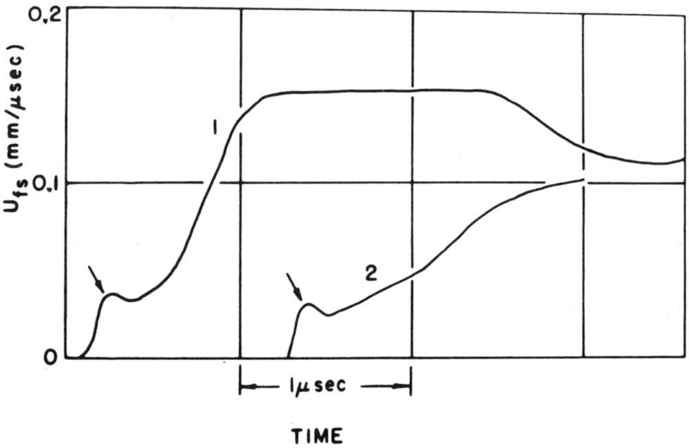

Figure 5
Free-surface velocity traces.
(From Ref.17.)

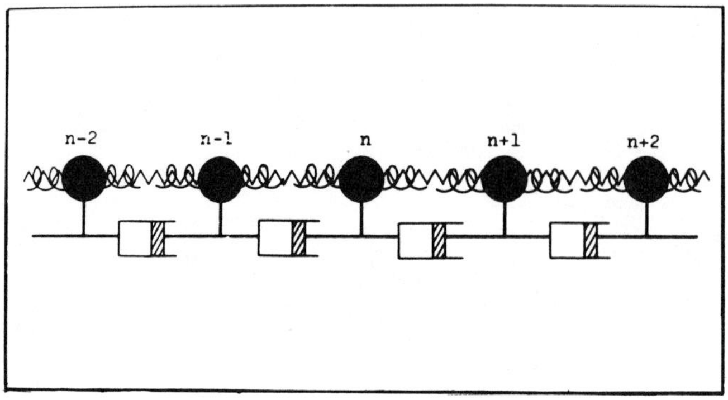

Figure 6
1D lattice with nonlinear elasticity and linear viscosity. (From Ref.23.)

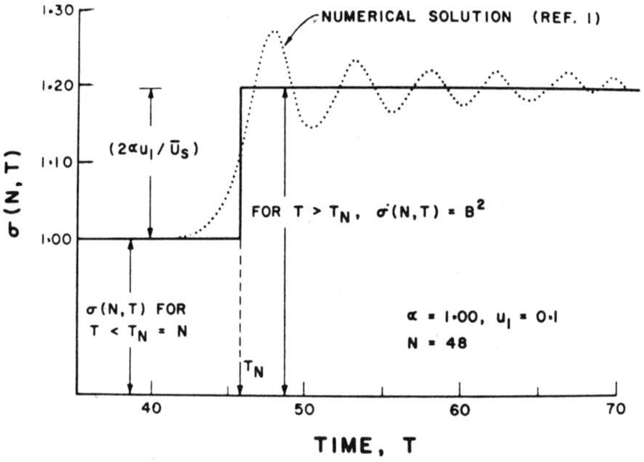

Figure 7
Acceleration of mass-point in linear-elastic 1D lattice. (From Ref.24.)

MEASUREMENTS OF INCIDENT-SHOCK TEST TIME AND REFLECTED SHOCK PRESSURE AT FULLY TURBULENT BOUNDARY-LAYER TEST CONDITIONS

ROBERT G. FUEHRER
Cornell Aeronautical Laboratory, Inc.
Buffalo, New York

SUMMARY

An experimental study has been made of incident-shock test time and end-wall, reflected-shock pressure at test conditions characteristic of high-pressure, high-enthalpy shock tunnels operating in the tailored interface mode. The driven gas in these experiments was either air or nitrogen and the driver gas was heated hydrogen. The incident-shock Mach number varied from 7.5 to 10.6 and the initial driven-tube pressure varied from 10 to 150 cm Hg. Measurements of the radiation intensity behind the incident shock and Flagg's interpretation of the end-wall "pressure dip" were used to infer the arrival of the interface region at three different axial stations in the driven tube. Incident-shock test time, for both air and nitrogen, was found to be considerably less than Mirels' turbulent boundary-layer test-time theory predicts at the test conditions commonly used in high-pressure shock-tunnel work. Combustion between the driver and driven gas was observed for the combination of hydrogen and air but this had no measurable effect on the usable test time. Mirels' theory appears to underestimate the mass flow in the boundary layer, particularly at the high Reynolds number test condition. Suitable modifications of this theory are suggested.

INTRODUCTION

Over the past several years it has been a common practice to interpret data and predict the performance of high-pressure shock tunnels in terms of driven gas slug-lengths computed from Mirels' turbulent boundary-layer, incident-shock test-time theory[1]. However, unlike the corresponding laminar boundary-layer theories (e.g. Refs. 2-8) which have been investigated experimentally by many researchers (e.g. Refs. 5-17), Mirels' turbulent theory has never been subjected to extensive experimental verification. A limited amount of turbulent boundary-layer test-time data has been reported[14-18]. However, in all cases these data were obtained either at low incident-shock Mach numbers (i.e. $M_s \leq 5$)[15-18] or at transitional rather than fully turbulent boundary-layer test conditions[14-17]. Data on incident shock test time, at test conditions where present-day high-pressure shock tunnels routinely operate, are virtually nonexistent in the open literature. The experimental investigation described in this paper was undertaken therefore to obtain new data on incident-shock test time for fully turbulent boundary-layer test conditions at relatively high shock Mach numbers ($7.5 \leq M_s \leq 10.6$) in air and nitrogen.

TEST FACILITY

The Cornell Aeronautical Laboratory's 96-inch Hypersonic Shock Tunnel (HST) test facility was used to obtain the data presented in this paper. The shock tube portion of this facility, depicted in Fig. 1, is a chambered configuration having a 5-inch ID driver and a 4-inch ID, 48.44 ft. long driven tube. The driver is externally heated and operated at a gas temperature of 775°F and pressures up to 30,000 psia. Hydrogen was used as the driver gas in these experiments. The theoretical tailoring shock Mach number corresponding to these driver conditions is 10.3 when air, at an initial pressure of 1 atm., is used for the driven gas. However, because of shock attenuation and other nonideal shock tube effects, the actual "tailoring" shock Mach number* for this facility is in the range of 8.0 to 8.5 depending on the driver pressure level. The data presented in this paper were obtained at near-tailored and over-tailored test conditions.

Diaphragm rupture was initiated by means of the double-diaphragm, firing cavity technique, shown in Fig. 1, to obtain reproducible test conditions. The firing cavity was vented to initiate diaphragm rupture so that the upstream diaphragm would always rupture first. This procedure is necessary to avoid the problem of multiple shocks in the driven tube which occur when the firing cavity is pressurized and the downstream diaphragm ruptures a finite time before the upstream diaphragm.

The nozzle throat configuration and centerbody valve (e.g. Ref. 18) normally used in the operation of the CAL 96-inch HST were not used in these tests. The end of the driven tube containing these components was replaced by an instrumented, constant-area tube section (4" ID) terminated by a flat end-plate.

INSTRUMENTATION

The data presented in this paper were obtained from side-wall measurements of the radiation intensity behind the incident shock at station 5 and 7-1/2 (see Fig. 1) and end-wall measurements of the

---

*The "tailoring" shock Mach determined experimentally corresponds to the shock Mach number for which the end-wall pressure has an average constant level prior to the arrival of the reflected head of the driver expansion fan. If the average pressure tends to increase with time this test condition would be referred to as being over-tailored.

pressure behind the reflected shock. In addition, side-wall measurements of the surface temperature behind the incident shock were obtained at station 10 using a thin-film heat transfer gage to provide data on boundary-layer transition.

Radiation Measurements

The incident-shock test time, determined by the radiation intensity behind the incident shock, was measured with an EG & G Model SGD-100 silicon diffused photodiode, sensitive in the spectral range from 0.5 to 1.05 microns.* The photodiode viewed the radiating gas through a 6-inch long, 1/16-inch ID orifice to assure good spatial resolution. Fused quartz windows (1/8-inch thick) mounted flush with the driven tube ID terminated the orifice of this optical viewport. No optical filtering was attempted in these experiments nor was any effort made to identify the radiating species.

The quartz windows were replaced after every run to avoid any question of window degradation affecting the shape or level of the radiation record. In addition, the bore of the driven tube was thoroughly cleaned before every run with a cloth dampened with acetone in an effort to reduce the level of radiation caused by gas contamination.

Pressure Measurements

A Kistler 603H pressure transducer was used for measuring the end-wall reflected-shock pressure. The output of this transducer was filtered by means of a specially designed "notch" filter having a 3 dB point at about 20 KHz, 37 dB maximum attenuation at 54 KHz and over 20 dB attenuation for all frequencies above 100 KHz. A thin coating (.025-inch thick) of G. E. RTV-102 was applied to the diaphragm of the transducer before every run to provide thermal insulation for the transducer.

Incident-Shock Velocity Measurements

Incident-shock velocity and trajectory was measured by means of five conventional ionization gages located at stations 4, 6, 7, 8, and 12 in the driven tube (see Fig. 1). High frequency electronic counters were used to record the elapsed time for passage of the shock wave between these stations (i.e. sta. 4 to 6, 6 to 7, etc.). The shock Mach number at the instrumentation stations 5 and 7-1/2 was computed directly from the elapsed time measurements obtained between stations 4 to 6 and 7 to 8, respectively. The end-wall Mach number was determined by linear extrapolation of the shock Mach number

---

*Optical transmission in this range is 50% or greater. In the spectral range from 0.35 to 1.13 micron (i.e. from the near ultraviolet to the near infrared) the transmission is 10% or greater. Peak sensitivity occurs at about 0.9 micron.

attenuation curve constructed for each run. Typical shock attenuation characteristics for the shock tube used in these experiments are given in Figure 2. The speed of sound used to calculate the shock Mach number was based on measured values of the initial gas temperature and the transport properties given in Reference 19.

Wall Surface-Temperature Measurements

Standard thin-film platinum resistance-thermometer gages of the type described by Vidal[20] several years ago were used to measure the side-wall temperature behind the incident shock. Boundary-layer transition time, τ_{tr} , representing the period that the wall boundary layer is laminar at a given station, was determined from these temperature records using the procedure described by Hartunian et al.[21] Data on τ_{tr} were of interest in establishing the extent to which the test conditions of these experiments satisfied the basic requirement in Mirels' theory that the boundary layer behind the incident shock be fully turbulent.

Considerable care was taken to assure that the response time of the wall-temperature instrumentation was adequate to measure the 1 to 10 μ sec transition times anticipated for these test conditions. The thermal response of various film-substrate combinations was examined using Kurzrock's[22] heat conduction analysis. This led to the selection of a platinum film on a substrate of vitreous alumina (Al_2O_3) for the high initial-pressure test condition (i. e. $P_1 = 10$ psia) and platinum on Pyrex 7740 for the low-pressure test condition of $P_1 = 10$ cm. Hg. Initially, no attempt was made to electrically insulate the platinum film from the ionized gas behind the incident shock because this would reduce the frequency response of the gage. Furthermore, some reduction in the laminar temperature step due to electrical shorting of the gage could be tolerated since heat flux measurements were not of interest. No problem was encountered with the noninsulated gages at the $P_1 = 10$ psia test condition. However at the $P_1 = 10$ cm. Hg. condition, severe electrical shorting of the film element was experienced. A thin coating of magnesium flouride ($\approx$ 0. 1 micron thick) had to be applied to the platinum film before data on τ_{tr} could be obtained at this test condition.

The output of the thin film gage was recorded on a Tektronix type 545 oscilloscope using a type M preamplifier plug-in unit to maintain the high frequency response capabilities of the entire system. With these precautions, signal rise times of 0. 3 μ sec were achieved with the uncoated alumina gage. This is the best that could be expected since the time for the incident shock to cross the width of the platinum film is about 0. 3 μ sec for the conditions of these tests.

DISCUSSION OF RESULTS

Data on incident-shock test time were obtained in air and nitrogen for shock Mach numbers ranging from 7. 5 to 10. 6 and initial driven tube pressures ranging from 10 to 150 cm. Hg. The driver pressures varied from 1900 to 26, 800 psi for the heated hydrogen driver used in these tests.

The primary source of data on incident-shock test time in this program was the side-wall measurements of the radiation intensity behind the incident shock obtained at station 5 and 7-1/2. However, the end-wall reflected-shock pressure proved to be an additional source of information regarding the location of the interface region near the end of the driven tube and also served to corroborate the radiation test-time data. These features are illustrated in Fig. 3 which shows typical oscillograms of the incident-shock radiation at stations 5 and 7-1/2 and the end-wall reflected-shock pressure as well as a wave diagram constructed from this data. The oscillograms shown are for hydrogen driving air at an initial pressure of 10 psia. The incident shock Mach number at station 7-1/2 is 8.17. The arrival of the interface region at station 5 is readily determined from the abrupt drop in radiation intensity seen in the radiation oscillogram (Fig. 3a) at a time interval "A" after arrival of the incident shock. The radiation record at station 7-1/2 (Fig. 3b) is similar to station 5 except that shortly after the radiation intensity begins to decrease (i. e. after time interval "B") an additional source of significant radiation passes the viewing station. This latter portion of the radiation record is the result of combustion between the driver and driven gas as a subsequent figure will clearly demonstrate. The "notch" in the radiation record of Fig. 3b, which occurs at the time interval "B" after passage of the incident shock, is a characteristic feature of the radiation records obtained at station 7-1/2 for all test conditions involving hydrogen driving air used in this program. The arrival of the interface region at this station is considered to occur at the beginning of this radiation "notch".

The end-wall reflected-shock pressure of Fig. 3c corroborates the interpretation of the radiation records by the following reasoning. Shortly after reflection of the incident shock from the end wall (i. e. at time interval "C") a sudden and significant drop in pressure is seen to occur in the pressure record of Fig. 3c. This easily recognized and very repeatable phenomenon, commonly referred to as the "pressure dip," is well known to operators of hydrogen-driven reflected shock-tunnels. Flagg [23], several years ago, correctly attributed this "pressure dip" disturbance to the interaction of the reflected shock with the combustion region behind the incident shock. Using this interpretation, the time interval "C" (Fig. 3c) represents the time for the reflected shock to encounter the combustion region and return an expansion wave to the end-wall. The location of the interface region near the end of the driven tube can therefore be determined from the "pressure dip" time interval "C", hereafter referred to as Δt "dip", and the theoretical reflected shock velocity and speed of sound (e. g. Ref. 24) by the method shown in the wave diagram of Fig. 3. The interface region located from the end-wall pressure data in this fashion is seen to be in good agreement with the radiation data identified as point "A" and "B" on the wave diagram. The interface region as determined by Mirels' theory[1] is also shown on the wave diagram for comparison. Incident-shock test time for this test condition is seen to be about 2/3 to 1/2 of that predicted by Mirels' theory. The results obtained for other test conditions are discussed in the next three sections.

Radiation Measurements

To answer the question of whether combustion between the driver and driven gas is responsible for the poor agreement of the experimental data with Mirels' theory (see Fig. 3), this test condition was repeated using nitrogen as the driven gas. The resulting nitrogen radiation data obtained at stations 5 and 7-1/2 are given in Fig. 4 along with the corresponding air radiation data for comparison. Incident-shock test time is seen to be essentially the same for nitrogen (Fig. 4a, b) as it is for air (Fig. 4c, d). Thus, although a combustion region definitely exists behind the incident shock (compare Fig. 4b to 4d) and even grows in intensity as it progresses down the driven tube (compare Fig. 4c to 4d), combustion does not contribute to a reduction in incident-shock test time at these test conditions. *

A summary of the incident-shock test time data obtained from the radiation measurements at stations 5 and 7-1/2 is given in Figs. 5 and 6. Theoretical incident-shock test times derived from Mirels' theory[1] are also given in these figures for comparison. ** Incident-shock test time for air is seen from Figs. 5 and 6 to be essentially the same as nitrogen for all the test conditions used in this program. Comparison of these data to Mirels' theory[1] shows that the agreement is poor at high initial pressure levels of the driven gas, P_1, and fair at low P_1's. In addition, the data exhibit substantially less sensitivity to P_1 than the theory would predict. A possible explanation for these particularly surprising and significant results is given in the "Data Analysis" section of this paper.

Wall-Temperature Measurements

The extent to which the data of Figs. 5 and 6 correspond to fully turbulent boundary-layer test conditions may be determined from Fig. 7 which shows typical oscillograms of wall-temperature records used to obtain boundary-layer transition times, τ_{tr}, and from Fig. 8

---

*It is interesting to note that about 20% lower driver pressures are required to produce a given shock Mach number in air than in nitrogen for the same initial pressure in the driven tube (e. g., Fig. 2, 4). This effect is very repeatable and is probably due to combustion behind the incident shock augmenting the driving energy of the expanding driver gas.

**All theoretical viscous shock-tube parameters given in this paper were obtained from a computer program written by the author which is based on the equations given by Mirels in Ref. 1 and the real-gas normal shock properties of Refs. 24, 25. This was necessary because a graphical presentation of the important viscous parameters is not given in Ref. 1 for nitrogen and that which is provided for air is based on real-gas normal shock properties corresponding to much lower initial pressure levels than were used in this test program.

which gives a summary of boundary-layer transition time data presented in terms of transition Reynolds number. The boundary-layer transition times given in Fig. 7 and in the table of Fig. 8 are seen to be of the order of a few microseconds as compared to the incident-shock test times given in Figs. 5 and 6 that are in the range of 120 to 240 μ sec. The laminar boundary-layer portion of the flow behind the incident shock therefore represents less than 3% of the entire incident slug length signifying that the data of Figs. 5 and 6 closely approximate a fully turbulent boundary-layer test condition.

The shock-tube boundary-layer transition data presented in Figs. 7 and 8 were obtained at a unit Reynolds number level that is higher than has ever been reported in the open literature, to the author's knowledge, for high wall-cooling rates (i.e. $M_s \approx 8$). For this reason it was of interest to compare these data, on a transition Reynolds number basis, to the data reported by Hartunian, et al.[21]. This result is given in Fig. 8. The agreement with the data of Ref. 21 is fairly good considering that the unit Reynolds numbers involved were 5 to 100 times greater than those of Ref. 21 and that the steel walls of the circular test section used in this test program were substantially rougher than the glass wall, rectangular test section used in Ref. 21. It appears, therefore, that reasonable estimates of shock-tube boundary-layer transition times may be obtained for present-day, high-pressure shock-tunnel test conditions by using the transition Reynolds number correlation generated by Hartunian et al.[21] many years ago at comparatively low initial-pressure levels in the driven tube.

End-Wall Pressure Measurements

In a previous section it was shown that the location of the interface region near the end of the driven tube may be determined from end-wall pressure data (see Fig. 3) when a combustion region exists behind the incident shock by using Flagg's[23] interpretation for the "pressure dip". It is not necessary, however, that the driver and driven gas be combustible in order to establish the location of the interface region in this fashion. A "pressure dip" similar to that obtained in air (Fig. 9 a) also occurs in nitrogen (Fig. 9 b) because a region of mixed driver and driven gas always exists behind the incident shock regardless of whether combustion takes place or not. The strength of the disturbance produced when the reflected shock encounters this mixed gas region is, of course, less for nitrogen than for air, as Fig. 9 shows and Dunn[26] working at much lower pressure levels also demonstrated, because combustion produces a greater impedance mismatch. Nevertheless, the "pressure dip" obtained with nitrogen is sufficiently well defined that inferring the location of the interface region from the nitrogen end-wall pressure data proved to be no more difficult than it was for the air data. A summary of these data expressed in terms of the time interval Δt "dip" (see Fig. 9) is given in Fig. 10 along with the corresponding theoretical time for this event computed as explained on this figure. As in the incident-shock test time results of Figs. 5 and 6, we see from Fig. 10 that (1) the air and nitrogen data are in good agreement (2) the data are in poor agreement with theory at high P_1's and in fair agreement at low P_1's and (3) the data exhibit substantially less sensitivity to P_1 than the theory would predict.

Data Analysis

In order to provide (1) a direct comparison of the radiation and end-wall pressure data in a way that takes into account differences in the axial location of the three instrumentation stations and (2) a convenient summary of the efficacy of Mirels' theory for various test conditions, the absolute test time and Δt "dip" results of Figs. 5, 6, and 10 are presented in terms of the ratio of the experimental to the theoretical time in Fig. 11. Compared in this fashion, the end-wall pressure data, for a given P_1, consistently corroborate the radiation data. However, a residual effect of the driven-tube length over diameter ratio, Ls/d, seems to exist since the low Ls/d instrumentation station tends to be in better agreement with theory than the higher Ls/d station in any given run.* The general agreement with Mirels' theory therefore depends not only on the driven-tube initial pressure level P_1, as was pointed out earlier and is certainly apparent in Fig. 11, but also on the driven tube geometry involved in a particular experiment.**

Thorough analysis of the data of Figs. 5, 6, 10 and 11 has uncovered a possible explanation for these unexpected results. However, before pursuing this, it is desirable to first review some of the properties of the important viscous shock-tube parameters given by Mirels' theory.[1]

Mirels' equation for the viscous, incident-shock test-time τ is given as

$$\frac{\tau}{\tau_i} = \left(\frac{1}{X}\right)\left(\frac{T}{1 + \frac{T^{4/5}}{W-1}}\right) \qquad (1)$$

where W is the density ratio across the incident shock (i.e. $W \equiv \rho_2/\rho_1$) and τ_i is the inviscid, incident-shock test time given by the equation below in which a_1 represents the initial speed of sound of the driven gas.

$$\tau_i = \frac{L_s/a_1}{M_s(W-1)} \qquad (2)$$

*It should be remembered, in interpreting the data of Fig. 11, that the shock Mach number at the end-wall is about 0.5 Mach number units lower than the shock Mach number at station 5 in any given run (see Fig. 2).

**For shock-tube configurations and test conditions commonly used in high-pressure shock-tunnel work, the incident-shock test time and hence slug lengths are only about one half of that predicted by Mirels' theory.

The viscous parameters X and T of Eq. 1 are defined by the following identities:

$$X \equiv \frac{\ell_i}{\ell_m} \tag{3}$$

$$T \equiv \frac{\ell}{\ell_m} \tag{4}$$

where ℓ_i is the inviscid-shock slug length (i.e. $\ell_i = Ls/W$) and ℓ is the viscous incident-shock slug length. The variable ℓ_m represents the maximum possible separation distance between the incident shock and the contact surface which occurs when all the mass flow entering the shock, in a shock fixed coordinate system, is in the boundary layer and the contact surface moves at the same velocity as the shock front.

Mirels shows that the parameter T is a function of the maximum separation distance parameter X according to the following implicit equation:

$$X = -\frac{5}{4}\left(\ln \frac{1 - T^{1/5}}{1 + T^{1/5}} - 2 \tan^{-1} T^{1/5} + 4 T^{1/5} \right) \tag{5}$$

From this equation it can be shown that when $X \rightarrow \infty$, $T \rightarrow 1.0$ signifying that virtually all of the mass flow entering the shock is in the boundary layer. Conversely, when $X \rightarrow 0$, $T \rightarrow X$ indicating that very little of the mass flow entering the shock is in the boundary layer and hence the viscous slug length ℓ is very nearly equal to the inviscid slug length ℓ_i.

It is apparent from Equations 1 and 5 that the viscous over inviscid test time ratio, τ/τ_i, depends simply on the maximum separation distance parameter X and the density ratio, W. Mirels[1] solves for the limiting slug length, ℓ_m, and therefore X by evaluating the mass flow in the boundary layer assuming a 1/7 power law for the velocity profile and an incompressible Blasius skin-friction law, applied to compressible flow by using Eckert's reference enthalpy technique for evaluating the fluid properties at a suitable temperature. Such analysis showed that ℓ_m was dependent on the driven tube diameter, d, and initial pressure P_1 according to the relation

$$\ell_m \sim p_1^{1/4} d^{5/4} \tag{6}$$

so from Eq. 3

$$X \sim \frac{L_s/d}{(p_1 d)^{1/4}} \tag{7}$$

Solutions for Mirels' maximum separation distance parameter, X, for air and nitrogen are presented in Fig. 12* in the normalized form suggested by Eq. 7. It is apparent from this figure that the maximum separation distance parameter, X, is a comparatively weak function of shock Mach number, M_s, and initial pressure, p_1, particularly in the range of M_s and p_1 covered in these experiments. The parameter, X, therefore is determined almost exclusively by the normalizing parameter $(p_1 d)^{1/4} / (Ls/d)$. As a result, τ/τ_i is similarly dependent on this parameter as can be seen be rewriting Eq. 1 in the form

$$\frac{\tau}{\tau_i} = (C_1)\left[\frac{(p_1 d)^{1/4}}{Ls/d}\right]\left(\frac{T}{1 + \frac{T^{4/5}}{W-1}}\right)^{-1} \tag{8}$$

where

$$C_1 = \left[\frac{(X)(p_1 d)^{1/4}}{Ls/d}\right] \simeq \text{constant (see Fig. 12)} \tag{9}$$

It is of interest now to present the data of Figs. 5, 6, 10 in the form suggested by Eq. 8 in order to evaluate the functional behavior of the data relative to the theory. This result is given in Fig. 13, which clearly demonstrates that the data can not be correlated by the parameter $(Ls/d) / (p_1 d)^{1/4}$ predicted by Mirels' theory. However, unlike previous presentations (i.e. Fig. 5, 6, 10, 11) the data do possess a definite coherency when viewed in this fashion for now the data organize themselves in three levels of the viscous over inviscid time ratio which are essentially independent of p_1 and vary linearly with the Ls/d of the corresponding instrumentation station. This observation suggested the correlation of data presented in Fig. 14 which proved very successful and provides insight into the probable reason that Mirels' theory fails to adequately predict incident-shock test time. The fact that the data correlate in the manner given in Fig. 14 strongly suggests that the mass flow in the boundary layer is much greater than is taken into account in Mirels theory. The basis for this conjecture is given in the discussion that follows.

---

*These results, obtained from the computer program referred to earlier, are based on the following numerical constants evaluated at an initial gas temperature, T_1, of 75°F which closely approximates the actual experimental test condition. $p_{st} = 1$ atm

Driven Gas	Specific heat ratio γ_1	a_1 ft/sec	Sutherland constant °K	Viscosity, lb_f-sec/ft^2 μ_1	Prandtl No.	$\frac{\gamma_1 p_{st}}{a_1 \mu_1}$, ft^{-1}
N_2	1.4	1152.87	104	3.7015×10^{-7}	0.713	6.942×10^{6}
Air	1.4	1133.79	110.4	3.8262×10^{-7}	0.709	6.829×10^{6}

If we assume that for these test conditions all rather than only a portion of the mass flow entering the shock is in the boundary layer, then T by definition is 1.0 and Eq. 8 reduces to the following form:

$$\left(\frac{\tau}{\tau_i}\right)\left(\frac{L_s}{d}\right) = (C_1)(p_1 d)^{1/4}\left(1 - \frac{1}{W}\right) \tag{10}$$

But now the exponent of the (P_1d) term must be less than 1/4 since when the theoretical mass flow in the boundary layer is made to increase at a given station, as a result of changes in the assumed velocity profile and skin friction law, the exponent of the (p_1d) term will decrease. This can be seen by comparing the functional behavior of l_m for a laminar boundary layer which is a low mass-flow sink to that of a turbulent boundary layer which is a relatively high mass-flow sink. Mirels shows in Ref. 1 that regardless of whether the boundary layer is laminar or turbulent l_m may be expressed as

$$l_m \sim (d)(p_1 d)^{\frac{n}{1-n}} \tag{11}$$

For a laminar boundary layer, n = 1/2 and for the turbulent boundary layer assumed by Mirels in Ref. 1, n = 1/5 so

$$l_m \sim (d)(p_1 d)\ , \quad \text{LAMINAR}\ , \quad n = 1/2 \tag{12}$$

$$l_m \sim (d)(p_1 d)^{1/4}\ , \quad \text{TURBULENT}\ , \quad n = 1/5 \tag{13}$$

The reduction in the exponent of the (p_1d) term as the mass flow in the boundary layer increases from the laminar to the turbulent case is clearly demonstrated by these last two equations (i.e. Eq. 12, 13).

Using the general expression of Eq. 11 in Eq. 10 then gives

$$\left(\frac{\tau}{\tau_i}\right)\left(\frac{L_s}{d}\right) = (C_2)(p_1 d)^{n/1-n}\left(1 - \frac{1}{W}\right) \tag{14}$$

where $n < 1/5$ and

$$C_2 = \left[\frac{(X)(p_1 d)^{n/1-n}}{L_s/d}\right]^{-1} \approx \text{CONSTANT} \tag{15}$$

If n were of the order of 1/10 to 1/20, which is likely since little or no sensitivity to p_1 can be detected in the test-time data of Figs. 5, 6 and 10, then the parameter (τ/τ_i) (Ls/d) of Eq. 14 would be essentially constant for the range of test conditions covered in these experiments. Within experimental accuracy, the data correlated as shown in Fig.14 are seen to behave exactly as would be theoretically expected from Eq. 14 which was derived assuming that the mass flow in the boundary layer was greater than predicted by Mirels' theory and that the limiting slug-length condition was achieved. This is a particularly significant result for it also suggests that limiting slug lengths and hence maximum incident-shock test times occur at much higher values of p_1 and lower values of Ls/d than was formerly believed.

In view of these results it appears that, although Mirels' theory is conceptually correct, the 1/7 power-law velocity profile and the incompressible Blasius skin-friction law used in developing that theory inadequately describe the high Reynolds number (i. e. $Re > 10^7$), high Mach number type flows to which this theory is frequently applied. Better agreement with experimental data might therefore be obtained by modifying Mirels' theory such that the solution of the shock-tube momentum integral equation is based on (1) the apparently universal velocity profile defined by Coles [27] well-known "Law of the Wall-Law of the Wake" equation and (2) a compressible skin-friction formula such as given by Nash and Macdonald[28].

CONCLUSIONS

Measurements of incident-shock test time and end-wall reflected-shock pressure have been presented for fully turbulent boundary-layer test conditions at shock Mach numbers ranging from 7. 5 to 10. 6 in air and nitrogen. Incident-shock test time and hence reflected-shock slug lengths were found to be considerably less than Mirels' theory predicts at the test conditions commonly used in high-pressure shock-tunnel work. Little or no sensitivity to p_1 could be detected in the experimental test-time data. The ratio of the experimental time to the inviscid time (i. e., τ/τ_i) multiplied by the Ls/d of the instrumentation station proved to be essentially constant for the range of test conditions reported herein. This is contrary to Mirels' theory which predicts a nonlinear dependence of τ/τ_i on the parameter $(p_1 d)^{1/4}$ / (Ls/d). Such results suggest that limiting slug lengths and hence maximum incident-shock test times probably occur at much higher values of p_1 and lower values of Ls/d than was formerly believed.

Comparison of the air and nitrogen data showed conclusively that combustion between the driver and driven gas, when hydrogen and air gas combinations are employed, is not responsible for the discrepancy in incident-shock test time observed. Combustion appears to occur in a region of mixed driver and driven gas that always exists behind the incident shock and for this reason combustion does not consume usable test gas. The end-wall pressure measurements, which provided an independent check both on the existence of a combustion region and on the location of the interface region near the end of the driven tube, consistently corroborated the radiation measurements for both air and nitrogen.

Measurements of shock-tube boundary-layer transition time showed that the test conditions of these experiments closely approximated Mirels' requirement that the boundary-layer behind the incident shock be fully turbulent. These data, which were obtained at a unit Reynolds number believed to be higher than has ever been reported in the open literature for high wall-cooling rates (i. e. $M_s \approx 8$), are in fairly good agreement with the transition Reynolds-number correlation of Hartunian, et al.[21]

The apparent reason for the discrepancy between the experimental and theoretical incident-shock test times is that the mass flow in the boundary layer is greater than Mirels' theory would predict. The problem probably stems from using a 1/7 power law for the velocity profile and an incompressible skin-friction law in developing this theory. However, definite proof of this conjecture awaits solutions based on a more reliable velocity profile such as defined by Coles [27] and a compressible skin-friction law such as given by Nash and Macdonald [28].

ACKNOWLEDGEMENTS

The assistance of W. Lorich and R. Kryszak in carrying out the experimental program is gratefully acknowledged. Thanks are also due to Dr. J. K. Dukowicz and Dr. M. G. Dunn with whom many informative discussions were held during the course of this work.

The research reported herein was supported by Cornell Aeronautical Laboratory internal research funding.

REFERENCES

1. Mirels, H. Shock Tube Test Time Limitations Due to Turbulent Wall Boundary Layer, Aerospace Rept. No. TDR-169 (3230-12) TR-3 May 6, 1963; also AIAA Journal Vol. 2, No. 1, Jan., 1964, pp. 84-93.
2. Mirels, H. Test Time in Low-Pressure Shock Tubes, Aerospace Rept. No. TDR-169 (3230-12) TN-5, Dec. 27, 1962; also The Physics of Fluids, Vol. 6, No. 9, Sept., 1963, pp. 1201-1214.
3. Brocher, E. F. Hot Flow Length and Testing Time in Real Shock Tube Flow, The Physics of Fluids, Vol. 7, No. 3, March, 1964, pp. 347-351.
4. Bertin, J. J. Analysis of Test Times and Boundary Layer Induced Property Variations in a Circular Shock Tube, NASA TN D-3759, Jan., 1967.
5. Roshko, A. On Flow Duration in Low Pressure Shock Tubes, The Physics of Fluids, Vol. 3, No. 6, Nov.-Dec., 1960, pp. 835-842.
6. Hooker, W. J. Testing Time and Contact-Zone Phenomena in Shock-Tube Flows, The Physics of Fluids, Vol. 4, No. 12, Dec., 1961, pp. 1451-1463.
7. Ackroyd, J. A. D. A Study of Running Times in Shock Tubes, Aeronautical Research Council Rept. C. P. No. 722, July 1963.
8. Camm. J. C. and Rose, P. H. Electric Shock Tube for High Velocity Simulation, Avco-Everett Research Laboratory, Research Report 136, July 1962.
9. Duff, R. E. Shock Tube Performance at Low Initial Pressure, The Physics of Fluids, Vol. 2, No. 2, March-April,1959, pp. 207-216.
10. Sandborn, V. A. Measurements of Flow Duration in a Low-Pressure Shock Tube, NASA TN D-1218, May, 1962.
11. Sandborn, V. A., Weisblatt, H., and Flagg, R. F. Test Time in a 1. 5-Inch Diameter High-Stagnation-Enthalpy Shock Tube, AIAA Journal, Vol. 1, No. 5, May, 1963, pp. 1236-1237.
12. Graber, B. C. and Nerem, R. M. Test Duration Measurements in an Arc-Driven Hypervelocity Shock Tube, The Ohio State University Research Foundation Rept. No. 1573-2, October 15, 1963.
13. Chapin, S. G. and Rumpel, W. F. Measurements of Uniform Flow Duration in a Chambered, Buffered Shock Tube, AIAA Journal Tech. Notes Vol. 2, No. 8, August, 1964, pp. 1499-1500.
14. Dunn, M. G. Experimental Shock-Tube Investigations of Conditions Behind Incident and Reflected Shocks In Air for Shock Mach Numbers Between 8. 5 and 16. 5, CAL Report No. AI-2187-A-3, October, 1966; also Fifth Hypervelocity Techniques Symposium, Denver, Colorado, March,1967.
15. Livingston, F. R. Experimentally Measured Effects of the Wall Boundary Layer on Shock-Tube Performance, Jet Propulsion Laboratory Tech. Report No. 32-714, March 1, 1965.

16. Roshko, A. and Smith, J. A. Measurements of Test Time in the GALCIT 17-Inch Shock Tube, AIAA Journal, Vol. 2, No. 1, Jan., 1964, pp. 186-187.
17. Appleton, J. P. and Musgrove, P. J. An Investigation of the Departure from Ideal Shock Tube Performance. Preliminary Results, Dept. of Aeronautics and Astronautics, University of Southampton, April, 1963.
18. Bird, K. D., Martin, J. F. and Bell, T. J. Recent Developments in the Use of the Hypersonic Shock Tunnel as a Research and Development Facility, Third Hypervelocity Symposium, Denver, Colorado, 1964.
19. Hilsenrath, J., Beckett, C. W., Benedict, W. S. et al. Tables of Thermal Properties of Gases, National Bureau of Standards Circular 564, November 1, 1955.
20. Vidal, R. J. Model Instrumentation Techniques for Heat Transfer and Force Measurements in a Hypersonic Shock Tunnel, Cornell Aeronautical Laboratory Rept. AD-917-A-1, February, 1956.
21. Hartunian, R. A., Russo, A. L. and Marrone, P. V. Boundary-Layer Transition and Heat Transfer in Shock Tubes, Cornell Aeronautical Laboratory Rept. AD-1118-A-3, December, 1959.
22. Kurzrock, J. W. Selection of Surface Thermometers for Measuring Heat Flux, Cornell Aeronautical Laboratory Rept. No. 124, February, 1963.
23. Flagg, R. F. Advances in Shock Tunnel Driving Techniques, Third Hypervelocity Techniques Symposium, Denver, Colorado, 1964, pp. 89-115.
24. Lewis, C. H. and Burgess, E. G. Charts of Normal Shock Wave Properties in Imperfect Air [Supplement: $M_S = 1$ to 10] AEDC-TR-65-196, Sept., 1965; also Numerical Tabulations of Air Normal Shock Properties provided by private communications with C. H. Lewis.
25. Lewis, C. H. and Burgess, E. G. Charts of Normal Shock Wave Properties in Imperfect Nitrogen [Supplement: $M_S = 1$ to 10] AEDC-TR-66-116, September, 1966; also Numerical Tabulations of Nitrogen Normal Shock Properties provided by private communications with C. H. Lewis.
26. Dunn, M. G. Experimental Study of Reflected-Shock Tunnel Flows in the Equilibrium Interface Region, Cornell Aeronautical Laboratory Rept. No. AM-1702-A-1, January, 1968.
27. Coles, Donald. The Law of the Wake in the Turbulent Boundary Layer, Journal of Fluid Mechanics, Vol. 1, Part 2, p. 191, July 1956.
28. Nash, J. F. and Macdonald, A. G. J. A Turbulent Skin-Friction Law for Use at Subsonic and Transonic Speeds, National Physical Laboratory, England NPL Aero Report 1206, A.R.C. 28 234, July 29, 1966.

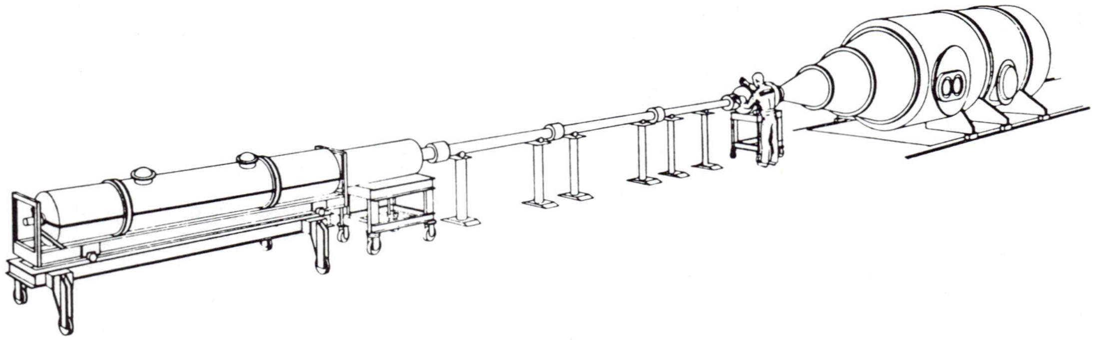

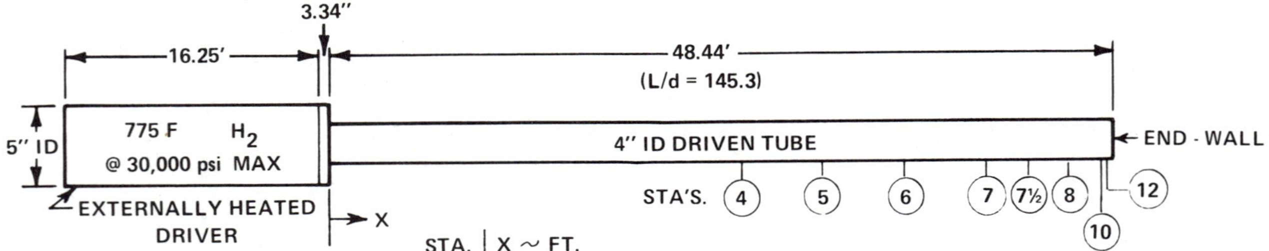

STA.	X ~ FT.
4	25.67
5	30.67
6	35.67
7	40.67
7-1/2	43.17
8	45.67
10	47.87
12	48.37

Figure 1 BASIC COMPONENTS OF THE CAL HYPERSONIC SHOCK TUNNEL 96″ HIGH ENERGY LEG

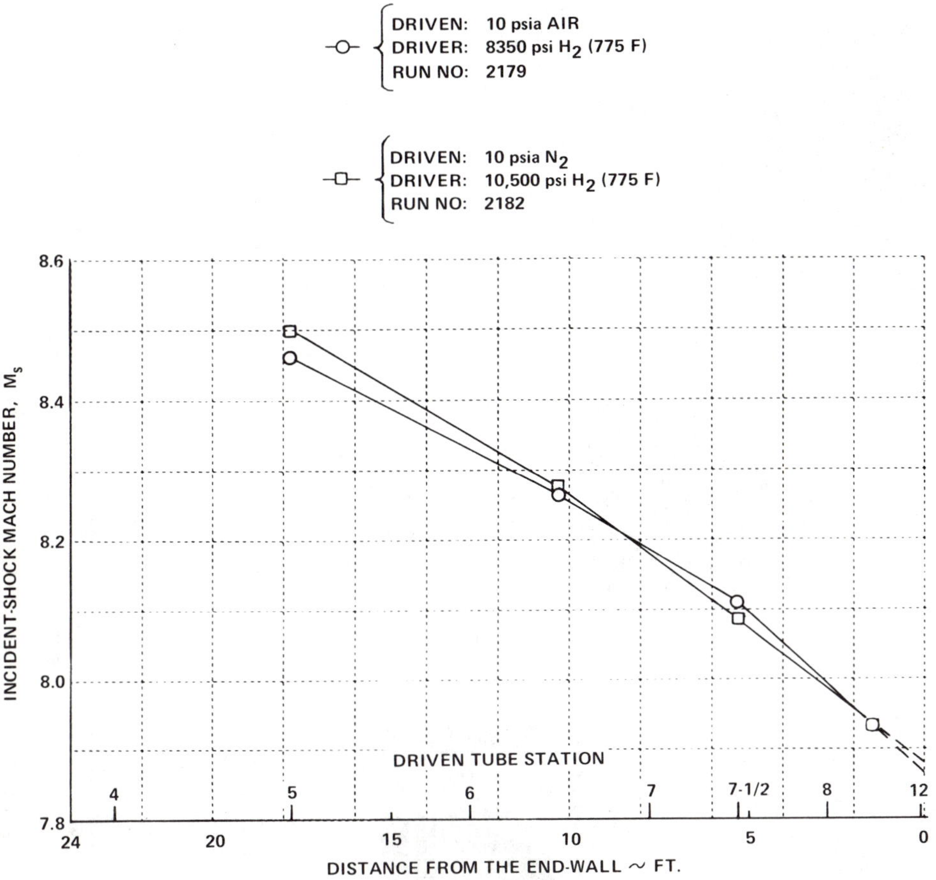

Figure 2 TYPICAL ATTENUATION CHARACTERISTICS OF THE INCIDENT SHOCK

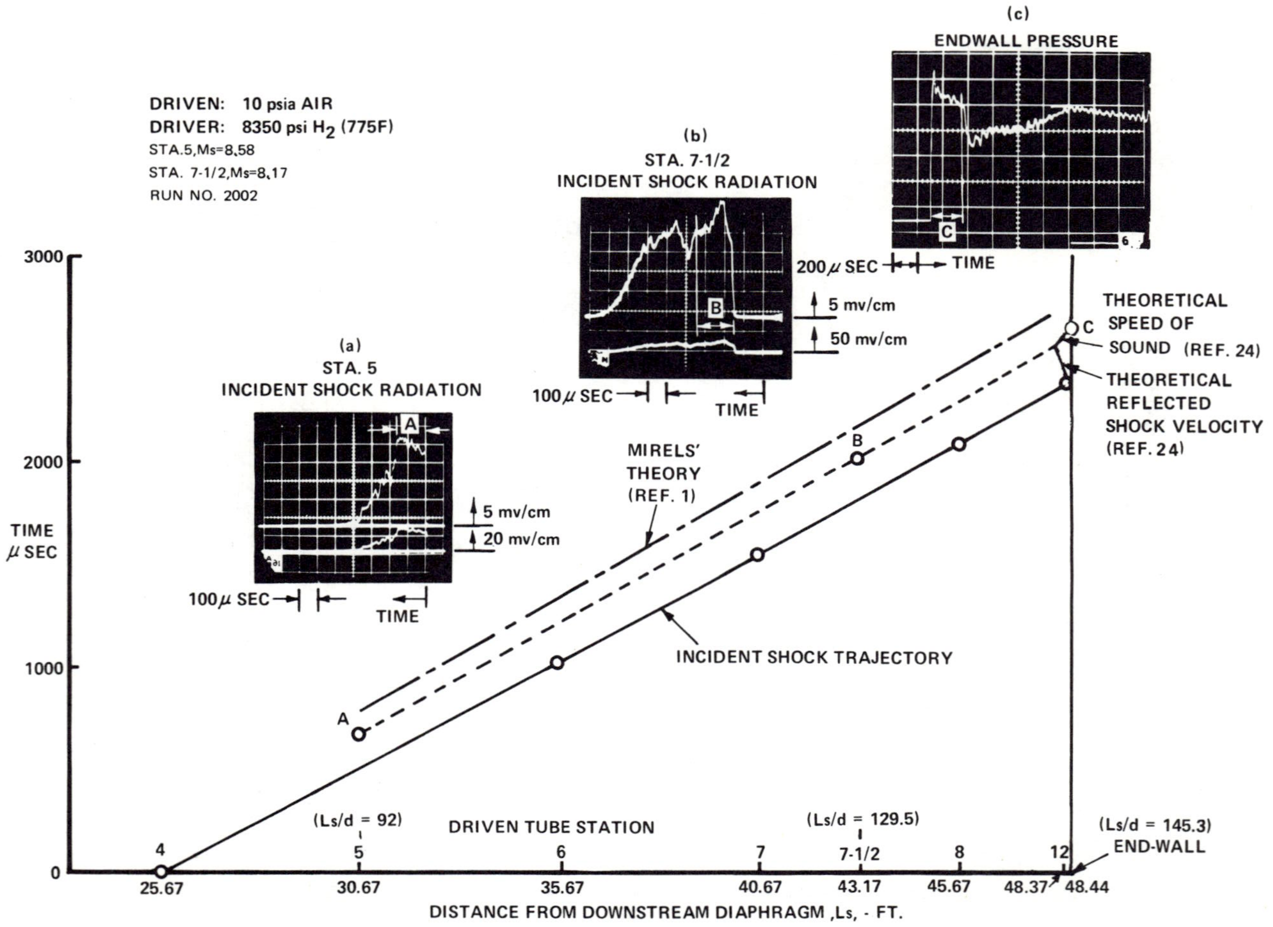

Figure 3 WAVE DIAGRAM CONSTRUCTED FROM EXPERIMENTAL DATA

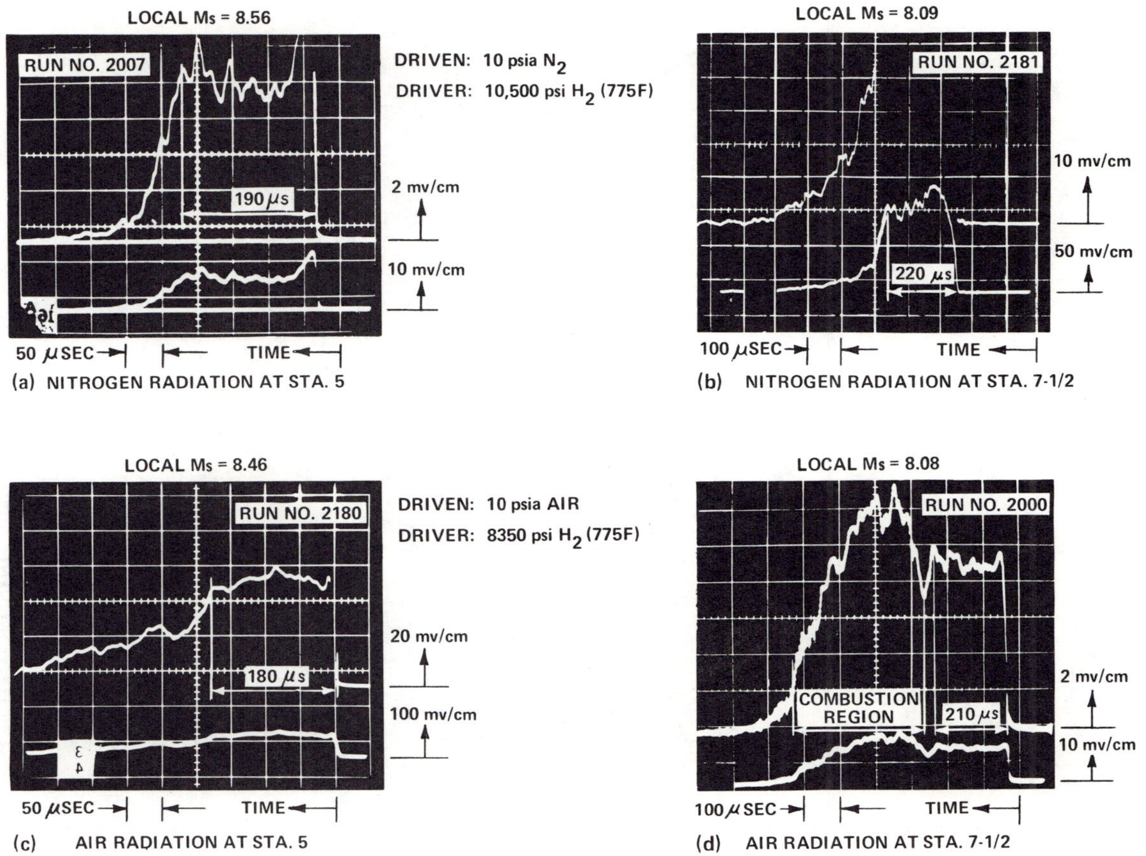

Figure 4 INCIDENT SHOCK RADIATION FOR NITROGEN AND AIR

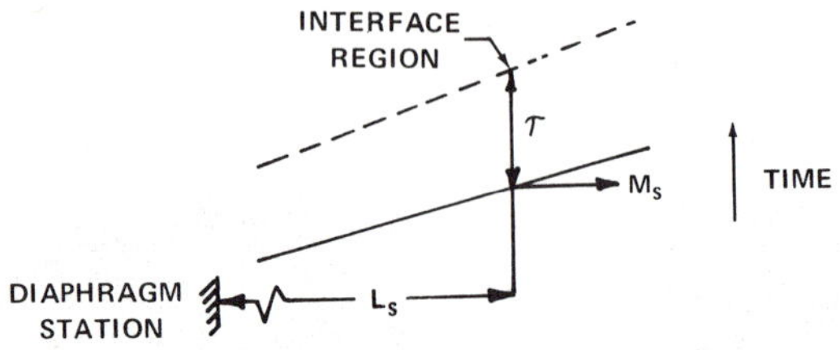

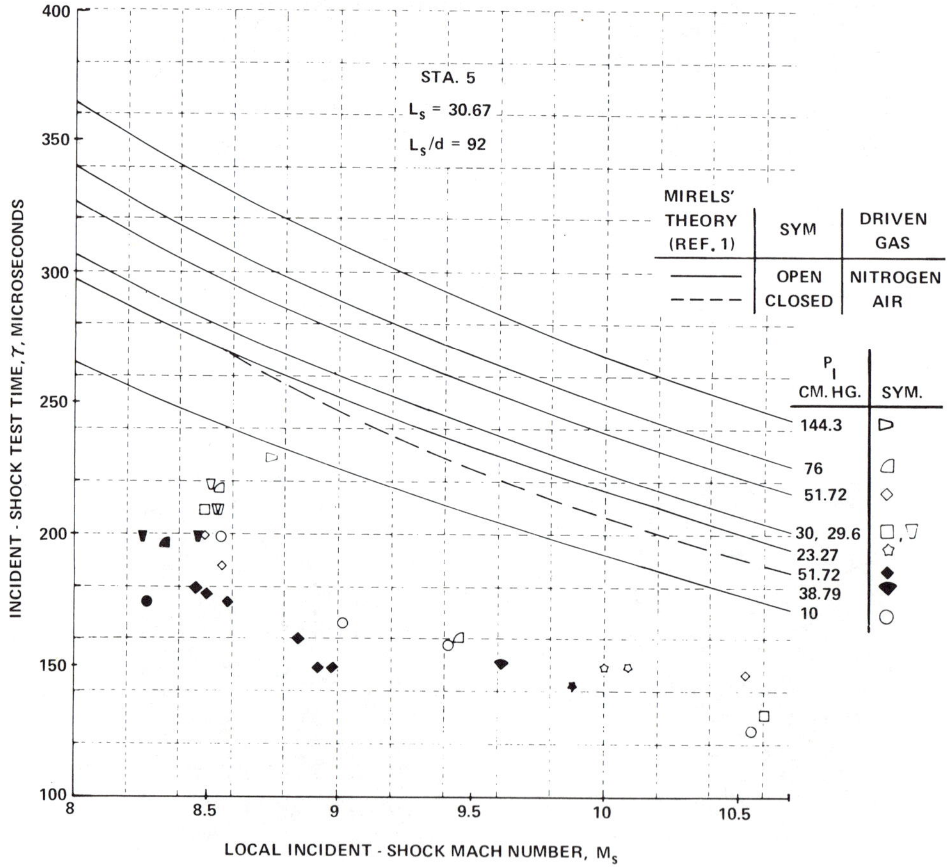

Figure 5 INCIDENT - SHOCK TEST TIME AT STA. 5

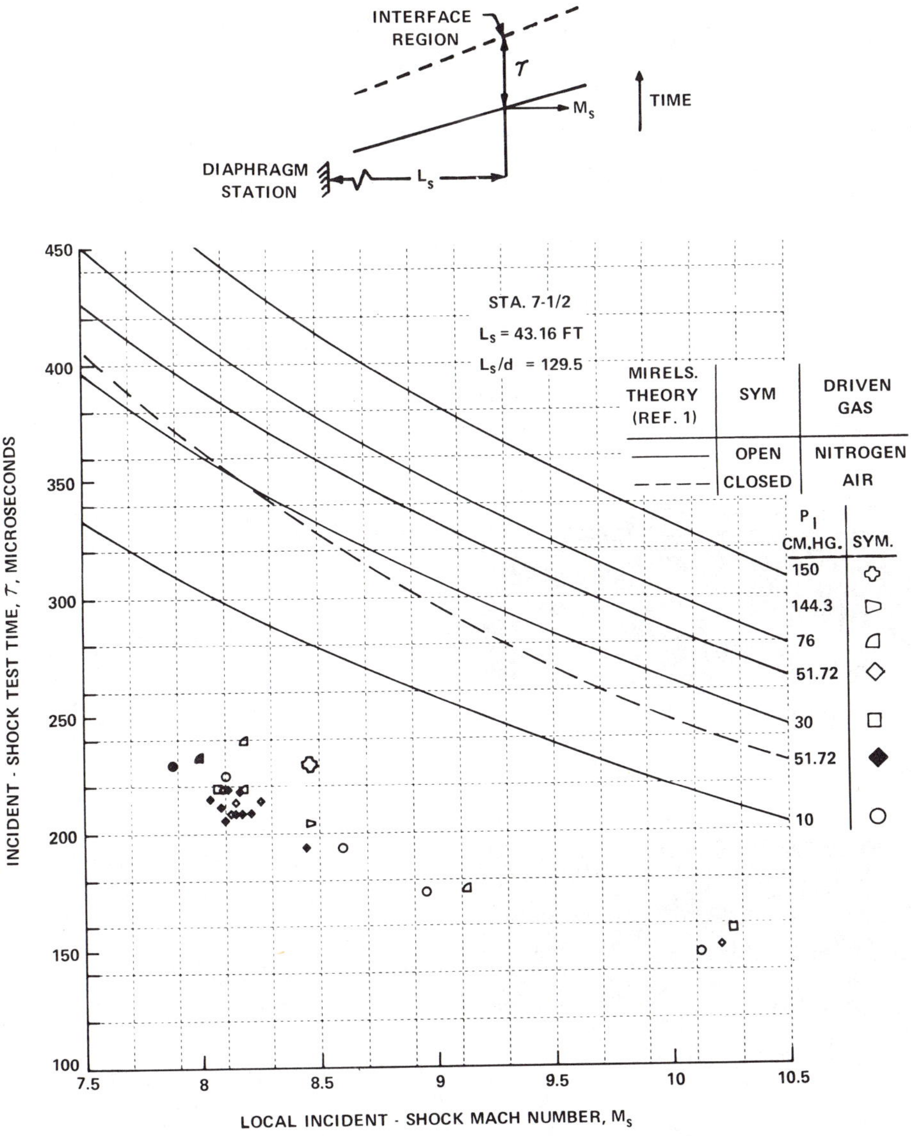

Figure 6 INCIDENT - SHOCK TEST TIME AT STA. 7-1/2

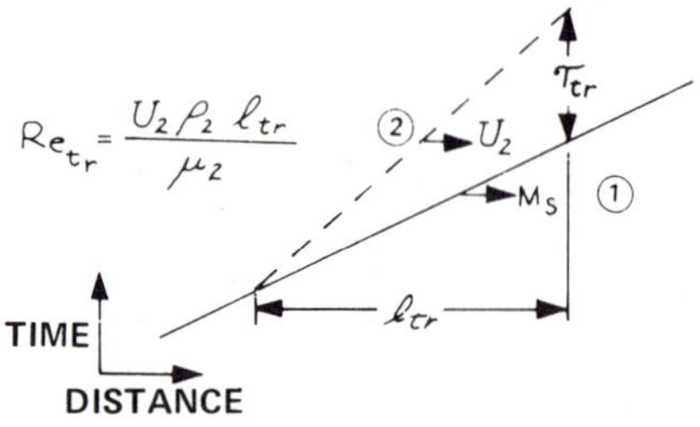

AIR

P_1 = 10 psia

STA 10

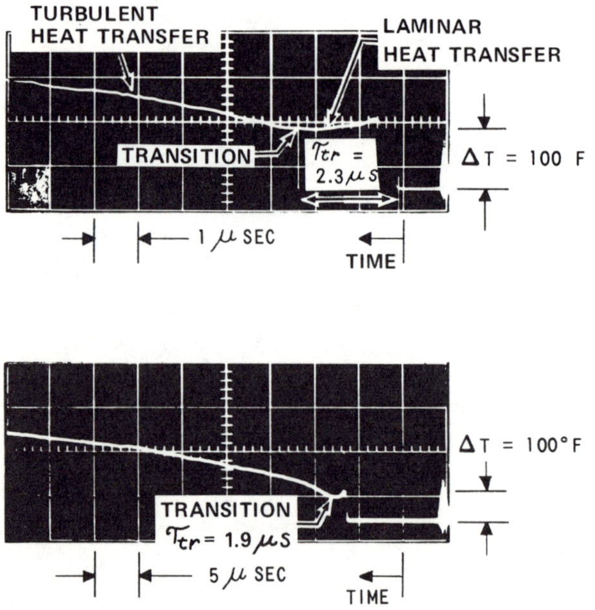

DRIVER: 8350 psi H_2 (775F)
LOCAL M_s = 7.86
Re/FT = 53 x 10^6
Re_{tr} = 6.76 x 10^6
RUN NO. 2180

DRIVER: 8200 psi H_2 (775F)
LOCAL M_s = 7.97
Re/FT = 53.6 x 10^6
Re_{tr} = 5.76 x 10^6
RUN NO. 2177

Figure 7 TYPICAL WALL - TEMPERATURE RECORDS SHOWING BOUNDARY - LAYER TRANSITION

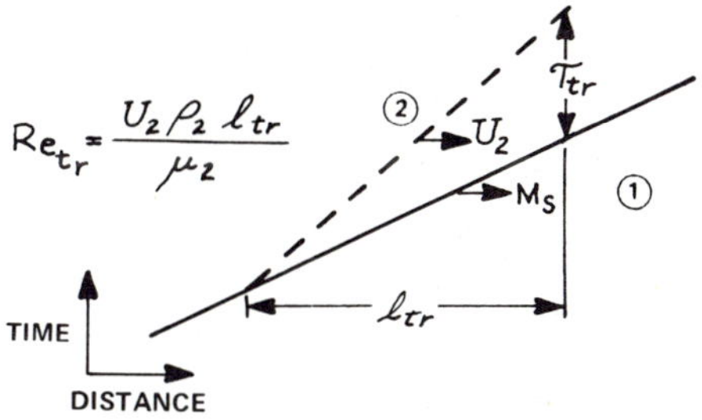

SYM	DRIVEN GAS	P_1 cm. Hg	M_s	UNIT REYNOLDS NUMBER Re/FT x 10^{-6}	τ_{tr} μ SEC
●	AIR	51.72	7.9	53	1.4 - 3.4
○	N_2	↓	7.96	48.6	2.5
■	AIR	10	7.6	10	1.6
□	N_2	↓	8.7	9.7	1.2

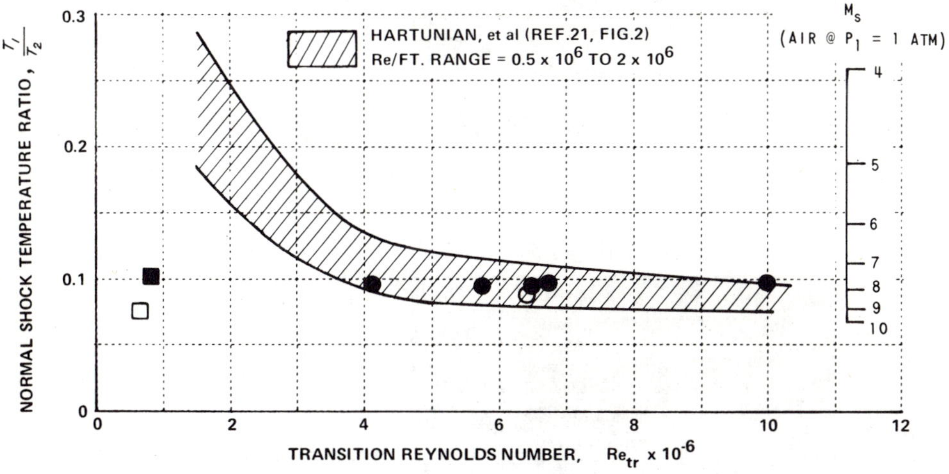

Figure 8 SHOCK-TUBE BOUNDARY-LAYER TRANSITION REYNOLDS NUMBER

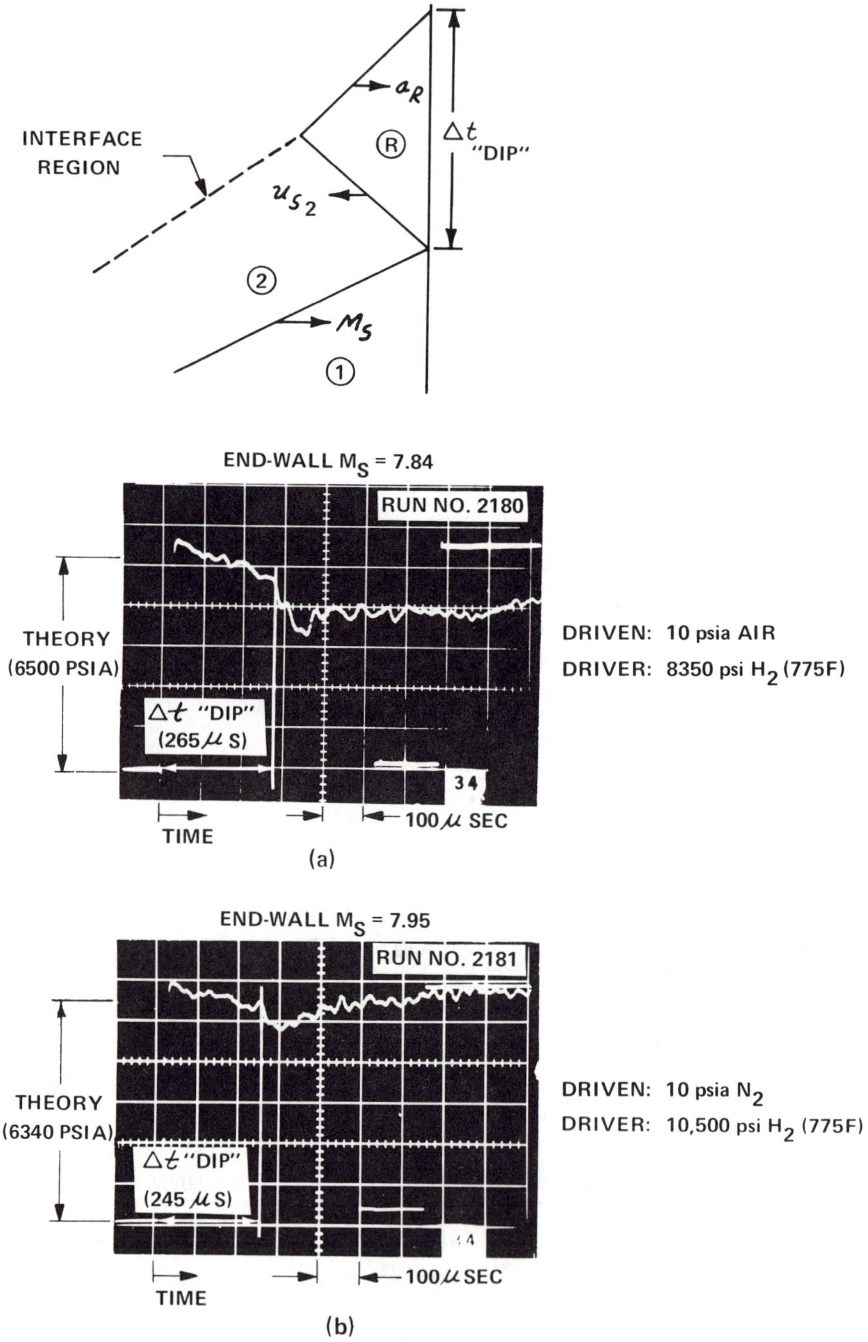

Figure 9 END-WALL REFLECTED - SHOCK PRESSURE

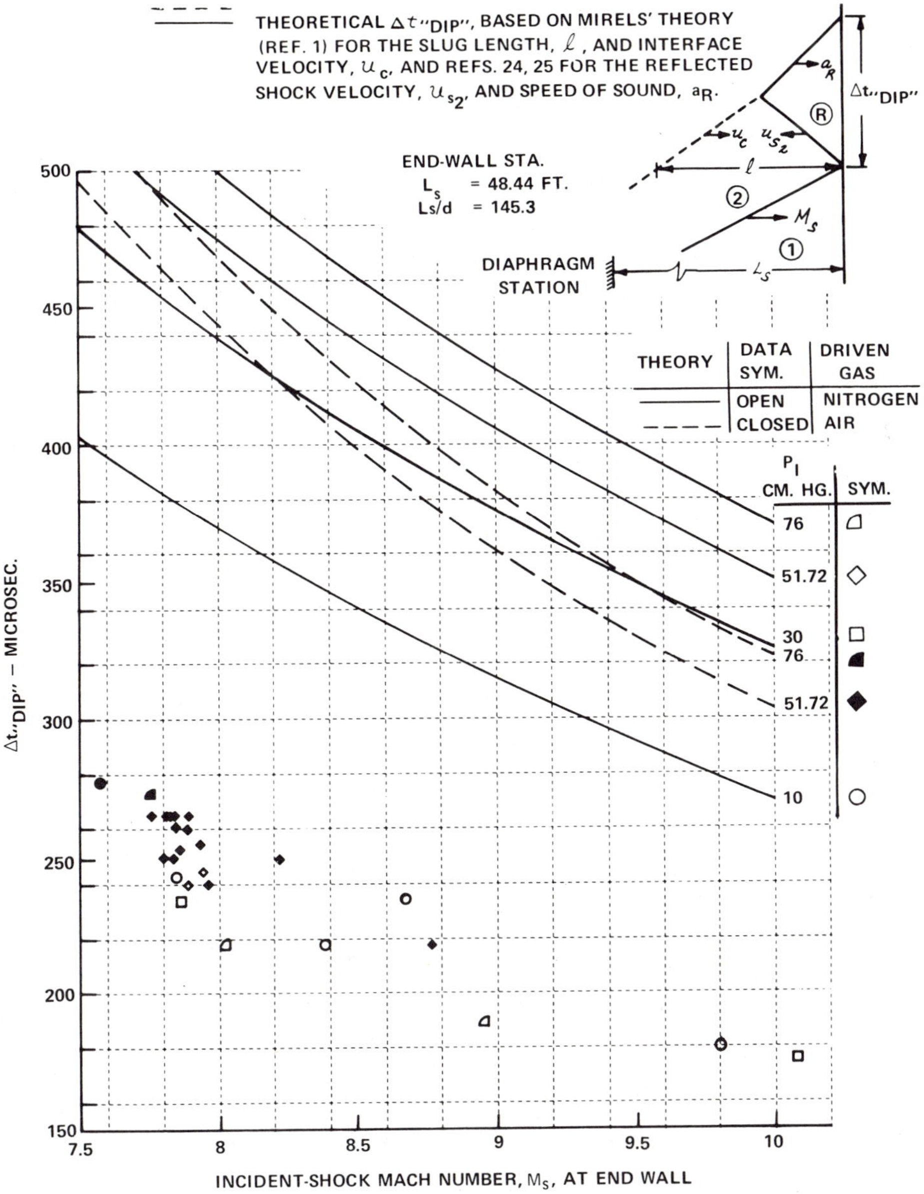

Figure 10 COMPARISON OF EXPERIMENTAL AND THEORETICAL TIME FOR THE "DIP" IN END WALL PRESSURE

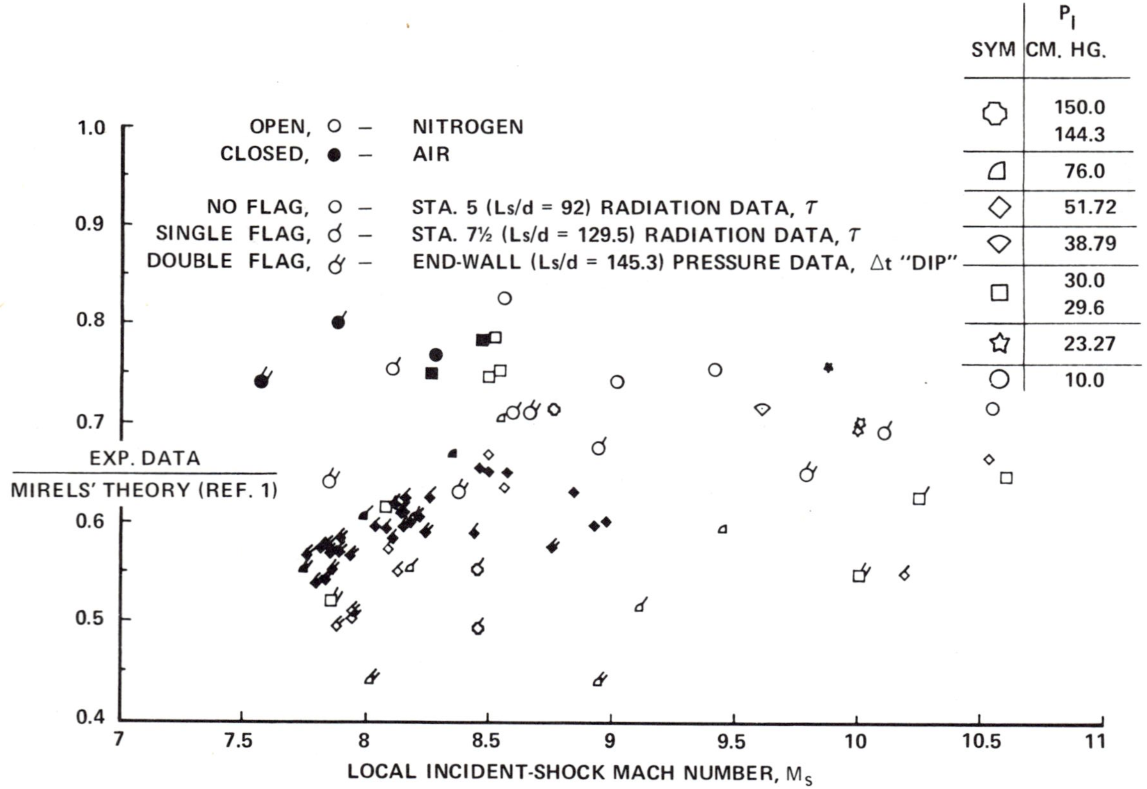

Figure 11 RATIO OF EXPERIMENTAL DATA TO MIRELS' THEORETICAL INCIDENT-SHOCK TEST TIME, τ, AND END-WALL Δt "DIP"

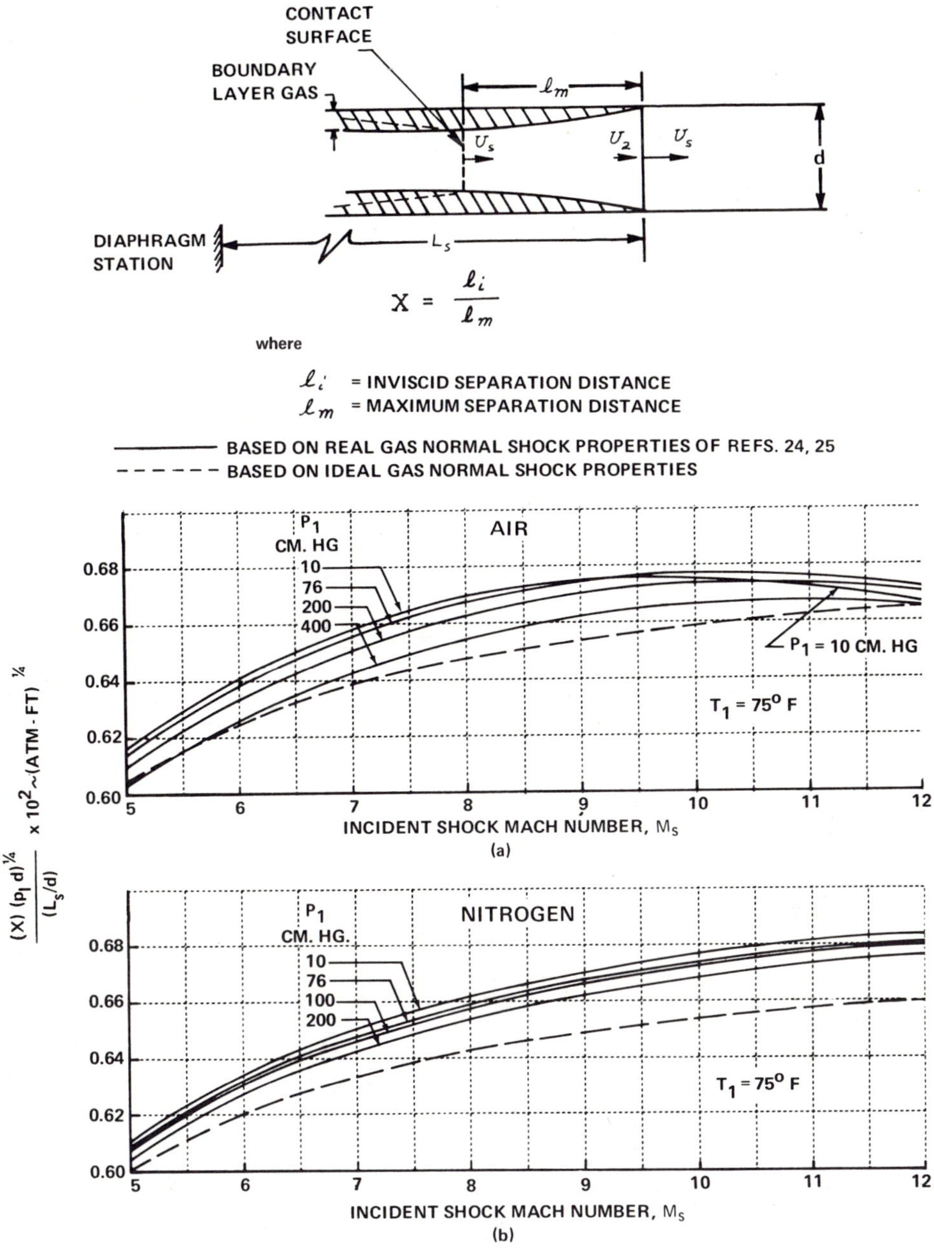

Figure 12 NORMALIZED FORM OF MIRELS' MAXIMUM SEPARATION DISTANCE PARAMETER, X (REF. 1)

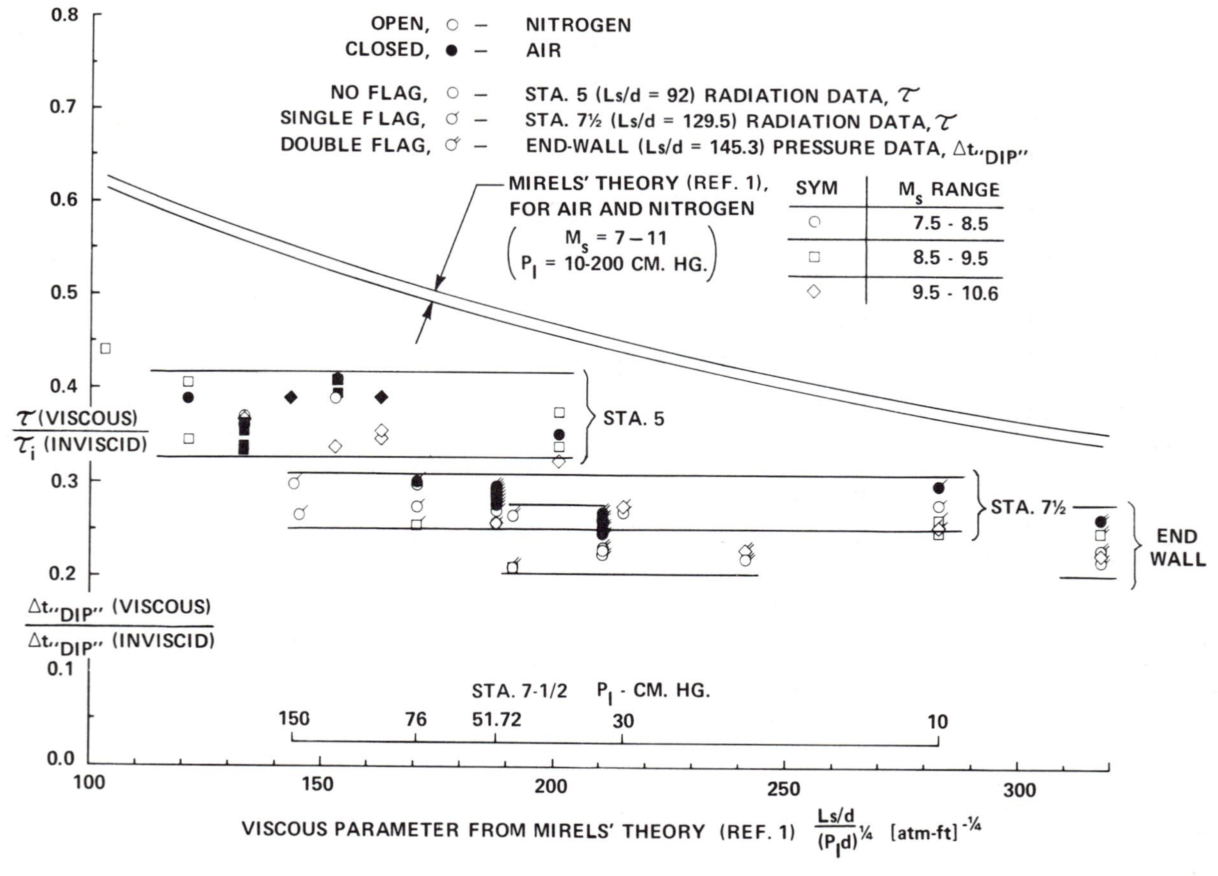

Figure 13 RATIO OF VISCOUS TO INVISCID INCIDENT-SHOCK TEST TIME AND $\Delta t_{\text{"DIP"}}$

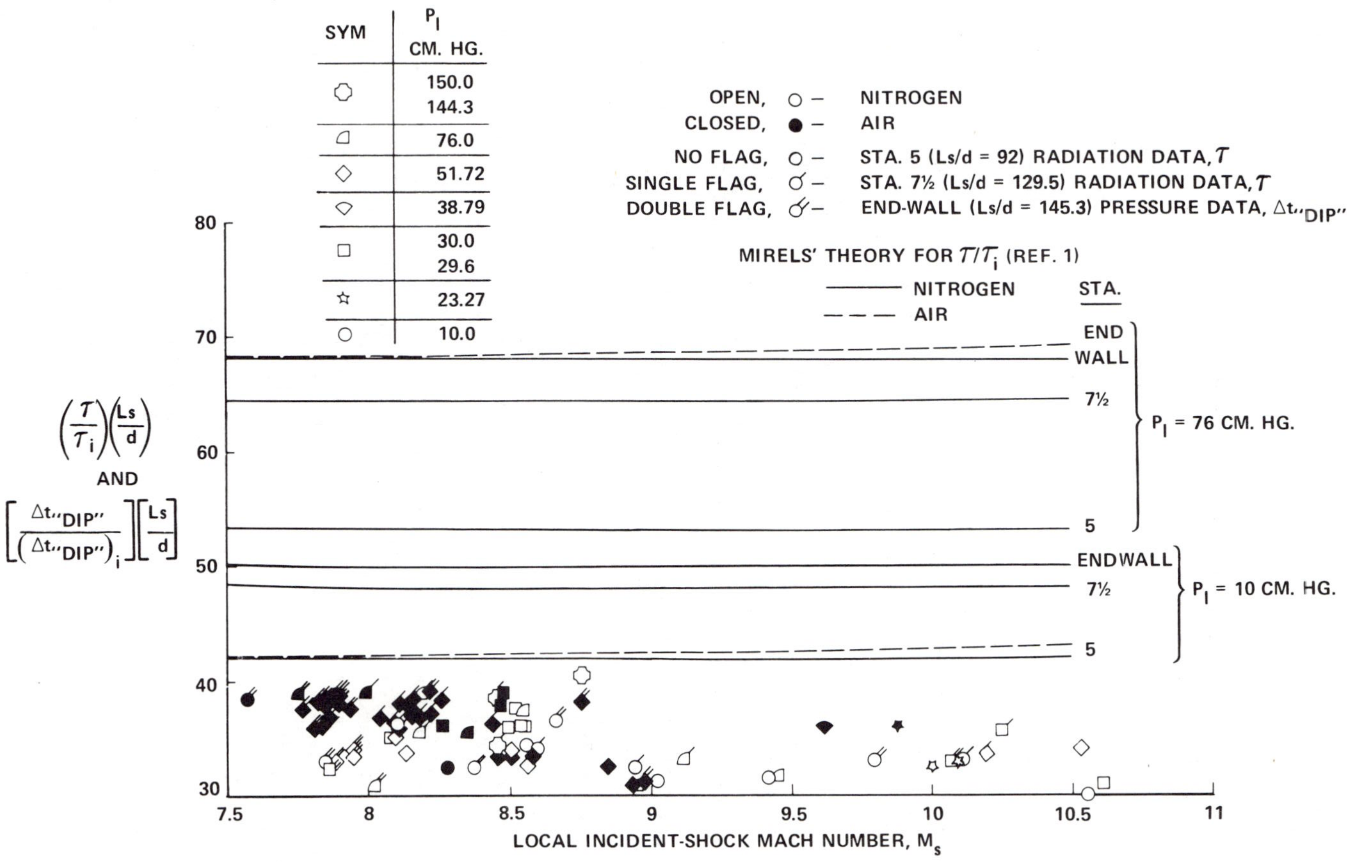

Figure 14 CORRELATION OF INCIDENT-SHOCK TEST TIME FOR TURBULENT BOUNDARY LAYER AND LARGE Ls/d TEST CONDITIONS

EXPERIMENTAL STUDY OF THE BOUNDARY LAYER EFFECTS ON THE DISTRIBUTION OF FLOW PARAMETERS BEHIND A SHOCK WAVE IN A SHOCK TUBE

T. V. BAZHENOVA I. M. NABOKO and R. G. NEMKOV
Institute for High Temperatures, Academy of Sciences of the USSR, Moscow

SUMMARY

According to the ideal theory of the shock tube, the flow parameters behind the shock front remain constant up to the contact surface, which moves with a velocity equal to the gas velocity behind the shock wave. A closer examination of the flow parameters behind the shock wave has revealed that these parameters vary from point to point in this region. In particular, the parameters determined from the conservation laws for an ideal gas are valid only at the shock front. Indeed, all along the length of this region, there is a boundary layer growth on the tube walls. This phenomenon gives rise to variations in the parameters in the nonviscous flow core as well.

The interaction of the nonviscous flow core with the boundary layer in a shock tube has been calculated theoretically by several authors. A most complete analysis of the problem is to be found in the known studies of Mirels. In particular, the problem of nonuniformity in the hot flow in a shock tube has been considered by this author in Ref. 7. The nonuniformity estimate given there is based on a few assumptions which are valid only in extreme cases. The first assumption is that the shock front velocity (U_s) is constant:

$$U_s = \text{const} \qquad (1)$$

Furthermore, U_s is assumed to be constant for such a long portion of the tube (up to the point where theoretical estimates are made) that the effects related to the history of the initial nonstationary wave propagation can be safely ignored. It is also supposed that the shock wave propagates with an induced "limiting length" when the contact surface velocity ($U_{2,c}$) is equal to U_s:

$$U_{2,c} \simeq U_s \qquad (2)$$

Even under these conditions, the distribution of flow parameters in the hot gas as obtained by Mirels is approximate. The accuracy of this prediction improves with increasing shock Mach number M_s, expressed by the shock wave velocity; that is, it is supposed that

$$1/M_s \ll 1 \qquad (3)$$

However, when the shock wave velocity is very high, this solution is no longer valid, as it was obtained under the assumption of constant gas heat capacity and gas composition behind the shock wave, whereas the assumptions of an ideal stationary state (Ref. 1 and 2) in real flows in the shock tube are satisfied only approximately. Thus the assumptions underlying the study of Mirels[1] in a real flow in the shock tube are fulfilled only approximately. Only a few studies have been devoted to experimental research on the nonuniformity of the hot flow due to the above interaction effect (Refs. 2 to 4). We therefore considered it desirable to carry out an experimental determination of the distribution of certain flow parameters behind a shock wave in nitrogen, with shock Mach numbers (M_s) ranging from 3 to 7.

EXPERIMENTAL TECHNIQUE

The experiments were performed in a square shock tube of 4x4 cm^2 cross-section. The distance between the test section and the diaphragm was 65 calibres. The flow in the test section was observed with a schlieren system.

The parameters to be measured were the flow velocity (U_2) and the flow Mach number (M_2). These parameters are quite sensitive to the presence of a boundary layer and are little affected by flow fluctuations. Therefore, a comparison of these quantities with calculations made according to Mirels' theory is quite meaningful. Here we have made use of the method of measuring the flow velocity by recording the propagation of an artificial density inhomogenity (schliere) produced by an electric discharge in the flow behind the shock. The discharge was produced by a breakdown of the gap between the electrode on the tube wall and the grounded surface of the semi-wedge, placed in the flow path within the shock tube 2 cm from the tube wall (Fig. 1a, b).

The optimum energy of the discharge was chosen (C = 50 pF, V = 50 kv), so as to visualize the nonuniform density of the hot gas and at the same time to produce a sufficiently weak density disturbance that the region including the inhomogeneity should move with practically the same velocity as the main flow. As is known, the region containing a density disturbance generally propagates with a speed equal to that of the main flow only at the initial moment. Later on, the nonuniformity may either lag behind or overtake the main flow, depending on the sign and the amplitude of the density disturbance. In the present experiments the observation time was short enough (20 μ sec) and, as shown by experiment, no schlieren velocity change was observed during this time (Fig. 2). Therefore, the error in velocity measurement was primarily due to the apparatus and included the inaccuracy in measuring the inclination angle of the schlieren traces in scanning and in determining the speed of the drum.

The Mach number was determined from the angle of inclination of the Mach line on the semi-wedge (Fig. 3) and particularly from the Mach line trace on scanning (Fig. 2). The accuracy of determination of the flow Mach number from the angle of inclination was $\pm$ 2 per cent, whereas that from scanning was $\pm$ 3 per cent.

RESULTS

Measurements were made of the values of U_2 and M_2 behind the shock wave front in N_2, with a shock Mach number (M_s) ranging from 3 to 7 and an initial pressure (P_1) in the low-pressure chamber of about 20 mm Hg. The values of U_2 and M_2 obtained were found to exceed those calculated from conservation laws ignoring the boundary layer effects. This is illustrated in Figs. 4 to 6, which give the experimental values of U_2, M_2, and C_2 the speed of sound ($C_2 = U_2/M_2$) corresponding to the first spark schliere (i.e. 35 μ sec after passage of the shock).

The calculated values in Figs. 4 to 6 correspond to the assumption of fully-frozen vibration[6]. Comparison between the "proper" time of a gas species (at the moment of measurement) and the values of vibrational relaxation as measured in Refs. 7 and 8 shows that, in the present experiments, the assumption of complete vibrational freezing holds true for most of the hot flow.

Figs. 7 and 8 give representative values of U_2 and M_2 measured all along the hot flow. The theoretical curves were calculated according to Mirels' method[1], where any physical flow parameter L is known to be a function of $X = x/\ell m$, where ℓm is the "limiting length" of the hot flow and x is the coordinate along the hot flow, that is,

$$L = L(X),\ X = x/\ell_m \qquad (4)$$

As in Ref. 5, we did not calculate the value of the limiting length ℓ_m but took the value directly from experiment and introducing for the finiteness of $\Delta_{sc}U$:

$$\Delta_{sc}U = (U_c - U_{2c}) \neq 0 \qquad (5)$$

That is, it is assumed here, just as it was in the original studies of Mirles (e.g. Ref. 9), that provided $\Delta_{sc}\ U/U_s$ is small, the profile of any L (in the shock reference frame) is the same as for the shock flow with the limiting length, the only difference being that because of (5) $\ell_c < \ell_m$, where ℓ_c is the contact surface coordinate. In the theoretical value of $U(X_c)$ (where $X_c = \ell_c/\ell_m$) is now equated to the contact surface velocity U_c, measured from experiment

$$U(X_c) = U_c \qquad (6)$$

then, given (U_c, X), it is possible to determine ℓ_m, which provides all we need to carry out our calculations. It goes without saying, however, that such a procedure of normalizing the U distribution to a known value of U_c is not equivalent to a theoretical calculation of ℓ_m.

The present study did not depend on whether the boundary layer was laminar or turbulent. The information available in the literature indicates that a considerable portion of the hot flow is expected to have a turbulent boundary layer (see e.g., Ref. 10). Two calculated curves have therefore been retained in each of Figs. 7 and 8, one corresponding to laminar (L) and the other to turbulent (T) boundary layers, beginning just behind the shock wave.

It may seem paradoxical that the calculated flow velocity increases more steeply with increasing x for a laminar boundary layer than for a turbulent one. In fact, however, it should be remembered that both curves refer to the same hot flow length, so that they are in a sense not comparable. For example, if the turbulent boundary layer develops, there can be no laminar boundary layer at all with such a short hot-flow region.

CONCLUSION

Comparison of the experimental values of U_2 and M_2 along a hot flow behind a shock wave in N_2, for $P_1 = 20$ mm Hg and $3 \leq M_s \leq 6$ with data calculated according to Mirels' theory[1] shows fair agreement between the two. That is, for most of the hot flow, except probably near the immediate vicinity of the contact surface, the experimental values (Figs. 7 and 8) agree with the calculations within an accuracy of $\pm$ 3% (i.e., the greatest departure from the calculated curves for laminar or turbulent boundary layer does not exceed 3%).

More accurate comparison with the theory could be made if the point of transition between the laminar and turbulent boundary layer were known. It may be supposed, for instance, that the distribution of experimental points in Fig. 8 can be attributed to just this transition. However, the present study,was not designed to determine the point of transition experimentally. Moreover, the hot flow length was also taken from experiment, not from calculations. Nevertheless, considering the approximations underlying the calculation in relation to the experimental accuracy of flow parameters (U_2, M_2), the agreement between theory and experiment may be considered satisfactory.

REFERENCES

1. H. Mirels, Flow Nonuniformity in Shock Tubes Operating At Maximum test times, Phys. Fluids 9, 1907 (1966).
2. J. N. Fox, T. L. McLaren, and R. M. Hobson, Test Time and Particle Paths in Low-Pressure Shock Tubes, Phys. Fluids 9, 2345 (1966).
3. M. P. Syshchikova, A. N. Semenov, and M. K. Berezkina, Measurement of Flow Parameters in Shock Tube. Collection of papers. "Aerophysical Studies of Supersonic Flows", Nauka, Moscow-Leningrad, 1967.
4. W. H. Christiansen, Measurement of Flow Uniformity in Shock Tubes Operating at Low Initial Pressure, Phys. Fluids, 11, 2279 (1968).
5. R. G. Nemkov, Flow Velocity as Measured in Shock Tube by Spark Technique. Collection of papers "Physical Gas Dynamics of Ionized and Chemically Reacting Gases", Nauka, Moscow, 1968.
6. I. M. Naboko, T. V. Bazhenova, and R. G. Nemkov, Experimental Studies of the Parameters of the Flow Behind a Shock Wave Moving in Relaxing Gas, First International Colloquium on Gas Dynamics of Explosions, Astronautica Acta, 14, 5, 497 (1969).
7. R. C. Millikan and D. R. White, Vibrational Energy Exchange Between N_2 and CO - The Vibrational Relaxation of Nitrogen, J. Chem. Phys. 39, 98 (1963).
8. I. R. Hurle, Line Reversal Studies of the Sodium Excitation Process Behind Shock Waves in N_2, J. Chem. Phys. 41, 3911 (1964).

9. H. Mirels, Test Time in Low-Pressure Shock-Tubes, Phys. Fluids, 6, 1201 (1963).
10. J. C. Breeze and C. C. Ferriso, Duration of Laminar Flow in a Shock Wave Boundary Layer, Phys. Fluids 7, 1071 (1964).

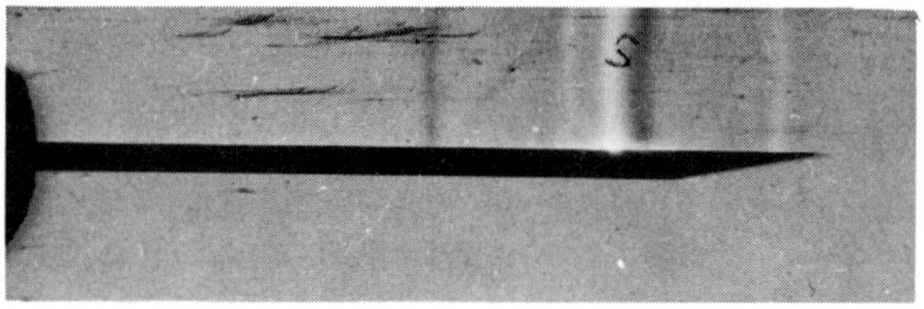

a) Photograph of spark schliere S in absence of shock wave

b) Photograph of spark schliere S behind the shock wave (P_2 = 1 atm)

Figure 1
Spark Photography

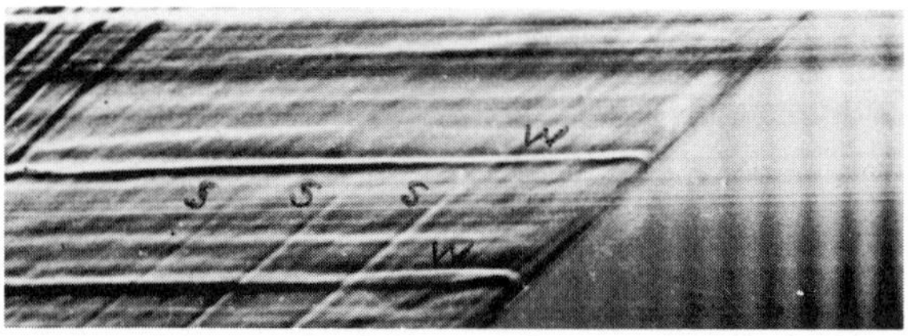

Figure 2
Scanning of traces of artificial nonuniformities (S_1-S_3) and the Mach line (M) behind the shock wave

Figure 3
Instantaneous photograph of the Mach line in flow streamlining over semiwedge behind the shock wave

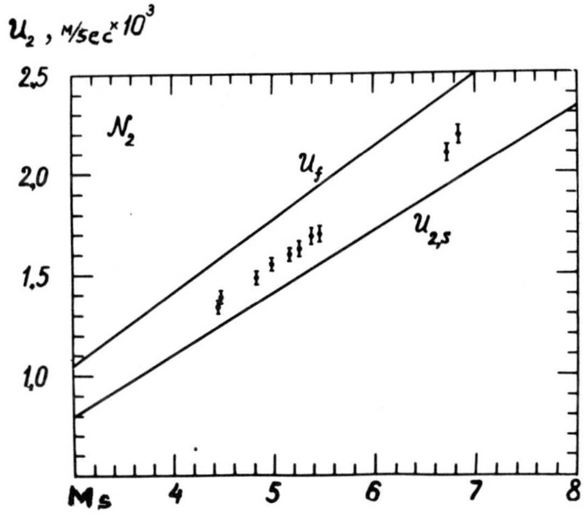

Figure 4
Flow velocity behind the shock wave

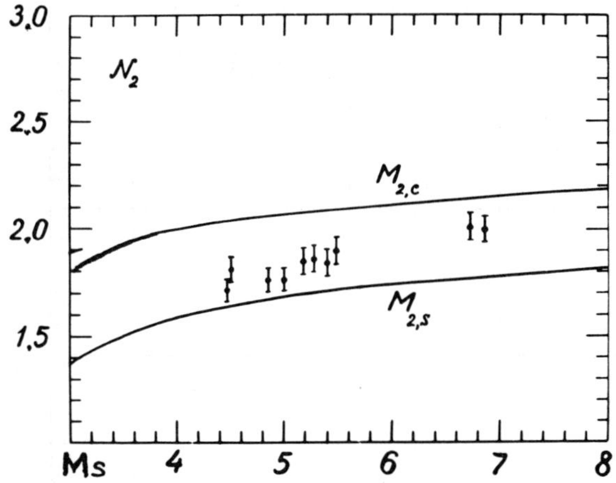

Figure 5
Flow Mach number behind the shock wave

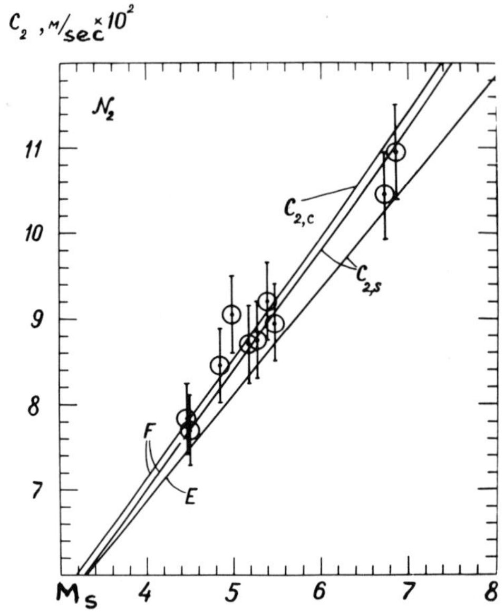

Figure 6
Sound speed behind the shock wave

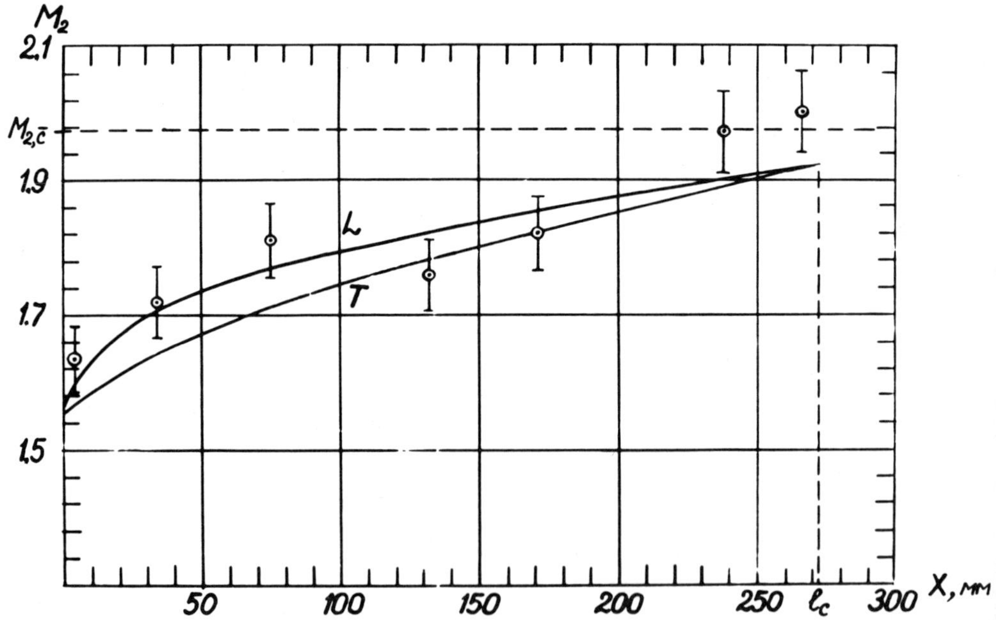

Figure 7
Mach number distribution along the hot flow (M_s = 4.09; P_1 = 27 mm Hg)

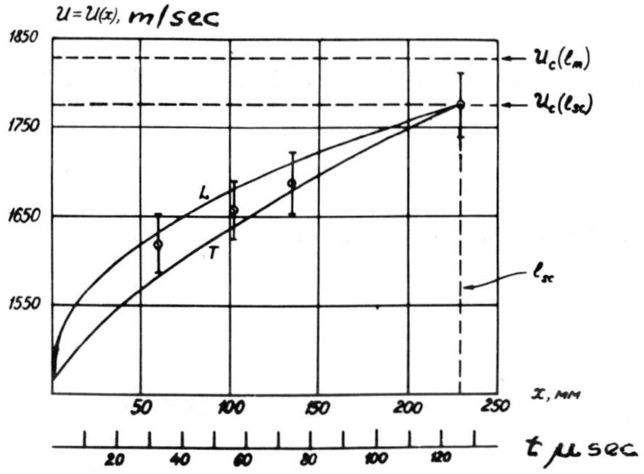

(a)

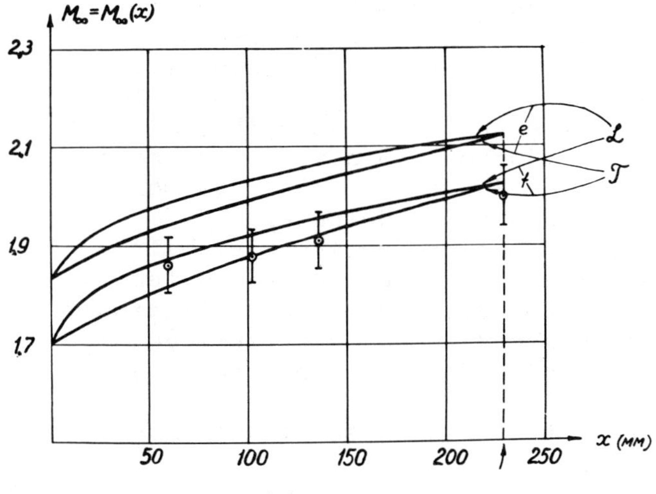

(b)

Figure 8 Velocity distributions along the hot flow (a) Flow Velocity (b) Mach number
(M_s = 5.18, P_1 = 20 mm Hg; e - complete equilibrium; f - frozen vibration)

SOME UNSTEADY AND NON-LINEAR EFFECTS IN THE SHOCK-WAVE REFLECTION PROBLEM

W. FISZDON, J. GOMULKA, H. PACZYNSKA
Institute of Basic Technical Problems, Polish Academy of Sciences, Warsaw

For the analysis of the flow behind a reflected shock-wave and of the heat transfer at the end wall of a shock-tube it is very useful to choose an adequate simple model and to formulate correctly the boundary conditions at the gas-solid interface.

Close to the center of the end wall of a shock-tube, for a period after reflection which is smaller than the time necessary for disturbances from the side walls to propagate to the considered region, whose dimensions are much larger than the mean free path, a one-dimensional continuum model seems appropriate. Limiting ourselves to the simplest case of thermodynamic equilibrium, without chemical reactions and radiation, when the pressure at the wall after shock reflection can be taken as constant, a boundary layer approximation is adequate.

This greatly idealised model gives the possibility of assessing some of the effects due to very rapid temperature variations and subsequent changes of transport properties of the gas as experienced in shock-tube measurements and of gas-solid interaction effects at the interface.

As shown in [1], [2], using the Mises transformation

$$\xi = \int_0^x \frac{\rho(x,t)}{\rho_5}\, dx, \qquad (1)$$

the gas-dynamic equations are reduced in this case to the heat-transfer equation

$$\frac{\partial T}{\partial t} = K_5 \frac{\partial}{\partial \xi}\left[\left(\frac{k\rho}{k_5\rho_5}\right)\frac{\partial T}{\partial \xi}\right] \qquad (2)$$

where T is the temperature; x is the distance along

the shock-tube axis measured from the end wall; k, ρ are the heat conductivity and density; $K_5 = \frac{k_5}{c_p \rho_5}$, the subscript 5 denotes conditions behind an ideal reflected shock-wave and 1 the conditions in front of the incoming shock-wave.

For the solid, the heat transfer equation is:

$$\frac{\partial S}{\partial t} = K_s \frac{\partial^2 S}{\partial x^2} , \tag{3}$$

where S denotes the temperature of the solid and the subscript "s" denotes the appropriate parameters of the solid, whose thermal properties are assumed constant.

The boundary conditions which are usually applied at the gas-solid interface are the heat flow continuity condition

$$- k_s \frac{\partial S(0,t)}{\partial x} = k \frac{\partial T(0,t)}{\partial x} + \rho_w u_w q_{ad} , \tag{4}$$

where the last term takes care of the heat of adsorption, u is the mean velocity at the wall surface and q_{ad} is the specific heat of adsorption.

The second boundary condition at the interface is the "temperature slip condition". Usually the temperature slip is assumed to be zero. In some cases [2], [3] the temperature slip is taken as proportional to the temperature gradient at the wall, according to the stationary Maxwell-Knudsen-Smoluchowski theory [4], [6]. As in the case of shock-tube experiments immediately after shock-reflection, the time and space derivatives of the temperature, at the interface, are very large, and the following relation for the temperature slip is assumed:

$$T(0,t) - S(0,t) = \sigma \frac{\partial T(0,t)}{\partial x} - \theta \frac{\partial T(0,t)}{\partial t} + \beta \frac{\partial^2 T(0,t)}{\partial t \partial x} , \tag{5}$$

where the first term is the usual "stationary" one, the characteristic slip length σ is of the order of the mean free path

$$\sigma = \frac{2 - \alpha}{\alpha} 2 \frac{k}{\rho c_v \bar{c}} , \tag{6}$$

where α is the temperature accommodation coefficient [2], [4], [5]; c_p, c_v are the specific heats at constant pressure and volume and $\bar{c}$ is the mean free particle velocity. The second term in (5) takes care of effects due to or influenced by rates of change of temperature at the interface. In the absence of any additional or chemical changes, considering only transport effects in the Knudsen layer, the temperature slip time is

$$\theta \cong \frac{\sigma}{\bar{c}} a_\theta . \tag{7}$$

The third term in (5) is due to the rate of change of the temperature gradient; the characteristic slip magnitude for this term is

$$\beta \cong a_\beta \, \sigma \cdot \frac{\sigma}{\bar{c}} = \frac{\sigma^2}{\bar{c}} a_\beta . \tag{8}$$

The coefficients a_θ and a_β are of the order of one. The boundary conditions at infinity for the gas and solid are respectively:

$$T(t,\infty) = T_5 \qquad S(t,\infty) = T_1 . \tag{9}$$

It is assumed that these rapid variations of temperature do not invalidate the Navier-Stokes equations from which (2) is deduced.

For the gas, all coefficients of the differential equations and boundary conditions are taken as functions of temperature according to the following relations:

$$\frac{k\rho}{k_5\rho_5} = \left(\frac{T}{T_5}\right)^{\epsilon} ; \quad \sigma = \sigma_5\left(\frac{T}{T_5}\right)^{\epsilon_\sigma} ; \quad \theta = \theta_5\left(\frac{T}{T_5}\right)^{\epsilon_\theta} ;$$

$$\beta = \beta_5\left(\frac{T}{T_5}\right)^{\epsilon_\beta} . \tag{10}$$

To determine analytically the influence of the different terms of the temperature slip relation and of the temperature dependence of the coefficients we will use the small parameter method and seek a solution of the form

$$\left.\begin{aligned} S &= S^{(0)} + \epsilon S^{(1)} + \epsilon^2 S^{(2)} + \dots \\ T &= T^{(0)} + \epsilon T^{(1)} + \epsilon^2 T^{(2)} + \dots \end{aligned}\right\} \qquad (11)$$

The superscripts in parentheses denote approximations of different order and the (0)-th solution corresponds to the linear case with constant coefficients ($\epsilon = \epsilon_\sigma = \epsilon_\theta = \epsilon_\beta = 0$).

Using the Laplace transformation method the following solution is obtained for the linear case:

$$\left.\begin{aligned} S^{(0)} - T_1 &= \frac{T_5 - T_1}{\frac{1+m}{m}}\,\Phi(x,t), \\ T^{(0)} - T_5 &= \frac{T_5 - T_1}{1+m}\,\Phi(\xi,t), \end{aligned}\right\} \qquad (12)$$

where

$$m = \frac{k_5}{k_s}\sqrt{\frac{K_s}{K}} = \sqrt{\frac{k_5 c_p \rho_5}{k_s c_s \rho_s}},$$

$$\begin{aligned} \Phi(x,t) = \operatorname{erfc}\frac{x}{2\sqrt{K_s t}} - \exp\left(-\frac{x^2}{4K_s t}\right)\Bigg\{ & \frac{q_2 q_3}{(q_1-q_2)(q_1-q_3)} \cdot \\ & \cdot \psi\left(\frac{x}{2\sqrt{K_s t}} - q_1\sqrt{K_s t}\right) \\ & + \Bigg[\frac{q_1 q_3}{(q_2-q_1)(q_2-q_3)}\,\psi\left(\frac{x}{2\sqrt{K_s t}} - q_2\sqrt{K_s t}\right) \\ & + \frac{q_1 q_2}{(q_3-q_1)(q_3-q_2)}\,\psi\left(\frac{x}{2\sqrt{K_s t}} - q_3\sqrt{K_s t}\right)\Bigg]\Bigg\} \end{aligned} \qquad (13)$$

$\psi(z) = \exp(z^2)\operatorname{erfc}(z)$,

and the q_i's are roots of the following equation:

$$\frac{\beta K_s}{m}\sqrt{\frac{K_s}{K}}\,q^3 + \frac{\theta K_s}{m}\,q^2 + \frac{\sigma}{m}\sqrt{\frac{K}{K_s}}\,q + \frac{1+m}{m} = 0. \qquad (14)$$

It should be noted that the physical magnitudes involved are such that q_2, q_3 are usually complex conjugate roots and it can be shown that the expression in square parenthesis of (13) is not complex.

Results of calculations of the temperature of the wall $S^{(0)}(0,t)$ for a particular case of helium reflected from a pyrex plate shown in fig. 1 indicate the influence of the different terms of the temperature slip relation (5) on the temperature-time history of the surface of the solid. These results are compared with experimental measurements made in 1967 at the Karman Laboratory of California Institute of Technology 17" Shock-Tube [3]. It is noted that the best agreement with theory is for the case when the temperature slip is taken as depending only on the temperature gradient and temperature rate of change. Two other examples for different parameters are given in figs. 2 and 3.

These results, although very encouraging, need further experimental verification, particularly in view of the noted adsorption effects reported in [7].

To obtain an indication of the effects of nonlinearities we will determine the first order correction:

$$\frac{\partial S^{(1)}}{\partial t} = K_s \frac{\partial^2 S^{(1)}}{\partial x^2}, \quad S^{(1)}(x,0) = 0, \quad S^{(1)}(\infty,t) = 0, \tag{15}$$

$$\frac{\partial T^{(1)}}{\partial t} = K_5 \frac{\partial^2 T^{(1)}}{\partial \xi^2} - K_5 \frac{\partial F_o(\xi,t)}{\partial \xi}; \quad T^{(1)}(\xi,0) = 0;$$

$$T^{(1)}(\infty,t) = 0, \tag{16}$$

where $F_o(\xi,t)$ is a known function obtained from the zeroth-order approximation:

$$F_o(\xi,t) = \frac{\partial T^{(0)}(\xi,t)}{\partial \xi} \ln \frac{T_5}{T^{(0)}(\xi,t)}. \tag{17}$$

The two following relations:

$$\left.\begin{aligned} S^{(1)}(0,t) &= \varphi(t), \\ T^{(1)}(0,t) &= \chi(t), \end{aligned}\right\} \tag{18}$$

can be considered as completing the set of relations (15), (16) necessary to determine $S^{(1)}$, $T^{(1)}$.

Using the transformation

$$S^{(1)}(x,t) = \exp\left(-\frac{x}{\sqrt{K_s}}\right)\varphi(t) + S^*(x,t), \tag{19}$$

the relevant boundary conditions for $S^*(x,t)$ become homogeneous, which enables the use of the known Green's function for the heat conduction equation in the region $x\geq 0$, $t\geq 0$, and we obtain:

$$S^{(1)}(x,t) = e^{-\frac{x}{\sqrt{K_s}}}\varphi(t) + \int_0^t\int_0^{+\infty} G_o(x,\bar{x},t-\tau)\Big[\varphi(\tau) - \varphi'(\tau)\Big] e^{-\frac{x}{\sqrt{K_s}}}\, d\bar{x}\, d\tau, \qquad (20)$$

where

$$G_o(x,\bar{x},t-\tau) = \frac{1}{2\sqrt{\pi K_s(t-\tau)}}\left\{\exp\left[-\frac{(x-\bar{x})^2}{4K_s(t-\tau)}\right] - \exp\left[-\frac{(x+\bar{x})^2}{4K_s(t-\tau)}\right]\right\}. \qquad (21)$$

A similar solution is obtained for $T^{(1)}(\xi,t)$.
The matching conditions for the first-order corrections are:

$$\left.\begin{aligned} T^{(1)}(0,t)-S^{(1)}(0,t) &= \sigma_5\frac{\partial T^{(1)}(0,t)}{\partial\xi} - \theta_5\frac{\partial T^{(1)}(0,t)}{\partial t} \\ &+ \beta_5\frac{\partial^2 T^{(1)}(0,t)}{\partial\xi\partial t} - \sigma_5\frac{\epsilon_\sigma}{\epsilon}F_o(0,t) \\ &+ \theta_5\frac{\epsilon_\theta}{\epsilon}G_o(0,t) - \beta_5\frac{\epsilon_\beta}{\epsilon}H_o(0,t), \end{aligned}\right\}(22)$$

$$-\frac{\partial S^{(1)}(0,t)}{\partial x} = \frac{K}{K_s}\left[\frac{\partial T^{(1)}(0,t)}{\partial\xi} - F_o(0,t)\right],$$

where

$$\begin{aligned} G_o(0,t) &= \frac{\partial T^{(o)}(0,t)}{\partial t}\ln\left[\frac{T_5}{T^{(o)}(0,t)}\right], \\ H_o(0,t) &= \frac{\partial^2 T^{(o)}(0,t)}{\partial t\partial\xi}\ln\left[\frac{T_5}{T^{(o)}(0,t)}\right]. \end{aligned} \qquad (23)$$

Substituting in the matching relations (22) $S^{(1)}$ from (20) and a similar expression for $T^{(1)}$ we obtain two integro-differential equations for the functions $\varphi(t)$

and $\chi(t)$. These equations are solved using the Laplace transformation method, which leads to the following relation:

$$\varphi(t) = S^{(1)}(0,t) = \int_0^t \Big[A(t-\tau)f(\tau) + B(t-\tau)g(\tau) + C(t-\tau)h(\tau)\Big]\, d\tau, \qquad (24)$$

where the functions $A(t)$, $B(t)$, $C(t)$ are known and $f(t)$, $g(t)$, $h(t)$ are expressed in terms of the zeroth-order approximation $T^{(o)}(\xi,t)$. To save space we will omit here these rather lengthy expressions and give only solutions obtained for two extreme values of the time t.

When the time tends to infinity we obtain:

$$\lim_{t\to\infty} S^{(1)}(0,t) = -\frac{\sqrt{\pi}\,(T_5-T_1)}{(1+m)^2}\int\Big[\frac{1}{\sqrt{\pi}}\exp(-\eta^2)-\eta\,\mathrm{erfc}\,\eta\Big]\cdot \ln\left[\frac{1}{1-\dfrac{(T_5-T_1)\,\mathrm{erfc}\,\eta}{(1+m)T_5}}\right]d\eta, \qquad (25)$$

and if it is assumed, as it occurs often, that

$$\ln\left[\frac{1}{1+\dfrac{(T_5-T_1)\,\mathrm{erfc}\,\eta}{(1+m)T_5}}\right] \cong \frac{(T_5-T_1)\,\mathrm{erfc}\,\eta}{(1+m)T_5}$$

the integral is easily calculated and the simple relation given below follows:

$$S^{(1)}(0,\infty) = -\frac{(T_5-T_1)^2}{2\sqrt{\pi}\,T_5(1+m)^3}, \qquad (26)$$

and

$$S(0,\infty) - T_1 = \frac{m(T_5-T_1)}{1+m} - \varepsilon\,\frac{(T_5-T_1)^2}{2\sqrt{\pi}\,(1+m)^3T_5} + O(\varepsilon^2). \qquad (27)$$

Hence, as would be expected, only the temperature dependence of $k\,\rho$ affects the asymptotic value of the wall surface temperature for large times.

In the other extreme case when the time tends to zero, the variation of the interface temperature with time depends strongly on the assumed temperature-slip relations (5) and the following three cases can be noted:

$\beta = 0,\ \theta = 0$:

$$\lim_{t\to 0} S(0,t)-T_1 = \frac{2(T_5-T_1)m}{G_5\sqrt{\pi}}\sqrt{K_5 t} - \epsilon\,\frac{(T_5-T_1)^2 m}{T_5\sigma_5^2}\cdot$$

$$\cdot\left(1 - \frac{\epsilon_\sigma}{\epsilon}\right) K_5 t + O(\epsilon^2), \qquad (28)$$

$\beta = 0,\ \theta \neq 0$:

$$\lim_{t\to 0} S(0,t) - T_1 = \frac{(T_5-T_1)m}{\theta_5}\, t - \epsilon\,\frac{(T_5-T_1)^2 m}{T_5\theta_5^2}\, t^2\left(\frac{3}{4} - \right.$$

$$\left.\frac{3}{4}J_1 - \frac{1}{2}\frac{\epsilon_\theta}{\epsilon}\right) + O(\epsilon^2), \qquad (29)$$

$\beta \neq 0,\ \theta \neq 0$:

$$\lim_{t\to 0} S(0,t)-T_1 = \frac{(T_5-T_1)m}{\beta_5}\,\frac{4}{3\sqrt{\pi}}\, t\sqrt{K_5 t} - \epsilon\,\frac{(T_5-T_1)^2 m}{T_5\beta_5^2}\cdot$$

$$\cdot\left(\frac{5}{12(1+m)} + \frac{5mJ_2}{8(1+m)} - \frac{1}{6}\frac{\epsilon_\beta}{\epsilon}\right)K_5 t^3 + O(\epsilon^2), \qquad (30)$$

where

$$J_1 = \int_0^1\!\!\int_0^\infty \frac{s}{(1-s)^{3/2}}\,\eta\,\exp\left[-\frac{2-s}{1-s}\eta^2\right]\left[\psi'^2(\eta) + \frac{\psi(\eta)\psi''(\eta)}{2}\right]d\eta\, ds = 0.526$$

$$J_2 = \int_0^1\!\!\int_0^\infty \frac{s^3}{(1-s)^{3/2}}\,\eta\,\exp\left[-\frac{2-s}{1-s}\eta^2\right]\left[\frac{3}{4}\psi''^2(\eta) + \frac{1}{6}\psi'(\eta)\,\psi'''(\eta)\right]d\eta\, ds = 0.406 \quad .$$

It should be remembered that the structure of the shock wave has not been taken into account. In the case considered, the above relations show that only when the temperature slip is assumed to depend only on the temperature gradient, and not on the rates of increase of temperature, the slope of the temperature-time curve becomes infinite for $t = 0$. The sign and relative magnitude of $\frac{\epsilon_\sigma}{\epsilon}$, $\frac{\epsilon_\theta}{\epsilon}$, $\frac{\epsilon_\beta}{\epsilon}$ have an appreciable influence on the first order correction, which is not negligible.

The existence of a region of constant slope in the end wall temperature-time slope may point to the advantage of using in some cases the hypothesis that $\beta \neq 0$ in (5), but further experimental evidence and analysis of the neglected effects of adsorption and shock-wave structure are necessary.

The first author would like to express his deep gratitude to Dr. H.W. Liepmann, Dr. A. Roshko and Dr. B. Sturtevant of GALCIT for the unforgettable discussions and help which led to work on this problem.

REFERENCES

1. Goldsworthy, F.A. The Structure of a Contact Region, with Application to the Reflection of a Shock from a Heat-Conducting Wall. J. Fluid Mech., Vol.5, 1959 .
2. Busing, S.R., Clarke, S.F. Some Reflection and Surface Effects in the Shock Tube. Recent Advances in Aerothermochemistry, Vol. 1 AGARD Conference Proceedings No. 12, 1967.
3. Fiszdon, W., Sturtevant B. Note on Temperature-Slip Effects on Reflected Shock-Wave Trajectories. Unpublished Note, 1967.
4. Chapman, S., Cowling T. The Mathematical Theory of Non-Uniform Gases. Cambridge University Press, 1952.
5. Schaaf, S.A., Chambre, P.L. Flow of Rarefied Gases, Fundamentals of Gas Dynamics. Princeton University Press, 1958.
6. Kennard, E.H. Kinetic Theory of Gases. McGraw Hill,1938.
7. Piva, H.O., Sturtevant, B. Electron-Beam Measurements of Density in Shock-Waves Reflecting from a Cold Wall. Sixth Rarefied Gas Dynamics Symposium, 1968.

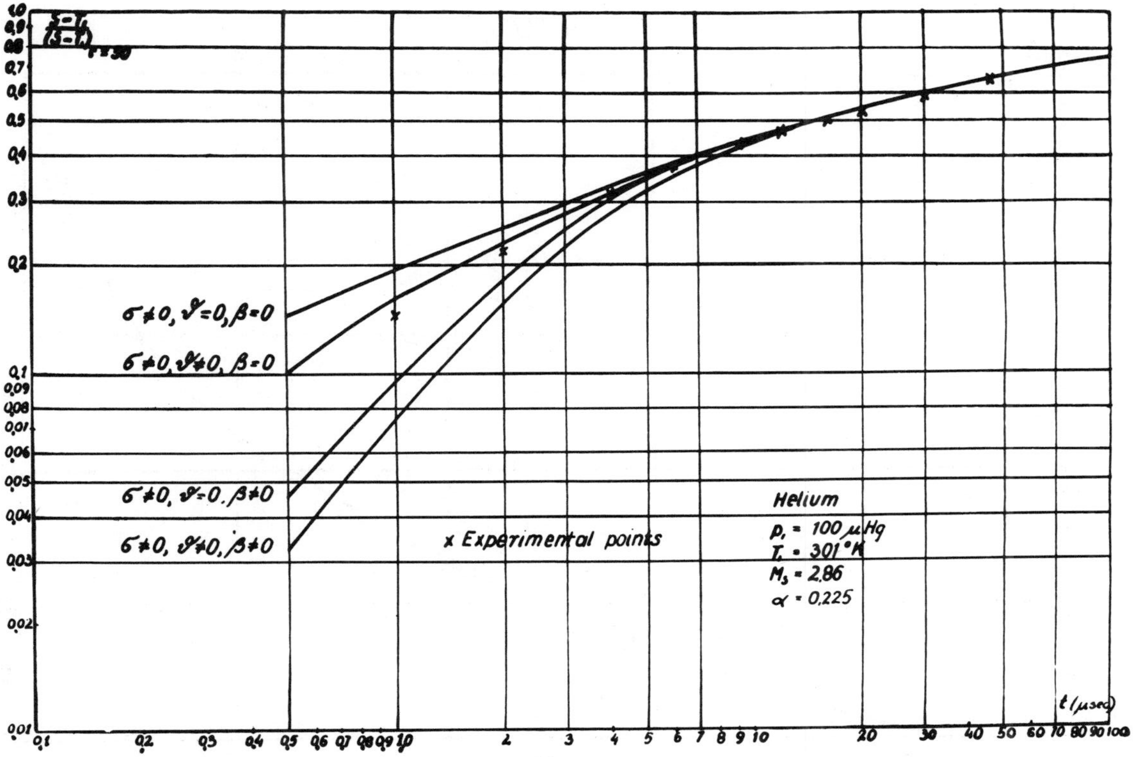
σ≠0, ϑ=0, β=0
σ≠0, ϑ≠0, β=0
σ≠0, ϑ=0, β≠0
σ≠0, ϑ≠0, β≠0
x Experimental points
Helium
p₁ = 100 μHg
T₁ = 301 °K
Mₛ = 2.86
α = 0.225
t (μsec)

Fig.1

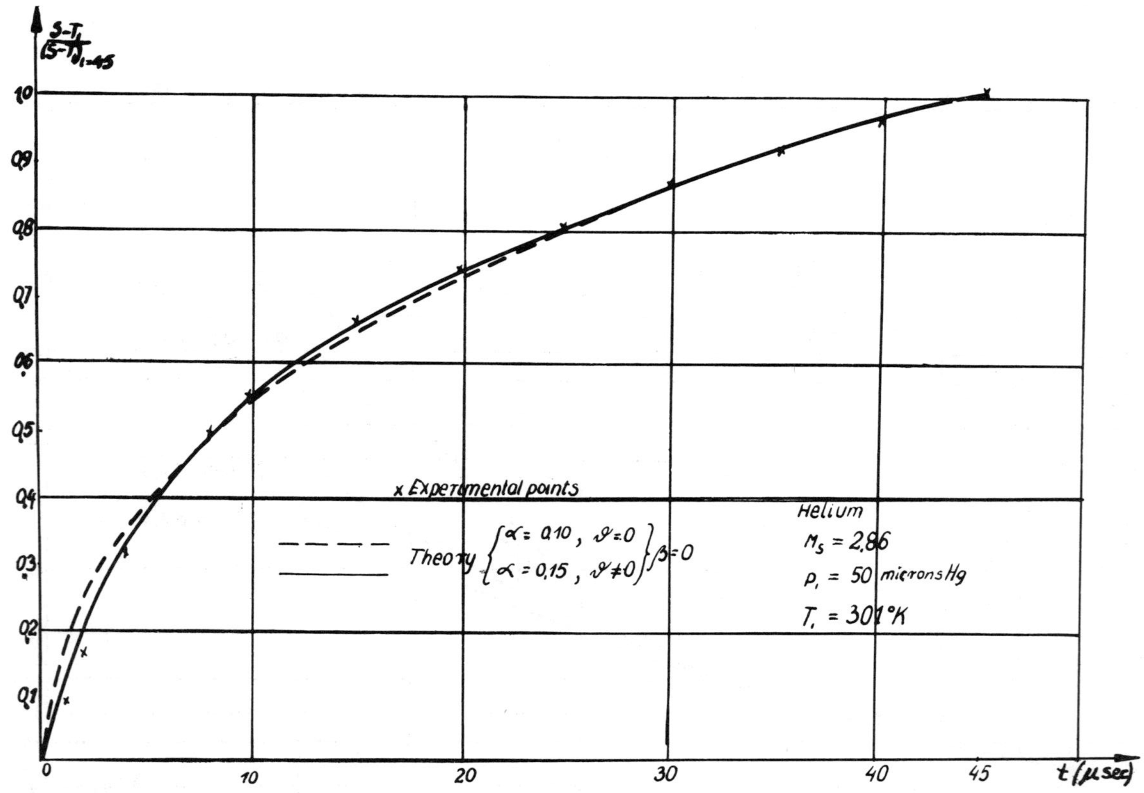

$\frac{S-T_1}{(S-T_1)_{t=45}}$
1.0
0.9
0.8
0.7
0.6
0.5
0.4
0.3
0.2
0.1
0
10
20
30
40
45
t (μsec)
x Experimental points
Theory $\left\{ \begin{array}{l} \alpha = 0.10, \ \vartheta = 0 \\ \alpha = 0.15, \ \vartheta \neq 0 \end{array} \right\} \beta = 0$
Helium
$M_s = 2.86$
$p_1 = 50$ microns Hg
$T_1 = 301$°K

Fig.2

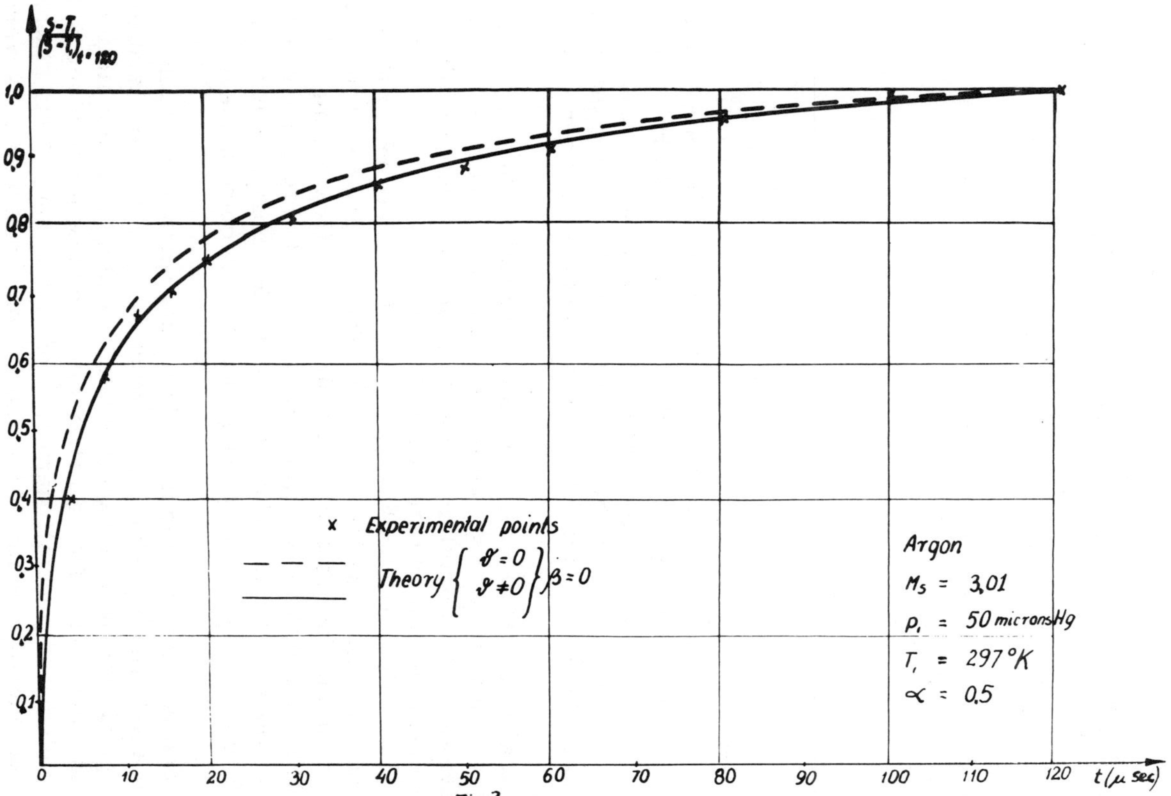

$\frac{S-T_1}{(S-T_1)_{t=120}}$
1,0
0,9
0,8
0,7
0,6
0,5
0,4
0,3
0,2
0,1
0
10
20
30
40
50
60
70
80
90
100
110
120
t (μ sec)
x Experimental points
Theory { $\vartheta = 0$ / $\vartheta \neq 0$ } $\beta = 0$
Argon
M_S = 3,01
p_1 = 50 microns Hg
T_1 = 297°K
α = 0,5

Fig.3

NON-REFLECTED SHOCK TUNNEL TEST TIMES

H. OERTEL
Institut Franco-Allemand de Recherches
de Saint-Louis (ISL), France

THE PROBLEM

Stationary hypersonic airflow behind the nozzle of a non-reflected shock tunnel represents an event of quite short duration within a long sequence of various flow phenomena. This is a well-known fact which has been recognized since the beginning of shock tunnel studies (1-4). As has been pointed out by B. D. Henshall in 1959 (5), there may even be no profitable running time at all when the driven tube is too short or too small or when the nozzle has been chosen too long.

Nevertheless, the non-reflected shock tunnel is quite a useful tool of hypersonic research, which has some basic advantages (6) in comparison with the reflected shock tunnel, the only disadvantage being the much shorter running time.

In the non-reflected shock tunnel of the French German Research Institute of Saint-Louis (Figure 1), for instance, good 3000 m/s airflow exists during 250 μs and good 5000 m/s airflow during 100 μs. Such test times are quite sufficient to take spark interferograms of stationary bow waves and boundary layers like those presented in Figures 2 to 9. They are also long enough to record pressures and heat flows.

But there is one very important difficulty when doing such measurements, which perhaps was not always fully recognized in the past: the good airflow is followed by bad, which is rather stationary as well and which only after a relatively long time gives place to the highly turbulent and nonstationary flow of driver gas. This stationary bad flow has unknown composition and is somewhat wavy. Nevertheless, bow waves and boundary layers look quite good as may be seen from Figures 10 to 14, and they are not very different from those in the good flow. Thus measurements made during the bad flow may, by error, be taken as made in the good one. This is especially true for studies of interactions like those depicted in Figures 15 to 17 which might only become stationary during the bad flow or not at all.

Thus there is the need to determine precisely the beginning and the end of good flow either by calculations or by special measurements. Calculations would be the easiest way if the required computer time were not too expensive relative to the cost of the experiments.

Therefore, it was the main object of the experiments reported below to determine whether simple calculations based upon the measured mean shock velocity u_s in the driven tube and done with the following assumptions would be appropriate:

1) The shortening of the running time Δt_2 at the nozzle entrance relative to that predicted by ideal shock tube theory is only due to the loss of driven gas into the shock-induced boundary layer (6, 7-17).

2) The shortening of the running time Δt_3 at the nozzle exit is only due to the different travel times t_1 and t_2 taken by two acoustic waves W_1 and W_2 to go through the nozzle as shown in Figure 18 (2, 6).

Certainly, these assumptions are not the best we can propose. Ahead of the nozzle the shock-induced boundary layer is by no means the only possible cause of shortened Δt_2 (6, 18 - 24). There are very complicated flow phenomena associated with the jet of driver gas blowing through the opening driver diaphragm. Moreover, turbulent spread of the contact zone as well as mixing and burning therein and within the boundary layer may contribute significantly.

Within the nozzle W_1 must surely be the head of an expansion wave traveling with local u(x) - a(x) whatever the preceding starting phenomena may be (25 - 32). But we observed that W_2 is a shock wave which might travel faster than the local u(x) + a(x). Moreover, u(x) and a(x) may be somewhat different for W_1 and W_2. Nevertheless, there was some hope that the afore-mentioned assumptions would not be too far from reality.

Considerable effort has already been put into comparing measured and calculated Δt_2 (8, 17, 33 - 44), with the general conclusion that the loss of driver gas into the shock-induced boundary layer is the dominant effect. Also some published comparisons of measured and expected Δt_3 (5) did not indicate that the assumption of acoustic W_1 and W_2 encountering equal u(x) and a(x) is too bad.

The problem was to find out whether the continued errors involved in the assumptions made ahead and within the nozzle can be tolerated. Inspection of the references revealed that the scatter of the previous results did not permit extrapolation to our flow conditions.

We found that the error cannot be tolerated. Then another question arose: Can some of the incertainties involved in the calculations based simply on measured shock velocities $u_s(x)$ be eliminated by obtaining records of the pressures p(t) at several stations x in the driven tube?

EXPERIMENTS

The basic dimensions of the shock tunnel were as follows:

Driver tube: Diameter 100 mm, length 1.5 m or 3 m.

Driver diaphragm: Flat scored, copper or stainless steel, with thickness up to 4 mm, clamped within a circular-quadratic-circular transition piece.

Driven tube: Diameter d = 100 mm, length l = 18.527 m, provided with 11 bores to receive thermo- or pressure probes as well as with four pairs of closeable 30-mm-diameter windows, 928 mm and 828 mm ahead of the nozzle diaphragm, to do optical measurements or to receive microwave antennas.

Nozzle diaphragm: Hostaplan with the smallest possible thickness down to 2.5 μ clamped 63 mm or 83 mm ahead of the nozzle entrance.

Nozzle: Conical, with sharp nozzle entrance, entrance diameter 20 mm or 30 mm, exit diameter D = 400 mm, length L = 772 mm or 752 mm, inner half-angle 14°, outer half-angle 26° or 36°.

Test section: Circular, diameter D = 400 mm, four flat windows with diameter 120 mm, the axis crossed on the test section axis 60 mm behind the nozzle exit.

Dump tank: Volume 9.9 m^3, 3.3 m^3 with diameter 1 m concentrically enveloping the nozzle and test section.

Hydrogen at room temperature was chosen as the driver gas and undried air or nitrogen at room temperature as the driven gas. The initial hydrogen pressure was varied over the range $40 < p_5/\text{bar} < 600$ and the initial driven gas pressure over the range $0.001 < p_1/\text{bar} < 1$. Most of the experiments were done with $300 < p_5/\text{bar} < 400$.

The following measurements have been performed under varied flow conditions (figure 19):

In the driver tube

1) 5 or 10 mean shock velocities $u_s(x)$ between 6 or 11 platinum surface thermometers on glass inserted flat to the wall (45).

2) 11 mean shock velocities $u_s(x)$ by means of a 2.25×10^9 cps microwave doppler apparatus exciting the H_{11} (TE_{11}) - mode with wavelength 21.33 cm by a loop fixed in the wall (46, 47).

3) Transversal microwave transmission 928 mm ahead of the nozzle diaphragm by means of the same microwave apparatus (48).

4) Wall pressures p(t) simultaneously at five stations x by means of quartz transducers Vibrometer 6 QP 500a clamped without vibration isolation (49).

5) Detachment distance d(t) of the bow wave of a sphere 10 mm or 12 mm in diameter 828 mm ahead of the nozzle diaphragm by means of streak interferography (44).

At the test section center

1) Detachment distance d(t) of the bow wave of a 150-mm-diameter sphere by means of spark or streak interferography (44).

2) Shock angle $\gamma(t)$ of the bow wave of a 30° wedge, 124 mm long, by means of spark or streak interferography (50).

3) Light phase shifts in the same bow wave of a 30° wedge taken from the same spark photographs or streak records (50).

4) Pressure $p_W(t)$ on the same 30° wedge by means of a quartz transducer Vibrometer 6 QP 500 with a fixed vibration isolated (50).

Moreover, some stagnation pressure and some stagnation heat flow measurements at various distances from the test section center have been made to ensure that the flow out of the nozzle was quite uniform, carrying quite plane waves W_1 and W_2. We don't need to discuss these further when dealing with running time at the test section center.

Of all these afore-mentioned measuring techniques streak interferography has given the most information about running times. The visualization as in the case of spark interferography has been achieved by means of different interferometers using polarized light and double-refracting prisms. Figure 20 displays the basic optical setup. (51 - 55). It can be adjusted either to small ray separation to look through smooth density gradients to visualize density jumps, as in Figure 4, or to large ray separation to visualize the densities themselves, as in Figure 6. In the latter case the fringe shifts in certain regions of the interferogram are like those observed with a Mach-Zehnder interferometer. To record streak interferograms, we simply added a mercury lamp, an obturator and a slit. Sparks then merely superimpose some slit pictures to document the orientation of the slit as well as the velocity of the film moving in a drum camera. Various orientations of the interference fringes, the slit and the motion of the film have been chosen to adapt the sensitivity to the phenomena to be observed (56). Some examples of streak records thus obtained are shown in Figures 21 to 23. The enlarged cut reproduced in Figure 24 gives the most direct proof of the fact that the wave W_2 is a shock wave. The shock can be seen coming and interacting with the bow wave of the 30° wedge. The angle of the trace as well as the direction of the fringe shifts therein correspond to this interpretation. This is also supported by the wedge pressure oscillograms shown in Figures 25 and 26, which have been obtained simultaneously with streak records. Here the shock W_2 produces a pressure "kick" which in nitrogen is somewhat less intense than in air. The height of the pressure "kick" reveals that, at least within the test section the shock is so weak that it may be used for diagnostics. Its velocity measured between two pressure transducers or taken from a streak record cannot be very different from u + a at the end of good flow.

Figures 27 and 28 present pressure oscillograms obtained in the driven tube. That in Figure 27 has shown that no perturbation by the reflected driver expansion wave need be feared. Records like those shown in Figure 28 together with simultaneous u_s-measurements served as the basis for a calculation done by F. Wecken and discussed below.

Two different techniques have been used to measure the shock velocities $u_s(x)$ all over the length of the driven tube, the thermoprobes being less reliable at shock mach numbers beyond 10 and the microwave reflection at the shock failing at shock mach numbers below 8. In figure 29 a comparison is made between simultaneous results.

Finally, the question arose as to whether the bad flow behind the interface is one of simply boundary layer air or nitrogen or one polluted with combustion products or hydrogen. This question has been settled by microwave transmission oscillograms like those presented in Figure 30. They clearly indicated that with nitrogen ionization disappears at the interface whereas with air ionization lasts for a rather longer time. There must be some cold hydrogen in one case and burning hydrogen in the other.

CALCULATIONS

Among the large number of published calculations of duration Δt_2 for given and assumed constant shock velocity u_s those of H. Mirels and those of P. J. Musgrove and J. P. Appleton have been chosen for comparison with experiment. From H. Mirels simply the working curves presented for wholly laminar (11) or for wholly turbulent (12) real air boundary layer have been used. From P. J. Musgrove and J. P. Appleton (17) we tried the expressions they used to fit calculated curves over the shock mach number range $2 < M_s < 8$. With our nomenclature these expressions may be written as follows:

For wholly laminar boundary layer:

$$- \ell n \left| 1 - \left(\frac{u_s \Delta t_2}{\ell_\infty}\right)^{1/2} \right| - \left(\frac{u_s \Delta t_2}{\ell_\infty}\right)^{1/2} = \frac{x}{2\left(\frac{\rho_2}{\rho_1} - 1\right)\ell_\infty} \quad (1)$$

$$\log \frac{p_1 d^2}{A\, \ell_\infty} = 0{,}0830\, M_s + 0.0147 \text{ with } A = 0.212 \text{ Torr cm} \quad (2)$$

For wholly turbulent boundary layer:

$$\ell n \frac{1 + \left(\frac{u_s \Delta t_2}{\ell_\infty}\right)^{1/5}}{1 - \left(\frac{u_s \Delta t_2}{\ell_\infty}\right)^{1/5}} + 2\tan^{-1}\left(\frac{u_s \Delta t_2}{\ell_\infty}\right)^{1/5} - 4\left(\frac{u_s \Delta t_2}{\ell_\infty}\right)^{1/5}$$

$$= \frac{4x}{5\left(\frac{\rho_2}{\rho_1} - 1\right)\ell_\infty} \quad (3)$$

$$\log \frac{p_1^{1/4} d^{5/4}}{B\, \ell_\infty} = 0.0349\, M_s - \frac{1{,}242}{M_s} + 0{,}580$$

$$\text{with } B = 5{,}65 \times 10^{-2} \text{ (Torr cm)}^{1/4} \quad (4)$$

Here, d is the diameter of the driven tube. ℓ_∞ would be the shock interface separation when in an infinitely long tube shock and interface finally travel at equal velocity.

If these relations were appropriate, then combined with an approximate expression for M_s (p_1/p_5) they would provide quite an easy tool for estimates.

A constant shock mach number has been assumed. Thus the question arose: Which of the measured M_s (x) has to be applied? We used the highest M_s, mean M_s and lowest M_s observed, and found that this was not an important problem. Moreover, for just one example the method proposed by F. Wecken (57, 58) has been used by itself to calculate the whole flow field between the shock and the air element coming from x = 4. 50 m for given shock velocities u_s (x) and pressures p(x, t).

The idea is quite simple. With $u_s(x)$, those states and flow velocities are given with which the air elements begin their paths. With $p(x,t)$, the accelerating $\partial p/\partial x(x,t)$ which the gas elements encounter on their paths, are known as well. Then the conservation equations for momentum and energy

$$\frac{\partial u}{\partial t} + u\frac{\partial u}{\partial x} = -\frac{1}{\rho}\frac{\partial p}{\partial x} \tag{5}$$

$$\frac{\partial h}{\partial t} + u\frac{\partial h}{\partial x} = \frac{1}{\rho}\frac{\partial p}{\partial t} + u\frac{\partial p}{\partial x} \tag{6}$$

$$\frac{\partial}{\partial t}\left(h+\frac{u^2}{2}\right) + u\frac{\partial}{\partial x}\left(h+\frac{u^2}{2}\right) = \frac{1}{\rho}\frac{\partial p}{\partial t} \tag{7}$$

are sufficient to calculate the flow outside the boundary layer if there is no turbulent or molecular transport and if the state equations are known. The conservation equation for mass and the knowledge of the mass lost into the boundary layer are not needed. The interpolations of measured $u_s(x)$ and $p(t)$ at several stations x must be consistent with the requirement that gas elements do not take over.

The fact that the measured $u_s(x)$ and $p(x,t)$ are just sufficient to replace the continuity equation was already pointed out by P.H. Rose and W. Nelson in 1958 (59). But their approach was quite different.

Finally, the relations

$$t_1 = \int_{x_1}^{x_3} \frac{dx}{u(x) - a(x)} \tag{8}$$

$$t_2 = \int_{x_2}^{x_3} \frac{dx}{u(x) + a(x)} \tag{9}$$

$$\Delta t_2 - \Delta t_3 = t_1 - t_2 \tag{10}$$

have been used to calculate the loss of running time within the nozzle. Here x_1 is the locus of the nozzle diaphragm, x_2 that of the nozzle entrance and x_3 that of the test section center. The one-dimensional approach has been used to calculate u(x) and a(x) with the assumptions of either equilibrium (e) or frozen (f) degrees of dissociation. The difference $x_1 - x_2$ is an important one, because the relatively slow motion of W_1 ahead of the nozzle entrance makes a significant contribution to the loss of running time.

For all but the F. Wecken calculations the driven gas outside the boundary layer has been assumed to be a Lighthill gas or mixture of Lighthill gases. Nitrogen oxide in air has been neglected. This is an assumption somewhat inconsistent with those of H. Mirels or P. J. Musgrove and J. P. Appleton, but we know from previous calculations that the error thus tolerated is small in comparison with experimental errors. The F. Wecken calculations used the real state equations of air.

RESULTS

The results obtained with hydrogen at initial pressures $300 < p_5/\text{bar} < 400$ and air at initial pressures $10^{-3} < p_1/\text{bar} < 1$ are presented as those of the most thoroughly studied example. The results found at other driver pressures were quite similar. Those obtained with nitrogen replacing air as the driven gas did not differ more than can be explained by the different state equations. No indication was found that burning in the air/hydrogen mixing zone had a significant effect upon the observed flow durations.

The results are reported as functions of the initial pressure ratio p_1/p_5. When interpreting, it must be kept in mind that with changing p_1/p_5 the shock mach numbers M_s in the driven tube were also changed. For a given value p_1/p_5 they vary with the distance x from the driver diaphragm as shown in Fig. 29. For varied p_1/p_5 the measured lowest shock Mach numbers $\check{M}_s$, mean $\bar{M}_s$ and highest $\hat{M}_s$ may be roughly approximated by (Fig. 31):

$$M_s + A = B \log \frac{p_5}{p_1} \qquad (11)$$

with the following constants:

Gas	A	$B(\check{M}_s)$	$B(\bar{M}_s)$	$B(\hat{M}_s)$
Air	1,30	2,60	2.70	2.75
N_2	1,20	2,50	2.60	2,65

In Fig. 32 a comparison is made between measured and calculated durations Δt_2 of good flow ahead of the nozzle. At $p_1/p_5 > 10^{-4}$ the measured Δt_2 are shorter than those calculated with the wholly turbulent boundary layer assumption. Within the domain $10^{-4} > p_1/p_5 > 5\text{x}10^{-6}$ they remain approximately independent of p_1/p_5 going from the calculated turbulent to the calculated laminar boundary layer curves. This corresponds to the statement made by H. Mirels that roughly a wholly turbulent boundary layer has to be expected for $p_1d > 130$ torr-cm, i.e. $p_1/p_5 > 5\text{x}10^{-5}$, and a wholly laminar one for $p_1d < 13$ torr-cm, i.e. $p_1/p_5 < 5\text{x}10^{-6}$. It seems that a wholly laminar boundary layer did not occur within the domain of the reported experiments. In the case of a wholly turbulent boundary layer the deviations from the Mirels curve are somewhat smaller than from the Musgrove/Appleton curve. These deviations might not look too serious, when presented on a double logarithmic scale, and as long as they are not contradictory to previous observations. But with regard to the problem under study they must be compared to Δt_3.

From the measured Δt_2 only the fractions Δt_3 plotted in Fig. 33 remained available behind the nozzle. The lost time interval $\Delta t_2 - \Delta t_3$ corresponds quite well to the calculated $t_1 - t_2$. Thus it seems that W_2 may in fact be considered to be an acoustic wave all over the length of the nozzle. The pointed part of the measured Δt_3 curve is somewhat uncertain. Here fragments of the nozzle diaphragm made it difficult to identify W_1 clearly. At higher p_1/p_5 these fragments passed with the starting flow.

When compared to Δt_3 most of the deviations of the calculated Δt_2 from the measured ones cannot be tolerated. For instance, at $p_1/p_5 = 10^{-3}$ the deviation - 125 μs between the measured and the Mirels Δt_2 amounts to 60% of the measured Δt_3 = 210 μs. At about $p_1/p_5 = 1.7 \times 10^{-5}$ the deviations from the two Mirels Δt_2 calculated with the assumption of either turbulent or laminar boundary layer are ± 105 μs, which must be compared to the measured Δt_3 = 100 μs. At $p_1/p_5 = 10^{-4}$ the experimental Δt_2 curve crosses the calculated one. For all initial pressures p_5 studied we always found such a crossing point at which the correspondence between the measured and calculated Δt_2 seems to be excellent. But this is not due to good correspondence between theory and experiment. It merely indicates that here the shock-induced boundary layer is a transitional one, the wholly turbulent boundary layer theory being used beyond its domain of validity.

When measurements are initiated by a shock detector ahead of the nozzle, then not only Δt_3 but also t_1 must be known. Therefore, the measured and calculated t_1 and $t_3 = t_1 + \Delta t_3$, with the calculated Δt_3 based upon measured Δt_2, are reported in Fig. 34. As can be seen from this figure, at $p_1/p_5 > 5 \times 10^{-5}$ an instantaneous measurement for instance by means of a single spark photograph, may be placed in the good flow with calculated $t_1 + \Delta t_3/2$, if u_s and Δt_2 at the nozzle entrance are known.

Moreover times of arrival t_p of the first detected pressure signal and t_ρ of the first detected interference fringe shift at the test section center are plotted in Fig. 34. A considerable scatter of those times was observed in spite of some effort to perform the experiments with equal initial pressures within the nozzle and dump tank.

Finally, the times t_4 of transition from somewhat wavy but stationary to highly turbulent and instationary flow are also reported in fig. 34. At low p_1/p_5 the duration $t_3 - t_1$ of good flow was much shorter than the duration $t_4 - t_3$ of that which by error may be taken to be good.

Figure 35 presents, in a somewhat restricted manner, results of the Wecken calculation done for the example H_2(p_5 = 300 bar)/air (p_1 = 0, 133 bar). Curves of constant density ρ and constant flow velocity u are given for the flow field between the path of the shock and the path of the air starting at distance x = 4.50 m from the driver diaphragm. Many other paths beginning at other x have been calculated. Here the 4.50 m air arrives at the end of the driven tube 674 μs after the shock. This may be compared to the theoretical Δt_2 = 340 μs or measured Δt_2 = 265 μs taken from Figure 32. Re-calculations with different interpolations of the measured p(t) at five stations x have shown that the discrepancy cannot be due to the errors involved in the determination of p(x, t).

This is a preliminary result which needs verification. If it is verified, then the conclusion is that the Wecken calculation can by no means give the correct Δt_2.

The explanation may be that quite an unexpectedly large portion of the air is not lost into the shock-induced boundary layer but by turbulent spread of the contact zone. According to the Wecken calculation, the air arriving at the end of the measured duration Δt_2 of good flow has begun its path at about x = 11 m. None of the air particles initially at smaller x arrive earlier than the interface. This is quite contradictory to the assumptions of the afore-mentioned Δt_2 calculations.

ACKNOWLEDGEMENTS

The present work has been assisted by Dr. G. Smeets, F. Gatau, A. Sanner, A. George, R. Resweber, G. Hantz, E. Stoffelbach. Moreover, the author would like to thank Dr. F. Wecken and A. Thierry for their calculation and M. König for his microwave measurements.

REFERENCES

1. A. Hertzberg
A shock tube method of generating hypersonic flow
J. aero. sci. 18 (1951) 803
2. A. Hertzberg, W. E. Smith, H. S. Glick, W. Squire
Modification of the shock tube for the generation of hypersonic flow
CAL-Rept. AD-789-A-2 (1955)
AEDC-TN-55-15 (1955)
3. H. S. Glick, A. Hertzberg, W. E. Smith
Flow phenomena in starting a hypersonic shock tunnel
CAL-Rept. AD-789-A-3 (1955), AEDC-TN-55-16 (1955)
4. A. Hertzberg
The shock tunnel and its applications to hypersonic flight
CAL-Rept. AD-1052-A-5 (1957)
5. B. D. Henshall
Experimental results from the NPL hypersonic shock tunnel
p. 1-44 in "Hypersonic Flow", Butterworths 1960

6. H. Oertel
Stossrohre - Shock Tubes - Tubes à choc
Springer 1966
7. G. F. Anderson
Shock tube testing time
J. aero/space sci. 26, 3 (1959) 184-185
8. R. E. Duff
Shock tube performance at low initial pressure
Phys. fluids 2 (1959) 207-216
9. A. ROSHKO
On flow duration in low pressure shock tubes
Phys. fluids 3 (1960) 835-842
10. W. J. Hooker
Testing time and contact zone phenomena in shock tube flows
Phys. fluids 4, 12 (1961) 1451-1463
11. H. Mirels
Test time in low pressure shock tubes
Phys. fluids 6, 9 (1963) 1201-1215
12. H. Mirels
Shock tube test time limitation due to turbulent wall boundary layer.
AIAA J. 2, 1 (1964) 84-93
13. E. F. Brocher
Hot flow length and testing time in real shock tube flow
Phys. fluids 7, 3 (1964) 347-351
14. H. Mirels, W. S. King
Series solutions for shock tube laminar boundary layer and test time
AIAA J. 4 (1966) 782-789
15. H. Mirels
Flow non uniformity in shock tubes operating at maximum test times
Phys. fluids 9, 10 (1966) 1907-1912
16. J. J. Bertin
Analysis of test times and boundary layer induced property variations in circular shock tube
NASA-TN-D-3759 (1967)
17. P. J. Musgrove, J. P. Appleton
The influence of boundary layer growth on shock tube test time
Appl. sci. res. 18, 9 (1967) 116-155
18. I. I. Glass, W. Martin, G. N. Patterson
A theoretical and experimental study of the shock tube
UTIA-Rept. 2 (1953)
19. D. R. White
Influence of diaphragm opening time on shock tube flows
Gen. Electric Res. Lab. Rept. 58-RL-1999 (1958)
20. D. A. Spence, B. A. Woods
Boundary layer and combustion effects in shock tube flows
p. 153-180 in "Hypersonic Flow", Butterworths 1960
21. D. A. Spence, B. A. Woods
A review of the theoretical treatments of shock tube attenuation
RAE-TN-AERO-2899 (1963), ARC-Rept. 25086

22. J. A. Copper
Effects of interface combustion and mixing on shock tunnel conditions
AIAA J. 2, 9 (1964) 1669-1671
23. V. D. Blankenship, P. M. Chung
Laminar boundary layer inviscid flow interactions in shock tubes
p. 969-1000 in NOL-AD-484 600 (1966)
Proc. 5 th Shock Tube Symp. 1965
24. C. S. Kim, C. C. Chang
Real flow effects on shock attenuation in shock tube flows
p. 879-914 in NOL-AD-484 600 (1966)
Proc. 5 th Shock Tube Symp. 1965
25. B. D. Henshall
Some notes on the flow durations occuring in hypersonic shock tubes
ARC-CP-290 (1956)
26. B. D. Henshall, G. E. Gadd
Factors affecting the performance of the nozzle of a hypersonic shock tube
ARC-CP-293 (1956)
27. L. L. Monroe
Investigation of the transmission of a shock wave through an orifice
CALCIT Memo 46 (1958)
28. C. E. Smith
An analytic study of the starting process in a hypersonic nozzle
PH. D. thesis Stanford Univ. 1962, Sudaer 135
29. C. E. Smith
An analytic study of the starting process in a hypersonic nozzle
p. 198-210 Proc. Heat Transf. Fluid Mech. Inst., Stanford Univ. Press 1964
30. C. E. Smith
The starting process in a hypersonic nozzle
J. Fluid mech. 24, 4 (1966) 625-640
31. P. Junghans
Theoretische Untersuchung des Anlaufvorganges in einer Lavaldüse
Dipl. Arbeit Univ. Karlsruhe 1967
32. H. O. Amann
Vorgänge beim Start einer Reflexionsdüse
Dissertation Univ. Karlsruhe 1968, EMI-Ber. 9/68
33. V. A. Sandborn
Measurements of flow dimensions in a low pressure shock tube
NASA-TN-D-1218 (1962)
34. V. A. Sandborn, H. Weisblatt, R. F. Flagg
Test time in a 1,5 inch diameter high stagnation enthalpy shock tube
AIAA J. 1, 5 (1963) 1236-1237
35. J. D. Appleton, P. J. Musgrove
An investigation of the departure from ideal shock tube performance. Preliminary results
ARC-Rept. 24, 721 (1963)

36. J. A. D. Ackroyd
A study on the running times in shock tubes
ARC-CP-722 (1964)
37. J. A. D. Ackroyd
Studies on the running time in a shock tube and shock tunnel
Ph. D. thesis Univ. London 1964
38. P. J. Musgrove
The influence of the boundary layer on shock tube running times
Ph. D. thesis Univ. Southampton 1964
39. A. Roshko, J. A. Smith
Measurements of test time in the GALCIT 17-inch shock tube
AIAA J. 2 (1964) 186-187
40. S. G. Chapin, W. F. Rumpel
Measurement of uniform flow duration in a chambered buffered shock tube
AIAA J. 2, 8 (1964) 1499-1500
41. T. A. Holbeche, D. A. Spence
Theoretical and experimental investigation of temperature variation behind attenuating shock waves
Proc. Roy. Soc. A 219 (1964) 111-128
42. J. N. Fox, T. I. McLaren, R. M. Hobson
Test time and particle paths in low pressure shock tubes
Phys. fluids 9 (1966) 2345-2350
43. J. A. D. Ackroyd
Some further notes on the laminar boundary layer development and running time in a shock tube
ARC-CP-966 (1967)
44. G. Smeets
Streakinterferometrische Untersuchungen von Kopfwellenabständen in Stossrohren
p. 285-298 EMI - Wiss. Ber. 4/68,
Proc. 6 the Shock Tube Symp. 1967
45. F. Gatau
Chaîne de mesure de la vitesse de propagation d'une onde de choc dans un tube à choc
ISL - N 27/63
46. B. Koch, M. König
Digitale Messung des Geschwindigkeitsverlaufes von Stosswellen mit Hilfe von Mikrowellen
p. 540-545 EMI - Wiss. Ber. 4/68,
Proc. 6 th Shock Tube Symp. 1967
47. B. Koch, M. König
Ein Verfahren zur Messung des Geschwindigkeitsverlaufes von Stosswellen mit Hilfe von Mikrowellen
ISL - T 26/65
48. M. König
ISL - Report in preparation
49. H. Oertel
Diagnostic d'écoulement au tube à choc hypersonique
ISL - N 27/65, 2ème AFITAE Colloque d'Aérodynamique appliquée, Toulouse1965
50. G. Smeets
ISL - Report in preparation

51. H. Oertel
Hypersonic research at the LRSL-St-Louis
p. 203-218 in "Hypersonic Flow"
Butterworths 1960
52. H. Oertel
Measurements in the hypersonic shock tunnel of the ISL
p. 305-316 in BRL-Rept. 1160 (1962)
Proc. 4 th Shock Tube Symp. 1961
53. H. Oertel
Ein Differentialinterferometer für Messungen im Hyperschall-stossrohr
ISL - T 17/61
54. H. Oertel
Bemerkungen zur Interpretation von Differentialinterferenz-aufnahmen
ISL - N 3/64
55. H. Oertel
Messungen im Hyperschallstossrohr
p. 759-848 in "Kurzzeitphysik - High Speed Physics"
Springer 1967
56. H. Oertel, G. Smeets
Schmieraufnahmen von Verdichtungsstössen mit Hilfe eines Differentialinterferometers mit streifenparallelem Spalt
ISL - T 35/65
57. F. Wecken
Zur Ermittlung des Strömungszustandes im Hyperschallstossrohr
ISL - T 16/61
58. F. Wecken
Ein Auswerteverfahren für Druckmessungen im Stosswellenrohr
ISL - N 15/62
59. P. H. Rose, W. Nelson
On the affects of attenuation on gas dynamic measurements made in shock tubes
AVCO Res. Lab. Rept. 24 (1958)

Figurc 1
The ISL-non-reflected shock tunnel.

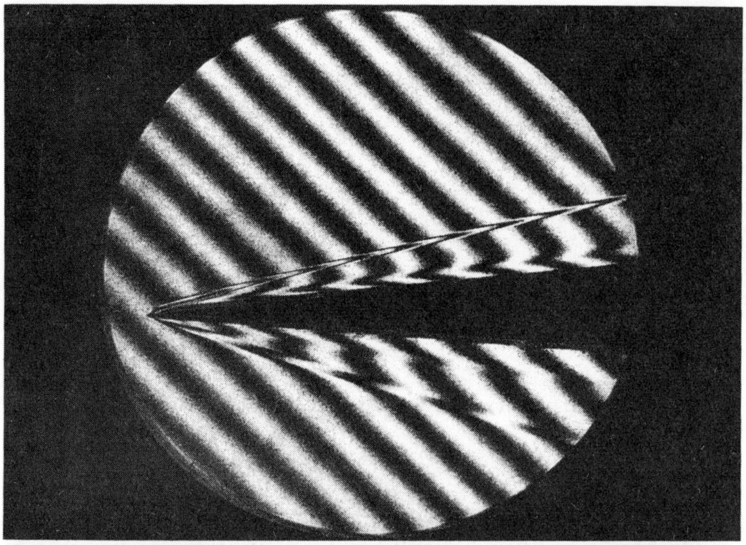

Figures 2, 3 and 4
Spark-interferograms of stationary bow waves and boundary layers in good test section flow.

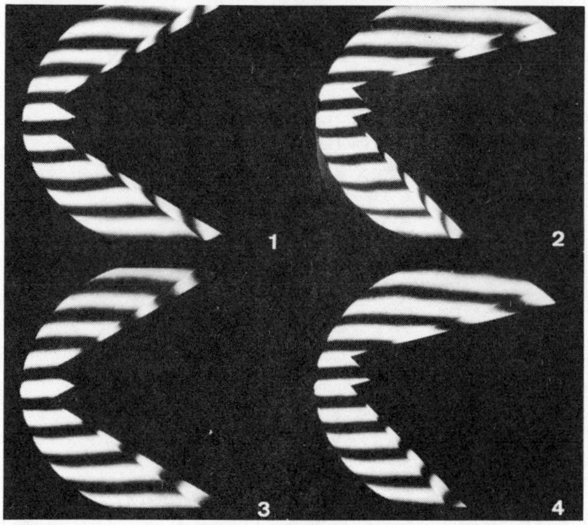

Figures 5, 6 and 7
Spark-interferograms of stationary bow waves and boundary layers in good test section flow.

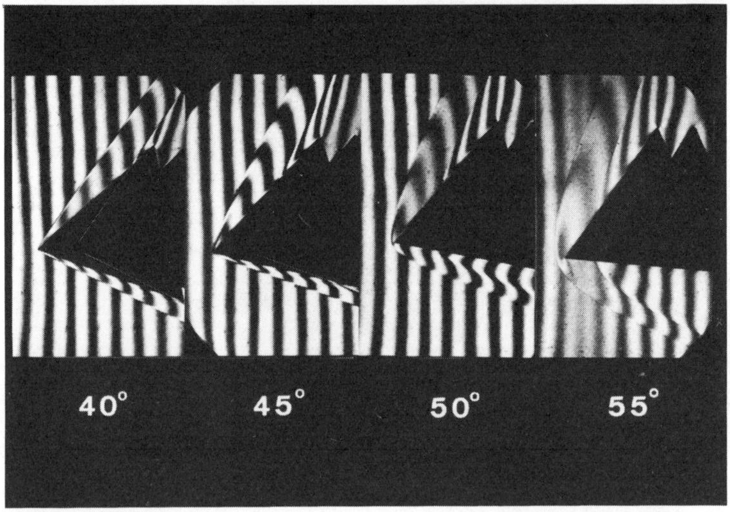

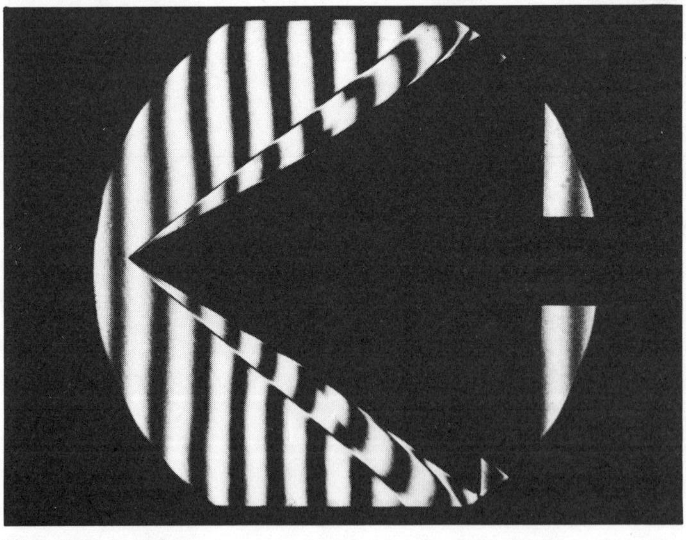

Figures 8 and 9
Spark-interferograms of stationary bow waves and boundary layers in good test section flow.

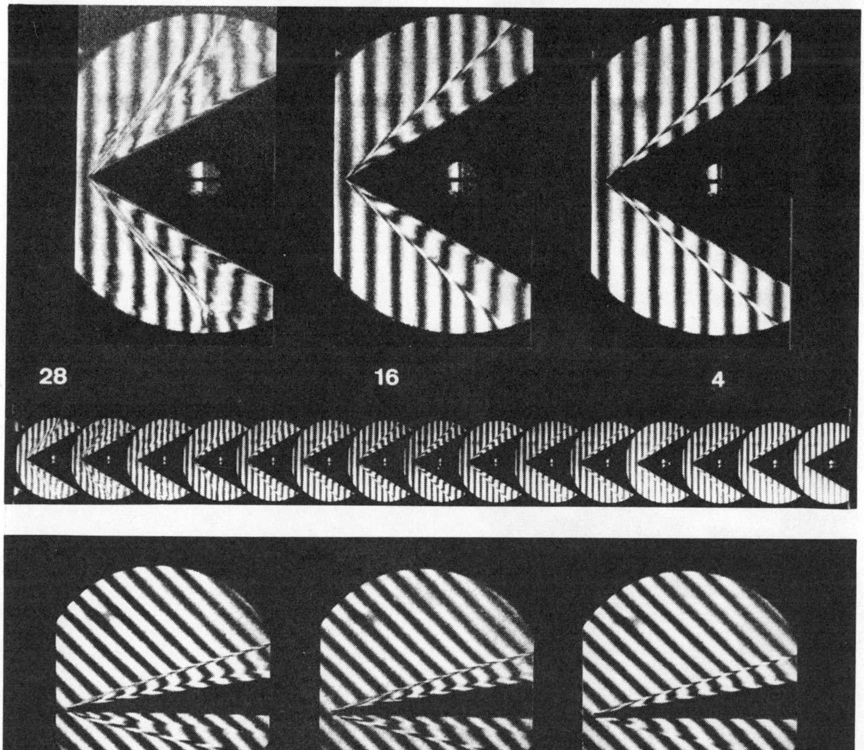

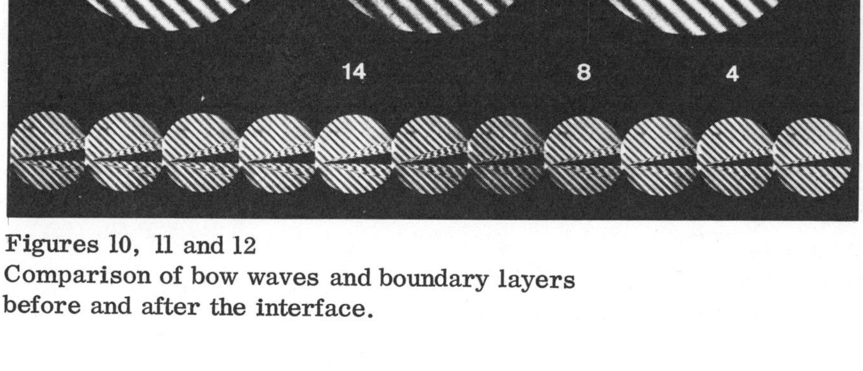

Figures 10, 11 and 12
Comparison of bow waves and boundary layers before and after the interface.

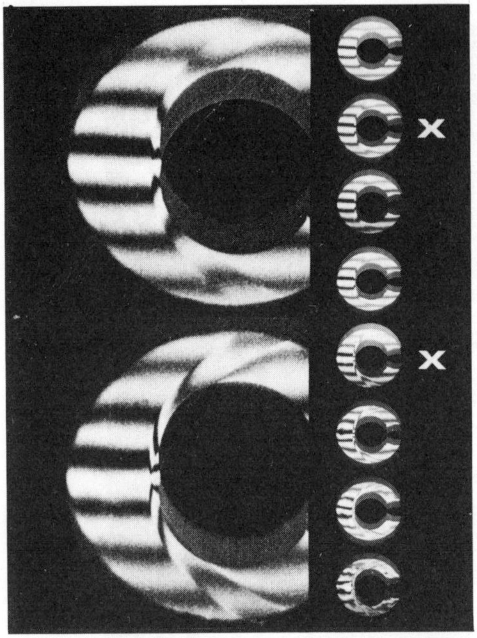

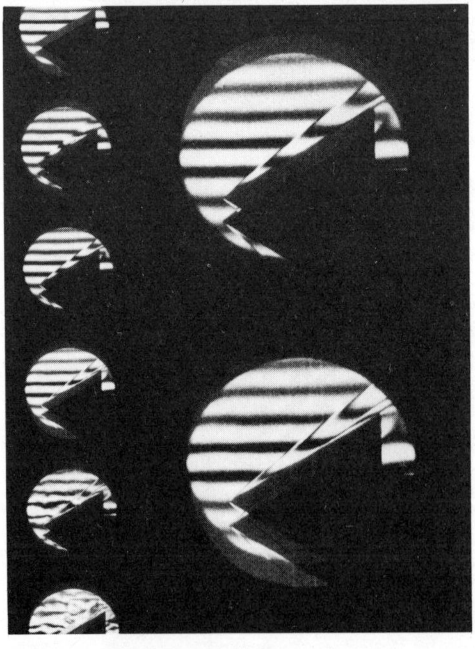

Figures 13 and 14
Comparison of bow waves and boundary layers before and after the interface.

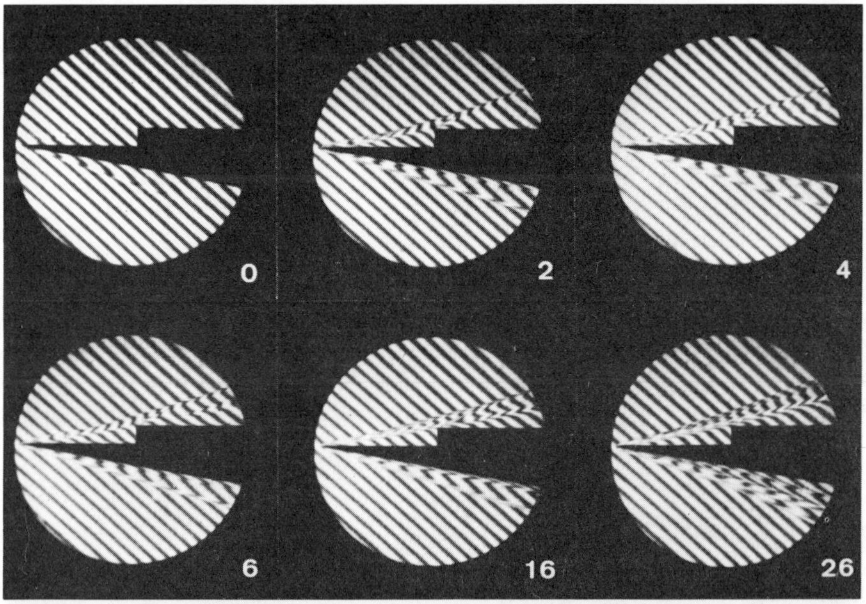

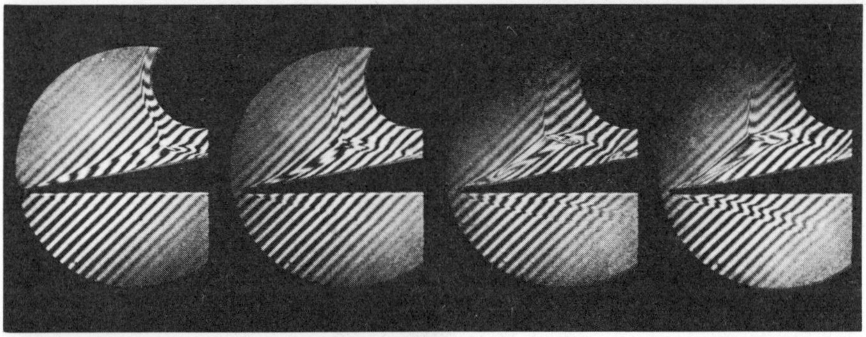

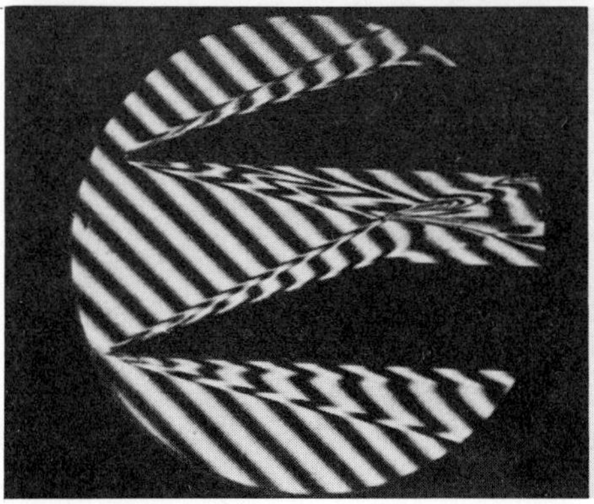

Figures 15, 16 and 17
Interactions which did not become stationary.

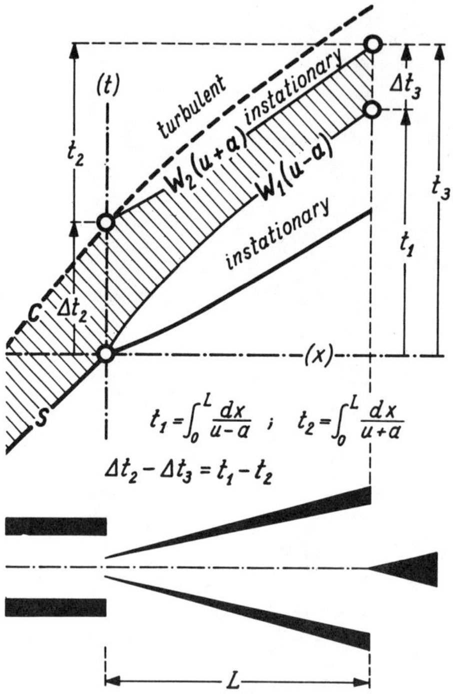

Figure 18
Loss of test time by waves W_1 and W_2.

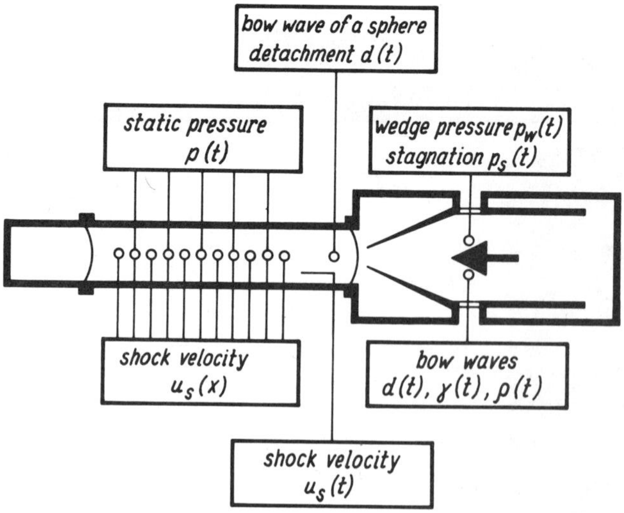

Figure 19
Measuring techniques used to study running times.

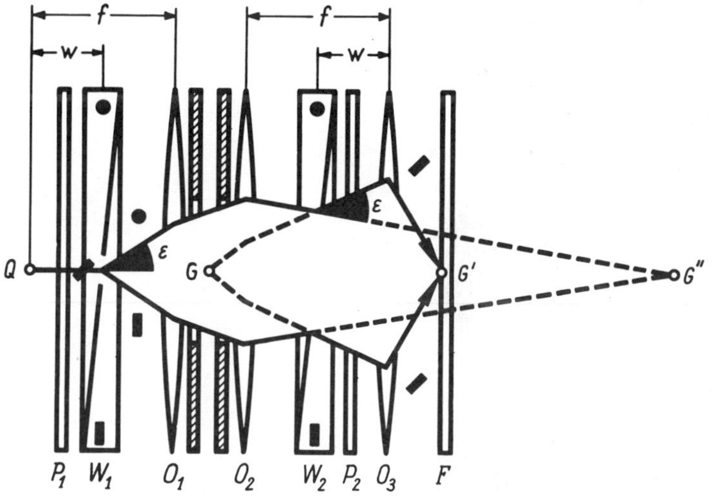

Figure 20
Differential interferometer

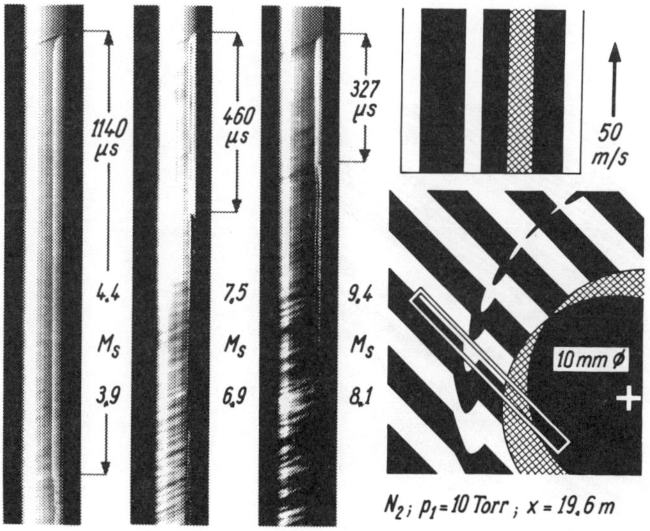

Figure 21
Streak-interferograms of bow waves in the driven section.

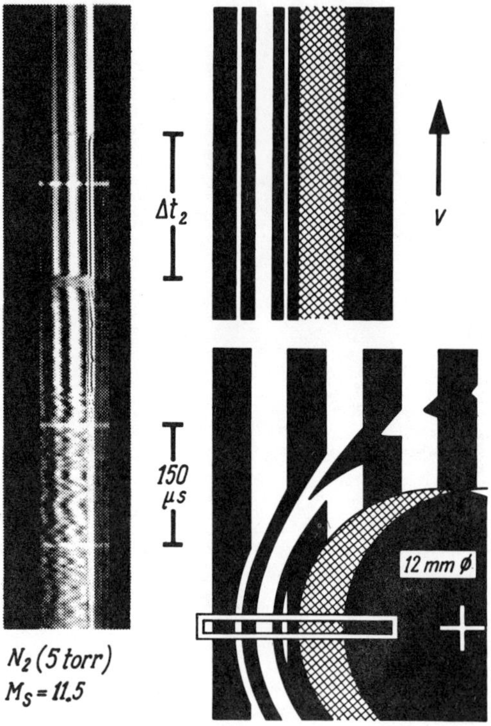

Figure 22
Streak-interferograms of bow waves in the driven section

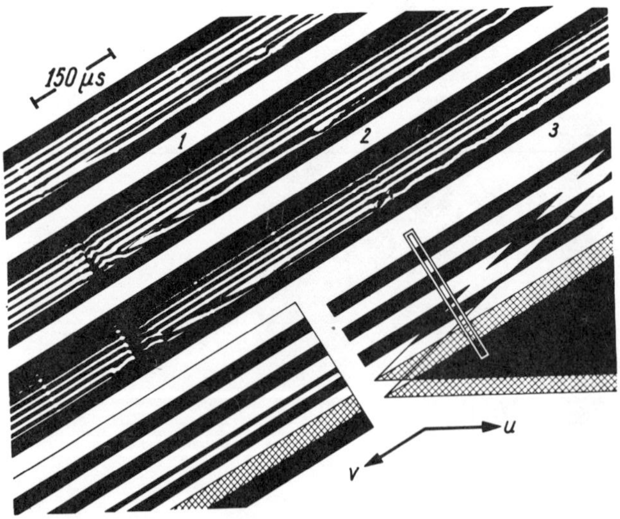

Figure 23
Streak-interferograms of bow waves at the test section center

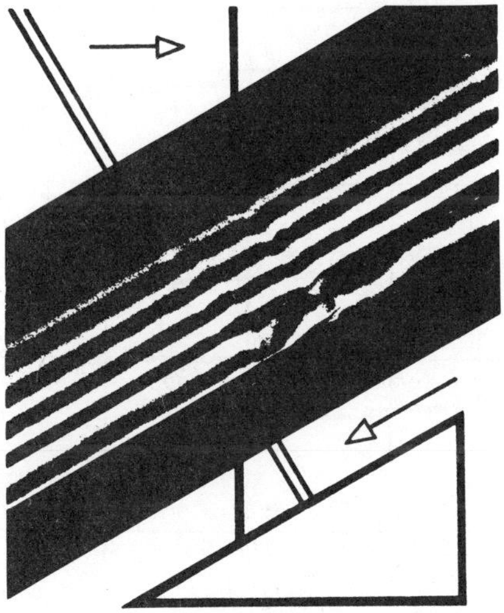

Figure 24
Traces of the shock W_2 and the interface

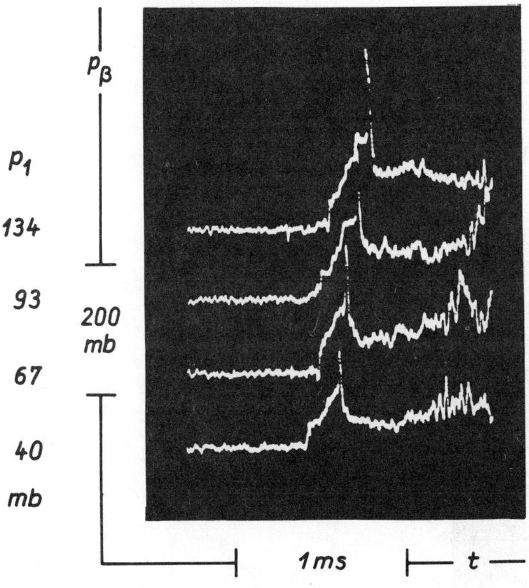

Figure 25
Pressures on a 30° wedge at the test section center.

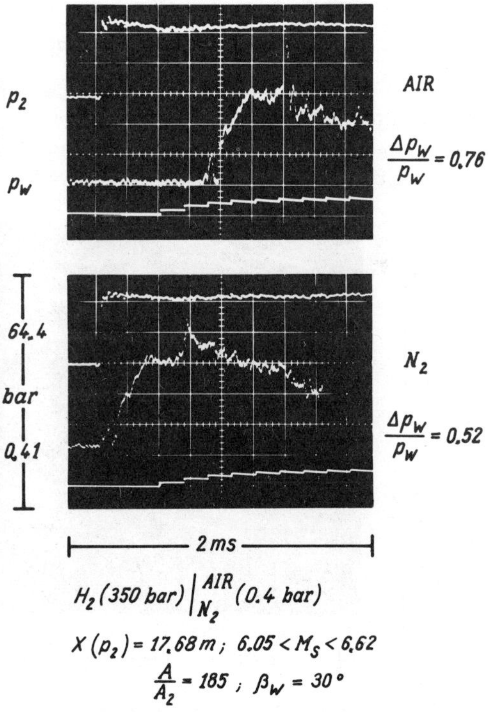

Figure 26
Pressures on the wall of the driven section and on a 30° wedge at the test section center.

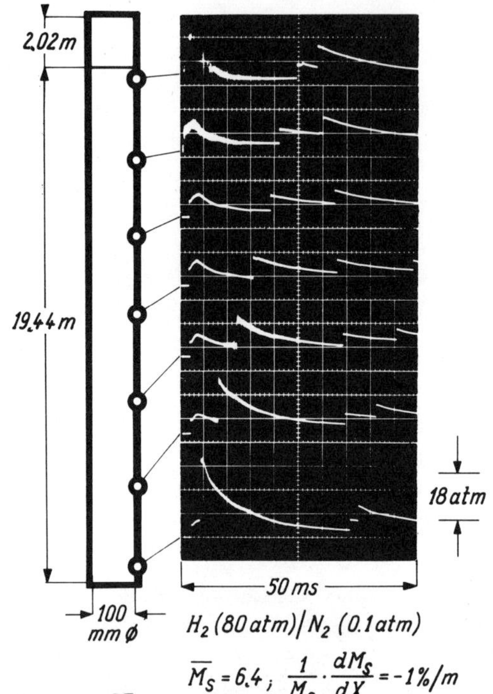

Figure 27
Pressures on the wall of the driven section when its end is closed.

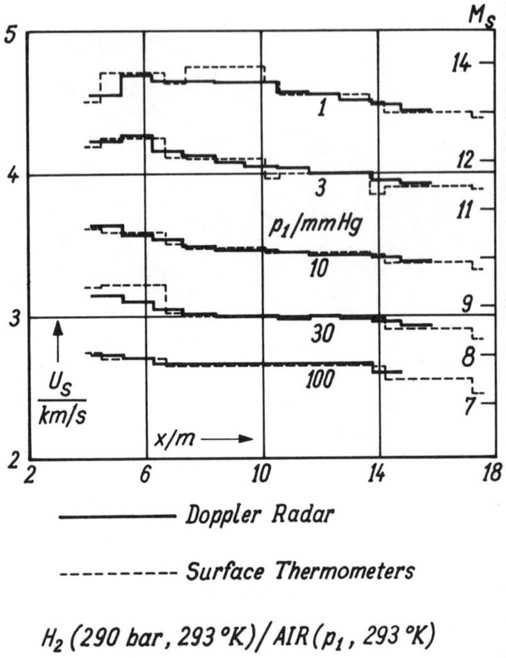

Figure 29
Shock velocities $u_s(x)$ measured by means of two different techniques.

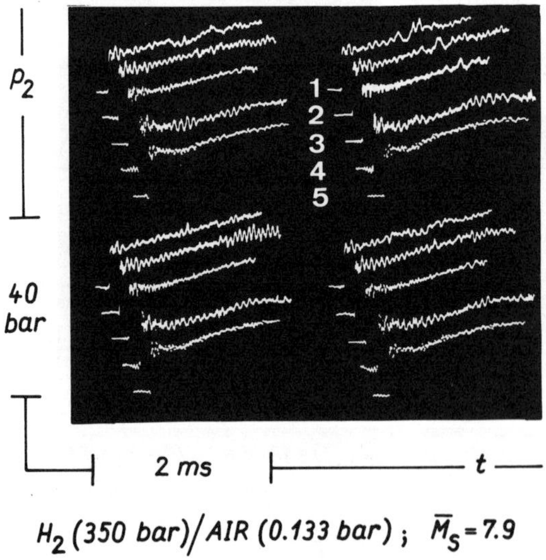

Nr.	1	2	3	4	5
x/cm	448	741	1059	1419	1768

Figure 28
Pressures on the wall of the driven section when its end is open.

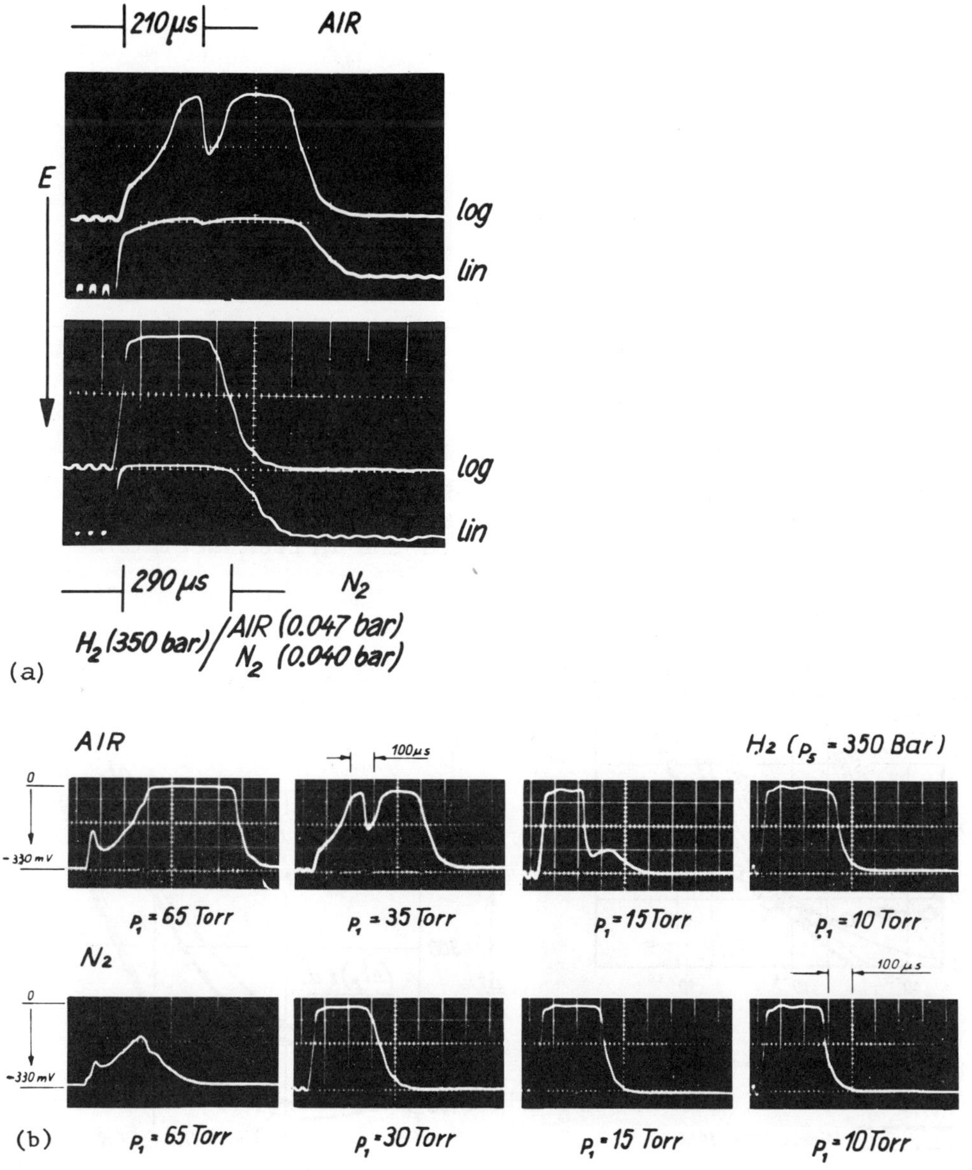

Figure 30
Transverse microwave transmission
ahead of the nozzle.

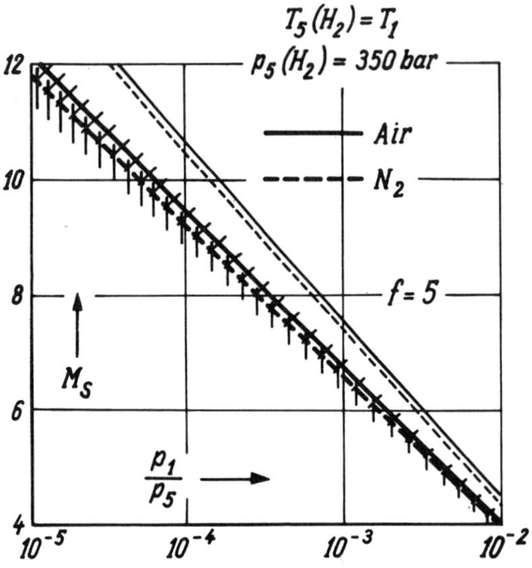

$$M_S + A = B_{10} \log \frac{p_5}{p_1}$$

Figure 31
Mean shock mach numbers in the driven section.

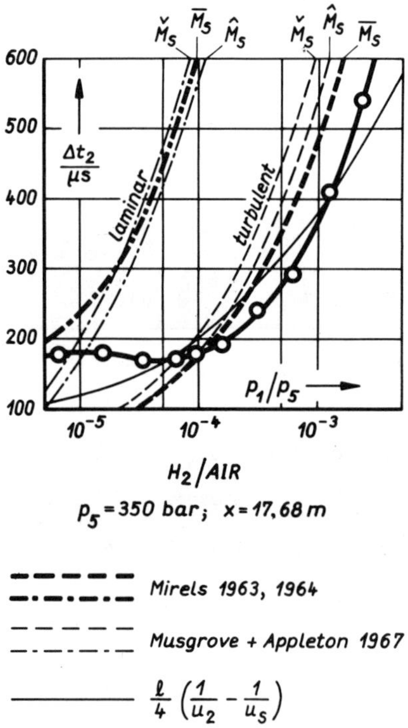

Figure 32
Measured and calculated durations Δt_2 of good flow at the nozzle entrance.

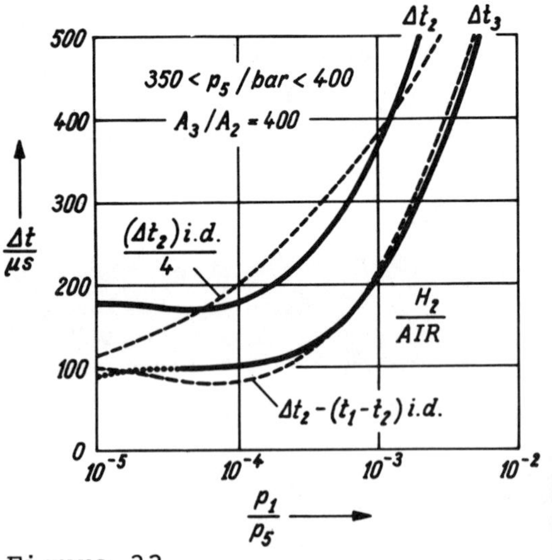

Figure 33
Measured and calculated durations Δt_3 of good flow at the test section center.

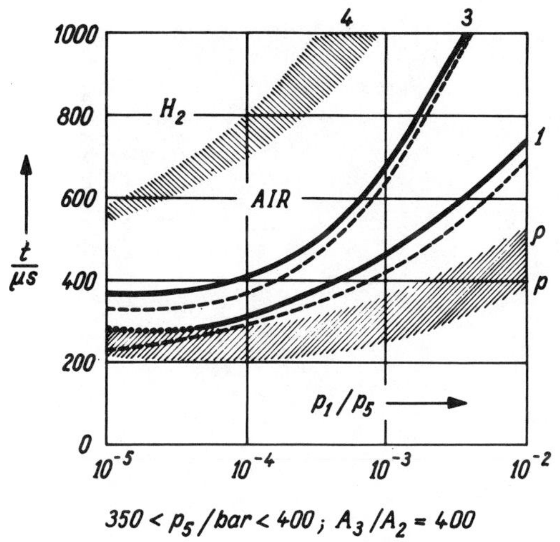

Figure 34
Measured and calculated times of arrival at the test section center.

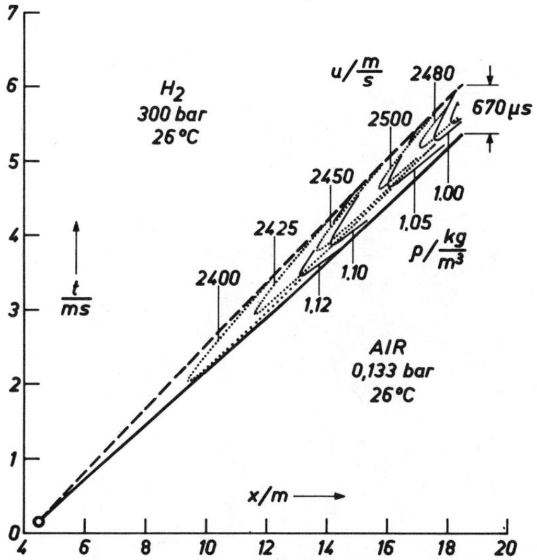

Figure 35
Flow field in the driven section calculated for measured shock velocities $u_S(x)$ and pressures $p(x, t)$.

Figure 36
The ISL-non-reflected shock tunnel.

FLOW FIELD BEHIND A SHOCK WAVE IN A LOW PRESSURE TEST GAS*

EDWARD W. HUBBARD** and P.C.T. de BOER
Cornell University

SUMMARY

A study is made of the two-dimensional flow field resulting from boundary layer growth in the driven gas of a shock tube. Attention is given to circular as well as to rectangular shock tubes. Formulas that describe the axial and transverse velocity components near the shock front are developed and are evaluated numerically. The properties of the two-dimensional flow field are discussed. It is found that the flow is nearly one-dimensional at distances greater than a tube radius from the shock front. In the region near the shock front, averaging of the perturbations gives results identical to those of Mirels' quasi-one-dimensional theory. An approximate description of the whole flow field can be obtained by taking the average velocity provided by Mirels' theory, and by superimposing the two-dimensional flow pattern stretched axially so as to conform to the prescribed average. The details of the flow field provided in the present paper can, therefore, be considered a complement to Mirels' work.

I. INTRODUCTION

The shock tube has become a popular tool for the study of high-temperature properties of gases. By applying the conservation relations across the shock wave, conditions in the high-temperature region between the shock wave and the contact surface can be found quite accurately. In principle, deviations from uniformity may be attributed to such real gas effects as vibrational or chemical relaxation or ionization, and studies of these deviations provide information about real gas processes. Unfortunately, the actual situation is not that simple because real gases also possess viscosity which leads to the development of a boundary layer at the shock tube wall. Acting

* Work supported in part by the U.S. Office of Naval Research.

** Present address: IBM Electronic Systems Center, Owego, New York.

as an aerodynamic sink, the boundary layer affects the uniformity of the shocked gas. The results obtained for the processes under study are of limited value unless the environmental conditions can be determined.

During the past decade considerable effort has been directed toward the theoretical description of the flow of a viscous gas in a shock tube. Excellent summaries of this work are given by Mirels, who has made many contributions to this subject[1-4]. He has been concerned mainly with the so-called "limiting condition," in which the total sink strength of the boundary layer exactly balances the mass flux through the shock wave. The contact surface moves at the same speed as the shock wave, so that in shock-fixed coordinates the flow is steady with the velocity decreasing from its Rankine-Hugoniot value behind the shock wave to zero at the contact surface. This case is of interest because the corresponding distance between shock wave and contact surface is the maximum test length obtainable[5]. Being a steady flow, it is also most easily amenable to theoretical treatment.

In reference 1 Mirels presents a theory for calculation of the maximum test length and non-uniformity of velocity in the presence of a laminar boundary layer. The boundary layer is assumed thin compared to the shock tube radius and is replaced by a sink distribution of appropriate strength. The inviscid core is assumed to be quasi-one-dimensional (Q-1-D) in the sense that flow variations are assumed to occur only in the streamwise direction. A similar treatment for a turbulent boundary layer is contained in reference 2. In later papers, Mirels presents empirical formulas based on his numerical solutions for wall shear stress, heat transfer and boundary layer thickness behind a moving shock[3], and an algorithm for estimating the non-uniformities in pressure, temperature, and other flow properties at the limiting condition[4].

The Q-1-D model becomes quite inaccurate in the region near the shock wave, where two-dimensionality becomes a significant feature of the flow. The purpose of this paper is to calculate the properties of the two-dimensional flow field. We limit our attention to pressures sufficiently low to ensure that no effects of the breaking of the diaphragm persist in the flow[6]. The formulation of the problem is the same as that used in reference 7 for the calculation of shock front curvature.

II. TWO-DIMENSIONAL CHANNEL FLOW

In shock-fixed coordinates the flow is presumed to move from left to right (Fig. 1) between two parallel infinite flat plates. The distance x is measured downstream from the shock wave, while y is measured vertically from the centerline of the channel. Consistent with boundary layer theory, the flow in the channel is taken to be inviscid. The effect of the boundary layer is taken into account by allowing the presence of a sink distribution of appropriate strength along the walls for positive x.

The flow in the region behind the shock wave is subsonic in shock-fixed coordinates. Furthermore, the flow behind the shock wave is vorticity-free to first order in shock angle ε [7] (see Fig. 1). Hence, the velocity field is characterized by a potential function $\varphi(x,y)$ that describes the velocity field $u_e(x,y)$, $v_e(x,y)$ as perturbations $u(x,y)$, $v(x,y)$ of the Rankine-Hugoniot velocity u_2. That is,

$$u_e(x,y) = u_2 + u(x,y) = u_2 + \varphi_x(x,y), \tag{1-a}$$

$$v_e(x,y) = v(x,y) = \varphi_y(x,y), \tag{1-b}$$

where the subscript on φ denotes partial differentiation.

The velocity potential satisfies the equation

$$(1-M_2^2)\varphi_{xx} + \varphi_{yy} = 0 \qquad (x > 0, \ -R < y < R), \tag{2}$$

where M_2 is the Rankine-Hugoniot value of Mach number. To lowest order in shock angle ε, the upstream boundary condition is

$$\varphi_x(0,y) = 0 \tag{3}$$

The wall boundary condition is chosen to account for the sink effect of the boundary layer. The velocity at the wall must be tangent to a streamline that represents the boundary layer displacement thickness relative to the wall. It is assumed that the boundary layer displacement thickness δ^* grows as

$$\delta^* = -A_1 x^s. \tag{4}$$

Then

$$v(x, \pm R) = \mp F(x) = \pm s A_1 u_2 / x^{1-s}. \tag{5}$$

Equation (5) may be taken as a definition of $F(x)$. The parameter s is $\frac{1}{2}$ for a laminar boundary layer; for a turbulent boundary layer with velocity profiles given by a 1/7 or 1/5 power law, s takes on the values 4/5 and 3/4, respectively.

Setting $\beta = (1-M_2^2)^{1/2}$, scaling the x-coordinate by $x = \beta R \xi$ and the y-coordinate by $y = R\eta$, one transforms equation (2) into Laplace's equation

$$\varphi_{\xi\xi}(\xi,\eta)+\varphi_{\eta\eta}(\xi,\eta)=0. \tag{2'}$$

The boundary conditions become

$$\varphi_{\xi}(0,\eta)=0 \tag{3'}$$

$$\varphi_{\eta}(\xi,\pm 1)=\mp R\,F(\beta R\xi) \tag{5'}$$

The solution of this boundary value problem was found by de Boer[7,8] as

$$\varphi(\xi,\eta)=-\frac{2}{\pi}\int_0^{\infty}\frac{\cosh(k\eta)\cos(k\xi)-1}{k\,\sinh(k)}\,dk\int_0^{\infty}R\,F(\beta R\alpha)\cos(k\alpha)\,d\alpha. \tag{6}$$

The α -integration in equation (6) may be carried out explicitly when F is given by equation (5). The result may be differentiated under the integral sign to obtain the non-dimensionalized velocity components

$$u_0(\xi,\eta)\equiv-\frac{u(\xi,\eta)\beta\sin(\pi s/2)\Gamma(1-s)(\beta R)^{1-s}}{s\,A_1\,u_2}=\int_0^{\infty}\frac{\cosh(k\eta)\sin(k\xi)}{k^s\sinh(k)}\,dk, \tag{7}$$

$$v_0(\xi,\eta)\equiv-\frac{v(\xi,\eta)\sin(\pi s/2)\Gamma(1-s)(\beta R)^{1-s}}{s\,A_1\,u_2}=\int_0^{\infty}\frac{\sinh(k\eta)\cos(k\xi)}{k^s\sinh(k)}\,dk. \tag{8}$$

We will also be interested in the rate of change of $u_0(\xi,\eta)$ in the x-direction obtained by differentiating equation (7) with respect to ξ:

$$u_{0,\xi}(\xi,\eta) \equiv -\frac{u_x(\xi,\eta)\,\beta\sin(\pi s/2)\,\Gamma(1-s)\,(\beta R)^{2-s}}{s\,A_1\,u_2} = \int_0^\infty \frac{\cosh(k\eta)\cos(k\xi)}{k^{s-1}\sinh(k)}\,dk. \tag{9}$$

Each of the ratios of hyperbolic functions in the integrands may be expanded in an infinite series of exponentials. The integrations may then be carried out term-by-term, which leads to

$$u_0(\xi,\eta) = \Gamma(s)\,\mathcal{I}m\sum_{n=0}^{\infty}\left[(2n+1-\eta-i\xi)^{s-1} - (2n+1+\eta+i\xi)^{s-1}\right], \tag{10}$$

$$v_0(\xi,\eta) = \Gamma(1-s)\,\mathcal{R}e\sum_{n=0}^{\infty}\left[(2n+1-\eta-i\xi)^{s-1} - (2n+1+\eta+i\xi)^{s-1}\right], \tag{11}$$

$$u_{0,\xi}(\xi,\eta) = \Gamma(s)\,\mathcal{R}e\sum_{n=0}^{\infty}\left[(2n+1-\eta-i\xi)^{s-2} - (2n+1+\eta+i\xi)^{s-2}\right]. \tag{12}$$

The three-dimensional flow field behind a shock wave in a rectangular shock tube may be obtained by superimposing the perturbations for the two plane channel cases corresponding to each pair of side walls[7,8].

III. FLOW IN A TUBE OF CIRCULAR CROSS-SECTION

Velocity profiles may also be derived for axisymmetric flow in a tube of circular cross-section of radius R. In that case the perturbation potential satisfies the equation

$$\varphi_{\rho\rho} + \varphi_\rho/\rho + \varphi_{\xi\xi} = 0 \qquad (\rho = r/R < 1,\ \xi > 0), \tag{13}$$

with the boundary conditions

$$\varphi_\xi(0,1)=0\ , \qquad \varphi_\varrho(\xi,1)=-RF\,(\beta R\xi) \tag{14}$$

The solution of this boundary value problem is[7,8]

$$\varphi(\xi,\varrho)=-\frac{2}{\pi}\int_0^\infty \frac{I_0(k\ )\cos(k\xi)-1}{k\ I_1(k)}\,dk\int_0^\infty RF(\beta R\alpha)\cos(k\alpha)\,d\alpha\,, \tag{15}$$

where I_n denotes the modified Bessel Function of the first kind and the nth order. Using equation (5), carrying out the η -integration in equation (15), and differentiating under the integral sign yields the velocity profiles

$$u_0(\xi,\varrho)=\int_0^\infty \frac{I_0(k\varrho)\ \sin(k\xi)}{k^s\ I_1(k)}\,dk\,, \tag{16}$$

$$v_0(\xi,\varrho)=\int_0^\infty \frac{I_1(k\varrho)\ \cos(k\xi)}{k^s\ I_1(k)}\,dk\,, \tag{17}$$

$$u_{0,\xi}(\xi,\varrho)=\int_0^\infty \frac{I_0(k\varrho)\cos(k\xi)}{k^{s-1}I_1(k)}\,dk. \tag{18}$$

No convenient series expansions were found for the ratios of Bessel functions in (16-18). For this reason it was not possible to integrate these equations analytically. Instead, determination of the properties of the axisymmetric flow required the use of numerical integration, which is described in the next section.

IV. EVALUATION AND DISCUSSION OF THE VELOCITY PROFILES

The velocity profile functions u_0, v_0, and $u_{0,\xi}$ are given by the Fourier integrals (7)-(9) and (16)-(18). In principle, the evaluation of (7)-(9) may be performed by summation of the series in equations (10)-(12). However, these series converge quite slowly and it is more convenient to evaluate the integrals (7)-(9) numerically. The numerical evaluation of the integrals is illustrated by considering (16). We limit ourselves to laminar boundary layers ($s = \frac{1}{2}$). The range of integration is split into three parts, the contributions of which are denoted by I_A (for $0<k<A$, $A<1$), I_B (for $B<\rho<\infty$, $B>1$), and I_{AB} (for $A<k<B$). The contributions I_A and I_B are obtained by analytic integration of the leading terms of an infinite series and of an asymptotic series, respectively. Provided $A<1$, the integral I_A is represented by

$$I_A = \xi A^{1/2}[4 + A^2(6\rho^2 - 4\xi^2 - 3)/30$$
$$+ A^4(15\rho^4 - 40\rho^2\xi^2 + 8\xi^4 - 30\rho^2 + 20\xi^2 + 10)/2160 + O(A)^6]. \tag{19}$$

Similarly, when $B\rho$ is large compared to one the asymptotic representation of I_B is

$$I_B \sim \rho^{-1/2}\int_B^\infty \frac{\sin(k\xi)}{k^{1/2}} e^{-k(1-\rho)}dk + \frac{1+3\rho}{8\rho^{3/2}}\int_B \frac{\sin(k\xi)}{k^{3/2}} e^{-k(1-\rho)}dk$$
$$+ \frac{3}{64\rho^{1/2}}\left(\frac{3}{2\rho^2} + \frac{1}{\rho} + \frac{11}{2}\right)\int_B \frac{\sin(k\xi)}{k^{5/2}} e^{-k(1-\rho)}dk + \ldots \tag{20}$$

Each of the integrals in equation (20) integrates into a complex expression involving $(1-\rho \pm i\xi)$ and incomplete Gamma functions whose second arguments are $(1-\rho \pm i\xi)B$. These expressions may be simplified by using the asymptotic expansion for the incomplete Gamma function, provided that $[(1-\rho)^2+\xi^2]^{1/2}B$ is large compared to one. After considerable algebra one obtains

$$I_B \sim \frac{e^{-B(1-\rho)}}{D(B\rho)^{1/2}}\left\{\sin(C+B\xi)+\frac{1}{B}\left[\frac{1+3\rho}{8\rho}\sin(C+B\xi)-\frac{\sin(2C+B\xi)}{2D}\right]\right.$$

$$+\frac{1}{B^2}\left[\frac{3}{64}\left(\frac{3}{2\rho^2}+\frac{1}{\rho}+\frac{11}{2}\right)\sin(C+B\xi)-\frac{3(1+3\rho)}{16D\rho}\sin(2C+B\xi)\right.$$

$$\left.\left.+\frac{3}{4D^2}\sin(3C+B\xi)\right]\cdots\right\}, \qquad (21)$$

where

$$C \equiv \tan^{-1}[\xi/(1-\rho)], \qquad (22)$$

$$D \equiv [(1-\rho)^2+\xi^2]^{1/2}. \qquad (23)$$

As stated before, this result is subject to the conditions

$$[(1-\rho)^2+\xi^2]^{1/2}B \gg 1, \qquad (24)$$

$$B\rho \gg 1. \qquad (25)$$

When $\rho=0$, the asymptotic series for $I_0(k\rho)$ must be replaced by $I_0(0)=1$, and the calculation must be repeated. The result, valid for $B \gg 1$, is

$$I_B(\rho=0) \sim \frac{(2\pi)^{1/2} e^{-B}}{F} \left\{ \sin(E+B\xi) + \frac{3}{8B} \sin(E+B\xi) \right.$$

$$\left. + \frac{1}{B^2} \left[-\frac{3}{8F} \sin(2E+B\xi) + \frac{33}{128} \sin(E+B\xi) \right] \right\}, \quad (26)$$

where E and F are C and D, respectively, evaluated for $\rho = 0$.

In the numerical computations A was taken to be 1/4. The expressions $[(1-\rho)^2+\xi^2]^{1/2} B$ and $B\rho$ are indicative of the accuracy of the asymptotic expansions and their values were kept in the neighborhood of 10. Thus B varied in value from 10 to about 100.

The third contribution to $u_0(\xi, \rho)$ is obtained by numerical evaluation of the integral

$$I_{AB} = \int_A^B \frac{I_0(k\rho)\sin(k\xi)}{k^{1/2}\, I_1(k)}\, dk. \quad (27)$$

The integrand oscillates with frequency ξ so that even at moderate values of ξ any integration scheme based on polygonal approximations to the integrand is rendered prohibitively difficult. Filon[10] has developed an algorithm in which the relation between accuracy and integration step size is governed only by the monotonic part of the integrand, thus facilitating the evaluation of sine or cosine integrals such as we have here. The most accessible discussion of Filon's algorithm is contained in reference 11. The algorithm itself is given concisely for both the sine and cosine integrals in reference 12.

By means of the method outlined above, the Fourier integrals of equations (7)-(9) and (16)-(18) were evaluated on the IBM 360/65 at Cornell University. The computational accuracy of the results obtained is estimated to be on the order of 0.1%. The results are plotted in Figs. 2-6. The profiles of u_0 and v_0 given in Figs. 2-5 indicate that the velocity field becomes approximately one-dimensional far downstream from the shock wave. The axial velocity component loses its dependence on the transverse coordinate, and the transverse velocity component becomes a linear function of the distance from the centerline. The idea of quasi-one-dimensionality, which forms the basis of much of Mirels' work, is thus shown to be a natural approximation in the region away from the shock wave. However, near the shock wave the flow field exhibits considerable two-dimensionality.

Mirels[1] has found that the axial velocity is well represented by

$$u_e/u_2 = 1 - (x/x_m)^{1/2}, \tag{28}$$

where x_m is the distance between the shock wave and contact surface. Equation (28) illustrates a serious difficulty with the Q-1-D theory: the slope of the axial velocity profile du_e/dx has a square-root singularity as $x \to 0$ for all values of y (or r). This difficulty arises because the Q-1-D approximation is not valid near the shock wave. In the two-dimensional treatment the leading edge singularity is relegated to its proper place at the foot of the shock wave (cf. Fig. 6). The presence of this singularity causes the transverse velocity component as a function of η to pass through a maximum, provided ξ is not too large (see Figs. 3 and 5). The influence of the singularity is also evident in the behavior of the streamwise velocity component (Figs. 2 and 4). However, it should be kept in mind that the linearized theory breaks down in the region near the singularity. It is not clear to what extent the behavior of the velocity components just discussed is real, or to what extent it is an artifact arising from the boundary layer approximation and from other assumptions made.

V. AVERAGE VALUES OF VELOCITY AND DENSITY

It is of interest to find average values of various perturbation quantities. Rather than proceeding from the detailed results obtained, we start from first principles. Applying the continuity equation to a volume encompassed by the shock tube walls together with the shock wave and a plane at distance x downstream of the shock wave, we have (see Fig. 1)

$$\rho_2 u_2 A = L\int_0^x (\rho_e v)_{y=R}\, dx + \int_A \rho_e u_e\, dA. \tag{29}$$

Here, A is the cross section of the shock tube, and L its perimeter. In the treatment given previously we have set

$$v(y=R) = -u_2\, d\delta^*/dx,$$

$$\rho_e = \rho_2 + \rho \quad ,$$

$$u_e = u_2 + u \, .$$

Substituting these relations in (29), and neglecting higher order terms gives

$$0 = -L\delta^*(x) + \int_A \left(\frac{\rho}{\rho_2} + \frac{u}{u_2}\right) dA. \tag{30}$$

The perturbation density ρ is related to the perturbation velocity u by the linearized Bernoulli equation and the isentropic relation between pressure and density. These give

$$\rho/\rho_2 = -M_2^2 \, u/u_2 \, , \tag{31}$$

and hence

$$\frac{\bar{u}}{u_2} \equiv \frac{1}{A}\int_A \frac{u}{u_2}\, dA = \frac{4\delta^*}{(1-M_2^2)\,d} \, , \tag{32}$$

$$\frac{\bar{\rho}}{\rho_2} \equiv \frac{1}{A}\int_A \frac{\rho}{\rho_2}\, dA = -\frac{4M_2^2\,\delta^*}{(1-M_2^2)\,d} \, , \tag{33}$$

where $d = 4A/L$ is the hydraulic diameter.

It is also possible to derive these results starting with the quasi-one-dimensional treatment given by Mirels[1]. The last integral in (29) then is replaced by $\rho_e u_e A$, where ρ_e is the density and u_e the velocity of the main flow at x. Setting $u_e = u_2 + \bar{u}$, $\rho_e = \rho_2 + \bar{\rho}$ gives

$$0 = -L\rho_2 u_2 \delta^*(x) + \left(\frac{\bar{\rho}}{\rho_2} + \frac{\bar{u}}{u_2}\right)\rho_2 u_2 A, \tag{34}$$

which is equivalent to (30). The rest of the derivation of (32) and (33) remains essentially unchanged. It can be concluded that under the assumption made for the boundary layer displacement thickness ($\delta^* = A_1 x^{1/2}$, A_1 independent of x), the linearized two-dimensional theory gives the same result for average quantities in the main flow as does the Q-1-D theory. The latter offers the advantage of providing a straightforward formalism for making corrections in δ^* arising from gradual changes in the free stream properties. This formalism consists of a step-by-step calculation starting at the shock wave. It cannot be used in the two-dimensional calculation. The agreement between the two theories is, therefore, limited to the region near the shock wave, where the corrections in δ^* still are negligible. In this region the Q-1-D theory is correct "on-the-average," while the details of the flow are given by the two-dimensional theory.

It should be kept in mind that in order to obtain the integrated density across the shock tube, the excess integrated density in the two boundary layers must be added to the integrated density of the main flow. This excess integrated density is given by the free stream density ρ_e times the boundary layer "density thickness"

$$\delta_\rho = \int_0^\infty (1 - \rho/\rho_e)\, dy .$$

VI. COMBINATION OF TWO-DIMENSIONAL AND Q-1-D DESCRIPTIONS

The two-dimensional theory provides a correct description of the flow near the shock wave, but breaks down for large x because the perturbations in velocity are no longer small compared to u_2, as was assumed. However, the breakdown is not severe. The reason for this is that the free stream Mach number decreases as x increases. Consequently, the flow becomes more and more incompressible, and thus becomes subject to the equation for incompressible flow

$$\Phi_{xx} + \Phi_{yy} = 0, \tag{35}$$

where Φ is the velocity potential. This equation must be compared to equation (2) for the perturbation potential φ. The two equations differ by a factor $1 - M_2^2$ in the first term. Provided the shock wave is not too weak, this factor is never very different from unity. To lowest accuracy, therefore, the results based on equation (2) could be used all the way up to the contact surface. Besides being based on an incorrect partial differential equation, the results obtained this way are also open to the objection that they are derived using the relation $\delta^* = A_1 x^{1/2}$ rather than a function δ^* properly calculated by taking account of the variations in free stream properties. These variations arise from two different effects. One effect finds its origin in the gradual loss of mass from the main flow to the boundary layer. This effect was shown by Mirels[1] to be quite important in the downstream part of the flow. The second effect is connected with leading edge corrections to the boundary layer. These corrections are necessary because the free stream perturbation velocities become quite large near the leading edge, and hence the free stream properties are not uniform as is assumed in the calculation of the boundary layer properties. Finding these corrections is a difficult problem in viscous flow theory. However, it is likely that they will not be of great importance for the main pattern of the flow, and we assume that they can be left out of account. In that case, Mirels' theory gives the correct average flow velocity everywhere. Furthermore, the two-dimensional theory then gives the correct flow pattern in the region near the shock wave, while it is not drastically incorrect further downstream, where it predicts the flow to be nearly uniform. An approximate solution for the whole flow field is obtained, therefore, by starting with the average velocity provided by Mirels' theory, and by superimposing the two-dimensional flow pattern stretched in the x-direction so as to conform to the prescribed average.

A criterion for the validity of the two-dimensional theory is the condition $u(x,y)/u_2 \ll 1$. Similarly, a measure for the degree to which the flow may be considered one-dimensional is given by the inequality $[u(x,R)-u(x,0)]/u_2 \ll 1$. Taking $y=0$ in the first condition, and adding the left hand side yields the quantity $u(x,R)/u_2$. The minimum of this quantity, which occurs at $\xi = x/\beta R = 0.55$, roughly separates an upstream region where two-dimensional effects are important from a downstream region where the flow is not far from one-dimensional. We have calculated the corrections in δ^* at $\xi = 0.55$ arising from flow non-uniformity according to Mirels' theory for some typical cases. These cases concerned air at initial pressures P_1 of 1 and 10 torr over a Mach number of 3 to 22. It was found that the corrections always were very small, and that leaving them out (i.e., using the relation $\delta^* = A_1 x^{1/2}$ with A_1 independent of x up to $\xi = 0.55$) had a negligibly small effect on the test parameter $x_m P_{st}/d^2 P_1$ $(P_{st} = 760$ torr$)$. Apparently, very little stretching of the two-dimensional flow pattern is required in the region upstream of $\xi = 0.55$. The stretching occurs mainly in the region downstream of $\xi = 0.55$, where the flow is nearly one-dimensional, and where as a consequence the stretching is justifiable. It can be concluded that the approximate picture proposed provides a quite accurate description of the whole flow in the cases considered and, by inference, for most cases of practical interest.

VI. CONCLUDING REMARKS

A study has been made of the effect of boundary layer growth on the non-uniformities of the flow field in the shocked-gas region of a shock tube. Formulas describing the two-dimensional details of the velocity field are presented in the form of Fourier integrals and are evaluated by means of numerical quadrature and series expansion near the end points. The results indicate that the flow field is significantly non-uniform in the region within about half a tube radius of the shock wave, while it rapidly approaches a one-dimensional situation farther downstream. The solutions obtained contain a singularity at the foot of the shock wave, implying a breakdown of the theory in the region near this foot. Apart from leading edge effects on the boundary layer displacement thickness, Mirels' Q-1-D velocity profile retains its validity as the profile of the average streamwise velocity component, a fact which justifies the use of Mirels' results to interpret shock tube data obtained by interferometry. The present model complements the Q-1-D result, and an approximate picture of the whole flow may be constructed by superposition.

In principle, the calculations could be improved by taking proper account of the interaction between the growth of the boundary layer and the development of the two-dimensional free stream. However, since the boundary layer parameter A_1 does not change very much in the region where two-dimensional effects are of importance, it is believed that the present results are sufficiently accurate for most cases in which two-dimensionality plays a role.

REFERENCES

1. H. Mirels, Phys. Fluids 6, 1201 (1963).
2. H. Mirels, AIAA J. 2, 84 (1964).
3. H. Mirels, Phys. Fluids 9, 1265, (1966).
4. H. Mirels, Phys. Fluids 9, 1907 (1966).
5. R. E. Duff, Phys. Fluids 2, 207 (1959).
6. H. W. Liepmann and R. M. Bowman, Phys. Fluids 7, 2013 (1964).
7. P.C.T. de Boer, Phys. Fluids 6, 962 (1963).
8. P.C.T. de Boer, Ph.D. Thesis, Univ. of Maryland (1962) (University Microfilms, Ann Arbor, Mich.).
9. I. S. Gradshteyn and I. M. Ryzhik, Tables of Integrals, Series and Products (Academic Press, New York, N. Y., 1965), p. 489.
10. L. N. G. Filon, Proc. Royal Soc. Edinburgh XLIX, 38 (1928-29).
11. G. J. Tranter, Integral Transforms in Mathematical Physics (John Wiley & Sons, Inc., New York, N.Y., 1951), pp. 67-72.
12. M. Abramowitz and I. Stegun, Handbook of Mathematical Functions (Dover Publications, Inc., New York, N.Y., 1965), p. 890.

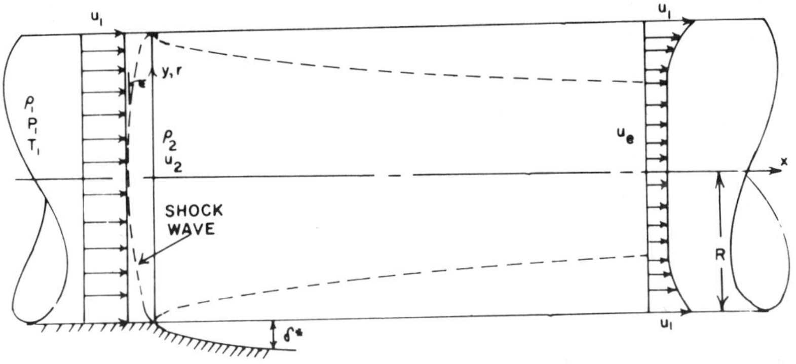

Figure 1
Geometry of flow in a shock tube.

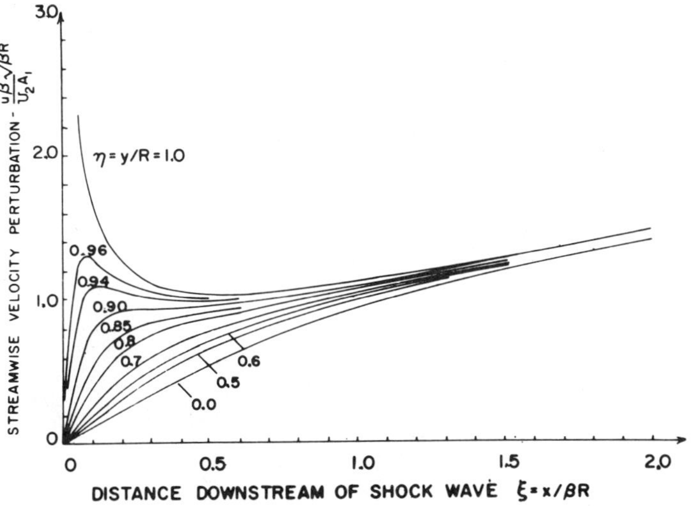

Figure 2
Streamwise velocity perturbation for channel flow.

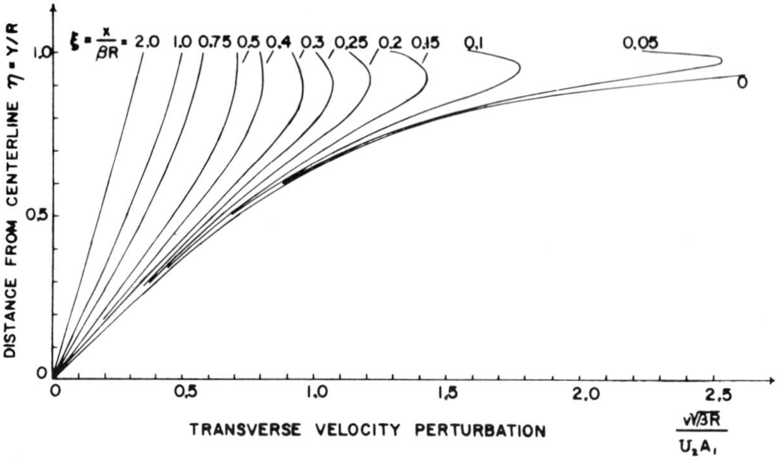

Figure 3
Transverse velocity perturbation for channel flow.

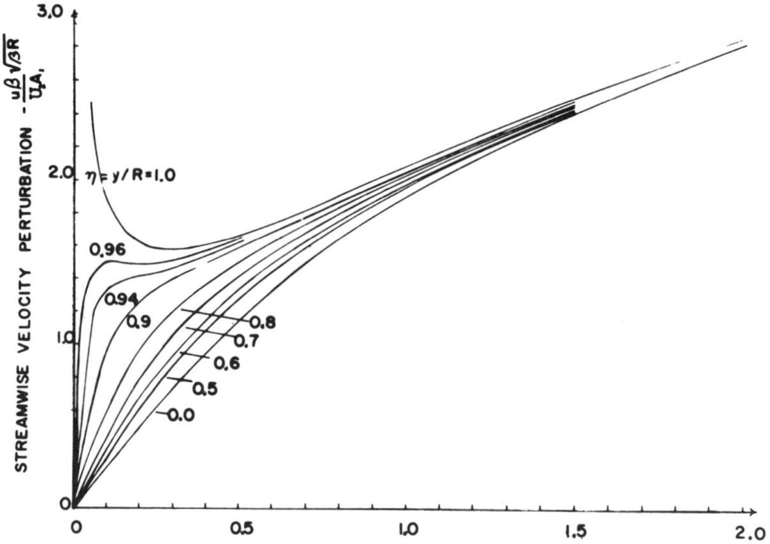

Figure 4
Streamwise velocity perturbation for flow in a tube of circular cross-section.

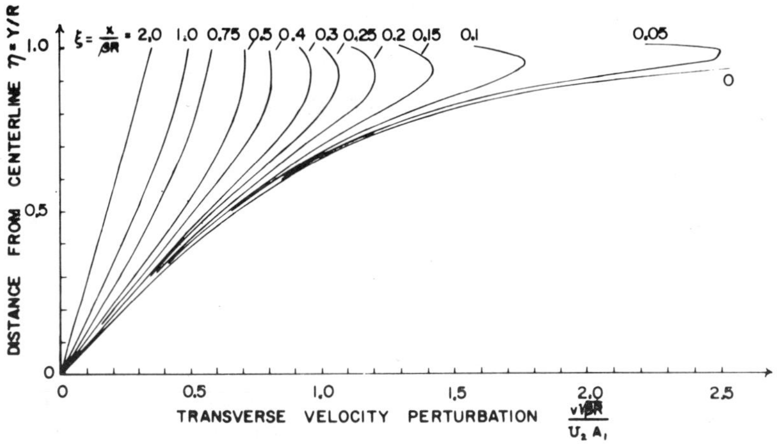

Figure 5
Axial velocity perturbation for flow in a tube of circular cross-section.

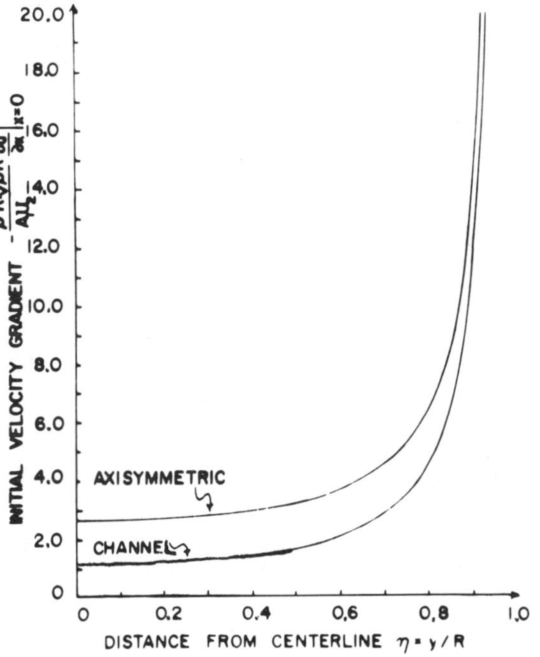

Figure 6
Velocity gradients at the shock front.

EXPERIMENTAL SHOCK TUBE TEST TIME - TURBULENT REGIME*

JOHN J. LACEY, JR.
ARO, Inc.
Arnold Air Force Station, Tennessee

SUMMARY

This report covers measurements of shock tube run time over the range of initial pressures from 100 to 0.5 psi, and shock Mach numbers from 2.1 to 10.5. Three different driver gases were used; N_2, He, and H_2. The driven gases used were air and nitrogen.

Pitot pressure measurements are used to infer shock tube run time and this run time is compared to that calculated by Mirels' turbulent theory. The data are correlated by a viscous parameter derived from Mirels' work, but for long driven tubes the experimentally determined run time is less than that predicted using Mirels' theory. Apparent interface burning is noted for the combination of hydrogen and air.

NOMENCLATURE

a	Speed of sound
A	Area
d	Diameter
J, J_I, J_x	Shock tube facility designation, see Equipment Section
ℓ_m	Maximum length of test gas, (Ref. 4)
L	Length of tube (see Fig. 1)
M_s	Shock Mach number, U_s/a_1
P	Pressure
P_o'	Pitot pressure
t	Time
T	Temperature
U_s	Shock speed
W	Density ratio, ρ_2/ρ_1
x	Distance from diaphragm
ρ	Density
τ	Shock tube run time, see Figure 1

Subscripts

1,2,3,4	Region of flow, see Figure 1

*The work reported herein was supported by the Arnold Engineering Development Center, Air Force Systems Command, under Contract F40600-69-C-0001 with ARO, Inc.

c	Calculated
i	Inviscid
P	Predicted using Mirels' theory (Ref. 4)
xy	Double subscript denotes ratio, e.g. $\rho_{21} = \rho_2/\rho_1$

INTRODUCTION

The test time of a shock tube determines, in many cases, the usefulness of the facility. Whether it is used to drive a shock tunnel or simply used by itself, the shock tube must have a test time long enough to obtain the desired data. In the conventional shock tube, the test gas is placed in the driven tube and heated and accelerated by the incident shock wave. This accelerated flow is called region 2 (see Figure 1 for wave diagram). The test time in region 2 increases directly as driven tube length increases if inviscid flow is assumed. However, viscous effects become more pronounced as driven tube length increases. Various authors have analyzed these viscous effects for the cases that have laminar boundary layers in region 2; see References 1-3. Mirels (Ref. 4) studied the viscous effects for a turbulent boundary layer and defined the conditions for which the boundary layer is turbulent.

Brabbs and Belles (Ref. 5) have published experimental test time results for argon with a turbulent boundary layer. Their data appear to be in fair agreement with predictions based on Mirels' theory (Ref. 4). Dunn presents test time data for air in Ref. 6; however, the conditions are predicted to be transitional and the run times are higher than calculations based on Mirels' theory. The present report presents test time data for air at fully turbulent conditions, as defined by Mirels.

DISCUSSION

The results of Mirels (Ref. 4) were derived in terms of the maximum separation distance (between shock and interface) ℓ_m and density ratio W (i.e. ρ_{21}). These parameters have been evaluated for real air using the shock crossing data of Lewis and Burgess (Ref. 7) and Figs. 3b and 5b of Mirels (Ref. 4). A viscous parameter $[(L_1/d_1)(d_1P_1)^{-1/4}]$ is the product of the ordinate and abscissa of Figs. 3b and 5b of Ref. 4 respectively and the density ratio. The ratio of predicted test time to inviscid test time decreases as this viscous parameter increases as shown in Fig. 2. Therefore, the shock tube conditions pertinent to test time predictions are driven tube length (L_1), diameter (d_1), and pressure (P_1).

Mirels uses the term d_1P_1 to determine if the flow will be laminar or turbulent. The value of d_1P_1 at which the boundary layer becomes turbulent is shown in Fig. 3, which is a cross-plot of Fig. 6a, Ref. 4. Note that higher values of d_1P_1 are required for turbulent boundary layer conditions at higher shock Mach numbers and that for each shock Mach number there exists a minimum value of d_1P_1 for fully turbulent flow behind the incident shock wave. The experimental conditions are shown on Fig. 3 and are above this minimum value;

therefore the region 2 boundary layer is turbulent for the conditions in this report.

The test conditions cover a wide range of the viscous parameter. Figure 4 presents the range of the viscous parameter for the shock tubes used in the present study and it is seen to vary from 45 to 250 $(\text{atm-ft})^{-1/4}$. The analysis based on Mirels' theory given in Fig. 2 leads to the estimate that the ratio of actual test time to inviscid test time will vary from 0.80 to 0.44.

The end of the useful test time is determined by the arrival of the interface. Ideally, there is a step-function change of flow properties across the interface. In particular, the pitot pressure changes and therefore a pitot probe can be used to detect the arrival of the interface. An analysis assuming one-dimensional inviscid flow can be used to calculate the pitot pressure ratio across the interface (P'_{o3}/P'_{o2}). Figure 5 shows the results of the calculation.

EQUIPMENT

Shock Tubes

The three shock tubes used in this study are described below.

<u>"J".</u> The "J" tube is shown in Fig. 6. The overall driven tube length is 43 feet and the inside diameter is 4 inches. Instrumentation ports are spaced at 3-foot intervals except for couplings at 18 and 36 feet. The pitot rake was installed at these coupling locations. The driver is 18 feet long and can be plugged to shorter lengths; for these tests the effective driver length was 5.3 feet. The driver internal diameter is 3.5 inches. Driver pressure for these tests ranged from 3600 to 18,000 psi with either hydrogen or helium as the driver gas. Two driver temperatures were used; 300°K and 500°K. External resistance heaters maintain the desired driver temperature.

<u>"J_I".</u> Shock tube J_I is shown in Fig. 7. The lengths of the driver and driven tubes are 2.1 ft and 7.4 ft, respectively, and the inside diameter is 2.5 inches for both tubes. The driver gas was helium or nitrogen at about 110 psia for all runs. The pitot probe is located 5.9 feet from the diaphragm. Ports at 3 stations (L_1 = 3.2, 4.2 and 5.2 ft) provide instrumentation locations for shock detection and side wall pressure measurements.

<u>"J_X".</u> A photograph and sketch of J_X are presented in Fig. 8. The driven tube is 21 feet long and 1.58 inches I.D. Couplings at 7 and 14 feet permit operation in the buffered mode so that driven tube length can be varied (7, 14, or 21 feet). The driver is 7 feet long and 2.25 inches I.D. Driver gas was helium at about 7000 psi and 300°K for all runs. The pitot probe is located at the entrance of a non-reflected-type, 7 deg half-angle nozzle. Wall pressures were measured at three driven tube locations (x = 6, 13, and 20 feet) and shock velocity was obtained from shock detectors over the last 5 feet of driven tube.

Instrumentation

Transducers used for shock detection and pressure measurement are shown in Fig. 9. The shock detectors were either ionization probes, thin-film resistance thermometers or pressure transducers. Piezoelectric transducers were used for pressure measurements. The operational principles of these types of transducers are summarized by Glass and Hall (Ref. 8).

Data records are photographs of oscilloscope sweeps (see Fig. 10). The pressure data are displayed on single-sweep oscilloscopes and the shock detector data are processed into an oscilloscope operating in the raster sweep mode. The sweep rate of each oscilloscope is verified by blanking pulses from a crystal-controlled time mark generator.

The pitot pressure models are shown in Fig. 11. In the "J" shock tube a 3-probe rake was used. The three probes were 3/8-inch-diam located on centerline, 3/4" below, and 1" above centerline. The "J_I" pitot pressure transducer was mounted in the end of a 3/8-inch-diam sting approximately 18" upstream from the endwall. The probe for shock tube "J_X" was a 1-inch-diam sting supporting a 3/8-inch-diam hemisphere-cylinder model. In all cases the piezoelectric transducer diaphragm was coated with a rubber-like thermal insulation to protect the transducer from heat during a run.

RESULTS

Pitot Pressure Measurements

Typical pitot pressure records are shown in Fig. 12. The rise time (from initial P_1 to P'_{o2}) is small compared to the test time and a reasonably level P'_{o2} is achieved behind the incident shock. The end of region 2 test time is indicated by a second rise (or dip) assumed to be the interface. The region of flow behind this second rise produces a second level, which is usually more noisy than the first. This second level is assumed to be P'_{o3}. Some traces (e.g. upper right hand record in Fig. 12) show a region between P'_{o2} and P'_{o3} levels, in which the pitot pressure increases gradually up to P'_{o3}. This increase is indicative of an interface mixing region. The gradual decrease following P'_{o3} is caused by the reflected driver fan.

Figure 13 shows a comparison of the pitot pressure ratios (P'_{o3}/P'_{o2}) with values obtained from Fig. 5. The experimental values are lower than the predicted values; however, the difference is not large. The general agreement indicates that regions 2 and 3 are properly described.

Test Time Measurements

The duration of the first level is the region 2 test time. Examples of this measurement are included in Fig. 12. The test time, measured from shock arrival to the interface arrival, is indicated on the records in Figure 12 by arrows. The inviscid region 2 test time (τ_i) is a function of M_s (except for real gas effects) and driven tube length (L_1) as given in Ref. 2.

$$\frac{\tau_i}{L_1} = \frac{1}{M_s a_1} \frac{1}{\rho_{21} - 1} \qquad (1)$$

Figure 14 compares the inviscid ratio of test time to tube length, with the experimental ratio obtained from the pitot pressure records. The experimental test time decreases as M_s increases, as in equation 1, and is less than the inviscid value. The observed test time was not affected by changes of the driver gas.

Figure 15a presents the comparison of experimental test time ratios and values calculated as in Figure 2. Most of the experimental test times are lower than the calculated value. The ratio of experimental to calculated test time is about 1.0 for tubes of small L_1/d_1, but it appears that the calculated time based on Mirels' theory is too high at large L_1/d_1. The ratio of experimental to calculated test time (τ/τ_c) decreases as L_1/d_1 increases as shown in Fig. 15b.

Interface Characteristics

Reference 9 reports reservoir pressure data that are influenced by hydrogen-air reaction at the interface. The present pitot pressure data also indicate a possible reaction of the hydrogen and air at the interface. Figure 16 presents the pitot records for He/Air and H_2/Air runs at L_1 = 18 feet. The runs are chosen for comparison because the pitot pressure ratios P'_{o3}/P'_{o2} are nearly equal. The comparison shows that the He/Air pitot pressure increases uniformly from P'_{o2} to P'_{o3} but the H_2/Air pitot pressure drops suddenly when the region 2 test time is over.

This decrease could be effected by heat addition to the local mixture. A heat addition lowers the Mach number compared to the Mach number without heat addition and a lower flow Mach number gives a lower pitot pressure for the same static pressure. The heat addition could be caused by burning of the H_2-Air mixture at the interface.

A similar analysis can be applied to the apparent difference for H_2/N_2 versus H_2/Air. Figure 17 presents pitot pressure records for H_2/N_2 and H_2/Air runs at L_1 = 36 ft. Note that the H_2/Air pitot pressure again drops suddenly as the mixing zone arrives, whereas the pitot pressure does not drop for the H_2/N_2 mixing zone.

CONCLUSION

Pitot pressure probes have been used to determine the test time in the 2 region of a shock tube. The ratio of experimental test time to inviscid test time correlates with a viscous parameter $(L_1/d_1)(d_1P_1)^{-1/4}$ and is less than calculated based on Mirels' theory.

The driver gas does not affect the region 2 test time. However, the mixing region that follows the 2 region shows some evidence of local heat addition when hydrogen drives air.

REFERENCES

1. Duff, R. E. "Shock-Tube Performance at Low Initial Pressure," Phys. of Fluids, Vol. 2, pp. 207-216, 1959.

2. Roshko, Anatol. "On Flow Duration in Low-Pressure Shock Tubes," Phys. of Fluids, Vol. 3, pp. 835-842, 1960.
3. Mirels, H. "Test Time in Low Pressure Shock Tubes," Phys. of Fluids, Vol. 6, pp. 1201-1214, 1963.
4. Mirels, H. "Shock Tube Test Time Limitation Due to Turbulent Wall Boundary Layer," AIAA Journal, Vol. 2, pp. 84-92, 1964.
5. Brabbs, T. A. and Belles, F. E. "Contact-Surface Tailoring in Real Shock Tubes," 5th International Shock Tube Symposium, U. S. Naval Ordnance Laboratory, Silver Spring, Maryland, 28-30 April 1965.
6. Dunn, M. G. "Experimental Shock-Tube Investigation of Conditions Behind Incident and Reflected Shocks in Air for Shock Mach Numbers Between 8.5 and 16.5," CAL No. AL-2187-A-3, October 1966. Also Fifth Hypervelocity Techniques Symposium, University of Denver, March, 1967.
7. Lewis, Clark H. and Burgess, E. G. III. "Charts of Normal Shock Wave Properties in Imperfect Air (Supplement: M_S = 1 to 10)," AEDC-TR-65-196, September, 1965.
8. Glass, I. I. and Hall, J. G. "Handbook of Supersonic Aerodynamics, Section 18, Shock Tubes," NAVORD Report 1488, Vol. 6, December, 1959.
9. Copper, J. A., Miller, H. R., and Hameetman, F. J. "Correlation of Uncontaminated Test Durations in Shock Tunnels," Fourth Hypervelocity Techniques Symposium, Arnold Air Force Station, Tennessee, November, 1965.

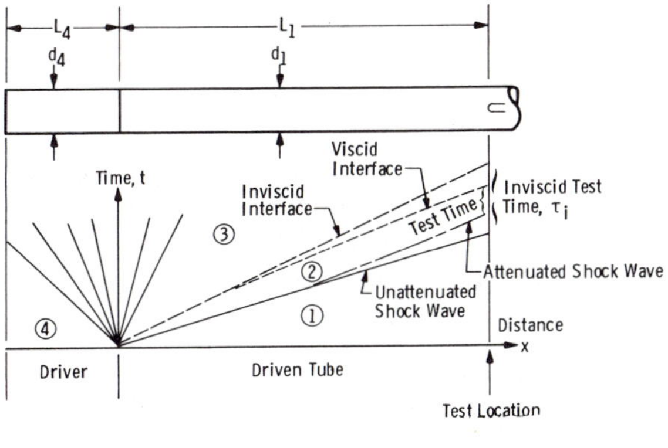

Figure 1
Shock tube wave diagram

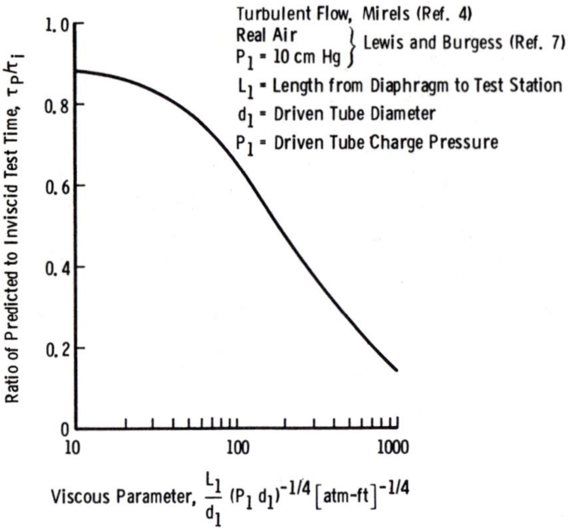

Figure 2
Predicted test time

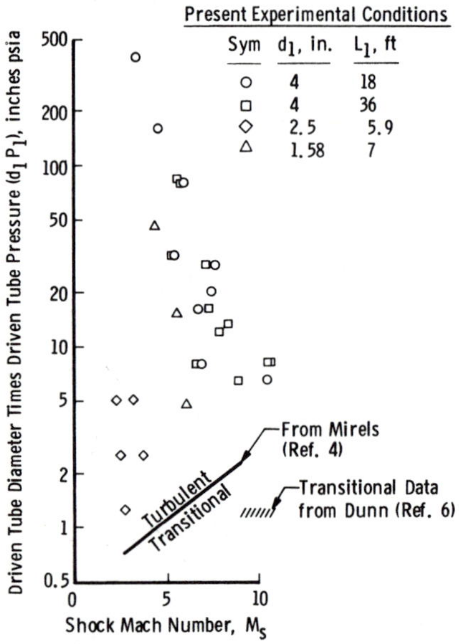

Figure 3
Parameter d_1P_1 for fully turbulent flow

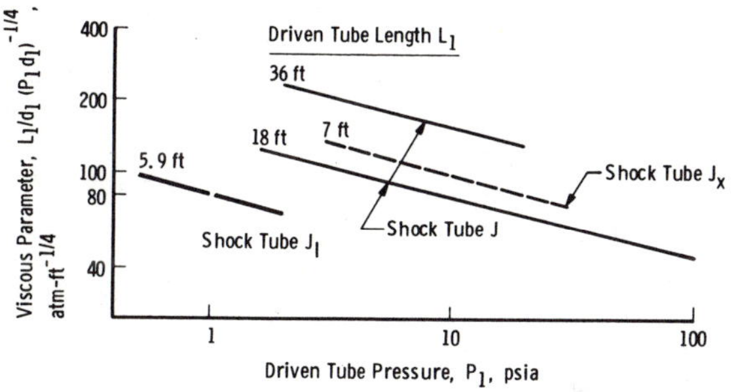

Figure 4
Ranges of viscous parameter and driven tube pressure tested

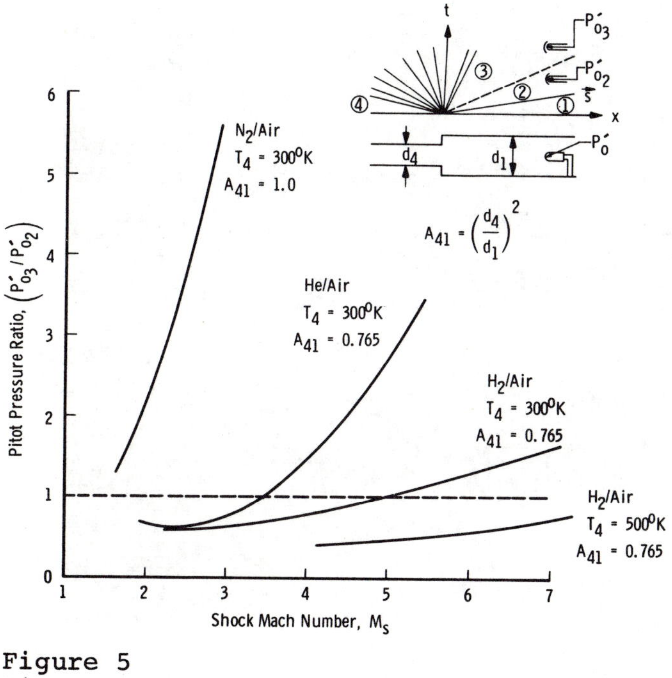

Figure 5
Pitot pressure ratio across the interface for various shock tube conditions

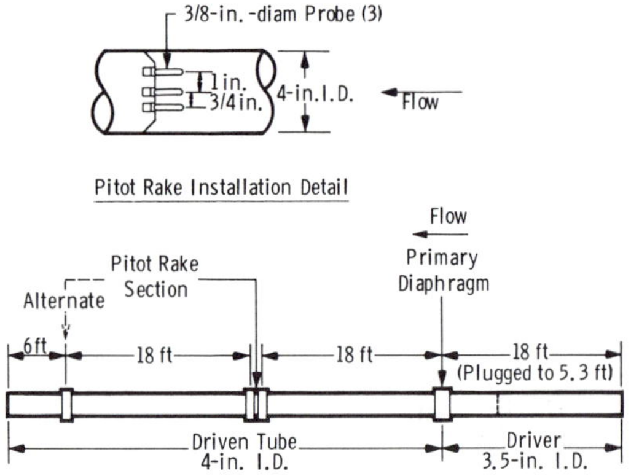

Figure 6a
Sketch of shock tube J

Figure 6b
Photograph of shock tube J

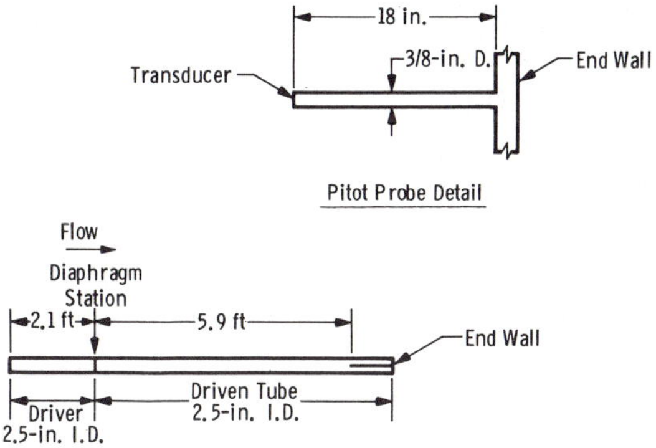

Figure 7a
Sketch of shock tube J_1

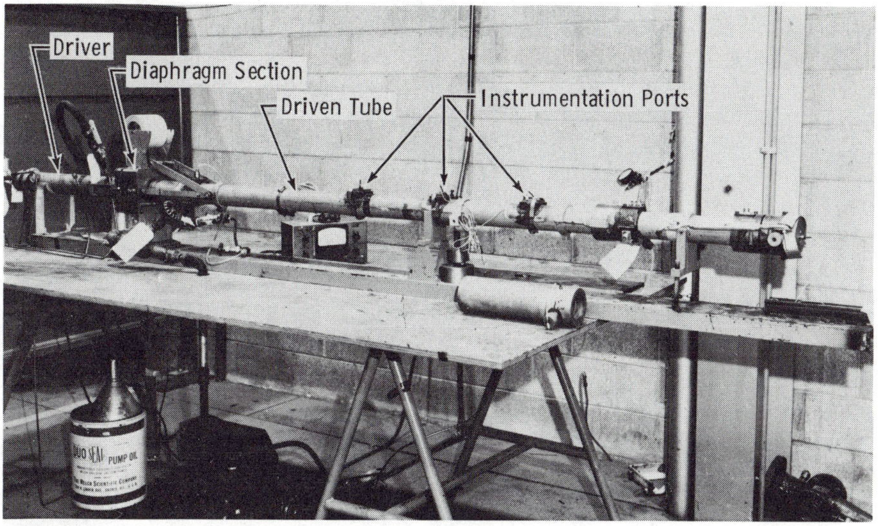

Figure 7b
Photograph of shock tube J_1

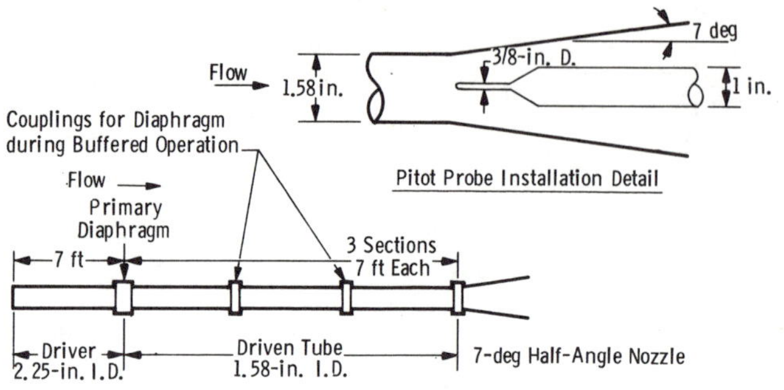

Figure 8a
Sketch of shock tube J_X

Figure 8b
Photograph of shock tube J_X

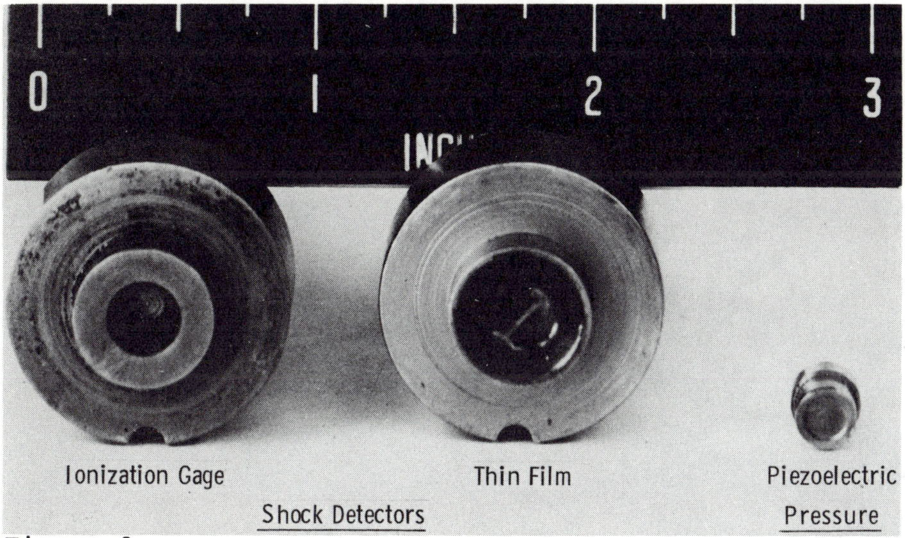

Figure 9
Transducers

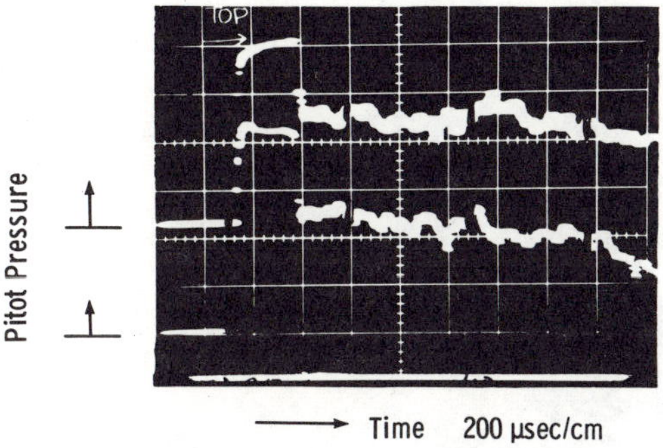

a. Single Sweep Pitot Pressure Record

Figure 10
Typical data records

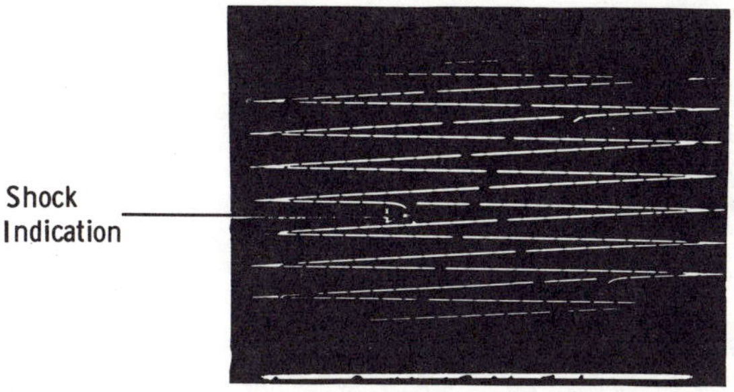

b. Raster Sweep Shock Passage Record

Shock Tube

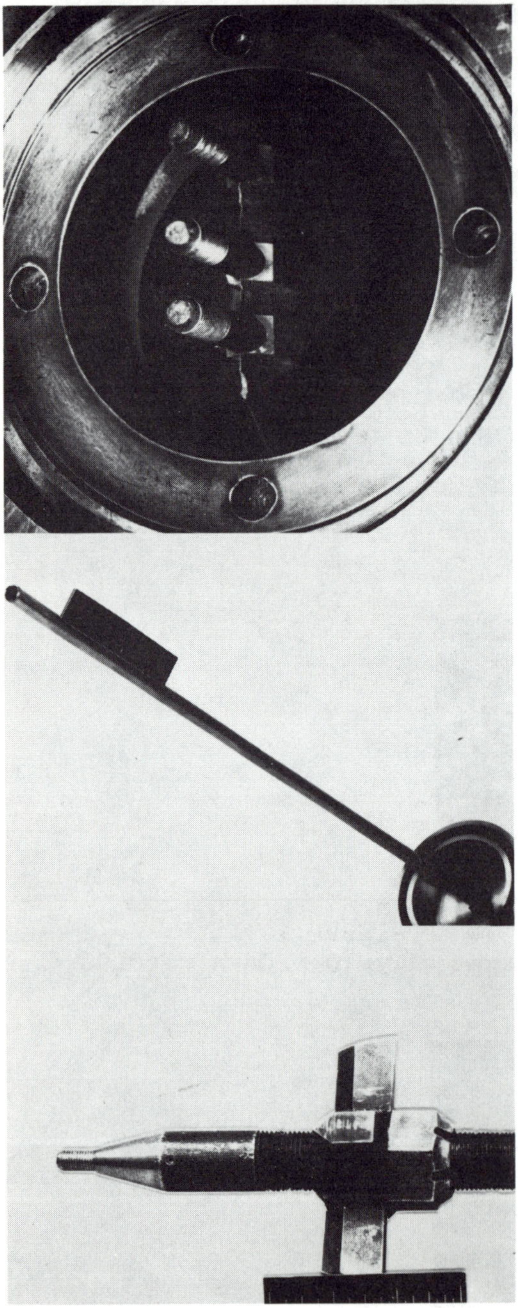

J

J_I

J_X

Figure 11
Pitot models

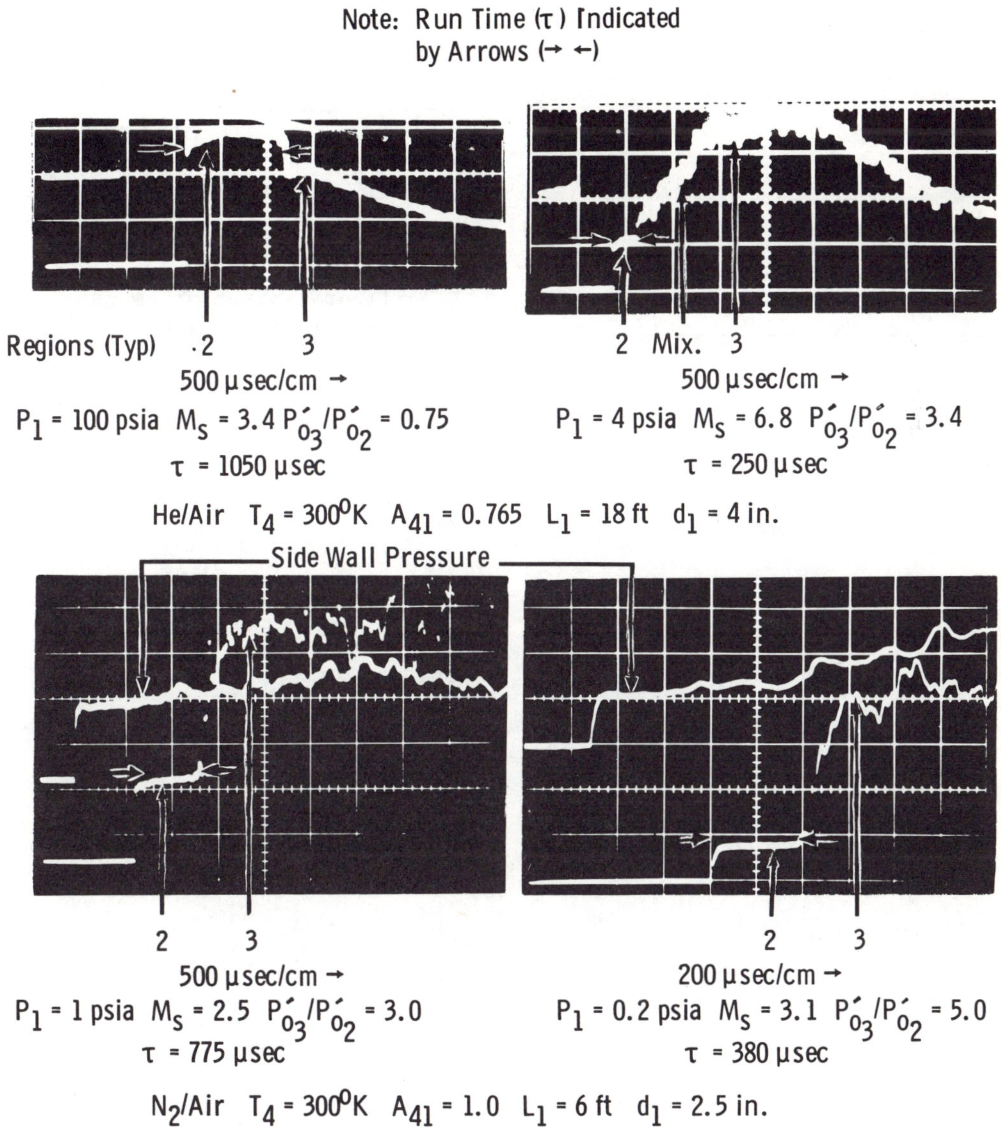

Figure 12
Pitot pressure records showing run time
and pressure levels in regions 2 and 3

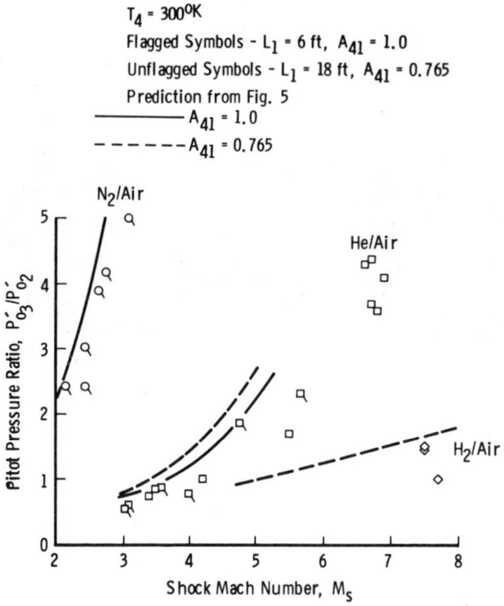

Figure 13
Experimental pitot pressure ratio across the interface

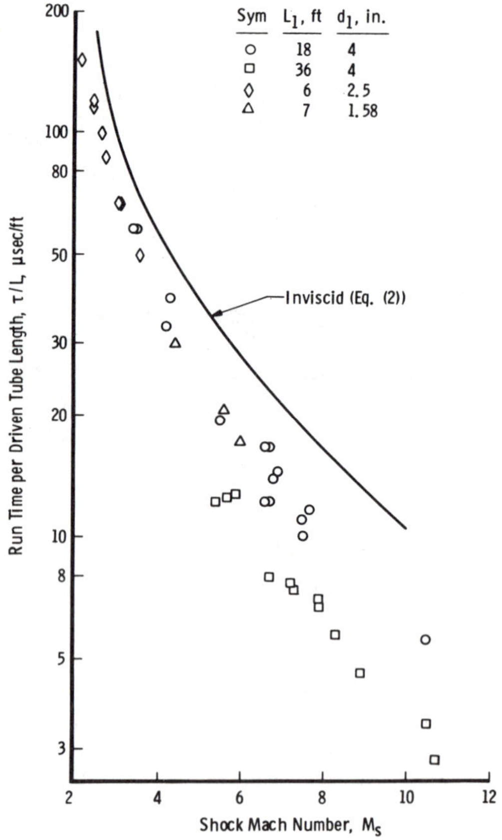

Figure 14
Experimental test time, t/L_1

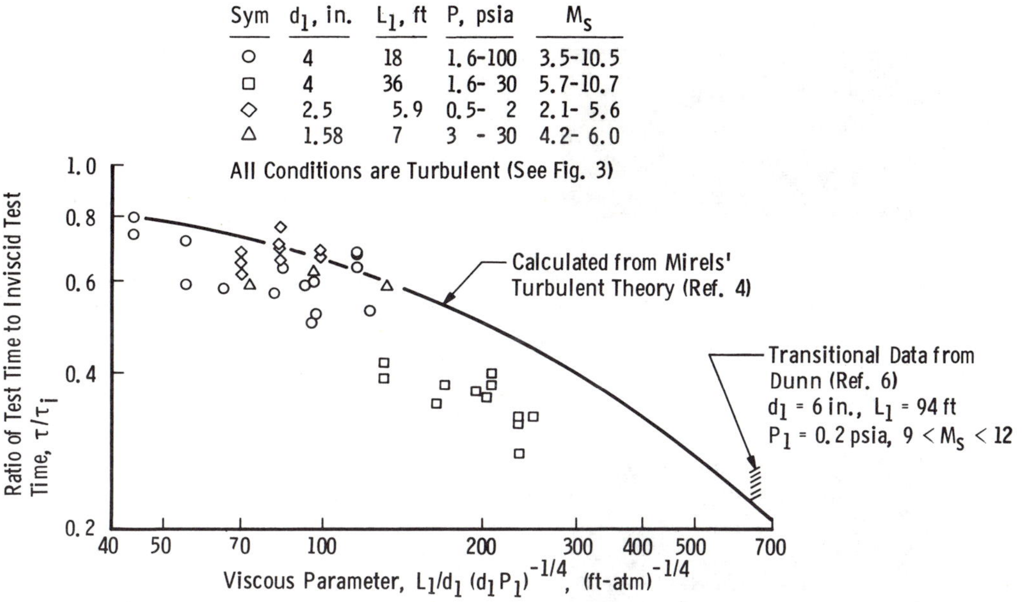

a. Correlation of Experimental Data

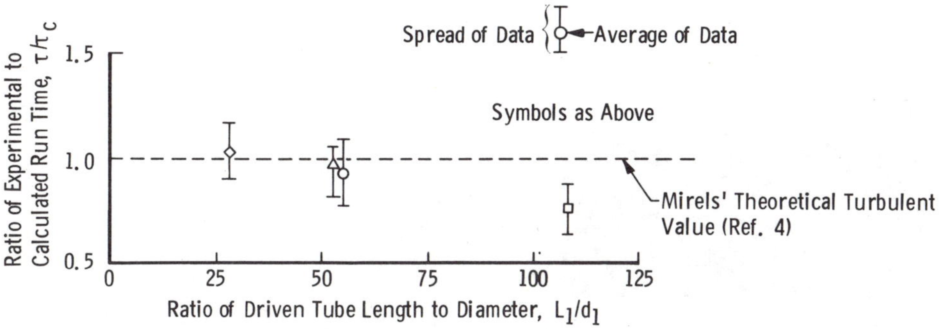

b. Effect of L_1/d_1 on Accuracy of Calculated Test Time

Figure 15
Experimental data

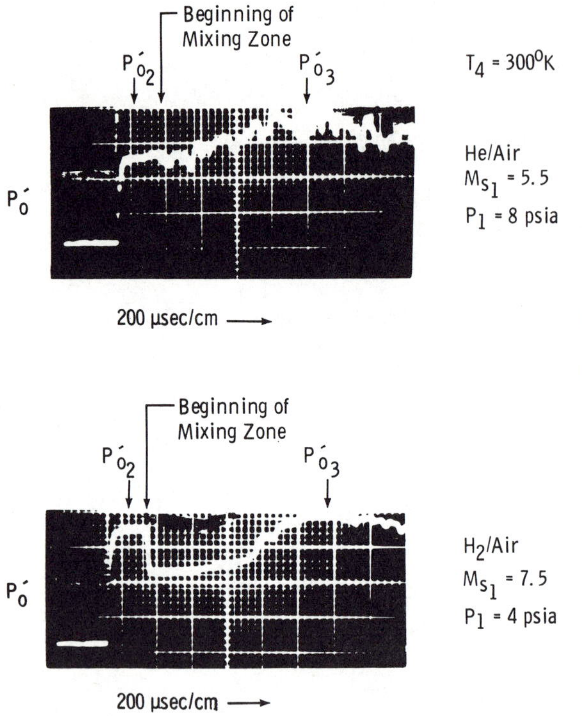

Figure 16
Pitot records showing H_2/air effects at interface

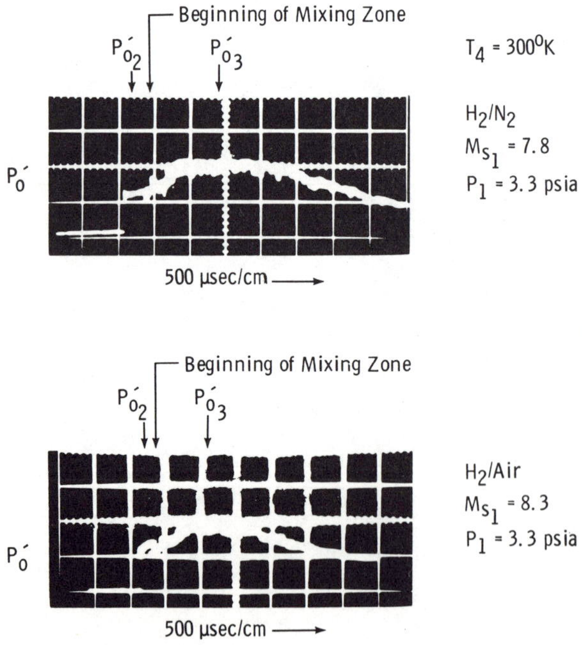

Figure 17
Pitot records showing H_2/air effects at interface

A CRITIQUE OF HIGH PERFORMANCE SHOCK TUBE DRIVING TECHNIQUES

W. R. WARREN
Director, Aerodynamics and Propulsion Research Laboratory
The Aerospace Corporation
El Segundo, California

C. J. HARRIS
Consultant, Fluid Dynamics Laboratory, Reentry Systems
General Electric Company
Valley Forge, Pennsylvania

SUMMARY

The various high performance shock tube driving techniques considered have been grouped as high pressure/high speed of sound conventional drivers, electric and magnetic field interaction drivers, detonation drivers, and those that derive their flow characteristics from the coupling of energy addition and wave processes in an unconventional manner. "High performance" can be taken to refer to any property of a shock tube flow that is important to a particular study: for example, shock velocity; driven gas temperature, pressure, and density levels; driven gas kinetic to static energy ratio; and uniformity and duration of the driven gas flow. Data from recent references are compared to determine the relative effectiveness of different driving techniques and the predictability of flow properties. Several fluid dynamic problem areas are outlined; requirements for their study in shock driven facilities are discussed and compared with capabilities resulting from the use of various driving techniques.

INTRODUCTION

During the past decade, several shock tube driving techniques have been conceived and developed with the general objective of increasing the ranges of operation of shock driven facilities. The shock tube as a generic facility is ideal for extension to extreme conditions because of its ability to deliver modest amounts of stored energy into confined regions at high power levels and to use the energy amplifying nature of unsteady flow processes in a number of ways. The intent of this paper is to evaluate the demonstrated performance of several high performance driving techniques primarily from the point of view of their usefulness in producing test flows for interesting fluid dynamic investigations.

DRIVING TECHNIQUES AND TEST FLOWS

To facilitate the description of the driver mechanisms we have grouped them somewhat arbitrarily into four categories (see Table 1). The first comprises conventional shock tubes in the sense that they function through the rapid removal of a diaphragm separating tubes of high and low pressure gases. Progress here has centered on increasing the pressure and temperature levels of a low molecular weight driver gas (usually hydrogen or helium) over those values previously available, for example, in cold hydrogen or combustion drivers. Techniques that have received considerable attention here are the rapid discharge of electrical energy through the driver gas (Fig. 1, Gruszczynski and Warren, 1963), the extension to extreme structural limits of relatively straightforward heating and pressurizing techniques, and the compression of a driver gas by a piston moving into the driver tube.

Second, we have grouped those drivers that propel the driver gas or the driven gas by body forces produced through electric and magnetic field interactions. Many of these techniques do not initially separate the driver and driven gases. Typical of these are the T tube, coaxial tubes, Hall and Faraday accelerators and variations on them. Also, we have included laser breakdown techniques here. In the third category, we consider those types of drivers which produce strong shocks as a result of a detonation process in a high explosive or solid propellant. Such drivers are self-destructive at least in part; however, the high power levels available in the detonation process make them attractive devices. In the fourth category, we have put techniques that derive their special flow characteristics from the coupling of the energy addition process with the wave processes in the driver and driven gases. The expansion tube, isentropic compression tube, continuous discharge driver, and the many variations of buffer techniques are examples of this type of driving mechanism. Sketches of several driver techniques listed in Table 1 are shown in Fig. 2. Figure 3 shows sketches of several shock driven facility test flows of interest in different studies.

DEMONSTRATED PERFORMANCE CHARACTERISTICS OF SEVERAL SHOCK TUBES

The development of a new driving technique is most frequently a response to a need to study an interesting fluid dynamic problem area. Therefore, the phrase "high performance" can refer to one or a combination of several properties of importance in a given study. Perhaps the most popular measure of high performance is the incident shock wave velocity. Figures 4a, 4b and 4c present representative experimental shock velocity, incident pressure data for several facilities as reported in the indicated references. Some of the pertinent physical characteristics of the shock tubes used to obtain the data in Fig. 4 are shown in Table 2. It is noted that the two highest velocity data sets (Schneiderman, et al., 1967 and Stalker, Plumb, 1968) were taken in relatively small devices.

Another important property is the duration of a test gas flow. Shock tube test time is usually considered to be the duration at a particular tube section of relatively uniform flow between the passage of the incident wave and the contact zone (see Fig. 3A). This is in itself a quite complex flow, but satisfactory analytical treatments of it exist

(Roshko, 1960; Mirels, 1963, 1964). Durations of other test flows shown in Fig. 3 are influenced by such phenomena as reflected shock interaction with the incident wave and contact zone flows, nozzle flow establishment, model flow establishment, uniformity of flow behind a contact zone, radiative losses, and focusing limitations. Reported incident wave test times for simple shock tubes using several types of high performance drivers are shown in Fig. 5 as functions of M_s and P_1 and in Fig. 6 compared to the appropriate prediction. The approximations in the theory are more questionable as test time values decrease below about 10 μs and this may be responsible for some of the scatter for the explosive and EM drivers; however, several of the data for these devices are approximately equal to or greater than the so-called ideal shock tube test time, which does not appear to be a possible result.

The shock tubes considered in Figs. 4, 5, and 6 are capable of producing incident shock velocities of up to approximately 20 km/sec. Several shock wave driving devices exist that have been reported to produce shock velocities of up to 10 times this value. Representative data from several of these are shown in Fig. 7. Essentially, the data shown are from two classes of shock tubes; the filled points correspond to electromagnetic drivers and the open points to detonation drivers. Figure 8 shows the incident shock wave attenuation performance for several high performance shock tubes. The characteristic of certain types of EM tubes in functioning as blast wave tubes (Muntenbruch, 1967) is evident here. The implosive/compression detonation driver, at least as developed by Voitenko, 1966, appears to have a similar blast wave behavior.

The question of test flow quality (steadiness, uniformity or prescribed variability, cleanliness) must be raised although it seldom has been demonstrated quantitatively what degradation in flow quality can be tolerated in high performance facilities. However, quality measurements can frequently be useful in determining test time or other properties of importance. Figure 9 summarizes several of the measurement techniques used. Frequently, single measures of flow quality will be insensitive to the changes in the property of interest and, therefore, many investigators rely on several concurrent measurements that depend on different properties of the flow. Figure 10 shows results of two techniques used to measure the duration and steadiness of the model test flow in an electrically driven tube. Image conversion camera photographs of the type shown in Fig. 11 permit us to assess the spatial uniformity of a test flow at several times during the test flow; in addition they provide us with quantitative data, such as radiating gas volume, needed in certain types of studies. Image conversion photography has frequently been used to sense the presence of driver gas behind the incident shock wave (for example, see Camm and Rose, 1962). Some recent photographs taken in a 12 inch id electrically driven tube at shock velocities of approximately 11 km/sec (L. Presley, NASA Ames Research Center, private communication) provide excellent definition of the apparent driver gas/test gas turbulent interface phenomena produced in that facility.

The shock driving ability of electrical discharge drivers should eventually be limited by losses in the driver. However, it is not clear that we have come close to their limit. Figure 12, which includes data from two facilities, shows that incident shock velocity is still increasing significantly with driver energy density at the highest values investigated, although the efficiency is dropping somewhat. A simple extrapolation of these data, and a limitation to a reasonable driver size, suggests that values of $u_s \approx 60$ kft/sec (or > 18 km/sec) at P_1 values of .25 to .5 torr would be possible with a capacitor bank of less than 10 mw. Also, it appears that considerably higher performance should be available at lower P_1 levels.

Let us now summarize with respect to the facilities discussed in this section. The satisfactory performance of electrically driven shock tubes has been demonstrated in the 10 to 15 km/sec regime with large sized, fairly uniform, and fairly well understood flows of sufficient duration to support several types of studies. Some growth in its capabilities (to $u_s \sim 20$ km/sec at $P_1 \geq .1$ torr) should be possible. Shock tubes driven by the detonation of high explosives and augmented by implosion and jetting phenomena appear capable of functioning in the 15 km/sec regime at fairly high P_1 values. Higher u_s values may not be available because of detonation velocity limitations, but P_1 values greater than the already demonstrated 5 torr may be possible. Test durations of 4 to 8 μs at the $P_1 = 5$ torr level seem to be reasonable, but reported $P_1 = .5$ torr test times of approximately the ideal test time are questioned, since such a result implies that wall boundary layer effects are unimportant. The annular EM driver accelerator (MAARC) appears to be competitive with the electrical driver, but results to date are limited to short driver tube lengths and it is not clear whether its performance will exceed or fall below that of an electrically driven tube of similar dimensions. Also, the test time situation is not clear, since the one reported high velocity result apparently exceeds the corresponding ideal test time prediction. Piston drivers are presently functioning at u_s values lower than electrical drivers but their flows appear to be well established and understood. The u_s performance of a single piston buffered driver is quite impressive; however, present results are for a very small device and test time data have not been presented. It remains to be seen whether a reasonably sized device will be competitive with the other high performance devices. Heated H_2 and He drivers of large scale have been developed for operation in the high pressure moderate u_s range (~ 5 km/sec). These devices provide relatively large test times and appear to be well understood.

At u_s values above about 20 km/sec the situation is less clear. While there are several devices capable of driving waves (perhaps even shock waves) to values as high at 200 km/sec, it is difficult to find finite test time values reported. In fact, Muntenbruch suggests that for EM blast wave type tubes there is no shock wave/current sheet separation until we go to u_s values below about 20 km/sec, a region where we have available several well understood devices. For the detonation driven implosion gas compression devices, waves with u_s > 40 km/sec have been produced at P_1 values of 1 atmosphere. The normal shock wave conditions corresponding to these u_s and P_1 levels are very impressive; however, no test time data have been reported and the blast wave nature of the velocity decay (Fig. 8) may limit the usefulness of this device.

Figure 13 shows the approximate u_s/P_1 performance regimes that have been demonstrated for several types of high performance shock tubes.

LIMITATIONS OF SEVERAL WAVE PROCESS DEVICES

In some studies it is important to restrict the static temperature of the test gas but to still reach a high level of total energy. While the simple shock tube partially meets this objective (if the flow is not reflected), interest in minimizing the static to kinetic energy ratio has led to a degree to the development of the expansion tube (Trimpi, 1962), isentropic compression tube (Stoddard, et al., 1965) and several MHD accelerator devices (Harris, et al., 1965; Norman and Chmielewski, 1965; Grabowsky, et al., 1968). Similarly, many studies require high pressure or density levels in the test flow and high shock velocity can be sacrificed to attain this objective; even if attainable, however, desired pressure levels are often so high that they exceed practical structural limits. This situation also has generated interest in wave process and MHD devices, since several of them can greatly increase the momentum flux level without increasing flow static pressure.

Figure 14 illustrates the type of limitations inherent on a stagnation or plenum chamber type of facility by test flow nonequilibrium and high pressure structural limitations. It appears that for practical structural reasons we will be limited to test flows of 20 kft/sec or less. Representative flow properties for several schemes proposed to eliminate these restrictions but provide about the same test flows (20,000 ft/sec at a density altitude of 100,000 ft) are shown in Fig. 15. It is seen that the isentropic compression and expansion tubes require maximum driven tube pressure levels of about 10^{-2} and 10^{-1} times those needed for the reflected and non-reflected tunnels, respectively. Also, the flow entropy levels are less which results in better test flow chemical properties. Therefore, a significant level of research has been directed toward their development. For the expansion tube Spurk, 1965, has found approximately 40 μs of test time at flow velocities near 7 km/sec. This data point is shown in Fig. 5; it is interesting to note that it is at the same level as several shock tubes operating at a shock velocity of about that flow velocity. Other investigators, looking primarily for higher velocity test conditions, have had difficulty in identifying a suitable test region with known flow properties. In experiments to date with one version of the isentropic compression tube, it has been reported (Stoddard, 1969) that test flow velocities of about 10 kft/sec have been measured.

Nozzle MHD accelerators (Fig. 16 illustrates the facility developed by Harris, et al., 1965) also are intended to provide high energy and momentum levels in the test gas flow without stagnating the gas, and it has been found by several investigators that they perform at close to simply predicted levels (for example, Grabowsky, et al., 1968). However, there are certain potential restrictions, illustrated in Figs. 17 and 18, that may limit their performance. These can be summarized as follows: if we decide that $S/R = 32$ is the highest value of test gas entropy we can allow because of free stream nonequilibrium (the nonequilibrium oxygen level as a function of entropy is shown in Fig. 17) and that $LB^2 \sigma = 4x10^5$ (Say L = 4m, B = $10w/m^2$, σ = 1000 mhos/m) is a maximum value of this parameter attainable in the near future, Fig. 18 shows us that an ultimate flow velocity of about 25,000 ft/sec is available with such a device.

STUDY OF FLUID DYNAMIC PROBLEMS IN HIGH PERFORMANCE SHOCK DRIVEN FACILITIES

We have grouped several fluid dynamics problem areas of current and future interest in the following categories: hypersonic aerodynamics, laminar to turbulent transition and turbulence, chemical and ionization kinetics, nonequilibrium flows, electromagnetic radiation interaction with high temperature flows, gas radiation and radiative/convective coupling in flow fields, noncontinuum flows, unsteady phenomena, MHD applications, and plasma production and properties. In the first two, density and Mach number simulation of flight tend to be important and we can frequently relax the high energy test gas requirements. It is possible then to use the power capabilities of a facility to provide large scale test flows. The further development of moderately heated high pressure H_2 and He conventional drivers to be used with shock tunnels is important here. In the other areas, and in some aspects of the first two, we are generally interested in high test gas energy content in addition to the other properties of importance. Many problems still require investigation to the energy level corresponding to about 20 km/sec, which is about the upper level of interest for vehicles returning to the Earth from deep space missions. Fortunately, several existing shock driving techniques are now performing near this level and can be expected to be extended to it in the future if the appropriate effort is applied. Other problems require good test gas simulation of flight flow parameters (for example, Mach number) as well as energy levels to the order of the Earth orbital level. Here, it appears that several of the wave process devices discussed, or similar concepts, will find usefulness with some future development. However, for certain atmospheric entry problems (for example, meteorite entry, Jupiter entry) we are interested in flight velocities to the order of 70 km/sec. While present techniques apparently can drive shock waves to such velocities, it is not clear that useful test flow durations have been obtained. This test time situation possibly will be worse as we go to still higher velocities (10^3 km/sec and greater) of interest in solar wind and fusion research, even if devices can be extended to drive shock waves in that region. At any rate, it seems safe to conclude that shock tube devices will not be capable of providing model test flows at such extreme velocities. However, as they have already done in several investigations, they may prove to be good short duration sources of high temperature plasmas.

CONCLUDING REMARKS

Several high performance shock tube driving techniques have been developed which perform well in the incident shock velocity region to about 20 km/sec. The choice between them may rest on such practical matters as initial cost, replacement costs, and run frequency.

High pressure conventional drivers and certain types of wave process devices are now reaching the point of development where they are useful for low and moderate temperature aerodynamic studies and for high energy aerodynamic studies at about the Earth orbital velocity level. Extension to useful applications near the Earth escape velocity level may be possible with one or more of these devices. We feel that the development of the continuous electrical discharge driver concept (Warren and Marston, 1969) should also be pursued, since it makes use of a well developed technology and avoids some of the problems of other wave process devices.

Studies at incident shock velocities of greater than 20 km/sec may be limited to plasma production and property studies by very short test times, at least in electromagnetically driven devices. (This tentative conclusion cannot yet be applied to detonation driven gas compression devices, since essentially no test time data have been reported.) Promising growth in this direction may be available with devices which do not give plane shock waves, but rather attempt to produce very high energy points or lines of plasma through the collapsing of waves (implosion) and the focusing of energy. Some interesting results have already been obtained in (1) a plasma focus discharge region of an EM coaxial device (Mather and Bottoms, 1968), (2) an imploding chemical detonation device where a temperature level of 1.9×10^5 °K has been reported (Knystantas, et al., 1967), and (3) a laser focus breakdown device in which a shock (or detonation) wave velocity of about 5 km/sec has been reported and in which "wave" velocities of up to 200 km/sec have been observed (Daiber, et al., 1967). In a later laser breakdown study the measurement of an average electron temperature of 1.5×10^6 °K was reported (Daiber, et al., 1969).

ACKNOWLEDGMENTS

The authors are indebted to the following people who provided some of the information and several of the references used in the paper: J. Daiber, R. Dannenberg, S. Gill, I. Glass, J. Grusczcynski, J. Jones, L. Presley, C. Wittliff.

BIBLIOGRAPHY

Bartle, E. R., and K. G. P. Sulzmann, Technical Report GD/C-DB E65-011, General Dynamics Convair Division Space Science Laboratory, June 1965.

Camm, J. C., and P. H. Rose, AFCO AERL Research Report 136 (AFCRL-62-568), July 1962.

Colin, C., et al., J. Appl. Phys., 39, (7), 2991-2993, June 1968.

Collins, D. J., et al., Report TR 32-620, Jet Propulsion Laboratory, Pasadena, Calif., June 1964.

Daiber, J. W., et al., Report 67-0069, USAF Air Research Laboratory, Washington, D. C., April 1967.

Daiber, J. W., R. G. Rehm, and H. M. Thompson, Report 69-0002, USAF Air Research Laboratory, Washington, D. C., January 1969.

Dukowicz, J. K., UTIA Technical Note 52 (AFOSR 2429), University of Toronto, Toronto, Canada, April 1963.

Dunn, M. G., Proceedings of the 5th Hypervelocity Technical Symposium, March 1967.
Fowler, R. G., G. W. Paxton, and H. G. Hughes, Phys. of Fluids, 4, (2), February 1961, p. 234.
Gill, S. P., and R. C. Goettelman, Report CR-950, National Aeronautics and Space Administration, November 1967.
Glass, I. I., and Heuckroth, L. E., Phys. of Fluids, 6, April 1963.
Glass, I. I., UTIAS Review No. 25, University of Toronto Institute for Aerospace Studies, January 1965.
Grabowsky, W. R., D. A. Durran, and H. Mirels, AIAA Paper GE-370, April 1968.
Gross, R. A., et al., Proceedings of 6th International Shock Tube Symposium, April 1967.
Gruszczynski, J. S., and W. R. Warren, AIAA Paper 63-450, August 1963.
Gruszczynski, J. S. and W. R. Warren, Final Report, General Electric Company, March 1964 (NASA Contract NASW-1303).
Gruszczynski, J. S., and D. A. Rogers, TIS Report R64SD67, General Electric Company, September 1964.
Gruszczynski, J. S., IEEE Transactions on Aerospace and Electronic Systems, AES-1, (1), August 1956, p. 43-56.
Gruszczynski, J. S., and W.R. Warren, AIAA Paper 66-103, January 1968.
Gruszczynski, J. S., AIAA Paper 68-665, June 1968.
Harris, C. J., et al., Proceedings of the 4th Hypervelocity Technology Symposium, November 1965.
Heywood, J. B., Fluid Dynamics Laboratory Publication 65-1, Massachusetts Institute of Technology, January 1965.
Holder, D. W., and D. L. Schultz, ARS Paper 1970-61, August 1961.
Hoppmann, R. F., and H. S. Glick, Proceedings of the 5th Hypervelocity Technical Symposium, March 1967.
Josephson, V., J. Appl. Phys., 29, (3), 1958.
Josephson, V. and R.W. Hales, Report TR-60-0000-19313, Space Technology Laboratories, Redondo Beach, California, September 1960.
Kaegi, E. M., and E. P. Muntz, Proceedings of the 3rd Hypervelocity Technical Symposium, March 1964.
Kamimoto, G., T. Mori, and T. Kimuna, Paper CP 20, Dept. of Aeronautical Engineering, Kyoto University, Kyoto, Japan, July 1967.
Kash, S. W., Magnetohydrodynamics, Stanford University Press, Stanford, California, 1957, pp. 92-98.
Knystantas, R., B. H. K. Lee, and J. H. Lee, Proceedings of the 6th International Shock Tube Symposium, April 1967.
Kolb, A. C., Magnetohydrodynamics, Stanford University Press, Stanford, California, 1957, pp. 76-91.
Koopman, D. W., Phys. of Fluids, 7, 1651, 1965.
Lin, S-C., and W. I. Fyfe, Phys. of Fluids, 4, (2), February 1961, p. 238.
Marston, C. H., and W. R. Warren, TIS Report R68SD1, General Electric Company, January 1968.
Mather, J. W., and P. J. Bottoms, Phys. of Fluids, 11, (3), March 1968.
Meehan, R., H. Presby, and R. A. Gross, APS Bulletin II, 13, (11), 1601, December 1968.

Mirels, H., Phys. Fluids, 6, (9), September 1963.
Mirels, H., AIAA Journal, 2, (1), January 1964.
Muntenbruch, H., Proceedings of the 6th International Shock Tube Symposium, April 1967.
Nagamatsu, H. T., R. E. Geiger, and R. E. Sheer, ARS Journal, 29, (5), May 1959.
Nerem, R. M., and B. C. Graber, Report 1573-2, Ohio State University Department of Aeronautics and Astronautics Engineering, October 1963.
Nerem, R. M., and R. A. Hopkins, AIAA Journal, October 1968.
Norman, W., and G. E. Chmielewski, AEDC-TR-65-37, USAF, February 1965.
Norman, W., AEDC-TR-66-25, USAF, May 1966.
Rose, P. H., and W. Nelson, AERL Research Report 24, Avco, April 1958.
Roshko, A., Phys. Fluids, 3, 835, 1960.
Schneiderman, A. M., E. R. Pugh, and R. M. Patrick, Proceedings of the 5th Hypervelocity Technical Symposium, March 1967.
Seay, G. E., L. B. Seely, and R. G. Fowler, J. Appl. Physics, 32, (11), November 1961.
Spurk, J. H., Ballistic Research Laboratories Report 1304, U. S. Army Materiel Command, Aberdeen Proving Ground, Maryland, October 1965.
Spurk, J. H., Proceedings of the 4th Hypervelocity Technical Symposium, November 1965.
Stalker, R. J., The Aeronautical Quarterly, November 1966.
Stalker, R. J., AIAA Journal, December 1967.
Stalker, R. J., and D. L. Plumb, Nature, 218, May 1968, pp. 789-790.
Stoddard, F. J., A. Hertzberg, and J. G. Hall, 4th Hypervelocity Technical Symposium, November 1965.
Stoddard, F. J., AIAA Paper No. 69-334, April 1969.
Trimpi, R. L., TR R-133, NationalAeronautics and Space Administration, 1962.
Veyrie, P. and F. Floux, AIAA Paper 68-678, June 1968.
Voitenko, A. E., Sov. Phys.-Doklady, 9, (10), April 1965.
Voitenko, A. E., Sov. Phys.-Doklady, 11, (1), July 1966.
Voitenko, A. E., I. Sh. Model', and I. S. Samodelov, Sov. Phys.-Doklady, 11, (7), January 1967.
Warren, W. R., D. A. Rogers, and C. J. Harris, TIS Report R62SD37, General Electric Company, April 1962.
Warren, W. R., et al., TIS Report R62SD56, General Electric Company, May 1962.
Warren, W. R., SSL Report, General Electric Company, September 1964.
Warren. W. R., and C. H. Marston, AIAA Journal, 7, (5), May 1969.
Willard, J. W. , Proceedings of the 5th Hypervelocity Technical Symposium, March 1967.
Willard, J. W., AIAA Paper 68-366, April 1968.

SYMBOLS

A	area
B	magnetic field strength
C	capacitance
D	diameter
E	energy
E_D	energy density
h	enthalpy
L	length
P	pressure
R	gas constant
R_N	model nose radius
S	entropy
T	temperature
t	time
t_t	test flow duration
u	velocity
V	voltage
σ	conductivity
ρ	density

Subscripts

1	initial conditions, driven tube
2	conditions behind incident shock wave
3	conditions behind contact zone
4	initial conditions, driver tube
5	conditions behind reflected shock wave or model flow conditions in expansion tube
e	test section
f	flight
m	model or maximum
o	reference
s	shock or stagnation
st	stagnation

Table 1
High performance shock tube driving techniques

	low, M_s (very high, mod., high)	P_1 (very low, mod., high)	t_t (very short, short, mod., long)
Conventional			
a. Electrical energy discharge	high	mod.	short
b. Heated hydrogen or helium			
(1) Internal	mod.	high	mod.
(2) External	mod.	mod.	mod.
c. Piston Compression	high	mod.	short
Electromagnetic			
a. T tube and variations, pinch devices	very high	very low	very short
b. Annular tubes, rail spark gap devices, plasma focus	very high	very low	very short
c. MHD			
(1) Driver	high	mod.	short
(2) Nozzle	high[1]	mod.[1]	short
d. Laser Breakdown	very high	high	very short
Detonation			
a. Linear Explosive driver	mod.	high	mod.
b. Implosion, jetting	high	mod.	short
c. Implosion, gas compression	very high	mod.	short
Wave Process			
a. Buffered, Double diaphragm	mod.	mod.	short
b. Large driven to driver area ratio	mod.	very low	mod.
c. Expansion tube	high[1]	mod.[1]	short
d. Isentropic compression tube	high[1]	mod.[1]	short
e. Conical tube	mod.	mod.	mod.
f. Continuous electrical discharge driver	mod.	mod.	mod.
g. Implosion (laser, chemical detonation, solid explosive detonation)	very high[2]	mod.	very short
h. Multiple driver			
(1) Wave Superheater	low	high	long
(2) Double driver	mod.	mod.	short
i. Opposed Flow			

[1] Effective, based on test flow properties

[2] Effective, based on maximum test gas plasma energy

Table 2
Summary of results given in Figure 4

SYM.	DRIVER			DRIVEN		INVESTIGATOR
	TYPE	L_4	D_4	L_1	D_1	
		ft.	ft.	ft.	ft.	
+	H_2, COLD	18.5	0.5	39	0.5	Holder, Schultz (1961)
×	H_2, 685°K	9.0	0.25	98	0.5	Dunn, et al. (1967)
⊙	COMBUST.	6.0	0.416	42	2	Lin, Fyfe (1961)
////	COMBUST.	22	0.67	112	0.5	Warren, et al. (1962)
\\\\	COMBUST.	5	0.22	30	.167	Kaegi, Muntz (1964)
□	COMBUST.	20	0.5	103	.33	Nagamatsu, et al. (1959)
◑	He, PISTON	1.64	0.42	24.6	.42	Kamimoto, et al. (1967)
◐	He, PISTON	3.28	0.42	24.6	.42	Kamimoto, et al. (1967)
▲	He, PISTON	0.73	.5	33	1.0	Williard (1967)
▲	He, PISTON	0.73	.375	33	1.0	Williard (1967)
◆	He, PISTON			6.5	.071	Stalker (1966, 1967)
▼	He, Pist. 2D	3.0	.08	3.5	.04	Stalker, Plumb (1968)
◆	ELECTRIC	1.42	.146	11.3	.167	Warren, et al. (1962)
●	" (He)	.88	.11	30	.5	Camm, Rose (1962)
■	" (He)	4.5	.21	17	.167	Harris, et al. (1965)
◣	" (H_2)	1.0	.084	15	.167	Hoppman, Glick (1967)
—	" (He)		..33	15	.33	Nerem, et al. (1963)
◆	" (He)	2.5	.25	31.5	.5	Grusczcynski, Warren (1966)
▼	" (He)	10	.33	37	.33	Dannenberg, Humphry (1968)
◢	" (He)	2.2	.12	43	.5	Collins, et al. (1964)
△	MHD, Nozzle			2.25	.125	Schneiderman, et al. (1967)
◇	Explosive			12.1	.12	Gill, Goettelman (1967)

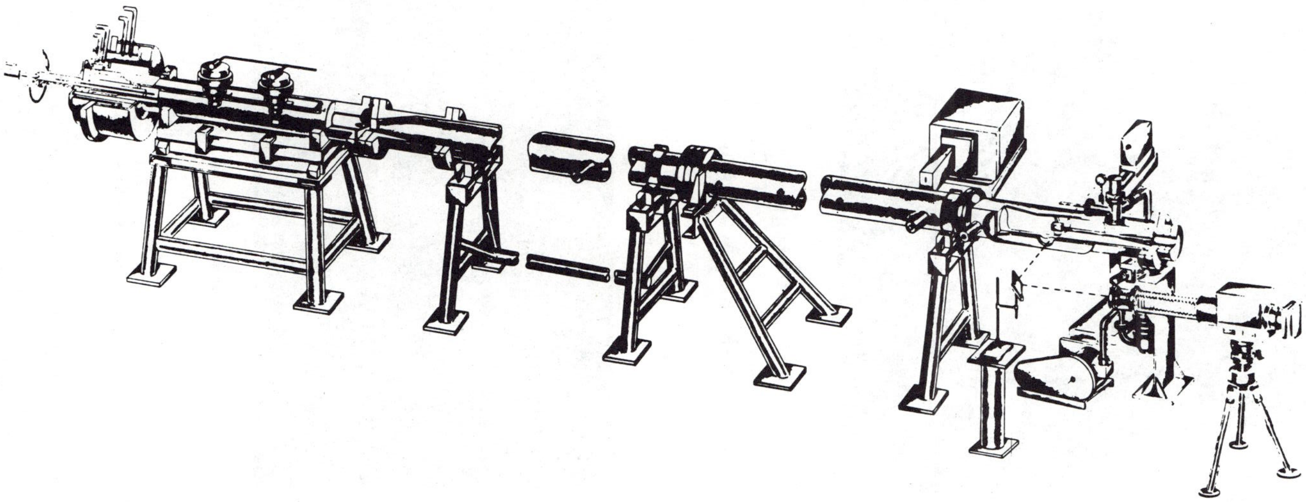

Figure 1a
Electrically driven shock tube: 6" id driver tube, 3" id driver (Gruszczynski and Warren, 1963)

Figure 1b
Photograph of electrical discharge driver for use with 40 kv capacitor banks (304 kj and 768 kj) (Gruszczynski and Warren, 1963)

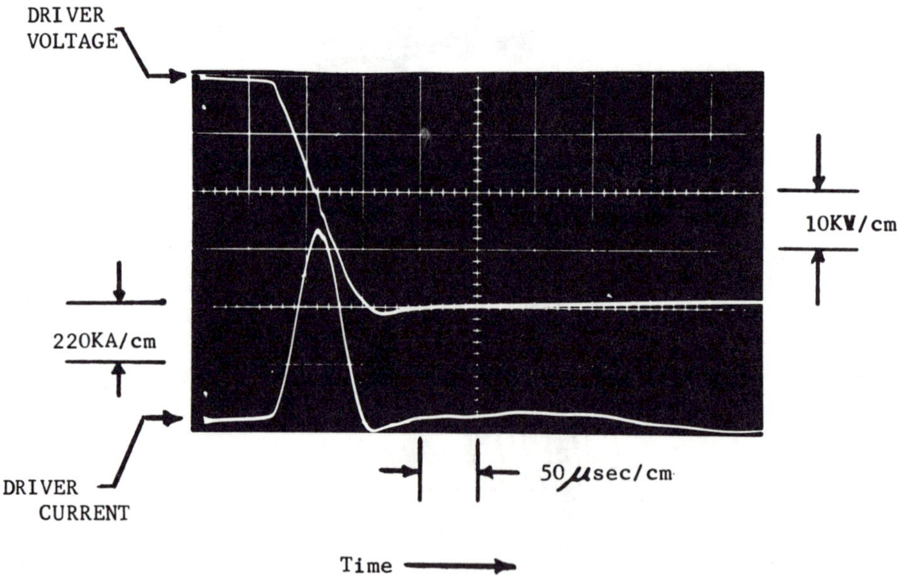

Figure 1c
Voltage and current traces for 6" electrically driven shock tube of Figs. 1a and 1b. Driver internal dimensions for this run: length 30", diameter 2.5". Other driver characteristics: C=960 μf, V_{nom}= 40 kv, Helium at initial pressure of 35 atm. Diaphragm ruptures 200-250 μs after initiation of discharge

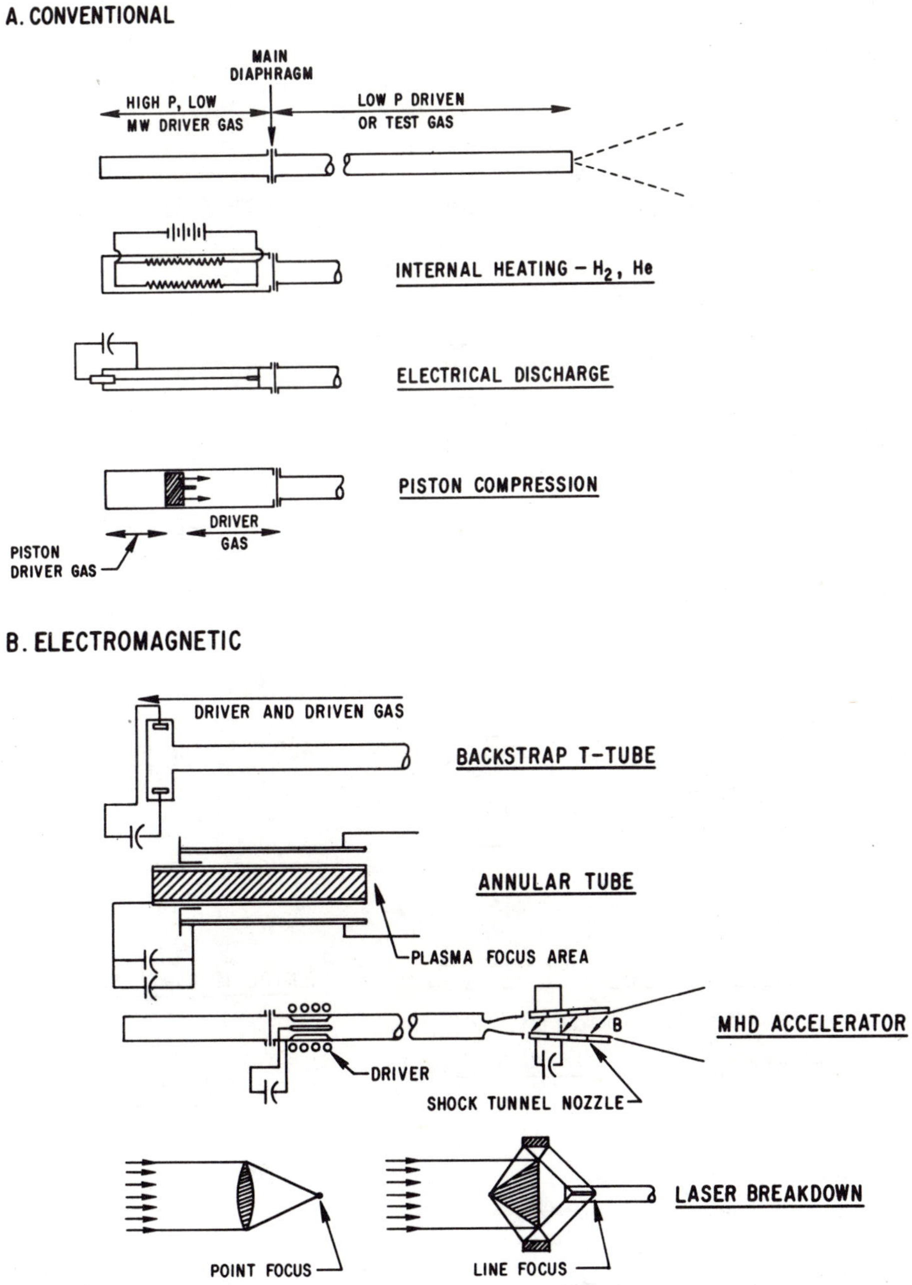

Figure 2
Some high performance shock tube driving techniques

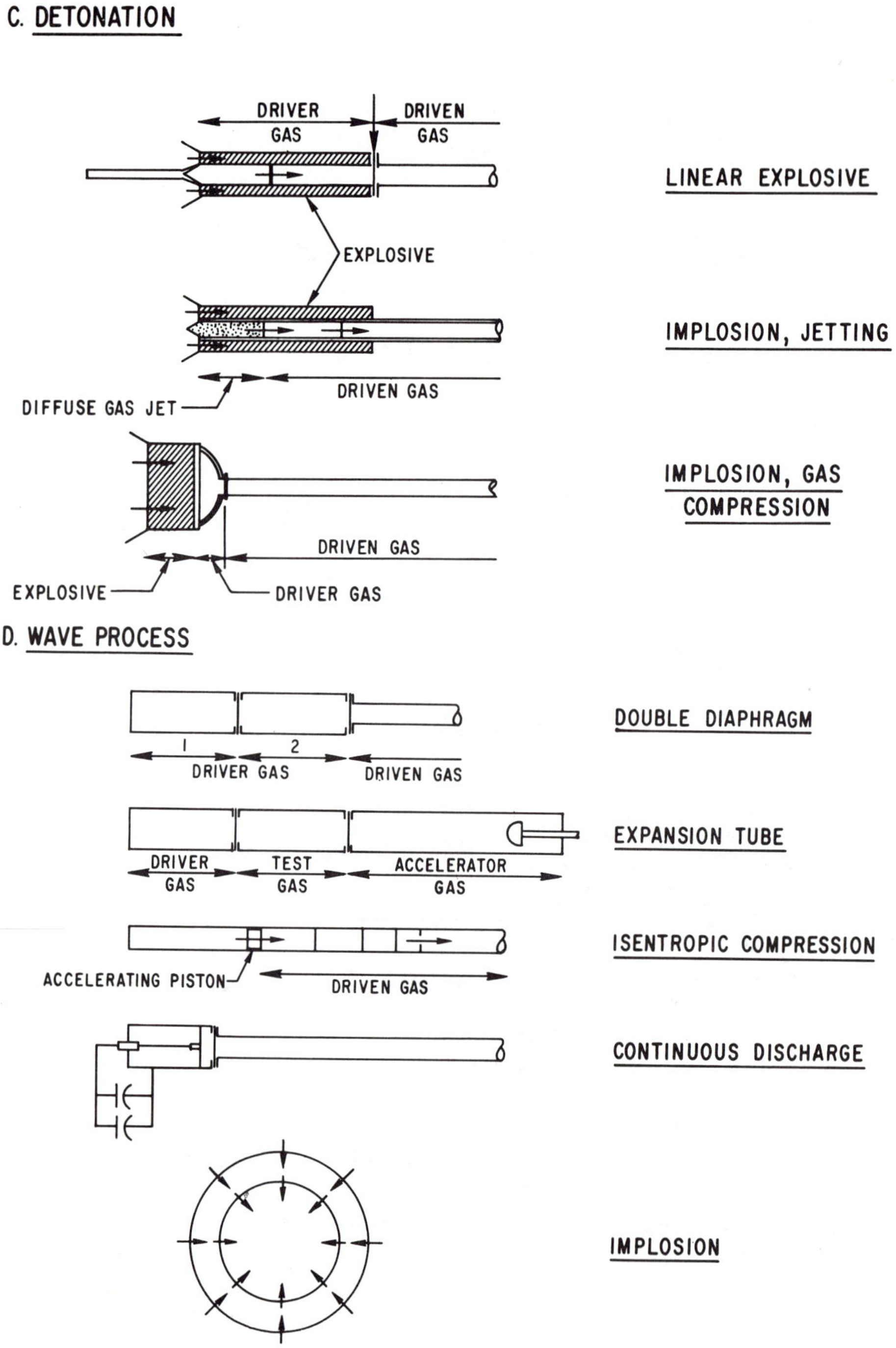

Figure 2 (concluded)

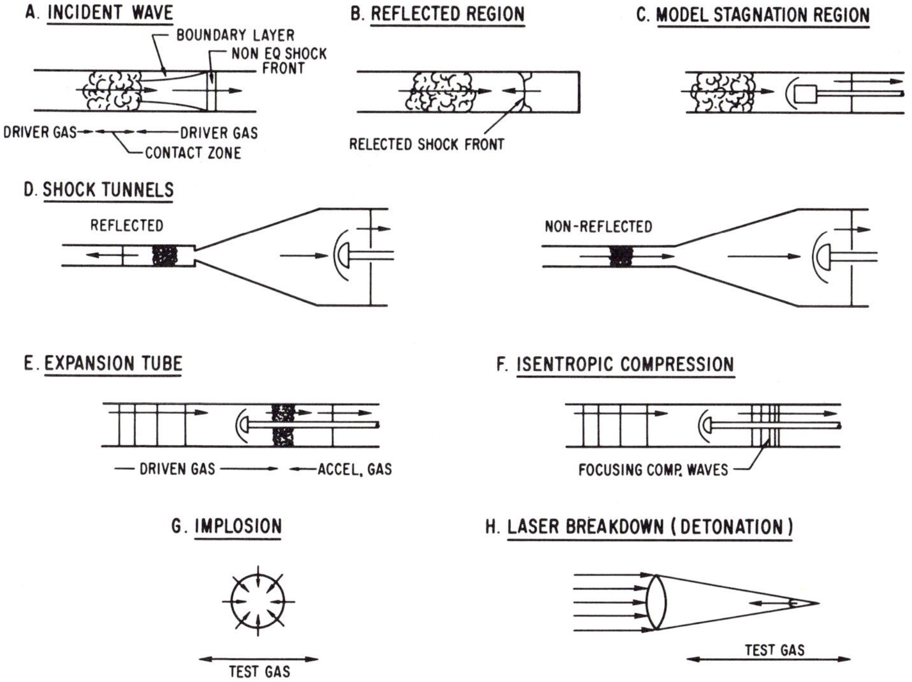

Figure 3
Test regions of interest

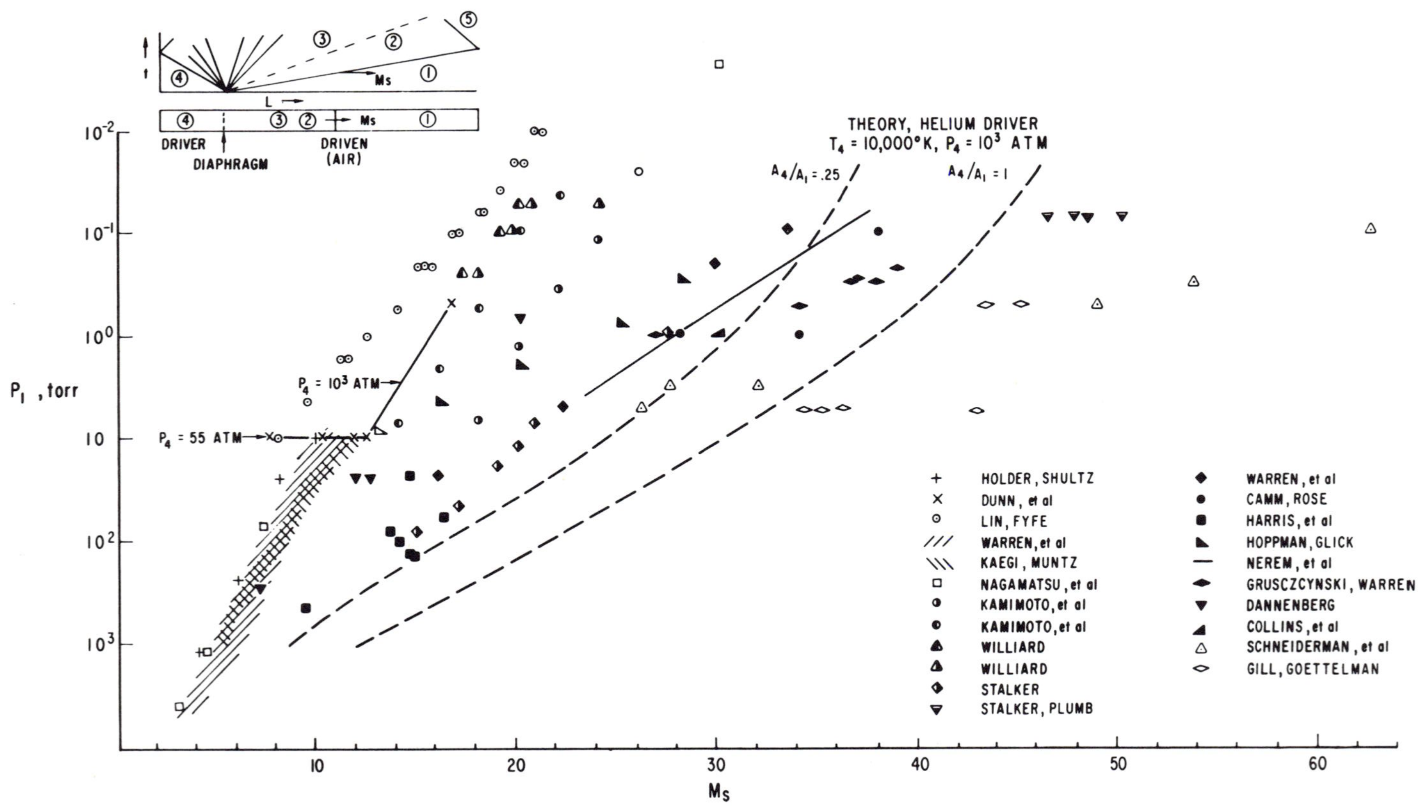

Figure 4a
Incident shock Mach number, initial pressure performance for several high performance shock tubes. See Table 2 for characteristics of different shock tubes.

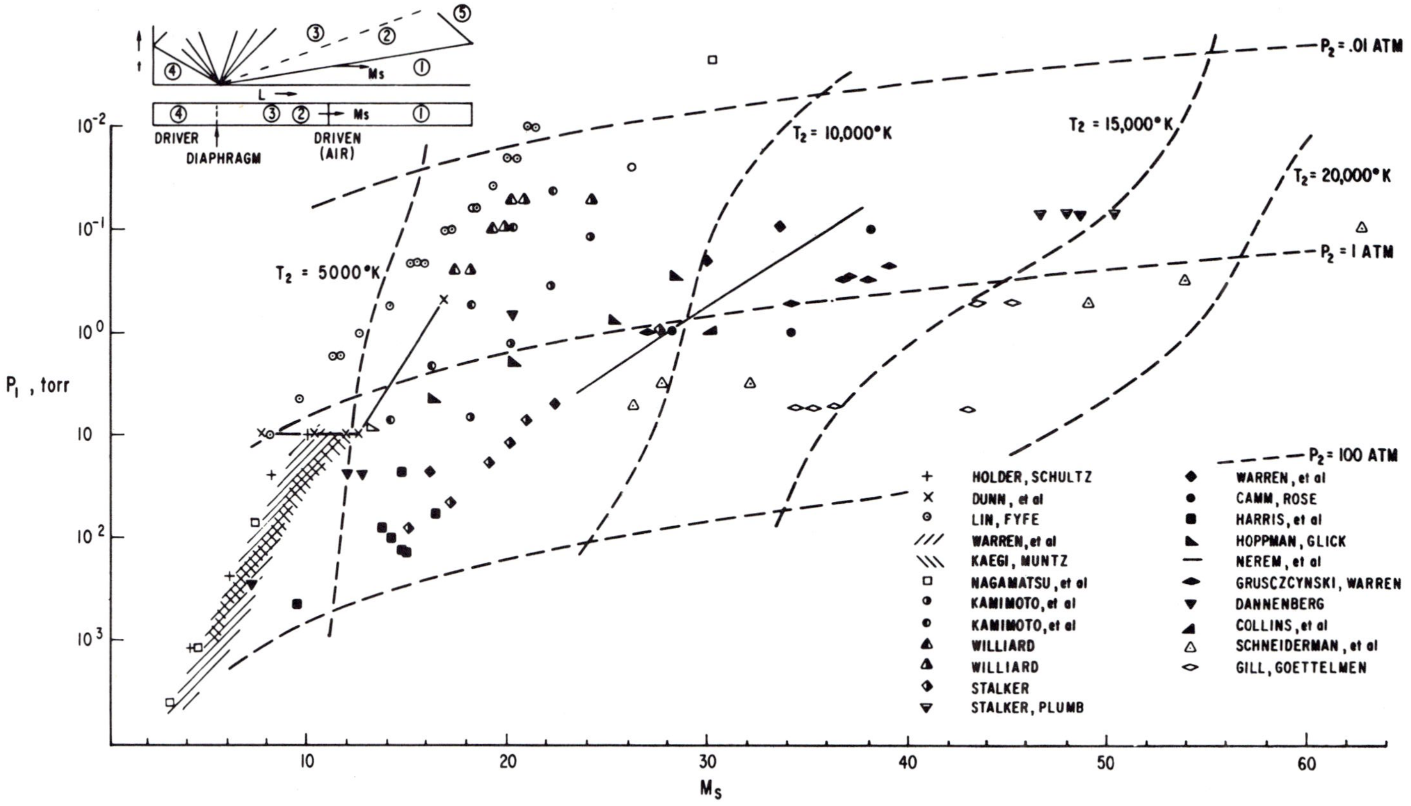

Figure 4b
Shock tube M_S/P_1 performance indicating equilibrium air properties behind incident shock wave

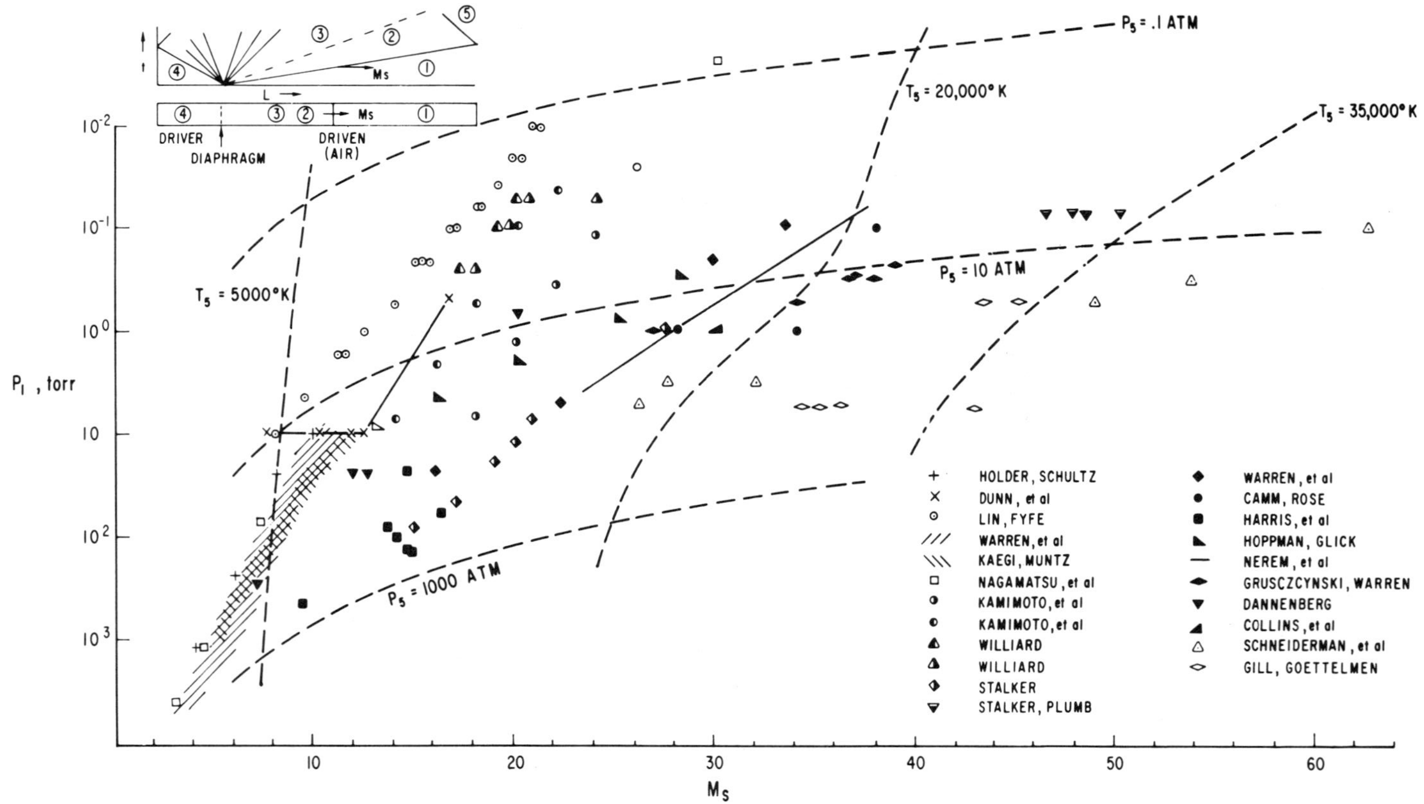

Figure 4c
Shock tube M_S/P_1 performance indicating equilibrium air properties behind reflected shock wave

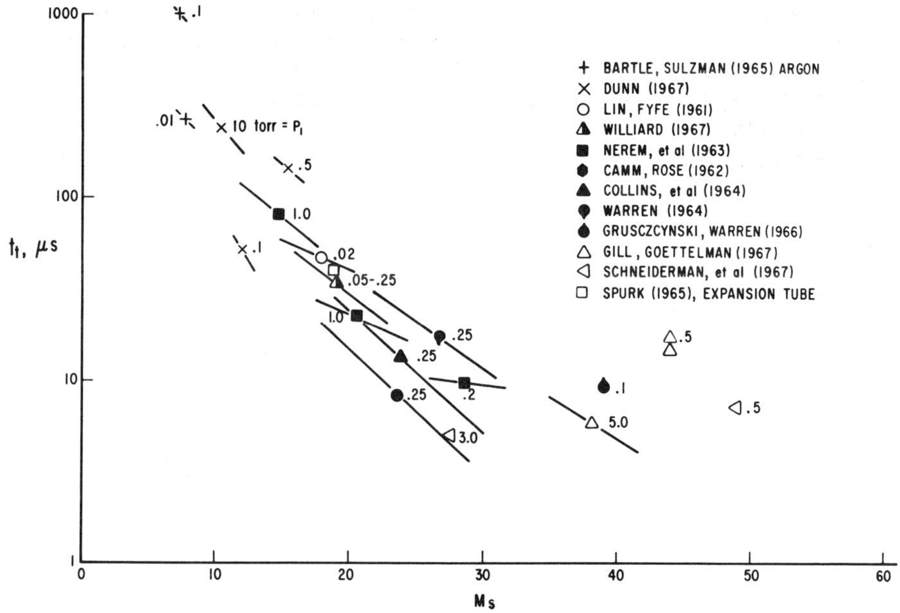

Figure 5
Reported approximate maximum experimental test time ranges as function of shock Mach number for several shock tubes and one expansion tube. Air is nominally the test gas.

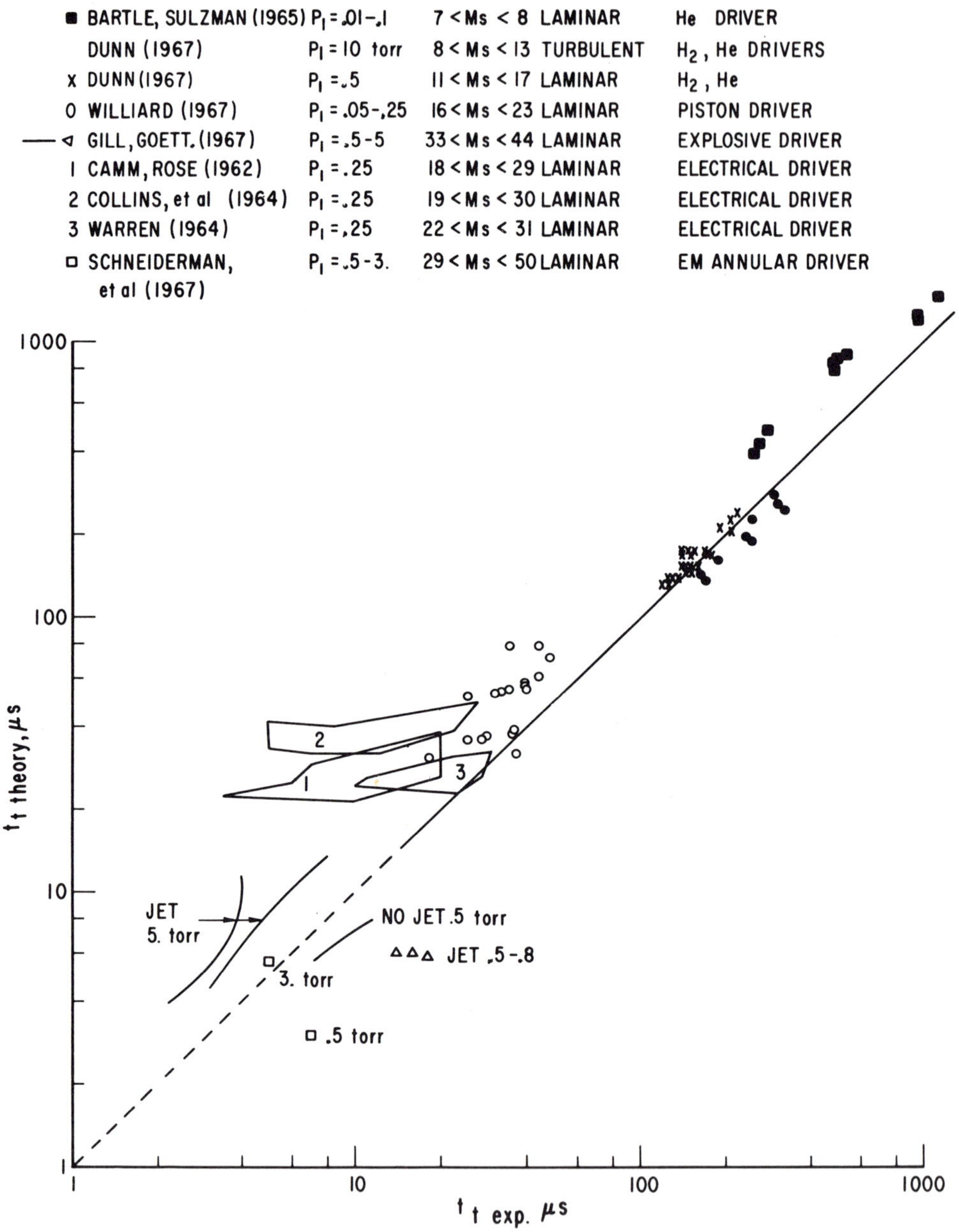

Figure 6
Reported experimental test times for several shock tubes compared to the theoretical predictions of Mirels (1963, 1964) and Roshko (1960)

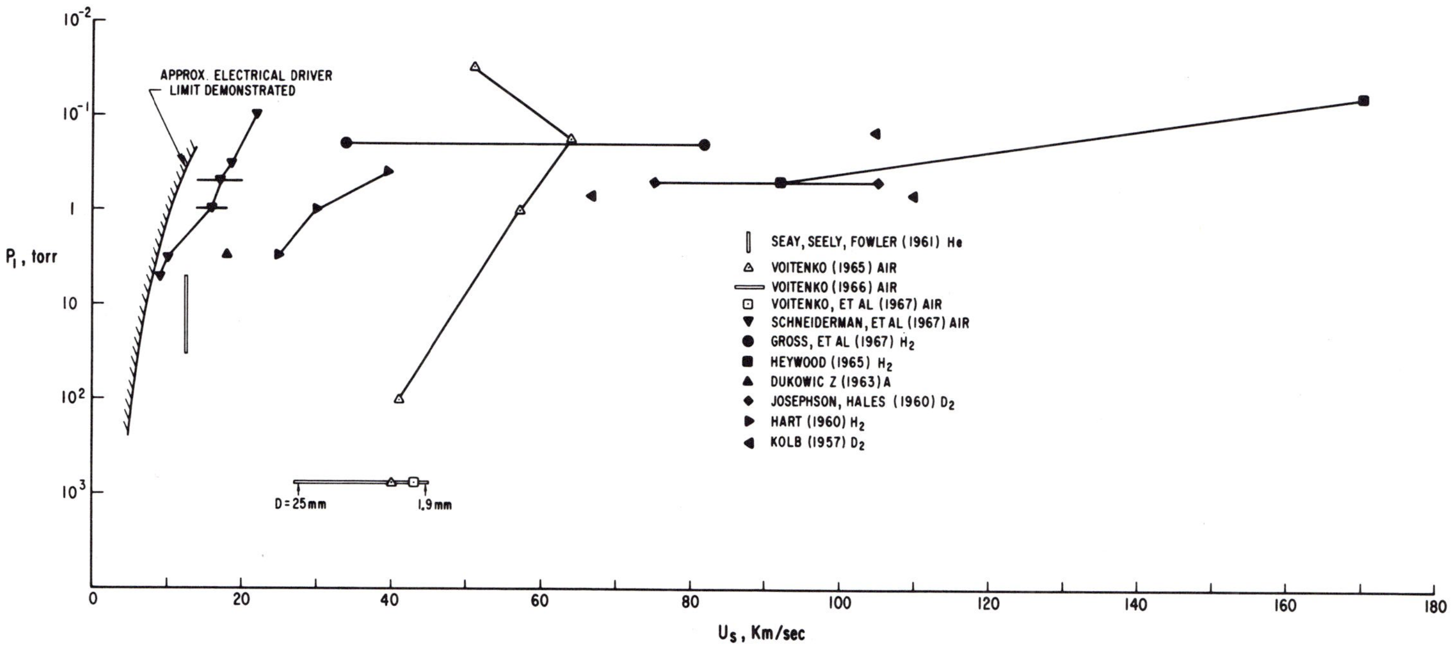

Figure 7
Incident shock velocity, initial pressure performance for several detonation and electromagnetically driven shock tubes

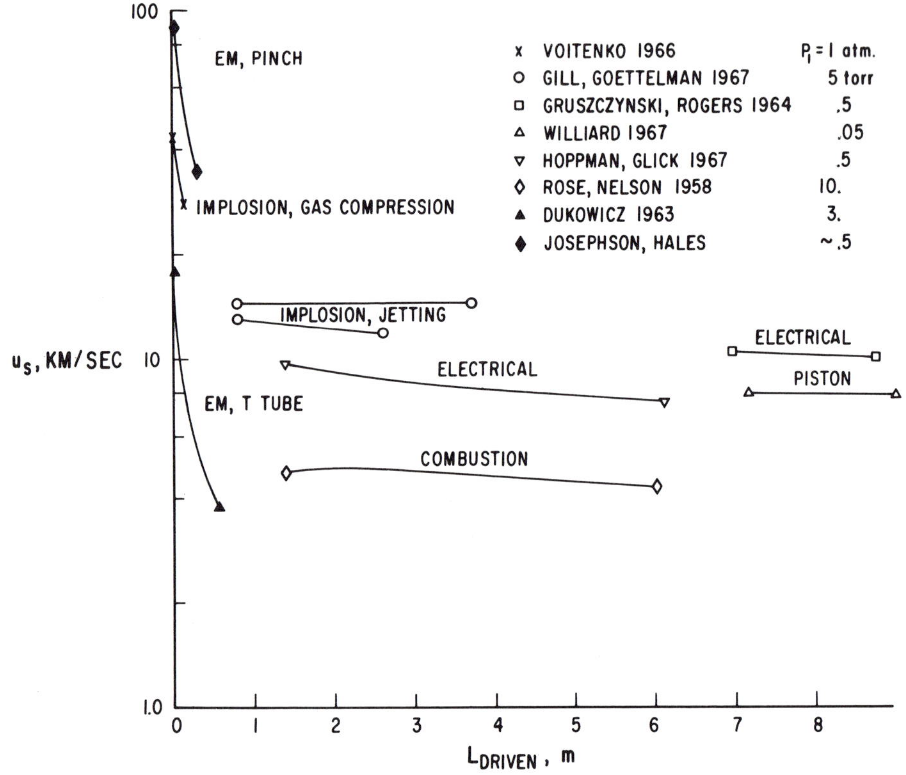

Figure 8
Incident shock velocity vs driven tube length

MEASUREMENT OBJECTIVES

INCIDENT SHOCK VELOCITY AND ATTENUATION

TEST TIME, $t_{\text{DRIVER GAS}} - t_{\text{INCIDENT WAVE}}$

FLOW STEADINESS (AND ESTABLISHMENT)

FLOW UNIFORMITY

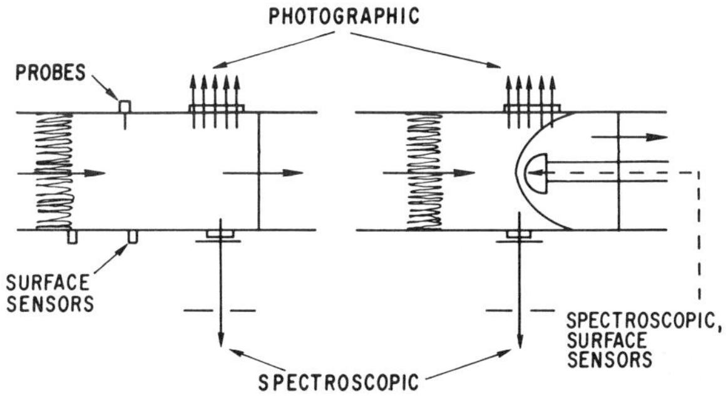

PHOTOGRAPHIC HIGH SPEED FRAMING, STREAK, IMAGE INTENSIFICATION AND CONVERSION, SPARK, LASER, INTERFEROMETER, SCHLIEREN, SHADOW

SPECTROSCOPIC EMISSION, ABSORPTION, MONOCHROMATIC AND WIDE BAND, WINDOW AND "WINDOWLESS" CAVITY GAGES, TIME RESOLVED SPECTRA

SURFACE SENSORS AND PROBES PRESSURE, THIN AND THICK FILM HEAT TRANSFER, ELECTROSTATIC AND CONDUCTIVITY

OTHER MICROWAVE, ELECTRON BEAM (LUMINOSITY, SCATTERING), LASER SCATTERING, GAS SAMPLING, HOLOGRAPHY

Figure 9
Several methods used to determine shock tube test flow properties and quality

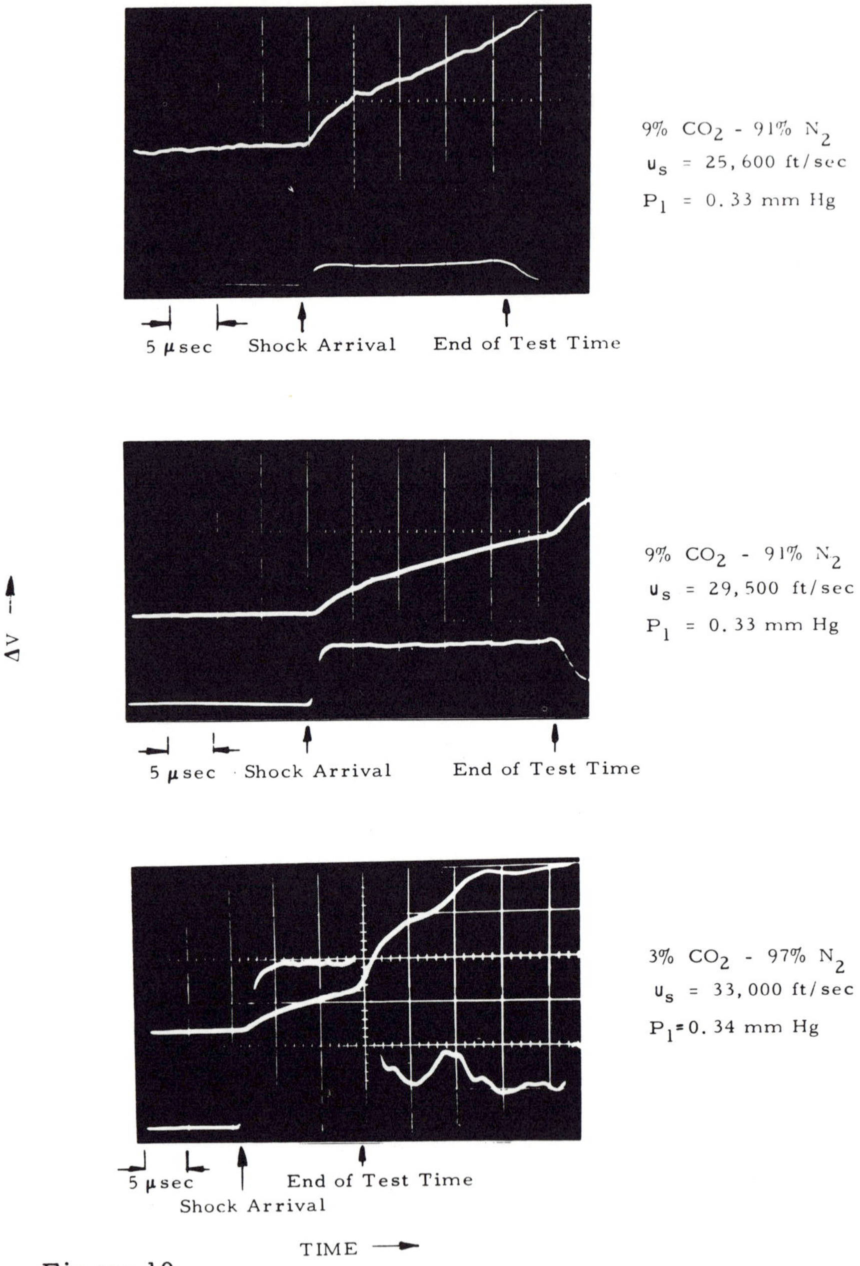

Figure 10
Oscilloscope trace showing thin film cavity gage response during test gas flow. Upper trace - model stagnation region cavity gage. Lower trace - sidewall photomultiplier sensitive in the red viewing stagnation region shock layer. (Gruszczynski and Warren, 1963)

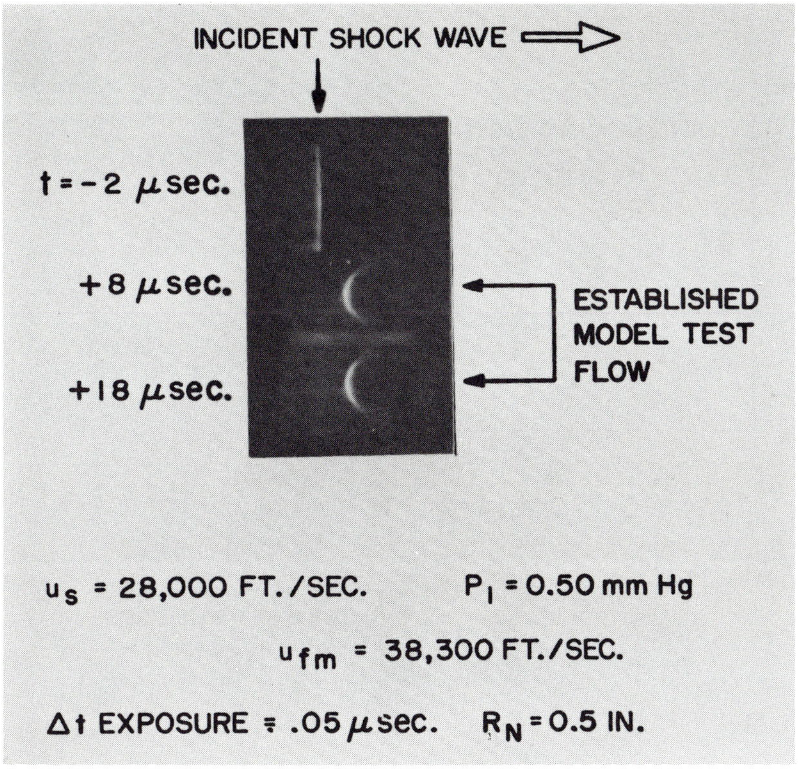

Figure 11
Image converter camera photograph of incident shock wave and model flow. Top frame shows incident shock approach model. Middle and bottom frames show established flow around a hemispherical model. (Gruszczynski and Rogers, 1964)

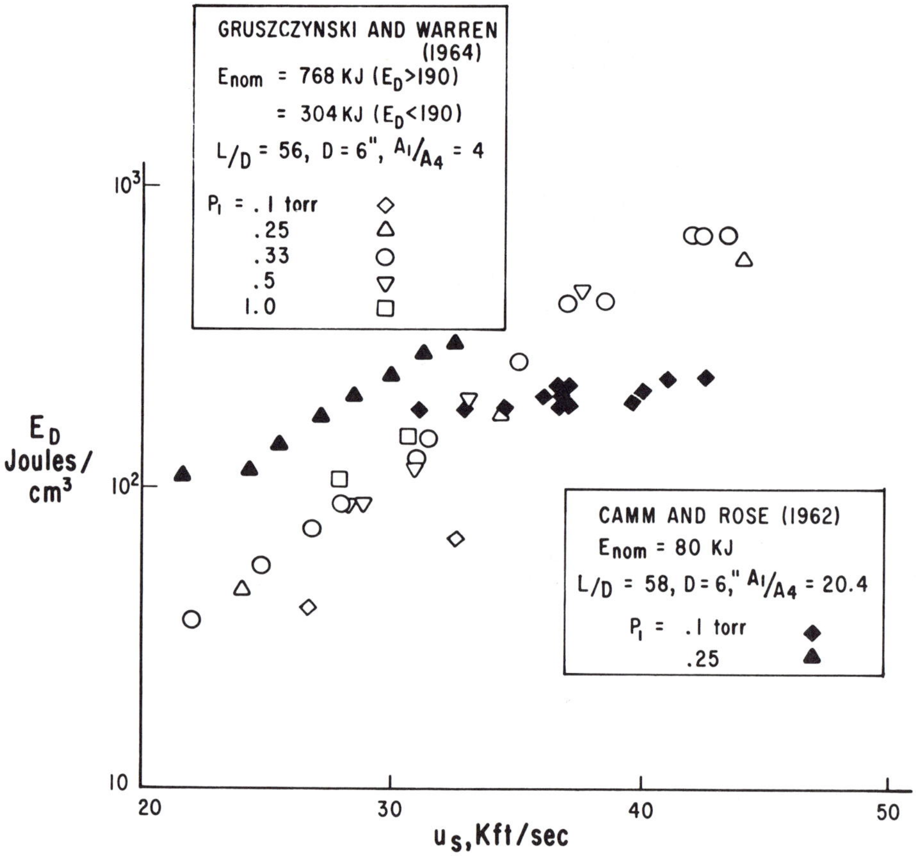

Figure 12
Shock velocity, driver energy density performance for two electrically driven shock tubes

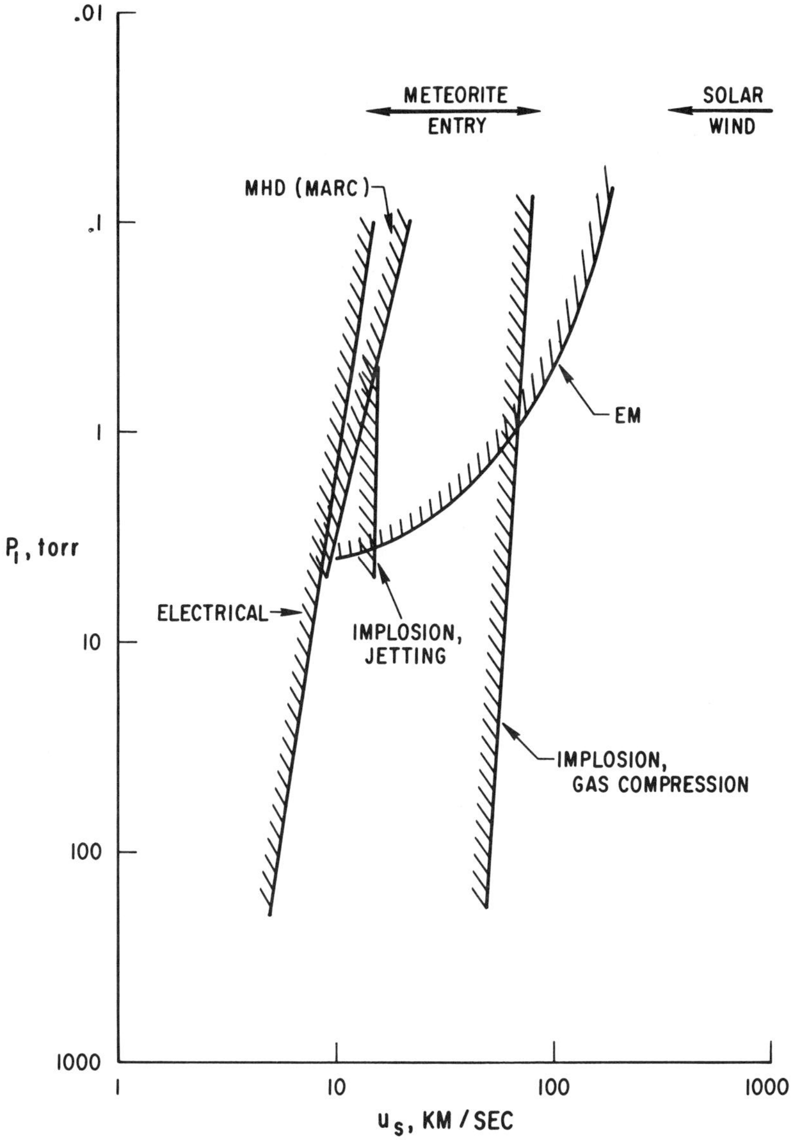

Figure 13
Demonstrated incident shock velocity/ initial pressure performance limits (approximate) for several types of shock tube drivers

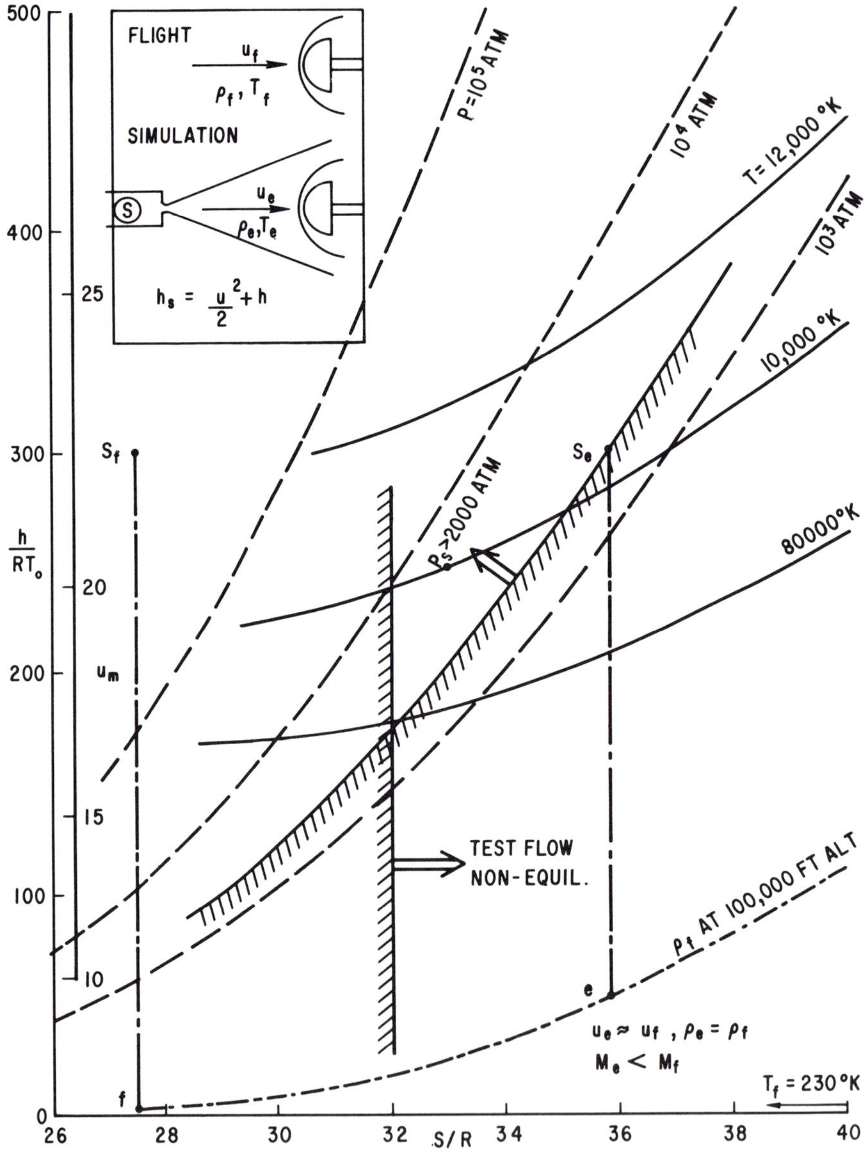

Figure 14
Pressure and chemical nonequilibrium limitations imposed on simulation of flight conditions with a stagnant-plenum-conditions test facility

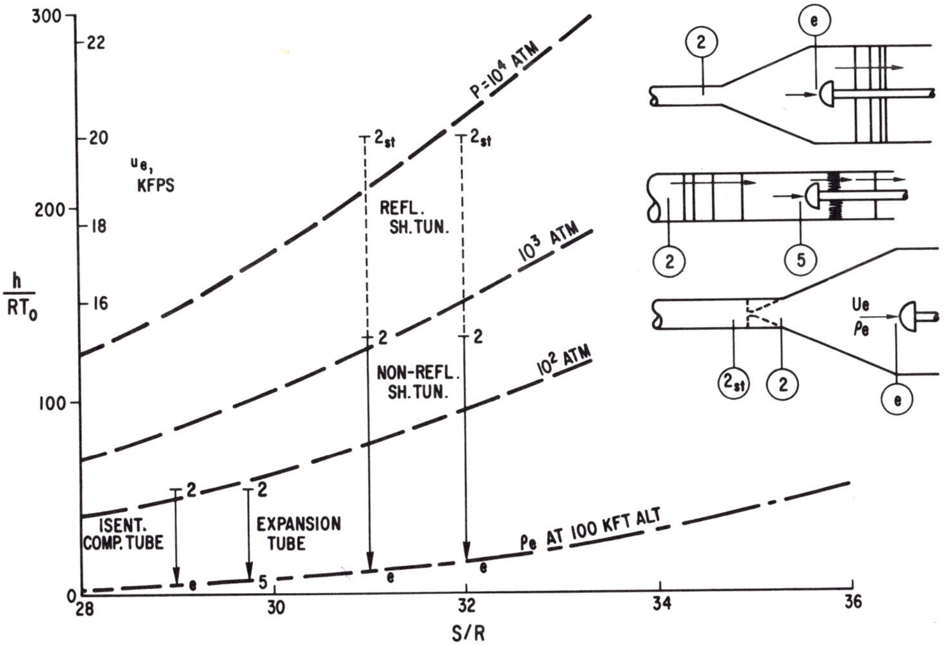

Figure 15
Comparison of shock tube flow properties for reflected and non-reflected shock tunnels, expansion tube, and isentropic compression tube (tunnel). Flight simulation is 20, 000 ft/sec at 100, 000 ft altitude

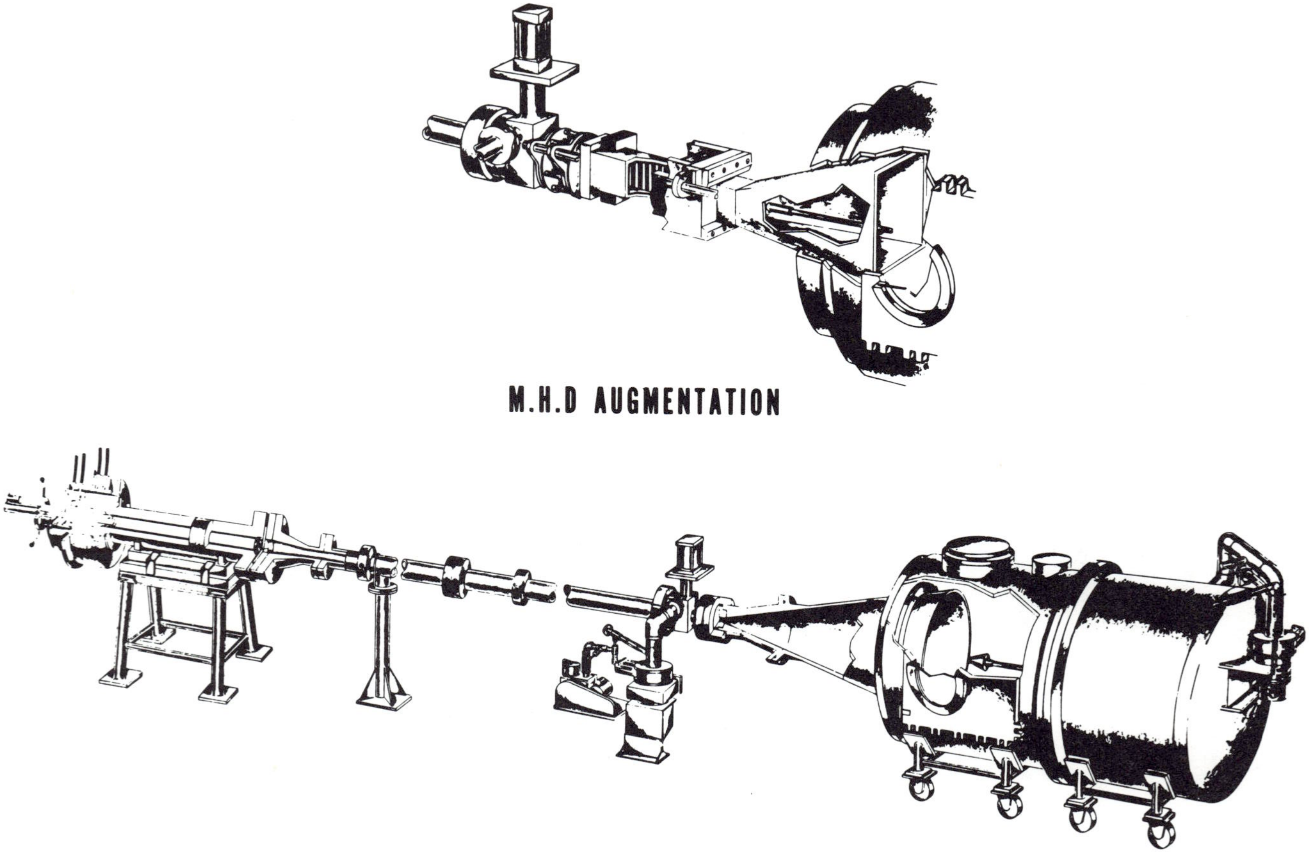

Figure 1.6
Electrically driven shock tunnel: 2'' id driven tube, 3'' id driver. Insert shows the facility modified with a Faraday type MHD accelerator located in the upstream part of the nozzle (Harris, et al., 1965)

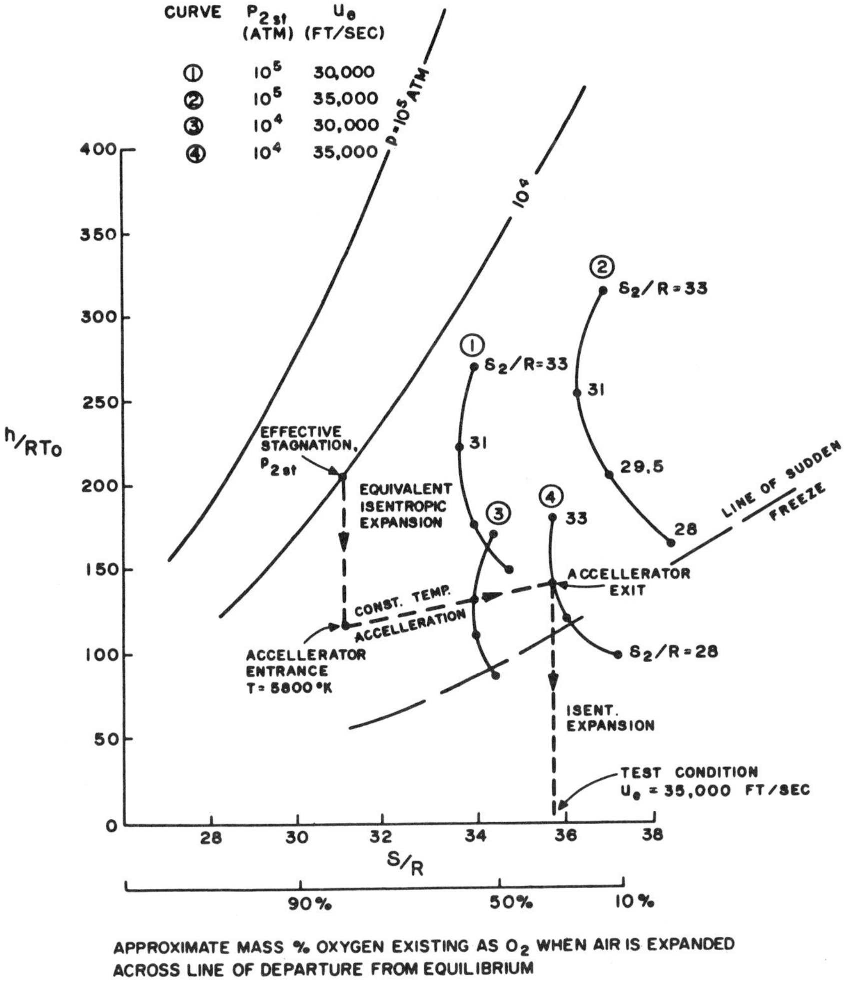

Figure 17
Locus of MHD accelerator exit states (for T = const., $LB^2\sigma = 4 \times 10^5$ in MKS units) from which isentropic expansion to a specified test velocity is possible for a specified initial stagnation pressure. The short dashed line indicates one typical expansion-acceleration-expansion path. (Marston and Warren, 1968)

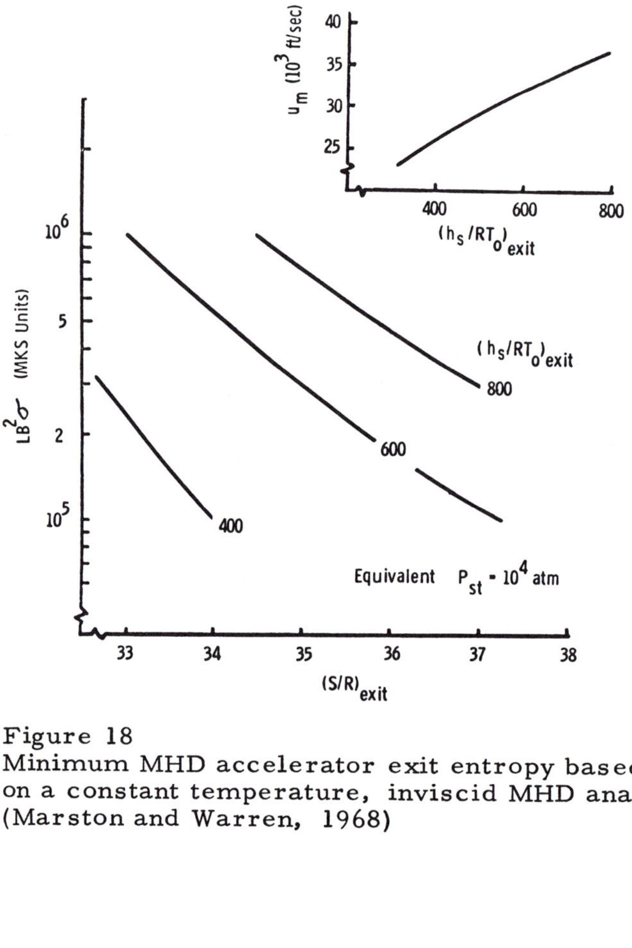

Figure 18
Minimum MHD accelerator exit entropy based on a constant temperature, inviscid MHD analysis (Marston and Warren, 1968)

EXPERIMENTAL STUDY OF SHOCK ATTENUATION IN A 10^6 JOULE ARC HEATED SHOCK TUBE USING HELIUM-ARGON MIXTURES AS THE DRIVER GAS

DAH YU CHENG
Mechanical Engineering Department
University of Santa Clara
Santa Clara, California, U.S.A.

SUMMARY

Shock wave attenuation in a high energy, high pressure shock tube can be delayed by using a helium-argon mixture as the driver gas. Species separation, caused by the difference in constituent molecular weights of this mixed driver gas under high acceleration (10^5 g) during the opening of the diaphragm, generates a high initial shock velocity in the driven tube (helium rich region) and a lower rarefaction wave velocity in the driver tube (argon rich region). This phenomenon elongates the "catch-up time" of the rarefaction wave with the moving shock. Different mechanisms proposed for attenuation thus can be separated. Velocities are measured as a function of distance along the shock tube. The small attenuation with a 30/70 per cent argon/helium driver gas is in good agreement with that predicted by boundary layer theory. The influences of the mean molecular weight and the temperature of the driver gas on attenuation are studied. Time histories of the pressure at the end of the shock tube, showing characteristic pressure bumps in the case of mixed driver gas, are presented as an evidence of species separation.

INTRODUCTION

Large shock wave attenuation in a high pressure (10,000 psi) and high temperature (15,000°R), electrical arc heated shock tube can be delayed by using a helium-argon binary mixture as the driver gas. Boundary layer theory can account for only relatively small shock wave attenuations. The expansion wave system in the tube contributes the major part of the observed attenuation; this is verified experimentally.

The present test program is conducted under the assumption that a binary mixture of noble gases, with a large difference in molecular weights, can be separated by the high accelerating force (10^5 g) that occurs during the opening of the main driver diaphragm. All noble gases have almost the same molar specific heats; hence, any mixture has nearly the same temperature and pressure, independent of its mixture ratio, after constant-volume arc heating with a given energy input. However, the sonic speed of each species of the driver gas mixture is different because of the difference in molecular weight. During the diaphragm opening, the mixture is accelerated into the low pressure gas in the driven tube to form a traveling shock wave. However, the lighter element in the mixture accelerates faster than the heavier element due to its higher sonic speed or, alternately, to satisfy the conservation of momentum; thus the species are separated. According to ideal shock tube theory, a stronger shock wave will also be formed by the gas composed of the lighter element. The steady-state shock structure in an inert binary mixture has been studied by Sherman (Ref. 1). He found that the species will separate in the shock zone due to pressure diffusion, molecular diffusion and thermal diffusion, but will recombine after the shock in order to satisfy the steady state boundary conditions.

The shock formation and subsequent attenuation are influenced by the way in which the diaphragm opens, of course. The diaphragms are scored stainless steel that tear open into four petals or lobes, and the opening process has been studied by a lobe contact technique and also by a high speed photographic technique (Ref. 2). The average total opening time is found to be about 150 μ sec, corresponding to the time required for the pressure load to accelerate a free hinge through 90° of arc.

The present paper studies high energy shock wave attenuation with variations in energy density, mixture ratio of the driver gas, and driver chamber length. The effects of boundary layer are also examined.

In a high energy shock tube, however, the separation of the driver gas encounters a transient boundary condition. The separation process can cause a "run-away" condition which will introduce a gradient in local sonic speed in the driver chamber during the expansion process. Rarefaction waves which travel toward the upstream end of the driver tube pass through a non-flowing mixture at a sonic speed determined by the local mean molecular weight. When this rarefaction wave is reflected from the end of the driver tube, it first passes through a gas of relatively high molecular weight and low local sonic speed. As it moves down the tube it is gradually accelerated toward the shock front, in passing through regions of decreasing molecular weight with higher local sonic speeds. Therefore, the "catch-up" time is increased. This allows us to separate the importance of different attenuation mechanisms.

It is obvious that this process will be influenced by the way in which the driver diaphragm opens. This has been studied by a lobe contact technique and, recently, by a high speed photographic technique (Ref. 2). The average total opening time is about 150 μ sec.

Variations in energy density, mixture ratio of the driver gas, and driver chamber length have been studied. Boundary layer theory for shock wave attenuation has been examined.

EXPERIMENT AND RESULTS

The experiments were conducted in the Electric Arc Shock Tunnel Facility at Ames Research Center, NASA (Fig. 1). The Facility has a 40 kv capacitor bank capable of storing 10^6 joules of electrical energy. The shock tube is of stainless steel construction, with an I.D. of 4 inches and a driven tube length of 37 feet. The driver tube is 4 inches in diameter and has a fiber glass reinforced R.T.V. rubber liner. The driver length can be varied from 30 to 114 inches. The discharge is triggered by the explosion of an 0.005 inch diameter tungsten wire. A stainless steel scored diaphragm is burst by high pressure. Noble gases are used in the driver; in these tests the driver chamber was prefilled to 272 psia. The driven gas is air, which is prefilled to 3.5 cm Hg absolute pressure. Timers are triggered by a delayed electrical pulse, initiated by the current flow of the capacitor bank discharge (Ref. 3). The arrival of the shock wave at different positions along the shock tube is detected by a series of twin-electrode probes mounted flush with the tube wall. The probe signals stop the timers. The average time between two stations determines the mean shock velocity. The probes are located 8 feet apart. When a hypersonic nozzle is not being used, the end of the shock tube is blanked off; here piezoelectric probes are mounted to detect the reflected shock pressure.

The shock velocity profile measured along the driven tube, as a function of driver-gas mixture, is presented in Fig. 2. Mach numbers of the incident shocks varied from 9.5 to 17. These results show that both the incident shock velocities and the percentages of total attenuation are decreased with increasing argon concentration. For argon concentrations above 30%, relatively flat velocity profiles extended to 72 diameters (24 feet) downstream of the diaphragm. For each of the mixtures, a sudden increase of attenuation is observed. It is believed that this sudden increase of attenuation is due to the "catch-up" of strong rarefaction waves.

Since the incident shock wave at the first measuring station (8 feet) should be only slightly affected by any boundary layer attenuation, the measured velocity should approach the prediction of the strong shock relations. If we match the measured and predicted velocities for 100%

helium, the variation of incident shock velocity with composition can be expressed as a function of the inverse square root of the driver gas mean molecular weight. A plot of incident shock velocity, for an ideal driven gas, versus argon concentration in the driver gas is given in Fig. 3. The measured velocities fall considerably above the theoretical, ideal, strong shock correction. If the air is assumed to be in equilibrium behind the shock wave (no relaxation is taken into account), the calculated incident shock speeds at 8 ft. station come close to the experiment data, although all the experiment data points are still higher than the calculated shock speeds. The largest deviation takes place between 20%-30% argon mixture. This seems to follow the trend reported in Ref. 1 for steady state, binary mixture shock waves, that the shock thickness (species separation) becomes a maximum between 20-30% argon.

To identify the effect of driver gas mean molecular weight, pure helium and 60/40 He/Ar mixtures were used to produce different strength shock waves by variations of the driver gas temperature. The results are presented in Fig. 4. The pure helium driver gas produces large attenuations even at very low shock speeds. In comparison, the He/Ar mixture driver gas allows little or no attenuation up to 72 diameters (24 feet) at relatively high shock speeds. For these results, the chamber temperature varies by a factor of five and the mean molecular weight varies by a factor of four and a half. The data clearly show that the predominating large attenuations are associated with the lighter gas only.

To identify the influence of the geometry of the driver and the associated boundary layer effect, a 70/30 He/Ar mixture was used as the driver gas, with driver lengths of 30, 54, and 114 inches. The energy density of the discharge for different driver lengths was kept the same. Velocity profiles are presented in Fig. 5. It can be seen the 30-inch driver has a large attenuation resembling that of the pure helium driver case, whereas the 114-inch driver reproduces the same velocity profile as the 54-inch length, except that the attenuation of the 114-inch driver has been delayed.

The boundary layer theory on shock attenuation by Mirels and Mullen (Ref. 4) is used for a theoretical comparison. The slope of the theoretical wave in Fig. 5 shows that surprisingly good agreement between this theory and the experimental data is achieved up to the point where severe attenuation sets in. This result suggests that previous data on strong shock attenuation does not separate the boundary layer effect from some rarefaction wave structure generated in the driver chamber.

It has been shown that the large attenuation is caused by some rarefaction wave originating in the driver chamber. The influence of diaphragm opening on the rarefaction wave could be of primary importance. Photographs show that lobes of the stainless steel diaphragm open uniformly and start to accelerate in an exponential fashion 85 microseconds after they are cracked (Ref. 2). To be fully opened requires a total time of 150 microseconds. During this time the strong shock wave is still in its formative stage and rarefaction waves have already started to propagate toward the upstream end of the driver chamber. The reflected rarefaction wave overtakes the shock at a much earlier time than that used for the design of the shock tube. Figure 6 shows schematic x-t diagrams for the cases discussed here. Case I is the idealized situation. Case II shows the modifications produced by the delay in diaphragm opening. Case III represents the situation where species separation in the driver gas has increased the shock strength and delayed the arrival of the rarefaction wave.

Since argon and helium have a molecular weight ratio of about ten, at the same temperature the sonic speed ratio will be 0.316 (Ar/He). During the opening of the diaphragm, the shock wave is formed as the driver gas rushes out at a speed close to its maximum speed, limited by Riemann's Invariant. For an ideal mixture, the result at some arbitrary time is visualized by the top pressure profile, at the left, in Figure 7. In the extreme case, if there were no mutual transport properties involved, such as a diffusion, the helium would propagate at a much higher speed than the argon to form a fast shock wave. However, the shock strength can only approach that for a pure helium driver at the helium partial pressure, as illustrated in the second profile in Figure 7, left. The argon and the rest of the helium, as a whole, will have to propagate as a pressure wave on top of the expanded helium driver gas, as in the fourth profile in the figure. The molecular diffusion between argon and helium through collisions tends to bring the mixture back to its equilibrium composition. Other diffusion mechanisms, such as pressure diffusion, tend to increase the separation (Ref. 1, 5), but if the total diffusion velocity is slow with respect to the shock velocity then there should be a pressure increase after the diffusion zone, as in the last pressure profile. Piezoelectric probes were used to measure the pressure-time history at the end wall of the shock tube; some typical results are shown at the right in Fig. 7. The scope traces show that the pressure at the end of the driven tube, using binary mixture driver gases, have pressure bumps some 2,700 microseconds after the shock passes, whereas the record obtained with the pure helium driver gas has a smoothly decaying trace. These bumps are interpreted as the arrival at the end wall of the successive pressure waves caused by species separation of the driver gas mixture.

ACKNOWLEDGEMENTS

The author is very grateful to Mr. Robert E. Dannenberg and the Fluid Mechanics Branch of Ames Research Center, NASA. Their help in making available their time and high energy shock tube equipment enabled the author to complete this work. Thanks are also extended to Mr. Howard A. Stine for his interested and challenging discussion of the work.

REFERENCES

1. Sherman, F. S., "Shock Wave Structure in Binary Mixtures of Chemically Inert Perfect Gases", J. Fluid Mech., No. 8, pp. 465-479, 1960.
2. Cheng, D. Y., Dannenberg, R. E., and Stephens, W. E., "A Novel Use of a Telescope to Photograph High Pressure Diaphragm Openings", J. A.I.A.A., June 1969.
3. Dannenberg, R. E. and Humphrey, E. E., "Microsecond Response System for Measuring Shock Arrival by Changes in Electrical Impedance in a Shock Tube", Rev. of Sci. Instr., Vol. 39, No. 11, Nov. 1968, pp. 1692-1696.
4. Mirels, H. and Mullen, J. F., "Small Perturbation Theory for Shock Tube Attenuation and Nonuniformity", Phys. Fluids, Vol. 7, No. 8, 1964, p. 1208.
5. Hirschfelder, J. O., Curtiss, C. F., and Bird, R. B., Molecular Theory of Gases and Liquids, John Wiley & Sons, Inc., 1964.

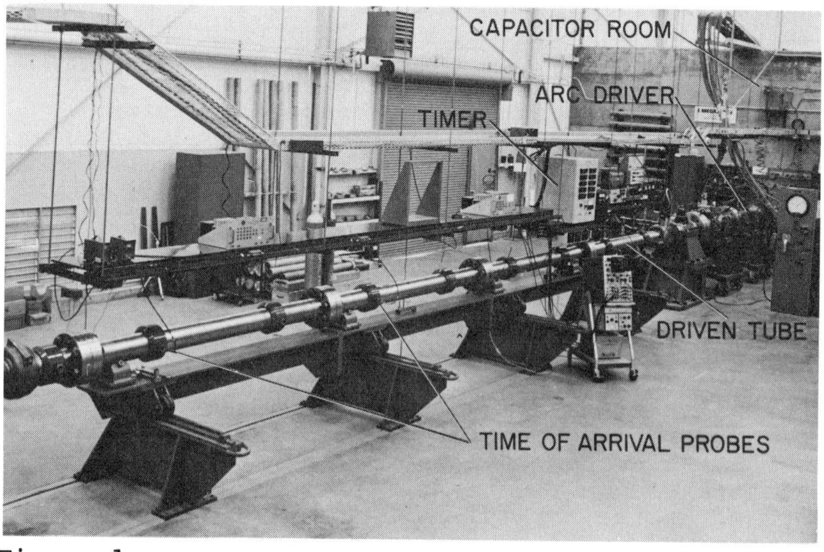

Figure 1
Ames Electric Arc Shock Tunnel

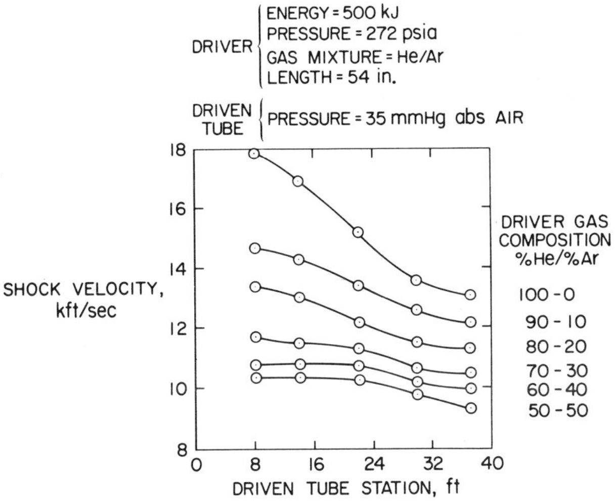

Figure 2
Velocity Profiles of He/Ar Driver Gas

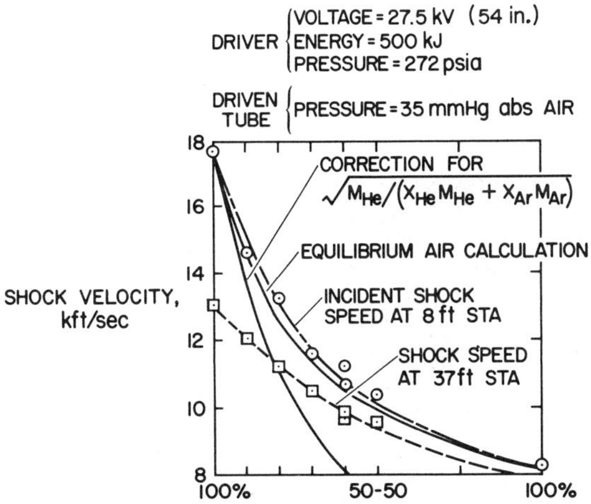

Figure 3
Shock As a Function of Concentration

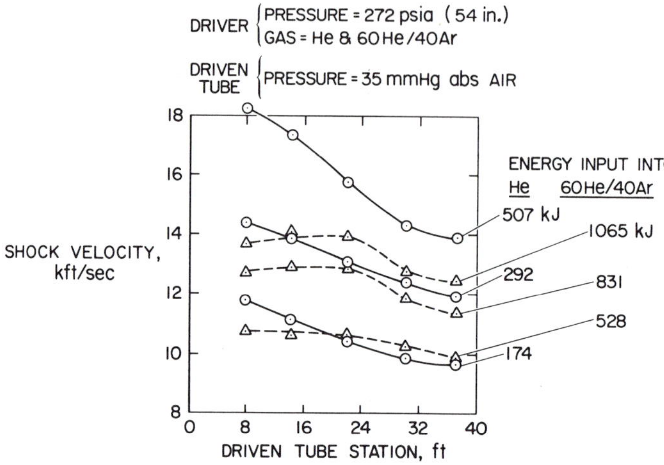

Figure 4
Effect of Driver Gas Temperature on Shock Attenuation

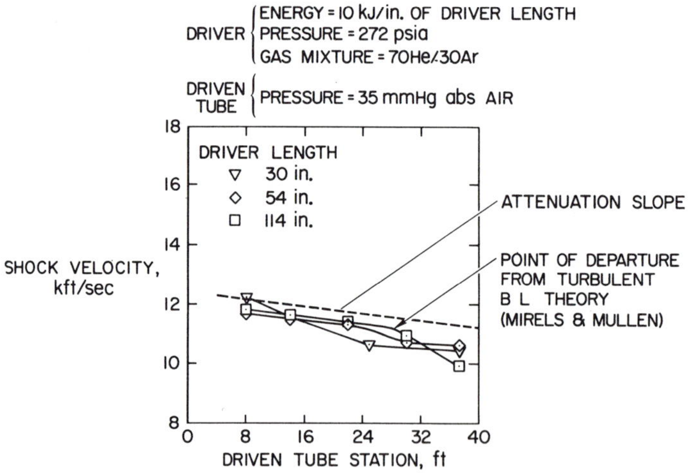

Figure 5
Influence of Driver Length and Boundary Layer Effect

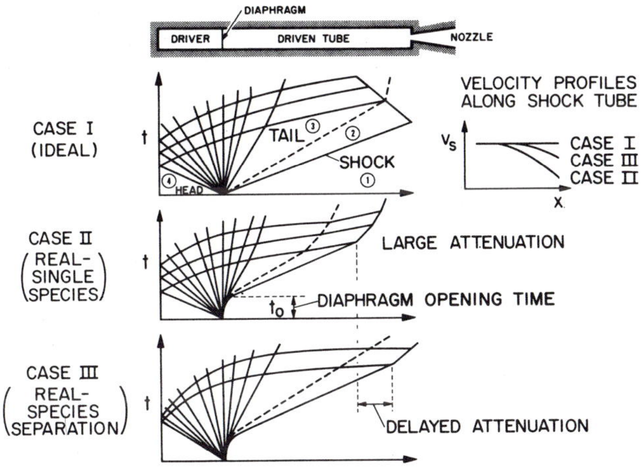

Figure 6
Influence of Diaphragm Opening on Shock Diagram

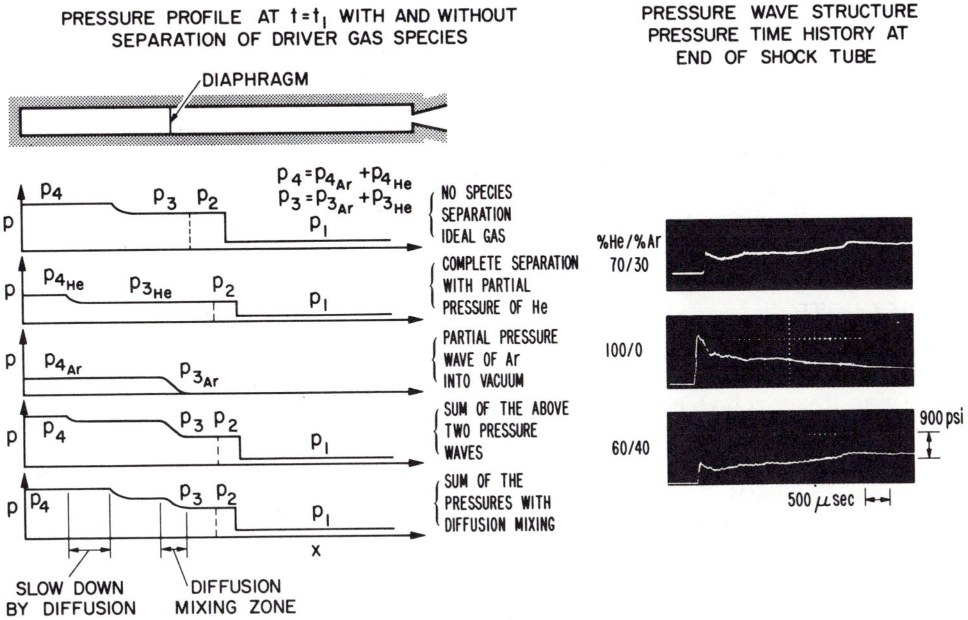

Figure 7
Pressure Profiles and Pressure Records

AN IMPLODING TRIGGER TECHNIQUE FOR IMPROVED OPERATION OF ELECTRIC ARC DRIVERS

ROBERT E. DANNENBERG
Ames Research Center
National Aeronautics and Space Administration
Moffett Field, California 94035

SUMMARY

A practical technique for achieving high shock velocities in a shock tube is to heat the driver gas with an electric-arc discharge. The greater the energy dumped into the driver chamber, the higher the shock velocity. Unfortunately, the driver efficiency falls off rapidly and is only about 30-40 percent at the high energy levels of current interest.

A new imploded wire concept has been developed to initiate transfer of capacitor bank energy into a driver. The mechanism is basically a coil of fine wire spiraled along the periphery of the insulated driver chamber. The imploded wire technique has been demonstrated on various drivers and the results indicate an increase in efficiency up to 80 percent.

A shock wave at a velocity of 48,700 feet per second has been driven into 1 torr of dry air. It is the highest speed known for this driven tube pressure level with an impulse type of facility.

INTRODUCTION

A variable length arc driver has been in operation for the past two years at the Ames Research Center. The driver length is easily changed to achieve proper matching of the driver to the driven tube for operation of a variety of test facilities used in gasdynamic and gas physics investigations. Energy for the driver is supplied by a 1 million joule, 40 thousand volt capacitor storage system. The driver has been operated with arc chamber lengths of 30, 54, and 114 inches, at energy storage levels from 350 to 1000 kilojoules, and with initial driver chamber pressures ranging from 100 to 500 pounds per square inch. Characteristic performance trends will be presented in terms of the discharge current and voltage waveforms, and driver efficiency. The production of high-speed shock waves is shown to be limited when a straight trigger wire is used in the driver chamber to initiate the arc discharge. This limitation can be removed by increasing the length (and inductance) of the path the arc must travel. By spiraling the trigger wire along the inner surface of the arc chamber, the driver and driven tube performances are improved, peak arc currents are reduced by factors of 2 to 3, and a new behavior is demonstrated in arc driver operation.

DRIVER DESCRIPTION

The electric arc driver system[1] was designed to provide a flexible and practical high-energy driver chamber to be used with different types of shock-driven facilities. The driver is composed of two 5-foot, high-pressure tube sections, one of which contains the current collector assembly. The latter section has an arc chamber 54 inches long and 4 inches in diameter. The chamber length can be reduced to 30 inches or less by means of high voltage electrode extensions and insulating plugs within the driver. When the two sections are connected, the arc chamber length is increased to a maximum of 114 inches. The system is quite versatile, and the different driver lengths are available for use at either of two parallel locations. In the arrangement shown in figure 1, the 54-inch driver is in position at a diaphragm test rig. A low-density shock tube is scheduled for installation at this station at a later date.

In the alternate position noted in figure 1, the driver powers the Ames 30-inch electric arc shock tunnel with a 4-inch I.D. driven tube that is 37 feet long. The 114-inch chamber is required for a tailored interface mode of operation.[2] With the collector assembly tube in this position, the system performance was evaluated for the different driver lengths. For a complete discharge of the stored energy from the capacitor bank into the arc chamber, the maximum obtainable energy density was 150, 87, and 55 J/cc for the 30-, 54-, and 114-inch driver, respectively.

The energy for the arc is supplied by a 1250 μF capacitor bank that is capable of storing 1 million joules when charged to a maximum of 40,000 V. The bank is charged in about 90 seconds. It can be discharged at maximum energy-storage capacity (1 MJ) or at a particular energy storage by varying the preset voltage. An overall view of the capacitor room is shown in figure 2. The capacitor bank consists of 180 capacitors, each rated at 27.5 μF for 20 kV DC. The capacitors are fused individually and are bussed into 10 sections of 18 units each. Each section is divided into two groups of nine cans each. The capacitors in each group are connected in parallel and the two groups are arranged in series. Thus, together, they form a series pair for operation at 40 kV. Two RG-17 coaxial cables connect each of the sections to the collector ring. In this type of bank design, the rack structure charges to one-half of the preset voltage - an arrangement first used by Mr. Dave Rogers of the General Electric Company.[3]

The high pressure driver chamber is insulated by an RTV (silicone rubber) coated, filament-wound, fiberglass liner. This liner has been found to be usable for at least 25 runs. After each firing, the liner and driven tube surfaces are wiped clean. The arc is initiated by means of a fine wire attached to the electrode terminal at the downstream end of the driver chamber and separated by at least 3 inches from the high voltage electrode with a piece of cotton string. The string is attached to a plastic rod that passes through the hollow core of the high voltage electrode. The space (occupied by the string) acts as a gas gap or switch and is the sole device preventing the capacitor bank energy from discharging since the bank is connected directly across the driver terminals. After the bank has reached its preset voltage, a pneumatically operated piston causes the firing rod to pull the trigger wire toward the main electrode until the breakdown voltage is reached and a high current is initiated to vaporize the wire and thus provide a path for the arc strike.

Flat scored diaphragms separate the driver tube from the driven tube. They are fabricated in different thicknesses (1/16 to 3/8 inch) and are scored from 20 to 50 percent of their depth to control their bursting pressure. The diaphragm opens into a square section that transforms smoothly into the circular tube. Diaphragm material is hot-rolled, fully annealed type 321 stainless steel.[4] This material has consistently produced lobe formations without fragmenting. For each test, the diaphragm is selected on the basis that the static burst value is about one-third of the expected equilibrium pressure of the driver gas after arc heating. With this mode of operation, the diaphragm starts to open about 150 μsec after the arc current has reached its maximum or peak value.

DRIVER PERFORMANCE

In evaluating performance, the system was examined on the basis of both the electrical characteristics of the arc discharge and the measured shock velocities and reservoir pressure in the driven tube. These data are taken for each test in the facility, and, to date, something like 350 firings have been made. Another performance monitor is a "first-light" signal which is generated when little more than a crack appears in one of the grooves of the diaphragm. This light (heated driver gas) is detected by a photomultiplier tube viewing upstream through a light pipe located in a port just downstream of the diaphragm. The first light signal also serves to reference the start of the diaphragm opening with the current records and other shock tube events.

The performance of the capacitor bank in terms of the energy available for an arc driver using helium is shown in Fig. 3. The bank or electrode voltage is presented as a function of the arc strike distance and the several parameters that limit the discharge capability are noted on the figure. For the short arc lengths, the onset of a critical or underdamped discharge restricts operation from the region indicated as underdamped. Operation in this mode is potentially dangerous to the capacitors as the units are rated for a damped or nonreversing type of discharge only. The upper limit of voltage is, of course, 40 kV.

Experimentally, it was found that there is a minimum voltage required to induce an arc strike. It corresponds to about 3.6 kV per foot of arc length.

For voltages below this minimum, the trigger wire was exploded but did not induce or result in an arc discharge. The shaded area noted in Fig. 3 represents the usable range of preset voltages capable of culminating in the discharge of stored energy from the capacitor bank. No difficulties were experienced in drawing the arc for any preset voltage within this range and it was found that the arc-out or voltage at which the arc was extinguished increased with arc length as indicated on the figure. Somewhat surprisingly, it was found also that the arc-out voltages for a given driver length were independent of the preset voltages. These arc-out voltages increased by about 1.8 kV per foot of arc length. The reduction in the available stored energy due to the residual arc-out voltage is small for both the 30- and 54-inch drivers.

In the case of the 114-inch driver, the arc-out voltages are high. Only about 70 to 80 percent of the bank energy is discharged into the driver chamber. In this case, the residual energy is about 250 kJ. Also, after the arc is extinguished, the residual voltage on the electrode (~20 kV) can be troublesome. The high temperatures within the arc chamber tend to carbonize the surfaces of the insulating materials surrounding the high voltage electrode, thus forming surfaces that become electrically conducting. If the dielectric strength of the seal fails between the end surfaces and the insulated liner, within the time period prior to an automatic dump through grounding resistors, a second electrical discharge can occur over to the driver tube. Such a discharge can cause mechanical damage to the liner, insulator plugs, and metal walls. This problem can be alleviated by good compression between adjoining insulator sections and by providing stress relief on all insulator surfaces exposed to the arc.

STRAIGHT TRIGGER WIRE

The voltage breakdown trends shown in Fig. 3 are characteristic of drivers with the arc initiated by a straight trigger wire generally located along the axis of the chamber. What happens to the characteristics if wires of various sizes or of different materials are used? Experimentation revealed that only certain types of wire material produced the arc discharge, although all wire materials tested were exploded within the chamber. Tungsten, titanium, chromel-C, and nickel wires initiated an arc. Copper, manganim, and 302 stainless steel did not. Aluminum was satisfactory only in the shorter driver. For the wires that culminated in an arc discharge, changing the ratio of the mass of the wire material between 0.4 to 10 percent of that of the mass of the helium gas in the driver did not significantly change the arc characteristics; nor did the different wires or wire sizes show any changes in the measured shock speeds. The performance of a driver was not changed by the trigger wire. It is only necessary that a proper material be used. Nor was the resistance of the solid wire pertinent to the wire-arc process. High or low resistance wires generated similar peak currents in the arc discharge.

It is interesting to note that the type of materials that induced a discharge are categorized as "high boiling point, high heat of vaporization" and fall into a classification of superfast exploding wires in which the magnetic pressures are believed to dominate the process. [5] In all the tests noted herein, the wires were exploded with an initial current conduction density of between 10^6 to 10^8 amperes/cm^2.

Analysis also indicated that for each arc length, the peak current recorded for its particular voltage drop (i.e., preset voltage minus arc-out voltage) followed a characteristic straight-line relationship. The slope of this curve represents the effective arc resistance in ohms. It was found, further, that the unit arc resistance was essentially the same for the three arc lengths tested. Simply stated, the arc acted as a conductor whose resistance increased directly with its length. For engineering purposes, a value of 15 mΩ per foot of arc length can be used in circuit calculations. This value is independent of the trigger wire material, thus the pressurized arc is a "gas" or "helium" discharge rather than a vaporized wire arc[6].

Figure 4 shows the measured shock speed in the driven tube as a function of the energy density discharged into the driver chamber. The driven tube loading was 35 torr of dry air. The data symbols represent the three driver lengths for an initial load pressure of 18.5 atmospheres of helium. The agreement of the experimental values for different driver lengths and the general trend of the data clearly show the predominant effect of energy density on shock speed. Also shown on the figure are the variation noted for two of the drivers over a range of initial pressure loadings. For the 114-inch driver, the temperature range after arc discharge corresponding to the indicated pressures is about 3 to 1. These results show that the driver-driven tube performance is not particularly sensitive to the initial load pressure of the driver gas and imply an increasing loss of energy as driver temperatures are raised.

Figure 4 also can be used to illustrate the rapid fall-off in the driver efficiency with increasing energy density. The dashed curve, designated as C-V, represents the shock speed calculated for an ideal gas with perfect energy transfer. It designates the upper limit or maximum shock speed that could be obtained for the particular driver-driven tube loadings. The ideal values of the temperature ratio, $(T_4/T_1)_{C-V}$, are shown along the top of the figure corresponding to the energy density values noted along the abscissa. The effective experimental values calculated for the particular shock speed are noted by the arrows in the figure. These temperature ratios were calculated using a shock-tube flow analysis developed by Dr. N. M. Reddy. The ratio of the experimental to the ideal C-V values represent a practical measure of the driver efficiency. On this basis, the efficiency is about 71 percent for the 114-inch driver at its highest energy density. The tabs along the abscissa denote the maximum value of energy density for each length of driver. The efficiency of the 54-inch arc length falls to about 57 percent, whereas with the 30-inch chamber the efficiency decreased to about 42 percent. If it were desired to decrease the chamber length and operate in the 200-300 J/cc range, the efficiency would be in the order of 20 to 30 percent. Other researchers have observed a similar trend.

It has been shown that the short arc is inefficient at high energy densities. In addition, a short arc generates high currents due to its low resistance, as illustrated on Fig. 5. In this figure are presented the current waveforms for a preset voltage of 40 kV for the three driver lengths using a straight trigger wire. It is interesting to note the large reduction in peak current for the longer arc. Operating at lower current reduces the internal mechanical forces of the capacitors and increases the life expectancy of the capacitor bank. The shape of the waveforms also are indicative of increased damping. Both the 54- and 114-inch traces are overdamped, with the longer arc the more overdamped. The 30-inch arc record shows a slight ring or underdamping as the arc was extinguished which is a detrimental factor in system operation as noted in connection with Fig. 3. The ringing was most likely caused by eddy currents within the cables and bus bars as opposed to a true oscillatory arc discharge. In any case, the ringing decreased (as did the peak current) as the preset voltage was reduced.

There is another trend to be noted from Fig. 5. While the trigger wire explodes in about 1 μsec, the arc does not strike immediately. There is a dwell or pause period between the wire explosion and the start of the main discharge of energy from the capacitor bank. This dwell period was surprisingly long, about 420 μsec for the longest arc. The duration of the dwell interval may be associated with the different magnitudes of the self inductance of the long trigger wire compared to that of the remainder of the system; the system inductance is only 0.25 μH compared to 5 μH for a 114-inch wire, 2.3 μH for a 54-inch wire, and 1.2 μH for a 30-inch wire length.[7] However, the mechanism that accounts for the long dwell period following the wire explosion is not fully understood. It is apparently a period of low conductivity and it is also consistent and reproducible. The interpretation of the dwell time following the exploding wire process could be an interesting subject for future research.[8]

Driver operation with a straight trigger wire can be summed up as follows: A long arc was stable (well overdamped), exhibited a high inductance, was relatively efficient, and had a high arc-out voltage. This latter characteristic tends to leave a sizable portion of the stored capacitor energy remaining in the bank. A short arc had a wider range of voltage operation, generated high peak currents, tended to be underdamped and was quite inefficient in producing high shock speeds. Furthermore, the high current flows induce self-constriction or magnetic pressures within the chamber that are large and vary as the square of the current. As the discharge current increases, the arc tends to become more constricted, a condition which promotes nonuniform heating and increases the core temperature of the plasma. In effect, the entire process of attempting to discharge more energy into a short arc initiated by a straight trigger wire appears to be self-defeating.

COILED TRIGGER WIRE

By coiling the trigger wire, and using the same driver environment, the driver performance can be substantially improved. As shown on Figure 6, the wire is spiraled along the inner surface of the insulating liner. It is installed by stapling it once at the upstream end of the liner and then winding the coil against the RTV surface. A second staple at the downstream end holds the helix in position as the wire is fastened to the ground or downstream electrode. The other end of the trigger wire is bent toward the axis of the chamber and tied to the cotton thread that passes through the hollow core of the high voltage electrode. (This is similar to the fastening arrangement used with the straight trigger wire.) A .007-inch-diameter tungsten wire was found to be an easily handled size. In operation, the string is drawn toward the electrode and the wire explodes leaving a helical path of ionized gases. After a dwell period, the arc discharge passes through this residual helical plasma column, heating the driver gas by ohmic heating. The self-magnetic field generated by the arc current flow, together with the wire core pressures, act to implode the heated gas toward the center of the driver chamber. The coil increases the overall length of the arc and its resistance is high.

Figure 7 illustrates the changes in driver/driven performance using a coiled trigger wire in the 54-inch driver. With a straight wire (Fig. 7(a)), the peak current is high and its waveform is marginally damped. With the coiled wire, the peak current was reduced by one half, the dwell period increased by a factor of 4, and the improvement to an overdamped discharge is evident. Of equal interest is the increase in shock tube performance, as can be noted by comparison of the tabulation on the figure. The driver and driven tube loading pressures were the same in both instances as was the energy density discharged from the bank. The driver efficiency was increased from 63 to 77 percent. The coil-loop spacing was about 6 inches. Also shown on Fig. 7 are the voltage measurements for both types of trigger wires. During the arc discharge period, the electrode voltages are characterized by a nearly straight line decay from the preset to the arc-out values.

The results shown in Fig. 7 were made at 3/4 MJ input to the driver chamber. The same coiled wire arrangement has also been tested at a 1 MJ discharge with a driven tube loading pressure of 1 torr dry air. While the dwell period and electrical waveforms were sensibly the same as those shown in Fig. 7(b), the peak current was increased to 336 kA. The arc extinguished at about 5600 volts. A shock speed of 36,100 ft/sec was measured in the driven tube at the station 24-tube diameters from the diaphragm.

The performance of the 30-inch driver for 1 MJ input with nitrogen driver gas is shown in Fig. 8. Nitrogen was used instead of helium as the driver gas in an effort to minimize underdamping, and, even so, the slight ringing accompanying the straight wire discharge caused localized spot welds between the breech and the driver tube. With the straight trigger wire, the peak current reached 863 kA. With the coiled wire discharge, the change in current waveform is quite evident. The peak current was reduced by almost 2/3 and the dwell period increased to 300 μsec. The circuit is well overdamped with none of the deleterious after-effects of the underdamped arc. The coiled wire raised the driver efficiency to 83 percent and increased the shock velocity by over 2000 ft/sec. Energy density was the same in both cases. The effective increase in driver temperature (T_4/T_1 from about 17 to 24) was noted on other instrumentation, particularly impedance probes[9] along the driven tube and the pressure transducer in the reservoir. Notice that the arc-out voltage with nitrogen is near zero potential, indicating a complete discharge of the stored energy from the capacitor bank. (Argon or helium-argon mixtures exhibited a similar low residual potential.)

Figure 9 presents the performance of the 30-inch driver with a coiled trigger wire for a full 1 megajoule input into helium. It must be emphasized that had a straight trigger wire been used, the resulting discharge would have been oscillatory and undoubtedly damaging to the equipment. With the coiled wire, the arc discharge is docile, there are no problems, and day-to-day operation is simple and straightforward. The effective driver temperature was about 20,000°R for an efficiency of 73 percent. The measured shock speed of 48,700 ft/sec driven into 1 torr of dry air is the highest velocity known for an arc driver and this driven tube pressure.

APPLICATION

The circuit damping restriction shown on Fig. 3 is removed with the use of a coiled trigger wire. Just how short a chamber can be and still benefit from a coiled wire is not known. Based on the test experience, it is anticipated that for a coiled wire to be effective the inductance of the single-layer coil should be at least double (or greater) than that of the system inductance. The inductances of the 4-inch-diameter coils used herein were about 0.5 μH. Also, and intuitively, it is probable that at least several coils are necessary and they must be spaced apart a sufficient distance to prevent a flashover between adjacent loops. The use of several coils assures that the arc is long enough so that its resistance is high. In an RLC series circuit, such as formed by the energy storage and driver system, the load resistance must be greater than the system resistance for a non-reversing, overdamped type of discharge. The calculated resistance of the system exclusive of the load is about 28 mΩ. Thus, in general, the results presented in this paper spanned a range of arc resistances from 2 to 5 times that of the system itself.

As noted previously, the coiled trigger wire was pinned in place against the silicon rubber coating on the fiberglass liner. There was no evidence of burn or wear on the liner surface from the wire explosion. This fact supports the contention that the arc is imploded. There is experimental evidence that the arc discharge is adversely affected if the coils are irregularly spaced or if the coil diameter is small (~2 inches) and the helix thus formed lies along the bottom of the chamber. In the future, another arrangement is to be tested. It consists of a thin metal film deposited on the inner surface of the liner to provide a cylindrical imploding pattern.[10]

It is expected that the imploded wire can offer a further improvement in driver operation. Generally, the attainment of a high driver gas temperature necessitates a high driver chamber pressure (~1000 atm). Internal chamber design and proper diaphragm opening are troublesome under these conditions. With the imploded wire and a concurrent diaphragm opening, the combination offers a driver with a high temperature but at a lower pressure compared to the existing mode of operation. More importantly, the driver arc discharge could be continued after the diaphragm opening and the net effect would be a continuous discharge driver[11]. Such a driver would be short, have a reasonable energy density requirement, and may improve shock wave attenuation, particularly at high driven tube pressure levels.

The coiled trigger wire technique used in the short chamber is suitable for driving the high speed shocks necessary for planetary return entry studies in the shock tube. Yet the system can be easily changed over to a long driver with a straight trigger wire for shock tunnel use without any limitations or changes to the capacitor system.

REFERENCES

1. Dannenberg, R. E. and Silva, A. F.: Exploding Wire Initiation and Electrical Operation of a 40-kV System for Arc-Heated Drivers Up to 10 Feet Long. NASA TN D-5126, 1969.
2. Loubsky, W. J. and Reller, J. O., Jr.: Analysis of Tailored-Interface Operation of Shock Tubes with Helium Driven Planetary Gases. NASA TN D-3495, 1966.
3. Gruszczynski, J. S. and Rogers, D. A.: Shock Tube Instrumentation Techniques for Study of Hypervelocity Entry Problems. G. E. Space Sci. Lab., Rept. R64SDG7, Sept. 1964.
4. Dannenberg, R. E. and Stewart, D. A.: Techniques for Improving the Opening of the Main Diaphragm in a Large Combustion Driver. NASA TN D-2735, 1965.
5. Webb, F. H.; Hilton, H. H.; Levine, P. H.; and Tollestrup, A. V.: The Electrical and Optical Properties of Rapidly Exploded Wires. Exploding Wires, Vol. 2., Plenum Press, New York, 1962, pp 37-75.
6. Mastrup, F. N. and Goldstein, R.: Radiation Measurements from a Pulsed High-Pressure Nitrogen Arc. TRW, Inc., Report ARL 68-D152, Aug. 1968.
7. Terman, F. E.: Radio Engineers' Handbook. McGraw-Hill, New York, 1943.
8. Hesselgreaves, J. E.; Nonweiler, T. R. F.; and Foord, T. R.: The Heating of Air by "Dark" Discharge, British Rand M, No. 3519, May 1966.
9. Dannenberg, R. E. and Humphry, D. E.: Microsecond Response System for Measuring Shock Arrival by Changes in Stream Electrical Impedance in a Shock Tube. Rev. Sci. Inst., vol 39, no. 11, pp. 1692-96, Nov. 1968.
10. Dennen, R. S. and Wilson, L. N.: Electrical Generation of Imploding Shock Waves. Exploding Wires, Vol. 2, Plenum Press, New York, pp 145-157.
11. Warren, W. R. and Marston, C. H.: A High Density, High Velocity Equilibrium Freestream Shock Tunnel Concept. AIAA 6th Aerospace Sciences Meeting, Jan. 22-24, 1968, New York, Paper N. 68-17.

Figure 1
Facility area of electric arc driver system.

Figure 2
40 kV, 1 megajoule energy storage bank.

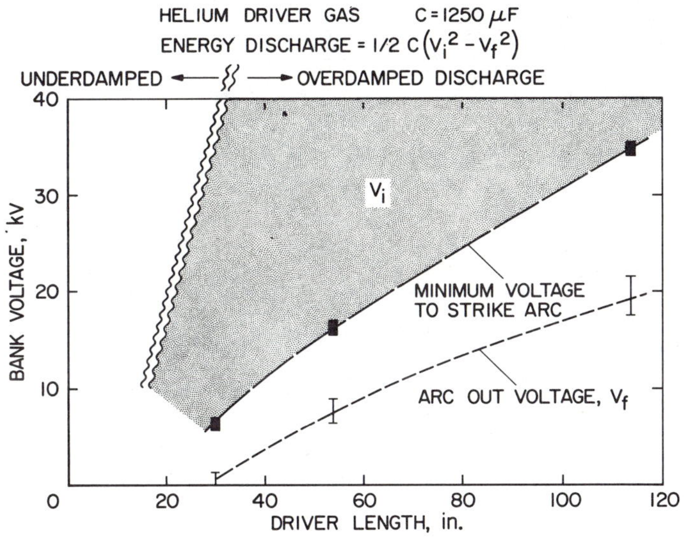

Figure 3
Characteristic voltage breakdown of system with straight trigger wire.

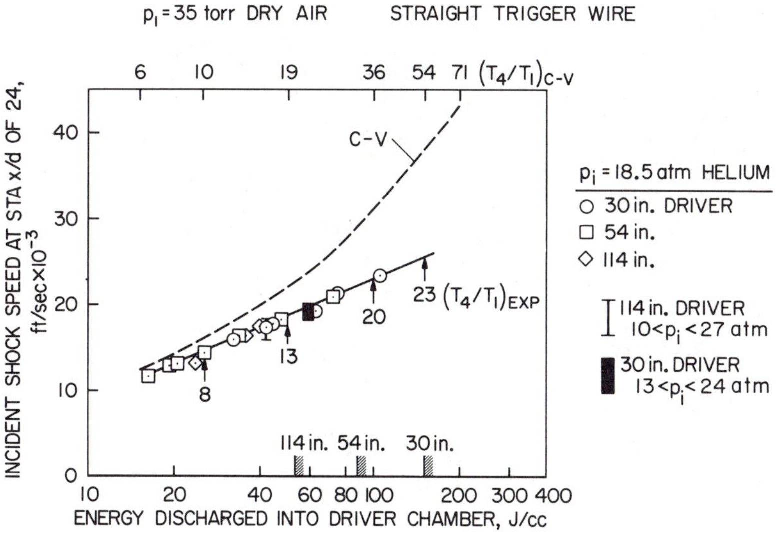

Figure 4
Variation in shock speed with energy density.

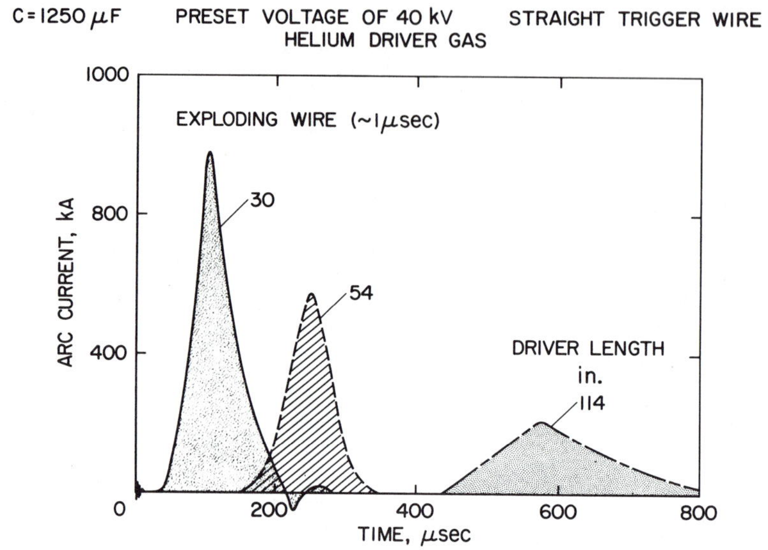

Figure 5
Typical current waveforms.

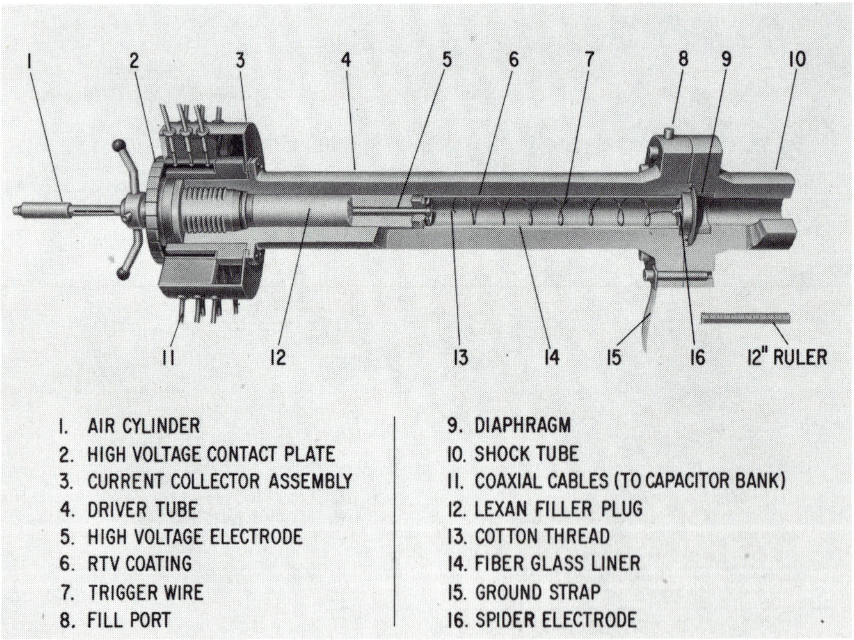

Figure 6
Schematic assembly of coiled trigger wire in
30-inch arc chamber.

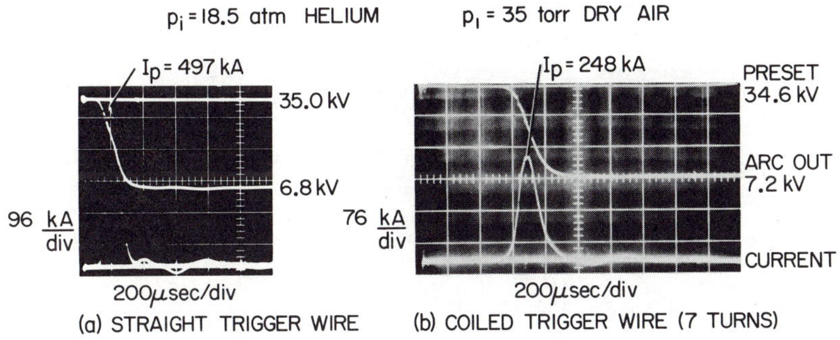

20,900 fps	V_s AT STA x/d = 24	21,850 fps
18.8	M_s	19.7
16.2	$(T_4/T_1)_{EXP}$	18.0
63	EFFICIENCY, %	77

ENERGY DENSITY = 71 J/cc LIMIT V_s = 25,420 fps

Figure 7
Performance characteristics with 54-inch arc driver for a 3/4 MJ helium discharge.

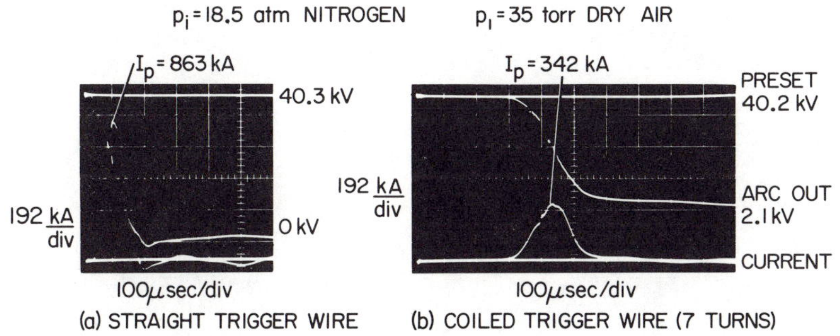

12,260 fps	V_s AT STA x/d = 24	14,530 fps
11.0	M_s	13.1
17.2	$(T_4/T_1)_{EXP}$	24.1
59	EFFICIENCY, %	83

ENERGY DENSITY = 144 J/cc LIMIT V_s = 15,800 fps

Figure 8
Performance characteristics with 30-inch arc driver for a 1 MJ nitrogen discharge.

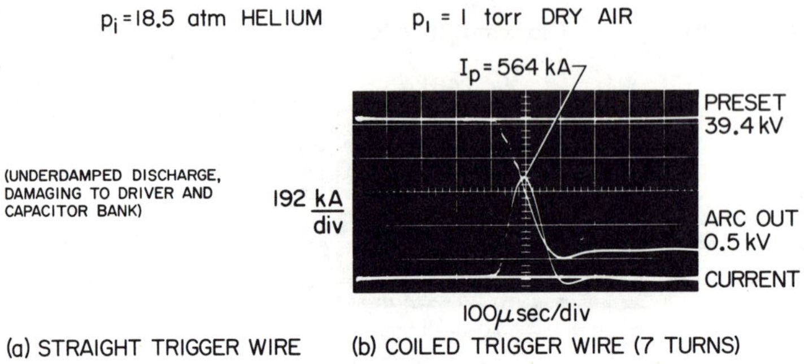

	V_s AT STA x/d = 24	48,700 fps
	M_s	43.7
	$(T_4/T_1)_{EXP}$	37.8
30-40 (Est)	EFFICIENCY %	73

ENERGY DENSITY = 153 J/cc LIMIT V_s = 54,100 fps

Figure 9
Performance characteristics with 30-inch arc driver for a 1 MJ helium discharge.

ELECTRICAL SHOCK TUBE PERFORMANCE WITH PREHEATING OF DOWNSTREAM GAS

S. LUDVIK and H.K. MESSERLE
School of Electrical Engineering
University of Sydney

SUMMARY

An electrically driven shock tube is described in which the ratio of the downstream to driver density can be varied by R.F. induction heating of the downstream gas. The results of an experimental study with such a system are discussed and it is shown that a uniform flow can be produced when the driver density is greater than the downstream gas density. The shock speed is shown to depend on the downstream pressure and the capacitor bank voltage and speeds of the order of 1.7 cm/μsec have been achieved.

INTRODUCTION

Since the appearance of the first electric shock tube[1] as a plasma source, various improvements in the performance of the electric shock tube have been attempted by changing the driver geometry. These have involved such developments as conical drivers[2] and coaxial drivers[3]. The emphasis of most of these investigations has been to increase the shock speed rather than to overcome the problems of poor reproduceability and the lack of a well defined flow region which are both significant objections to the value of the electric shock tube as a plasma source. Only recently, there has been a reexamination of the uniformity problem and this study by Scott and Anderson[4] showed that a symmetrical conical driver could lead to a steady flow by reflection off a surface. The incident shock however was subject to the same characteristic non-uniformity of electric shock tubes.

The purpose of this paper is to bring attention to the role of the driver to downstream density ratio on the uniformity of flow. Experimentally it is possible to alter the density difference between these regions by R.F. heating of the downstream gas and a driver system has been developed by the authors[5] for this situation. The approach has been found to lead to a well behaved flow, and the results discussed here present some of the characteristics of such a system.

EXPERIMENTAL ARRANGEMENT

One of the major aims of the experiment was to allow for flexibility and provide for the use of different driver gases. Most of the work has concentrated on a study in argon but helium and hydrogen are also being used as drivers. The design of the driver system is based on a multiple flow technique as first suggested by Cloupeau[6] and the general experimental arrangement is shown in figure 1.

The main section of the shock tube consists of a 25mm inner diameter vycor glass tube which is about 80cm long and it is situated inside an induction heating coil of 60cm length. The energy for the heating coil is supplied by a radio frequency generator rated at 50kW. The glass tubing is cooled with the aid of a water jacket and the temperature rise of the water flow is used to measure the power delivered to the downstream plasma. Attached to the right hand end of the main section is the driver which is constructed from two concentric vycor tubes of inner diameters 25mm and 36mm. The larger tube is fused to the main section while the inner tube acts as an extension of the main tube. The driver subsection is located outside the heating coil and consequently the ambient or pre-firing conditions of the shock tube assembly take the form of a hot downstream gas with a cold driver.

Gas flow in the system is arranged so that gas may be introduced both into the downstream end of the main section as well as the inner driver tube. The system is evacuated through the annulus between the driver tubes by a 500 litre/min backing pump. This technique provides scope for increasing the driver density dynamically by adjusting the gas flow and pumping rates. Another advantage of the gas flow arrangement is to permit the use of a driver gas different to the downstream gas. The effectiveness of the gas separation produced by the technique can be qualitatively demonstrated by using argon as the downstream gas and hydrogen in the driver. The plasma formed downstream exhibits the normal white colour of a medium pressure argon plasma (10 torr) and only 1-2cm of the plasma near the driver section show any sign of the characteristic redness of the hydrogen plasma.

The driving energy for the system is obtained from a parallel bank of five 8.5μF capacitors which can be charged to 20kV. Triggering of the bank is achieved through a conventional thyratron/ignitron circuit using a GL5550 ignitron. Typical performance of the LC transmission line for the bank produces a current pulse of 25μs duration and with the capacitor bank voltage at 8kV the peak current is about 25,000A.

The electrode geometry provides two modes for coupling the capacitor bank energy to the shock tube. The simplest mode involves an axial discharge down the length of the tube where most of the energy is dumped into the two cold gas regions outside the heating coil. The presence of two cold regions results in shock waves emanating from either end of the tube and eventually colliding near the centre of the induction plasma. Preliminary studies[5] of the shock tube were made using the axial discharge mode. It was found that the shock speed reached a maximum with increasing energy input, due to an apparent equalization of energy distribution throughout the tube at higher discharge energies. Another disadvantage of this mode from the point of view of quantitative measurement was the fact that the induction plasma was disturbed by the axial discharge and as a consequence the initial state of the plasma before the arrival of the shock was uncertain.

The other mode, referred to as the single-ended mode, is operated by discharging the capacitor bank into only the right hand driver region. In the single ended configuration no speed limitation is encountered as in the case of the axial mode and, in argon at pressures of 0.1mm, speeds of 1.7cm/μs have been obtained with five 8.5μF capacitors charged to 14kV. In both modes streak photographs have shown a separated shock front and contact surface with a well defined flow.

The speed characteristics of the shock tube have been determined by means of streak photography and by a photomultiplier technique in which a series of equally spaced slits along the tube are focussed on to the photocathode of the photomultiplier. The latter diagnostic method is a simple means of determining the speed and attenuation along the tube and it has the advantage of requiring only one photomultiplier. The decay in the intensity of the shock can be compensated by using smaller slits near the driver region.

Some preliminary spectral investigation has also been carried out to confirm the structure of the shock wave development indicated by streak photographs. Two spectroscopes operated as monochromators were set up by means of a half-reflecting mirror on to a 1mm wide slit at a point approximately 25 cm downstream of the driver section and the output of one spectroscope was aligned to a characteristic line of the downstream gas (Argon II, 4806Å) while the other spectroscope was set to a line of the driver gas (Helium I, 5876Å). The spectroscopes employed in the study were a Jarrell-Ash, Model 82-000 (0.5 meter focal length, f/8.6, 16Å/mm) and a Hilger & Watts Constant Deviation Prism Spectrometer D186. Photomultipliers (E.M.I. 9558B) attached to each spectroscope were used to trace the time history of the argon and helium spectral lines and it was possible to infer shock heating of the downstream gas.

EXPERIMENTAL RESULTS

General Performance

Some of the factors which were used in assessing the performance by the system were the range of shock speed with pressure and various capacitor bank voltages. Also considered was the uniformity and attenuation of the gas flow along the tube. The shock speed was measured by the photomultiplier technique described earlier with a slit spacing of 10cm. Figure 2 shows the dependence of the shock on capacitor bank voltage for a number of different pressures. For each measurement the R.F. power input to the downstream gas was held constant at about 10kW. As can be seen the shock velocity increases with discharge energy and maximum velocities attained so far appear only to be limited by available equipment. The general pattern of behaviour of the shock speed observed is similar to the cold downstream shock tube, however, the characteristic turbulence observed in cold tubes does not occur.

An indication of the reproduceability of the shocks can be seen from the scatter of points in figure 2. The reproduceability observed is basically a long term measure since many measurements were done on different days and with a replacement tube after fracture of the vycor glass. Nevertheless, a figure of 10% can be assigned to the system's overall reproduceability. Investigation of the short term reproduceability was also carried out with conditions kept constant while a number of consecutive shots were fired. It was found that the reproduceability was within 3% although sometimes a spurious shock was obtained.

Streak photographs taken of the plasma sample show a clearly defined shock front followed by a steady flow region. This observation contrasts with the highly turbulent flow generally encountered with electric shock tubes[3,7]. Figure 3 shows a typical streak record for a shock moving at 0.5cm/μs into argon at 30 torr. The shock region shows up clearly as the brighter and hotter front preceding the colder driver.

Attenuation

An important contributing factor to the inhomogeneity of the plasma sample generally found with electric shock tubes has been the very severe attenuation of the shock. The mechanism producing turbulence has been suggested[8,9,10] to be similar to the Rayleigh-Taylor instability of the interface of two immiscible fluids when subject to a deceleration. Perhaps the simplest analogy for the instability can be found in the case of a heavy fluid on top of a lighter one under the action of gravity. Streak photos of the conventional cold downstream shock tube indicate[7] that turbulence commences behind the shock front and it is considered therefore that the interface which initiates the instability is the contact surface rather than the shock heated gas/downstream gas interface as envisaged by Smy[8].

The attenuation characteristic obtained with the shock tube described here is shown in figure 4. The x-t diagram was derived using argon both as the driver and downstream gas at a pressure of 7.6 torr and with a capacitor bank voltage of 8kV. The attenuation of the shock speed over the useful length of the tube is about 0.5% per/cm and this figure is considerably better than the usual performance of electric shock tubes[10,11] where typical attenuation is about 6% per cm. The broken line represents a theoretical attenuation curve which has been fitted using the 'blast wave' relation [$x=x_o\ (t-t_o)^{2/3}$] that is generally used to describe the attenuation of conventional electric shock tubes. It is worthwhile to compare the attenuation performance of the present shock tube with another low attenuation electric shock tube[8] for which the attenuation was reported as 0.2% per cm. The construction of this tube was such as to raise the driver density above the downstream gas by means of a thin nylon film in a manner similar to conventional mechanical shock tubes. The greater density of the driver appeared to give rise to a low attenuation shock which corresponds with the observations made here. The diameter used in that study, however, was twice that used in our experiments which could account for the lower attenuation.

Effect of Varying the Driver Gas

The advantage of using a lighter gas such as hydrogen or helium in the driver is well known for mechanical shock tubes. It was possible to study a similar effect for the electric shock tube employing the gas separation technique described earlier. The curves in figure 5 indicate the speed characteristics using argon and helium in the driver. Under the same energy input to the driver the shock speed with helium can be seen to be about 60 percent higher. Relating this to the simplified theory of driver expansion (i.e. where the limiting speed of the expanding driver surface is proportional to the speed of sound) one would anticipate a maximum speed increase in the ratio of $\sqrt{(M_A/M_{He})}\approx 3$, where M is the atomic weight.

The discrepancy is probably due to the fact that the driver temperature for identical energy inputs will be higher in helium and that the greater ionization in argon will reduce the effective density ratio. There is also some initial mixing of downstream and driver gas at the interface region before the capacitor bank is discharged and this will also reduce the effectiveness of the lighter driver.

Contact Surface

The study of the shock wave development has been performed on the system using a helium driver and argon as the downstream gas. Typical traces in figure 6 show the intensity of the helium and argon spectral lines as function of time for a shock speed of 0.8cm/μs (p=5.0 torr) and it is apparent that heating of argon has taken place in a narrow region near the front of the shock. Some helium is also present in the shock region but this is probably due to the relatively crude gas separation technique. The contact surface is identified as the discontinuity just behind the shock front and it can be seen to appear at the same point in both traces. The length of the shock heated sample is shown as a function of shock speed and the curve in figure 6 indicates how the length decreases with shock speed.

DISCUSSION AND CONCLUSION

The low attenuation and the uniform flow generated by the shock tube used in these investigations appears to confirm the need for a higher density driver in order to achieve well behaved shock heating with electric shock tubes. The stability of the contact surface, implied by the spectroscopic investigation, indicate that the Rankine-Hugoniot equations should apply to this situation. Theoretical calculations have been carried out for argon to determine the shock heated temperature as a function of shock speed, taking into account preheating of the downstream gas. The curves plotted in figure 7 indicate general trends for an initial pre-discharge pressure of 7.6 torr and a number of initial downstream temperatures. For an initial temperature of 10,000°K the broken lines show also the behaviour at pressures of 30 and 50 torr. At 7.6 torr the maximum speed so far achieved is 1.2 cm/μs which corresponds with a shock temperature of 23,500°K in argon with an initial temperature of 10,000°K.

The purpose of this report has been to describe the improvement in behaviour achieved by preheating the downstream gas of an electrical shock tube and to demonstrate the value of a higher driver gas density. The good reproduceability of the system also establishes the usefulness of the device as a means of studying basic plasma properties such as wave structure in an ionized gas.

ACKNOWLEDGMENTS

The authors would like to acknowledge the assistance in the development of these experiments provided by Dr. A.D. Stokes and to thank Mr. B. Campbell for taking the streak photograph. They also wish to acknowledge the financial support provided by the Australian Research Grants Committee and the Australian Institute of Nuclear Science and Engineering.

REFERENCES

1. Fowler, R.G., Goldstein, J.S., Clotfelter, B.E. Phys.Rev., 82, 879 (1951).
2. Josephson, V., J.Appl.Phys., 29, 30 (1958).
3. Koopman, D.W., Phys. of Fluids, 10, 1651 (1964).
4. Scott, D.S., Anderson, T.P., 9th Symposium on Engineering Aspects of MHD (Tennessee) April (1968).
5. Campbell, B., Messerle, H.K., Ludvik, S., Stokes, A.D., A.I.A.A. Journal, 6, 1579 (1968).
6. Cloupeau, M., Phys. of Fluids, 6, 679 (1963).
7. Campbell, B., George, D.W., Messerle, H.K., Aust.J.Phys., 22, 23 (1969).
8. Smy, R.M., Rev.Sci.Inst., 36, 1334 (1965).
9. Barach, J.P., Vermillion, R.E., Phys. of Fluids, 8, 1976 (1965).
10. Camm, J.C., Rose, P.H., Phys. of Fluids, 6, 663 (1963).
11. Thornton, J.A., Cambel, A.B., J.Quart. Spectrosc. Radiat. Transfer, 2, 249 (1962).
12. Barach, J.B., Mayes, T.W., Phys. of Fluids, 11, 89 (1968).

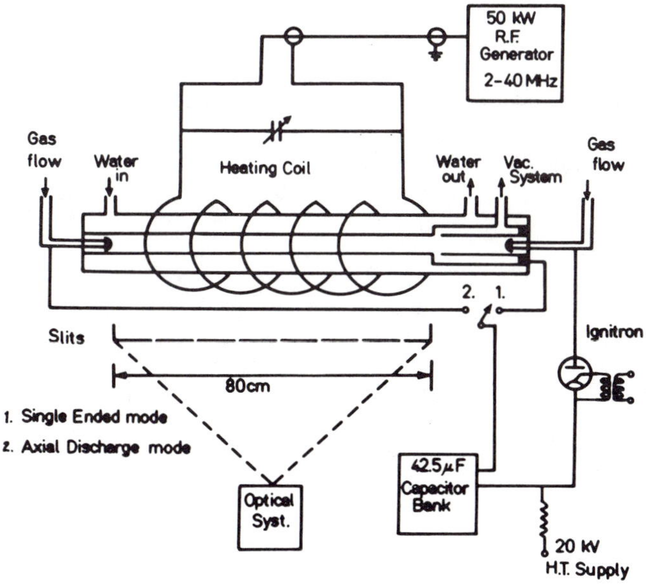

Figure 1
Schematic diagram of shock tube.

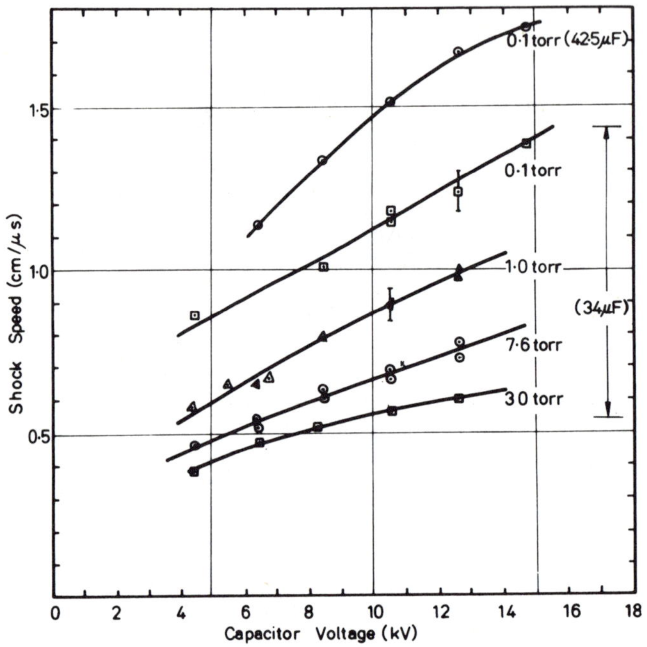

Figure 2
Shock speed as function of capacitor bank voltage, pressure and capacitor bank size.

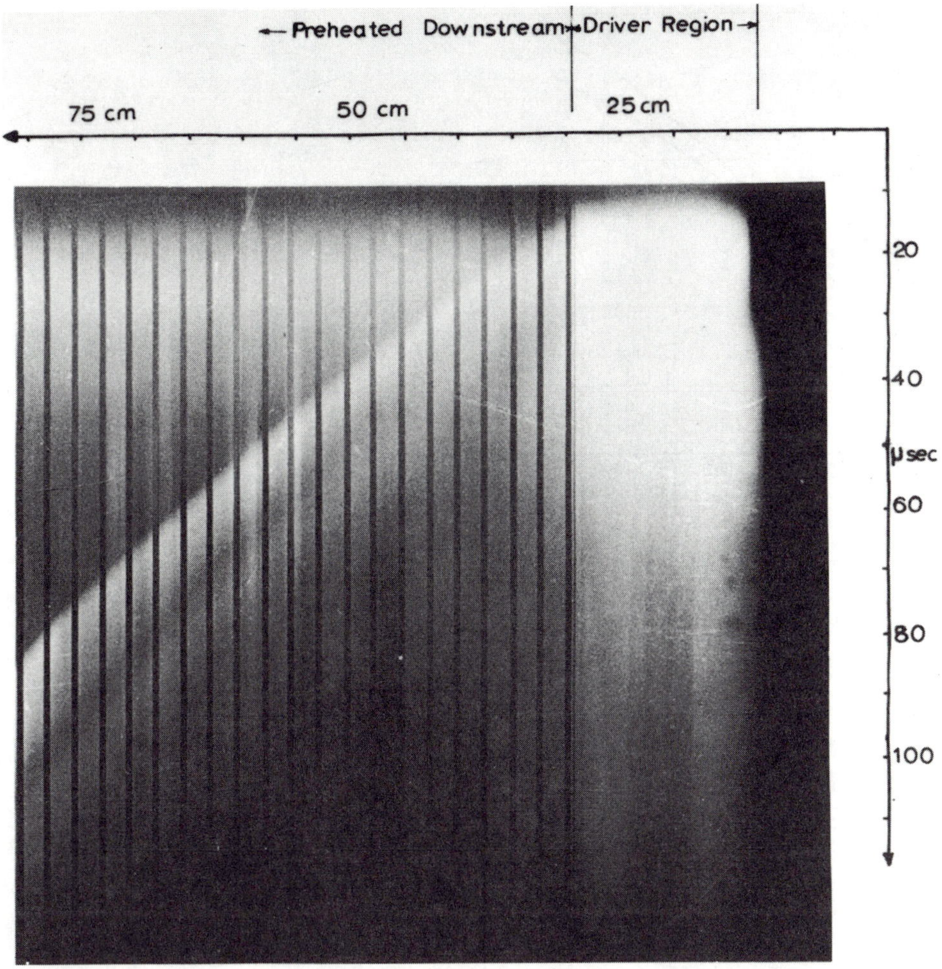

Figure 3
Streak photograph for an argon driver.
Discharge conditions: Voltage = 8kV; p = 7.6 torr;
Capacitor bank size = 34μF.

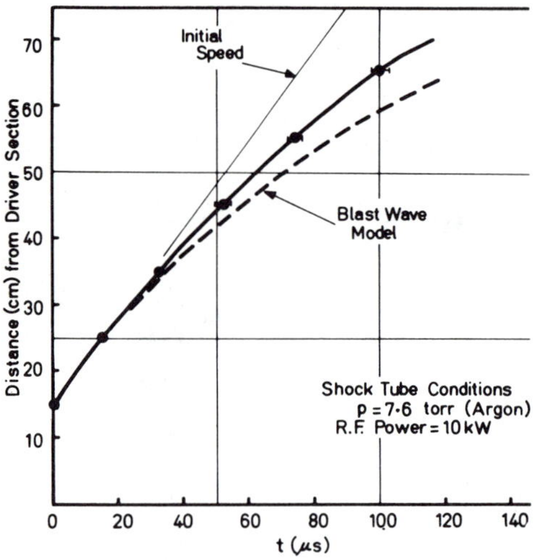

Figure 4
x-t diagram showing the shock wave attenuation.

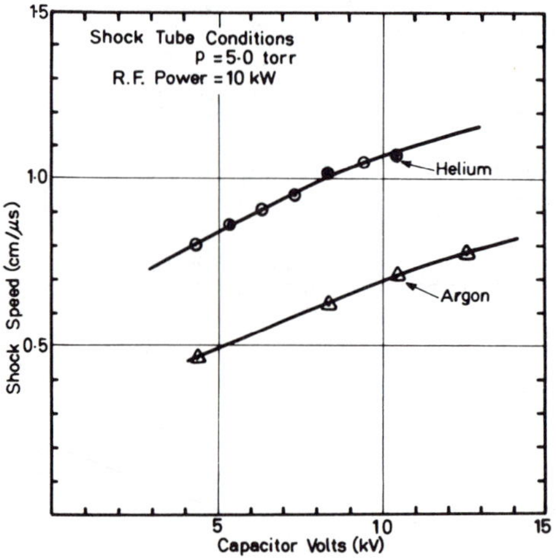

Figure 5
Plot of θ^{-1} for LiBr and CsBr

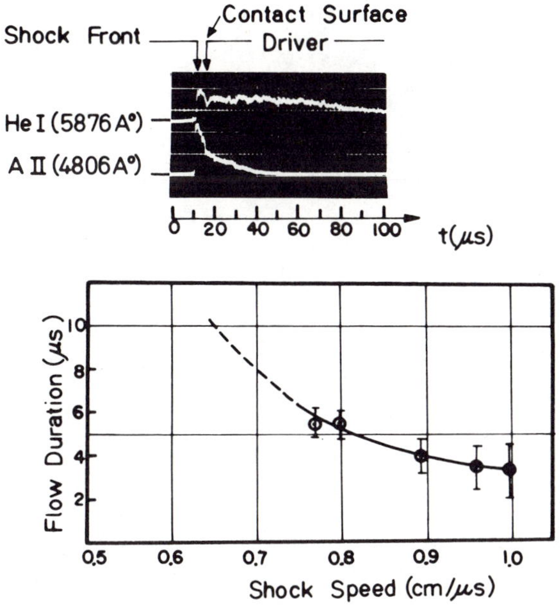

Figure 6
Structure of shock wave. Flow duration time of shock-heated gas shown as a function of shock speed.

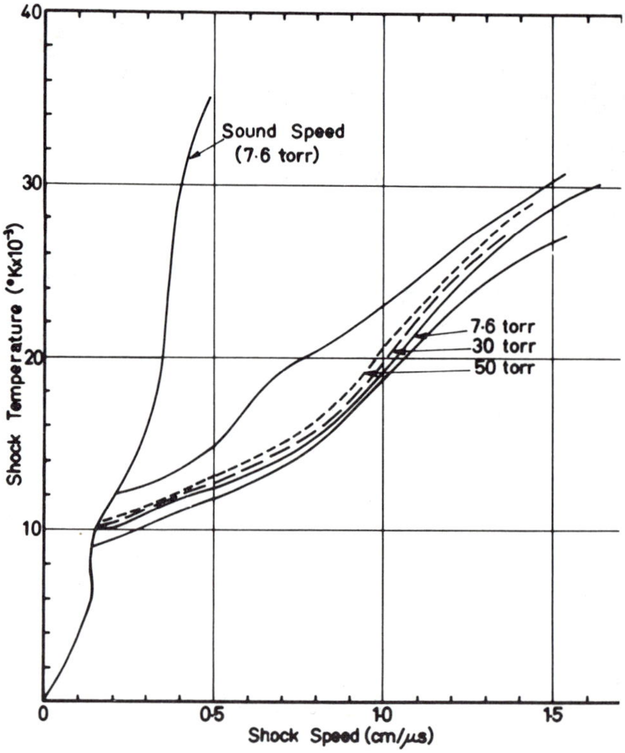

Figure 7
Curves showing the dependence of the shock-heated gas temperature on shock speed for a number of initial temperatures.

STUDY OF A BYPASS PISTON SHOCK TUBE

STELLAN KNÖÖS*
Aeronautical Research Institute of Sweden (FFA),
Stockholm, Sweden

SUMMARY

The bypass piston tube has been further developed and used for driving a shock tube. Using the pressure energy of room-temperature nitrogen or helium as the energy source, peak temperatures higher than 9000 °K and pressures of 300 to 600 atm in the helium driver gas were measured; in nitrogen, shock velocities were above 12 km/s. Phenomena of mixing between the accelerating and driver gases were observed and found to limit the overall performance of the BPF-heating. A piston retardation insert in the end of the compression chamber was tested and found extremely useful. The experiments add to the understanding of the gas dynamics of the bypass piston tube and support the original theory. It now seems possible to build shock-tube drivers with internal energies in the megajoule regime with temperatures above 2×10^4 °K and pressures above several thousand atmospheres.

INTRODUCTION

At present, most high-performance pressure-driven shock tubes employ electric arc-heating of the driver gas.[1] Several serious limitations are inherent with the use of this technique. For example, contamination of the driver gas from evaporated wall material generally limits the speed of sound of the driver gas to 5-6 km/s and the attainable shock speed in the driven gas to approximately 12 km/s. Also, the storage capacity of electrical energy cannot be

*Visiting scientist. Present address: Northrop Corporate Laboratories, Hawthorne, California. The support given to this study by the Swedish National Science Foundation (Styrelsen for Teknisk Utveckling) is gratefully acknowledged.

raised above a few hundred thousand joules without considerable technical effort and very high cost. A well-established alternative way of heating the driver gas is in a free piston compressor, e.g., as described by Stalker.[2,3] Generally, high pressures can be obtained with the free piston compressor, but generated temperatures rarely exceed 5000°K because of practical limitations to the volumetric compression ratio.

The bypass piston (BPP) tube,[4,5] first used in 1966, offers a promising alternative and efficient way of heating a shock-tube driver gas with possibilities of achieving helium driver gas temperatures above 10^4 °K and shock velocities higher than 12 km/s. Basically, in the BPP tube, energy is transferred from a so-called accelerating gas to the driver gas (to be heated) in two separate steps. Thermodynamically speaking, the first step is an entropy-producing shock-heating process and the second is a nearly isentropic compression. Before the second heating step, the gas undergoes an entropy-producing free expansion with small temperature change. In practice, shock heating of the driver gas (test gas) takes place in a shock-tube section of the BPP tube which is located upstream of a heavy piston (Figure 1). Subsequently, a controllable amount of the heated gas, bypassing the piston, enters the evacuated compression chamber. The piston is brought into motion, before or during the bypass process, by the expanding accelerating gas. For shock-tube driver applications, useful thermodynamic conditions of the helium gas are obtained when the motion of the ballistic piston has been retarded and the piston transfers its temporarily gained kinetic energy to the driver gas ahead of it. The major advantage of the BPP tube over conventional (isentropic) free-piston compressors is in shock-tube driver applications where the entropy in the driver gas during its overall compression is significantly increased. Hence, for a given temperature of the driver gas and given driver volume, a BPP shock-tube driver could be made short and compact, a feature not possessed by the isentropic free-piston compressor.

In the early exploratory experiments by Knöös,[5] peak temperatures of the order $T_f \sim 12000$°K and peak pressures $p_f \sim 350$ atm were obtained in argon, in agreement with the simple theory. From these experiments and later theoretical work, increased understanding of the gas dynamics and associated phenomena of the BPP tube has been gained. Experiments have been conducted with a bypass piston tube used as a shock tube driver, and shock waves have been produced with velocities higher than 12 km/s in a small shock tube. The apparatus includes some novel features and concepts. Experimental results for the compression processes of helium within the BPP tube compare favorably with simple theory.

FACILITY AND EXPERIMENT

The bypass-piston shock tube used in the present experiments is shown in Figure 2. The accelerating gas chamber (denoted by 5), for nitrogen or helium, had a volume of approximately 40 liters. A helium buffer tube (denoted by 4) 2.8 m in length and 40 mm i.d. and a primary helium shock tube 2.1 m long, separated from this chamber by a diaphragm (denoted M1 in Figure 2a), followed. The bypass channel was located behind the third diaphragm (M3) and was followed by a compression tube of approximately 0.83 m length and 65 mm i.d. The most frequently used high-pressure section had a length of 0.12 m and an i.d. of 40 mm and was directly connected to the compression tube. Separated from the compression tube by a thick stainless steel diaphragm (denoted M4 in Figure 2b) was a secondary shock tube of 3.0 m length and 40 mm i.d. in which the high-speed shock waves were generated.

Several different bypass channels were used (Figures 3a, 3b). A particularly simple bypass configuration employed four (90° spaced) 20-by-10-mm rectangular channels machined into a 65-mm-i.d. honed hydraulic tube, giving perfect guiding of the free piston (Figure 3a). The initial location of the piston was usually within the bypass channel, and not in front of it as reported in the original experiment.[5] By changing the initial location of the piston the amount of gas bypassing the piston could readily be controlled. Three different types of steel pistons (Figure 3c) were used with masses varying from 0.39 to 1.92 kg. Two of the pistons had conical, rather than truncated flat afterbodies, to permit smooth flow and small heat transfer during the bypass process. The third piston had a wedge-like afterbody; this piston was used in conjunction with only two diagonally located bypass channels.

Since premature rupture of high-pressure diaphragm M4 could cause destructive piston impact into the high-pressure section, a piston retardation insert was used as is shown in Figure 4. Carefully designed, this insert throttled the flow of heated driver gas into the high-pressure section to such an extent that the piston performed a bounce in front of the insert, avoiding metal-to-metal impact. Subsequently the piston smoothly docked into the insert with its small cone-cylinder forebody (used on the two larger pistons shown in Figure 3c) and entirely closed the throat of the insert during the final few millimeters of motion.

Diaphragms M1, M2, and M3 were made from aluminum 1.5 to 3.0 mm thick and diaphragm M4 from stainless steel 1.0 to 2.0 mm thick. All diaphragms were cross-scribed to a depth of 0.3 mm. Folding of the diaphragm flaps against the tube walls was generally good, with the exception of the upper M1 diaphragm (used for firing) where the diaphragm flaps were not fully folded and the accelerating gas flow was possibly choked.

The flow in the bypass piston shock tube was analyzed with help of piezoelectric pressure transducers (Kistler 603A and 601F) that could be inserted at ten axial positions along the tube (P1 through P10 in Figure 2b). Silicon-diffused photodiodes (EG&G SGD-100), viewing into the secondary shock tube at small solid angles through fiber-optic light guides and small-diameter glass windows, were used for the measurement of shock velocities and the time-history of spectrally integrated emission of radiation behind the shock front.

Each of the five tubes that comprise the BPP shock-tube had adequate means for evacuation and gas filling. Several gas valves had requirements to operate in the wide pressure regime of $< 1\ \mu$ Hg to 300 atm. These requirements were handled by inexpensive home-made valve units, each basically consisting of one high-pressure back-flow valve, one safety rupture disk, and one low-pressure diaphragm valve. The vertically mounted back-flow valve was fully open during the evacuation and gas-filling processes owing to the gravitational force of the sealing piston, but was automatically closed during the test by exposure to the high-pressure gas.

The driver gas used in the experiments was helium. Because of its high cost in Sweden, helium could not at first be used as the accelerating gas; and combustion gases were not considered for safety reasons. Instead, nigrogen was chosen as in the reported first experiment with BPP tube.[5] Sizable entropy production through shock waves in the helium driver gas in the first compression step was achieved by using a helium buffer gas. With such an arrangement, expanding nitrogen compressed the helium buffer gas nearly isentropically after rupture of M1 and raised the gas pressure until diaphragm M2 ruptured. The heated helium buffer gas subsequently expanded into the primary shock tube, acting as a usual shock-tube driver, and caused the irreversible shock-heating of the helium driver gas, with resulting higher temperature and lower density than would have been obtained without a helium buffer.

Initial values for the pressure of the accelerating gas were typically $100 < p_5 < 200$ atm, for the helium buffer gas $1 < p_4 < 12$ atm, and for the helium driver gas in the primary shock tube $0.5 < p_1 < 2$ atm.

During the later phase of the experiments the use of nitrogen as accelerating gas was abandoned in favor of helium. In order to save helium and bring the consumption to an acceptable low level, a helium recovery system was built. The use of a buffer tube did not then offer great advantages and it could, in fact, be entirely omitted. Alternative schemes were tested in which no gas was placed initially in the buffer tube or in which the helium driver gas was located in the buffer tube with the primary shock tube initially under vacuum.

The experiments aimed towards further understanding of the basic gas dynamics of the BPP-heating of a light gas driver such as helium and the optimization of the initial gas loading conditions and bypassing processes. As is inherent in new types of facilities, a large amount of the experimental work became developmental in nature, e.g., with the introduction of novel piston designs, piston retardation units, and bypass channels. The shock-heated gas in the secondary shock tube was not subject to any detailed study, although interesting radiation profiles were recorded. The strong shock waves were merely produced in order to demonstrate the general capability of the BPP tube as a high-performance shock-tube driver, and recorded velocities of the shock waves were used as an independent measurement of the heating of the helium driver gas achieved in the BPP tube.

RESULTS AND COMPARISON WITH THEORY

A. Initial Heating

During the early course of the experiments nitrogen was used as accelerating gas. Typical pressure records of the flow in the buffer and primary shock tubes are shown in Figure 5. Here, the piezoelectric transducer in the buffer tube at position P2, located 0.32 m upstream of diaphragm M2, showed the presence of only weak shocks in the helium buffer gas. The total duration of the initial buffer gas compression was typically 5.5 ms. It was estimated that convective energy losses to the walls were small for the initial pressures employed. Oscilloscope trace $P2_A$ indicated that the buffer gas pressure increased almost linearly with time during the latter part of the compression. This indicated that the density of the helium gas varied with time in accordance with $\rho \sim t^{1/\gamma}$, where γ is the ratio of specific heats, and hence the velocity u_c of the fictitious contact surface between the accelerating and buffer gases decreased with time in accordance with $u_c \sim t^{-(\gamma+1)/\gamma}$. In most experiments the rupture pressure of diaphragm M2 was $p_{M2} \sim 120$ atm, and the pressure ratio of the buffer gas compression was $p_{M2}/p_4 \sim 30$. This would ideally result in a temperature $T \sim 1140^{\circ}K$ for the compressed helium, assuming the initial gas temperature to be $T = 293^oK$. The associated ideal volumetric compression ratio would take the value 7.7 and the heated helium would occupy a length $L = 0.37$ m of the buffer tube, at the time of rupture of diaphragm M2. The incident and reflected shock waves, and the subsequent practically shock-free compression of the helium driver gas, are also shown in Figure 5, recorded at transducer position P3, located 0.31 m from diaphragm M3 in the primary shock tube. The pressure ratio across the incident shock at P3 was $p_2/p_1 \sim 8.3$, indicating the local Mach number of the shock to be $M_{s1} \sim 2.6$, which was considerably lower than the value calculable from ideal shock-tube theory. However, the P3 pressure records clearly indicate that the pressure profile behind the incident shock wave is decreasing with time and the shock wave is attenuating. Obviously the length of the heated buffer gas is much shorter than predicted by the ideal shock-tube theory. Figures 6a and 6b show values of the shock Mach numbers M_{s1} calculated from measured pressure ratios across the shock waves at transducer location P3. The initial pressure of the nitrogen accelerating gas was held constant at approximately $p_5 = 140$ atm, and the helium buffer gas pressure p_4 as well as the initial pressure of the helium driver gas p_1 varied over wide ranges. In Figure 6b, the experimental curve for M_{s1} exhibited a maximum near $p_4 \sim 5$ atm. At lower pressures, all pressure records show evidence of an attenuating shock wave.

From the experiments described it is concluded that serious mixing is most likely present between the accelerating and buffer gases during a few milliseconds preceding the rupture of diaphragm M2. Mixing is pronounced, particularly for small values of the initial buffer gas pressure p_4, when the density difference across the contact surface and the deceleration of the contact surface are both large and favorable in sustaining Taylor-instability mixing. Also, the finite opening time of diaphragm M2 of more than 100 μs could partially explain why the measured shock velocities were considerably lower than the theoretical, but it does not explain the observed attenuation of the shock wave.

In later experiments, when helium was used as the accelerating gas, shock waves were naturally present in compression of the buffer gas. As shown in Figure 7a, measured shock Mach numbers M_{s4} of the incident shock wave in the buffer gas were in close agreement with Mach numbers predicted from ideal shock-tube theory. Measured shock Mach numbers of the incident shock wave in the driver gas (1) were slightly higher than those obtained with nitrogen as accelerating gas, but substantially lower than predicted from ideal shock-tube theory, as visualized in Figure 7b. The shock waves were certainly attenuating at transducer location P3 for buffer gas pressures $p_4 < 4$ atm and when mixing was still present. The use of helium as an accelerating gas, instead of nitrogen, was certainly to be favored since it offered a more rapid compression of the driver gas and a reduction of additional Taylor-instability mixing in the primary shock tube.

With nitrogen as accelerating gas, the shock wave reflected at closed diaphragm M3 was reflected again when it reached the ideal contact surface to the buffer gas. Repetitive reflections caused the measured pressures of the driver gas to show a temporary local maximum (Figure 8a). Thereafter, the driver gas suffered an intermediate expansion. The slowly expanding nitrogen accelerating gas, which continuously compressed the helium buffer and driver gases, subsequently caused the expanding helium driver gas to be recompressed, until the rupture of diaphragm M3 initiated the by-pass process. The compression time of the driver gas was measured to be typically 7 ms.

With helium as accelerating gas, the contact surface between the accelerating and buffer gases moved faster because of the more rapid expansion of the helium gas, and the shock-heated helium driver gas did not suffer any intermediate expansion (Figure 9a). The time required for compression of the driver gas until rupture of M3 was typically 2.5 ms. As predicted, the slope immediately before rupture of diaphragm M3 of the measured pressure-time profile decreased with increasing time. Since the density of the buffer and accelerating gases was then higher than the density of the driver gas, favorable conditions for Taylor-instability mixing were present.

Mixing effects limited efficiency of the initial heating of the helium driver gas. To obtain optimum overall performance for the given geometrical dimensions of the present bypass piston tube, part of the buffer gas in tube (4) had to be injected into the compression tube, along with the driver gas in tube (1). The calculation of accomplished initial heating before bypass of the part of the buffer gas to be injected was difficult, and was at the time not performed with great accuracy. Better accuracy would have required more detailed measurements of the spatially varying strengths of the entropy-producing shock waves passing through the buffer gas; particularly measurements of the strengths of those shock waves passing through the gas after the reflection at diaphragm M3 of the incident shock wave in the driver gas would have been necessary. The helium temperatures before bypass were calculated from the perfect-gas relation

$$\frac{T_{Bp}}{T_o} = \left(\frac{p_{Bp}}{p_o}\right)^{\gamma - 1/\gamma} \exp\left(\frac{\gamma - 1}{\gamma} \; \frac{\Delta s}{R}\right) , \qquad (1)$$

where subscript o refers to the initial thermodynamic condition of the appropriate gas and Bp to the condition before bypass. The pressure p_{Bp} was taken as the rupture pressure of diaphragm M3. The non-dimensional average entropy change parameter $\Delta s/R$ for the driver gas was calculated from the measured strengths of the incident and reflected shock waves at P3, neglecting the effects of succeeding weaker shock waves, as well as a decrease in entropy from possible heat transfer to the walls. With helium as the accelerating gas, contributions to the average entropy-change parameter $\Delta s/R$ for the helium buffer were taken from the incident and reflected shock waves in the buffer tube before the rupture of diaphragm M2, and finally from the shock wave reflected from diaphragm M3 and penetrating the driver-buffer gas interface. Typical values for parameters of interest in the estimated initial heating to 1700 - 3100°K are listed in Table 1. This table indicates that both the entropy production and the final temperature before bypass of the driver gas were larger than in the buffer gas.

Computer calculations using idealized gas dynamic flow models have been carried out in order to determine optimum gas loading conditions. Results show that it is advantageous in maximizing the initial heating to place initially only the amount of gas that is to bypass the piston in the buffer and primary shock tubes, the division of this gas in one tube or the other being relatively unimportant. If part of the helium accelerating gas could be allowed also as a driver gas, schemes of having initially both the buffer and primary shock tubes completely evacuated would also give high initial temperatures before bypass. The gas injected into the compression tube would then be the front portion of the twice freely expanded and recompressed helium accelerating gas. Because of the high impurity level acquired during repeated recovery of the accelerating helium gas in the present set of experiments, this particular scheme was not followed.

B. Bypass Process and Final Compression

Theoretically, for the optimum operation of the bypass piston tube, the driver gas injected into the evacuated compression tube should be portioned such that whatever the time-history of the pressure accelerating the piston, the bounce would take place immediately in front of the piston retardation insert, and hence the heated driver gas would all be located inside the high-pressure section. To make an accurate theoretical prediction of the corresponding initial location of the piston in the bypass channel was difficult. Bypass calibration experiments were performed for different initial piston locations with room-temperature argon, initially occupying joint buffer and primary shock tubes and with diaphragm M2 removed; after the bypassed argon in the closed-end secondary shock tube (known volume) was cooled to room temperature, the gas pressure was measured. Using these pressure measurements and the known volume, the amount of bypass gas was determined and found to agree with predictions from first-order theory.[5] However, when heated helium was actually bypassed, simple scaling laws of the bypass parameters did not apply. Most likely this was due to existing nonuniformities in the state of the gas upstream of the piston and the associated reflections of unsteady expansion waves generated in the gas during the bypass process. In practice, the initial piston position giving the desired amount of bypassing gas was determined in an experimental iteration procedure involving estimations of the piston bounce position for a closed diaphragm M4.

Thermodynamic conditions of the gas prevented the direct determination of the piston-bounce location; for example piezoelectric transducers at the very end of the compression tube probably would have been destroyed or given erroneous readings. Therefore, the location was ascertained with corresponding peak gas pressure determined indirectly from the measurements of gas pressures in front of and behind the piston and of the piston velocity generally when the piston was passing transducer locations P5 and P6 (located 0.48 m and 0.28 m, respectively, from the piston retardation insert). In this way, the bounce location (for closed diaphragm M4) was found using an isentropic compression law and an energy conservation equation involving the gaseous volumetric compression ratio to be calculated $\tilde{V}_r = V_r/V_f$, from the measured reference condition "r" to the bounce condition "f". The energy equation could be written in the following implicit form:

$$\frac{\gamma - 1}{2} \frac{M u_p^2}{p_r V_r} + (\gamma + 1) \frac{\langle p_a \rangle}{p_r} \left(1 - \frac{1}{\tilde{V}_r}\right) = \tilde{V}_r^{\gamma - 1} - 1, \tag{2a}$$

and the corresponding peak pressure ratio as

$$\frac{p_f}{p_r} = \tilde{V}_r^{\gamma} \tag{2b}$$

Here $\langle p_a \rangle$ is an appropriate average gas pressure acting on the piston back surface during the piston stroke from "r" to "f", M is the piston mass, and u_p is the piston velocity corresponding to the measured reference condition "r" of the driver gas under compression.

Equation (2) applied with better accuracy to the experimental situation when the reference condition "r" was taken to correspond to the piston passage over the P6 transducer location during a later phase of the compression rather than an earlier, e.g., immediately after closing the bypass channel. Large discrepancies between predicted and actual volumetric compression ratios could arise from nonisentropic changes in the state of the gas due to shock waves present immediately after bypass and possibly large heat transfer to the walls.

The bypass process was analyzed from pressure records mainly from the transducer locations P3, upstream of diaphragm M3, and P6 in the compression tube. Typical records are shown in Figures 8 and 9. As can be seen from these figures, the pressure at P3 exhibits a rapid decrease by the rupture of diaphragm M3, when the heated helium driver gas flows through an unsteady expansion into the compression tube and the volume between diaphragm M3 and the piston. The piston was generally started with its rear surface at a nearest distance $0.05 < L_o < 0.20$ m from diaphragm M3. It is most clearly visualized (e.g., in Figure 8b, where the piston starts upstream of the bypass channel) that the pressure recorded at P3 (after filling of the evacuated volume on the upstream side of the piston) was raised to an almost level plateau. At a later time, the initiation of the bypass could be seen from the abrupt increase in pressure of the P6 pressure transducer. Figures 8 a to c (where the piston started upstream of the bypass channel) show injection to take place 1 or 2 ms after rupture of M3. Figures 8 c and d and Figures 9 a to c (corresponding to a piston start inside the bypass channel) show injection almost immediately after rupture of M3. A shock generated at the closing of the bypass channel is propagated into the buffer and accelerating gases and recorded in some of the P3 oscilloscope traces (Figures 8 a and c). The strength and time of arrival of this shock at the transducer location P3, measured from the rupture of M3, varied considerably, particularly for the nitrogen accelerating gas, with the mass of the piston. With a lighter piston the shock strength was smaller and the time of arrival of the shock at P3 was delayed because of the associated higher velocities of the piston and the gas accelerating the piston.

With the use of nitrogen accelerating gas, the pressure-work energy transfer to the piston was found to be comparatively efficient only when the heavier pistons with $M \sim 2$ kg were used. With the use of a light piston of $M = 0.39$ kg, the pressure accelerating the piston was measured to be much lower than the initial pressure p_5

of the nitrogen gas. This was obvious and calculable when it was taken into account that the expanding nitrogen accelerating gas had to flow through the choking buffer and primary shock tubes into a compression tube of larger cross-sectional area. However, with the use of a heavy piston the energy transfer efficiency was increased. Typically, the piston stroke time, measured from the closing of the bypass channel to bounce, was then 14 ms and the average piston accelerating pressure $< p_a > \sim 70$ atm for an initial nitrogen pressure of $p_5 \sim 180$ atm.

The transfer of pressure-work to the piston from the helium accelerating gas was efficient, even with the use of the light piston. Figures 9 a through c show from corresponding P3 traces how the shock produced at the closing of the bypass channel arrived at the P3 position after a time lag not visible on the millisecond scale of the oscillograms. For heavy pistons, and with helium-accelerating-gas pressures of $p_5 \sim 140$ atm, measured piston stroke times (from closing the bypass channel) were of the order of 10 ms, the average piston accelerating pressure $< p_a > \sim 90$ atm, the average piston velocity between P5 and P6 locations $u_p \sim 120$ m/s, and the associated piston kinetic energy $E \sim 14000$ J. For a light piston, corresponding values were for the stroke time 4.5 ms, the average piston back pressure $<p_a> \sim 80$ atm, the average piston velocity between P5 and P6 $u_p \sim 235$ m/s, and the piston kinetic energy $E \sim 11000$ J.

With the comparatively short compression chamber used in the present experiments, the maximum possible volumetric compression ratio with the high-pressure section of 0.12 m length was $\tilde{V} \sim 18$. With the use of the piston retardation insert it was possible to operate the BPP tube with final compression ratios near the maximum allowable of $\tilde{V} \sim 18$. In some experiments a smaller amount of gas was injected, which ideally would have resulted in a volumetric compression ratio of $\tilde{V} \sim 25$ and mechanical impact without the use of the insert. With the insert, no piston impact was observed, and the function of the piston retardation insert was considered satisfactory.

The possible increase in the average energy content during injection of the bypassing gas from pressure-work performed by the gas which did not bypass is explained in Ref. 5. The ideal peak temperature T_f achieved after completion of the bypass and final compression processes could be estimated. Assuming a perfect gas and neglecting energy losses, the temperature for $\tilde{V} = 18$ was found to be $T_f/T_{B_p} = 6.9(1 + \xi(\gamma - 1))$, where ξ is a

dimensionless parameter,[5] $0 \leq \xi \leq 1$. Still lower theoretical values for the temperature ratio T_f/T_{Bp} would apply if the piston motion were rapid and the injected gas had inadequate time to equilibrate before the piston has traveled a large fraction of its stroke. Judging from the P6 pressure records in Figures 8 and 9 (where no shock waves are visible in the injected gas shortly after completion of the bypass process) this situation does not apply to the present experiments with helium. This is also predicted by theory for a large ratio of piston mass to the mass of bypassed gas, as employed here, but is in sharp contrast with the experimental results in the BPP-heating of argon reported in Ref. 5 where this mass ratio was smaller.

The diaphragm M4 was ruptured at a pressure lower than the ideal bounce pressure, usually between 300 and 600 atm, before the piston had reached its theoretical bounce position. From early rupture alone, the peak gas temperatures were 10 to 25% lower than those predicted from the bounce.

Energy losses during the bypass and final compression processes were found to be important. From comparisons of measured pressures in the compression tube for known locations of the piston, e.g., at the closing of the bypass channel and when the piston was passing over P5 and P6 transducer locations, the deviation from the isentropic compression law could be determined and some knowledge obtained of the magnitude of the entropy-decreasing energy transfer to the walls. Preliminary results for convective energy losses estimated from experiments were in agreement with those calculated in a simplified laminar boundary layer model (to be published). Typically, the peak temperature would be reduced less than 10% from the ideal value owing to convective losses. Radiative losses would typically be important above 5000°K and the dominant loss above 10,000°K. Assuming black-body radiation as a close approximation valid for the peak thermodynamic conditions of helium obtained in the present experiments, calculations indicate that the radiative losses alone lower the peak temperature 20% for peak temperatures of the order $T_f \sim 10^4$ °K and peak pressures near $p_f \sim 500$ atm.

C. Shock Velocity Measurements

Typical oscilloscope records from the photodiodes in the secondary shock tube are shown in Figure 11. At the time of arrival of the shock wave the emitted radiation from the shock-heated nitrogen gas increased from the zero level to (uncalibrated) large values with almost infinite slope, except for lower initial pressures $p_{11} <$ 0.5 mm Hg, when the transition was smoother. Typically, from 8 μs after passage of the radiative shock front the observed radiation level decreased rapidly, indicating the arrival of expanded cooler helium driver gas possibly in conjunction with large absorption in the boundary layer. As is seen in Figure 11, the recorded radiation level was subsequently smoothly increased because of the arrival of helium of higher temperature and most likely also radiation from contaminant gases and metal dust from the diaphragms and tube walls.

Recorded shock velocities, obtained with a light piston of mass 0.39 kg and a given set of initial gas pressures in the accelerating, buffer and primary shock tubes, are collected in Figure 12 and plotted as a function of the corresponding overall pressure ratio across diaphragm M4 (before rupture). Shock velocities higher than 11.0 km/s for $p_{11} \sim 1.0$ mm Hg, and higher than 9 km/s for $p_{11} \sim 5.0$ mm Hg. For the plotted data, the amount of bypassed gas was varied, and hence the calculed closed-end volumetric compression ratios V are different. Corresponding values are not indicated, but for the fastest recorded shock waves, calculed ideal compression ratios were in the range $15 < \tilde{V} < 20$, but with the rupture at $\tilde{V} \lesssim 18$. Also shown in parametric form in Figure 12 is the influence of helium driver gas temperatures on theoretical shock velocities. These curves were calculated using ideal shock-tube theory, assuming a perfect driver gas with $\gamma = 5/3$, negligible opening time of the diaphragm M4, and for simplicity, but applicable to the nitrogen gas, a very high density ratio across the shock wave.

Measured shock velocities of Figure 12 indicate an equivalent driver gas temperature of $T = 9000^{\circ}K$ for the fastest shocks. Note that the opening time of the stainless-steel diaphragm M4 was large (~ 100 μs) and during that time a sonic wave in the heated helium driver gas could generally travel a distance more than four times the 0.12 m length of the driver tube. Hence, the assumption of negligible diaphragm opening time used in the calculations of the parametric curves for the equivalent driver gas temperatures T_o is certainly invalid, and the temperature actually achieved in the helium driver should be considerably higher than the corresponding equivalent temperature T_o found from Figure 12.

Measured attenuation in shock velocities in the secondary shock tube between P7 and P10 transducer locations was small. This is in agreement with simple theory for attenuation caused by reflection of the expansion wave at the end of the driver. This theory predicts attenuation after 40 driver-lengths from the diaphragm M4, for the typical pressure expansion ratio of 300 for the driver gas.

IV. CONCLUDING REMARKS

Independent shock-velocity measurements in the secondary shock tube and pressure measurements of the two-step heating processes in the basic BPP tube indicate achieved peak temperatures higher than $9000^{o}K$ in the helium shock-tube driver gas. The BPP tube used in the experiments permitted only a moderate volumetric compression of $\widetilde{V} \sim 18$ or less in the compression tube. With the use of higher volumetric compression ratios (e.g., $\widetilde{V} \sim 100$-200, as is employed in free piston compressors of conventional type), the attainable temperatures with helium in a BPP-tube would reach the $20,000^{o}K$ level. Higher values would be restricted by associated radiative energy losses.

Correlation between the simple theory of the gas dynamics of the BPP tube and the results from the present experiments show satisfactory agreement. However, several loss mechanisms, not accounted for in the simple theory, were found to be present and should be incorporated in a second-order theory of the BPP-tube. Thus, mixing effects were found to be very important, particularly when a gas of high molecular weight, such as nitrogen, is used as the accelerating gas and a light gas, such as helium, is used as the driver gas. Convective energy losses, prominent particularly during an early phase of compression in the compression tube, and radiative losses in the final compression are important and should be analyzed further. It is proposed that further experiments be conducted, using, for example, as originally proposed in Ref. 5, a combustion accelerating gas. Also, new schemes should be tested, in which no separate buffer or driver gases are used but part of the helium accelerating gas is used as driver gas and allowed to freely expand, under entropy production, into an evacuated part of the BPP tube before bypass. Using such techniques it seems possible, given the current state of the art, to build BPP shock-tube drivers at a reasonable cost, with energies in the megajoule regime.

REFERENCES

1. Camm, J. C. and Rose, P. H., "Electric Arc-Driven Shock Tube," Physics of Fluids, Vol. 6, 1963, p. 663.
2. Stalker, R. J., "Isentropic Compression of Shock Tube Driver Gas," ARS Journal, Vol. 30, 1960, P. 564.
3. Stalker, R. J., "Diaphragm-Type Shock Tube for High Shock Speeds," Nature, Vol. 218, 1968, No. 5143, P. 789.
4. Knöös, S., "Sätt Och Anordning Att Astadkomma Extremt Hoga Temperaturer," Swedish patent application 11275/65, U.S. application 690 172.
5. Knöös, S., "Bypass Piston Tube, a New Device for Generating High Temperatures in Gases," AIAA Journal, Vol. 4, 1968, p. 632.

Table 1
Typical measured and estimated parameters for the initial heating process of helium in the BPP-tube with helium as accelerating gas

Gas loading	
Helium accelerating gas pressure	p_5 = 140 atm
Helium buffer gas pressure	p_4 = 3.0 atm
Helium driver gas pressure	p_1 = 1.0 atm
Diaphragm M_2 rupture pressure	$p_{M2} \sim$ 120 atm
Diaphragm M_3 rupture pressure	$p_{M3} \sim$ 110 atm
Buffer gas parameters	
Pressure ratio (incident shock)	p_{42}/p_4 = 4.2
Incident shock Mach number	M_{s4} = 1.9
Reflected shock Mach number	M_{sr} = 1.6
Estimated Mach number of penetrating shock in primary shock tube	M = 2.2
Associated total entropy change (calculated)	exp $(\Delta s/R)_{tot}$ = 2.1
Temperature ratio (calculated)	T_{Bp}/T_o = 5.7
Temperature before bypass (calculated)	$T_{Bp} \sim 1700^oK$
Driver gas parameters	
Pressure ratio (incident shock)	p_2/p_1 = 11
Incident shock Mach number	M_{s1} = 3.0
Reflected shock Mach number	M_{sr} = 2.2
Associated total entropy change (calculated)	exp $(\Delta s/R)_{tot}$ = 3.4
Temperature ratio (calculated)	T_{Bp}/T_o = 11
Temperature before bypass (calculated)	$T_{Bp} \sim 3100^oK$

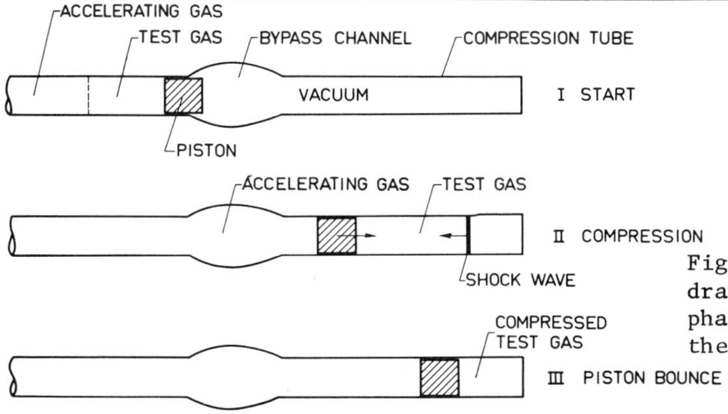

Figure 1. Schematic drawings showing basic phases of operation of the bypass piston tube.

Figure 2a. Photograph of the bypass piston shock tube. To the left is visible part of the gas handling system and a reinforced-concrete bunker for personnel protection.

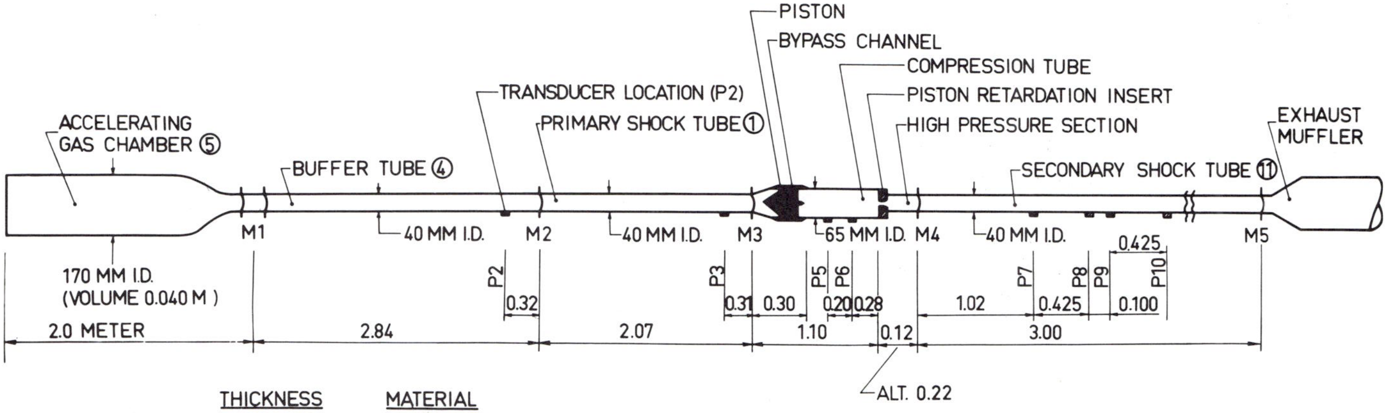

DIAPHRAGMS	THICKNESS	MATERIAL
M_1	2 -3	SOFT ALUMINIUM
M_2	2	SOFT ALUMINIUM
M_3	1.5-3	SOFT ALUMINIUM
M_4	1.0-2.0	STAINLESS STEEL
M_5	0.1	MYLAR

TRANDUCER LOCATIONS P2-P10 ARE SHOWN

Figure 2b. Schematic outline of the bypass piston shock tube, showing dimensions as well as diaphragm and transducer locations.

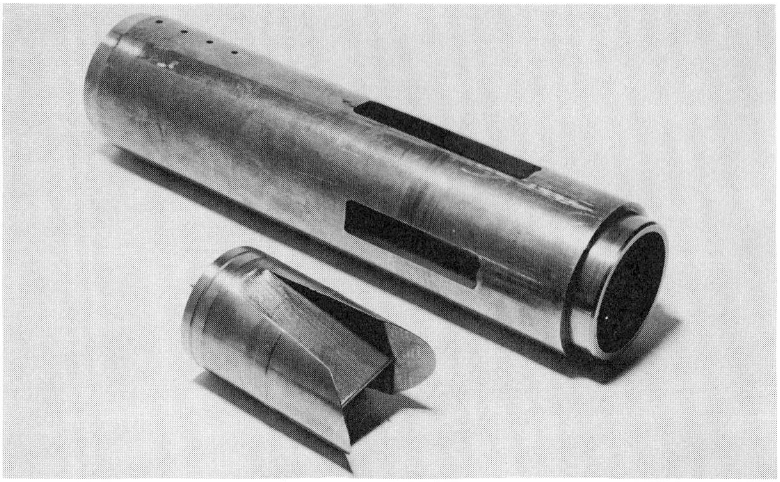

Figure 3a. Disassembled bypass channel and a heavy piston (piston type III, mass M ≃ 1.90 kg). Only the rear surface of the piston is visible. The direction of motion of the piston in the tube is to the upper left.

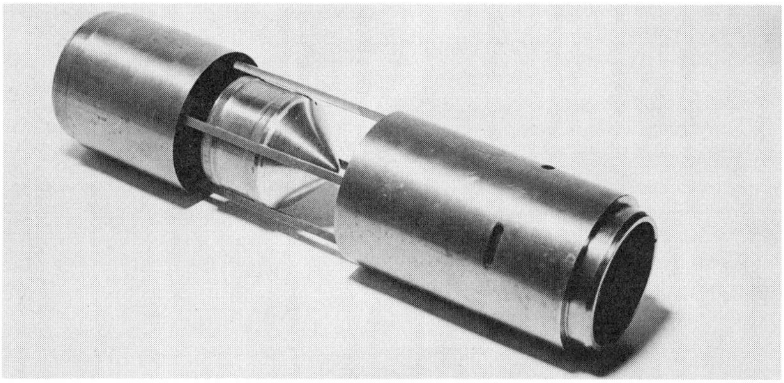

Figure 3b. Disassembled bypass channel (earlier version for large amount of bypass) and a light piston (piston I with mass M ≃ 0.29 kg). The piston is hollow and has a retractable aluminimum cone as a rear body. The direction of motion of the piston in the tube is to the upper left.

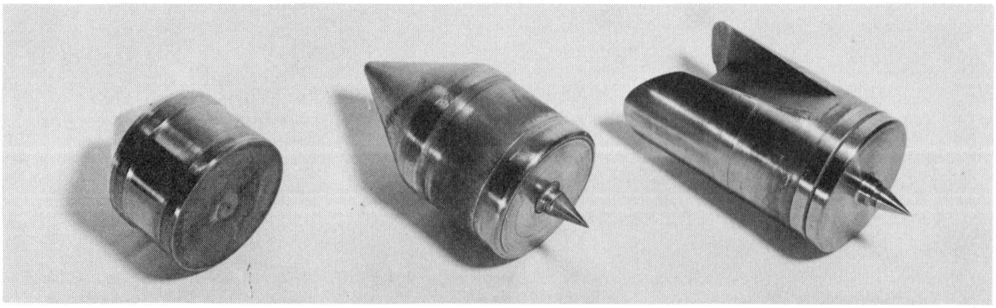

Figure 3c. Three different pistons. From left to right: piston I with mass M ≃ 0.39 kg, piston II with mass M ≃ 1.92 kg, and piston III with mass M ≃ 1.90 kg. The front surface is shown and the direction of motion is to the lower right.

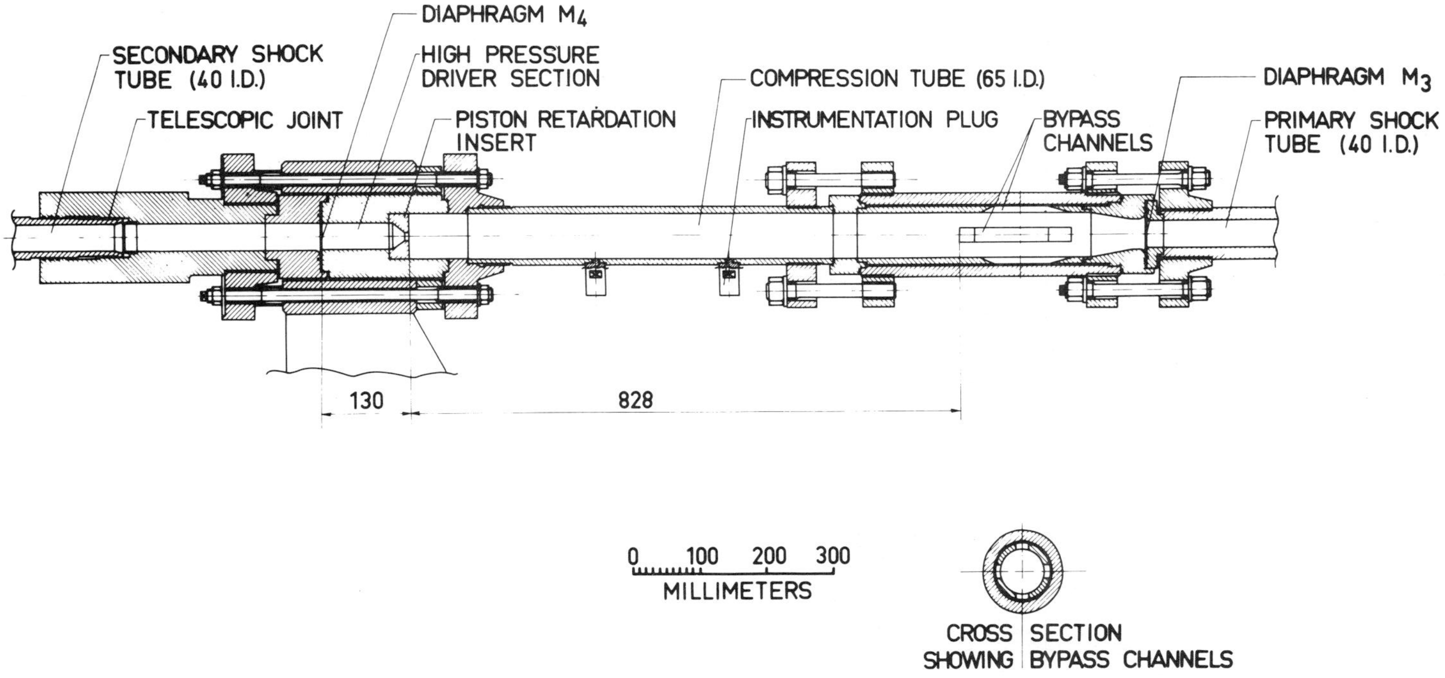

Figure 4. Drawing of the bypass channel, compression tube, and high pressure section. Diaphragms M3 and M4 are shown as well as the telescopic joint to the secondary shock tube.

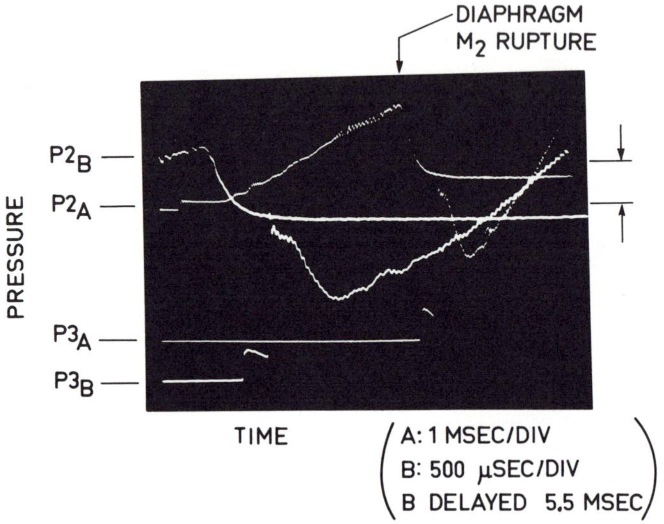

Figure 5. Oscilloscope traces showing pressure-time history of the initial heating processes of helium in the buffer and primary shock tubes with nitrogen as accelerating gas. Upper traces are from the P2 transducer location in the buffer tube, vertical sensitivity 45.4 atm/div. Lower traces are from the P3 transducer location in the primary shock tube; vertical sensitivity 10.3 atm/div. Initial pressures: p_5 = 138 atm (N_2), p_4 = 4.0 atm (He), p_1 = 0.99 atm (He).

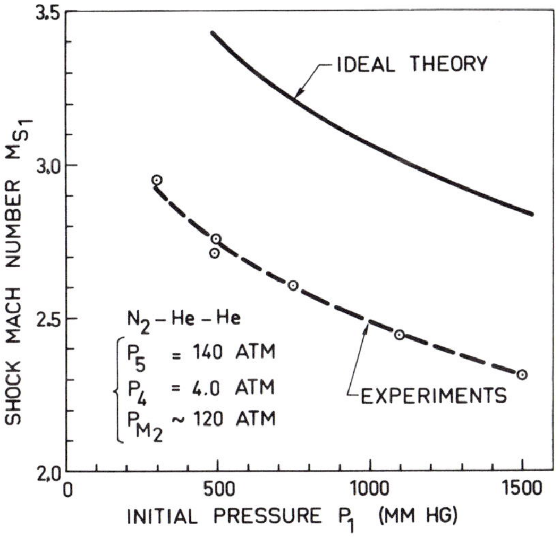

Figure 6a. Experimentally measured shock Mach numbers in the primary shock tube with nitrogen accelerating gas, showing the effect of variation of the initial pressure p_1. The theoretical curve is calculated from ideal shock-tube theory, neglecting mixing effects and heat losses in the buffer gas.

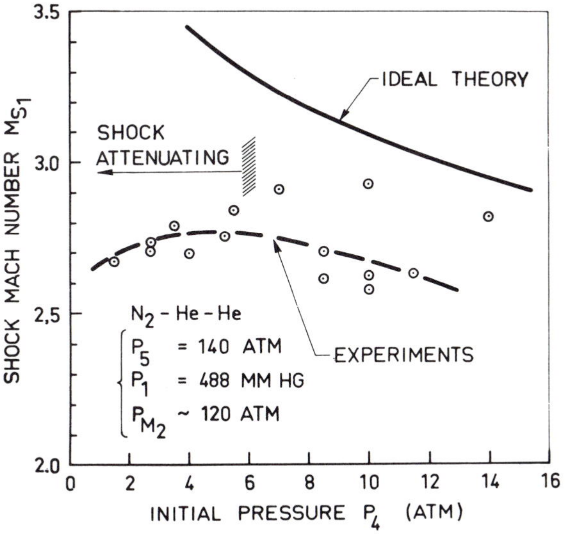

Figure 6b. Experimentally measured shock Mach numbers in the primary shock tube with nitrogen as accelerating gas showing the effect of the initial buffer gas pressure p_4.

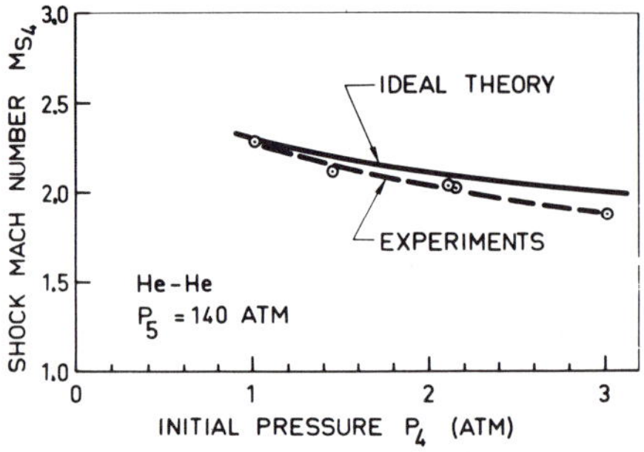

Figure 7a. Experimentally measured shock Mach numbers in the buffer tube with helium as accelerating gas.

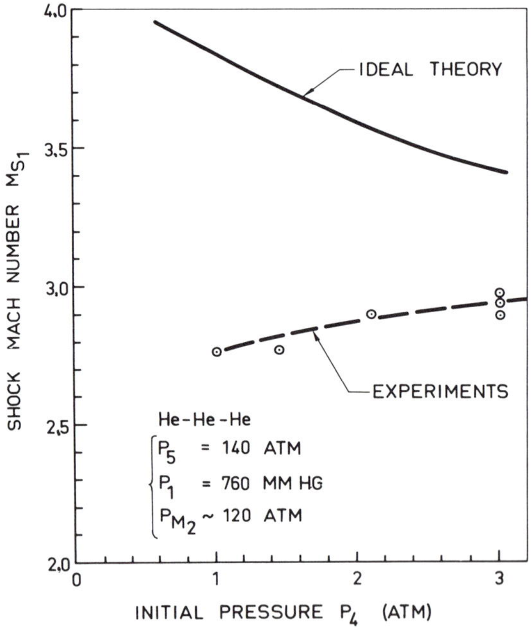

Figure 7b. Experimentally measured shock Mach numbers in the primary shock tube with helium as accelerating gas. The pressure records indicate that the shock waves are attenuating. The theoretical curve is calculated from ideal shock-tube theory, neglecting mixing effects and heat losses in the buffer gas as well as any attenuation.

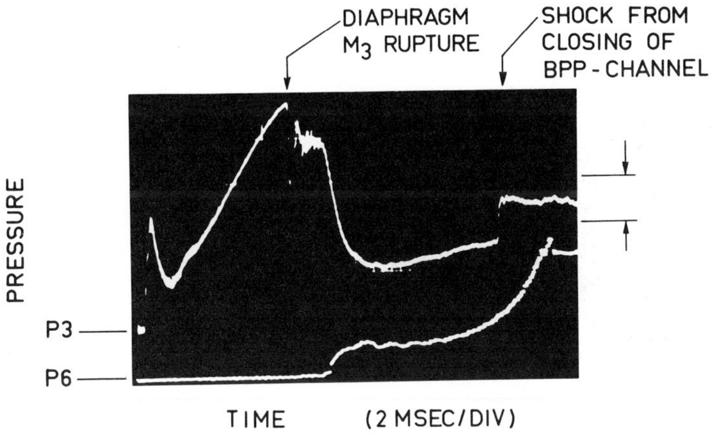

Figure 8. Typical oscilloscope traces of pressure - history before, during, and after the bypass process in the primary shock tube (P3 location) and in the compression tube (P6 location) with nitrogen as accelerating gas.

a. Vertical sensitivities, P3: 20.6 atm/div, P6: 22.7 atm/div. Piston II with mass M = 1.92 kg. Initial piston location upstream of bypass channel at a distance L_o = 35 mm from diaphragm M3. Initial pressures, p_5 = 147 atm (N_2), p_4 = 5.8 atm (He), and p_1 = 0.40 atm (He).

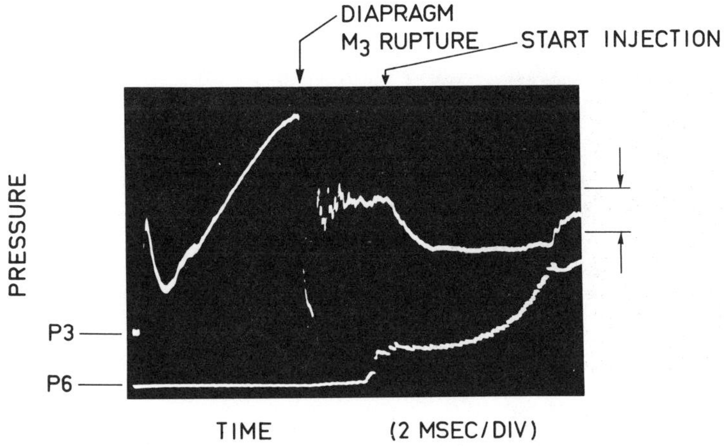

Figure 8b. Vertical sensitivities, P3: 20.6 atm/div, P6: 22.7 atm/div. Piston II with mass M = 192 kg. Initial piston location upstream of the rotated bypass channel (i.e., not in the position indicated in Figure 4) at a distance L_o = 160 mm from diaphragm M3. Initial pressures, p_5 = 141 atm (N_2), p_4 = 5.8 atm (He), and p_1 = 0.40 atm (He).

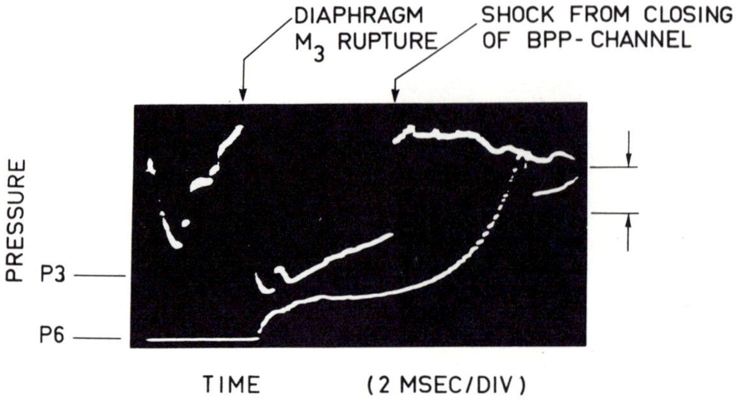

Figure 8c. Vertical sensitivities, P3: 20.6 atm/div, P6: 22.7 atm/div. Piston II with mass M = 1.92 kg. Initial piston location within the bypass channel at a distance L_o = 130 mm from diaphragm M3, giving a large amount of injected gas. Initial pressures, p_5 = 176 atm (N_2), p_4 = 11.6 atm (He), and no gas in the primary shock tube.

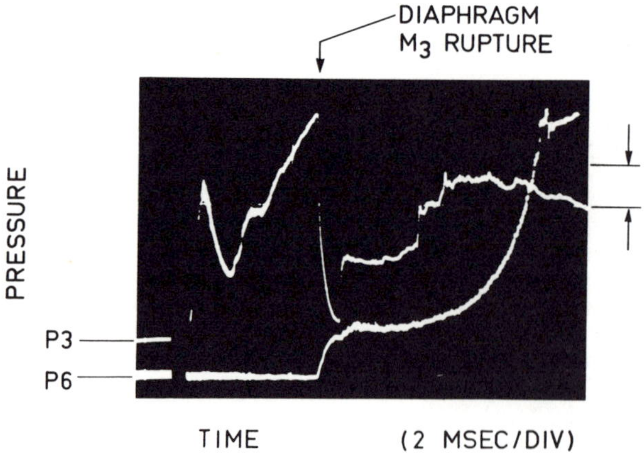

Figure 8d. Vertical sensitivities, P3: 20.6 atm/div, P6: 11.4 atm/div. Piston II with mass M = 1.92 kg. Initial piston location within the bypass channel at a distance L_o = 143 mm from diaphragm M3, giving less amount of injected gas than Fig. 8c. Initial pressures, p_5 = 171 atm (N_2), p_4 = 11.6 atm (He), and no gas in the primary shock tube.

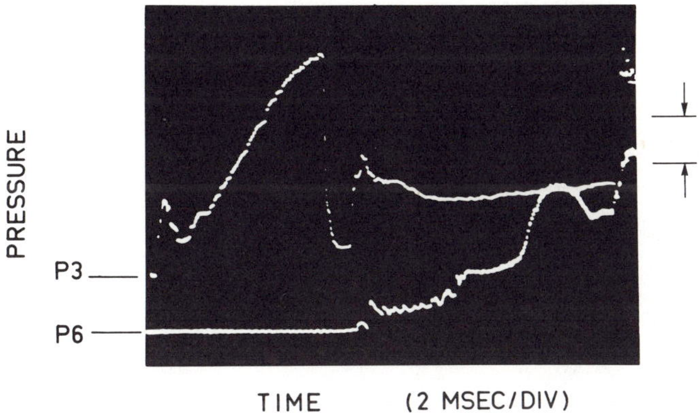

Figure 8e. Vertical sensitivities, P3: 20.6 atm/div. P6: 22.7 atm/div. Piston I with mass M = 0.39 kg, different from previous figures. Initial piston location immediately upstream of the bypass channel. Initial pressures, p_5 = 137 atm (N_2), p_4 = 4.0 atm (He), and p_1 = 0.40 atm (He).

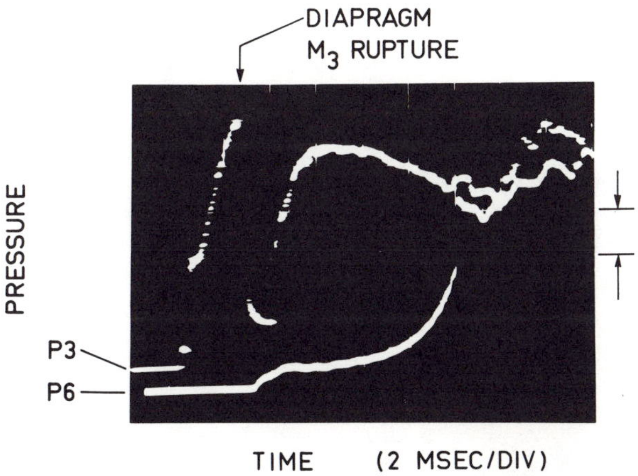

Figure 9. Typical oscilloscope traces of pressure-history before, during, and after the bypass process in the primary shock tube (P3 location) and in the compression tube (P6 location) with helium as accelerating gas. Vertical sensitivities, P3: 20.6 atm/div, P6: 22.7 atm/div.

a. Piston III with mass M = 1.90 kg. Initial piston location within the bypass channel at a distance L_o = 168 mm from diaphragm M3. Initial pressures, p_5 = 135 atm (He), p_4 = 3.0 atm (he), and p_1 = 1.0 atm (He). Measured shock velocity in secondary shock tube, u_s = 10.7 km/s with p_{11} = 0.5 mm H_g (N_2).

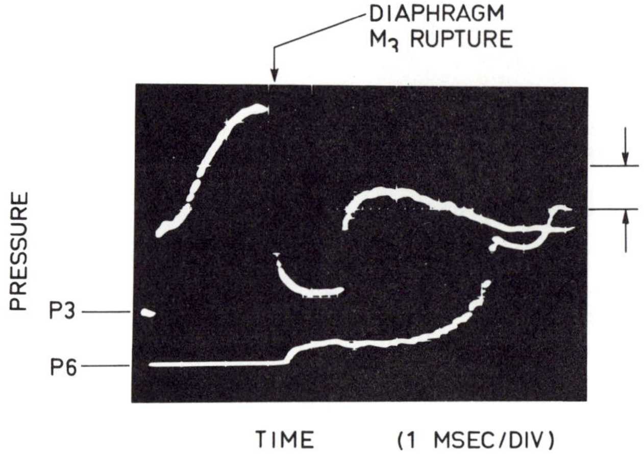

Figure 9b. Piston I with mass M = 0.39 kg. Initial piston location within the bypass channel at a distance L_o = 175 mm from diaphragm M3. Initial pressures, p_5 = 129 atm (he), p_4 = 3.0 atm (He), and p_1 = 1.0 atm (He). Measured shock velocity in the secondary shock tube, u_s = 11.0 km/s with p_{11} = 0.5 mm Hg (N_2).

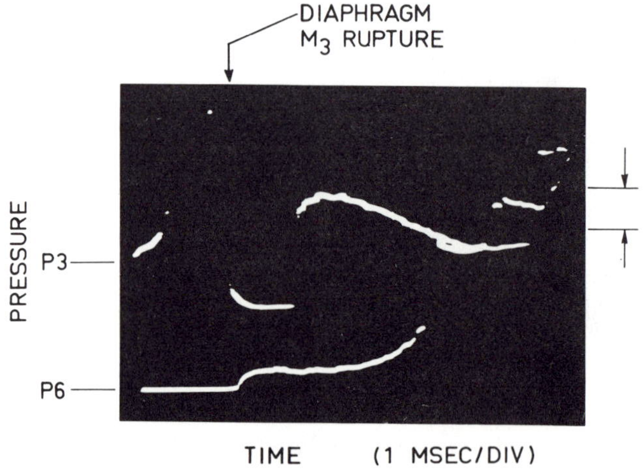

Figure 9c. Piston I with mass M = 0.39 kg. Initial piston location within the bypass channel at a distance L_o = 179 mm from diaphragm M3. Initial pressures, p_5 = 135 atm (He), p_4 = 3.0 atm (He), and p_4 = 1.0 atm (He). Measured shock velocity in the secondary shock tube u_s = 11.5 km/s with p_{11} = 0.2 mm Hg (N_2).

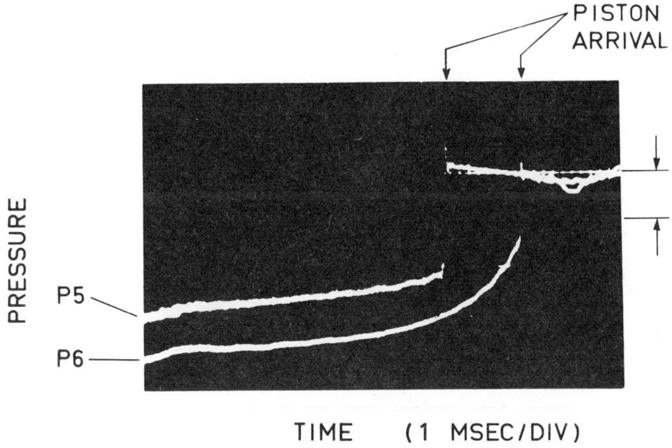

Figure 10. Typical oscilloscope traces of the pressure-history of a piston passage over the transducer locations P5 and P6, used for the measurements of the piston velocity. Piston III with mass M = 1.90 kg. The measured average piston velocity between P5 and P6 was u_p = 123 m/s. Initial pressures, p_5 = 135 atm (He), p_4 = 3.0 atm (He), and p_1 = 1.0 atm (He).

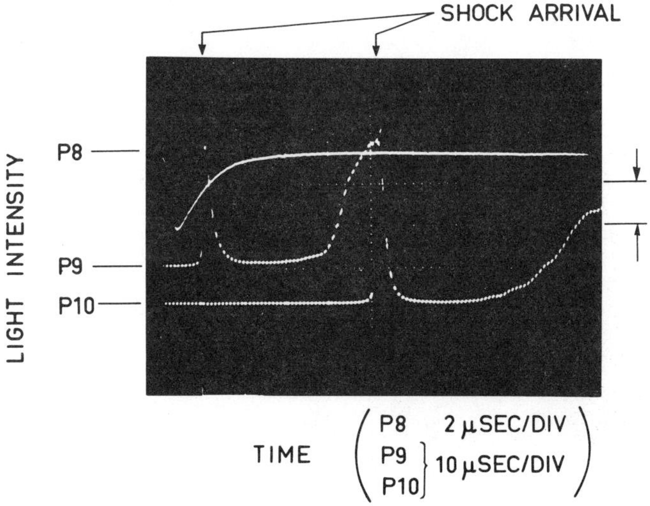

Figure 11. Oscilloscope traces of the uncalibrated output from three photodiodes at the locations P8, P9, and P10, with an axial separation of 0.100 m and 0.425 m, respectively.

a. Shock velocity u_s = 10.7 km/s with p_{11} = 0.5 mm Hg.

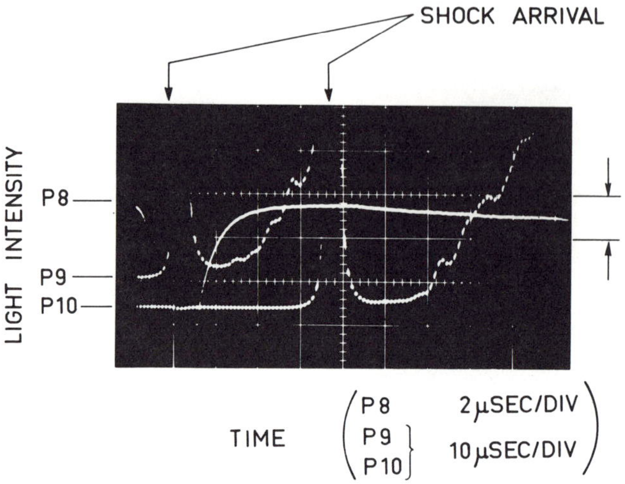

Figure 11b. Shock velocity u_s = 12.1 km/s with p_{11} = 0.2 mm Hg.

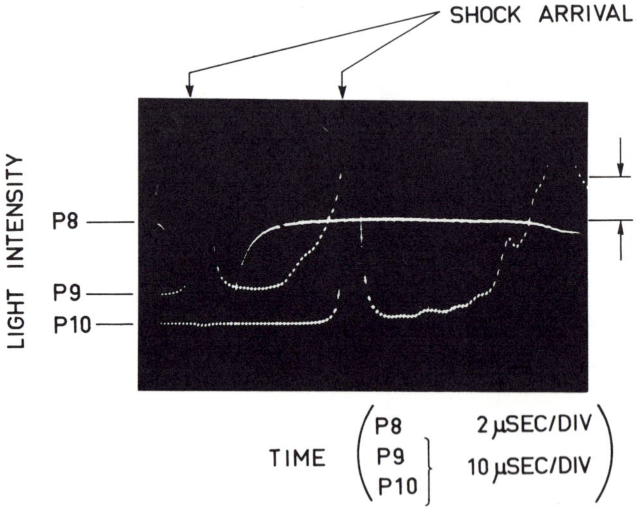

Figure 11c. Shock velocity u_s = 12.2 km/s with p_{11} = 0.3 mm Hg.

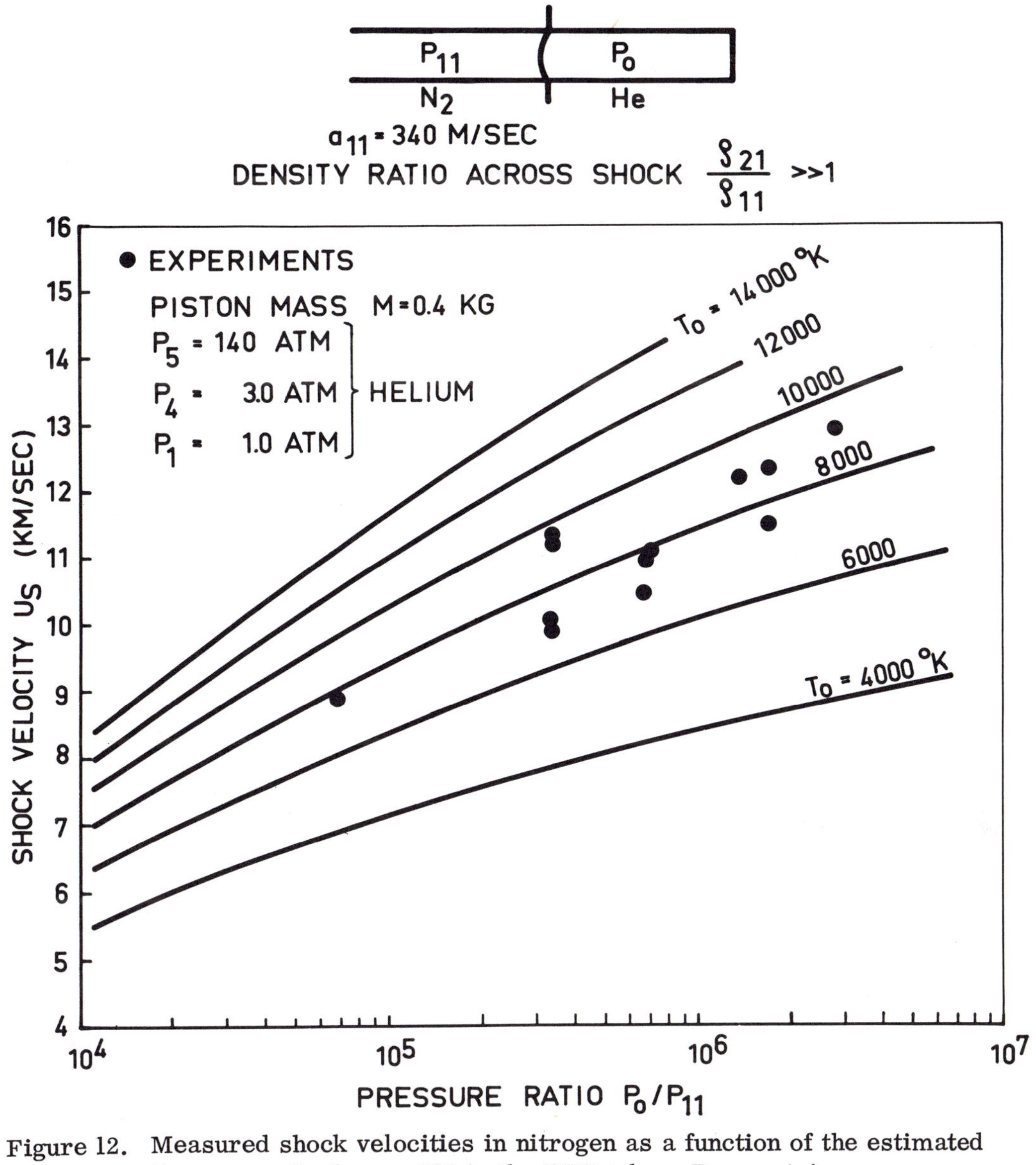

Figure 12. Measured shock velocities in nitrogen as a function of the estimated pressure ratio across diaphragm M4 in the BPP tube. Parametric curves are shown for an equivalent driver gas temperature T_o, calculated from idealized theory.

TWO DEVELOPMENTS WITH FREE PISTON DRIVERS

R. J. STALKER and H. G. HORNUNG
Physics Department
Australian National University

SUMMARY

Two developments employing the free piston driver technique are discussed, and are used to explore features of the "plateau pressure reduction" effect. This effect is characterized by a substantial reduction in the value of the plateau pressure after shock reflection at the end of the shock tube, without a corresponding reduction in the primary shock speeds. It increases with increase in performance. The first development concerns initial testing of a relatively large free piston shock tunnel, and is used to show that the plateau pressure reduction is not sensitive to scale effect. The second development concerns exploitation of the effect to produce high shock speeds in a shock tube, and involves operation of a double diaphragm shock tube attached to a free piston driver. Shock Mach numbers of 50 in air have been measured in a shock tube which employs this technique.

INTRODUCTION

In the early stages of their development, free piston drivers were employed for shock tubes [1-5]. To analyze the manner in which they operated was a simple matter. Heating of the driver gas was effected by a compression process, using a single stroke of a free piston moving in a relatively large tube. As the compression process was slow in comparison with the time scale of operation of the shock tube, it could be assumed that, when the diaphragm ruptured to initiate the shock tube flow, the compressed gas in the driver section of the shock tube was spatially uniform in pressure and temperature. The subsequent flow in the shock tube was therefore expected to behave in a classical manner, with the shock speed determined by the initial conditions in the shock tube, and the pressure and temperature of the driver gas at diaphragm rupture. Taking the driver pressure to be the diaphragm rupture pressure, and the temperature as that determined by isentropic compression, from the initial pressure in the compression tube to the diaphragm rupture pressure, reasonably accurate predictions of shock speed could be made, at least up to peak driver temperatures in helium of 8,800^{o}K [4]. It appeared that, for practical purposes, the operation of a shock tube using a free piston driver was satisfactorily understood.

Approximately six years ago, studies began of shock tunnel operation with free piston drivers. The cross sectional area of the shock tube was made much smaller than that of the compression tube, and the diaphragm was placed at the entrance to the shock tube. The moment at which the diaphragm ruptured, and initiated the shock tube flow, was chosen to occur just before the piston completed its compression stroke. The subsequent motion of the piston face then approximately compensated for the flow of gas from the driver volume into the shock tube, and maintained nearly constant conditions in the driver for a period sufficient to allow satisfactory shock tunnel operation. Again, a simple analysis of the process of operation could be formulated. For the same reasons as above, the operation of the shock tube could be considered independently of the compression process. Classical shock tube theory (with a contraction of infinite area ratio at the entrance to the shock tube), could be used to calculate the primary shock speed, together with the initial variation of the pressure at the end of the shock tube after shock reflection, and the value of the quasi-steady "plateau" of pressure which follows (see fig.1). Termination of the period of quasi-steady pressure could be calculated by reference to the expansion waves generated at the driver end of the shock tube as the piston slowed down.

The analysis was tested in a relatively small free piston shock tunnel, which embodied a compression tube 10ft. long and 3" in diameter, with a shock tube 6.5ft. long and 0.85" in diameter [6]. Not surprisingly, it was found that the primary shock speeds measured were satisfactorily predicted by the theory, thus repeating the experience with the free piston shock tubes. However, this was not so with the other parameters relevant to shock tunnel operation. As the performance was increased, by raising the driver gas volumetric compression ratio, it was found that the values of the plateau pressure after shock reflection were reduced considerably below those expected. Since the shock speeds were as predicted, this implied that tailored interface operation occurred at shock speeds higher than the expected value. It was also found that the period of quasi-steady pressure, which determines the maximum test time in a nozzle located at the end of the shock tube, persisted for longer than expected.

The plateau pressure reduction has important consequences for free piston shock tunnel operation. The plateau pressure constitutes the reservoir pressure for the test section flow, and, for a high performance shock tunnel, this should be kept as high as possible. Use of the free piston compression process is most advantageous in this regard, as the pressure at which the shock tube diaphragm bursts may be many times that prevailing anywhere in the apparatus when the compression process began. The plateau pressure reduction implies that some of this advantage is lost.

This paper describes two developments with free piston drivers, in which the plateau pressure reduction effect features prominently. Whilst these two developments were undertaken separately, and for largely independent reasons, they are presented together here in order to summarize what has been learned in this laboratory about the effect. The first development concerns operation of the compression tube-shock tube combination with a free piston shock tunnel much larger than that described above. After some discussion of mechanical and general performance details, attention is directed towards the plateau pressure reduction effect. This effect has not yet been related satisfactorily to a theoretical model, and therefore it was important to determine if it was affected by a substantial increase in scale. It was found that the effect persisted with increase in scale, displaying slight, if any, modifications in magnitude.

The second development concerns exploitation of the plateau pressure reduction to increase the shock speeds available in a free piston shock tube. Essentially, a double diaphragm shock tube is employed, with helium in the intermediate tube. The enhanced tailored interface shock speed resulting from the plateau pressure reduction allows shock reflection in the intermediate tube to produce a sample of helium at a considerably higher temperature than the original driver gas, leading to higher shock speeds. It is found that use of such a configuration does indeed lead to higher shock speeds in practice.

FREE PISTON SHOCK TUNNEL-SCALE INCREASE

The compression tube-shock tube assembly of the larger free piston shock tunnel is shown in fig.2. For convenience, this is referred to as shock tunnel T3, whilst the smaller facility, the leading dimensions of which were given in the introduction, is referred to as T2. A stainless steel forging 30ft. long is used for the compression tube of T3, but because shock tunnel operation requires substantial quantities of stored energy, it was necessary to use 10ft. of this length for the piston driving reservoir. At an air pressure of 3,500 p.s.i, which is the maximum pressure available from the air supply system, the stored energy in the reservoir is 9×10^6 joules. The remaining 20ft. of the forging is bored and honed to an internal diameter of 12", and it is this portion which constitutes the effective compression tube for performance calculations. A stainless steel piston is used, of mass 200lb. The high pressure end of the compression tube is reinforced by a steel sleeve, of 30" outside diameter, which is shrunk onto the stainless steel forging. This allows peak pressures of 2,000 atmos. to be generated in the compression process, without exceeding the yield stress of the stainless steel at the inner walls of the compression tube.

The shock tube is 10ft. long and 3" in internal diameter. Five shock timing stations, with a distance of 3ft. between each pair, are located to time the shock wave downstream of a point 7ft. from the main diaphragm. Piezo-electric pressure transducers are used for shock timing. An 0.5" diameter throat is placed at the end of the shock tube, and during these tests, the test gas expanded through a short conical nozzle to the atmosphere.

The shock tube is attached rigidly to the compression tube in a conventional manner, using a single large nut at the diaphragm station. Large axial forces are generated during operation, and it is necessary to ensure that these are not transmitted to the supporting structure and the surrounding building. The entire compression-tube-shock tube assembly is therefore mounted on rollers (Rogers Chain Co. "Easimove" skates) which permit free axial movement. As the piston is the only other moving element of substantial mass, and the centre of gravity of the complete system must remain stationary, the motion of the piston can be deduced by monitoring the movement of the compression tube.

Piston Recoil

It is obvious that an essential requirement of free piston shock tunnel operation, is that damaging impact between the piston and the end of the compression tube should be avoided. Satisfying this requirement is more difficult than may first appear. If the wave motions in the shock tube are ignored, then it is clear that the ideal situation occurs when the kinetic energy of the piston at diaphragm rupture is entirely utilized in pushing the driver gas into the shock tube, and the piston comes to rest with its forward face in contact with the end of the compression tube. In practice, the piston velocity at rupture may be expected to be either too high or too low to satisfy this condition. If it is too high, then the piston will impact at the end of the tube on first approaching it. On the other hand, if it is too low the piston will rebound, and the driver gas will continue to drain into the shock tube. Then, driven forward once more by the reservoir air behind it, the piston will approach the end a second time and, in the absence of the "cushion" afforded by the driver gas, will impact there. Thus some impact must be accommodated, and accordingly, the downstream end of the compression tube is fitted with four nylon buffers, each 2" in diameter and 0.5" thick. In the tests conducted to the present, these buffers have almost invariably suffered permanent deformation.

The movement of the compression tube is monitored using a potentiometer, the output of which is displayed on an oscilloscope. A typical record is shown in fig.3. The "dip" in the trace immediately following the first peak* indicates that the piston rebounded after the first approach towards the end of the tube. It was found that this always occurred, and that the piston impacted at second approach. However, the energy involved in the impact was not large. Noting that the slope of the trace in fig.3 provides a measure of the piston velocity, it is clear that the velocity with which the piston approaches the end is of the order of 0.1 times the maximum velocity achieved in the initial compression stroke. A typical maximum piston kinetic energy in these tests was 4×10^5 joules, indicating that the energy involved in piston impact was of the order of 4×10^3 joules. This was confirmed by measuring the deformation of the buffers, and

---

* The long term decay of the trace is due to the action of the recoil mechanism, which automatically returns the assembly to its original position.

estimating the strain energy required to produce this deformation. The kinetic energy of the piston at diaphragm rupture was typically of 10^5 joules.

Thus, although the ideal situation could not be produced, with all the piston kinetic energy transferred to the driver gas, it is clear that a close approximation can be achieved, and that it is necessary to dissipate only a few percent of the piston energy by mechanical means.

Performance

Shock speeds were measured by displaying the output of the pressure transducers on oscilloscopes with pre-calibrated time bases. Shock speed attenuation was less than could be consistently detected with the oscilloscopes, and was therefore less than 3% over the instrumented portion of the shock tube. Low shock attenuation was also observed in shock tunnel T2 [6], and is probably due to a weak compression wave, generated by slight overcompression of the driver gas by the moving piston. The variation of pressure at the end of the shock tube after shock reflection was measured by a Kistler type 6201 piezoelectric pressure transducer, located in the side wall of the tube, 2" upstream of the downstream end. The shock speed for tailored interface operation was determined by using measured values of shock speed and initial pressure in the shock tube to plot the variation of shock reflection pressure with shock speed. When this pressure matched the measured value of the plateau pressure, it was concluded that tailored interface conditions prevailed.

Tailored interface shock speeds are plotted against driver gas volumetric compression ratio in fig.4(a). A scale of driver gas temperature is shown below the figure. Both volumetric compression ratio, and temperature, were obtained by assuming isentropic compression from the initial conditions in the compression tube to the diaphragm rupture pressure. For comparison, results obtained with the small shock tunnel T2 are shown, together with the variation of tailored interface shock speed expected theoretically. It has previously been established [6] that measured values of tailored interface shock speed in T2 exceed theoretical predictions, and it can be seen that the same trend is followed in T3. The highest tailored interface shock Mach number shown is 24. This yields a test section stagnation enthalpy corresponding to a flight velocity of approximately 40,000 ft/sec. It has not yet been established that useful flow with such a stagnation enthalpy is realized at the end of the nozzle. However, it has been shown that a useful test section flow at a stagnation enthalpy corresponding to 30,000 ft/sec. is realized in shock tunnel T2 [6], and this leads to some optimism concerning the outcome of tests which will be undertaken to assess the characteristics of the nozzle flow in the larger facility T3.

The plateau pressure reduction, which is of particular interest in this paper, is shown by fig.4(b). Here the ratio of measured plateau pressure to the diaphragm bursting pressure is plotted as a function of the volumetric compression ratio. The expected theoretical variation of this pressure ratio is also shown, together with results obtained in T2. It can be seen that the values of the plateau pressure fall well below those expected, and that there is not a significant consistent difference between results obtained in T2 and T3. The magnitude of the effect, and the fact that it is not sensitive to variations in scale, implies that it must always be considered in estimating the performance of a free piston shock tunnel.

Estimates of the performance of shock tunnel T3 illustrate this point. In fig.5, the expected maximum test section reservoir pressure (i.e. the plateau pressure) is plotted against the flight velocity corresponding to the test section stagnation enthalpy. This is referred to on the figure as "equivalent flight velocity". In explaining the trends shown on the figure, it first should be noted that, as an approximation, the diaphragm bursting pressure may be taken to be directly dependent on the driver gas volumetric compression ratio, and the pressure in the piston driving reservoir. Thus, at relatively low equivalent flight velocities, where low volumetric compression ratios are used, the maximum diaphragm bursting pressure is limited by the pressures available for driving the piston (maximum value 3,500 p.s.i), and this in turn limits the test section reservoir pressure. At high equivalent flight velocities, the driver gas volumetric compression ratios used are such that the maximum allowable internal pressure in the compression tube is achieved, but the test section reservoir pressure is limited by the plateau pressure reduction effect. Thus, the useful pressures achieved in the facility are greatly reduced. In the absence of the plateau pressure reduction, the maximum design pressure of 2,000 atmos. in the compression tube would allow test section reservoir pressures of approximately 3,000 atmos. to be produced. In fact, it seems unlikely that values of 1,000 atmos. will be exceeded.

Discussion

In seeking an explanation for the plateau pressure reduction, it is natural first to consider the possibility of a "vena contracta", or aerodynamically induced flow constriction, at the entrance to the shock tube. This idea is encouraged not only by noting that the reduction is greatest when the piston face is closest to the end of the compression tube at diaphragm rupture, but also by the fact that the plateau pressure reduction is accompanied by an increase in the period of approximately constant plateau pressure, suggesting that the driver gas is not escaping from the driver into the shock tube as rapidly as expected. If such a constriction occurs, it is possible to estimate its magnitude from the ratio of the plateau pressure to the diaphragm bursting pressure [1]. For T3, with a driver gas volumetric compression ratio of 60, the measured value of the plateau pressure, shown in fig.4(b), yields an effective area at the entrance to the shock tube which is 0.13 of the actual area. This may be expected to increase the period of constant plateau pressure by a factor of 7.5, whereas the increase which has been observed represents a factor of approximately 3. Further, it is possible to calculate the effect the constriction would be expected to have on measured shock speeds. Fig.6 shows the results of such a calculation (see curve for A/A_* = 7.5), and it is seen that the predicted values are much less than those actually achieved. Thus, although the concept of a flow constriction at the entrance to the shock tube perhaps may constitute a part of the explanation of the results, it is not adequate in itself.

In fig.6 the expected variation of shock speed when there is no constriction at the entrance to the shock tube is also compared with the results, and confirms that the primary shock speeds are predicted with reasonable accuracy by classical shock tube theory. The figure also shows that the same conclusion follows from comparing similar predictions with results obtained in T2 at a driver gas volumetric compression ratio of 66. If, for purposes of discussion, it is assumed that the shock speeds are correctly predicted by theory, then it follows that the pressure and velocity of the driver gas at the contact surface are also as predicted. Now, the plateau pressure may be looked upon as the pressure achieved behind a shock wave which, reflected upstream through the driver gas following the contact surface, brings it to rest. This pressure rise depends only upon the local pressure and the Mach number of the driver gas flow, whence the plateau pressure reduction implies that the Mach number is reduced below its predicted value. Since the velocity is fixed, a reduction in Mach number must be interpreted as an increase in the speed of sound, and therefore the temperature, of the driver gas. Thus it appears that a mechanism purporting to explain the plateau pressure reduction must incorporate an additional source of energy in the driver gas, or a means of increasing its entropy, in a fashion which does not substantially affect the primary shock speeds. The authors are unable to suggest such a mechanism at present.

DOUBLE DIAPHRAGM FREE PISTON SHOCK TUBE

Principle of Operation

Although the plateau pressure reduction effect cannot be explained, the fact that it appears to imply additional heating of the driver gas suggests that it may be exploited to increase the shock speeds produced in a free piston shock tube.

A double diaphragm shock tube configuration can be used for this purpose. Such a configuration is shown diagrammatically in fig.7(a). An intermediate shock tube is placed between the driver and the test shock tube. In operation, the free piston driver is used to produce a sample of shock heated driver gas in the intermediate tube, which then acts as driver for the test shock tube. For some years, it has been appreciated that this method theoretically offers a means of improving the performance of conventional shock tubes [8,9]. However, its use offers practical difficulties. If it is operated at the tailored interface condition in the intermediate shock tube, the matching conditions at the interface require that conditions in the shock heated gas be identical with those in the driver gas immediately upstream of the contact surface. Because of this, no substantial performance gain is realized. It must therefore be operated at shock speeds in excess of the tailored interface value. The finite opening time of the secondary diaphragm (diaphragm B in fig.7(a)) then allows the contact surface to move towards the secondary diaphragm before the test shock tube flow is initiated, and also permits mixing of the relatively cooler driver gas with the shock heated gas. The former effect reduces the effective driver length for the test shock tube, leading to early shock attenuation, and the latter reduces the temperature of the shock heated driver gas. In view of this, it is not surprising that the double diaphragm technique has not been used extensively with conventional shock tubes.

In employing the technique with a free piston driver, these considerations are modified favourably by the plateau pressure reduction effect. This may be expected to allow tailored interface operation at shock speeds considerably in excess of those predicted by classical shock tube theory, thereby producing considerable gains in the temperature of the shock heated gas. Further, because conditions in the shock heated driver gas must be identical with those in the driver gas upstream of the contact surface, it follows that the contact surface should cease to exist as a meaningful feature of the flow, and the effective length of the driver for the test shock tube should then occupy a substantial fraction of the length of the intermediate shock tube. The situation is shown in the wave diagram in fig.7(b). The finite opening time of the secondary diaphragm becomes an advantage in this case, as it allows the reflected shock wave to travel a substantial distance upstream into the driver before the expansion waves generated by rupture of the diaphragm overtake and reflect from it.

An indication of the benefit conferred by the plateau pressure reduction effect may be seen by estimating the effect which the double diaphragm modification may be expected to have on the contact surface velocity for given values of the overall driver gas expansion ratio. This is defined as the ratio of the bursting pressure of the primary diaphragm (i.e. diaphragm A) to the pressure at the contact surface in the test shock tube. Fig.8 shows this effect, in terms of the ratio of the contact surface velocity to that which would be achieved by a conventional free piston shock tube (i.e. without the intermediate shock tube). The curves were calculated by assuming that the primary shock speeds in the intermediate tube were in accord with the ideal shock tube theory, and that the plateau pressure followed the experimental correlation shown in fig.4(b). For low values of the compression tube driver gas volumetric compression ratio, where the conditions for tailored interface operation accord with classical shock tube theory, only a slight augmentation of the contact surface velocity occurs. However, as the volumetric compression ratio increases, the part of the increase in the intermediate tube tailored interface shock speed which is due to the plateau pressure reduction, produces augmentation of the contact surface velocity which amounts to as much as 50% at high pressure ratios. Combined with the relatively long effective driver lengths which may be expected to prevail for the shock tube flow, this makes operation of a free piston shock tube in this mode appear very attractive.

Experiment

Tests have been made with a small double diaphragm free piston shock tube. The results have been reported previously [10], and here emphasis is placed on relating them to the plateau pressure reduction effect. Details of the apparatus and techniques used are given in reference [10], so only a brief description is required here. A compression tube 10ft. long and 3" in diameter was employed, with an intermediate tube 3.3ft. long and 1" in diameter. The apparatus was originally constructed with a test shock tube 1" in diameter. However, this was destroyed on the first test, and it was necessary to replace it by a test shock tube 0.5" in diameter, which was 4ft. long. Because of the resulting contraction at the entrance to the test shock tube, it was difficult to study the attenuation characteristics associated with the driver. Streak photographs of the motion of the shock heated test gas were taken with an image converter camera, and were used to measure shock speeds. The tests were conducted at a driver gas volumetric compression ratio of 85, and a primary diaphragm rupture pressure of 5,800 p.s.i.

Prior to the experiments with the double diaphragm configuration, tests were conducted in shock tunnel T2 with helium as test gas. It was confirmed that primary shock speeds were approximately predicted by classical shock tube theory, and that the plateau pressure reduction was as shown in fig.4(b) for air. The tests were conducted only at a driver gas volumetric compression ratio of 38, but it is reasonable to regard the results as a general confirmation of the effect for helium. Thus, from fig.4(b), the expected value of the plateau pressure in the intermediate tube during the double diaphragm test will be approximately 2,000 p.s.i. Using this value, together with primary shock speeds calculated by classical shock tube theory, the expected value of the helium temperature after tailored interface shock reflection will be 24,000°K.

Measured values of the shock speed in the test shock tube are shown plotted in fig.9. These were obtained with intermediate shock tube filling pressures somewhat in excess of the tailored interface value*, but spot checks with the filling pressure at the tailored interface value indicated that speeds were the same, or slightly lower. A curve is also shown which displays the shock speeds expected with the above driver pressure for the test shock tube (i.e. 2,000 p.s.i), but with a temperature of 15,000°K. It can be seen that this represents an upper limit for the effective test shock tube driver gas temperature. The difference between this value and the

---

* Early data had indicated that the filling pressure used, i.e. 3.5 p.s.i, would be the tailored interface value, and curves for predicted shock speed have been presented in reference [10] which were based on these results. Subsequent work has shown this data to be in error.

expected value is consistent with the trend, explored by other investigators [11], for radiation losses in helium drivers to become large at temperatures approaching 20,000°K. It also implies that, at these high shock speeds, the gains in performance will not be as large as those predicted. Thus, where fig.8 predicts a gain in contact surface velocity of 50%, comparison of the two temperatures above indicates that a gain of 25% is more representative of that achieved. This remains sufficiently interesting that the present work, which is preliminary in nature, is being followed by a more thorough and comprehensive investigation.

CONCLUSION

A study of two developments with free piston drivers has demonstrated the practical importance of the plateau pressure reduction effect, but has not given rise to an explanation of its occurrence. Tests with a relatively large free piston shock tunnel have shown that it is not sensitive to scaling and, unless it is understood and controlled, may be expected to persist in any free piston shock tunnel of practical size. Its consequences are partly beneficial, in that it increases tailored interface shock speeds and available test times, but are also detrimental in that the reservoir pressure for the test section flow is reduced. For example, in a facility designed to withstand peak pressures of 2,000 atmos., it seems unlikely that test section reservoir pressures will exceed 1,000 atmos.

Consideration of the characteristics of the effect suggested that it may be exploited to produce high shock speeds by using a double diaphragm configuration with a free piston shock tube. By operating in the tailored interface mode in the intermediate shock tube, a shock heated high temperature driver could be produced with a driver length which was not limited by the presence of the contact surface. The experimental measurements of the plateau pressure reduction were used to show that substantial performance gains could be expected with this method. Model tests indicated that these gains could be only partially realized at high performance levels. Nevertheless, it was possible to produce an effective driver temperature approaching 15,000°K, and a shock Mach number of 50 in air, with an estimated peak driver temperature in the compression tube of only 5,700°K.

ACKNOWLEDGEMENTS

The authors wish to thank Mr. R. French for his help in performing the experiments. Construction and operation of the free piston shock tunnel T3 was made possible through support provided by the Australian Research Grants Committee.

REFERENCES

1. Stalker, R.J. Isentropic compression of shock tube driver gas. ARS J. 30, 564, 1960.
2. Stalker, R.J. An investigation of free piston compression of shock tube driver gas. Nat. Res. Council, Canada. Report MT-44, 1961.
3. Greif, R. The free piston shock tube. Ph.D. Thesis, Harvard University, 1962.
4. Stalker, R.J. The free piston shock tube. Aero. Quart. 17, 351, 1966.
5. Williard, J.W. Design and performance of the J.P.L. free piston shock tube. Fifth Hypervelocity Techniques Symposium. Denver, Colorado, 1967.
6. Stalker, R.J. A study of the free-piston shock tunnel. AIAA J. 5, 2160, 1967.
7. Gore, M.M. and Stalker, R.J. Shock tunnel operation with an expansive area change at the main diaphragm. To be published.
8. Bernstein, H.J. A double-diaphragm shock tube to produce transient high Mach number flows. J. Ae. Sci. 20, 790, 1953.
9. Henshall, B.D. Use of multiple diaphragms in shock tubes. ARC 18062 Aero. Res. Council (HMSO London, 1955).
10. Stalker, R.J. and Plumb D.L. Diaphragm-type shock tube for high shock speeds. Nature, 218, 719.
11. Schneiderman, A.M., Pugh, E.R. and Patrick, R.M. Magnetic annular arc-driven shock tube. Phys. Fluids, 11, 278, 1968.

NOTATION FOR FIGURES

A	cross-sectional area of shock tube
A_*	cross-sectional area at entrance to shock tube
λ_p	driver gas volumetric compression ratio (= $(P_R/P_a)^{\frac{1}{\gamma}}$)
P_a	initial driver gas pressure in compression tube
P_c	pressure at contact surface in shock tube
P_R	diaphragm rupture pressure
P_{RA}	primary diaphragm rupture pressure - double diaphragm configuration
P^o	plateau pressure
P^o_{He}	plateau pressure in intermediate tube - double diaphragm configuration
P_1	initial pressure in shock tube
M_s	shock Mach number in shock tube
M_{st}	tailored interface shock Mach number
u_c	contact surface speed - double diaphragm configuration
u_c'	contact surface speed - conventional free piston shock tube

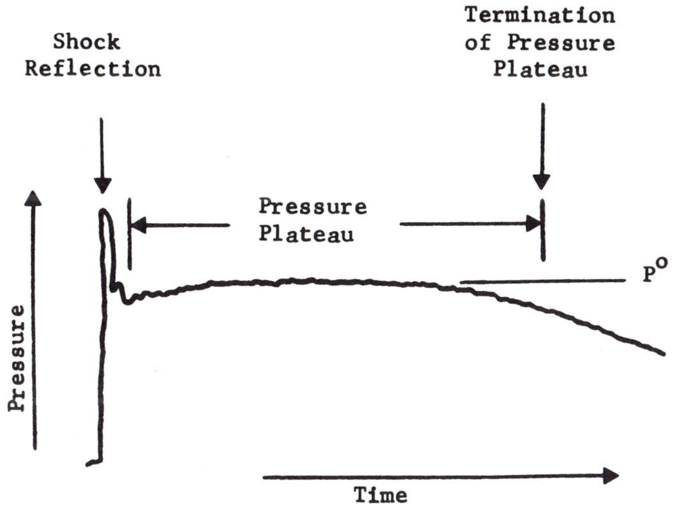

Figure 1
Typical pressure record after shock reflection

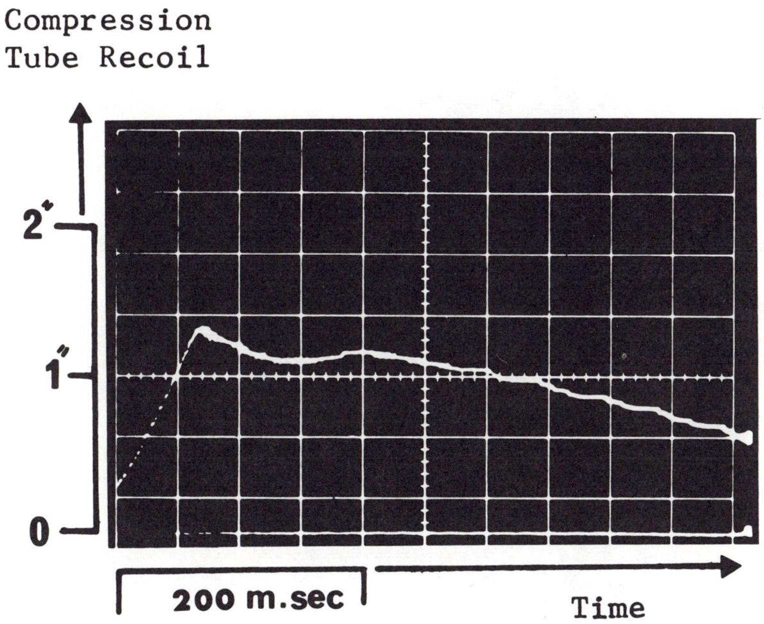

Figure 3
Piston trajectory

Figure 2
Free piston shock tunnel T3

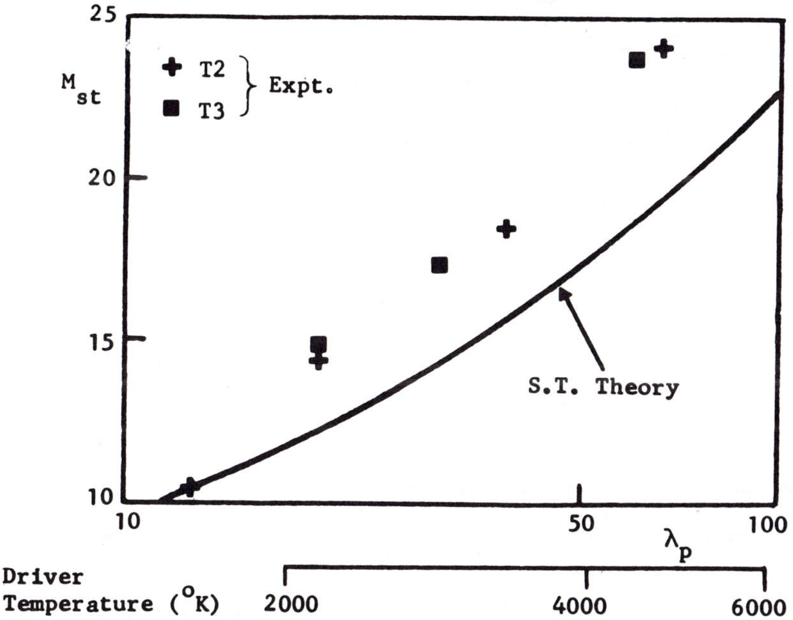

(a) Shock speeds

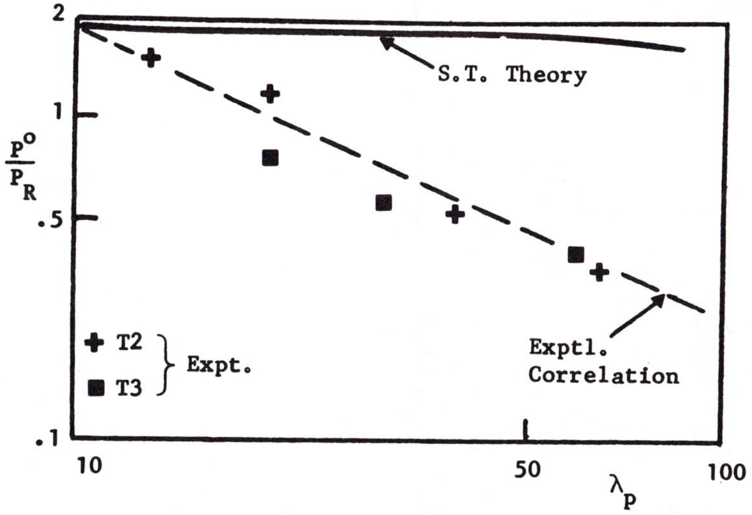

(b) Plateau pressure

Figure 4
Tailored interface operation (He/Air) - free piston shock tunnels

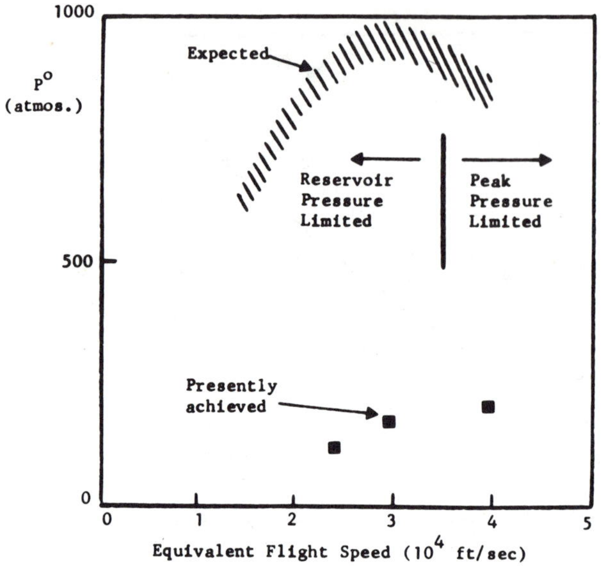

Figure 5
Expected performance - tailored interface operation (He/Air) - shock tunnel T3

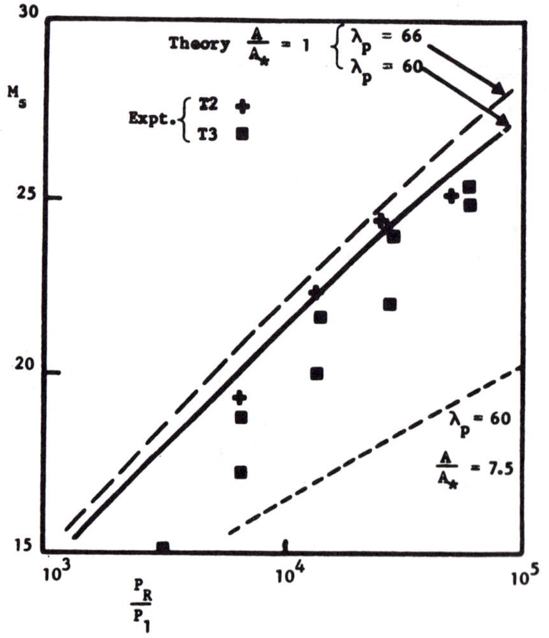

Figure 6
Shock speeds in free piston shock tunnels (He/Air)

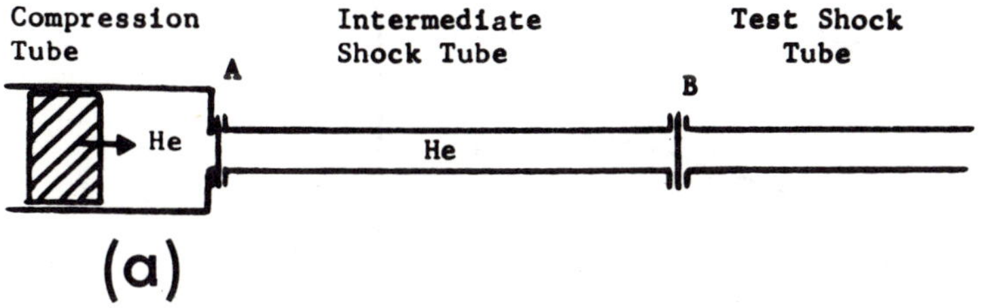

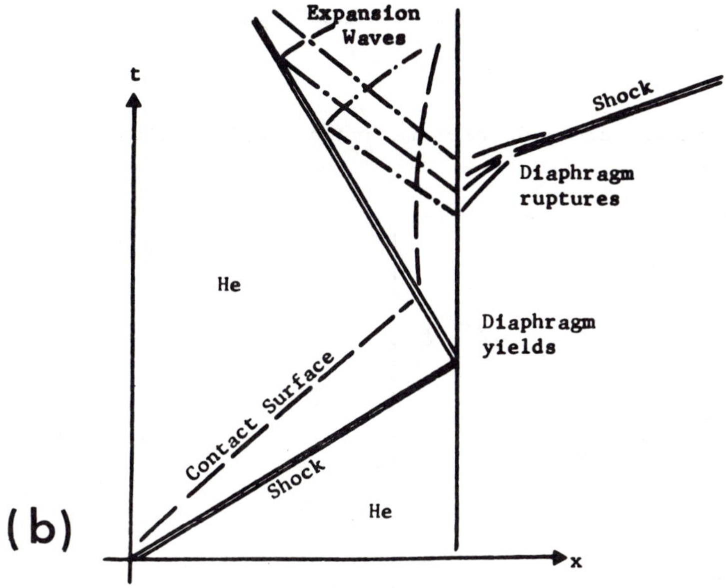

Figure 7
Double diaphragm free piston shock tube
(a) configuration
(b) wave diagram showing operation

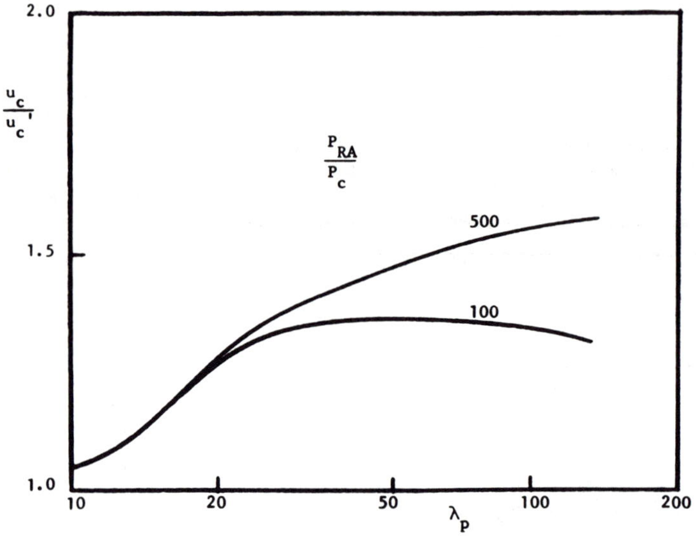

Figure 8
Expected performance gain - double diaphragm configuration (He/He/Test Gas)

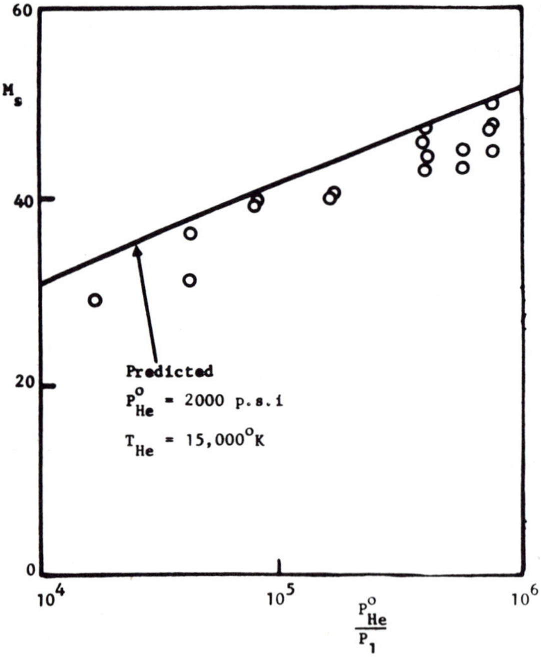

Figure 9
Shock speeds - double diaphragm free piston shock tube (He/He/Air)

"SLINGSHOT" - AN ADVANCED AERODYNAMIC TEST FACILITY

M. H. BLOOM, R. J. CRESCI, G. MORETTI, and J. LIBRIZZI
Gas Dynamics Laboratory
Polytechnic Institute of Brooklyn

SUMMARY

A conceptual view is presented of the SLINGSHOT method of mechanically setting fluids or other materials into motion with respect to models. In the special context of gas-dynamic testing, unpreprocessed test gas is made available at high Reynolds numbers and high Mach numbers with fixed models. Performance in regimes now inaccessible with fixed models is theoretically possible; however, practical implementation may require significant effort. Exploratory experiments made with the simplest conceivable apparatus show agreement with estimated model pressures and running times. The basic setup now consists of a gas-gun accelerated capsule containing the test gas, 2" dia. x 12" in length, which has been driven to 3600 fps. An analysis of the internal wave behavior agrees with the experimental observation indicating the absence of shocks within the test gas. Among the main problems are running-time limits, booster simplicity, and disposal of capsule debris. Examination of somewhat higher performance conditions appears warranted.

I. INTRODUCTION

This paper describes a conceptually simple device for fluid mechanical and gas dynamic experimentation with several novel features. Some results of exploratory gas-dynamic experiments with a rudimentary prototype apparatus are presented. In particular, pressure measurements are seen to be reasonably predictable at a moderate performance level. Particular attention is paid, analytically and experimentally, to transient behavior and ultimate uniformity of the test gas.

The SLINGSHOT concept consists of raising the velocity of an encapsulated mass of gas (or a mass of some other material, liquid or solid) by mechanical means, employing any of a number of available methods of propulsion (gun-type, rocket, etc.). The momentum and energy of the mass are thus increased, and the mass may then be passed over a model or otherwise processed. Numerous variations of this procedure and combinations with other processes, such as expansion and counter-motion, are possible.

The SLINGSHOT is discussed here in the particular context of aerodynamic testing. In this case it is possible to provide the following features:

1. High Reynolds numbers at high Mach number.
2. Minimum or controlled thermochemical preprocessing of test gas.
3. Fixed models.
4. Wide flexibility of test gas composition and density.

Although simple forms of SLINGSHOT provide short test times (on the order of capsule length divided by final velocity), the use of long capsules or of multitudes of coordinated capsules, in the spirit of the Cornell Aero. Lab. wave superheater, can be envisaged to provide longer run times or even time-varying test conditions. The present paper will deal only with the single-capsule mode, and acceleration by means of a light-gas expansion driver.

Even this simple form of SLINGSHOT* provides substantial improvements in high-speed wind tunnel performance, in comparison with those involving compression and expansion of the test gas.

---

*An elemental form of SLINGSHOT was conceived independently at the Polytechnic Institute of Brooklyn in September 1965. The technique was proposed by Prof. R. Perry in the course of discussions with Prof. M. H. Bloom, who initiated the study of staged devices such as a gun-accelerated shock tube to move gas relative to a model. It was subsequently found to be similar to a proposal for a novel dense molecular beam facility made to the USAF Cambridge Research Laboratory in June 1963. The two applications differ in that more mass is required for wind tunnel purposes, and more importantly, in that complete "stripping" of the capsule, a formidable problem, was required for the crossed-molecular-beam experiment proposed.

The further development of the SLINGSHOT concept, including the use of a deflector nozzle, multiple capsules, other types of propulsion and some experiments, were carried out by Bloom and Cresci. It was brought to the current stage of development by the authors of this paper.

The initial theoretical work was sponsored by the Mathematical Sciences Division of ONR under Contract Nonr 839(25), monitored by Mr. Morton Cooper. The subsequent work was conducted under Contract Nonr 839(38) for PROJECT STRATEGIC TECHNOLOGY and was made possible by the support of the Advanced Research Projects Agency under Order No. 529 through the Office of Naval Research.

II. DESCRIPTION OF "SLINGSHOT" CONCEPT

In essence, a SLINGSHOT test resembles a conventional wind tunnel test in which the entire test section is projected past the model. This procedure has advantages and disadvantages in comparison with other types of test methods. A very significant advantage is the possibility of attaining a high degree or even complete simulation of Mach number, Reynolds number, flow composition and real gas effects. This is accomplished by virtue of the test fluid having the same static conditions (i.e., temperature, density, etc.) as in the free flight counterpart, and by the test gas flowing over the model at the same relative velocity as experienced in flight. Limitations on model size can be alleviated by utilizing an auxiliary expansion-nozzle technique to be discussed further below.

A disadvantage of the method may be the limited test duration, which is approximately equal to the ratio of capsule length to capsule velocity. For high test velocities, long capsules may be required to provide useful test times. It is observed that the possibility of the appearance of shock-wave non-uniformities in the test gas is enhanced by increases in capsule length for a given capsule acceleration (the main determinant of driver length), since compressions from the rear face are not as readily counteracted by expansions from the front face. This consideration of gas uniformity, in the main, sets the capsule length for a given driver and test condition. On the other hand, decreases in acceleration level and alterations in acceleration history, which imply increased driver lengths, permit increases in capsule length for a given test condition. In this regard the chief limitation is space available for the facility and appropriate propulsion arrangements.

This can be depicted roughly by considering the analytically simple case of constant capsule acceleration. For this case it is known that compression waves from the rear face of the capsule will coalesce at a point $x=(c^2/\alpha)(2/1+\gamma)$ after an elapsed time $t=x/c$; where x is measured from the rear face, c is the sonic velocity of gas within the capsule, γ is the specific heat ratio of the gas in the capsule, and α is the capsule acceleration. Since the distance traversed by an expansion wave from the front face is likewise ct, if one wishes to utilize a compression-expansion interaction to achieve shock-free test gas, a capsule length roughly equal to 2x would be employed. Clearly more detailed analysis along this line is required to take into account the development of coalescence and interaction, and the effects of other kinds of acceleration history. Indeed, initial results of such an analysis are shown in Section IV of this paper for conditions employed in the present experiments utilizing a light-gas expansion driver which produces a time-wise diminishing acceleration. This computerized analysis of one-dimensional unsteady flow including shocks and interactions is rather sophisticated, and will be the subject of an additional paper.

However, it can be seen from the elemental considerations above, that for air as a test gas and a capsule length $L_c=2x$, the estimate $\alpha L_c \simeq 2 \times 10^6$ (units in ft and sec) gives an idea of the order of permissible acceleration below which shock formation is avoided for a given capsule length. For a capsule length of 1 ft., accelerations as high as 60 thousand g's can be tolerated. For constant acceleration, the driver length L_D required for test velocity u_T is given by $\alpha L_D = u_T^2/2$, so that the ratio of driver to capsule length is $L_D/L_c \simeq (U_p \times 10^{-3}/2)^2$. For $U_p = 2 \times 10^4$ fps, a value $L_D/L_c \simeq 100$ is estimated.

It is of interest to note that the use of light gases such as helium or hydrogen for the test gas within the capsule permits an order of magnitude increase in the capsule length $L_c \sim c^2$ for a fixed acceleration α, without significant change in the driver length L_D. This may be of use for certain special purposes wherein increased test time is significant.

These rough considerations indicate that shockless flow may be attainable with realizable driver lengths for non-trivial test times, and that additional analyses and experiments are of interest.

Figure 1a presents a schematic of the simple SLINGSHOT. The capsule acceleration may be accomplished by several known driver techniques such as high-pressure light gas, combustion, electromagnetic fields, etc. The resulting acceleration history, which is somewhat different for each of these techniques, has implications with respect to the flow processes within the capsule, as previously indicated. A simple light gas expansion, as used in the present exploratory experiments, produces a high initial acceleration which subsequently diminishes. A variable-thrust rocket motor may provide constant acceleration over an extended period of time, a mode which may be quite attractive with respect to test-gas behavior, driver-length, and simplicity.

In the rudimentary setup shown in Fig. 1a, a thin diaphragm (e.g., mylar) covers the forward end of the capsule; it is sufficient when the capsule pressure is low and the launch tube is evacuated. Upon reaching the model, the diaphragm is broken, either by the model or otherwise. When the capsule speed is high, the disturbance of the test gas due to the breaking diaphragm is small. However, the mechanical behavior of the diaphragm under conditions of high performance must be given careful attention.

The test gas then passes over the model, which is expendable in this arrangement and is destroyed when the rear of the capsule hits it. Remotely mounted instruments are unaffected, whereas model-mounted sensors may be expendable. To minimize or avoid such damage a stripper-deflector in the form of an asymmetric expansion nozzle may be employed as shown in Fig. 1b. This scoop, located at the end of the driver, ruptures and deflects the diaphragm material and strips off the capsule, whereas the test gas passes over the model which is mounted in an offset position. With proper design, debris impingement on the model should be minimal.

The expansion nozzle may serve a second purpose, previously cited, which is to expand the flow to a higher Mach number and a larger flow area. Test Reynolds number per unit length may be preserved by a degree of initial pressurization of the capsule test gas. The expansion permits the use of larger models with corresponding increases in model Reynolds number.

Figure 2 shows a performance plot in the form of test Reynolds number versus Mach number. For a given capsule pressure, the test Reynolds number produced by SLINGSHOT increases with test Mach number as a result of the direct dependence of both parameters on capsule velocity. This is in contrast to the behavior of other methods which are based on compression and expansion and wherein Reynolds numbers decrease with increasing Mach number. A subsequent expansion of SLINGSHOT gas decreases the Reynolds number per unit length as the Mach number increases. This unit Reynolds number reduction is roughly inversely proportional to the expanded flow area ratio or to the square of the increased diameter ratio. If the model size is increased in proportion to the enlarged diameter, the Reynolds number based on model size is only diminished in inverse proportion to the diameter of the flow area, rather than the diameter-squared. Of course, this same relation applies to the aforementioned expansion methods as well.

As a specific example, Fig. 2 shows that a capsule pressurized to 3 atmospheres and boosted to Mach 8.5 (end velocity of 9500 fps) yields a unit Reynolds number of 2×10^8. A subsequent expansion to Mach 10 reduces the unit Reynolds number by a factor of about 3 to 7×10^7. This compares with the indicated shock tunnel Reynolds numbers of about 10^7 at Mach 10. The static temperature is reduced about 30 percent during the SLINGSHOT expansion.

III. TEST APPARATUS

This section describes initial experiments to explore the feasibility of translating into practice the concept of propelling an encapsulated slug of gas in order to produce unpre-processed, high Mach number, high Reynolds number test flow. The simplest conceivable apparatus is employed for this purpose.

A schematic of the existing PIBAL SLINGSHOT facility is shown in Fig. 3. The 4.5 inch I.D. driver section was designed by means of characteristics in conjunction with the internal ballistics equations for the available barrel length. This yields the effect of an infinite driver and therefore maximum muzzle velocity. The driver is designed for a working pressure up to 10,000 psi. Adjacent to the driver is the double diaphragm section which is used to initiate the test. This section is terminated at either end by two scored stainless steel diaphragms. The test procedure consists of pressurizing the driver and double diaphragm section simultaneously to approximately half the test pressure. At this point a valve separating the driver and double diaphragm is closed and the driver is pressurized to the desired test pressure. Venting of the diaphragm section causes the pressure differential across the diaphragm to exceed the breaking pressure.

The gun barrel has a 2.0 inch I.D. and a honed chrome finish which provides an extremely smooth and uniform surface. The effective length of the barrel is 62 ft. At the end of the barrel there is an instrumentation port to which the model support is attached. The end of the barrel is sealed with a .005 inch mylar diaphragm which allows the barrel to be evacuated and also allows the projectile to pass freely out of the barrel into an armor plate "catcher". The model support is broken off at the wall and the model is carried out of the barrel and destroyed. This fact exemplifies the utility of the expansion nozzle described previously. Break wire switches are placed at various points along the barrel to determine position and velocity of the projectile.

The capsule, which contains the test gas, is shown in Fig. 4a. The capsule is made of Lexan, a polycarbonate, and fabricated in three pieces. Lexan tubing is used and closed off at the ends with caps that are machined out of solid Lexan. The end caps are fused to the tubing with methylene chloride and the entire assembly is turned down in a lathe in order to maintain a uniform diameter of 1.98 inches. The front end has a screw-on cap which holds in the .015 inch cellophane diaphragm which is sealed with an O-ring. When the capsule is completely assembled, the fused joints and diaphragm seal are checked under vacuum for leaks. The capsule is 14 inches long and weighs 250 gm. A neoprene O-ring seal is used between the barrel walls and the back end of the capsule to prevent leakage of driver gas past the capsule. A teflon O-ring is used to provide alignment of the front of the capsule and to prevent contact of the Lexan with the barrel. This O-ring has a small section missing (as seen in Fig. 4a) in order to allow the region between the capsule and barrel to be evacuated. In order to insure against shear failure of the rear end cap due to the driver gas pressure, a 1/16 inch aluminum plate is glued to the back of the capsule.

To evaluate the feasibility of aerodynamic testing in a facility such as the SLINGSHOT, two types of pressure models were used. The pitot pressure and cone pressure models are shown in Figs. 4b and 4c, respectively. Both models are 30 degree cones and constructed of brass. To reduce the time lag due to convecting tube length, the models are constructed to house the sensor. The cone pressure model simply has four static holes which lead to the pressure gage. The pitot pressure model, however, provides a direct passage for diaphragm fragments to the sensor and had to be designed in a double chamber arrangement. That is, the first chamber is used to stop the diaphragm fragments and allow the pressure to pass into the second chamber where the pressure is measured.

The models are constructed to contain either a piezoelectric crystal transducer or a diaphragm supporting a strain gage. The crystal transducer was developed by PIBAL and yields extremely reliable pressure data. The use of these transducers is somewhat costly since they are destroyed after each test. The strain gage device was made expressly for these tests to provide a more economical and fairly reliable means for measuring the pressure. The characteristics of these gages, response and sensitivity, proved to be ideally suited for these initial tests. The device consists of a strain gage glued to a circular spring steel diaphragm which is constrained within the model.

Each model is fully assembled, as shown in Fig. 4d, and then calibrated in the range of the expected pressure.

A static pressure tap is located in the instrumentation port. This tap records the pressure history in the vicinity of the model in order to detect any leakage of driver gas past the capsule or possible buckling failure of the capsule. The break wire switches mentioned earlier consist of a wire imbedded into a 1/8 inch plastic cylinder which protrudes slightly into the barrel. The passage of the capsule breaks off the tip of the plastic and breaks the circuit formed by the wire. The switches are constructed in this manner so as not to interfere with the capsule diaphragm.

Typical test data for the cone pressure and pitot pressure are shown on the oscilloscope traces in Fig. 5; the sequence of events indicated on these traces are as follows: The capsule accelerates down the barrel to the position of S-2, the breaking of which triggers the oscilloscope and starts an electronic counter. The breaking of S-3 stops the counter and indicates the impingement of the capsule on the model. The model pierces the cellophane diaphragm and traverses the length of the capsule. The breaking of S-4 indicates the termination of the test, i.e., the collision of the model with the back of the capsule. The second trace from the bottom represents the model pressure, and the bottom trace the time history of static pressure in the barrel.

The test data was obtained using both air and helium as driver gases. The air was used at 2,000 psi and the helium, for the higher velocities, was used in the 2,000 to 3,000 psi range. Test times ranged from 330 to 650 microseconds and the velocity ranged from 1,800 to 3,600 feet per second. The results of these tests are shown in Fig. 6. The theoretical cone and pitot pressure curves are obtained using 14.7 psi as the free stream static pressure. The experimental data show fairly good agreement with the theoretical predictions.

The above-mentioned values of driver pressure for the air and helium represent practically optimum conditions for test velocity. Significant increases in pressure above these values yield very little performance increase. For instance, an increase of pressure to 3,000 psi for air yields only a 4.0 percent increase over the velocity at 2,000 psi. For helium, an increase of pressure from 3,000 psi to 4,000 psi yields a velocity increase of 6.0 percent. Therefore, for the present facility, an increase in driver pressure will not yield an appreciable performance increase. A very effective method for increasing the performance of this type of facility is the heating of the driver gas whereby a marked increase can be obtained without approaching unrealistic pressures. As an example, the heating of helium to 2,000^{o}R produces a test velocity of 12,000 feet per second. It can be seen (cf. Fig. 2) that this simulates sea level flight at a Mach number of 11 for a capsule initially at one atmosphere; for pressurized capsules, the simulated Mach number is considerably higher.

IV. INTERNAL FLOW ANALYSIS

Determination of the ultimate feasibility of the SLINGSHOT and its limitations can be assisted by a theoretical study of the dynamics of the test gas contained in the capsule. In principle, this is a straightforward problem since the gas can be assumed to be inviscid and perfect, its motion one-dimensional, and the geometry (length) of the capsule invariant. If shocks build up, however, the computational method must be capable of following their evolution, their reflections on the end-walls of the capsule, and their mutual interactions. In Ref. 1, reasons have been given against using the method of characteristics for such a calculation. In the same paper, a method which combines the advantages of a second order accurate, finite-difference technique with a special treatment of boundaries and discontinuities is presented; the reliability of the results is also indicated.

A number of calculations have been made in an attempt to find a practical operational limit to the device, as a function of capsule-to-barrel length ratio, and initial acceleration. These results will be reported elsewhere; it may be stated here that no serious limitations as to feasibility or high performance have been observed as yet. Here we show only the case which has been achieved experimentally, at the present stage of the project. The capsule is accelerated according to:

$$U_p = \frac{2a_o}{\gamma-1}\left\{1-\left[1+\frac{\gamma+1}{2a_o}\left(\frac{p_oA}{m}\right)t\right]^{-\frac{\gamma-1}{\gamma+1}}\right\}$$

where U_p = capsule velocity

a_o = initial sonic velocity of driver gas

p_o = initial pressure of driver gas

A = cross-sectional area of barrel

m = capsule mass

γ = specific heat ratio of driver gas

t = time

This is the usual velocity history resulting from the expansion-driver interior ballistics analysis and has been found to give extremely good agreement with the experimentally measured capsule velocity for both the helium and air drivers.

The results of the analysis for the air driver at an initial pressure of 2,000 psia are shown in figures 7 and 8. The velocity of the test gas is shown in figure 7 for various times after initiating the acceleration. It can be seen that in a very short time the gas within the capsule achieves a uniform value equal to the capsule velocity. A practically constant pressure is also reached quickly (cf. figure 8)

---

[1] Moretti, G., "Complicated One-Dimensional Unsteady Flows". To appear as a PIBAL Report, 1969.

and the gas in the capsule then moves as a rigid body until the capsule reaches the test section. The analysis was also applied to the Mach 3.5 test case with a helium driver and the same qualitative results were obtained.

V. CONCLUDING REMARKS

Exploratory theoretical and experimental studies of the SLINGSHOT concept have been depicted. The main experimental results have involved measurements of transient cone pressures and pitot pressures produced by gun-accelerated capsules containing air initially at standard atmospheric conditions. Nominal capsule dimensions were 2" diameter by 12" long. With expansion-gun propulsion, capsule velocities up to 3600 fps have been produced. The measured pressures and their duration, and the capsule ballistics, are consistent with theoretically-based expectations. A sophisticated computer calculation of the transient wave behavior within the capsule indicated the absence of shock formations under the relatively modest performance conditions of the present tests. The computer program is now available for studies of advanced nature; however, rough estimates in the constant-acceleration case, and experience gained with the calculations already made, provide an indication that shock-free flow may be possible with the present apparatus up to capsule velocities of at least 10^4 fps, with air as the test gas. This will be one objective of the next phase of the study. Ideas have been generated for dealing with a number of problems which are anticipated in connection with operation at the next performance level.

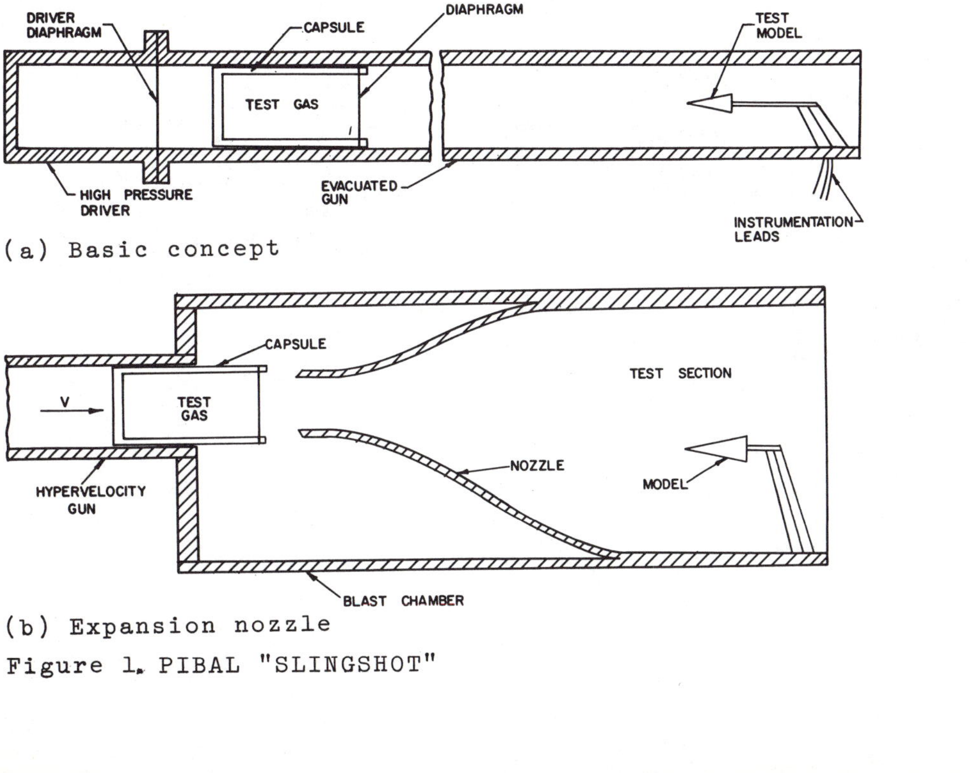

Figure 1. PIBAL "SLINGSHOT"

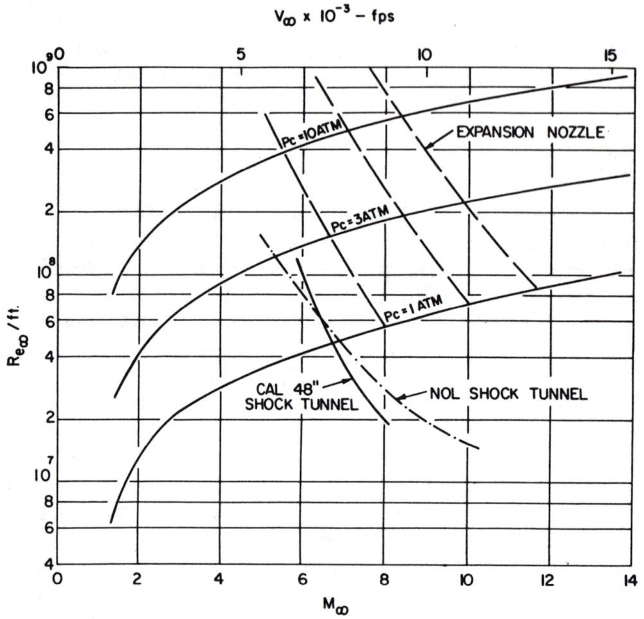

Figure 2
"SLINGSHOT" performance

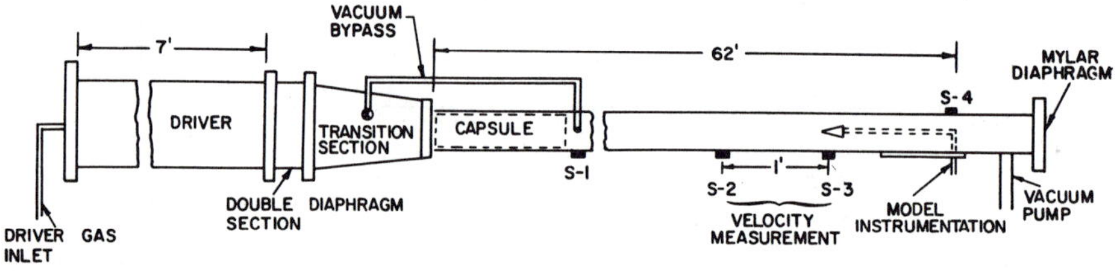

Figure 3
Schematic of facility

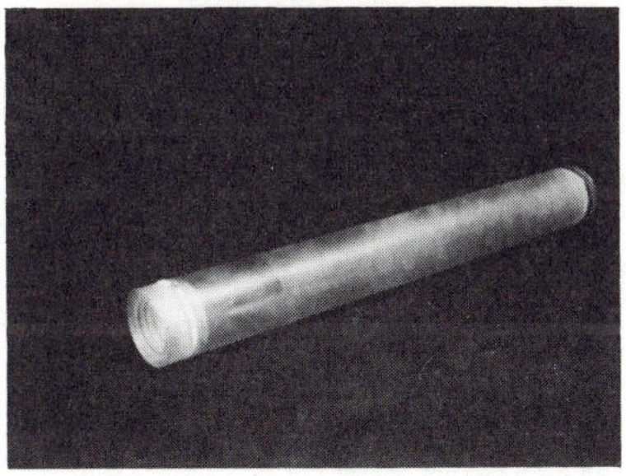

(a) CAPSULE

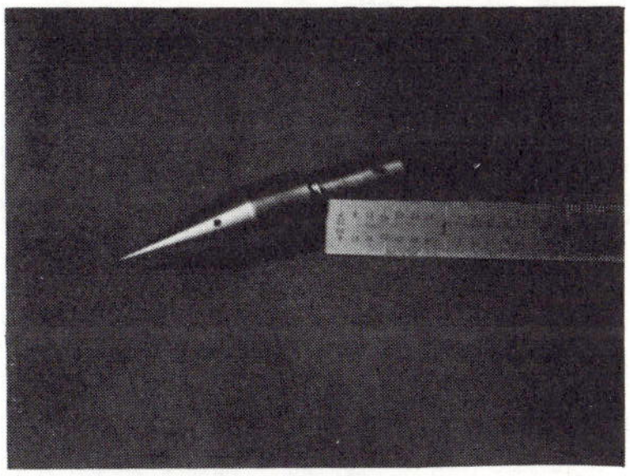

(c) CONE PROBE

(b) PITOT PROBE

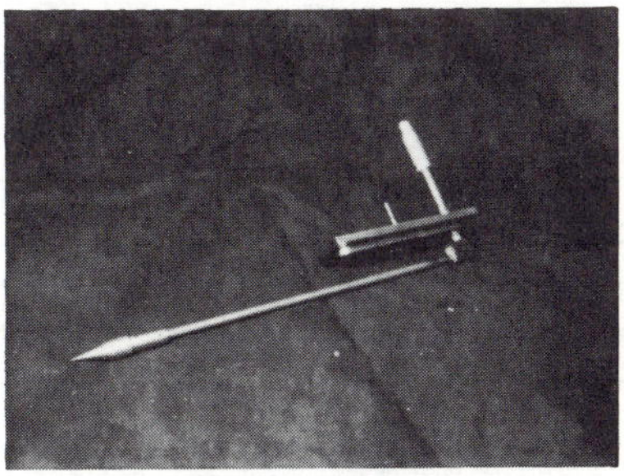

(d) MODEL ASSEMBLY

Figure 4
Photographs of test components

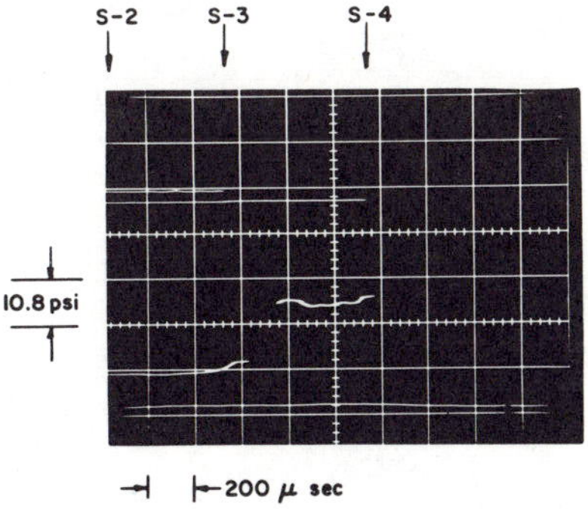

(a) Cone pressure

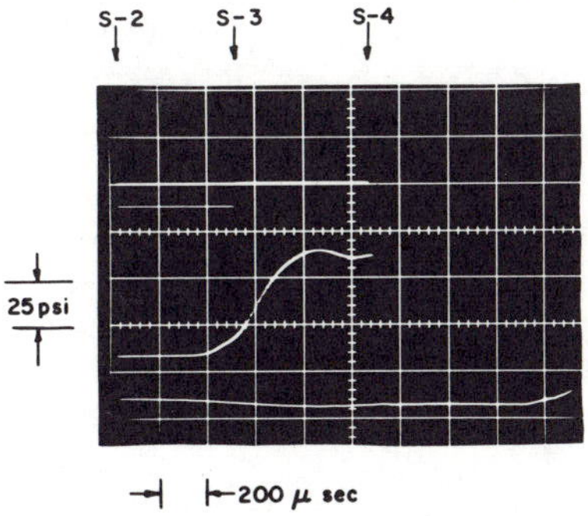

(b) Pitot pressure

Figure 5
Typical oscilloscope trace

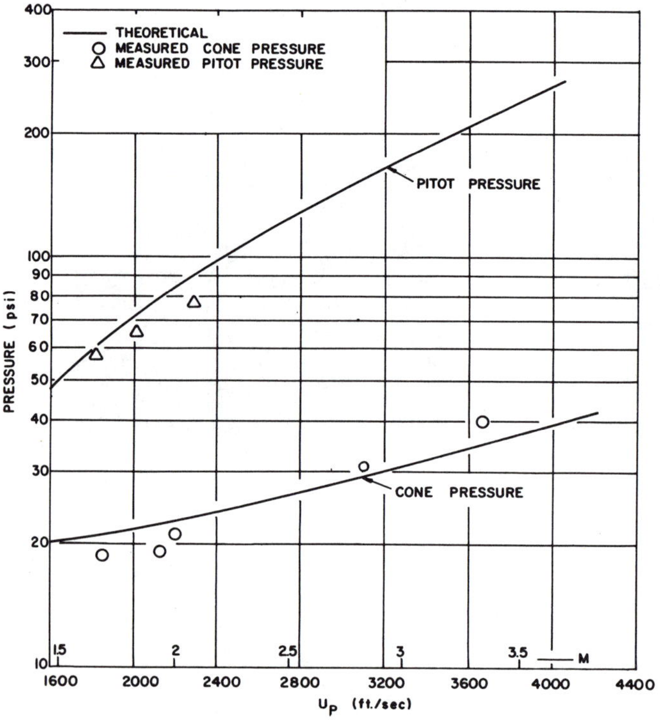

Figure 6
Test data

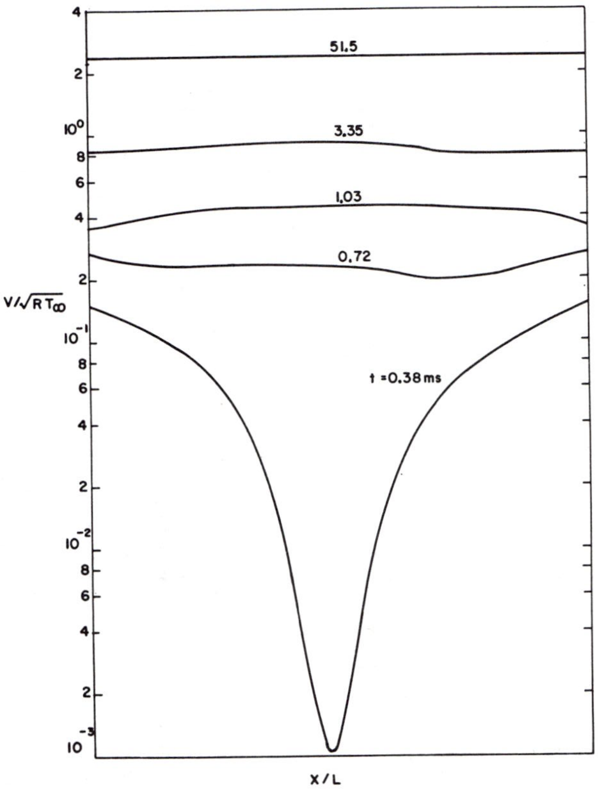

Figure 7
Theoretical velocity distribution
in capsule

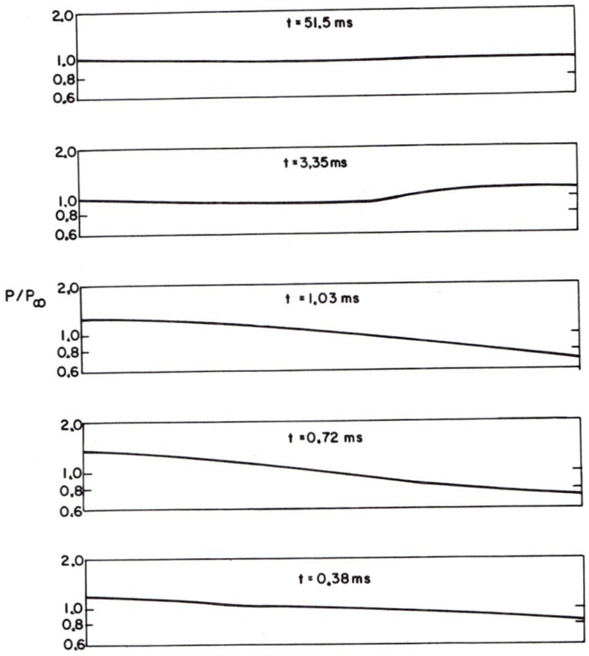

Figure 8
Theoretical pressure distribution
in capsule

AN INVESTIGATION OF A DOUBLE DIAPHRAGM SHOCK TUBE WITH A DETONABLE BUFFER GAS

H. L. GIER
University of Colorado
Boulder, Colorado
Presently Univ. of Wyoming
Laramie, Wyoming

THOMAS G. JONES
University of Colorado
Boulder, Colorado
Presently Boeing Scientific
Research Laboratories
Seattle, Washington

SUMMARY

The double diaphragm shock tube with a hydrogen-oxygen mixture in the buffer section is shown both analytically and experimentally to produce higher Mach numbers than a cold hydrogen driven shock tube. Detonation on shock reflection from the second diaphragm is shown to produce higher Mach numbers in the test gas than detonation on rupture of the first diaphragm. Analysis shows that the Mach number attainable increases with buffer gas pressure and performance is optimum for buffer gas mixtures between $3H_2 + O_2$ and $5H_2 + O_2$.

Successful tests were run with hydrogen to oxygen mixture ratios of 3 and 5 and with buffer gas pressures from 100 to 750 mm.Hg. Mach numbers of over 15 in argon at 5 mm.Hg. pressure were obtained with driver pressures of less than 50 atm. Testing time and attenuation are comparable to those observed in standard shock tubes and the Mach numbers produced are 20 per cent higher for the same driver pressure.

INTRODUCTION

In recent years several modifications, directed at increasing the incident Mach number, of conventional pressure driven shock tubes have been described.[1, 2, 3, 4] The most popular experimental arrangements employ heated drivers, multiple diaphragms, combustion drivers, variable cross-sections, or various combinations of these. Significant increases in Mach number, however, can usually be realized only at the expense of massive and sometimes expensive modifications of existing facilities.

A relatively simple and inexpensive device, in which a detonable gas mixture is used in a section of the shock tube situated between the driver and the test gas, has been used by Coates and Gaydon.[5] Because of the ease with which this idea can be incorporated into a conventional shock tube, it is felt that a further study of its operation at higher initial pressures and different mixture compositions is desirable.

The work by Coates and Gaydon[5] on the double diaphragm detonation shock tube treated initial conditions for a given driver pressure and the overall shock tube performance. The present study includes a much wider range of initial conditions and examines the conditions for the detonation on reflection from the second diaphragm which will produce satisfactory test conditions. The equipment used in the present study is shown in Figure 1.

ANALYTIC TECHNIQUES

The analytic results were obtained by programming the idealized problem on the University of Colorado Graduate Computing Center's CDC 6400 computer. This allowed the solution of many more cases of the problem than would have been possible by hand calculations.

The problem separated easily into three sections for purposes of calculation. The first was the breaking of the first diaphragm and the resulting shock and expansion waves. The second was the reflection of the initial shock wave from the second diaphragm, and the third that of breaking the second diaphragm with the expansion of the gases from behind the reflected shock and the formation of the shock in the test gas.

All gas properties used in the program were taken from the JANAF tables[7]. The thermodynamic properties of He, A, H_2, O_2, H, O, H_2O, OH, and N_2 were stored in tabular form and linear interpolation was used to obtain intermediate values.

The input data included the pressure, temperature, and the mole fractions of each gas present in each section of the shock tube. Section number designations are shown in Figure 2. Except for comparison purposes, the initial conditions were limited to a hydrogen driver (4), a hydrogen-oxygen mixture in the buffer chamber (5), and argon in the test section (1). All of these were initially at room temperature.

The first problem, which was the solution for the conditions resulting from the breaking of the first diaphragm, existed because the shock strength is an implicit function of the initial pressure ratio across the diaphragm. A simple iteration was therefore necessary to match the velocities and pressures at the contact surface. All pertinent quantities were then calculated from either the perfect gas expansion or the normal shock relations.

Calculation of the properties of the detonation on reflection was performed from the conservation equations using the properties of the gas behind the initial shock wave as determined in the previous section. The values of both the detonation velocity and heat release were assumed and a double iteration was performed. The gas was assumed to reach equilibrium behind the detonation and to have zero velocity relative to the shock tube.

It is obvious that the above computations produce waves which satisfy the conservation equations but do not necessarily produce a Chapman-Jouquet detonation on the reflection from the second diaphragm. However, after two computations at the same buffer gas pressure and composition but different driver pressures, it was possible to write a $(\Delta v_7 / \Delta p_4)_{p_5,\, m_{i5}}$ term with which the driver pressure required to produce a C-J detonation could be calculated. Seldom were more than four pressures required to be run before the C-J condition, $v_7 = a_7 = D$, was obtained. The velocity of the gas in region seven (Fig. 2) relative to the detonation front, v_7; the speed of sound behind the detonation, a_7; and the detonation velocity relative to the shock tube D; must be equal to match the boundary and C-J conditions.

The breaking of the second diaphragm was computed in exactly the same manner as for the first diaphragm. This resulted in the dual assumption that the gas expanding from behind the reflected wave was frozen and that the test gas remained perfect.

EXPERIMENTAL APPARATUS AND TECHNIQUE

The shock tube in which these experiments were performed was assembled from five and ten foot long sections of type 304 stainless steel tube with a 3.070 inch bore. 0-rings were mounted between the tube ends to insure a vacuum seal. The driver section was built of a five foot length of 3/4 inch wall mild steel. The section including the schlieren window and second diaphragm holder was machined from a piece of 304 stainless steel, 14 inches long, with a wall thickness of 3/4 inch.

The gases used in the experiments were all commercially available grades of hydrogen, oxygen, and argon. Mixtures of these gases were prepared at least 12 hours before use to insure proper homogeniety. The mixing was done by partial pressures in high pressure gas bottles which were first evacuated to a pressure of less than 20 microns and then filled to a total pressure of 800 mm. Hg. The attachment of the storage and mixing bottles to the shock tube is shown in Figure 1.

Several materials were used for the second diaphragm. Mylar was used on only a few tests because of a tendency to cause pitting on the aluminum end plate of the shock tube. Soft aluminum sheet of

.012 thickness and scored in an X pattern was also tried in some experiments but the petals always separated with resulting damage in the test section. Most tests were run using Reynolds Heavy Duty aluminum foil which had a thickness of .001 inch. Several layers were used for higher initial pressures. The aluminum foil always melted and left a fairly easily removable deposit on the shock tube walls.

Schlieren streak photographs were taken of the shock wave trajectories upstream of the second diaphragm position. The shock tube windows were made of $\frac{1}{2}$ inch thick optical glass set into the wall of the shock tube. The camera used was an AVCO Type MC300 Model 2 rotating drum camera. The camera mirror rotational speed was measured on a Tektronix 555 or 585A oscilloscope as the picture was being taken. The light source was an E. G. & G. Xenon flash tube operated with an energy input of 32 Joules and a light intensity remaining constant to within 10 per cent of the peak value for 220 μsec. Pictures were taken through a .025 inch wide slit.

Pressure and shock wave data on the reflected shock wave and the detonation on reflection were obtained by replacing the second diaphragm with a plug type end wall to bring the reflecting surface within the optical window. The end wall was designed to mount either a Baganoff type or a piezoelectric pressure gauge.

The piezoelectric pressure bar gauge as described by Edwards, et al[8] was modified and used to get a longer time of valid pressure measurement on the end wall than was possible with the Baganoff gauge. The composition of the bar gauge was aluminum-X cut quartz-lead of $\frac{1}{4}$ inch diameter with the quartz at $x/d = 3.5$.

Platinum film gauges were used both upstream of the second diaphragm position and in the test section to measure shock front arrival times. The gauges were formed by applying platinum paint to the end of a 4 mm. quartz rod which had been ground and fire polished; the rod was then baked at 600-800°C. These gauges have raise times of less than one μsec. Lifetime of these film gauges is adequate even under detonation conditions. Two of these gauges were mounted in the window section of the shock tube, flush with the inner wall and 22.58 cm. apart; the first was used to trigger the instrumentation and the time elapsed until the second responded provided a check on the photographic value of the incident Mach number. Two more platinum film gauges were mounted in the test section at distances of 162 cm. and 297 cm. from the second diaphragm. The output of these were recorded on synchronized sweeps of a Tektronix 555 oscilloscope to give the test gas shock Mach number.

Another method used to measure the shock velocity was by microwave reflection. The microwave apparatus was the same as that used by Glick and Jones[9] except for the antenna. The guide wavelength for the shock tube was 20.84 cm. for the microwave frequency of 2671 megacycles. In the Mach number range of the experiments the argon will be partially ionized and the reflected microwave signal undergoes one cycle in the time that the plasma moves half of a guide wavelength, 10.42 cm. The guide wavelength was checked by moving a short along the shock tube and agreed with the calculated value to better than .01 cm. The antenna used was a straight dipole which could be moved along the axis of the tube to obtain a maximum response amplitude.

Simultaneously with the microwave measurements the testing time was measured. The piezoelectric gauge described above was mounted in the side wall of the test section at a distance of 137 cm. from the diaphragm. A quartz rod, 4 mm. diameter by 18 mm. long, was mounted in the wall at 145 cm. A photoelectric tube was mounted to observe this window through a pinhole which was 5 cm. from the window; thus achieving some collimation of the light. The sole light source in the shock region was assumed to be the hot gas between the shock and the contact surface. Because there was some photo scattering effect both ahead of and behind the shock, the shock arrival time was determined by the pressure gauge and the test time was judged to be the time elapsed until the light intensity dropped to the intensity which was measured at shock arrival.

RESULTS

The first segment of this experiment was the analytical work to calculate the properties behind the detonation on reflection and the test gas conditions. The experiment confirmed the assumptions involved in the calculations by measuring the velocities of the shock waves incident to the second diaphragm, reflected from the second diaphragm, and in the test gas; measuring the pressure of the gas behind the detonation; and checking on usable initial conditions, testing time, and shock attenuation.

The performance of a double diaphragm shock tube with detonation occurring on rupture of the first diaphragm, mentioned by Coates and Gaydon[5], was calculated for comparison purposes. The calculated performance was little better than that of a standard shock tube. The x-t diagram for the double diaphragm shock tube with detonation on reflection from the second diaphragm (hereafter abbreviated DDDST) shows the parallel reflection and expansion waves from the second diaphragm in Figure 2.

The overall performance of the detonation tubes is shown in Figure 3 together with that of hydrogen driven shock tubes with driver gas temperatures equal to, and four times the test gas temperature. In addition the performance of the detonation driven tube of Lee[6] is shown to be by far the most efficient. The DDDST is much more efficient than the standard shock tube and is comparable to the heated hydrogen driver shock tube. The curve for the DDDST also shows that there is little difference caused by a factor of ten variation in the buffer gas pressure.

The first two steps of the calculations led to the conditions for a C-J detonation on reflection of the shock wave from the second diaphragm or an end wall. The theoretical pressure ratio across the first diaphragm is shown in Figure 4. Because the breaking of the diaphragm is never without losses, the incident Mach number needed for a C-J detonation on reflection is also shown in Figure 4. The ratio of specific heats and the pressure behind the reflected wave are shown in Figure 4. Due to the energy addition in the detonation the recovery pressure is greater than the driver pressure.

It would be expected that there might be an optimum mixture ratio in the buffer gas due to the decrease in both temperature and molecular weight and the increase in specific heat ratio of the combustion products with an increase in percentage of hydrogen in the initial gas. The optimum mixture ratio for pressure ratios from 100 to 50,000 is plotted in Figure 5. The shaded area shows how widely the mixture ratio can vary without causing a detectable decrease in the shock Mach number. Because of the small variation with mixture ratio no attempt was made to verify the optimum mixture ratio experimentally.

One condition for the operation of the detonation on reflection shock tube is that the shock wave in the combustion mixture must be slow enough that the mixture does not detonate before reflection and fast enough that ignition is virtually instantaneous on reflection. Fortunately, the calculated Mach numbers fall within the range in which the desired effect is possible. Schott and Kinsey[10] reported a linear dependance of $\log_{10}([O_2]t_i)$ on 1000/T so that an increase in M_5 will decrease the induction time, t_i, by increasing both the oxygen molal concentration, $[O_2]$, and the temperature behind the reflected wave. They also report the explosion peninsula near 800° K; therefore, the incident Mach number should be kept below about 3.2.

Most experimental tests were made with $3H_2 + O_2$ and $5H_2 + O_2$ mixtures and 100 mm. Hg pressure in the inner chamber. Additional tests were made with the same mixtures and buffer chamber pressures of 250, 500, and 750 mm. Hg which was the highest pressure possible in the apparatus.

For the mixtures and pressure ratios used, detonation of the incident wave before reflection was very rare. The detonations on diaphragm rupture which did occur were at 500 and 750 mm. Hg in $3H_2 + O_2$ and at 750 mm. Hg in $5H_2 + O_2$ and constituted only a few of the tests run at these conditions. Based on this experience the shock tube may be operated at buffer gas pressures up to at least one atmosphere, without special valves or buffer gases, if normal care is taken.

The measurement of the velocities of the incident shock wave, the reflected shock wave, and the detonation wave was performed from the schlieren streak photographs as shown in Figure 6. Time increases to the right and the end wall is at the bottom in the pictures. In Figure 6a the incident Mach number is low enough that the induction time is greater than 20 μsec. The corresponding pressure reading in Figure 6d shows an initial pressure rise to the reflected shock wave pressure and then a wildly fluctuating pressure after the detonation waves reach the end wall. The detonation starts in the gas from what is apparently a deflagration. Figure 6b shows a higher incident Mach number with a correspondingly shorter induction time. In these cases the detonation appears to start at the wall. The pressure fluctuations in Figure 6e are much reduced from the longer induction time condition of Figure 6c. Figure 6c shows pictures for a relatively high incident Mach number and very short induction time. The corresponding pressure trace in Figure 6f shows an initial pressure peak and a subsequent fairly steady pressure. Figure 6f shows the conditions at which it was later attempted to operate the double diaphragm shock tube.

The pressures measured behind the reflected waves are shown in Figure 7 for a gas mixture of $3H_2 + O_2$. The solid symbols designate the reflected shock pressure whenever the induction time was long enough to make a measurement. The bars on the detonation pressures indicate the maximum and minimum pressures in approximately 100 sec after detonation. On this figure the fluctuations are large at low incident Mach number and decrease at about $M_5 = 2.9$ to a much lower level. Although this was a higher initial Mach number than was calculated for a C-J detonation on reflection, it was judged better to operate the shock tube at conditions which would minimize the disturbances in the test gas than at the calculated C-J conditions.

The detonation on reflection from an aluminum foil second diaphragm was seen to occur much as it did from the flat end wall despite the curved surface and the short time of existence of the diaphragm. The diaphragm rupture was observed to be on the order of 50 μsec which is an order of magnitude longer than the induction time.

The experimentally measured performance of the DDDST is given in Figure 8. The pressure probe data was obtained by reading the elapsed time from the initial deflection of the microwave curve to the pressure probe initial response. This was the first 137 cm. of the test section. The platinum film gauge data was taken starting at 162 cm. from the diaphragm. The difference in measuring locations tend to give higher Mach numbers for the pressure probe data. No distinction is made between different pressures or mixtures in the buffer section because no trends were discernible. The major cause for the Mach numbers being lower than predicted was a rather severe loss in effective pressure at the first diaphragm. The agreement of data and theory in Figure 8 is satisfactory when the large degree of ionization which exists at the higher Mach numbers is considered.

A representative trace of the side wall mounted pressure probe and the test gas light emission is shown in Figure 9a. The pressure is seen to be steady throughout the test period as determined by the period of light emission. The testing time measured is shown in Figure 9b. The method used to measure the testing time was not intended to give exact values, but the experimental data obtained shows a good approximation to the value given by Roshko[11] of one-half of the simple calculated testing time for an initial pressure of 5 mm. Hg.

Since the microwave output exhibits a maximum for every 10.42 cm. of shock wave travel, it is possible to estimate the wave attenuation to a high degree of accuracy. It was noticed that for all high pressure ratio shocks the microwave signal indicated that the shock wave was arriving at the pressure probe position long before the pressure probe recorded the shock. From this data it was concluded that a precursor wave of electrons existed for shock waves with an initial Mach number greater than about 11. Figure 10 shows the velocity variation along the shock tube for runs which gave no evidence of a precursor wave; i.e. the microwave and pressure probe both show the shock wave to be at the same point at the same time.

CONCLUSIONS

The double diaphragm shock tube with detonation on reflection from the second diaphragm promises performance on par with a hydrogen driver heated to four times the test gas temperature. Only the detonation shock tube proposed by Lee[6] gives a better performance and that advantage disappears when the maximum shock tube pressure is considered.

Although the analytic work showed that the shock tube would perform better at higher buffer gas pressures and mixture ratios ranging from $3H_2 + O_2$ to $5H_2 + O_2$, the variations were so small that no attempt was made to verify these trends experimentally.

The combined streak schlieren photographs and pressure measurements in Figure 6 show that for the pressure to be reasonably constant behind the detonation the induction time for the reaction must be less than about 5 μsec. The incident Mach number must be kept above 2.9 for an initial pressure of 100 mm. Hg to keep the pressure fluctuations to a minimum. At higher initial pressures the initial Mach number could be reduced because of the inverse dependance of induction time on the molal concentration of oxygen.

Overall performance of the DDDST was about 20 per cent below the calculated value as seen in Figure 8. This was still 20 per cent better than theoretical for a standard shock tube. About half of the discrepancy was due to losses from the first diaphragm. Therefore, the shock tube performs much as it would be expected to if ionization were taken into account. Resler, Lin, and Kantrowitz[3] predict a degree of ionization of 0.12 for an initial pressure of 10 mm. Hg and a shock Mach number of 15 which is obtainable with the present apparatus.

In general the shock tube seems to bear out its promise of better performance than a standard shock tube with small increases in operating difficulty. The aluminum contamination of the test section and instrumentation may prove to be the major limiting factor; particularly for operations which need ultra-clean conditions.

ACKNOWLEDGEMENTS

The work is based on a thesis submitted by one of us (H. L. G.) in partial fulfillment of the requirements for the degree of Doctor of Philosophy at the University of Colorado and was supported by the Advanced Research Projects Agency under Contract No. DA 31-124-ARO-D-139.

Our thanks to Mr. R. O. Bartsch, Mr. V. R. Buonadonna, and the Natural Resources Research Institute, University of Wyoming for their assistance.

REFERENCES

1. Wurster, W. H., Glick, H. S. and Treanor, C. E.; Cornell Aeronautical Lab. Report. QM-987-A-1 (1957).
2. Henshall, B. D.; The Use of Multiple Diaphragms in Shock Tubes, A. R. C. 18, 062, N. P. L. (1955).
3. Resler, E. L., Lin,S. C., and Kantrowitz, A.; J. Appl. Phys. 23, 1390 (1952).
4. Alpher, R. A., and White, D. R.; J. Fluid Mechanics, 3, 457 (1958).
5. Coates, P. B. and Gaydon, A. G.; Proc. Roy. Soc. A 283 18, (1965).
6. Lee, B. H. K.; AIAA Journal, 5, 791 (1967).
7. Stull, D. R., et al; JANAF Thermochemical Tables, U. S. Dept. of Commerce, PB 168470, (1965).
8. Edwards, D. H., Davies, L., and Lawrence, T. R.; J. Sci. Instrum., 41, 609, (1964).
9. Glick, H. S. and Jones, T. G.; AIAA Journal, 7, 4 (1969).
10. Schott, G. L. and Kinsey, J. L.; J. of Chem. Phys., 29, 1177, (1958).
11. Roshko, A.; Phys. of Fluids, 3, 835, (1960).

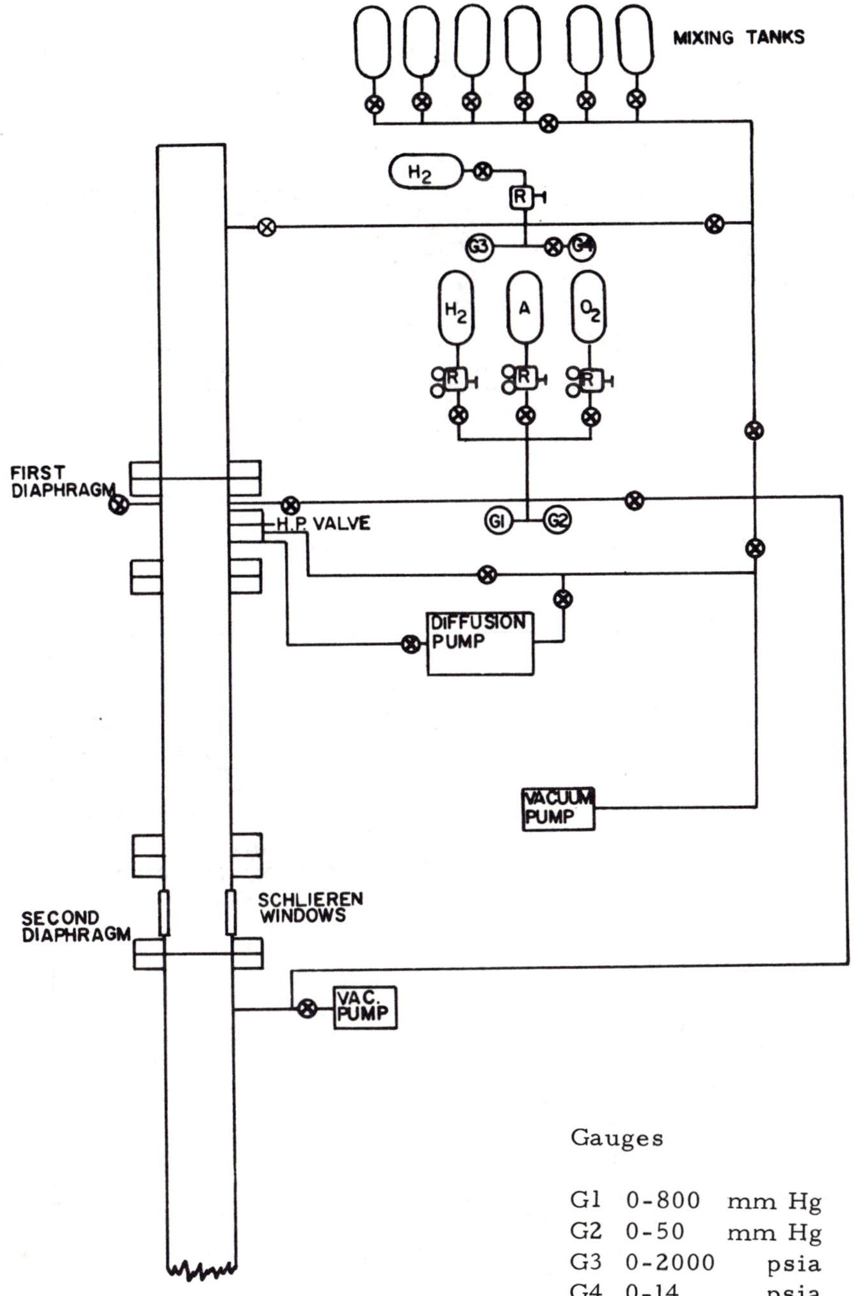

Gauges

G1	0-800	mm Hg
G2	0-50	mm Hg
G3	0-2000	psia
G4	0-14	psia

Figure 1
Shock tube schematic

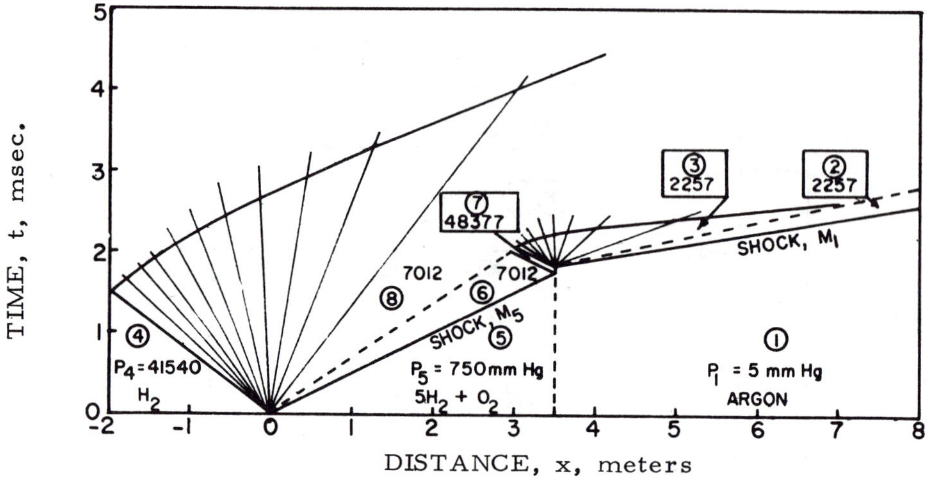

Figure 2
x-t diagram for a double diaphragm shock tube with detonation on reflection from the second diaphragm.

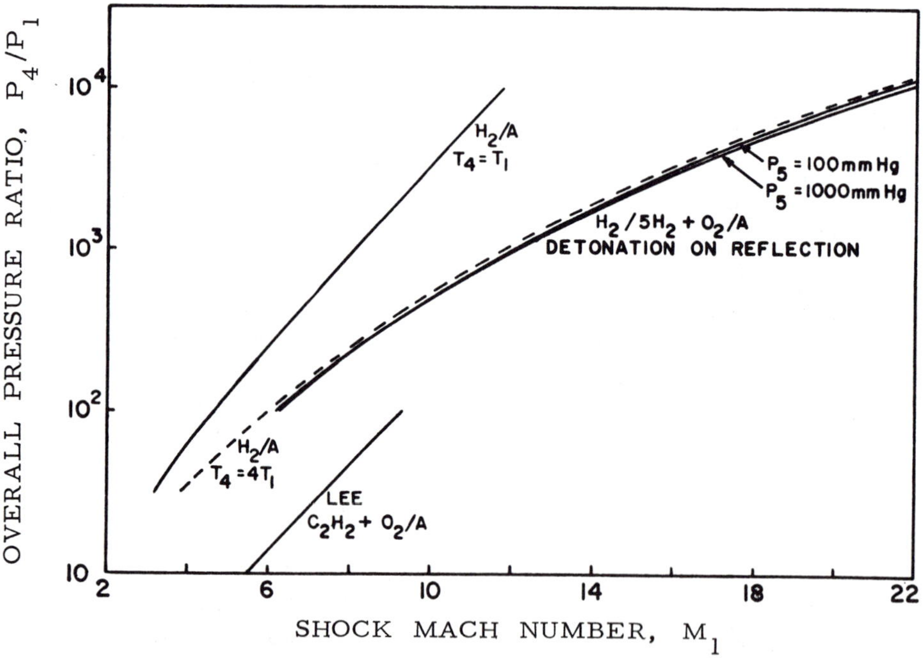

Figure 3
Driver to test gas pressure ratio as a function of Mach Number for various shock tubes

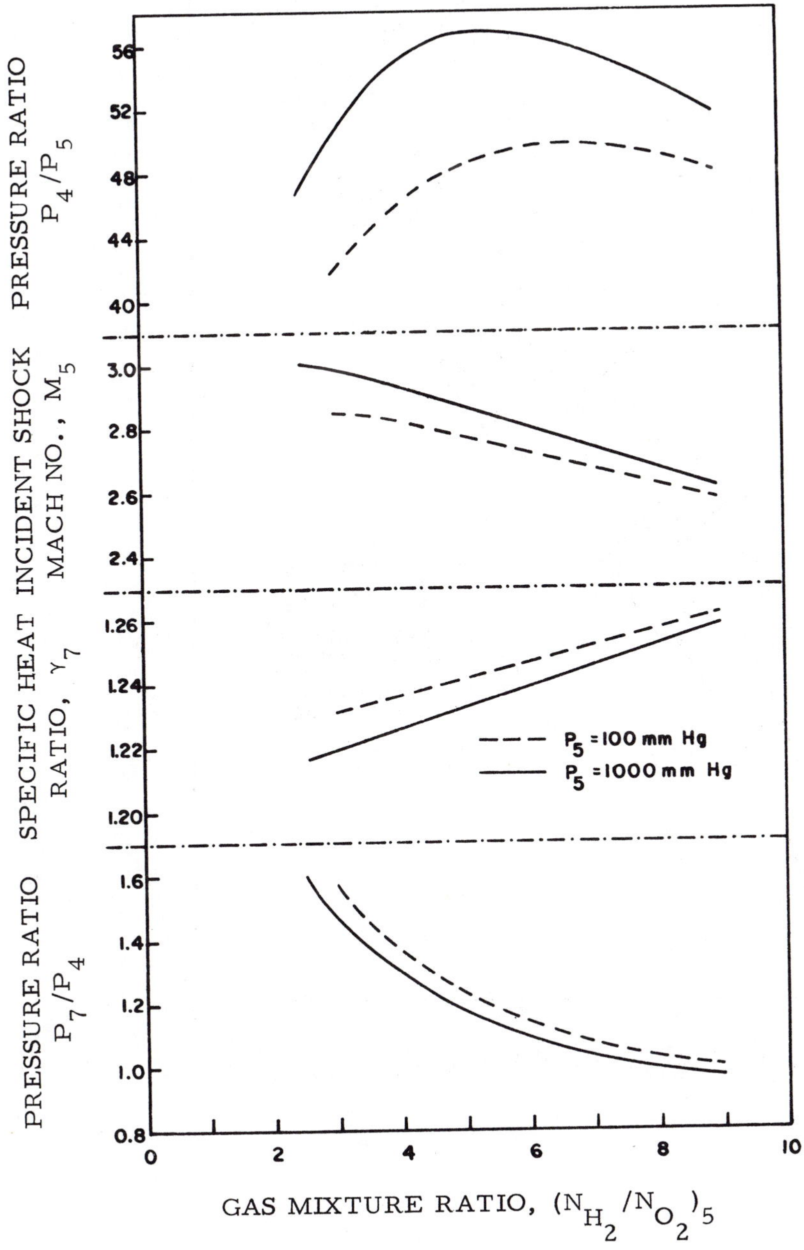

Figure 4
Shock tube conditions for a C-J detonation on reflection

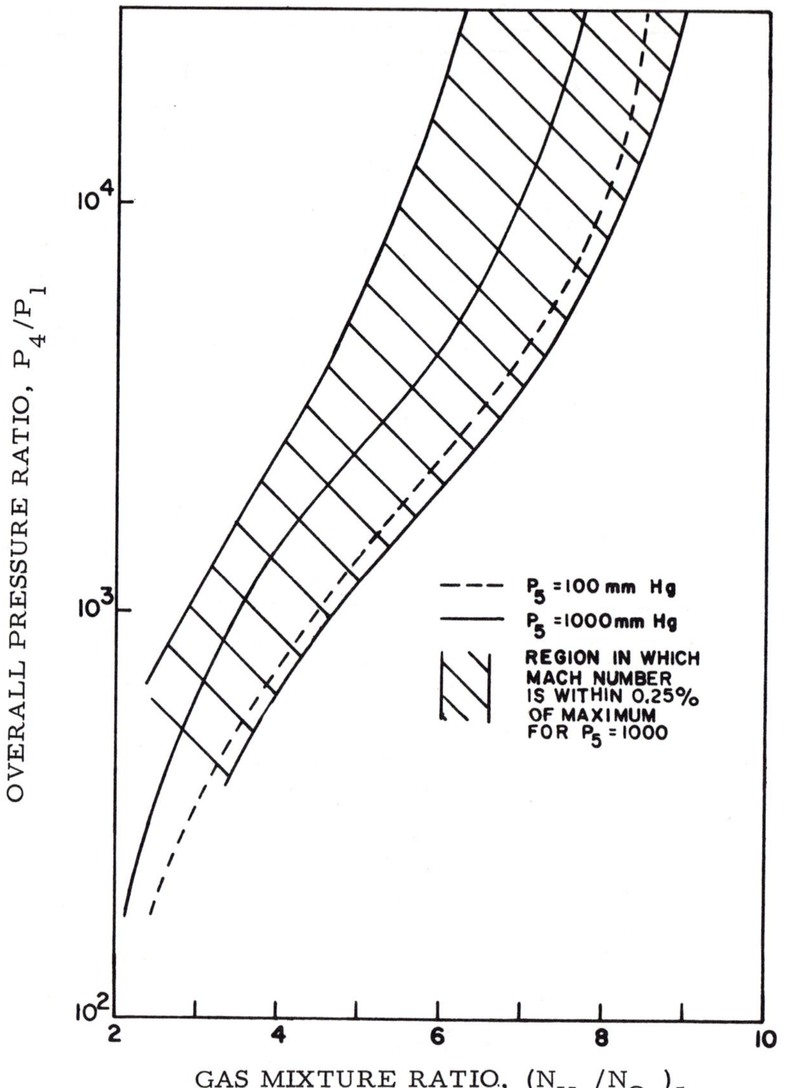

Figure 5
Optimum buffer gas mixture ratio for high test gas Mach No. as a function of overall pressure ratio.

a. Incident Mach No. $M_5 = 2.55$

b. Incident Mach No. $M_5 = 2.81$

c. Incident Mach No. $M_5 = 3.10$

Figure 6
Schlieren streak photographs of shock wave reflection and detonation from the shock tube end wall; gas: $3H_2 + O_2$; pressure: $P_5 = 100$ mm Hg.

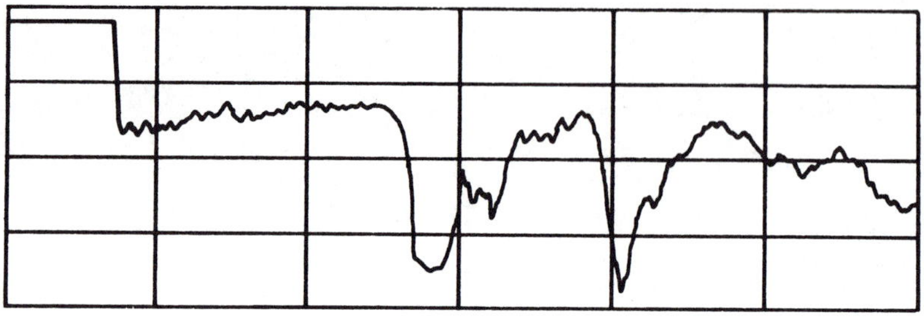

d. Trace for Figure 6a.

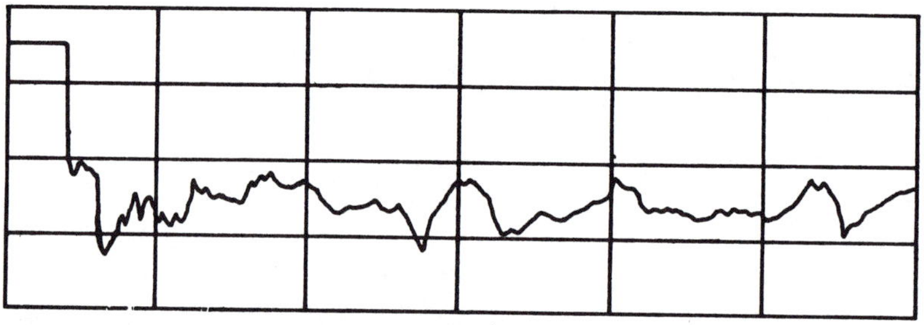

e. Trace for Figure 6b.

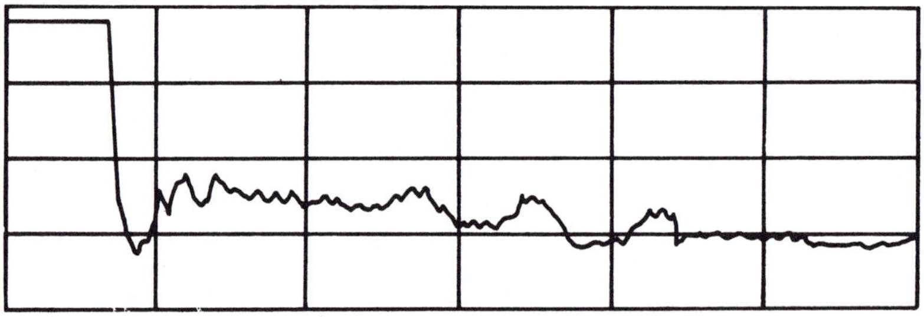

f. Trace for Figure 6c.

Figure 6
Oscillogram tracings of end wall pressures.
Piezoelectric gauge sweep expanded 2:1; 20 μs/div. ;
60 psi/cm.

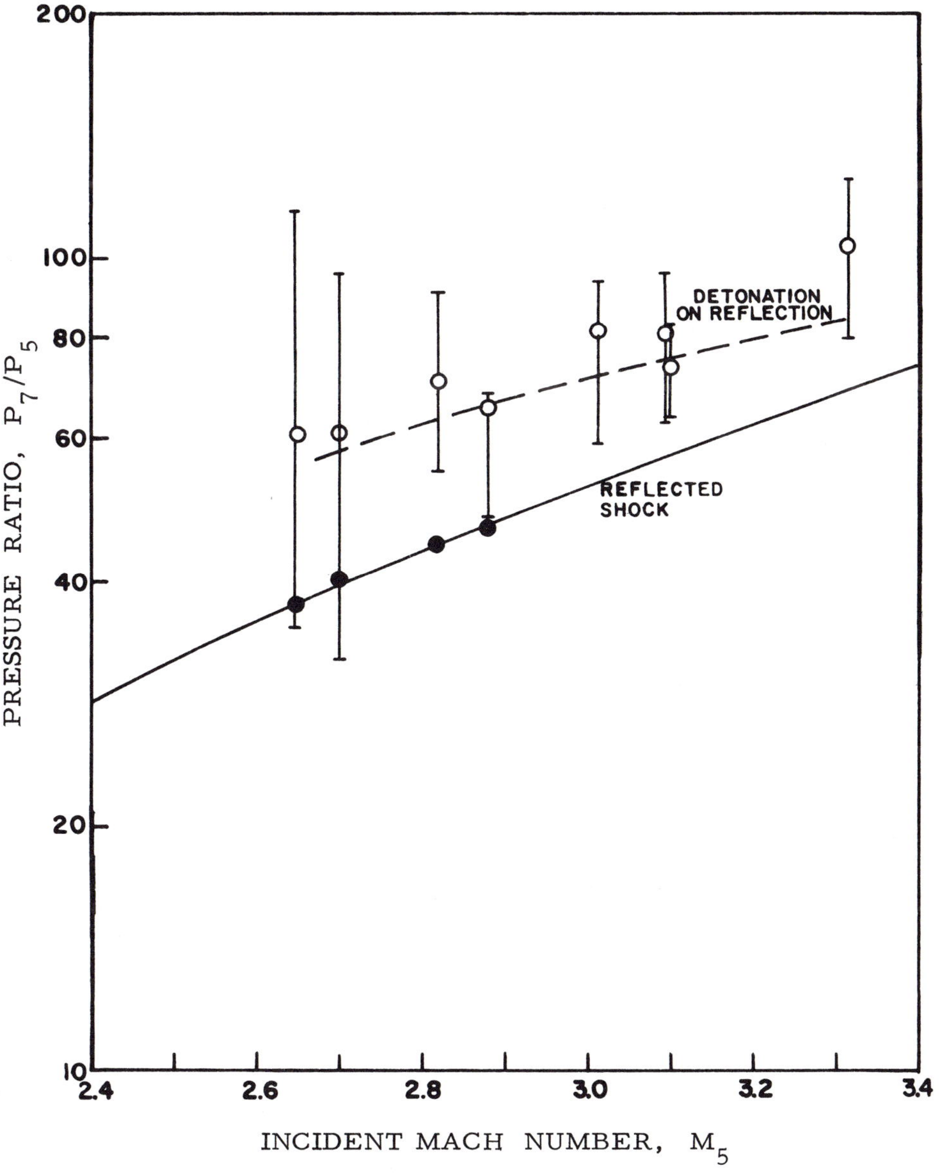

Figure 7
Pressures on the shock tube end wall from the reflected shock wave and the detonation on reflection in a $3H_2 + O_2$ mixture. P_5 =100 mm Hg

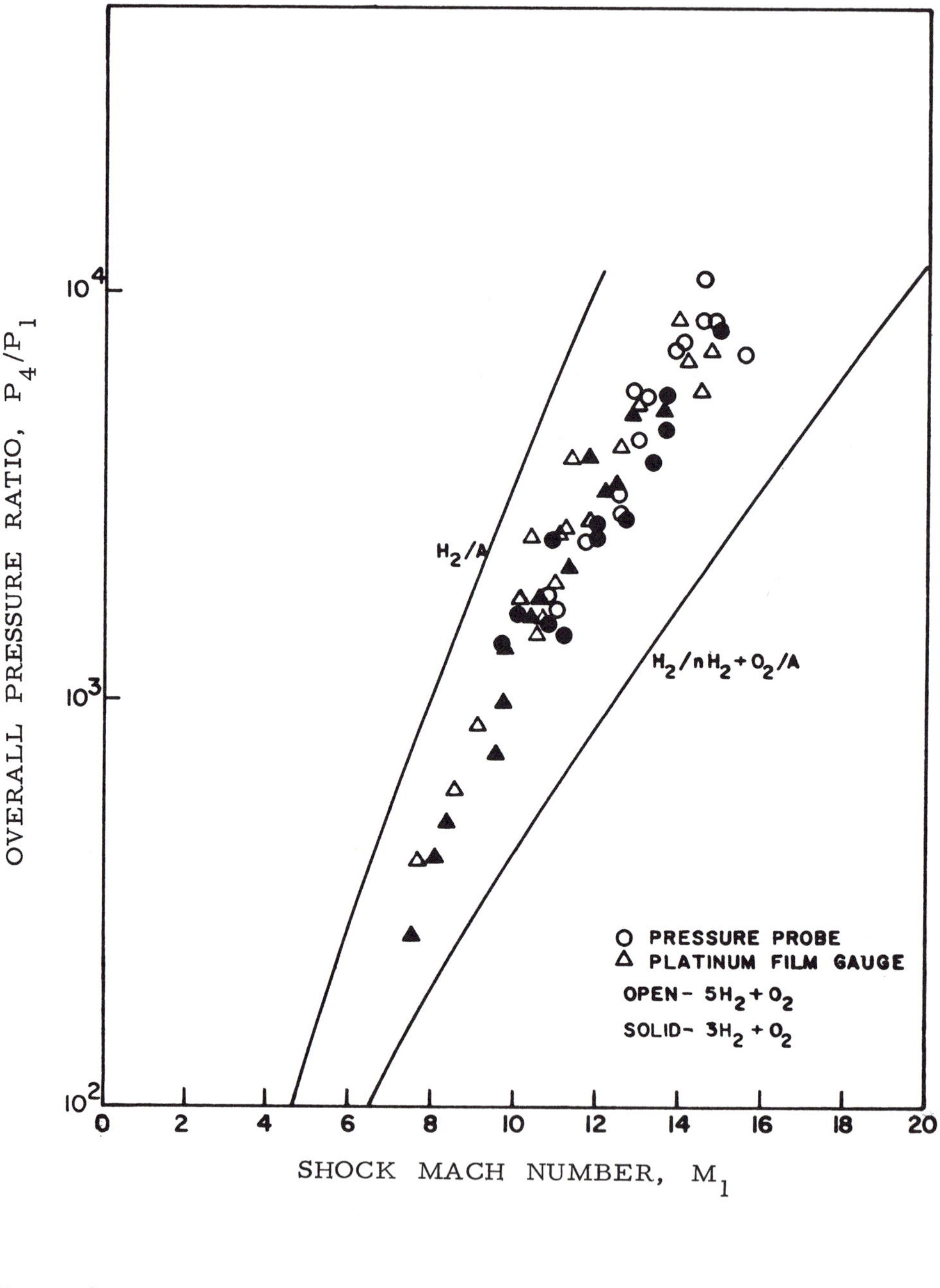

Figure 8
Comparison of experimental and analytical performance of the DDDST.

Pressure gauge

Light intensity

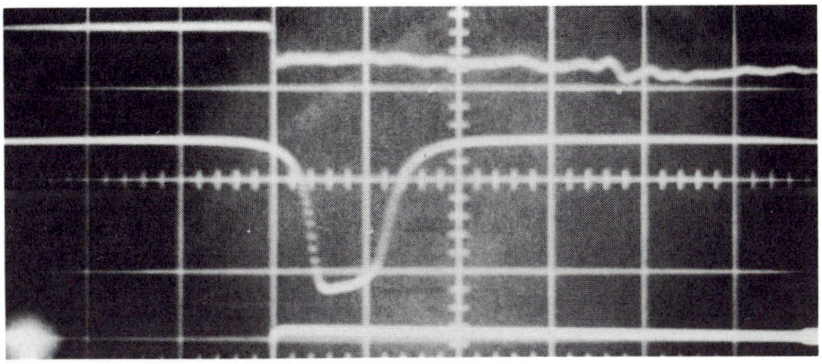

Figure 9a
Sidewall pressure probe and photoelectric tube output. Initial shock tube conditions: $H_2/3H_2 + O_2/A$, 14920/250/6.1; $M_1 = 12.1$.

Pressure sensitivity 6000 mm Hg/cm.

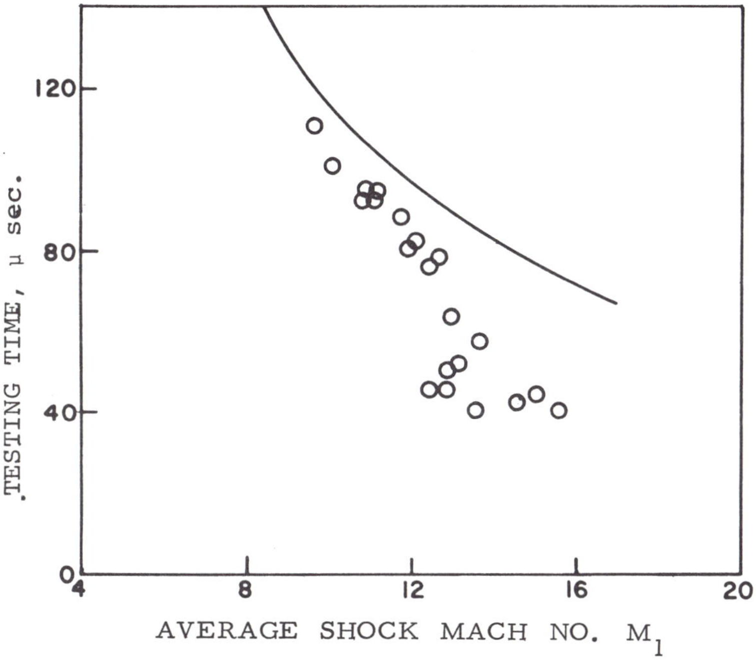

Figure 9b
Comparison of measured and simple theory testing time.

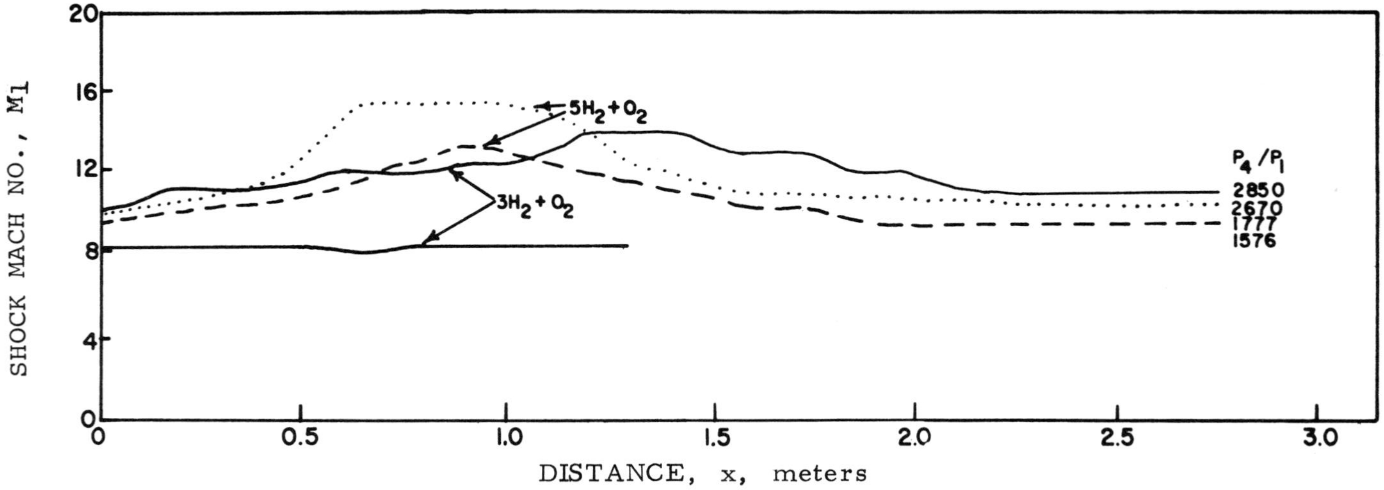

Figure 10
Shock wave velocity with no apparent precursor wave.

HIGH EXPLOSIVE DRIVERS FOR SHOCK TUBES

RUSSELL E. DUFF
Systems, Science and Software
La Jolla, California

SUMMARY

The author of an introductory paper should try to cover his subject broadly without anticipating the original contributions to be made by the following speakers. This author has attempted to bring together work done over many years on a spectrum of subjects, some of which may seem unfamiliar to you. The choice of examples is a personal one. Without doubt this selection has ignored much important research which the serious student of this technology will want to investigate. Nevertheless, I hope this discussion will broaden our view of shock tubes and help to weave a unifying thread through several areas of physics, chemistry and engineering.

High explosive has been used in several quite different ways to drive shock waves in facilities that can be called shock tubes. Three main objectives can be identified in the various investigations. Included are more or less basic studies of gases, similar studies of condensed phases, and the development of blast simulation facilities.

Before we proceed to discussions of specific technical work, it would be well to define a shock tube. For present purposes, a shock tube will be considered as a facility in which a basically one-dimensional flow is established involving a shock wave driven by a piston. This piston may be either a discreet, identifiable object which is forced into a fluid with known velocity, or it may be a fluid-dynamic flow generated by the well-known rarefaction process.

In the first case, it is a relatively simple task to describe a shock state to be found in front of a given piston. In the more interesting second case, the flow situation is usually described in the pressure-particle velocity plane as shown in Figure 1. In the conventional shock tube the two initial states in the expansion and compression chambers are indicated along the ordinate at different initial pressures and zero particle velocity. The locus of all states which can be reached from the low pressure state, the so-called shock Hugoniot, is indicated in the figure. The locus of all states which can be reached through a rarefaction wave from the upper state is also indicated. At the intersection of these two lines the particle velocity and the pressure in the two flows are equal, and this is a physically realizable flow situation. It is the shock tube solution, and the familiar Taub equation is nothing more than an analytical statement of the above construction.

If we now consider a shock tube situation in which a high explosive is used as a driver, we introduce a new element into this consideration as shown in Figure 2. Here, the high pressure state is no longer characterized by an initial velocity of zero. There may or may not be an intersection between the shock Hugoniot from a given initial state and a rarefaction from the explosive, Chapman-Jouget detonation state. It may be that the initial state of the explosive lies to the right of the shock Hugoniot; then a reflected shock wave must be sent back into the shock tube driver in order to develop an acceptable solution to the shock tube problem. This situation is also indicated in Figure 2. Thus, since a high explosive simultaneously generates a high pressure and a significant particle velocity, a shock tube driven by such an explosive can produce flow configurations in which the compression chamber flow is characterized by either a rarefaction wave as in the conventional shock tube or a shock wave as outlined above.

Another introductory comment should be made before we move on. High explosives are unique substances. They provide a very high energy density at a very modest cost. For instance, the energy of explosion of one gram of high explosive is of the order of 1,000 calories. The density of a good high explosive is 1.7 gm/cc, so one cubic centimeter of explosive contains approximately 7,000 joules of available energy. This energy is comparable to that which can be stored in a large capacitor costing as much as $2,000.00. The explosive cost is a fraction of a cent.

Explosive assemblies can be tailored to meet many specific requirements and the technology of safely working with such hazardous material is well developed. On the other hand, large quantities of explosive cannot ordinarily be used in the proverbial basement of the physics department. There are certain rather dramatic and disruptive effects associated with the detonation of explosives, and relatively elaborate facilities are usually required to assure their safe and effective use in research work.

One ordinarily considers a shock tube as a device in which a flow of gases is studied. However, the availability of high explosive allows us to generalize this concept to the study of fluids. Since the pressures generated by high explosives are measured in hundreds of thousands of atmospheres while the strengths of most common materials are only a few thousand atmospheres, any material may be considered a fluid at shock wave conditions characteristic of high explosives. We can therefore imagine a very simple shock tube to consist of a piece of high explosive in intimate contact with a piece of material, a metal for instance. When the explosive is detonated so that a plane wave is incident on the interface, the early part of the resulting flow is properly described as a shock tube flow. It is one-dimensional, it is described by the pressure-particle velocity argument presented above, and it is absolutely free of boundary layer peturbations which often influence conventional shock tube work. This independence of boundary disturbances is assumed by making the length to diameter ratio of the configuration so small that no disturbances reach the central area from the boundaries during the time of measurement.

Obviously, the advent of high explosive driven shock tubes which can be applied to the study of various characteristics of fluid and solid materials has strengthened and broadened the inter-disciplinary relationships between fluid mechanics, solids mechanics, and the physics of solids. This development has its counterpart in the better-known inter-disciplinary relationships between fluid mechanics and chemistry which have become so familiar to all of us in the last 15 years of shock tube research.

GENERAL CONSIDERATIONS

A brief consideration of the well-known conservation equations for mass, momentum and energy will be instructive.

$$\rho_o U_s = \rho(U_s - U_p)$$

$$p_o + \rho_o U_s^2 = p + \rho(U_s - U_p)^2$$

$$H_o + \frac{1}{2} U_s^2 = H + \frac{1}{2} (U_s - U_p)^2$$

where p, ρ, and H are the pressure, density and specific enthalpy, respectively, U_s and U_p are the shock and shock particle velocity, and the subscript o refers to conditions in front of the wave.

The usual next steps in the development of shock tube theory involve the introduction of the ideal gas relationship for H as a function of p and ρ and the elimination of the velocities from the three equations. One is left with the familiar Rankin-Hugoniot relation between pressure and compression.

It is instructive to take a more general view and manipulate the conservation equations without making any assumptions about the equation of state. It is easy to show that

$$p - p_o = \rho_o U_s U_p$$

and

$$\frac{\rho}{\rho_o} = \frac{U_s}{U_s - U_p} \quad .$$

These equations demonstrate that the measurements of shock wave and shock particle velocities are sufficient to determine the pressure, density and specific enthalpy of an arbitrary substance compressed by a shock wave. This is equation of state information.

The important question is, under what conditions can shock wave measurements contribute new information not available by more familiar techniques. Logarithmic differentiation of the equations shows that the percentage uncertainty in pressure or enthalpy is comparable to the uncertainties in the velocities, but

$$\frac{\delta\rho}{\rho} = \frac{U_p}{U_s - U_p}\left(\frac{\delta U_s}{U_s} + \frac{\delta U_p}{U_p}\right) = \left(\frac{\rho}{\rho_o} - 1\right)\left(\frac{\delta U_s}{U_s} + \frac{\delta U_p}{U_p}\right) \quad .$$

It is clear that the percentage error in compression is proportional to the percentage errors in the measurements of the two velocities multiplied by a coefficient which involves the compression across the shock wave. Thus, in the case of a gas, where compressions in strong shocks are of the order of 10, one must make extremely accurate measurements of shock and particle velocity in order to determine the compression to one percent. And, one percent is not a particularly interesting accuracy for many applications because the ideal gas law is an excellent first approximation. It would be incorrect to assume that shock measurements of this sort have no place in the study of gases, but that place must be carefully chosen as will be illustrated shortly.

On the other hand, in the case of solids and liquids, where the compression which can be produced with explosive pressures is rarely as high as two, the coefficient is less than one, and the percentage uncertainty in the measured density is comparable to the uncertainty in the measurements of the velocities.

The development of shock tube techniques for the study of condensed materials is of particular interest to the physicist because very important material properties can be determined simply by measuring the characteristics of the shock waves themselves. Elaborate special instrumentation from other technical domains is not required. The equations of state of condensed substances are poorly known, and measurements precise to a few percent make significant contributions to our knowledge.

However, material velocity is not easily measured directly. Fortunately, the compression produced in a conventional metal by pressures of several hundred thousand atmospheres is relatively small. Walsh and Christian[1] have shown that the shock particle velocity generated when the material is raised to such a shock pressure is very nearly identical to the particle velocity imparted by a rarefaction wave which returns the material to zero pressure. In other words, the velocity of the free surface of a substance upon which the shock is normally incident is, to a good approximation, twice the shock particle velocity. Therefore, the measurement of shock wave and free surface velocities makes possible the determination of pressure, compression and the specific internal enthalpy by simple use of the conservation equations.

This technique for determining equation of state parameters has been extensively used to investigate the properties of condensed substances at pressures from a relatively few kilobars up to 10 megabars by a number of investigators in many nations.

Now we turn from general to specific considerations. The remainder of this paper will present examples of high explosive shock tubes used either to study gases or condensed materials or to serve as blast simulation facilities.

EXPERIMENTS IN GASES

High explosive driven shock tubes have been used to study several features of the shock process in gases, including the development and characteristics of precursor signals, general radiation phenomenology and equations of state. Examples of these investigations follow.

Perhaps the most dramatic example of an explosive shock tube is also one of the earliest published. I refer to the 1954 paper of Shreffler and Christian[2] in which very prominent radiation-induced precursors were examined. In these experiments a 20 cm square by 5 cm thick block of composition B explosive was used to drive a gas filled 12.5 x 12.5 cm square by 40 cm long shock tube. The shock and precursor phenomena were observed by high speed framing cameras. A sketch of a typical configuration is shown in Figure 3.

An example of the self-luminous flow configuration observed in argon at an initial pressure of 0.8 atmosphere is shown in Figure 4. Note that prominent precursor signals are observed at the top and the bottom of the tube in front of the shock wave which is propagating from left to right at a velocity of 8 mm/μsec. Similar disturbances were found in other gases which became prominently luminous when driven by high explosives. Included are chlorine and argon-nitrogen, 50-50 mixtures. No boundary disturbances were observed in shock tubes filled with helium, air, or butane-protane mixtures. The precursor disturbance was observed in tubes as small as 1 inch in diameter. The disturbance was also found to propagate along wires or rods extending along the axis of the tube. The velocity and energy content of the precursor are independent of the axial material. Whether it be a copper wire, a threaded brass rod, a steel rod or a birch dowel makes very little difference. The energy associated with the precursor, as measured by its ability to penetrate a metal plate, is roughly comparable to that of a 0.22 cal long rifle bullet fired at close range. Note that Figure 4 also shows a luminous shock generated at the end wall in the shock tube before the main shock gets there. Clearly, this wave must be supported by radiation.

It was demonstrated that these prominent and very energetic precursor disturbances are associated with radiation emitted by the plane shock front. Their possible existence and importance in other configurations involving strong shocks should be a matter of concern for the investigator.

A simple explosive driven shock tube has also been used to generate extreme conditions in helium for the study of electron-induced energy level shifts. Seay, Seely, and Fowler[3] used the configuration shown in Figure 5 to produce reflected shock temperatures of 1.8 ev and electron densities of 10^{18}/cc in an initial helium pressure of approximately 25 torr. Note that there is no evidence of boundary disturbances of the type discussed above because the incident shock is not luminous.

Time-resolved spectrograms of the reflected shock region similar to the one shown in Figure 6 were used to determine the quadratic Stark shift of individual spectral lines. Both red and blue shifts were observed, and the measured shifts of individual lines lay between approximately 2 and 26 Å. This investigation demonstrated that the line shift produced by electrons can be a major part of the total Stark effect in helium.

The Soviet investigators, Voitenko, Model' and Samodelov,[4] have used a more elaborate shock tube system to study the apparent brightness temperature of shock waves driven into xenon or air. Their experimental configuration is sketched in Figure 7. Note that a block of solid explosive is used to propel a metal plate (piston) into a hemispherical chamber of hydrogen at an initial pressure of one atmosphere. This compressed and heated hydrogen breaks a diaphragm and expands into the shock tube. This apparatus can generate shock velocities as high as 4 centimeters per microsecond in air or xenon. The temperature of the shocked gas was estimated from the optical density of a photograph of the shock wave viewed through the end of the tube. Figure 8 shows the results. Note that for xenon, the apparent brightness temperature increases to a maximum of approximately 50,000°K and then falls as shock velocities increase. In the case of air, the observed temperature is below that predicted but does not show a well-defined maximum and subsequent decrease. The authors interpret these results as indicating the development of a luminous region in the gas ahead of the shock front. The gas absorbes soft x-rays radiated from the front and re-radiates the energy at a lower characteristic temperature.

High explosives have also been used to push pistons at various velocities into an atmosphere of nitrogen in order to quantitatively determine the equation of state of this gas. Figure 9 shows the results of Christian, Duff and Yarger.[5] Here, the two theoretical curves are predicted Hugoniot relations for this gas assuming the dissociation energy to be 7.4 or 9.7 ev. Note that the data are in excellent agreement with predictions based on the higher value of dissociation energy. This work was one of the contributions which settled the controversy concerning the binding energy of two nitrogen atoms which was finally resolved little more than ten years ago.

The shock density was determined by wave and particle velocity measurements as described above in spite of the fact that large compressions were observed. Clearly such measurements were significant only because 5 percent accuracy was adequate to resolve this uncertainty. Few instances have been found where such relatively coarse measurements could be used to answer significant questions.

EXPERIMENTS IN CONDENSED SUBSTANCES

As mentioned above, a unique and unfamiliar class of high explosive driven shock tubes can be used to investigate the thermodynamic properties of condensed substances. This possibility was recognized approximately 15 years ago in research centers around the world. For obvious reasons, most of these centers are connected with the atomic energy or the military research programs of various governments. Particularly worthy of note, are the pioneering investigations of Walsh[6] and his co-workers at the Los Alamos Scientific laboratory in the United States and Al'tshuler[7] and his colleagues in the Soviet Union.

The two main techniques which have been developed for equation of state measurements of solid and liquids are the electrical and optical techniques illustrated in Figure 10. In the one case, pins or contact switches are used to indicate the arrival of a shock wave or a free surface at a given place. In the other, this arrival is indicated by a flash of light generated in a thin layer of highly compressed argon or xenon. Ingeneous variations of these schemes have been developed by investigators to meet special requirements. For instance, the optical technique has been modified to permit simultaneous measurements to be made on a half dozen samples in a single experiment, and pin techniques can be used in geometric situations where optical visibility cannot be assured.

The early workers in this field found that the simple shock tube could be used at pressures up to approximately 500,000 atmospheres. At still higher pressures the explosive could be used to drive a metal plate or piston, which upon impact with a substance would produce pressures as high as two million atmospheres. Figure 11 shows an example of early equation of state measurements[8] on bismuth compared with static data by Bridgeman.

It is also possible to measure the equation of state of liquid substances by making more or less obvious variations of the experimental technique. Figure 12 shows the equation of state of water as deduced from dynamic shock tube measurements.[9]

In order to investigate still higher pressures more elaborate techniques must be employed. Al'tshuler has most extensively developed convergent high explosive systems to generate very high material velocities which can be used to generate very high shock pressures. Figure 13 illustrates equation of state measurements on copper and lead at pressures up to nine million atmospheres.[10] It is interesting to note that the very high pressure measurements seem to lie along a simple extrapolation of the low pressure measurements in the shock velocity-shock particle velocity plane.

A particularly interesting investigation by Krupnikov, Brazhnik, and Krupnikova[11] shows the shock compression of porous tungsten to go through a maximum at pressures of approximately 1 megabar and then to decrease at still higher pressures. This apparently anamalous result is quite consistent with the compression maximum to be found in dissociating gases.

It takes little imagination to realize that once a capability exists to measure thermodynamic variables at extreme conditions, interest would rapidly grow in other classes of measurements of interest to the solid state physicist. For example, it has been possible to augment the equation of state measurements with investigations of properties such as the electrical conductivity, magnetic properties, and the phase transitions which occur at very high pressures.

Mitchell and Keeler[12] have reported measurements of the electrical conductivity of carbon tetrachloride over the pressure range of 65 to 160 kilobars. The conductivity varies by more than 6 orders of magnitude over this range as shown in Figure 14. Such measurements are useful in determining changes in electronic configurations associated with significant compression of the fluid.

It is known that compression changes the magnetic properties of substances slightly. Royce[13] has investigated the shock induced magnetic effects in nickel ferrite. He made a thin sample part of a magnetic circuit. When the specimen was shocked to a high pressure, the change in saturation magnetization was measured as an induced voltage in coils surrounding the circuit caused by a change in magnetic flux in it. Figure 15 shows the measured "demagnetization" as a function of compression. Experimental data are the circled points and the solid lines are theoretical demagnetization predictions based upon either a constant Néel temperature for this material or a Néel temperature which increases at the rate of 1.16°C/kbar as indicated by static measurements.

It is interesting to note that there is no correlation between the experimental results and the theoretical predictions. This surprising result has been ascribed to the fact that shock compression of a solid substance is not purely hydrostatic. The elastic strength, though usually relatively small, does introduce a slight anisotropy to the compression. This anisotropy develops a crystal field which turns the magnetization of the ferrite from parallel to the shock front to perpendicular to it. Such a change is sensed in the external circuit as a demagnetization. In fact it is only a change in magnetic orientation. This explanation has been verified by further experiments in which the coercive strength of the magnetic field is increased by an order of magnitude. Under these circumstances the coercive field dominates that associated with crystallographic anisotropy, and the expected small demagnetization is all that is observed.

Many substances undergo phase transitions at high pressures. In fact, one of the more interesting classes of investigation under static high pressures has been the investigation of these transformations. Similarly, a number of phase transitions have been observed under dynamic conditions. The most noteworthy of these is the new iron phase at 130 kbar which was discovered by dynamic techniques.

It has become second nature for those of us interested in shock tubes to think of shock waves as stable compression processes. However, as pointed out many years ago by Bethe and Weyl[14], the condition for shock stability is that the second deriative of the pressure with respect to volume be positive. This condition is met for the great majority of substances in nature, and it is uniformly true for gases away from their critical points. At a phase transition, however, the isentrope of a substance can have a discontinuity in slope, and this discontinuity can cause violation of the stability condition. Then a shock wave between two states will break up into two waves. The first running at the higher speed carries material from the initial pressure up to the transition pressure. The second wave which continues the compression to the upper state is associated with the phase transformation process. The sharpness of the second front is related to the kinetics of the phase transformation. This situation is analgous to the well-known situation in shock waves in gases where excitation processes lead to domains of widely differing pressure gradient. The first and largest is associated with the equilibration of the translational and rotational degrees of freedom and the second, slower process, involves the kinetics of vibrational excitation or molecular dissociation.

Returning to solids, the pressure of the first shock wave is the phase transformation pressure. If the characteristics of this wave are measured in our high explosive shock tube, information about the transition has been determined. Figure 16 shows results from two laboratories, LASL and LRL, on the measurement of the phase transition pressure for several iron-nickel and iron-chromium-nickel alloys.[15]

You realize, of course, that these few examples of investigations made with high explosive driven shock tubes provide a most cursory introduction to the subject. They have been presented only as a brief indication of the kinds of work which can be done with these unique, but very powerful instruments.

BLAST SIMULATION FACILITIES

We have seen examples of the use of several kinds of high explosive driven shock tubes to study fundamental properties of materials in either the gas, liquid or solid phase, and also to investigate the influence of intense radiation fields upon shock phenomena and observables. These investigations may be characterized as rather fundamental ones in which basically scientific questions have been answered.

High-explosive-driven shock tubes can also be used as blast simulation facilities. Two examples will be given. Almost 20 years ago the need was recognized for a facility in which relatively low level but long duration blast pulses characteristic of distant explosions could be simulated. This need was met by a 2m diameter, 60m long shock tube driven by explosive. This shock tube has been used to investigate many problems of the interaction of blast waves with structures. The facility[16] is an almost conventional shock tube in the sense that high explosive, in this case primacord, is used to provide a volume of high pressure gas which can drive a shock in air. The explosive simply obviates the necessity of installing a diaphragm across the large shock tube, and it provides an inexpensive and easily varied source of energy to drive the facility.

The primacord charge for the tube is suspended along a 6m closed chamber at one end of the tube. A loading of 19 kg of PETN in primacord will produce a blast wave overpressure of about 7 atm, and a total positive phase duration of approximately 0.2 sec.

High explosives can also be used to simulate much stronger blasts by increasing the amount of material used and by shortening the overall length of the tube. Such a facility was described by Duff and Blackwell[17] at the Shock Tube Symposium three years ago. Figure 17 is a sketch of this facility. The 2m diameter steel pipe is set vertically into the ground and its length may vary between 3 and 15m. The bottom of the tube is filled with a low density high explosive, usually TNT. Objects to be tested are suspended in the mouth of the shock tube.

This facility also can best be described as a simple shock tube in which the high explosive detonation products act as a conventional driver. However, it differs from most in that the primary test fluid is the driver gas itself, and not the initial gas of the expansion chamber. Stagnation pressures in excess of 700 atm. have been measured about 10m from a 10 gram per square centimeter charge of low density TNT.

A spectacular demonstration of the performance of this explosive driven shock tube was made as shown in Figure 18. Here we see one frame from a motion picture record of an experiment in which an array of six, 80 kg steel spheres were located in a plane at the mouth of the tube. Each sphere carried a smoke flare so that its trajectory could be observed photographically. The six spheres can be seen near the horizon. The average apogee was 130m. This experiment proved that the impulse imparted to an object by the shock tube is primarily axial and quite uniform across the flow. It is interesting to note that two of the objects which flew the highest came from near the outer tube wall.

Perhaps this figure illustrates why workers at the Sandia Laboratory have dubbed this shock tube facility Thunderwell. Approximately 225 kg of explosive are used in a representative test. It follows that a particular shock tube can be used only once.

SUMMARY

The objective of this paper was to provide an introduction to high-explosive-driven shock tubes. I hope I have succeeded in this, and I hope you also recognize that this discussion is by no means the exhaustive or definitive statement of the use of high explosives as shock tube drivers. In fact, the speakers that follow in this session will talk of other high explosive shock tube applications which are quite different and distinct from anything which has been discussed so far. Nevertheless, I hope that several points have been adequately demonstrated. The very high pressure and specific energy density available in high explosives at an economical price makes these materials uniquely suited for shock tube applications where extreme conditions or large size are important considerations. It is clear that unique and interesting investigations of physical and chemical problems can be made with high-explosive-driven facilities. Finally, blast simulation facilities can be designed in which high explosives are used to simulate the amplitude and duration of various blast phenomena. Explosive facilities have been shown to be adaptable and economical for many applications.

REFERENCES

1. J.M. Walsh and R.H. Christian, Phys. Rev. 97, 1544 (1955).
2. R.H. Christian and R.G. Shreffler, J. Appl. Phys. 25, 324 (1954).
3. G.E. Seay, L.B. Seely, Jr., and R.G. Fowler, J. Appl. Phys. 32, 2439 (1961).
4. A.E. Voitenko, I.Sh. Model', and I.S. Samodelov, Dok. Akad. Nauk SSSR, 169, 547 (1966) (English translation Sov. Phys. Dokl. 11, 596 (1967)).
5. R.H. Christian, R.E. Duff, and F.L. Yarger, J. Chem. Phys. 23, 2045 (1955).
6. M.H. Rice, R.G. McQueen and J.M. Walsh, in Solid State Physics, edited by F. Seitz and D. Turnbull (Academic Press Inc. New York, 1958), Vol. 6.
7. L.V. Al'tshuler, Uspek. Fiz. Nauk (USSR) 85, 197 (1965). (English translation, Soviet Phys. Uspekhi 8, 52 (1965)).
8. J.M. Walsh, M.H. Rice, R.G. McQueen, and F.L. Yarger, Phys. Rev. 108, 196 (1957).
9. M.H. Rice and J.M. Walsh, J. Chem. Phys. 26, 824 (1957).
10. L.V. Al'tshuler, A.A. Bakanova, and R.F. Trunin, J. Exptl. Theoret. Phys. (USSR) 42, 91 (1962) (English translation, Sov. Phys. JETP 15, 65(1962)
11. K.K. Krupnikov, M.I. Brazhnik, and V.P. Krupnikova, (J. exptl. Theoret. Phys. (USSR) 42, 675 (1962) (English translation, Sov. Phys. JETP 15, 470 (1962)).
12. A.C. Mitchell and R.N. Keeler, Rev. Sci. Inst. 39, 513 (1968).
13. E.B. Royce, J. Appl. Phys. 37, 4066 (1966).
14. Bethe, H. Office of Scientific Research and Development, Division B, Report No. 545 (1942).
Weyl, H. Shock waves in arbitrary fluids. National Defense Research Committee, Applied Mathematics Panel Note No. 12 (Applied Mathematics Group - New York University No. 46) (1944).
15. W.H. Gust and E.B. Royce, Shock Induced Phase Transition Pressures in Fe-Cr and Fe-Cr-Ni Alloys, Lawrence Radiation Laboratory Report UCRL 50581 (1968).

16. R.E. Holt and R.A. Crist, Calibration of a Six-Foot and a Two-Foot Diameter Shock Tube, Air Force Special Weapons Center, Report AFSWC - TDR-63-5 (1963).
17. R.E. Duff and A.E. Blackwell, Rev. Sci. Inst. 37, 579 (1966).

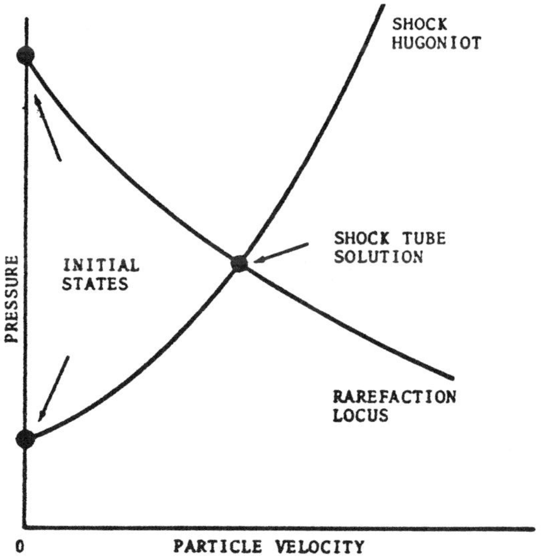

Figure 1
Pressure-particle velocity diagram for a conventional shock tube

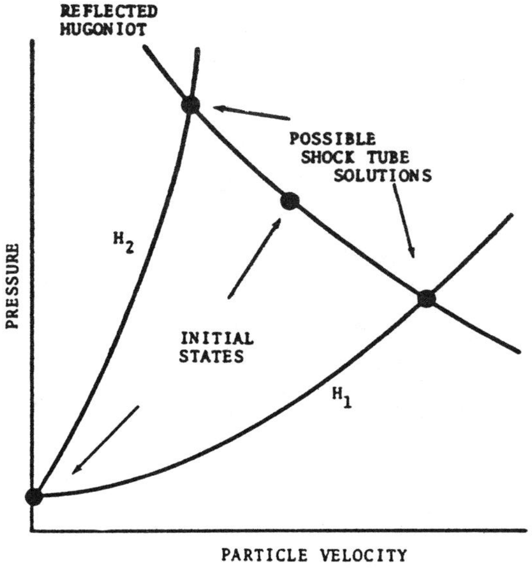

Figure 2
Pressure-particle velocity diagram for a high-explosive-driven shock tube. H_1 and H_2 represent Hugoniot curves for materials with shock impedances less than and greater than the high explosive.

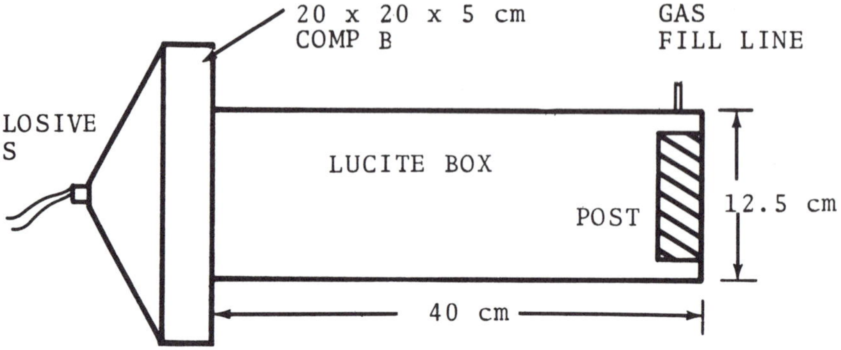

Figure 3
Experimental configuration used to investigate radiation-induced precursor disturbances

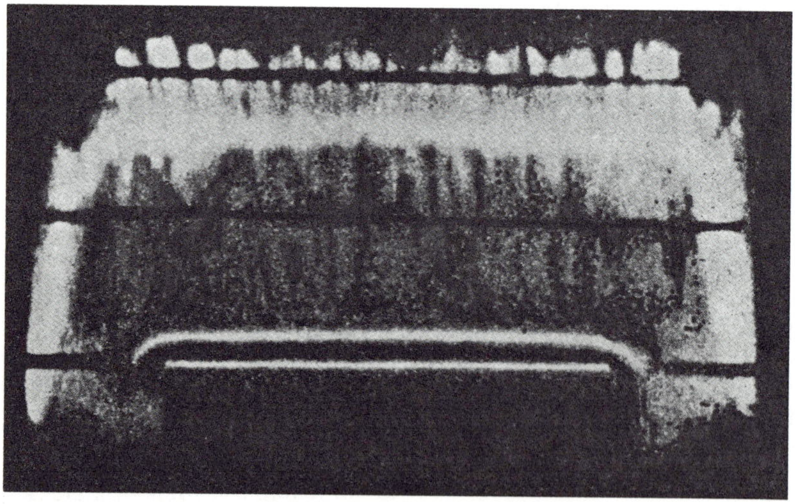

Figure 4
Self-luminous shock configuration showing precursors on the top and bottom tube walls and on the end wall

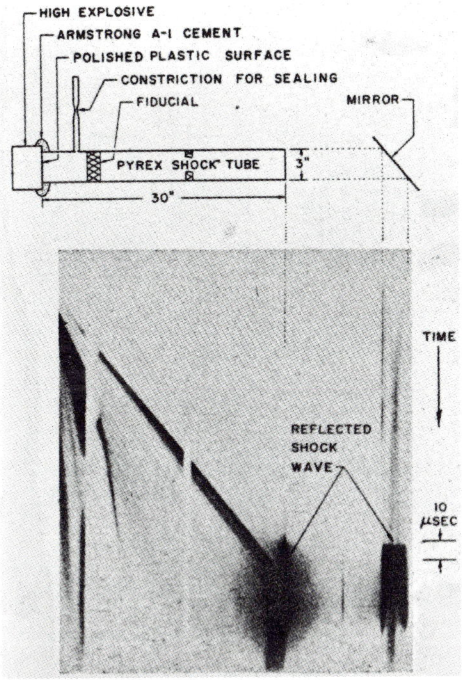

Figure 5
Experimental configuration used to investigate the quadratic Stark effect in helium

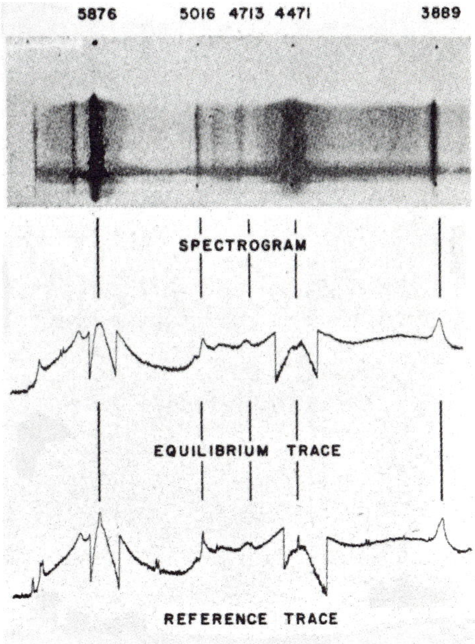

Figure 6
Spectrogram and photo-densitometer traces from shock heated helium

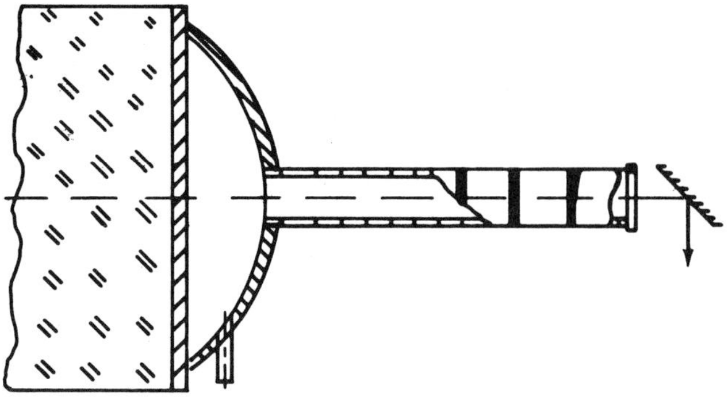

Figure 7
Experimental configuration of the high-explosive-driven shock tube developed by Voitenko

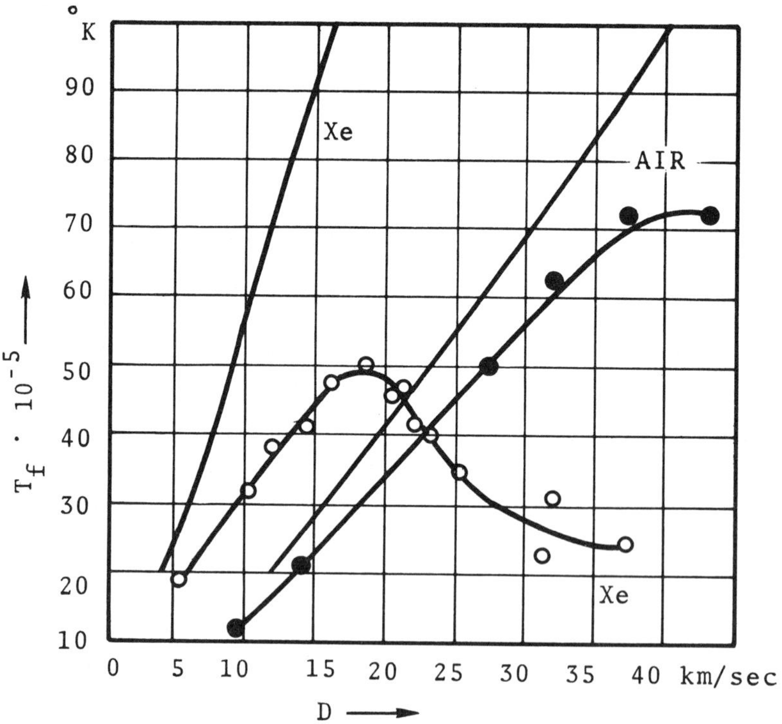

Figure 8
Calculated and observed shock wave temperatures of xenon and air

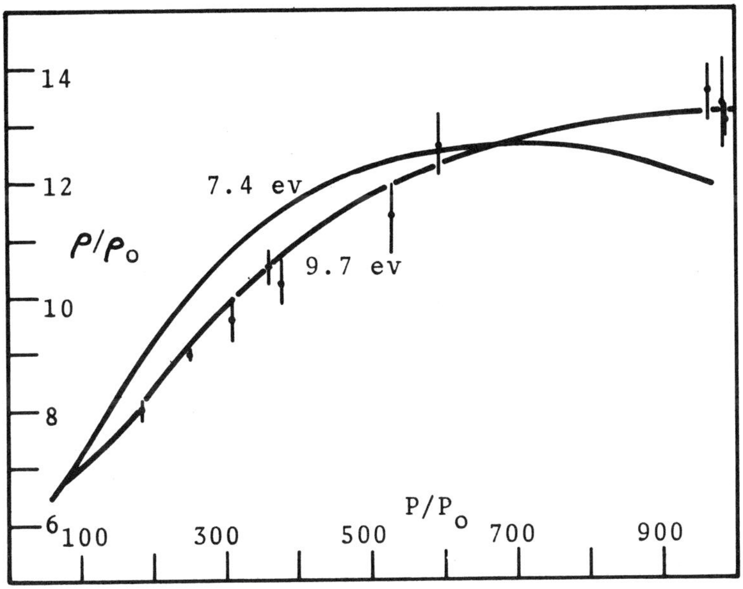

Figure 9
Theoretical and experimental values of density as a function of shock strength in nitrogen

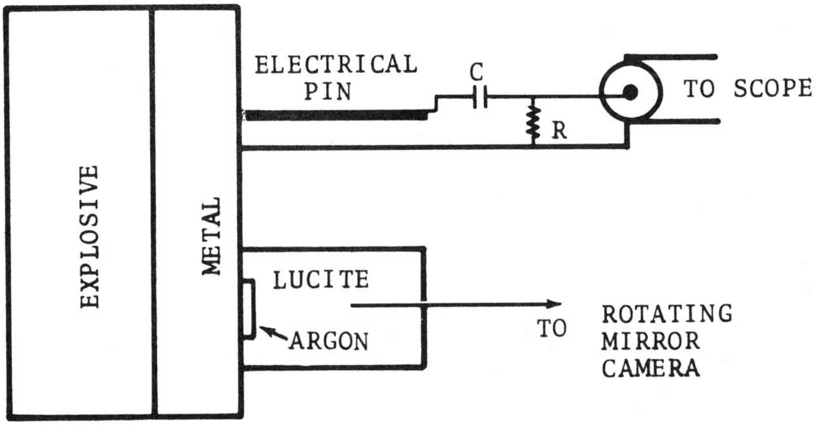

Figure 10
Sketch of electrical and optical techniques widely used to measure shock and free surface velocity

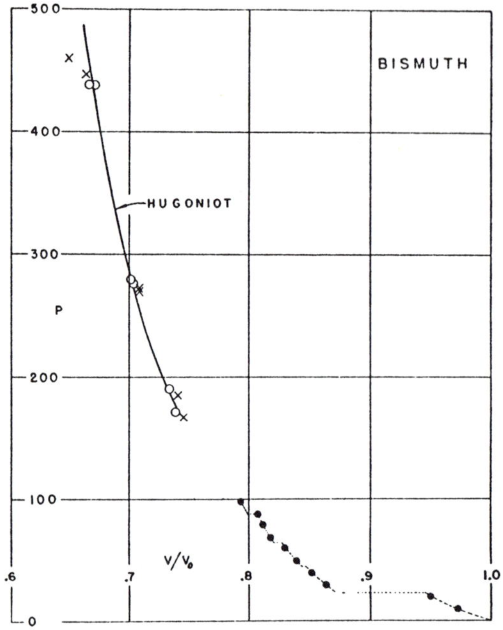

Figure 11
Dynamic equation of state measurements on bismuth compared with low pressure static data

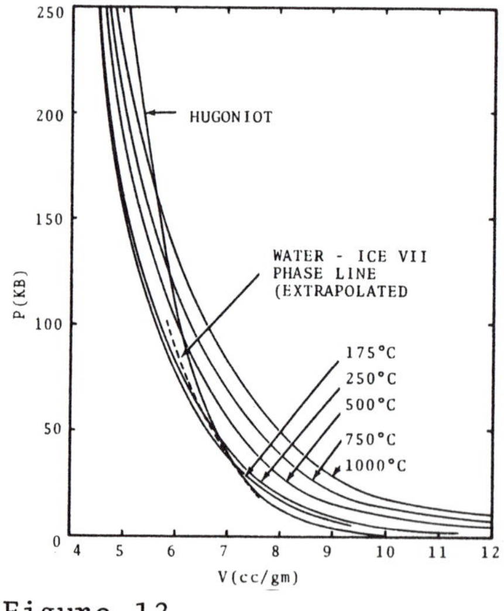

Figure 12
Equation of state of water determined from shock wave measurements

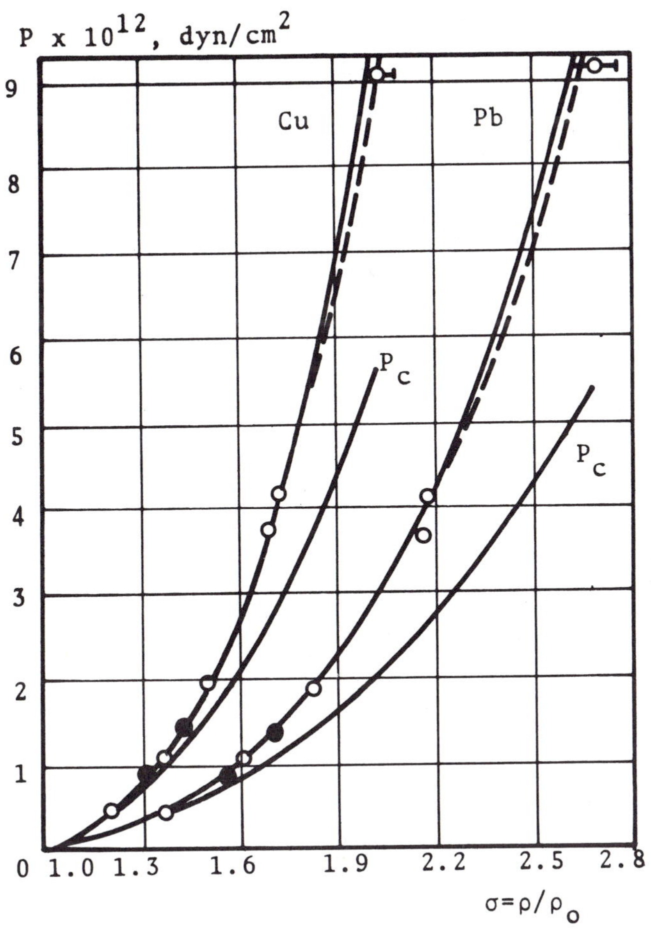

Figure 13
High pressure measurements of the equation of state of copper and lead

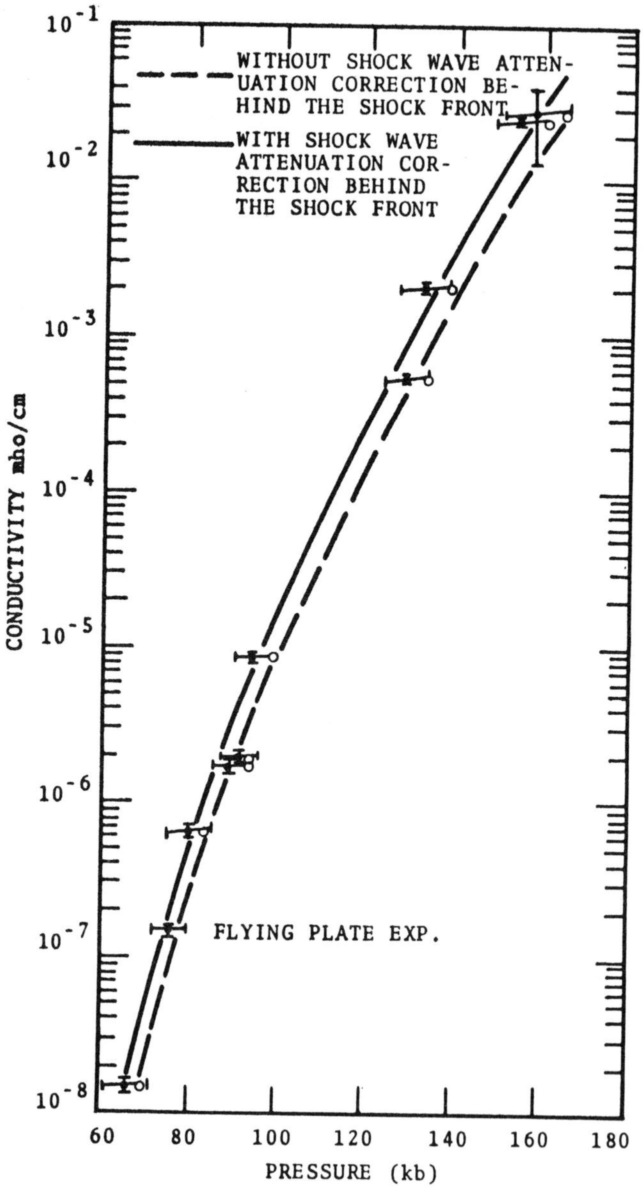

Figure 14
Electrical conductivity of shock-compressed CCl_4

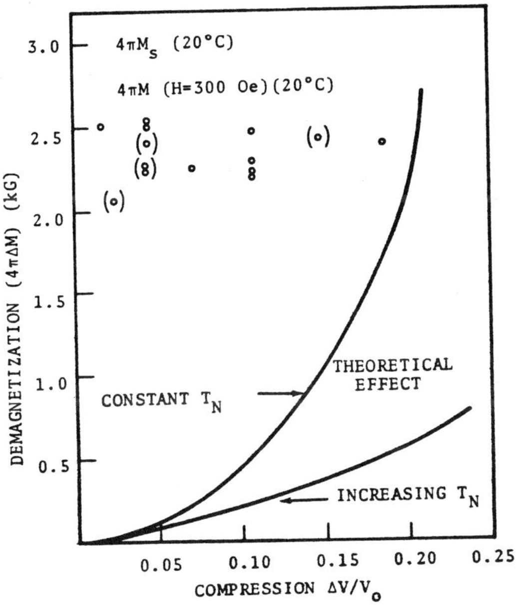

Figure 15
Predicted and measured shock demagnetization of nickel ferrite

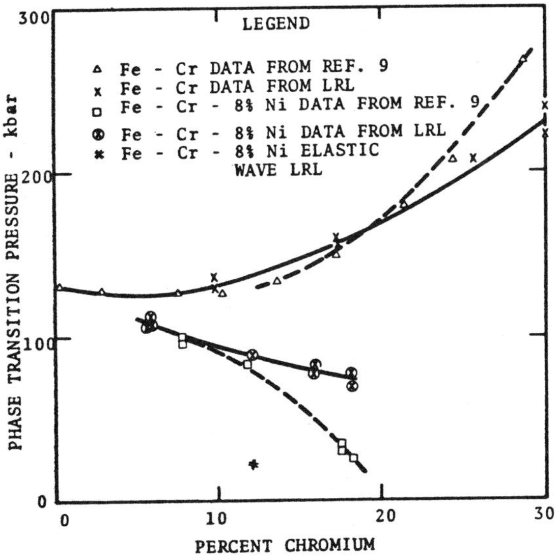

Figure 16
Phase transition pressures measured in the Fe-Cr-Ni system

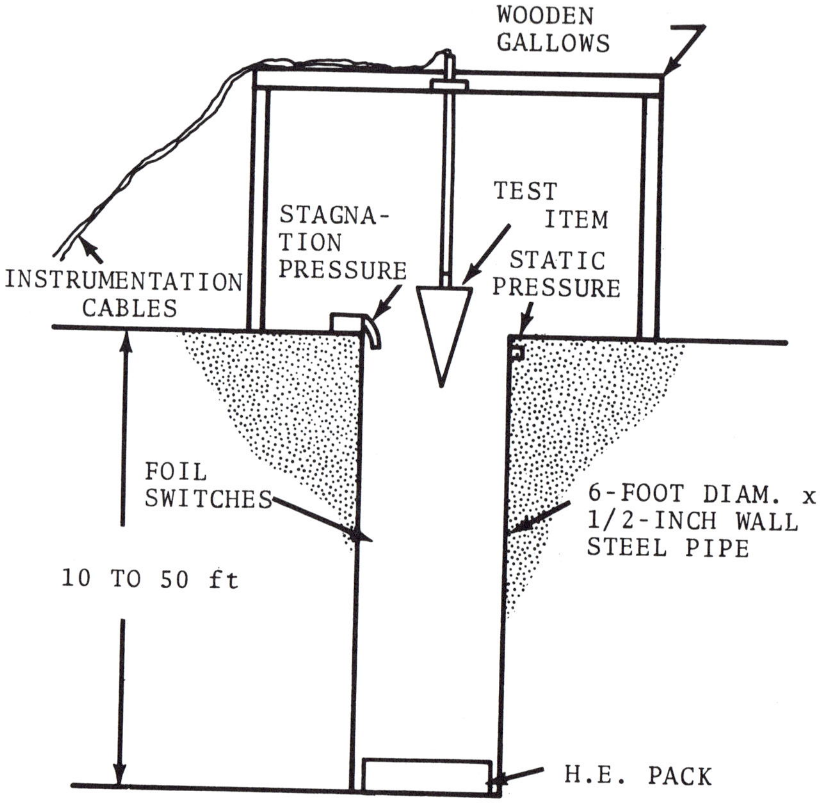

Figure 17
Sketch of large explosive-driven shock tube designed to produce high dynamic pressures

Figure 18
Test of a vertical explosive-driven shock tube

NUMERICAL SIMULATION OF A HIGH-ENERGY (Mach 120 TO 40) AIR-SHOCK EXPERIMENT*

BARBARA K. CROWLEY and H. DAVID GLENN
Lawrence Radiation Laboratory, University of California
Livermore, California 94550

SUMMARY

A shock tube experiment was performed which used a Voitenko compressor to deposit approximately 8×10^{13} ergs of energy into a driver gas consisting of 29 gm of air. The high energy driver gas exited into a 2-cm-i.d., 142-cm-long steel exit pipe containing the driven gas, atmospheric air. The shock was experimentally monitored in the exit pipe where it initially attained a velocity of approximately 4.0 cm/μsec and attenuated to 1.5 cm/μsec at the end of the pipe. A numerical simulation of this experiment was performed which calculated the air shock generation, propagation, and attenuation. This paper briefly reviews the experimental arrangement and results; the numerical codes and models used in the simulation are described; and the calculational results are discussed and compared with the data. During the 57-μsec interval required for the shock to transverse the 142-cm pipe, the calculations indicated that approximately 17 gm of the steel pipe was ablated. The largest fractional mass increase occurred in zones near the driver-driven-gas contact surface. The calculations indicated that mass entrainment was the predominant factor in attenuating the shock and contact surface in this experiment.

INTRODUCTION

To obtain both higher energy and greater than 1 cm/μsec velocity shocks into air, shock tube development in recent years has turned to driver sections which use H. E. (high explosives) in a variety of ways. Among the H. E. shock tubes in current use, the jet-augmented devices are capable of producing the highest energies and velocities.[1]

In 1966 Voitenko reported on a compressor device which uses high explosives to drive a steel plate into a rapidly converging chamber containing a driver gas.[2,3] The basic compressor design has the advantages of: (1) being able to impart enough energy to the driver gas to generate shocks of 4 cm/μsec into a driven gas of atmospheric air, (2) having a known mass (tens of grams are possible) of driver gas; and (3) being adaptable to numerical simulation by existing codes. Since the compressor has a known driver mass and is

*Work performed under the auspices of the U. S. Atomic Energy Commission.

adaptable to numerical simulation, the time and space histories of the flow variables can be reproduced calculationally. This calculational reproducibility enables a more complete analysis of high-energy gas flow than is obtainable by analytical techniques. In addition, the driver masses and internal energies attainable with the compressor are an order of magnitude larger than those obtainable with the jet devices; hence, the compressor also extends the range of achievable experimental conditions.

A Voitenko compressor was modified and used by Sawle et al. to accelerate foils.[4] Using a compressor similar to Sawle's, an experiment which used 29 gm of air as a driver gas was performed.[5] A steel exit pipe, 142-cm long with a 2-cm i.d., was designed as an integral part of the compressor. With atmospheric air in this exit pipe, a shock with mean velocity of 4.0 cm/μsec was attained over the first 10 diameters. This shock was monitored as it propagated down the pipe and attenuated to approximately 1.5 cm/μsec at the end of the pipe.

This paper briefly reviews the data from the above experiment. The numerical codes are briefly discussed; the numerical models are presented; the calculations performed are described; and agreement between experimental data and the calculations is shown. This paper demonstrates that experimental data combined with accurate numerical simulation can provide considerable information not accessible by purely experimental or analytical means.

EXPERIMENTAL ARRANGEMENT AND RESULTS

The compressor (Fig. 1) and its operation, which is discussed below, were essentially the same as reported by Sawle.[4] The first section of H. E. on the compressor was a plane wave lens which generated a plane shock that traveled through the thin aluminum plate. The plane shock ignited uniformly, over the cross sectional area, the straight cylinderical section of H. E. This second section of H. E. provided the force necessary to move the stainless steel driver plate into the compressor chamber. A previous unreported compressor study, by Sawle, indicated that the steel driver plate maintained a plane uniform shape (Fig. 2) while compressing the driver gas into a small volume before the aluminized diaphragm burst. The shock then propagated down the 142-cm-long 2-cm-i.d. steel outlet pipe, which contained atmospheric air.

An experimental zero-time reference was provided by the electrical signal which initiated detonation in the plane wave lens. The sequence of events following detonation are briefly summarized below, and those experimental results pertinent to the calculations are noted. Details and accuracy of the experimental results are given in ref. 5.

H. E. Detonation to Diaphragm Burst Time

The time at which the shock passed through the diaphragm was determined with a light pipe which was directed to view the center of the diaphragm. The elapse time between detonation of the plane wave lens and diaphragm bursting was measured as 59.7 ± 0.15 μsec. The calculations described here use the detonation of the straight section of H. E. as the zero time reference. Subtracting 29.0 ± 0.1 μsec[6] for the burn time of the plane wave lens and 0.5 ± 0.15 μsec[7] for shock transit through the thin aluminum plate results in 30.2 ± 0.25 μsec. This is the elapse time between detonation of the straight section of H. E. and diaphragm bursting

Shock TOA Data

Shock propagation was recorded by monitoring the TOA (time of arrival) by means of: light pipes (luminosity associated with the shock front); sigma pins (impact and electronic technique); pressure transducer (pressure of the shock front); and finally, recording the time at which the shock emerged from the end of the pipe with a framing camera. Fig. 1 illustrates the light pipes, sigma pins, and transducer in the exit pipe. The luminosity measurements were accomplished by positioning 3-foot-lengths of light pipes at fifteen preselected positions in the steel wall. Each light pipe viewed the inside of the outlet pipe through a 0.12-cm circular hole in the steel wall. Each light pipe was brought radially out of the wall and tied into a display board. As the shock propagated down the outlet pipe, the luminosity associated with the shock front arrival was transmitted to the board and recorded with two streaking cameras. Eight sigma pins were centered in the bore of the outlet pipe. Upon impact, the pins gave sharp electrical signals on raster scopes. The steepness of the shock front was indicated by the pressure transducer signal which gave a full scale deflection (>1.0 kbar) of the scope trace in less than 0.2 μsec. The transducer was a piezoelectric quartz gage factory calibrated to ~4.5 kbar. The framing camera provided a visual record of the shock emerging from the end of the outlet pipe, and the TOA was determined by optical timing fiducials that appeared on the film. Agreement between sigma pins, pressure transducer, and light pipe measurements were generally within the experimental error limits. The TOA data are discussed later and are compared with the calculations.

Motion of the Stainless Steel Driver Plate

Motion of a driver plate in the compressor chamber was recorded by Sawle et al.[4] and is discussed in the calculations section. After this experiment, the outlet pipe beyond 15 cm from the diaphragm was recovered intact. A spark spectrograph analysis on samples from the bore of the outlet pipe wall showed that the center segment of the stainless steel driver plate traveled the full length of the outlet pipe. However, no dynamic data exists which could determine the velocity of that segment of the driver plate which traveled down the exit pipe.

Radial Expansion of Exit Pipe

A framing camera focused on the compressor and first 25 cm of outlet pipe gave no photographic evidence of exterior wall motion for 60 μsec after the diaphragm burst. However, only that portion of the outlet pipe beyond 15 cm from the diaphragm was recovered intact after the experiment. Neglecting end effects, diameter measurements on the recovered pipe indicated the bore had undergone expansion from 2.0 cm to values centered around 2.30 cm at positions between approximately 20 and 80 cm and to values of 2.20 cm at 120 cm.

Ablation of the Walls

No direct experimental attempt was made to quantitively measure ablation in this experiment. Fig. 3 qualitatively illustrates the effect of erosion for the pin holder and light pipe apertures at the 20- and 100-cm positions. Erosion in the pressure transducer emplacement at 100 cm is also shown. To minimize such erosional effects, the pin holders and pressure transducer implacements were machined to very close tolerances (±0.0025 cm).

DESCRIPTION OF THE NUMERICAL CODES

The PUFL,[8] CEL,[9] and TENSOR-PUFL[10] codes which are used in the numerical simulation are all finite difference calculations which, with respect to time, increment the partial differential conservation equations of continuum dynamics. The conservation equations of mass, energy, and momentum together with appropriate equations of state (EOS) are considered.[11] These codes are based on the "artifical-viscosity" approach which was first introduced by Von Neumann and Richtmyer[12] for the calculation of hydrodynamic shocks.

PUFL

The conservation equations in PUFL are cast in the "almost-Lagrangian" form for quasi-one dimensional, or pipe geometry flow. In this geometry, only axial velocity and gradients are calculated. Zones in the PUFL code are allowed to have mass sources or sinks; hence, the equations are said to be cast in the almost-Lagrangian form. PUFL is used in this analysis to simulate burn of the cylindrical H. E. section and to start the driver plate into the chamber. It is used in the TENSOR-PUFL link to simulate flow in the exit pipe with friction, heat transfer, and mass entrainment. The basic PUFL equations which are solved for each zone each time cycle are:[8]

Conservation of mass:

$$\frac{\partial \rho}{\partial t} = \frac{1}{V}\left(-\rho\,\frac{\partial V}{\partial t} + \dot{m}S\right) \qquad (1)$$

The second term on the right describes the change in density, ρ, of a zone due to the mass flux $\dot{m}$ (= mass/area-time). The zone has a volume, V, and is in contact with the pipe walls over a surface area S.

The momentum equation:

$$\frac{\partial u}{\partial t} = \frac{1}{m}\left\{\dot{m}S\,(u_w - u) - V\frac{\partial p}{\partial x} - \tau_w S\right\} \tag{2}$$

On the right, the first term describes the change of the axial particle velocity, u, due to the entering mass whose axial velocity is u_w. The second term is the axial pressure gradient. The last term describes frictional stresses from the pipe surface where $\tau_w = 1/2\, C_f\, \rho u^2$; and C_f, the dimensionless coefficient of friction, may be constant or calculated from auxiliary equations.

Conservation of energy:

$$\frac{\partial e}{\partial t} = \frac{1}{m}\left\{\dot{m}S\left[(u-u_w)^2/2 + e_w - e\right] - p\frac{\partial V}{\partial t} + \tau_w S\,|u| + \dot{H}\right\} \tag{3}$$

The first term on the right modifies the specific internal energy, e, by the entering mass whose specific internal energy is e_w and velocity is u_w. The second term describes the effects of volume change; the third term accounts for irreversible dissipative work done on a zone by the pipe surface stresses; and $\dot{H}$ (= energy/time) is an energy source or sink.

Equation of state (EOS):

$$p = f(e, \rho) \tag{4}$$

The function may be any equation or table, and each material present may use its own EOS.

Ablation

The energy sink, $\dot{H}$, can be due to a turbulent convective heat flux, q_c, and a radiative heat flux, q_r. These heat fluxes act over the surface area, S, and are calculated by[8,13,14]

$$\dot{H} = -S\,(q_c + q_r) = -S\left(\frac{C_H \gamma\, p\, u}{\gamma - 1} + (\alpha\sigma)T^4\right) \tag{5}$$

where: p is the pressure; u particle velocity; γ the ratio of specific heats; C_H the dimensionless coefficient of turbulent heat transfer (the Stanton number); σ the Steffan Boltzman constant; α the absorptivity; and T the temperature of the gas which is determined from additional EOS information.

The mass flux which enters the flow, $\dot{m}$, is related to $\dot{H}$ by

$$\dot{m} = \frac{q_c + q_r}{q^*} = -\frac{\dot{H}}{Sq^*} \tag{6}$$

q^* is the energy required to ablate a unit mass of wall material and is calculated by

$$q^* = E_v + \eta\, h \tag{7}$$

where E_v is the total specific heat required to vaporize wall material, $E_v \approx 1.25 \times 10^{11}$ ergs/gm for steel; h is the specific total enthalpy of the gas flow in the pipe; and η is the turbulent transpiration coefficient, $\eta \approx 0.2$ for steel.[14]

Boundary conditions

Boundary conditions are specified for the pipe radii and for the ends of the pipe. Generally, the radii may be prescribed as a function of time and axial position. The ends of the pipe may use one of a variety of boundary conditions; pressure, velocity or position, flow in, or flow out, as functions of time.

CEL

CEL is a two dimension calculation which is used in this analysis in the all Eulerian, single fluid, cylindrical geometry, form. It is primarily used here to investigate the two dimension effects of chamber convergence which are not considered by PUFL. The CEL equations are:[9]

Conservation of mass:

$$\frac{\partial \rho}{\partial t} + \underset{\sim}{\nabla} \cdot \rho \underset{\sim}{U} = 0 \tag{8}$$

where $\underset{\sim}{\nabla}$ is the divergence in cylindrical coordinates.

The momentum equation:

$$\frac{\partial(\rho u)}{\partial t} + \underset{\sim}{\nabla} \cdot (\rho u)\, \underset{\sim}{U} = -\frac{\partial p}{\partial x} \tag{9}$$

$$\frac{\partial(\rho v)}{\partial t} + \underset{\sim}{\nabla} \cdot (\rho v)\, \underset{\sim}{U} = -\frac{\partial p}{\partial y} \tag{10}$$

where $\underset{\sim}{U} = (u, v)$.

Conservation of energy:

$$\frac{\partial \epsilon}{\partial t} + \underset{\sim}{\nabla} \cdot \epsilon \underset{\sim}{U} = - p \underset{\sim}{\nabla} \cdot \underset{\sim}{U} \qquad (11)$$

where ϵ is the internal energy per unit volume.

Equation of state:

$$p = f(\epsilon, \rho) \qquad (12)$$

The function f may be any arbitrary equation or table.

Boundary conditions

The two types of boundary conditions used in the following analysis consisted of rigid walls and an impermeable boundary (like a piston) which moved as a specified function of position and time.

TENSOR-PUFL

The TENSOR-PUFL code is actually two codes, TENSOR[10] and the previously mentioned PUFL which are linked on an overlapping cycle by cycle basis on the CDC-6600 computer. TENSOR is a two-dimensional code which solves the Lagrangian conservation equations in cylindrical geometry. The inner boundary of TENSOR is the inner radius of the pipe and is the outer boundary of PUFL.

The inner boundary of TENSOR uses a pressure vs. time and axial position, boundary condition applied to it. PUFL calculates this pressure profile on each staggered cycle. The pipe radius vs. time and axial position boundary condition used by PUFL is supplied by TENSOR. Generally, TENSOR calculates new radii each cycle for each of its two dimensional array of zones by solving the momentum equation. The radii of TENSOR's inner boundary may change due to a radial pressure gradient existing between the pressure profile applied by PUFL and the pressure in the first row of TENSOR zones along the boundary.

Linking the two codes, each cycle enables the simultaneous calculation of two regions (the high-energy gas flow and the pipe walls) whose predominant directions of motion are at right angles (axial and radial) and whose flow properties are quite different. In addition, the advantages of a Lagrangian formulation can be used for both regions.[15] In this analysis, the TENSOR-PUFL calculations are started when the CEL calculations indicate that the gas flow is no longer two dimensional, that is, radial gradients are insignificant.

TENSOR considers the following:

Conservation of Mass

Since TENSOR zones are Lagrangian, the mass of these zones is held constant; and densities are modified by volume changes. For the

TENSOR-PUFL link, it is thus assumed that the vaporized wall mass is negligible compared to the mass of a TENSOR boundary zone. However, the vaporized wall mass which enters the PUFL zones may not be negligible compared to the mass of a PUFL gas zone. PUFL uses a time step control to insure stability[8] which requires that the mass added per cycle is less than 10% of the current mass.

Conservation of Momentum

$$\rho \frac{\partial \underset{\sim}{U}}{\partial t} = \text{div}\ \tau \tag{13}$$

where $\underset{\sim}{U} = (u, v)$; two momentum equations, in the axial and radial directions are solved; τ is the symmetric stress tensor in cylindrical symmetry and consists of an isotropic and a deviatoric part; and div is the divergence in cylindrical coordinates.

Conservation of Energy

$$\frac{\partial \epsilon}{\partial t} = \tau \cdot \text{grad}\ \underset{\sim}{U} \tag{14}$$

where ϵ is internal energy per unit volume, and grad is the gradient in cylindrical coordinates.

Equation of State

To simulate the steel exit pipe, equation of state

$$P = f\ (\rho) \tag{15}$$

(discussed in the section on calculations) was used. TENSOR's use of material properties such as bulk and rigidity moduli is discussed in ref. 10.

Boundary Conditions

In TENSOR-PUFL, the inner radius TENSOR boundary condition is always a pressure profile supplied by PUFL. The outer radius and sides of the cylinder may use either free or rigid wall boundary conditions.

MODEL FOR NUMERICAL CALCULATIONS

The straight cylindrical section of H. E. shown in Fig. 1 was lit uniformly over its cross sectional area and provided the force necessary to move the stainless steel driver plate into the compressor chamber. Preliminary calculations indicated that the 13.2-cm length of this section was sufficient to appear the same as an infinite length to the stainless steel driver plate. Thus, the plane wave lens and aluminum plate did not affect the time history of the steel driver plates; and they are omitted from the numerical models.

Fig. 4 shows the initial conditions in the model of the compressor which is used for the first PUFL calculation. All dimensions and physical parameters shown in Fig. 4 are "as-build" conditions except for the diaphragm region.

Preliminary tests indicated that with 11.5 atmospheres, 29 gm of air, in the chamber, the 10-mil mylar diaphragm bulged ~0.175 cm. This bulge is equivalent to the volume of a cylinder of about 0.085-cm length in the 1-cm radius outlet pipe. Using a measured density of $\rho = 1.39$ for the mylar and spreading the mass in the 10-mil diaphragm over 0.085 cm of the outlet tube gives a density of 0.415 gm/cm^3. This is shown for the diaphragm region in Fig. 4. To numerically contain the 11.5-bar chamber pressure, a pressure hold of 50 bars is placed on the diaphragm. This pressure hold keeps the front surface of the diaphragm at its initial position (+0.085 cm) until the pressure in the diaphragm zones reach 50 bars. The first shock traveling through the chamber has a sufficiently large pressure gradient associated with it that the 50-bar pressure hold has no affect in retarding its propagation.

Equations of state used for PBX 9404 H. E. and iron in the driver plate were obtained from refs. 16 and 17. For the air in the compressor chamber, the air in the exit pipe, and the mylar diaphragm, the equation $p = e(\gamma-1)\rho$ was used. For the mylar, $\gamma = 1.2$ (a constant) was used.

To determine the effect of the equation of state of air on the calculations, two calculations were performed using $\gamma = f(e, \rho)$ for real air[18] and $\gamma = 1.2$, a constant. These two calculations gave nearly identical results. The reason for the similarity is that $\gamma \approx 1.2$ over most of the range of e and ρ achieved in the experiment. Hence, a constant $\gamma = 1.2$ was used for air in all of the calculations described.

As indicated in Fig. 4, the origin of the coordinate system used in the calculations is placed at the chamber side of the diaphragm zones. The zones representing the diaphragm and air in the exit pipe extend in the positive direction; zones representing the chamber air and H. E. extend in the negative direction.

CALCULATIONS

Shock Generation in the Chamber Region

The first part of this section describes calculations which simulate the cylindrical section of H. E. burn and generation of the high-energy shock in the compressor chamber. First, the PUFL code is used to simulate the cylindrical section of H. E. and the initial driver plate motion into the compressor chamber. Then, in order to consider the important two-dimensional effects of the highly convergent section of the compressor chamber, the two dimensional CEL calculation is used.

The first calculation is a PUFL calculation which uses the initial conditions shown in Fig. 4. The time of ignition of the second or cylindrical section of H. E. starts the PUFL calculation and is taken to be t = 0.

From this PUFL calculation, the position of the interface between the steel driver plate and chamber air is plotted versus time in Fig. 5. As seen in Fig. 5, the motion of this interface is detectable at slightly more than 15 μsec after ignition of the H. E. Since the H. E. length is 13.2 cm and the PBX 9404 detonation velocity is about 0.88 cm/μsec,[16] a 15-μsec time interval is reasonable before plate motion is detected.

Fig. 5 also shows experimental arrival data[4] for the front of the driver plate as it passes through the chamber. The PUFL calculation agrees nicely with the first experimental data point located -5.8 cm from the diaphragm where the chamber is nearly cylindrical. However, at -1.95 cm from the diaphragm, where the chamber is sharply convergent (see Fig. 4), the PUFL calculation differs from the experimental data. This calculation shows the plate continuing to accelerate through the convergent section and arriving in the exit pipe before the air shock was experimentally observed in the exit pipe.

In the one-dimensional PUFL calculation, no radial gradients or radial velocities are present. Hence, pressure increases induced by chamber convergence are immediately present throughout the zone, even at the axis. This immediate increase in pressure at the axis causes an unrealistic increase in the axial velocity of both the driver plate and the shock.

Because of significant two-dimensional effects, the two-dimensional CEL code was used to calculate the chamber region. The model used for the CEL calculations is like Fig. 4, but the H. E. and the driver plate are omitted. To simulate the front of the driver plate moving into the chamber, the CEL calculations used a radially straight impermeable boundary whose axial coordinates were specified as a function of time. Three CEL calculations (C1, C2, and C3) were performed which differed only in the driver plate vs. time boundary condition that was specified (see Fig. 5). The three boundary condition curves are all the same as the PUFL calculation, up to the -5.8 cm data point. The three curves are all linear between the -5.8 and -1.95 cm data points and give a plate velocity of 0.55 cm/μsec. They all gradually decrease slope until 31 μsec when they are located at -0.05, -0.15, and -0.25 cm for C1, C2, and C3, respectively. After 31 μsec, the last slope is continued in each calculation.

As the boundary moved into the chamber at 0.55 cm/μsec, a nearly plane, relatively low-energy air shock was formed. This shock moved through the chamber with a nearly constant velocity of 0.6 cm/μsec, as would be calculated from a 0.55 cm/μsec piston using plane one-dimensional shock theory.[19] The velocity of this shock through the chamber was the same in the three calculations and was unaffected by the variations in plate position shown in Fig. 5.

Axial velocities are imparted to the gas behind the 0.6 cm/μsec shock. However, due to the radial convergence of the chamber, large radial velocities soon develop. Fig. 6 shows the total axial and radial momentums plotted vs. time from the C2 CEL calculation. A negative radial momentum in Fig. 6 implies an average velocity pointed toward the axis of symmetry.

At about 29.5 μsec, when the 0.6 cm/μsec plane shock reaches the diaphragm region, a tremendous radial convergence of the air in the chamber takes place (Fig. 6). The stagnated radial velocities induce high pressures which act to greatly accelerate the shock in the region of the diaphragm. Fig. 7 shows the position of the shock through the chamber and exit pipe as a function of time from the three CEL calculations and the experimental data.

The total amount of energy deposited in the chamber air is quite sensitive to the plate position vs. time curve which is used between 27 and 31 μsec. This is illustrated in Fig. 8, where the total energy in the chamber air from the three CEL calculations is plotted vs. time.

Fig. 7 indicates that by 30 μsec the high velocity shock is at the diaphragm. This 30 μsec compares favorably with the 30.2 $\pm$ 0.25 μsec obtained from experiment (H. E. Detonation to Diaphragm Burst Time). Shortly after 30 μsec, pressures in the chamber start to relieve due to expansion down the exit pipe; and, as Fig. 8 indicates, the plate is no longer capable of doing significant work on the chamber gas.

Fig. 7 indicates that the 4.0 cm/μsec velocity in the exit pipe from the C2 CEL calculation closely approximates the experimental data. Hence, the C2 calculation, with a total energy of approximately 8×10^{13} ergs in the 29 gm of chamber air, is used in the following calculations of the exit pipe region.

Shock Propagation and Attenuation in the Outlet Pipe

The C2 CEL calculation was continued, and the shock position vs. time for C2 is shown in Fig. 9 along with the experimental data. Fig. 9 indicates that the 4.0 cm/μsec shock velocity calculated in C2 agrees nicely with the experimental data over the first 20 cm. However, the shock calculated by CEL maintains its 4.0 cm/μsec velocity and does not attenuate as the experimental data indicates. In the following, a number of phenomena that are not considered in the C2 calculation which could attenuate the shock are investigated.

Since the flow is virtually one-dimensional at 31 μsec (see Fig. 6), the pressure, density, and velocity results from the C2 CEL at 31 μsec are used as initial conditions in the PUFL part of the following TENSOR-PUFL calculations. Three preliminary TENSOR-PUFL calculations were performed with these initial conditions from the C2 CEL. These calculations used $C_f = \dot{H} = \dot{m} = 0$ in eqs. 1, 2, and 3. When (1) the driver plate boundary condition was given the same constant velocity in the exit pipe as in the C2 CEL (see Fig. 5); and (2) the exit pipe walls in TENSOR were forced to remain rigid, the flow variables from PUFL agreed with the radially averaged results from the C2 CEL between 31 and 65 μsec. The shock position vs. time from this TENSOR-PUFL calculation also agreed with the C2 CEL results shown in Fig. 9. This agreement between the C2 CEL and the TENSOR-PUFL calculation indicates that the mapping of flow variables from CEL to TENSOR-PUFL was performed correctly.

As discussed in the first section of this paper, it was found that portions of the driver plate traveled the full length of the exit pipe. However, no data exist which could give the velocity of the driver plate in the exit pipe. The constant plate velocity shown in Fig. 5 is undoubtedly an upper limit. Two additional TENSOR-PUFL calculations were performed. The first held the boundary (that is the plate position) stationary after 31 μsec. The second applied a constant 1 bar pressure to the boundary which allowed the plate to move backwards at later times. The shock TOA results from these two calculations are indistinguishable from those shown in Fig. 9 for C2 and the TENSOR-PUFL with a constant velocity plate. In addition, the TOA of the contact surface down the exit pipe was constant and the same in the three TENSOR-PUFL calculations. This constant contact surface velocity was 3.6 cm/μsec; the value expected from ideal shock tube theory with a 4.0 cm/μsec shock velocity.[19]

Hence, propagation and attenuation of the shock down the exit pipe was independent of the velocity of the driver plate. These results are not too surprising since Fig. 8 shows that after about 30 μsec the plate no longer imparts significant energy to the gas. The remaining TENSOR-PUFL calculations all used a zero plate velocity boundary condition.

The TENSOR-PUFL calculations considered below all used TENSOR to calculate the radial expansion of the exit pipe walls. These calculations summarized in Table 1 all use the following description in TENSOR for the "Ledloy A" exit pipe. The two-dimensional TENSOR cylinder extended over a radial region from 1.0 to 8.25 cm with an initial density of $\rho_0 = 7.81$ gm/cm^3. A pressure, p, vs. μ ($\mu = \rho/\rho_0 - 1$ = compression - 1) curve for the equation of state was used which matched shock Hugoniot data[20] for mild steel at pressures above a megabar. The curve thru the high-pressure Hugoniot data was smoothly joined to a straight line through the point ($p = 0$, $\mu = 0$) with an initial slope corresponding to a constant bulk modulus of 1.65 megabars. This bulk modulus was obtained by using a Poisson ratio of 0.3 and a Young's modulus for "Ledloy A" of 2 megabars.[21]

As mentioned earlier, photographs of the exit pipe at 60 μsec after diaphragm burst showed no visible movement of the outer radius. Using a free-surface boundary condition in TENSOR with a large static, 10 kbar, yield strength resulted in significant radial expansion of the outer pipe walls well before 60 μsec. Currently, no experimental data are available for the dynamic yield strengths of Ledloy; but dynamic yield strengths are credited with being greater than the static yield strengths. Hence, the outer radius boundary condition in TENSOR was treated as a rigid wall in the following calculations which considered periods up to only 60 μsec after diaphragm burst. The calculations and experimental evidence indicate a dynamic yield strength in excess of 25 kbars for durations of up to 40 μsec.

Table 1
TENSOR-PUFL calculations performed on the exit pipe

Calculation number	Mass of sigma pins	Friction and heat transfer: Stanton no. $C_H \approx 1/2\ C_f$	Friction and heat transfer: Absorptivity coefficient, α	Mass addition: Transpiration coefficient, η
TP1	No	0	0	No mass addition
		No heat transfer		
TP2	Yes	0	0	No mass addition
		No heat transfer		
TP3	Yes	0.002	1	No mass addition ($\eta = \infty$)
TP4	Yes	0.002	Independent	.2
TP5	Yes	0.002	Independent	.1

The first TENSOR-PUFL calculation, TP1, used $C_f = \dot{H} = \dot{m} = 0$ in eqs. 1, 2, and 3. The results of this calculation are shown in Fig. 9, where it is seen that radial expansion has only a slight effect in attenuating the shock velocity.

The previous calculations used air only at atmospheric conditions in the exit pipe. Because of the sigma pins present in the exit pipe, an attempt was made to calculationally simulate their effect by adding their mass to certain zones in the exit pipe. The pins extended radially 1 cm, from the pipe wall to the center line, and along the axis of the pipe for 1.2 cm. This 2.2-cm length of the pin had a mass of 0.02 gm which was inserted into PUFL zones at each of the eight axial positions where sigma pins were located. In all other respects, TP2 was exactly like TP1. The total added mass of the pins, 0.16 gm, had a nearly indetectable affect in attenuating the general motion of the shock down the pipe. A small localized retardation of the shock velocity was, however, observed both experimentally and in this calculation at the pin positions.[5] The shock arrival from TP2 is so close to the TP1 results that it is not detectable in Fig. 9.

In the next calculation, TP3, the effects of friction and energy lost by heat transfer to the walls is considered. For the turbulent flow present in the smooth exit pipe, a constant friction coefficient of $1/2\ C_f = 0.002$ is used.[13] (To provide a smooth surface, the inside of the outlet pipe was machined and honed to a 12-μin. finish.) By invoking Reynold's analogy, which assumes that the same mechanism causes the exchange of momentum and of heat transfer in turbulent flows, $C_H \approx 1/2\ C_f = 0.002$ is obtained.[13] To maximize the energy lost by radiation, the absorptivity, α, is used as unity. Eq. (5) is used to calculate the heat fluxes to the walls in TP3; but no mass addition is considered, ($\dot{m} = 0$ by setting $\eta = \infty$). In all other respects, TP3 is like TP2.

Shock arrival from the TP3 calculation is shown in Fig. 9, and a slight amount of attenuation is noticeable. During the 42 μsec it took the shock to travel down the exit pipe in this calculation, 1.1×10^{13} ergs (15%) was lost by heat transfer to the walls of the exit pipe. TP3 shows that even <u>maximized</u> thermal energy loses without mass addition do <u>not</u> account for the observed attenuation. During the TP3 calculation, the convective heat flux was generally 5 to 10 times larger than the radiative flux. It should be noted that this calculation did not include opacities for the air and used $\alpha = 1$ in eq. (5). A more realistic calculation of the radiative flux would result in even lower values. Additional calculations that varied the value of α were performed, which indicated that at all times and positions the solution was virtually independent of the radiative heat flux.

To consider the effects of mass addition due to ablation, eqs. (6) and (7), which link mass addition and energy losses, are used in the next calculation, TP4. This calculation uses a turbulent transpiration coefficient of $\eta = 0.2$ and a specific energy of vaporization of 1.25×10^{11} ergs/gm.

The results of shock and contact surface arrivals from TP4 are also given in Fig. 9 where it is seen that mass addition has a tremendous effect. The calculation indicates that 1.2×10^{13} ergs of thermal energy was transferred to the pipe walls during the 56 μsec of the calculation. Fig. 10 shows the convective heat flux at various axial positions as a function of time.

Fig. 11 shows the mass flux entering the flow at various axial positions. A total of 17 gm was ablated from the walls of the exit pipe during the calculation. The vaporized mass was assumed to enter the flow with a specific internal energy of $e_w = 1.3 \times 10^{11}$ ergs/gm and no kinetic energy, $u_w = 0$. By integrating the fluxes shown in Fig. 12, the thickness of steel ablated during the calculation can be obtained. Such a computation indicated that during the 56 μsec of the TP4 calculation, approximately 0.003, 0.0025, 0.0013, and 0.0004 cm thickness of the steel wall was ablated at the 10, 80, 100, and 120 cm axial positions, respectively.

Fig. 12 shows pressure vs. time from the TP4 calculation at a number of axial positions. As mentioned above, the pressure transducer located at 100 cm had a range of 1 kbar but it failed about 0.2 μsec after shock arrival. Fig. 12 is consistent with this minimal information.

During the TP4 calculation, 1.1×10^{13} ergs of energy were imparted to TENSOR by the gas doing work on the pipe walls. The pipe walls started to expand shortly after the air shock arrived at any particular axial position, and a shock traveled to the outer boundary of the exit pipe, where it reflected. The radial shock and its reflection soon interacted with pressures from similar shocks up and down the pipe. Such complex pressure histories resulted in an initial expansion of the inner radius of the pipe and, at later times, a contraction. Thus, a plot of inner radius vs. axial position is not an expanding cone with respect to time. At 56 μsec, the TP4 calculation showed that the radii at positions beyond about 60 cm were still undergoing their initial expansion. Between 20 and 60 cm, the radii had attained their maximum values at earlier times and, subsequently, oscillated between values of approximately 1.10 and 1.20 cm. These results do not appear to be inconsistent with the information in the section on post-shot measurements of the exit pipe radii.

Mass addition acts directly in the momentum equation, eq. (2), to decrease particle velocity and consequently to decrease the shock and contact surface velocities. After 56 μsec in the TP4 calculation, approximately 14 gm of the entrained mass entered the region which contained the 29-gm driver gas. The remaining 3 gm mixed with the 0.6 gm of driven gas. The average fraction β = (mass added)/(original mass) for the driver and driven regions is thus $\beta \simeq 0.5$ and 5, respectively. In the driver gas region, the zones near the driver plate have $\beta \approx 1.1$, while near the contact surface, $\beta \approx 3.0$. In the driven gas region, $\beta \approx 12$ for zones near the contact surface, and decreases to $\beta \approx 3$ for zones approximately half way between the shock (where $\beta = 1$) and the contact surface. The large fractional mass increase in the zones near the contact surface is the primary factor in attenuating the contact surface and shock velocities.

The calculations presently assume instantaneous mixing of the wall mass with the gas. A small lag time which is proportional to the radius of the pipe and inversely proportional to the velocity would be more realistic. However, with such a lag time, the same effects of a large fractional increase in mass near the contact surface would occur.

To demonstrate the sensitivity of the calculations to the ablation parameters, a TP5 calculation is considered. TP5 is exactly the same as TP4, but a turbulent transpiration coefficient $\eta = 0.1$ is used.

The shock and contact surface of TP5 are also shown in Fig. 9. During shock transit of the exit pipe in the TP5 calculation, 32 gm of wall material was entrained in the gas.

CONCLUSIONS AND DISCUSSION

A high-energy air shock experiment using the Voitenko compressor and an associated exit pipe is shown to be capable of accurate numerical simulation with existing finite difference calculations. It is demonstrated that these calculations give considerable information about the flow which is not currently accessible by analytical or purely experimental means.

The calculations indicate that in the energy range of this experiment (Mach 120-40) the turbulent convective heat flux is of greater importance than the radiative flux. By using the sum of these heat fluxes, ablation mass fluxes are obtained. The ablated mass enters the gas flow such that the largest fractional increase of mass occurs in zones of the calculations which are near the driver-driven gas contact surface. The TENSOR-PUFL calculations performed with mass entrainment are shown to agree with the experimental data.

The calculations also indicate that the mass flux, shock arrival, and other flow variables are sensitive to the value used for the turbulent transpiration coefficient, η. Hence, experiments using available techniques, combined with calculations, may provide valuable information about the time-dependent ablation process. Future experiments are being designed which will provide pressure-time data at a number of positions in a longer exit pipe. In addition, an effort is being made to identify propagation of the contact surface in the exit pipe.

ACKNOWLEDGMENTS

The authors wish to express their sincere appreciation for the many discussions and helpful comments contributed by J. B. Knox, D. R. Sawle, and E. R. Horton; the modification and running of the CEL calculation by W. F. Noh and the CEL group; the programming and running of the PUFL and TENSOR-PUFL calculations by L. K. Barr and E. A. Reed; and for the typing of this manuscript by C. L. Myers.

REFERENCES

1. Gill, S. P., Implosively Driven Shock Tubes, Stanford Research Institute, Technical Rept. 001-66, December 1966.
2. Voitenko, A. E., Generation of High Speed Jets, Soviet Physics-Doklady, 9, 860, April 1966.
3. Voitenko, A. E., Strong Shock Waves in Air, Soviet Physics-Technical Physics, 11, 128, July 1966.
4. Sawle, D., Characteristics of the Voitenko High Explosive Driven Gas Compressor, First International Colloquium of Gas Dynamics of Explosions, Sept. 18-21, 1967, Brussels.
5. Glenn, H. D. and Crowley, B. K., Optical Technique for Monitoring High Energy (Mach 35-130) Shocks in Steel Pipes Containing Ambient Atmospheric Air, Lawrence Radiation Laboratory, Livermore, Rept. UCRL-71007, 1968.
6. Mullins, R. K., Research Operators Handbook, Lawrence Radiation Laboratory, Livermore, M-013, 1968.
7. Wilkins, M. L., The Use of One- and Two-Dimensional Hydrodynamic Machine Calculations in High Explosive Research, Lawrence Radiation Laboratory, Livermore, Rept. UCRL-14531, 1965.
8. Crowley, B. K., PUFL, an Almost-Lagrangian Gasdynamic Calculation, J. of Comp. Physics, Vol. 2, 61, 1967.
9. Alder, B., Fernbach, S., and Rotenberg, M. (eds.), Methods in Computational Physics, Vol. 3, Academic Press, New York, 1964. Noh, W. F., CEL: A Time-Dependent Two-Space-Dimensional, Coupled Eulerial-Lagrangian Code.

10. Alder, B., Fernbach, S., and Rotenberg, M. (eds.), Methods in Computational Physics, Vol. 3, Academic Press, New York, 1964. Maenchen, G. and Sack, S., The TENSOR Code.
11. Crowley, B. K., Numerical Simulations of Shock Tubes, Conference on Military Applications of Blast Simulators, Suffield Experimental Station, Ralston, Alberta, Canada, July 18-21, 1967.
12. Von Neumann, J. and Richtmyer, R. D., A Method for the Numerical Calculation of Hydrodynamic Shocks, J. Appl. Physics, 21, 232, 1950.
13. Schlichting, H., Boundary Layer Theory, McGraw-Hill Book Co., New York, 1960.
14. Rose, P. H., and Offenhartz, E., Ablation Measurements in Turbulent Flow, Avco-Everett Research Lab., AVCO-PR-114, 1959.
15. Crowley, B. K., A Comparison of Analytical and Numerical Solutions for the Shock Tube Problem, Lawrence Radiation Laboratory, Livermore, Rept. UCRL-14198, 1965.
16. Wilkins, M. L., Squier, B., and Halperin, B., The Equation of State of PBX 9404 and LX 04-01, Lawrence Radiation Laboratory, Livermore, Rept. UCRL-7797, 1964.
17. Tillotson, J. H., Metallic Equation of State for Hypervelocity Impact, General Atomics Rept. 3216, 1962.
18. Fenter, F. W., The Thermodynamic Properties of High Temperature Air, Chance Vought Research Center, RE-1R-14, 1961.
19. Courant, R. and Friedrichs, K. O., Supersonic Flow and Shock Waves, Interscience Publishers Inc., New York, 1963.
20. vanThiel, M., Compendium of Shock Wave Data, Lawrence Radiation Laboratory, Livermore, Rept. UCRL-50108, Vol. 1, 1966.
21. Ryerson, Joseph T. Inc., Ryerson Data Book for Steel and Aluminum, 1958.

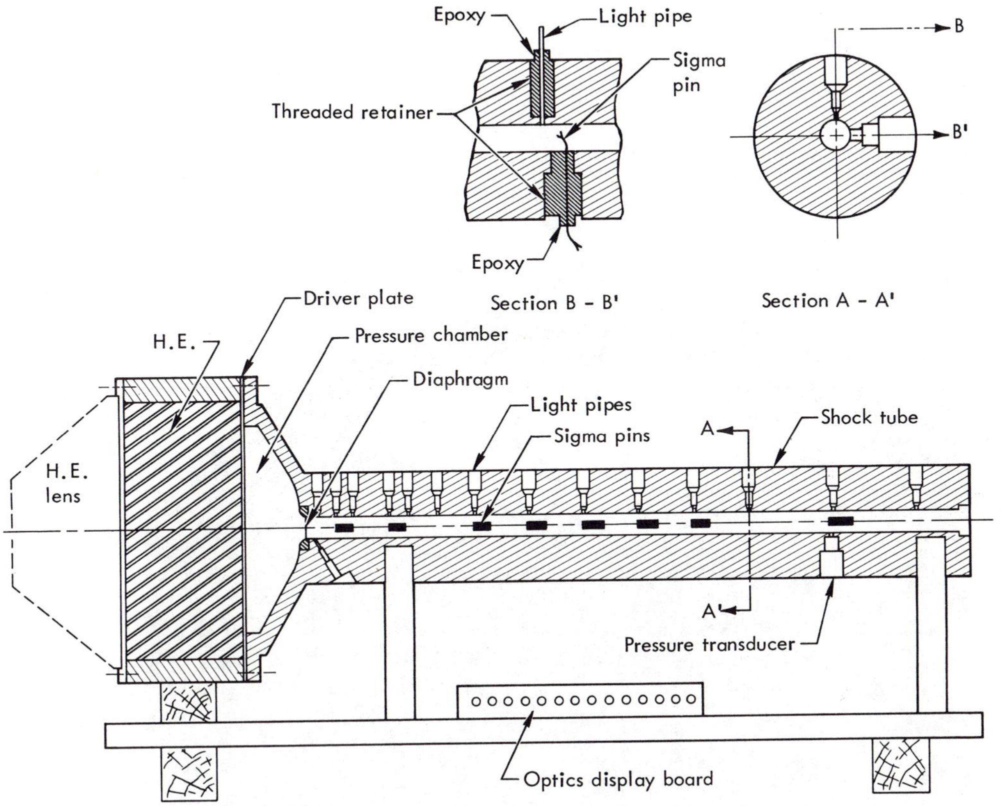

Figure 1
Voitenko compressor, outlet pipe and diagnostic systems

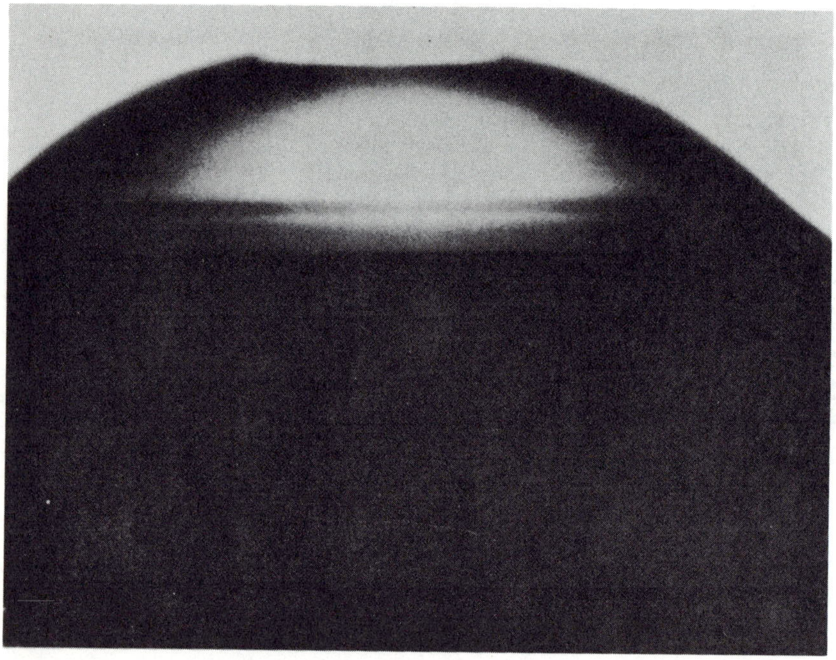

Figure 2
X-ray photograph showing uniform planar motion of the driver plate at the time the plate has moved ~3.2 cm into the chamber. The x-rays were generated with 35-MeV linac sending a pulsed (0.25-μsec beam width) 2-amp electron beam current into a Platinum-Iridium Target (Courtesy D. Sawle, Lawrence Radiation Laboratory, Livermore)

Figure 3
Post-shot cross-sections of the steel outlet pipe showing erosional effects of the high-temperature gas flows in the pipe

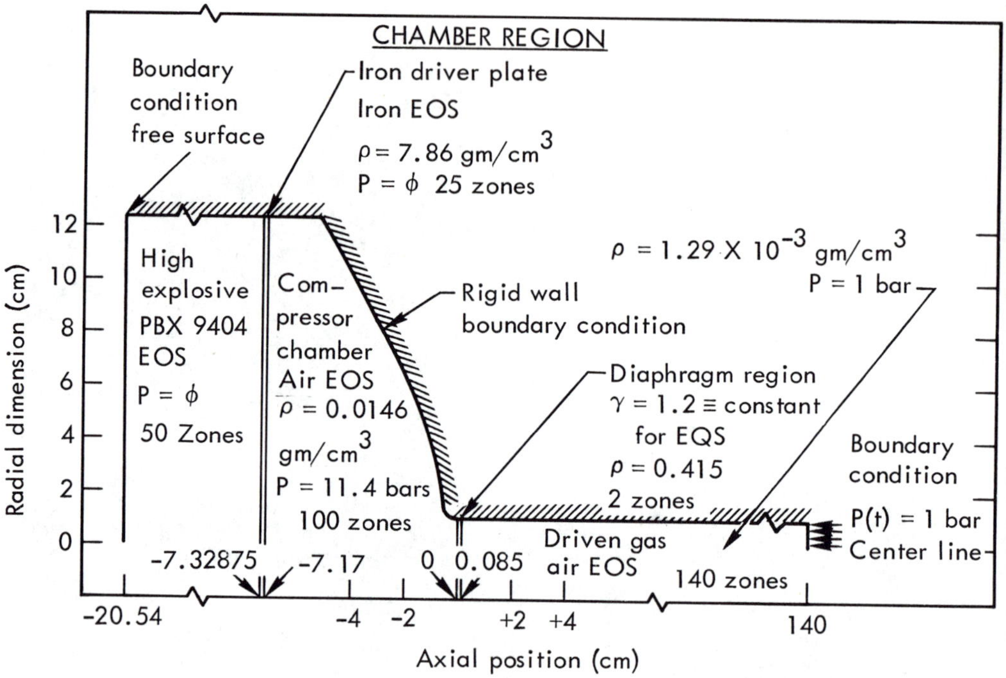

Figure 4
Initial conditions in the model of the compressor used in the one-dimensional PUFL calculation

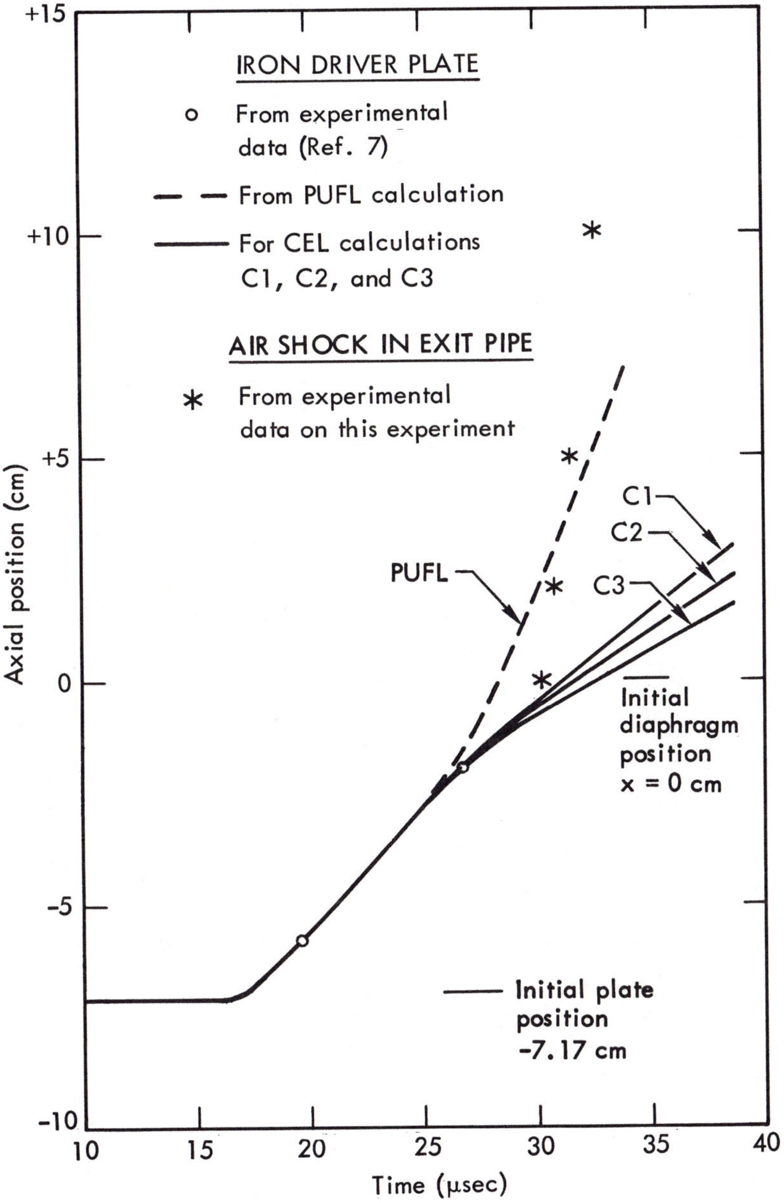

Figure 5
Position of the iron driver plate — chamber air interface from the PUFL; CEL calculations of the chamber C1, C2, and C3; and the experimental data

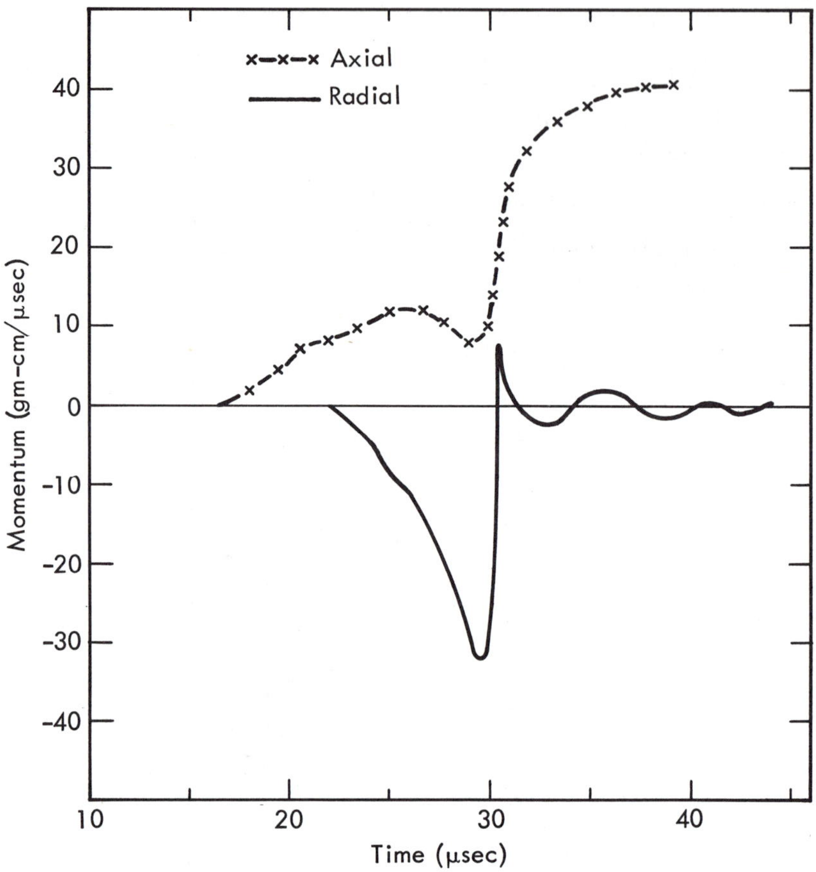

Figure 6
Total axial and radial momentum of driver gas vs. time from C2 CEL. Negative radial momentum is toward axis.

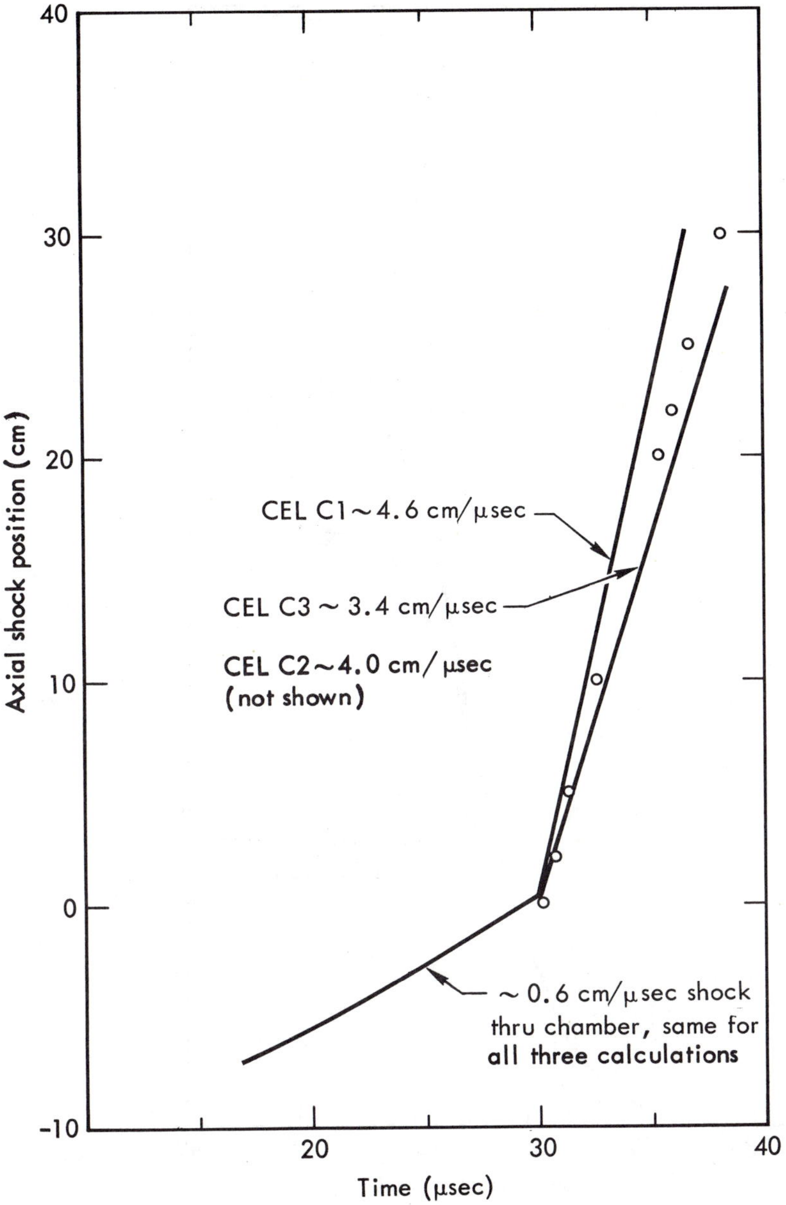

Figure 7
Shock arrival through chamber and start of exit pipe from CEL C1, C2, and C3 calculations and experimental data

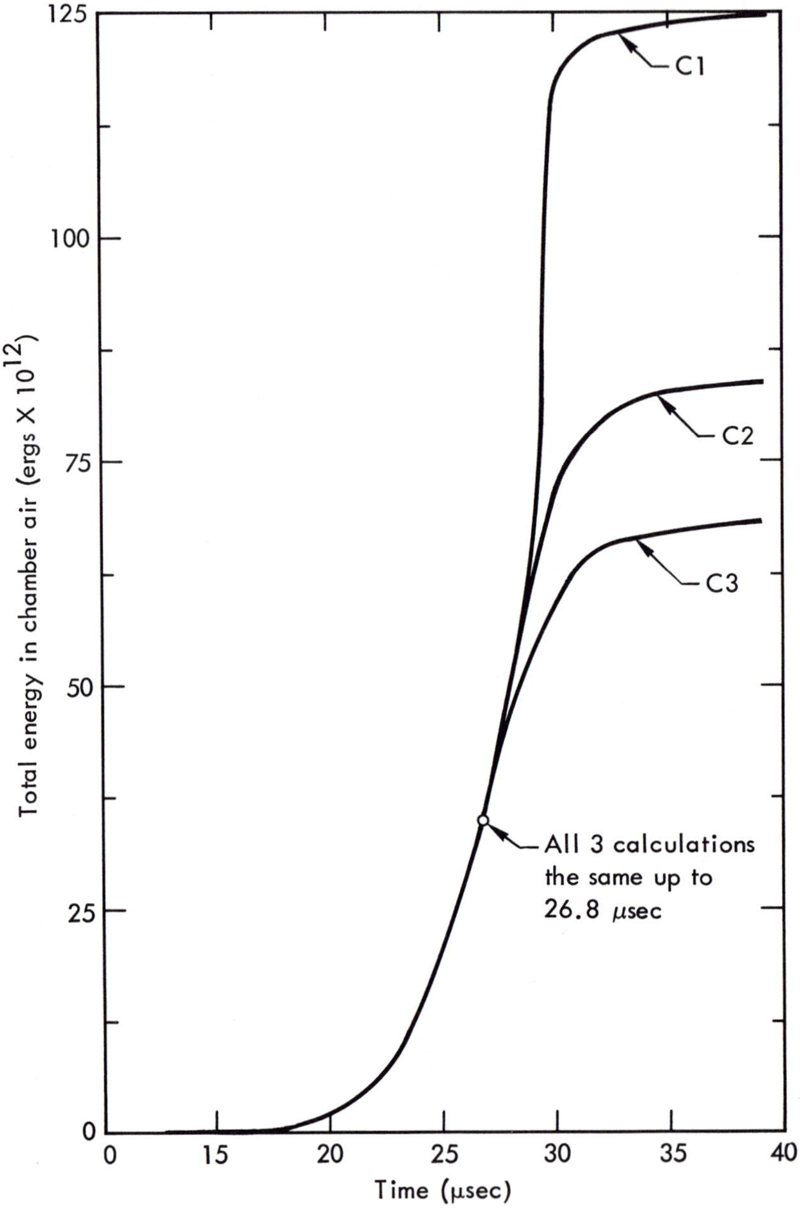

Figure 8
Total energy vs. time from CEL's. Calculations C1, C2, and C3.

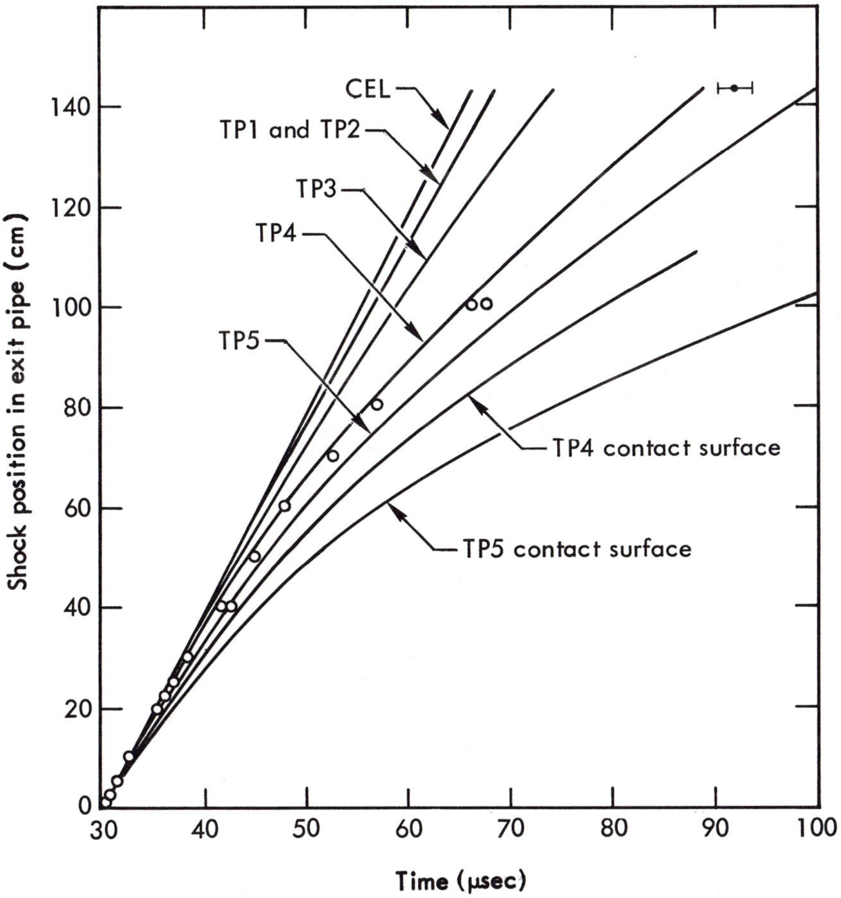

Figure 9
Shock position vs. time from the TENSOR-PUFL calculations (Table 1) and experimental data

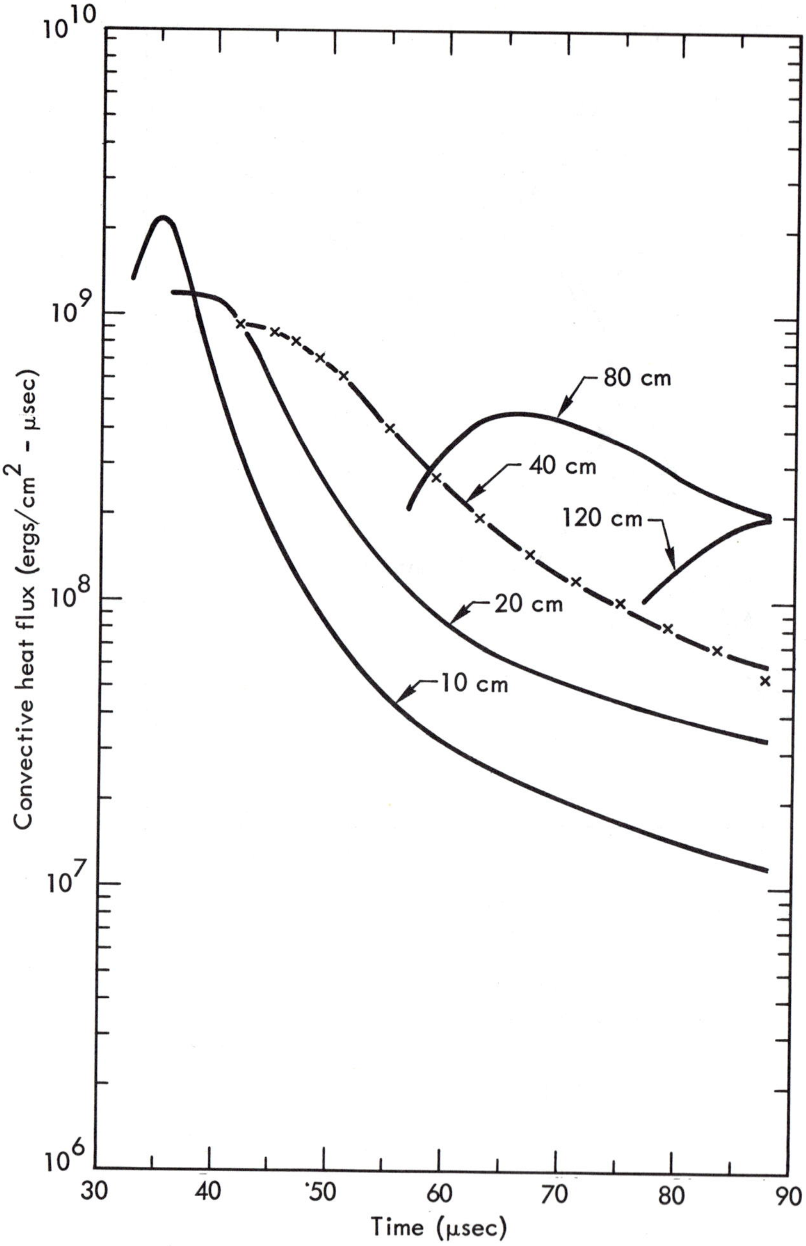

Figure 10
Convective heat flux vs. time at the 10, 20, 40, 80, and 120 cm positions in the exit pipe from the TP4 calculation

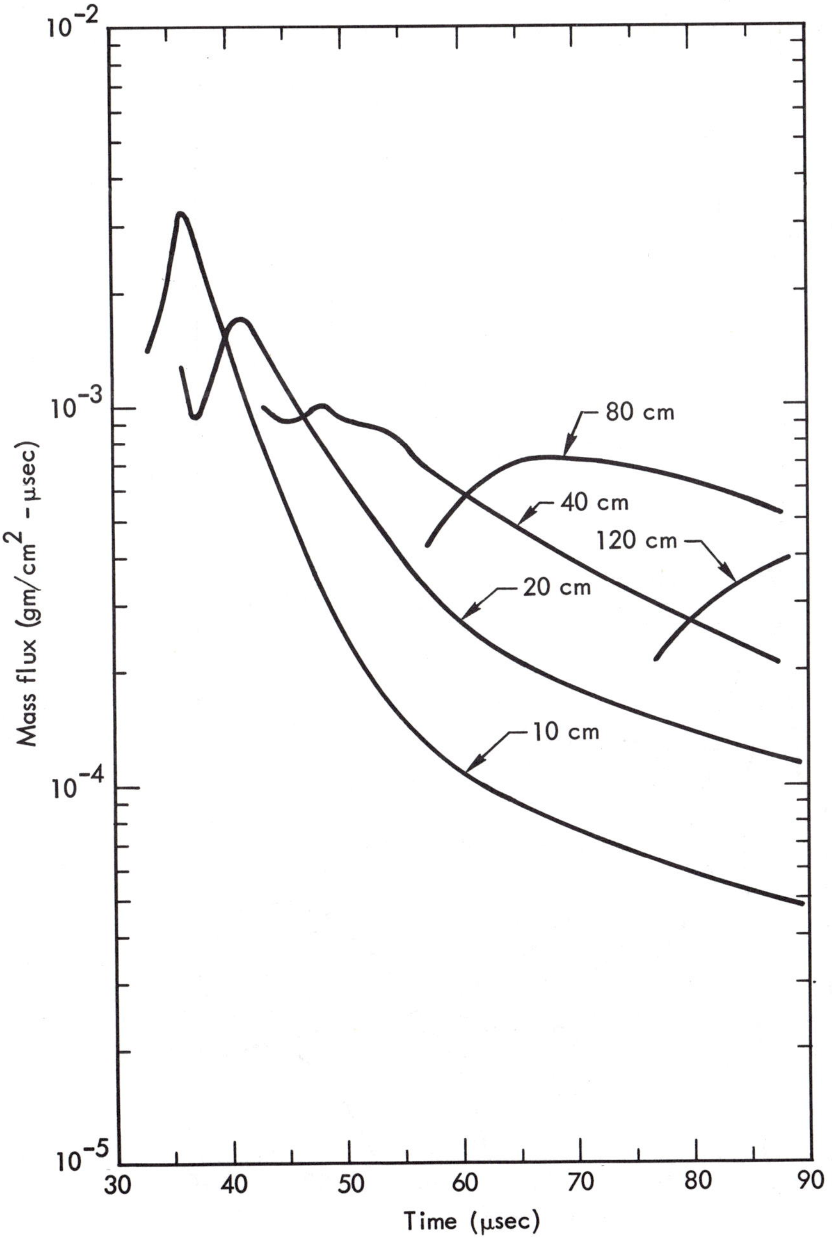

Figure 11
Mass flux vs. time at 10, 20, 40, 80, and 120 cm positions in the exit pipe from the TP4 calculation

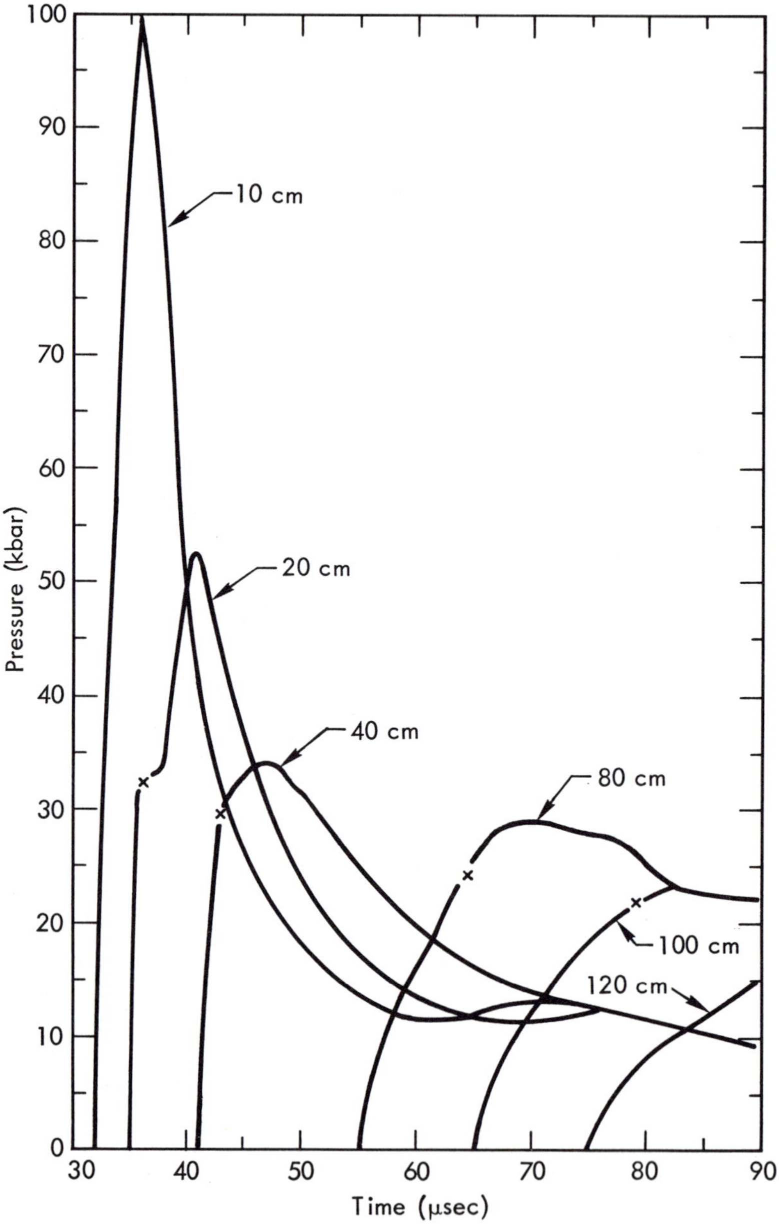

Figure 12
Pressure vs. time at 10, 20, 40, 80, 100, and 120 cm positions in the exit pipe from the TP4 calculation; X denotes arrival of contact surface

IMPLOSION-DRIVEN SHOCK TUBE

I. I. Glass and J. C. Poinssot
Institute for Aerospace Studies
University of Toronto
Toronto,Canada.

SUMMARY

Stoichiometric mixtures of hydrogen-oxygen at 200, 400, and 600 psi initial pressure were detonated at the geometric centre of an 8 in dia hemispherical chamber by a very short exploding wire. The detonation wave in the mixture moves out towards the periphery where it can strike a shell of PETN explosive to produce an explosive-driven implosion, or it may be allowed to reflect from the steel surface only, without an explosive liner, as an implosion wave moving into a highly preheated gas. As expected, for the gas runs only, the shock strength increases nonlinearly with initial pressure. For 600 psi initial, a shock Mach number $M_s = 40$ was obtained in a 8 mm dia channel filled with air at 1 torr. The shock decayed to $M_s = 11$ over about 180 cm. Blast-wave decay and viscous and ablative losses undoubtedly cause the high attenuation. A numerical analysis predicts shock Mach numbers $M_s \sim 100$, using a modest loading of 200 g PETN and 200 psi $2H_2 + O_2$. The small barrel bore was used to investigate the limiting case when the facility is used as a massless projectile hypervelocity launcher or shock tube. Explosive runs will be made in 8 mm and 25 mm dia channels, using the 8 in dia driver chamber, in the near future.

1. INTRODUCTION

The controlled generation of strong, symmetrical implosion waves in a reusable facility by using safe, secondary explosives is described in Ref.1. Basically (see Fig.1), an 8 in dia hemispherical chamber is filled with a mixture of $2H_2 + O_2$ to several hundred psi. The mixture is detonated by a short exploding copper wire (1/16 in long x 0.002 in dia), as shown in Fig. 1A. The detonation wave races at about 2.9 km/sec to impact on a hemispherical shell of PETN weighing several hundred grams (2 to 6 mm thick), in order to detonate it impulsively and simultaneously, and send an implosion wave into the highly preheated gas (about 3,000°K) to further heat and compress it as it collapses and reflects at the origin (see Fig.1B).

The pocket of gas at the origin can be used as a driver for a hypervelocity launcher or a shock tube (see Fig. 1C), or for many other physical experiments requiring such a pocket of concentrated high-pressure, high-temperature gas.[2,3,4] Implosions produced in this manner have been used to drive plastic projectiles (7.9 mm dia x 7.9 mm long, that is, 5/16 in dia single-calibre cylinders), weighing 356 mg, to velocities up to 17,600 ft/sec. Although numerical analyses showed that velocities which are factors greater than the above should ideally have been achieved, it was concluded that problems of projectile integrity and those arising from viscous and ablative losses would limit the actual velocities which could be obtained.

The numerical analyses[5,6,7] consist of solving the set of nonlinear partial differential equations of mass, momentum and energy in Lagrangian form, with the aid of equations of state. The set of equations is solved by numerically integrating a set of corresponding finite difference equations, using an artificial viscosity technique, subject to prescribed initial and boundary conditions [8,9,10,11] fo hypervelocity launchers and shock tubes.

It is important to note that the shock tube can be considered as a special case of the hypervelocity launcher, where the projectile mass is reduced to zero, or the case of the massless projectile. In view of the actual loss mechanisms that exist to limit projectile velocity, it was considered very worthwhile examining the massless projectile case, since it does not suffer from projectile disintegration, friction, ablation, and leakage problems[1].

2. NUMERICAL ANALYSIS

Both the hypervelocity launcher and the shock tube have been studied analytically, as noted above. The numerical analysis consists of solving the set of lossless (no friction, heat transfer or ablation) nonlinear partial differential equations of mass, momentum and energy, in Lagrangian form, with the aid of the equations of state, as follows:[5,]

$$\text{mass:} \quad V = \frac{1}{\rho} = \frac{1}{\delta}\frac{\partial R^{\delta}}{\partial m} \qquad (1)$$

$$\text{momentum:} \quad \frac{\partial U}{\partial t} = -A\frac{\partial (P + Q)}{\partial m} \qquad (2)$$

$$\text{energy:} \quad \frac{\partial E}{\partial t} = -(P + Q)\frac{\partial V}{\partial m} \qquad (3)$$

$$\text{state:} \quad E = E(P, V), \qquad (4a)$$

for gas mixtures $2H_2+O_2$ and air

$$P = P(E, V), \qquad (4b)$$

for the explosive, PETN.

The artificial-viscosity pressure[7], Q, is restricted to compression waves only, and is expressed as

$$Q = \frac{(C\Delta x)^2}{V} \cdot \left(\frac{\partial U}{\partial x}\right)^2 \qquad (5a)$$

for $\frac{\partial U}{\partial x} < 0$

and $Q = 0$ (5b)

for $\frac{\partial U}{\partial x} \geq 0$

where
A = area of elemental interface
C = artificial viscosity constant
$m = \frac{1}{\delta} \rho_o R_o^{\delta}$, mass per steradian
δ = 1, 2, 3, for planar, cylindrical, and spherical geometries, respectively
ρ_o = initial density
R_o = initial position of Lagrangian surface
U = particle velocity
V = specific volume
R = radial distance
P = pressure
Q = artificial viscosity pressure
E = specific internal energy
t = time
Δx = zone width in difference scheme

The system is divided into three regions (see Fig.2): the explosive PETN, the gas mixture $2H_2 + O_2$, and the air in the channel. Each has its own equation of state. Each region is divided into zones, and mass points containing one half of the mass of two adjacent zones are assumed to be at the interface of these zones.

Equations 1 to 5 are then solved by numerically integrating a set of corresponding finite difference equations, using the artificial viscosity technique subject to the prescribed initial and boundary conditions.[8,9,10,11] Using the artificial viscosity, which spreads the shock over a specified number of zones, avoids treating the shock wave discontinuities running through discrete mass points by spreading them over about 3 to 6 zones.

The two independent variables, time and distance, are represented by the number of the cycle, N, in the computational scheme which is being calculated, and J, the label assigned to a mass point (Fig. 2).

The finite difference equations can be written as follows:

mass:
$$V^N_{J-1/2} = \frac{v_{j-1/2}}{m_{j-1/2}} \qquad (6)$$

momentum:
$$\frac{DU_J}{Dt} = [(P+Q)_{J-1/2}-(P+Q)_{J+1/2}]\frac{A^N_J}{1/2(m_{j-1/2}+m_{j+1/2})} \qquad (7)$$

energy:
$$E^{N+1}_{J-1/2} = E^N_{J-1/2} - [1/2(P^N_{J-1/2}+P^{N+1}_{J-1/2})+Q^{N+1/2}_{J-1/2}]\,[V^{N+1}_{J-1/2} - V^N_{J-1/2}] \qquad (8)$$

state:
$$E^N_{J-1/2} = E(P^N_{J-1/2}, V^N_{J-1/2}) \qquad (9a)$$

for $2H_2 + O_2$ and air

$$P^N_{J-1/2} = P(E^N_{J-1/2}, V^N_{J-1/2}) \text{ for PETN} \qquad (9b)$$

artificial viscosity:
$$Q^{N+1/2}_{J-1/2} = C^1 \frac{(U^{N+1/2}_J - U^{N+1/2}_{J-1})^2}{V_{J-1/2} + V_{J-1/2}} \cdot \frac{\rho_o}{2} \qquad (10a)$$

for $U_J < U_{J-1}$

$$Q^{N+1/2}_{J-1/2} = 0 \qquad (10b)$$

for $U_J > U_{J-1}$

The quantity A^N represents the area of a particular interface where DU/Dt is the acceleration of the interface. The constant C^1 in the artificial viscosity is related to the usual constant C by the equation $C^1 = C^2/4$ (for further details see Ref.10).

A particular solution is obtained by integrating the momentum equation, **Eq.2,twice**, to get new velocities and positions for the next step. From the new position a new specific volume is calculated for each zone and a simultaneous solution of the energy and state equations, Eqs. 8 and 9, gives the new pressure and energy. A new time step is then calculated subject to the Courant and the artificial viscosity stability conditions[10], and the process is then repeated.

During the course of the analysis, several difficulties were found to exist near the origin, where there is a discontinuity in going from a spherical to a planar geometry. In reality, the flow around the origin cannot be considered as one-dimensional, as it is actually a radially-symmetric, two-dimensional phenomenon.

The use of the artificial viscosity technique may be strongly criticized when it is applied in the region surrounding the origin. In the spherical geometry this technique is valid only as long as the shock thickness is small compared with the radius of the shock, and this condition is clearly violated near the origin.

In addition, there are very severe zoning problems. The effect of a discontinuity on zone mass has been studied and it had been found that to obtain precise results, there should not be significant discontinuities in zone mass, particularly at the interface between the $2H_2 + O_2$ mixture and the air. Consider the movement of the interface between the gas mixture $2H_2 + O_2$ in the chamber and the air in the barrel, at the time when the diaphragm breaks. This movement is governed by the momentum equation, Eq(7):

$$\frac{dU_J}{dt} = \frac{A_J^N \left[(P + Q)_{J-1/2} - (P + Q)_{J+1/2} \right]}{1/2 \ (m_{J-1/2} + m_{J+1/2})}$$

For a given pressure difference across the interface, the acceleration depends on the values of the masses $m_{J-1/2}$ and $m_{J+1/2}$. But, due to the very low pressure in the barrel, $m_{J+1/2}$ is very small and is negligible compared with $m_{J-1/2}$, so that the acceleration of the interface is almost inversely proportional to the mass of the last zone in the chamber. However, the continuity in zone mass was impossible to maintain, since the density of the combined gas is very high (in some cases $\sim$ 1.0g/cm^3). It was therefore necessary to compromise, such that the formation of the planar shock in the channel was not instantaneous, and the final velocity was approached exponentially in a time that depended on the zone mass. That is, the dense mass zone behaves like a projectile. To improve the precision of the calculations, many more zones must be taken in the highly dense gas in order to have continuity in zone mass across the interface of the gases, $2H_2 + O_2$-Air. But, if very fine zones are used, then the computing time becomes too costly.

A computed wave-speed diagram of the shock-tube case having a 5/16 in dia(8 mm) launcher barrel or shock tube channel appears on Fig.3. It is seen that for an 8 in dia(20.3 cm) hemispherical driving chamber with $2H_2 + O_2$ at 200 psi(13.6 atm) initial, a shell of PETN weighing 200 g, and a counterpressure of air of 1.0 torr, produces a shock velocity of 32.8 km/sec, or a shock Mach number of 95. The above is by no means an optimum driving condition and stronger shocks could be produced with optimized higher explosive loading (beyond a maximum explosive charge, the shock strength will be decreased by wave interactions) (see also Ref.8) and initial conditions.

The wave diagram shows some other major features. The detonation wave in the $2H_2 + O_2$ mixture is initiated instantly at the origin ($t = 0$, $x = 10^2$ cm). It races at 2.9 km/sec to initiate the PETN liner when it strikes the surface and reflects as a shock to overtake the previous shock wave. A single imploding shock wave (as well as reflected rarefaction wave and a new contact surface--the details have been omitted on the diagram for simplicity throughout) is thereby generated. The imploding shock wave strikes the diaphragm (which is assumed to break instantly) at the origin station, producing a reflected hemispherical shock wave in the driving chamber and the primary planar shock wave followed by a contact surface in the channel. The process is repeated many times, as shown. However, the other shock waves that enter the channel are too slow to overtake the primary shock wave.

It is of interest to note that the computed results do not show any shock wave decay, and the computed pressure profiles (see Fig.4) show a uniform flow behind the planar shock wave. In practice there is a severe decay, probably arising from blast wave effects (indicated by the motion reversals of the contact front between the explosive gases and the hydrogen-oxygen driver gas after 50 μsec), viscous losses, and ablative losses from the chamber and channel walls.

The wave motions and the pressure histories in the chamber and channel, as a function of time, are shown in Fig.4. The reflection of the rarefaction wave at the periphery of the chamber, and the wave motions about the origin are clearly shown. The development of the primary shock wave with time is readily seen and an ideal uniform flow appears to develop. Nearly 200 cm of this hot gas is available at t = 407 μsec. However, as noted above, this idealization is not achieved.

3. EXPERIMENTAL RESULTS

Some experimental work was done with the actual 8 in dia chamber. It was not meant to be an extensive study of the performance of this type of shock tube. Rather, it was aimed to check the results of the computations and to give an estimate of the losses, as the computer code did not take account of them. However, this preliminary work has been extremely helpful in assessing the deficiencies of the code and in the design of a new 25 mm dia shock tube that will not suffer from as many viscous effects.

The results were obtained in an 8 in dia chamber. The barrel was a stainless steel, high-pressure tube having a 5/16 in (8 mm) internal diameter. It was approximately 4 m long. A stainless-steel, scribed diaphragm, 0.015 in (0.4 mm) thick, previously calibrated, was used to separate the channel from the hemispherical driver near the origin.

The velocities of the shocks were measured using three different methods. Five ionization probes were placed along the barrel 14 in (36 cm) apart to detect the arrival of the ionization front. These were corroborated by using small optical windows and photomultipliers to detect the luminous front at the same stations. The agreement was excellent. Additional substantiating velocity data for the 400 psi case, as shown on Fig.5, were obtained with a 34.4 Ghz (12 mm wave length) microwave Doppler system[12]. The agreement was again satisfactory.

A summary of the experimental results that have been obtained to date appears in Fig.5. It shows the maximum average velocities and their decay over a distance from 10 to 225 cm using 200, 400 and 600 psi $2H_2 + O_2$ initial into 1.0 torr air, without explosive liners. It is seen that maximum average velocities of 5.3, 8.2 and 13.7 km/sec, respectively, are obtained near the origin, giving shock Mach numbers of $M_s \sim 6$, 24 and 40. The rate of decay of the shock is very high. For example, it drops from $M_s \sim 40$ to $M_s \sim 11$, in a distance of about 180 cm, or an average Mach number decay rate of approximately $\Delta M_s/l \sim 5$/ft. Such high rates of decay can be expected if we consider the very small diameter of the channel, the prevailing high shock Mach numbers and a detonation driver that relies on focussed implosions that can give rise to several loss mechanisms[13]. The present results are also in reasonable agreement with the recent work of Ref.14, on the Voitenko launcher.

4. CONCLUSIONS

Considerable additional work will have to be done in order to evaluate the influence of radiative, convective, ablative and frictional losses; to optimize performance; to assess the future possibilities of this unique facility. Nevertheless, when one considers the sophisticated electrical condenser banks and switching equipment that are required even to produce $M_s \sim 40$ for electromagnetic drivers, this facility offers some worthwhile possibilities, especially with explosive liners when shock Mach numbers in the hundreds appear possible even with low density PETN ($\rho \sim 0.6$ g/cc) and very much greater with high-density PETN ($\rho \sim 1.5$ g/cc).

The existing facility will be used in the very near future to extend the present results by employing PETN with the $2H_2 + O_2$ driver into 8 mm and 25 mm channels with various gas counterpressures. Very much higher shock Mach numbers can be expected. At the same time, numerical analyses will be done to keep pace with and to guide the experimental program.

ACKNOWLEDGEMENTS

The financial support received from the Canadian National Research Council and the USA National Aeronautics and Space Administration under NASA Grant NGR 52-026-023 is acknowledged with thanks.

REFERENCES

1. Robert F. Flagg and I. I. Glass, "Explosive-Driven, Spherical Implosion Waves", The Physics of Fluids, Vol.11, No.10, pp.2282-2284, October 1968.
2. I. I. Glass, "Shock and Combustion-Wave Dynamics in an Implosion-Driven Hypervelocity Launcher", UTIA Review No.25, 1965 (also available as an AIAA preprint 25-108).
3. Robert F. Flagg, The Application of Implosion Wave Dynamics to a Hypervelocity Launcher. UTIAS Report No.125, December 1967.
4. I. I. Glass, "Research Frontiers at Hypervelocities", CASI Jl, Vol.13, No.8 and Vol.13, No.9, 1967, pp.348-366 and 401-426.

5. H. L. Brode, "Numerical Solutions of Spherical Blast Waves" J. App. Phys. 26, 6. p.766 (1955).
6. H. L. Brode, "Blast Wave from a Spherical Charge", Phys.Fluids, 2, 2, p.217, 1959.
7. J. Neumann and R. P. Richtmyer, "A Method for the Calculations of Hydrodynamic Shock",J. Appl.Phys. 21, p.232, (1950).
8. P.A.L.Sevray, Performance Analysis of UTIAS Implosion-Driven Hypervelocity Launcher. UTIAS Technical Note No.121, January 1968.
9. Robert F. Flagg and Gregory P. Mitchell, An Optimization Study of the UTIAS Implosion-Driven Hypervelocity Launcher Mk II, UTIAS Technical Note No. 130, 1968.
10. J.C. Poinssot. A Preliminary Investigation of a UTIAS Implosion-Driven Shock Tube, UTIAS Technical Note No.136, January 1969.
11. A. Elsenaar, A Numerical Model for a Combustion-Driven Spherical Implosion Wave, UTIAS Technical Note No. 144, 1969.
12. A. Elsenaar, Microwave Measurements of Projectile Motion in the Barrel of the UTIAS Implosion-Driven Hypervelocity Launcher, UTIAS Technical Note No.145, 1969.
13. D. E. Roberts, A Spectroscopic Investigation of Combustion-Driven Spherical Implosion Waves, UTIAS Technical Note No.140, May 1969.
14. B. K. Crowley and H. D. Glenn, "Numerical Simulation of the Generation, Propagation and Attenuation of a High-Energy (Mach 120 to 40) Air-Shock Experiment", Proceedings, 7th International Shock Tube Symposium, University of Toronto, 1970.

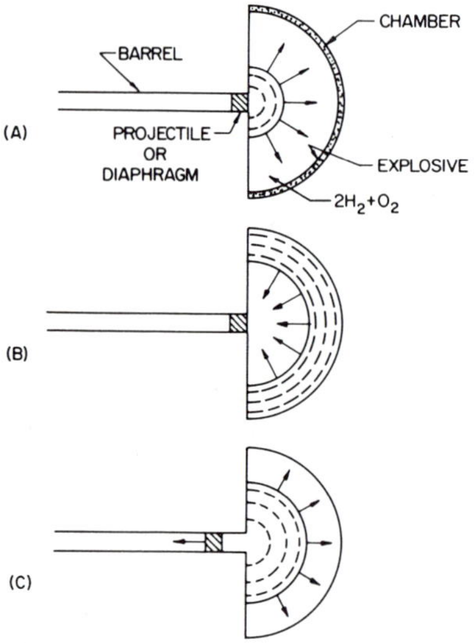

Figure 1
Schematic of principle of operation of UTIAS hypervelocity launcher or shock tube

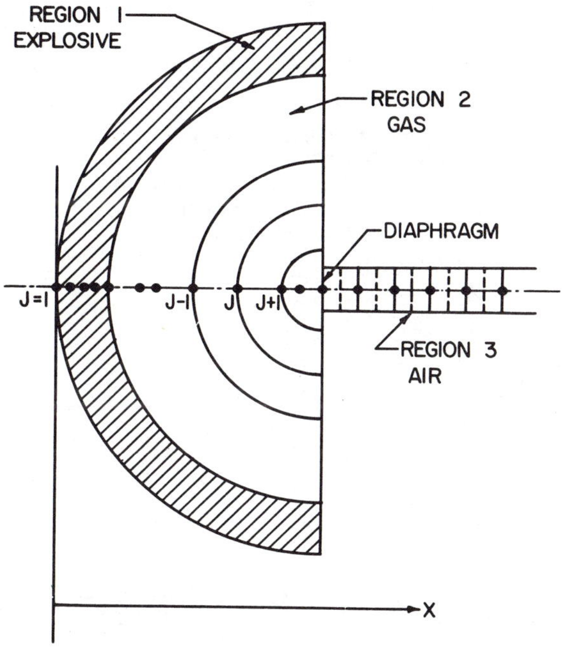

Figure 2
Distribution of mass points

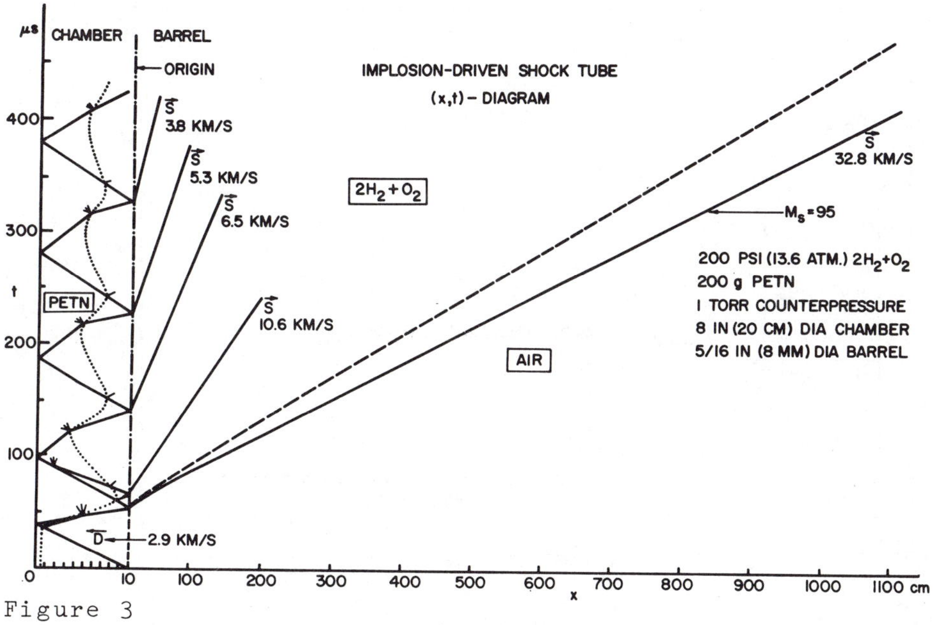

Figure 3
Wave-speed diagram in the (x,t) plane
for a UTIAS implosion-driven shock tube

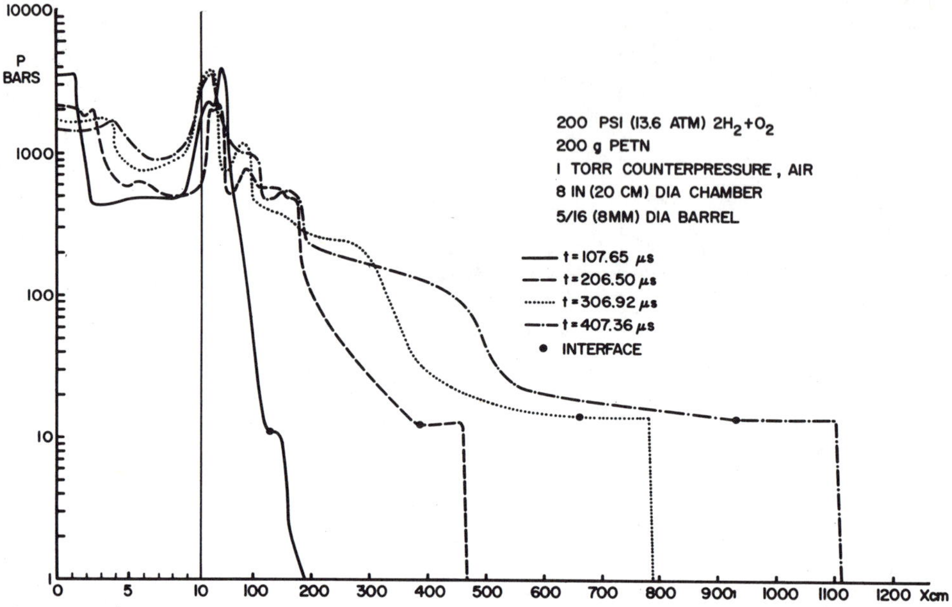

Figure 4
Shock tube case, pressure profiles

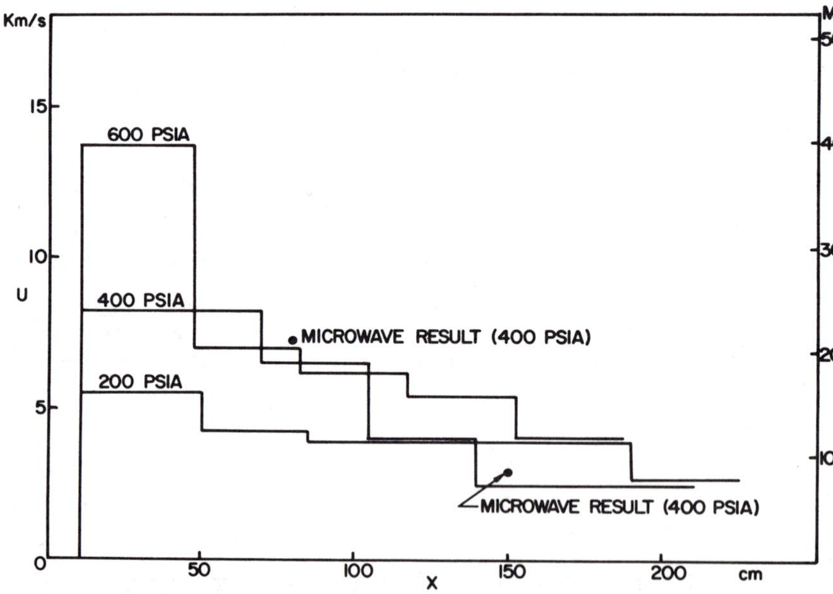

Figure 5
Experimental shock Mach number M_s vs. distance. 8 in (20.3 cm) dia chamber 5/16 in (8mm) dia barrel; 1 torr counterpressure, air; no explosive

AMES' HIGH-EXPLOSIVE SHOCK-TUBE FACILITY

ROBERT E. BERGGREN, DALE L. COMPTON,
THOMAS N. CANNING, WILLIAM A. PAGE
National Aeronautics and Space Administration,
Moffett Field, California

SUMMARY

This paper is one of two* presented at this symposium, describing a shock-tube facility in operation at the Ames Research Center of NASA, that utilizes a high-explosive driver to generate shock-wave speeds higher than achieved with more conventional shock tubes. This facility was developed primarily to simulate the flow conditions encountered at speeds from about 10 to 20 km/s during earth reentry of manned planetary flights. In its present configuration, a glass shock tube, 7.3-m long by 47-mm I.D., is imploded at one end by an explosive charge which produces a high-speed glass-jet piston. This piston, in turn, shock-heats the gas to the required test conditions. The explosive-shock-tube operation is fully confined by containment chambers which, with proper precautions, allow safe, quiet, and convenient operation within an existing laboratory building.

Initial tests in this facility have been made to determine the uniformity, purity, and usable length of shocked test gas (air). A relatively short explosive charge has been found adequate to drive a planar shock for extended distances. The shock speeds produced are approximately twice the detonation speed - about 16 km/s with the explosive presently used. These shock speeds, together with the appropriate shock-tube loading pressures, permit duplication of stagnation-region temperatures and pressures on flight vehicles. Test times to date have been less than desired for proper simulation of stagnation-region stand-off distances, but are adequate for the study of radiative cooling of the test gas.

*The other is "Instrumentation for Measuring Flow Properties in an Explosively Driven Shock Tube," by W. J. Borucki, D. M. Cooper, and W. A. Page.

SIMULATION REQUIREMENTS AND CAPABILITIES

It is the objective of this laboratory facility to duplicate shock-layer conditions produced during earth entry of vehicles at hyperbolic speeds. The requirements for simulating the thermodynamic conditions generated during return of a typical manned vehicle are given on an altitude-velocity plot in Fig. 1, taken from Ref. 1. This figure shows the corridor in which representative vehicles are constrained to fly after pullout for initial entry speeds in the range from about 10 to 18 km/s. The upper boundary of the corridor is imposed by the requirement for atmospheric capture of the vehicle and the lower boundary by a limitation of 10 g's on the maximum deceleration. Superimposed on this plot is a chart of thermodynamic properties behind a normal shock wave. These are representative of stagnation conditions on actual vehicles and it is these pressures and temperatures which must be duplicated in the test facility. It is seen that for the typical vehicle chosen the pressure range of interest, particularly at the higher entry velocities, is relatively narrow, with an average value of approximately 0.8 atm. The gas temperature, which is a strong function of vehicle speed, is 18,000° K at the maximum velocity shown, 18 km/s. In addition to the thermodynamic requirements shown in Fig. 1, the test-gas sample must be of sufficient size to allow reasonable simulation of the non-adiabatic effects expected during the high-heating-rate portion of the vehicle entry. It is desirable that the length of the test-gas sample be comparable to full-scale stagnation-region shock-layer thicknesses which range typically from 1 to 20 cm.

A shock tube is an effective device for producing these conditions, for at least a portion of the flight region of interest. By generating incident shock waves at speeds equal to the actual flight speeds and into air at the actual free-stream pressure, the gas properties and kinetics, which are primarily functions of speed and pressure, can be acceptably simulated. Further, if the region of test gas behind the incident wave can be made sufficiently long, the effects of finite-rate chemistry and the non-adiabatic effects of radiative energy loss at high speeds can also be simulated reasonably well.

The requirements of shock-wave speed, pressure, and flow length (i.e., test time) have been partially met by conventional arc-driven shock tubes. For example, Fig. 2 shows the capabilities of the Ames 12-inch arc-driven shock tube for duplicating part of the altitude-velocity corridor of interest. The explosively driven shock tube described in this paper provides a complementary facility with a test regime at the higher speeds of interest. The altitude-velocity range covered in this facility to date is also shown in Fig. 2. The arrows indicate that its velocity capability can be extended to both higher and lower speeds by proper choice of explosives. Shock-tube loading pressures are also given on this figure to show the range of that variable for proper simulation. Further studies are necessary to determine if adequate test times can be obtained with the present facility over the full velocity and pressure range of interest.

EXPLOSIVE DRIVER SYSTEM

The concept of the particular explosive driver chosen for use in the present facility evolved from a study conducted for NASA Ames Research Center by G. R. Fowles, et. al. (Ref. 2) of Stanford Research Institute to develop an explosively driven light-gas gun. During this work it was suggested by T. N. Canning that one of the explosive drivers being tested be also considered as a driver for a high-performance shock tube. The feasibility of this driver, which utilizes a "jetting piston" to generate a shock in a single-gas system, was then investigated for NASA Ames, by Gill and Goettelman (Ref. 3) of Stanford Research Institute. Gill and Goettelman also briefly examined a non-jetting explosive driver to shock-compress helium, used as a driver gas, in a two-gas system similar to a conventional shock tube. The study showed the jetting-piston driver to be particularly promising, with the result that this scheme was chosen for initial use in the present facility.

The manner in which the shock is generated is shown in Fig. 3. The shock tube is a Pyrex-glass tube, 7.3-m long, with a 47-mm I. D., and 51-mm O. D., and is operated as a single-gas system with a uniform initial pressure throughout the tube length, i.e., without a diaphragm. A cylindrical charge of high explosive, (presently 2.25 kg of Composition C-4 explosive, about 28-cm long and 10-cm O. D.), is positioned concentrically about the outside diameter of one end of the tube. The explosive is detonated at its base and a plane detonation front proceeds through the charge at the characteristic detonation velocity of the explosive. This detonation front crushes the glass inward as it proceeds with the result that a small fraction of the imploded glass is accelerated forward in a jet of glass particles. This jet then acts as a piston to drive a shock wave through the test gas. As indicated by the theory of Birkhoff, et. al. (Ref. 4) for jetting produced by collapsing plates and by the work of Koski, et. al. (Ref. 5) on jetting from collapsing metallic cylinders, the resulting jet velocity for this charge geometry is approximately twice the detonation velocity. A wide range in shock velocity (approximately as shown on Fig. 2) may be obtained by using explosives of different detonation speeds. In the present facility, using Composition C-4 explosive having a detonation velocity of about 8 km/s, shock speeds from 14.7 to 16.9 km/s have been measured. This variation in velocity may be due (1) to differences in the effectiveness of the glass jet which are caused by inconsistencies in charge construction, and/or (2) to variations in the detonation velocity which may result from differences in charge density and the age of the explosive (the oldest explosives used showed loss of plasticity and signs of deterioration).

It was found in the initial tests by Fowles that when a glass cylinder is used, the jet produced consists of a stream of glass vapor and small particles (estimated by Gill and Goettelman to have a characteristic dimension of the order of 10^2 to 10^3 microns) which spread out over the cross section of the tubing forming a diffuse piston that is sufficiently uniform to produce a planar shock front. Although this piston is porous and there is some leakage of the test gas through it, the piston density is adequate to drive a shock for extended distances beyond the relatively short charge. It was concluded by Gill, from measurements of test time (Ref. 3), that the leakage or mass flow of test gas past the piston at initial gas pressures of 5 torr was large, with the result that the actual test time was very much shorter than ideal, and the flow rapidly reached steady-state conditions where test time did not increase with shock travel distance. At pressures below 1 torr Gill reported test times that approached those predicted by ideal shock-tube theory. This result has not been substantiated in the present facility, where the measured times are, in fact, much shorter than ideal as will be discussed in a subsequent section.

FACILITY DESIGN

The complete facility is located within a reinforced-concrete laboratory building and is operated from a remote control room. The shock tube itself is enclosed within four steel containment chambers, which are: an explosion chamber, a containment tube, a test section, and a catcher (see Fig. 4). The chambers are fastened together with quick-connect clamps for rapid assembly and disassembly and, except for the explosion chamber, are separately moveable on a common track to provide easy access to the inside for operation and maintenance. This provides a versatile arrangement, in that individual components may be separately modified so that changes in the geometry and mode of operation of the basic shock tube can be readily made.

The explosion chamber is a cylindrical pressure vessel, approximately 3-m in diameter by 7-m long with a 2.5-cm-thick wall, that was designed to contain the explosive detonation. Access to this chamber is gained through a large door located at one end. The glass shock tube is positioned along the chamber axis with the relatively short explosive charge located at the center. The design of this chamber was based upon the empirical relation of Loving (Ref. 6), which gives the shock pressures developed in a closed chamber as a function of chamber volume, and quantity and type of high explosive. Chamber shape, which may influence the stresses developed, is not, however, accounted for in this relationship. The pressure-time history experienced at the walls by successive reflections of the initial pressure wave is determined, in part, by the shape of the chamber. In certain cases the stresses produced by these multiple pressure reflections can combine with those due to vibration of the structure at its natural frequency to produce peak stresses that are higher than that induced by the initial pressure pulse. Thus to obtain a more precise measure of the maximum safe explosive charge for this particular vessel, the stresses at several points on the chamber wall are monitored with strain gages during operation. The measurements have shown that in this chamber the stress created by the first pressure wave from the detonation is the highest but that continued chamber vibration and/or multiple pressure reflections act to sustain a relatively high-stress level for many cycles so that the fatigue life of the structure may be important. (The pressures from the initial blast wave are reduced by evacuating the chamber to approximately 1 torr.) The measured stress levels appear to be substantially higher than that predicted by Loving's relation but additional measurements will be made to confirm this.

The glass shock tube extends out one end of the explosion chamber and through the smaller steel containment tube, 20-cm I. D. by 3-m long, containing small observation ports at nine equally spaced stations. Photomultiplier tubes view the shock tube through six of these ports to detect the passage of the shock. The remaining length of glass shock tube extends into a test chamber containing eight 20-cm-diameter ports through which a variety of observations and measurements are made. Additional measurements are made just past the end of the glass tube through slits in a splitter-plate assembly which extends slightly into the tube and provides a means of excluding the wall boundary layer from the field of view. The facility assembly is terminated with a thick-wall steel catcher that serves to arrest the high-velocity jet. With the present glass-tube configuration however, no significant damage has resulted to the catcher from the impact of the glass jet particles. The larger craters formed in the catcher have been typically 1 to 2 mm in diameter, which would correspond to a particle size of 10^2 to 10^3 microns, in agreement with the size estimated earlier by Gill and Goettelman. These larger craters are relatively few however, which, coupled with the very minor cumulative damage that has occurred from many firings indicates that the mean particle size may be considerably smaller than the above estimate.

The shock tube is constructed of welded and mechanically joined sections of Pyrex-glass tubing. Each tube is chemically cleaned, evacuated to a pressure of less than 10^{-2} torr and held at this pressure in an oven at 400° K for several days to reduce contamination of the test gas. It is then transferred to the facility while evacuated and charged with the test gas without exposing its interior to the atmosphere. Initially it was found that impurities were introduced during the bakeout process from deterioration of the neoprene O-ring pressure seals at the mechanical joints. This was alleviated by replacing the neoprene with teflon seals.

During operation the entire length of glass shock tube is shattered but there is essentially no damage to the permanent enclosing structures. The test-section observation windows are protected from the impact of glass fragments with replaceable plastic disks (plexiglas, 6-mm thick) mounted in front of the windows. In the main explosive chamber, light-weight plastic materials are used for support of the shock tube and charge to minimize any damage from shrapnel. An exception has been the metal cases of the detonators, which, although small, have produced significant cratering on the vessel walls. After firing, the noxious gases, carbon, and other particulate matter are partially removed from the explosive chamber with a positive-pressure blower system. Further purging is then accomplished with the vacuum system.

INSTRUMENTATION

The facility instrumentation presently consists of a variety of fast-response photographic and electronic instruments to measure shock-wave speed and gas properties behind the wave front (see Fig. 5). Those used to observe and measure gas properties include an image-converter camera, a Kerr-cell-shuttered camera, a spatially-resolving Kerr-cell-shuttered spectrograph, and time-resolving multichannel radiometers, all of which are described in the companion paper to be presented at this symposium (Ref. 7). A description of instrumentation for the measurement of shock speed, the basic quantity required for all tests, is given below.

Six shock detectors are spaced along the shock-tube axis, (Fig. 5). A diagram of one of these shock-detector systems is shown in Fig. 6. The passage of the shock wave at each station is detected by a photomultiplier tube having an S-20 photocathode surface that views the shock tube through a series of 1-mm-diameter collimating orifices. The photomultiplier output is amplified (details of the electronic circuitry are described in Ref. 8) and the resulting signal is twice differentiated and further amplified in a trigger amplifier to provide the pulse required to trigger a 100-MHz electronic chronograph. Double differentiation of the signal produces a fast-rise pulse and gives a more accurate measure of the precise time of passage of the shock front. High-pass filters are included in the trigger-amplifier circuitry to reject any relatively slowly changing signals preceding the shock, which might otherwise provide an erroneous trigger pulse to the chronographs. The photomultiplier signal is also recorded on an oscilloscope to measure the variation of radiative intensity with time. Thus, the shock detectors also serve as broad-band radiometers to measure changes in the shock-tube flow from station to station; a typical record is shown in Fig. 6. These shock detectors have proved to be reliable and allow determination of shock speed to an accuracy better than 1 percent. A measured time-distance history for a typical test is shown in Fig. 7. A plot of the variation of shock speed with distance, as determined from the time-distance data, is also included in this figure, and it can be seen that the velocity change of the shock front is small (about 1 percent/meter) over the measured distance.

SHOCK-TUBE TEST TIME

Tests to date in this facility have been made primarily to determine the uniformity, purity, and usable length of the shocked test gas. The shock tube has been operated at initial loading pressures of 0.2, 0.4, and 0.8 torr (as measured with a McLeod gage) with a test gas of dry, high-purity air, to which 1 or 2 percent of hydrogen was added as a diagnostic aid in spectrographic studies. A description of the gas properties behind the shock, as determined from these tests, is contained in the companion paper (Ref. 7). A discussion of the useful length of test gas for the particular explosive-driver configuration used in these tests is given below.

Test time is defined here as L/U_s, where U_s is the shock-wave speed and L is the length of the column of useful gas behind the shock front in which the gas is free of contamination and the observed flow is uniform. The two observables which signaled the end of useful test time for this test series were: (a) the first appearance of radiation from an impurity, sodium, as determined from radiometric and spectrographic data, and (b) the observance of irregular luminous structure seen in photographs of the flow, which is interpreted to be large-scale turbulence or mixing of the flow. The test data indicate that there is a relationship between these two factors, in that sodium is generally first observed at about the same position behind the shock that the irregular structure is first noted. Spectra and photographs are shown in Ref. 7. Experimental evidence suggests that the sodium radiation results from glass vaporized during the initial implosion process rather than from subsequent heating of the walls of the tube by the shocked test gas (see Ref. 7). It appears, therefore, that this turbulent region contains, in addition to the test gas, piston material that has been swept forward.

The test time, as determined by the above criteria, is plotted in Figure 8 for the three initial-loading pressures. At each pressure the measurements made by spectrographic, radiometric, and photographic means are separated slightly for clarity. There is no more than one data point per instrument for each run, but data from all three instruments were not always obtained. The data show that at a given loading pressure there is an appreciable variation in test time for different tests but no systematic difference between the times of appearance of the sodium and the flow irregularities. The test time predicted by Mirels' laminar-flow theory (Ref. 9) is also plotted on this figure as a reference for comparison, although it does not strictly apply in the present case because it does not consider either the mass flow through the porous glass piston or radiative cooling. Furthermore, as mentioned above, photographic evidence suggests that test time is determined by the turbulent mixing process rather than by mass entrained in a laminar boundary layer or by flow through the piston. The large scatter in observed test time also suggests a random rather than an ordered mechanism. The first appearance of irregular structure is influenced by test pressure; at the lowest pressure (0.2 torr) it was observed to be very close to the shock and in one case actually produced distortion of the shock front. In contrast, at the higher pressures (0.4 and 0.8 torr) the shock fronts were consistently planar and the start of the region of turbulence was appreciably farther behind the front.

As mentioned previously, the earlier tests of Gill and Goettelman (Ref. 3) showed much longer test times than measured here, but the discrepancy may be primarily due to a difference in the criteria used for determining this time. In the earlier studies the length of test gas was taken to be the length of the highly luminous region behind the shock front, but Gill's spectrographic data showed the presence of sodium much nearer the shock, at approximately the same position (or time) as that observed in the present facility.

Although the test times at the higher pressures in the present facility are long enough to provide information on the approach to equilibrium in the flow behind the shock wave and the effects of radiative cooling, the times are not sufficient to duplicate representative stagnation-region shock-standoff distances. The test results reported have all been made with the same explosive-driver geometry. Work is now being directed, in part, towards exploring the effects of changes in explosive-charge geometry, and shock-tube length and diameter, in an effort to substantially increase the useful test time for this type of jetting driver.

REFERENCES

1. Page, William A.; Compton, Dale L.; Borucki, William J.; Ciffone, Donald L.; and Cooper, David M.: Radiative Transport in Inviscid Nonadiabatic Stagnation-Region Shock Layers. AIAA 3rd Thermophysics Conference, Los Angeles, June 24-26, 1968. AIAA Paper No. 68-784.
2. Fowles, G. R.; Isbell, W. M.; and Shipman, F. H.: Feasibility Study of an Explosive Gun. Stanford Research Institute Final Report No. 1, Contract NAS2-1361, July 1964.
3. Gill, S. P.; and Goettelman, R. C.: Implosively Accelerated Shock Tube Driver. SRI Final Report, Contract NAS2-3129. NASA CR-950, November 1967.
4. Birkhoff, G.; MacDougall, D. P.; Pugh, E. M.; and Taylor, G.: Explosives with Lined Cavities. J. Appl. Phys., vol. 19, 1948, pp. 563-582.
5. Koski, W. S.; Lucy, F. A.; Shreffler, R. G.; and Willig, F. J.: Fast Jets from Collapsing Cylinders. J. Appl. Phys., vol. 23, 1952, pp. 1300-1305.
6. Loving, F. A., Jr.: Noise-Reducing Structure. U. S. Patent No. 3,165,916, Jan. 19, 1965.
7. Borucki, William J.; Cooper, David M.; and Page, William A.: Instrumentation for Measuring Flow Properties in an Explosively Driven Shock Tube. Seventh International Shock Tube Symposium, Toronto, Canada, June 23-25, 1969.
8. Craig, R. A.; and Davy, W. C.: Absolute Radiometers for Use in Ballistic-Range and Shock-Tube Experiments. Prospective Ames publication.
9. Mirels H.: Test Time in Low-Pressure Shock Tubes. The Physics of Fluids, vol. 6, 1963, pp. 1201-1214.

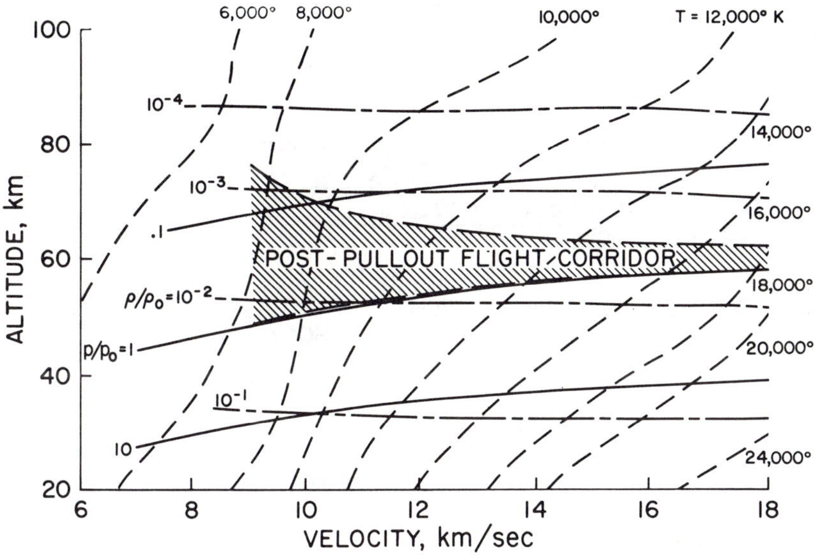

Figure 1
Thermodynamic properties behind normal shock wave. Corridor is for a vehicle with $W/C_DA = 500\ kg/m^2$, $L/D = 0.5$.

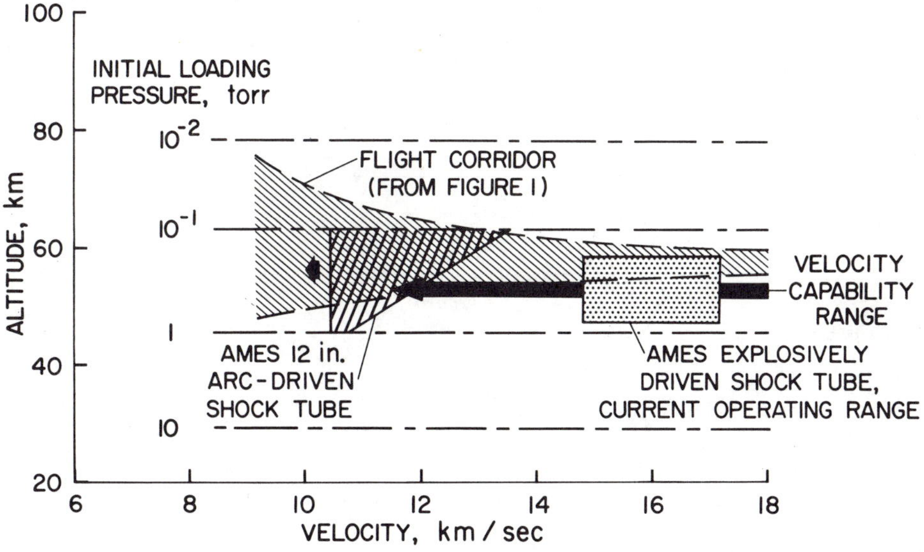

Figure 2.
Simulation of flight conditions by two NASA - Ames shock tubes.

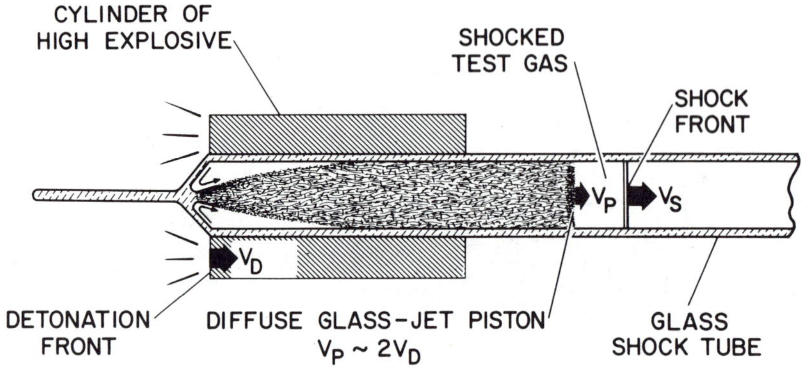

Figure 3
Shock-tube driver.

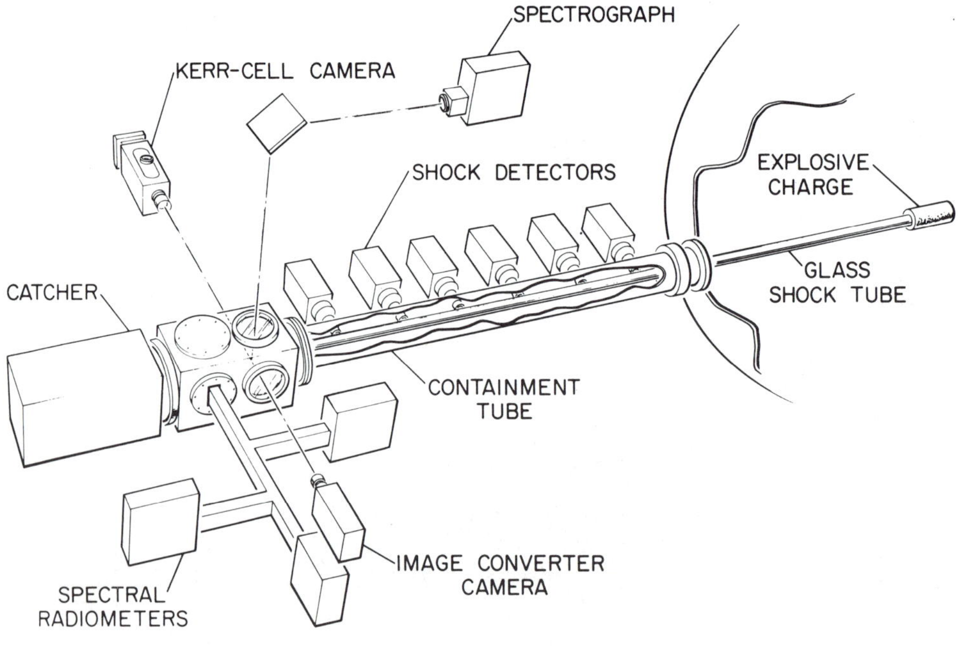

Figure 5
Facility instrumentation.

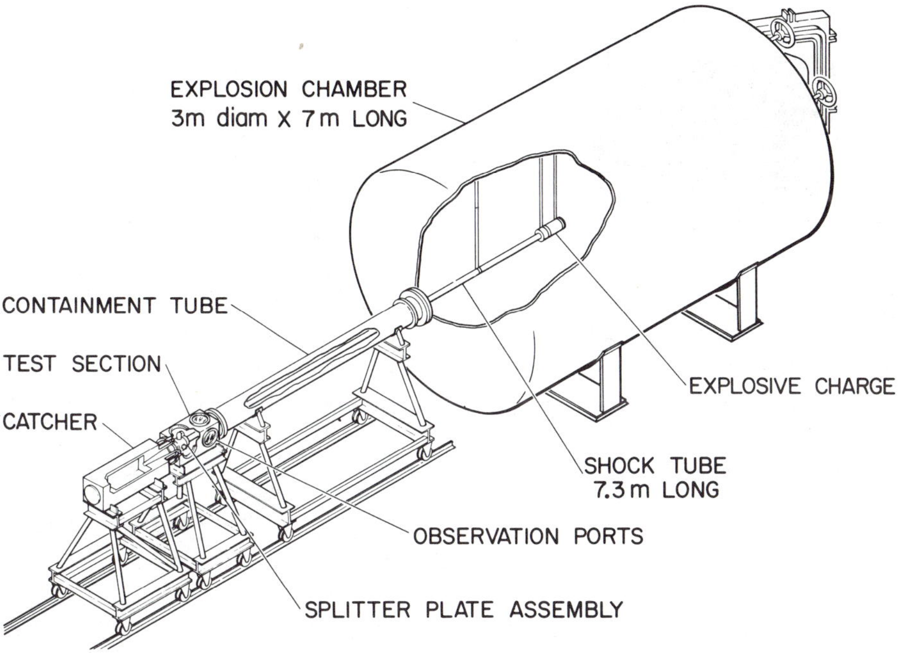

Figure 4a
Explosive-shock-tube facility.

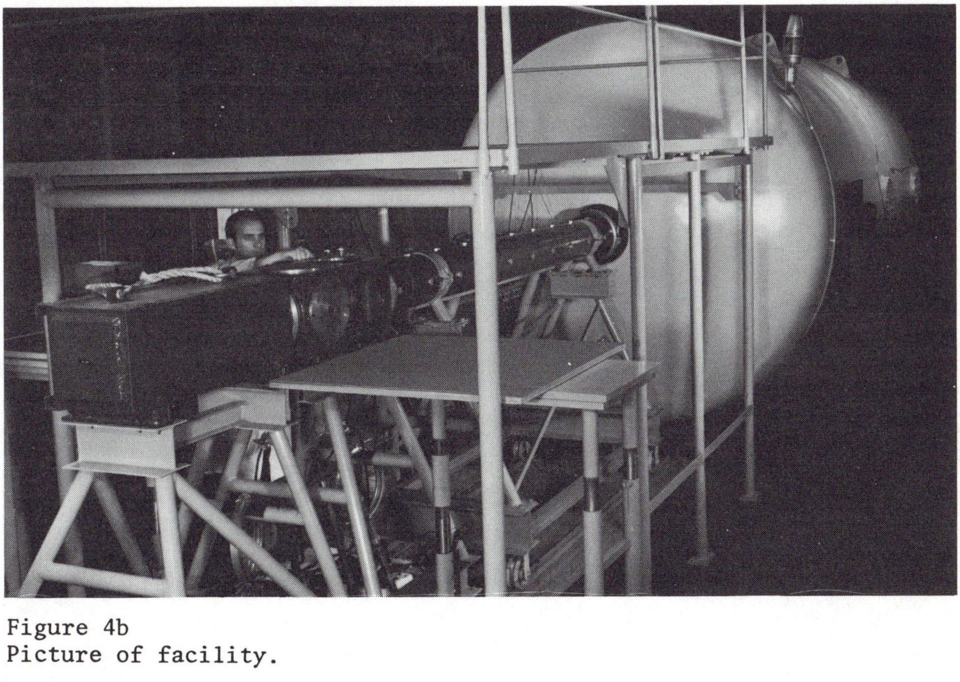

Figure 4b
Picture of facility.

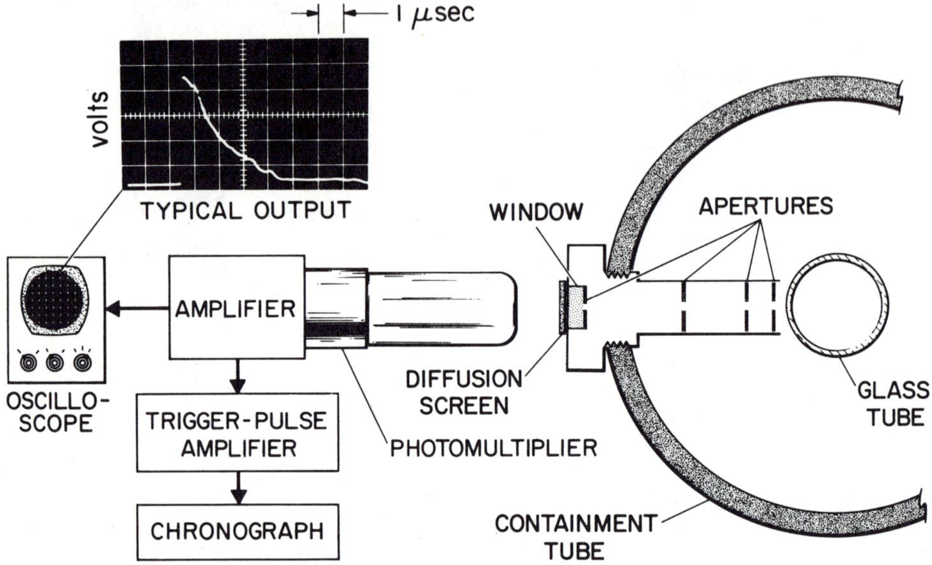

Figure 6
Shock detector system.

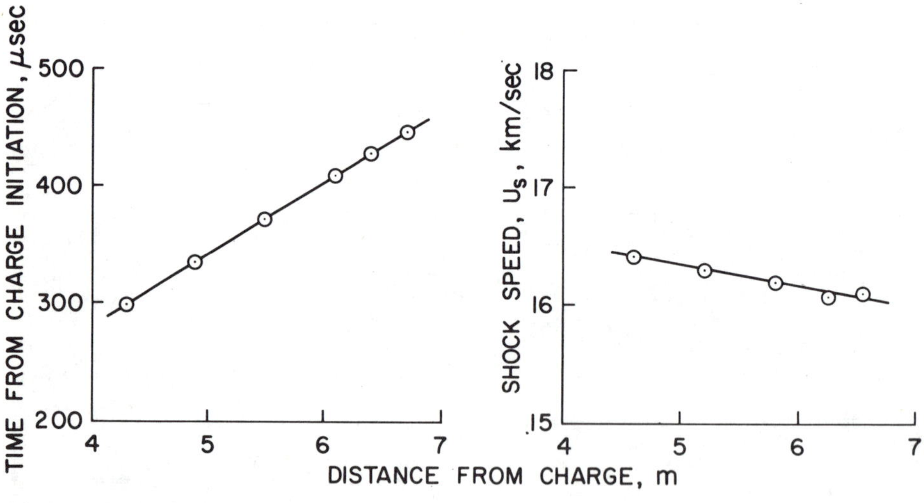

Figure 7
Time-distance and velocity-distance histories
of shock wave, initial pressure = 0.4 torr.

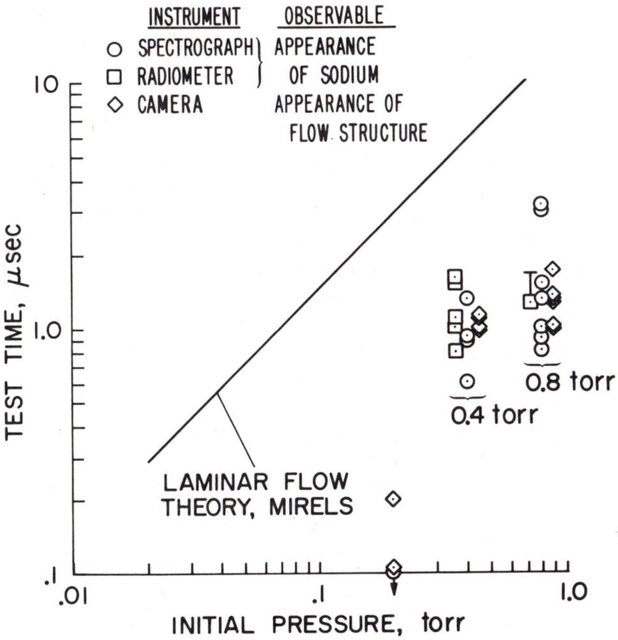

Figure 8
Test-time results, U_s = 14.7 to 16.9 km/sec.

ANALYSIS AND PERFORMANCE OF AN EXPLOSIVELY DRIVEN SHOCK TUBE*

S. P. GILL and W. V. SIMPKINSON
Physics International Company
San Leandro, California, USA

SUMMARY

An explosive shock tube driver is described in which helium initially at 42 atm is shock compressed to 3.3 kbar by an implosively formed piston traveling at 6.3 km/sec. An idealized theory of explosive driver operation is presented, and the observed nonideal effects of pressure tube radial expansion, boundary-layer growth, liner jetting, preinitiation of the driver explosive, and explosive gasdynamics are described and theoretically analyzed. Performance of an explosively driven shock tube containing air at initial pressures in the range 200 to 500 μHg is discussed. Experimental data were obtained from the shock-heated air by means of high-speed rotating mirror cameras, photodiodes, time-of-arrival pins, and pressure transducers. Experiments were also conducted to determine the extent and degree of precursor ionization. Observed shock velocities were in the range 14 to 16 km/sec.

INTRODUCTION

The potential performance capability of explosively driven hypervelocity shock tubes, shock tunnels, and projectile launchers far exceeds the existing capability of conventional energy sources. In recognition of this potential a number of experimental programs have been conducted in recent years to determine the principal technical problems in the way of realizing this potential.

The primary difficulty in developing explosively driven systems is effectively controlling and directing the violent energy release of an explosive detonation. Various methods are currently being developed. I. I. Glass and his colleagues (References 1 and 2) utilize a spherical implosion generated by the detonation of a hemispherical liner of solid explosive. In this method the products of the explosive reaction drive an implosion wave toward the center of a hemisphere to yield essentially a point source of very high energy gas. This gas is subsequently used to accelerate a hypervelocity projectile. Seay, et al. (Reference 3), have used explosives to generate shock waves in a gas. In this method the explosive is in contact with gas in a shock tube, and a plane detonation wave in the explosive expands at its surface to drive a strong shock in the gas.

* This work was supported in part by Contracts No. NASw-1769 and No. NAS2-4903 with the National Aeronautics and Space Administration

The USSR has been particularly active in explosively driven devices. Voitenko (References 4, 5, 6) has developed a device in which an explosively accelerated metal plate compresses gas in a cup-shaped cavity. Extremely high energy densities are obtained in the trapped gas, which is used to drive strong shocks in a shock tube. Zatsepin (Reference 7), Titov (Reference 8), and others have used a "cumulative charge" to drive high-velocity shock waves and projectiles. One of the authors (Reference 9) has developed an explosive shock tube in which a cloud of glass fragments jetted forward at twice detonation velocity by implosive tube collapse is used to drive a shock in low-pressure air.

The purpose of this paper is to describe a particular type of explosive driver system and its application to a hypervelocity shock tube. In its simplest form, the linear explosive driver (Figure 1) consists of a thin-walled tube containing the driver gas, usually helium. This pressure tube is surrounded by an annulus of explosive which, when detonated, progressively collapses the tube. The collapse region, traveling at detonation velocity of the explosive, completely seals the tube and acts like a high-velocity piston driving a strong shock wave into the helium driver gas.

This simple mechanical system offers great potential in controlling the release of explosive energy, and transforming the energy into a form appropriate for research in high-energy gasdynamics. Over the past few years considerable effort has been spent in investigating the detailed behavior of linear explosive drivers, and the present investigation is representative of research currently being conducted on applications of the driver to the study of hypervelocity shock wave phenomena.

LINEAR EXPLOSIVE DRIVER

Ideal Driver Operation

In the idealized model of explosive driver operation (Figure 1) a conical piston traveling at detonation velocity is formed by the implosively collapsed pressure tube walls. This piston is assumed to seal the tube completely, so that the driver gas is forced forward to form a strong shock wave.

For the range of thermodynamic conditions achieved in explosive driver systems the helium driver gas can accurately be considered as an ideal gas with polytropic exponent $\gamma = 5/3$. The ideal driver can therefore be analyzed in terms of an ideal piston-driven shock tube with an ideal gas.

Ideally, the explosively formed piston travels at the constant detonation velocity, D, and the flow velocity of the gas is equal to the piston velocity. The Hugoniot shock jump conditions for a strong shock in ideal gas relate the shock velocity and shock pressure to the known piston velocity:

$$p_s = \frac{\gamma+1}{2} \rho_o D^2 \tag{1}$$

$$U_s = \frac{\gamma+1}{2} D \tag{2}$$

For helium driver gas $U_s = 4/3\ D$. Assuming typical values of D = 6.3 km/sec (nitromethane explosive) and $\rho_o = 7.0 \times 10^{-3}\ g/cm^3$ (42 atm He), the ideal shock velocity is 8.4 km/sec and the shock pressure is 3.7 kbar.

Operation of the driver is most conveniently represented in terms of a wave diagram; i.e., a distance-time trajectory of the detonation and shock waves (Figure 2). Since the driving piston is conical, a startup process occurs before the shock front emerges ahead of the detonation. Normalizing the coordinates with respect to the diameter, d, and the detonation velocity, D, so that

$$\bar{x} = \frac{x}{d} \quad \text{and} \quad \bar{t} = \frac{Dt}{d},$$

the entire wave diagram including the startup process is ideally independent of size scaling.

A typical wave diagram for the explosive driver, including experimental data points, is shown in Figure 2. It is interesting to note that the detonation velocity increases by a factor of approximately 1.06 at a normalized time of 10; this detonation speedup is due to precompression of the explosive by the driver shock. Ideal detonation and shock wave trajectories are shown in Figure 2. If the ideal shock wave trajectory is corrected to account for the detonation speedup, the ideal shock wave trajectory coincides exactly with the experimental data. If the wave diagram is drawn from shock breakout and normalized to the pressurized nitromethane detonation velocity, the observed wave diagram is very nearly ideal. Experimental results are shown for drivers differing in size by a factor of 6; the close coincidence of the normalized data justifies the idealized concept of driver operation as a simple mechanical system.

Driver efficiency is defined to be the fraction of chemical energy in the explosive converted into kinetic and internal energy of the driver gas. The rate at which the explosive energy is released is CDE, where C is the explosive mass per unit length and E is the specific chemical energy. Similarly, the rate at which energy of the driver gas is increased is $GU_s(e_s + \frac{1}{2}D^2)$, where G is the gas mass per unit length in the unshocked gas and e_s is the specific internal energy of the shocked gas. From the shock jump conditions for a strong shock wave $e_s = \frac{1}{2}D^2$ and $U_s = \frac{\gamma+1}{2}D$. The efficiency, ϵ, of energy conversion is therefore given by

$$\epsilon = \frac{\gamma+1}{2}\frac{G}{C}\frac{D^2}{E} \tag{3}$$

in a helium driver using nitromethane explosive $D^2/E = 8.85$ and $\epsilon = 11.8\ G/C$.

Energy conversion efficiency is limited by a number of nonideal phenomena which affect driver performance, not the least of which is that at sufficiently high shock pressures the explosively formed piston fails to seal the tube. The explosive driver used in the hypervelocity shock tube experiments described in this paper has an ideal efficiency of 12% and a measured efficiency in excess of 7%. Efficiencies of up to 20% have been obtained in some advanced designs. It should be noted, however, that efficiency of energy conversion is not necessarily a dominant factor in explosive driver systems. For example, nitromethane, with an energy density of 4.5 MJ/kg, is a commonly available industrial solvent that can easily be obtained in large quantities. For small-scale experiments the quantity of explosive is not a significant factor, so that efficiency is of secondary importance. Only in large-scale experiments is it necessary to optimize efficiency in order to reduce the total amount of explosive.

Nonideal Phenomena

The ideal explosive driver is characterized by an explosive collapse process that acts as an ideal piston. In this simple view details of the driver gasdynamics and explosive collapse process are neglected. In the past few years Moore, (Reference 10); Waldron, et al. (Reference 11); and one of the authors (Reference 12) have experimentally identified and analyzed several phenomena that profoundly influence the performance of explosive driver systems. For the efficient high-pressure drivers found to be most useful as energy sources, the most important phenomena have been identified (Figure 3) as pressure tube expansion, boundary-layer growth in the driver gas, preinitiation of driver explosive, and jetting phenomena. Considerable additional observations and analyses have been carried out, but are too extensive for the purposes of this paper.

Pressure Tube Expansion. To increase efficiency of energy transfer between the explosive and the driver gas, shock pressure is maintained at a significant fraction of explosive detonation pressure. Typical driver pressures, 3 to 15 kbar, exceed the yield strength of the driver tube, and dynamic radial expansion occurs. A flash radiograph (Figure 4) of an explosive driver indicates the expansion of the pressure tube by the driver shock, and subsequent collapse of the tube by the explosive detonation. The amount of radial expansion of the pressure tube depends on the duration and amplitude of the pressure pulse, on the geometry of the pressure tube and explosive containment tube, and on the properties of the confining material. Tube rupture has been observed experimentally to occur between 20% and 40% radial expansion. Rupture normally causes severe disturbances in the shock trajectory, and the confinement is usually chosen to limit the maximum radial expansion to less than 20%.

Details of the pressure tube expansion are complex and difficult to estimate analytically. Transit time for the shock-induced pressure disturbance through the pressure tube, explosive layer, containment tube, and back again is the same order of magnitude as the duration of the pressure pulse. Calculations based on the limiting cases of short transit times (quasi-steady analysis) and long transit times (infinite confinement) are grossly in error. We have found it necessary to perform detailed computations of the process using a one-dimensional (radial symmetry), time-dependent computer code in which the pressure tube and confinement tube materials are accurately represented by tensor constitutive equations. These computations have been checked where possible with experimental results (Figure 5), and have been found to be of sufficient accuracy to justify confidence in the results.

Expansion of the pressure tube generates rarefactions in the shocked driver gas which modify its properties and weakens the driver shock. The complex, time-dependent effects of the pressure tube expansion on the driver performance have been calculated numerically using the quasi-one-dimensional stream tube approximation. As the wall expands, radial rarefactions are created throughout the shocked-gas region that cause the pressure to decrease nearly uniformly in the gas slug. This drop in pressure causes the shock velocity to decrease, which in turn causes the particle velocity behind the shock front to decrease. The particle velocity at the piston is of course constant and is equal to the piston velocity. Briefly, the results of these calculations indicate that radial expansion of the pressure tube generates a quasi-static reduction in pressure of the shocked-gas slug, and a subsequent weakening of the shock strength.

Boundary-Layer Effects. Another gasdynamic phenomenon that has an influence on driver performance is boundary-layer growth. As a result of relative motion between the shock-accelerated driver gas and the pressure tube surface, a boundary layer will be formed behind the shock front. The effect of this boundary layer on the performance of the explosive driver is similar to that observed in conventional shock tubes. The deceleration of the gas near the tube walls caused by the boundary-layer growth generates rarefactions that tend to lower the shocked-gas pressure and velocity and to attenuate the shock front. In a nonjetting driver, the boundary-layer gases appear to receive little axial acceleration in the collapse region and, as a result, these gases are trapped by the implosive collapse. This process continues until a steady state is attained and the mass flow trapped in the collapse process equals the mass flow across the leading shock. At this time the shock velocity has attenuated to the detonation velocity and the slug of shocked gas has grown to its maximum length.

The effect of boundary-layer growth on the performance of an explosive driver is shown in Figure 6. For $\overline{x} > 20$ the measured shock trajectory coincides with the ideal shock trajectory, since the boundary-layer growth is negligible. The measured shock trajectory is seen to parallel the detonation trajectory at a distance of $\overline{x} > 100$. The mass flow entering the shock is equal to the mass trapped by the liner collapse process, and the shocked-gas length remains stationary.

Mirels (Reference 13) has investigated the growth of shocked-gas length in a conventional pressure-driven shock tube. The basic assumption underlying his work is that boundary-layer gas behind the shock front flows around the driver gas interface. The resulting loss diminishes shocked-gas length. The same analysis is applicable to an explosive driver if the assumption is made that boundary-layer gas is trapped in the collapse process.

Mirels' calculation of maximum shocked-gas length--the length at which mass flux in the boundary layer equals mass flux through the shock front--exceeds the observed maximum length by a factor of two. This discrepancy is probably due to inadequate data on the transport properties of helium at high temperatures and pressures. However, when the theoretical steady-state value is adjusted to fit experimental data, the predicted growth rate greatly underpredicts the observed rate of growth of shocked-gas length (Figure 7). This discrepancy appears to come primarily from Mirels' assumption of quasi-steady flow.

In order to improve the boundary-layer predictions, a calculation was performed using a quasi-one-dimensional computer code. The explosive piston was assumed to be ideal except for a boundary-layer-induced leak rate given by

$$V = V_m \left(\frac{\ell_x}{\ell_m} \right)^{4/5} \qquad (4)$$

Here ℓ_x is the shocked-gas length at shock position x, and V_m is the leak velocity corresponding to the maximum shocked-gas length ℓ_m. In this calculation rarefaction waves caused by the leak overtake and attenuate the shock. Results of these calculations agree well with experiment (Figure 7).

Preinitiation of Driver Explosive. Expansion of the pressure tube by driver shock pressures in the range 3 to 15 kbar exposes the explosive to transient high pressures, and under certain circumstances the explosive can detonate prematurely. Nitromethane has been the preferred explosive in explosive drivers for several years, primarily because it is classed as an industrial solvent and is easily available in large quantities. Nitromethane drivers were used for some time without any apparent problem arising from exposure to high pressure, but as driver systems were scaled to larger sizes, apparent irregularities in the late-term behavior of the detonation trajectory were noticed. Ionization pins used to detect passage of the detonation wave reported early, and indicated a possible acceleration of the detonation wave to a velocity roughly equal to the shock velocity in the driver. The ionization pin readings were erratic, but clearly implied some energy release when the nitromethane was exposed to pressures of approximately 5 to 10 kbar for periods in excess of 75 μsec.

Initiation of homogeneous nitromethane at shock pressures in excess of 80 kbar is at least fairly well understood at present (Reference 14). The initiation process consists of three phases: (1) compression and shock heating of nitromethane to temperatures in excess of 1000°K, (2) a chemical induction period governed by an Arrhenius reaction rate leading to runaway thermochemical decomposition of the explosive, and (3) a rapid increase in pressure and temperature behind the shock front and acceleration of the shock into a selfsustaining detonation. Calculations of the chemical induction time for homogeneous nitromethane compressed to 10 kbar indicates an expected induction time of 10^7 sec, far in excess of the observed times on the order of 10^{-4} sec. It was clear from this analysis that the observed initiation phenomenon was based on a physical mechanism substantially different from that observed at pressures in excess of 80 kbar.

After an experimental and analytical investigation too extensive to report in this paper, the initiation mechanism was identified as the compression of minute air bubbles trapped in the liquid nitromethane.

The initiation process proceeds by compression-heating of the trapped air to temperatures on the order of 3000°K. The compression temperature depends on the equation of state, and it was necessary to use real-air thermodynamic tables to agree with experiment. The temperature of the nitromethane at the liquid-gas interface rises to temperatures on the order of 1000°K by heat conduction, and a complex thermochemical heat transfer process begins. For bubble dimensions smaller than approximately 10^{-4} cm, heat loss by conduction exceeds heat input by chemical decomposition, and the disturbance is quenched. For larger bubble dimensions the high surface temperature leads to runaway thermal decomposition with a characteristic chemical induction time. Calculations of chemical induction time for real air bubbles are shown in Figure 8, and indicate an induction time at 10 kbar on the order of 10^{-4} sec.

Dependence of the initiation mechanism on temperature rise in trapped air bubbles implies that the observed preinitiation phenomenon can be suppressed by prepressurization of the explosive. Temperature rise in a compressed gas depends on the initial pressure, and increasin the initial pressure decreases the final temperature. The induction time is dependent on temperature, and a small change in nitromethane interface temperature greatly increases initiation time. Calculations for an initial pressure of 40 atm (Figure 8) indicate an increase in induction time by five orders of magnitude.

Prepressurization of the explosive driver has been experimentally verified as a means of suppressing preinitiation phenomena, and at present there appears to be no known limit in scaling nitromethane explosive driver systems to very large sizes for aerodynamic applications.

Jetting Phenomena. At sufficiently high implosive collapse velocities, the imploding pressure tube walls can jet forward into the driver gas (Figure 3). Jetting is essentially a dynamic phenomenon similar to splashing of a liquid. Birkhoff's classic analysis of jetting (Reference 15) assumes that impact pressures are sufficiently great to neglect material strength, so that flowing metal is treated as a liquid. In a frame of reference moving with the collapse, the process is steady state and Bernouilli's law holds:

$$h + \frac{1}{2}u^2 = \text{constant} \tag{5}$$

where h is the enthalpy and u the flow velocity in the metal pressure tube walls. This equation is essentially a statement of the First Law of Thermodynamics, and is independent of most of the complex processes occurring within the flowing metal. If the density variations induced by the flow pressures are small, then $h = p/\rho$. Substituting this expression in Equation 5, we obtain the classical Bernouilli equation for an incompressible material. At a free surface exposed to constant pressure, the velocity is constant. In the steady-state frame of reference the imploding pressure tube is initially at the detonation velocity, D, so that the emerging jet also has the velocity D. In the laboratory frame of reference the forward jet has the velocity 2D.

Following Birkhoff, the mass of the jet can be calculated by applying conservation of axial momentum. Suppose that pressure of the expanded detonation gases and the compressed driver gas are negligible. Then the axial momentum of the pressure tube wall material flowing into the collapse region must equal the sum of the flow of vector momentum flowing out of the collapse region:

$$MD^2 \cos \alpha = M_s D^2 - M_j D^2 \tag{6}$$

Here M, M_s, and M_j are the mass per unit length of the imploding wall, the emerging slug, and the jet; α is the cone half-angle of the imploding pressure tube wall. Combining Equation 6 with conservation of mass ($M = M_s + M_j$), we obtain Birkhoff's equation for the jet mass:

$$\frac{M_j}{M} = \frac{1 - \cos \alpha}{2} \tag{7}$$

Since the cone half-angle α is typically 5° to 10°, Birkhoff's equation implies that as much as 0.8% of the wall material is jetted forward.

We have observed that jet mass can be significantly diminished by irreversible processes occurring in the flowing metal as it traverses the impact region. If an irreversible process occurs, such as plastic flow of the metal or shock deceleration, energy is transferred (according to Equation 5, expressing the First Law of Thermodynamics) from kinetic to internal energy. If the irreversible processes are minor, the velocity change is small, and the equations governing conservation of mass and momentum can be written in terms of a velocity decrement through the impact region:

$$\text{Mass:} \qquad MD = M_s D(1-\epsilon_s) + M_j D(1-\epsilon_j) \tag{8}$$

$$\text{Momentum:} \quad MD^2 \cos \alpha = M_s D^2 (1-\epsilon_s)^2 + M_j D^2 (1-\epsilon_j)^2 \tag{9}$$

Solving for jet mass, M_j, we find:

$$\frac{M_j}{M} = \frac{1 - \cos \alpha - \epsilon_s}{(2 - \epsilon_s - \epsilon_j)(1 - \epsilon_j)} \tag{10}$$

Even though ϵ_s and ϵ_j may be small, total jet mass can be substantially modified. For a 10° cone half-angle, the jet mass vanishes for $\epsilon_s = 0.015$. The relatively small irreversible effects, such as plastic flow and shock wave phenomena in the metal, can therefore have a substantial effect on jet mass.

Growth of a boundary layer in the driver gas also acts to minimize jetting. The mass of gas moving at wall velocity acts as an extension of the wall, and the metal jet is diminished by the mass of boundary-layer gas jetted forward. Jetting has been found to be a problem for explosive driver systems only for low driver pressures (less than 2 kbar), and it can usually be eliminated by reducing the velocity of the imploding pressure tube walls.

Explosive Driver Performance

Observed nonideal phenomena in the linear explosive driver limit its performance. At driver pressures in excess of 15 kbar the problems of pressure tube expansion, boundary-layer effects, and explosive preinitiation are excessive. At low pressures (less than approximately 2 kbar) efficiency of the driver is poor.

Explosive drivers using nitromethane explosive have been found to be particularly effective as gasdynamic energy sources for operating pressures of approximately 3 to 10 kbar. Detonation velocity ranges from 6.3 to 6.9 km/sec (depending on driver pressure), and ideal efficiency ranges from 5% to 15%.

For explosive driver systems designed to minimize the effects of boundary-layer growth and pressure tube expansion, the measured performance closely matches the ideal performance (Figure 2), and the measured efficiency is equal to the ideal efficiency. As boundary-layer effects begin to dominate, measured efficiency decreases by the amount of gas loss.

The explosive driver configuration chosen for the hypervelocity shock tube experiments is shown in Figure 9. It consists of a 5-cm i. d. steel pressure tube containing 42 atm helium, a 6-mm annulus of liquid nitromethane explosive, and a 9.5-mm steel containment tube backed by 30 cm of concrete. The ideal driver pressure is 3.7 kbar, and ideal efficiency of the driver is approximately 12%.

The length of the driver is approximately 40 tube diameters, which indicates that boundary-layer effects are present, but not dominant (Figure 7). Calculated maximum pressure tube expansion is 15%, a value leading to slight expected degradation of performance.

Experimental performance of this explosive driver design is presented in the next section.

EXPLOSIVELY DRIVEN SHOCK TUBE

The purpose of the experimental investigation reported in this paper was to investigate the possibility of using a linear explosive driver to generate hypervelocity shock waves in air. A second objective of the program was to investigate the practical limitations of sophisticated diagnostic measurements in an explosive experiment. The experiments were conducted in a permanent shock tube laboratory specially designed to protect delicate instrumentation from the explosive blast wave.

Preliminary experimental results have been encouraging. Shock velocities from 14 to 16 km/sec in air at initial pressures of 200 to 500 mTorr have been achieved in a 10-cm shock tube, and the measured shock velocities compare well with theoretical performance based on an ideal driver. Instrumentation was found to be well protected from the explosive blast, and measurements of shock properties were made with a variety of instruments.

Ideal Shock Tube Operation

Operation of an explosively driven shock tube is similar to a conventional pressure-driven shock tube, except that the driver gas is traveling at a high velocity. The shock-compressed driver gas breaks a diaphragm and expands through a nozzle into the driven section, driving a strong shock into the low-pressure test gas.

The wave diagram of an explosively driven shock tube is shown in Figure 10. Regions of constant state are identified in a manner consistent with conventional shock tube notation: Region 1 identifies the unshocked test gas (usually low pressure); Region 2 is the shocked test gas; Region 3 is expanded driver gas; and Region 4 is shocked driver gas. In addition, Region 4' stands for the unshocked driver gas, and Region 4D for the driver gas after expansion from the nozzle.

Details of explosive shock tube operation differ from conventional pressure-driven shock tubes in a number of ways. Since flow behind the strong driver shock is supersonic, the unsteady rarefaction fan caused by bursting of the diaphragm is swept downstream. The first wave disturbance in Region 2 identifying the end of constant-state test time is due to a rarefaction generated by the sudden stop of the explosive piston. In a conventional shock tube the first disturbance is caused by reflection of the diaphragm opening rarefaction. Details of the calculation of rarefaction fans and test times in an explosive shock tube therefore vary considerably from a conventional shock tube, although the basic principle of operation remains the same.

For the range of conditions achieved in an explosive driver, helium driver gas can accurately be characterized as a perfect gas. Calculations of the unsteady rarefaction fans, steady expansion, and trajectory of the terminating characteristics can all be carried out analytically in terms of known driver conditions.

Conditions in the test gas are far too severe to use a perfect gas approximation. It is not necessary, however, to obtain a complete equation of state. For a strong shock wave, the Hugoniot shock relations give, in general,

$$p_2 = \frac{\rho_1 u_2^2}{1 - \frac{1}{\eta}} \tag{11}$$

where η is the density ratio across the shock. Even though η varies between 10 and 15 for air at high shock velocity (10 to 20 km/sec), the quantity $1 - \frac{1}{\eta}$ only varies between 0.90 and 0.933; i.e., the uncertainty in shock pressure is less than 4%.

For given driver conditions, the flow velocity u_2 in the driven section is obtained by determining the isentropic p_3-u_3 expansion relationship. Across the contact surface pressure and velocity are constant, so that $p_3 = p_2$ and $u_3 = u_2$. The isentropic p_3-u_3 expansion curve, combined with the p_2-u_2 shock relationship (Equation 11) suffices to determine the theoretical shock conditions in the driver tube.

The isentropic p_3-u_3 relationship can be calculated by assuming the expansion process to consist of a steady expansion through the nozzle followed by an unsteady expansion to the driven tube pressure. The perfect gas relationships for the steady expansion are (References 16 and 17):

$$\frac{u}{a_t} = M\left[1 + \frac{\gamma-1}{2}(M)^2\right]^{-\frac{1}{2}} \quad (12)$$

$$\frac{p}{p_t} = \left[1 + \frac{\gamma-1}{2}(M)^2\right]^{-\frac{\gamma}{\gamma-1}} \quad (13)$$

$$\frac{A}{A*} = \left(\frac{\gamma+1}{2}\right)^{-\frac{\gamma+1}{2(\gamma-1)}} M^{-1}\left[1 + \frac{\gamma-1}{2}(M)^2\right]^{\frac{\gamma+1}{2(\gamma-1)}} \quad (14)$$

For the unsteady expansion

$$\frac{p}{p_t} = \left[1 - \frac{\gamma-1}{2}\frac{u}{a_t}\right]^{\frac{2\gamma}{\gamma-1}} \quad (15)$$

$$\frac{u}{a_t} = M\left[1 + \frac{\gamma-1}{2}M\right]^{-1} \quad (16)$$

Here M is the local Mach number, p the pressure, u the velocity, a the sound speed, and A the cross-sectional area of the flow. A* denotes the area for unity Mach number, and the subscript t identifies total conditions (i. e., conditions that would exist if the gas were initially at rest in the given process). Total conditions for the steady expansion process and the unsteady expansion are different, and should not be confused.

It is useful to normalize the shock tube calculations with respect to the driver velocity, u_4, and the driver shock pressure, p_4.

The normalized p_2-u_2 shock relationship (Equation 11) becomes

$$\frac{p_2}{p_4} = K\left(\frac{u_2}{u_4}\right)^2 \quad (17)$$

where

$$K = \rho_1 u_4^2 / [p_4(1 - \frac{1}{\eta})] \quad (18)$$

From the strong shock relations it can easily be shown that

$$M_4 = \left(\frac{\gamma(\gamma-1)}{2}\right)^{-\frac{1}{2}} \quad (19)$$

Equations 19 and 14 can be used to relate M_{4D} to the nozzle area ratio:

$$\frac{A_{4D}}{A_4} = \left(\frac{\gamma(\gamma-1)}{2}\right)^{-\frac{1}{2}} \left(1 + \frac{2}{\gamma}\right)^{-\frac{\gamma+1}{2(\gamma-1)}} M_{4D}^{-1}\left(1 + \frac{\gamma-1}{2} M_{4D}^{2}\right)^{\frac{\gamma+1}{2(\gamma-1)}} \qquad (20)$$

Given the ratio A_{4D}/A_4, the Mach number M_{4D} in the expanded flow can be iteratively calculated.

Next, Equations 12, 13, 15, 16, and 20 can be manipulated to yield

$$\frac{p_3}{p_4} = \left\{\frac{\left(\frac{\gamma+1}{\gamma}\right)^{1/2}\left(1 + \frac{\gamma-1}{2} M_{4D}\right)}{\left(1 + \frac{\gamma-1}{2} M_{4D}^{2}\right)^{1/2}} - \sqrt{\frac{\gamma-1}{2\gamma}}\ \frac{u_3}{u_4}\right\}^{\frac{2\gamma}{\gamma-1}} \qquad (21)$$

This equation is the desired p_3-u_3 relationship for the isentropic expansion of the driver gas.

The normalized pressure-velocity curves for the expansion of helium driver gas (Equation 21) and the driven tube shock (Equation 17) are plotted in normalized coordinates in Figure 11. Intersection of these curves represents a simultaneous solution of the two equations.

Operating point in a given explosively driven shock tube is obtained by first determining the value of K (Equation 18) from the known density ρ_1 and an assumed value of η. Since η cannot be exactly determined unless the operating point is known, K must be approximated at first and then refined by iteration. However, since K varies by only a few percent over the total expected range of η, this iterative correction is usually unnecessary.

The value of K determines a particular shock pressure-velocity curve (Figure 11), and the intersection of this curve with the driver expansion curve determines the p_2-u_2 operating point. The exact value of the shock density ratio of η can then be determined from shock tables, and K updated if necessary. Shock velocity in the driver section is given by the shock relation:

$$u_s = u_2 \, / \, (1 - \frac{1}{\eta}) \qquad (22)$$

Determination of ideal test time from the intersection of the terminating characteristic with the contact surface requires an integration of the characteristic trajectory through the steady nozzle expansion and the unsteady rarefaction fan in the driven section. This integration is readily carried out either numerically or graphically, but the process is tedious. Sufficient accuracy for some design purposes can be obtained by assuming that the characteristic velocity through the expansion regions is equal to the average of the velocities in adjacent regions of constant state. This analysis yields the following equation for the distance from the diaphragm at which the terminating rarefaction overtakes the contact surface:

$$\frac{L_o}{T_D} = u_3 \frac{(u_4 + a_4)(u_3 + a_3 - u_4 + 3a_4)}{a_4(u_4 + a_4 - u_3 + 3a_3)} \qquad (23)$$

where L_o is the overtake distance and T_D is the driver run time.

Experimental Program

A short experimental program has been carried out to investigate the performance of a hypervelocity shock tube driven by an explosive driver. Three experiments were conducted on a 10-cm-diam. shock tube with initial air pressures ahead of the shock of 250 and 400 mTorr. The experimental program also served to investigate the possibility of decoupling the explosive blast from the shock tube and its associated instrumentation.

Of the three experiments conducted, one was performed with air at 250 mTorr initial pressure and two with pressures of 400 mTorr. These three firings resulted in shock velocities of 14.2, 15.4, and 16.4 km/sec. The shock velocities were constant to within the precision of the measurements for the entire length of the 8-m shock tube. The observed velocities were slightly less than predicted for an ideal explosive driver.

Photographs of the shock-heated air showed the shock front to be plane. The homogeneous luminous region behind the shock front extends at least 3 cm.

Measurements of the electron conductivity ahead of the shock wave indicated appreciable ionization to a distance of 65 cm ahead of the shock front. An electron density in excess of 10^9 cm^{-3} was measured 16 cm ahead of the shock wave.

This work has demonstrated the ability of the explosive driver to generate plane shock waves moving at velocities in excess of 15 km/sec into air at densities corresponding to altitudes of approximately 60 km. The work has also demonstrated that diagnostic instrumentation, employing delicate apparatus, can successfully be applied for a permanent explosive-driven shock tube.

Explosive Shock Tube Laboratory. Explosively driven shock tubes present a number of experimental problems not ordinarily associated with shock tube research. One of the major problems is the necessity of decoupling the shock tube instrumentation from the destructive blast wave associated with detonation of explosive. The two simplest approaches to this problem are containing the explosive in a large steel tank, or protecting the instrumentation in a blast-proof bunker. The first method has been used with success for a variety of explosive devices, but it is practically limited to a few kilograms of explosive. If a field test site is available, the second approach allows more flexibility in explosive driver design.

Physics International Company has constructed an explosive shock tube facility at its Tracy Test Site (Figure 12) specially designed to protect sensitive shock tube instrumentation from explosive blast effects. The shock tube passes through a steel-sheathed, reinforced concrete wall into a blast-proof bunker containing the shock tube instrumentation. The shock tube and instrumentation setup are shown in Figure 13.

Personnel and remote control and recording instruments are located in a second bunker (Figure 14). The purpose of maintaining two bunkers is to eliminate the danger of shock tube rupture at the very high shock pressures possible in an explosively driven shock tube.

Explosive Driver Performance. The driver shock and detonation trajectories (Figure 15) were monitored by contact and ionization pins. The contact pins are closed by the shock-induced expansion of the pressure tube, and the ionization pins are short circuited by the ionized detonation products. The pins discharge capacitors on the passage of the shock front and detonation front, and these discharge pulses are monitored by a raster oscilloscope in the instrumentation bunker.

The observed detonation velocity of 6.8 km/sec is greater than ideal detonation velocity of 6.3 km/sec because of precompression of the nitromethane by the driver shock. The observed shock velocity, 7.6 km/sec, is less than either the ideal velocity of 8.4 km/sec based on ideal detonation velocity, or 9.1 km/sec based on observed detonation velocity. The difference is attributable to the nonideal effects of pressure tube expansion and boundary-layer growth discussed previously.

The initial helium pressure in the driver was 42 atm, which corresponds to a pressure of 3.7 kbar in the ideal driver. Calculated pressure based on the observed shock velocity is 3.1 kbar.

An independent measure of driver performance was obtained by monitoring driver pressure with a piezoelectric pressure transducer. The transducer was calibrated at low pressures with a static test device before and after the shot, but it has been our experience that dynamic accuracy of the calibration is no better than $\pm 10\%$. The oscilloscope record (Figure 16) indicates an initial pressure step of approximately 2.7 kbar, followed by a slow rise over 20 μsec to a steady value of approximately 3.3 kbar. The most probable explanation for the slow rise is that driver gas in the expanded pressure tube is recompressed when it flows into the shock tube coupling section.

Observed run time in the explosive driver is 60 μsec to the start of the terminating rarefaction. Assuming an average pressure of 3.3 kbar during this period, the calculated thermodynamic efficiency of the driver is slightly greater than 7%.

Shock Wave Trajectory. The shock trajectory in the shock tube was determined with ion pins similar to those used in the explosive driver. The measured trajectory in Shot 121-2 (Figure 15) indicates a shock velocity of 16.4 km/sec, constant to within the accuracy of the measurements.

Three shots of the explosive shock tube were fired during the experimental program. The results are tabulated in Table I, and are compared with calculations based on an ideal driver. The agreement between the ideal calculations and experimental observations for the 400 mTorr shots are of course fortuitous, as there are a number of interacting nonideal effects present which happen to cancel each other. It is not clear at this time why the 250 mTorr shot has an unusually low velocity; driver data were not recovered on this shot, and there is the possibility of a driver malfunction.

Table I
Comparison of Ideal and Observed Shock Velocities

a. Ideal shock velocities (u_s)

Initial Shock Tube Pressure (mTorr)	Velocities for Area Ratio ∓ 4 (km/sec)
250	16.0
400	15.7

b. Observed shock velocities (u_s)

Shot Number	Initial Shock Tube Pressure (mTorr)	Measured Shock Velocities (km/sec)
121-1	250	14.2
121-2	400	16.4
121-3	400	15.4

It is interesting to note the absence of diaphragm break time usually associated with shock tubes. The measured shock trajectory in the shock tube starts no later than a few microseconds after arrival. The explosively driven shock tube has the advantage that the diaphragm can be quite thin, as it is necessary to withstand only the low initial pressure (42 atm) of the driver gas. The reflected driver pressure is 36 kbar, which is sufficiently high to destroy almost instantaneously the thin diaphragm.

Evaluation of Shocked-Air Conditions. One of the purposes of the experimental program was to investigate the possibility of using sophisticated instrumentation in our explosively driven shock tube. The design of the blast-proof shock tube bunker proved to be adequate in reducing explosive blast effects, and even delicate high-speed photographic equipment was successfully used.

A photographic view of the self-luminous shocked air in Shot 121-3 was taken with a high-speed rotating mirror camera (Figure 17). A clear plastic ring was inserted between flanges of the shock tube, and formed a slit-like window for viewing the shock wave. The optical image of this slit was reflected from the rotating mirror, and was focused onto 35-mm film. The speed of the rotating mirror was adjusted so that the slit image traveled at the rate of 2 km/sec on the film. The streaked slit image generates a picture of the shocked gas as it flows past a point; assuming the shock velocity to be constant, this time-resolved image is equivalent to a direct photograph of the shocked gas.

Within the resolution of the optical system, the photograph (Figure 17) indicates that the shock front is plane. It is not clear where the contact surface is located. The luminosity of the gas diminishes rapidly behind the shock due to radiative cooling, so that reduction of film exposure is not a reliable indication of the contact surface. The observed luminous region extends for at least 3 cm behind the shock front, and establishes a lower bound of approximately 2 μsec for the test time.

Photodiodes were also used to monitor the luminous plasma. The photodiodes were mounted at several stations on the shock tube wall, and were collimated by means of two 1-mm apertures separated by 43 mm. The diodes were saturated by the light output, but indicated a test time of approximately 5 μsec at a distance 6.4 m from the diaphragm.

Mirels' theory of shock tube test time (Reference 13), extrapolated to the experimental shock Mach number, predicts a test time of approximately 4 μsec. This value is well within the experimental uncertainty concerning the location of the contact surface.

Spectroscopic measurements were also conducted with a high-speed spectrograph. The measurements were exploratory in nature and fairly crude, but they did indicate the presence of metallic emission lines. The source of these metallic impurities is thought to be the boundary layer, judging from experimental data in other programs. However, it is possible that the contamination is due to metallic impurities in the helium driver gas diffusing into the shocked-air slug. Theoretical predictions based on estimates of driver contamination indicate that this is an unlikely possibility, but it is a question that can be resolved only by obtaining additional experimental data.

Precursor Conductivity Measurements. A four-wire conductivity probe based on Brown's design (Reference 18) was used to measure precursor conductivity in Shot 121-3. This probe (Figure 18) consists of four equally spaced bare wires exposed to the conducting plasma. A voltage is applied across the two outer conductors. Measurement of the current, i(t), through the outer conductors and the voltage, e(t), across the two inner conductors is sufficient to determine conductivity of the plasma. The probe was calibrated by measurements in an ionic solution of soluble salts in water.

Measurable conductivity (approximately 0.02 mhv/m) was observed as much as 40 μsec before arrival of the shock. At the measured shock velocity of 15.4 km/sec, this time corresponds to a distance of 62 cm ahead of the shock front. Conductivity values were converted to approximate electron density values using the well-known equations of a weakly ionized plasma (Reference 19). Electron temperature was assumed to be approximately 1 eV.

CONCLUSIONS

The purpose of the experimental program described in this paper was to investigate the performance of a hypervelocity shock tube driven by a linear explosive driver, and to determine the feasibility of performing sophisticated diagnostic measurements in a permanent shock tube facility protected from the explosive blast. The shock tube facility proved to be adequately protected, and measurements of shock tube performance were made with such delicate instruments as a high-speed rotating mirror camera.

The present stage of investigation into the performance of the explosively driven shock tube is encouraging but not conclusive. Observed shock velocities were in the range 14 to 16 km/sec in air at initial pressures of 0.25 to 0.4 Torr. From the streak camera records the shock appears to be plane, and test time is approximately the maximum time predicted by Mirels' boundary-layer theory. Preliminary spectroscopic results indicate the presence of metallic contamination, but this is thought to be confined to the boundary layer. The basic performance potential of the explosively driven shock tube has been demonstrated in this program, but additional research is necessary to determine thermodynamic conditions in the shocked gas.

It is interesting to compare the performance potential of the explosively driven shock tube described in this paper with the performance of existing arc-driven and combustion-driven shock tubes. Camm, et al. (Reference 20), have described the performance of a very high performance, 80-kJ, arc-driven shock tube and a large combustion-driven shock tube. These data are compared to performance calculations for the explosively driven shock tube in Figure 19.

Total explosive energy in the explosive driver described above is approximately 11.2 MJ. Observed energy in the driver gas, both internal and kinetic, was approximately 0.8 MJ, so that the overall thermodynamic efficiency of the explosive system is slightly greater than 7%.

ACKNOWLEDGMENTS

The shock tube research presented in this paper was possible only because of extensive previous research on the linear explosive driver. The authors are particularly appreciative of the contributions of Dr. C. S. Godfrey, Mr. E. T. Moore, Jr., Mr. J. D. Watson, Mr. G. B. Steel, and Dr. H. F. Waldron in developing the explosive driver system, and in identifying the principal nonideal phenomena affecting its performance.

REFERENCES

1. I. I. Glass, "Shock and Combustion Wave Dynamics in an Implosion-Driven Hypervelocity Launcher," UTIAS Review No. 25, University of Toronto, Toronto, Ontario, Canada (January 1965).
2. J. D. Watson, "Implosion-Driven Hypervelocity Launcher Performance Using Gaseous Detonation Waves," UTIAS Technical Note No. 113, University of Toronto, Toronto, Ontario, Canada (May 1967).
3. G. E. Seay, et al., J. Appl. Phys. 32, 2439 (1961).
4. A. E. Voitenko, Sov. Phys. Doklady 9, 860 (1965).
5. A. E. Voitenko, Sov. Phys.-Tech. Phys. 11, 128 (1966).
6. A. E. Voitenko, I. Sh. Model', and I. S. Samodelov, Sov. Phys. Doklady 11, 596 (1967).
7. Yu. A. Zatsepin, E. G. Popov, and M. A. Tsikulin, Sov. Phys. JETP 27, 63 (1968).
8. V. M. Titov, Yu. I. Fadeyenko, and N. S. Titova, Letters of the Acad. of Sci. USSR 180, 1051 (1968)(in Russian).

9. S. P. Gill, "Shock Tube With Implosive Jet Driver," Phys. Fluids Supplement, to be published in 1969.
10. E. T. Moore, Jr., "Explosively Driven Shock Tubes," AEDC-TR-66-238, Arnold Engineering Development Center, Arnold Air Force Station, Tennessee (November 1966).
11. H. F. Waldron, E. T. Moore, Jr., G. B. Steel, and C. S. Godfrey, "A Mechanism for the Conversion of the Chemical Energy of Explosives to the Kinetic and Internal Energy of a Gas," AIAA Paper No. 67-178, presented at the AIAA Fifth Aerospace Sciences Meeting, New York, New York (January 1967).
12. S. P. Gill, "Implosively Driven Shock Tubes," Poulter Laboratories Technical Report No. 001-66, Stanford Research Institute, Menlo Park, California (December 1966).
13. H. Mirels, "Shock Tube Test Time Limitation Due to Turbulent-Wall Boundary Layer," AIAA J. 2, 84 (January 1964).
14. J. W. Enig and F. J. Petrone, "An Equation of State and Derived Shock Initiation Criticality Conditions for Liquid Explosives," 4th Detonation Symposium, U.S. Naval Ordnance Laboratory, White Oak, Silver Spring, Maryland (October 12, 1965).
15. G. Birkhoff, D. P. MacDougall, E. M. Pugh, and G. I. Taylor, "Explosives with Lined Cavities," J. Appl. Phys. 19, 563 (June 1948).
16. "Equations, Tables, and Charts for Compressible Flow," Report No. 1135, National Advisory Committee for Aeronautics, Washington, D. C. (1953).
17. R. Courant and K. O. Friedricks, Supersonic Flow and Shock Waves, Interscience (1948), p. 104 ff.
18. R. T. Brown and M. Mitchener, "Measurements in a Two-Temperature Plasma Boundary Layer," AIAA Paper No. 69-692, AIAA Fluid and Plasma Dynamics Conference, San Francisco, California (June 1969).
19. See, for example, D. J. Rose and M. Clark, Jr., Plasmas and Controlled Fusion (MIT Press, 1961) p. 147.
20. J. C. Camm and P. H. Rose, Phys. Fluids 6, 663 (1963).

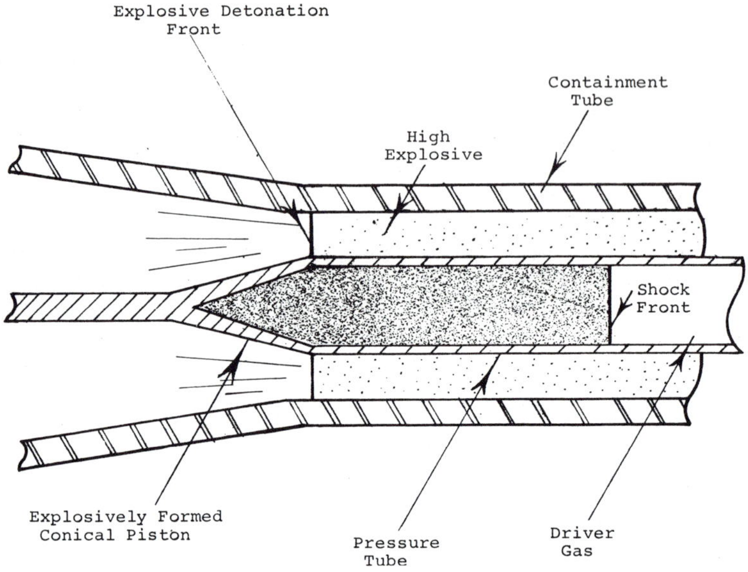

Figure 1
Idealized schematic of linear explosive driver

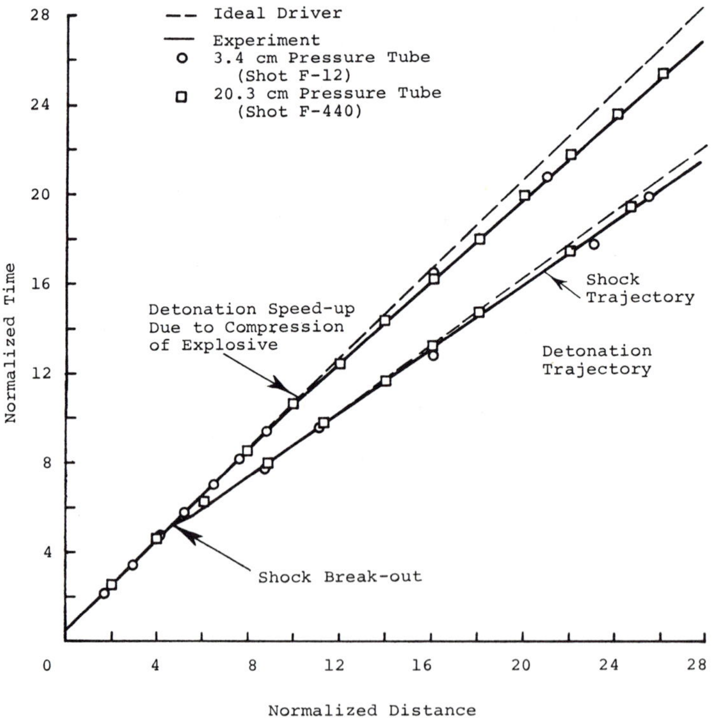

Figure 2
Explosive driver wave diagram

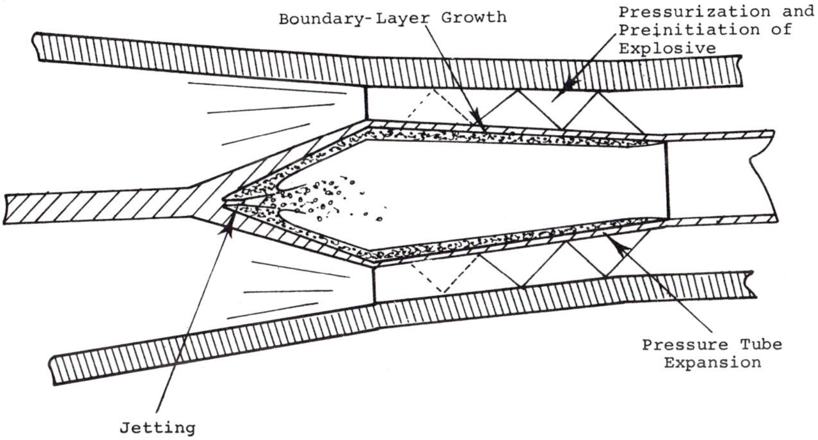

Figure 3
Schematic of nonideal phenomena in explosive drivers

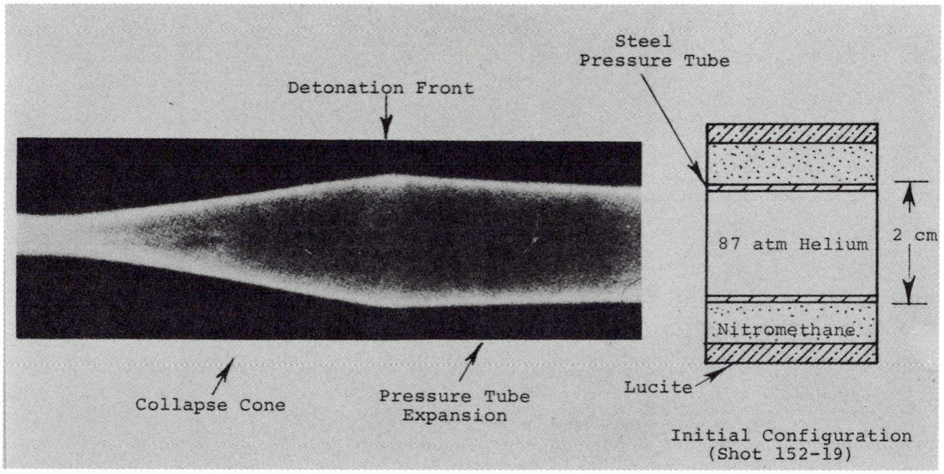

Figure 4
Flash radiograph of explosive driver operation

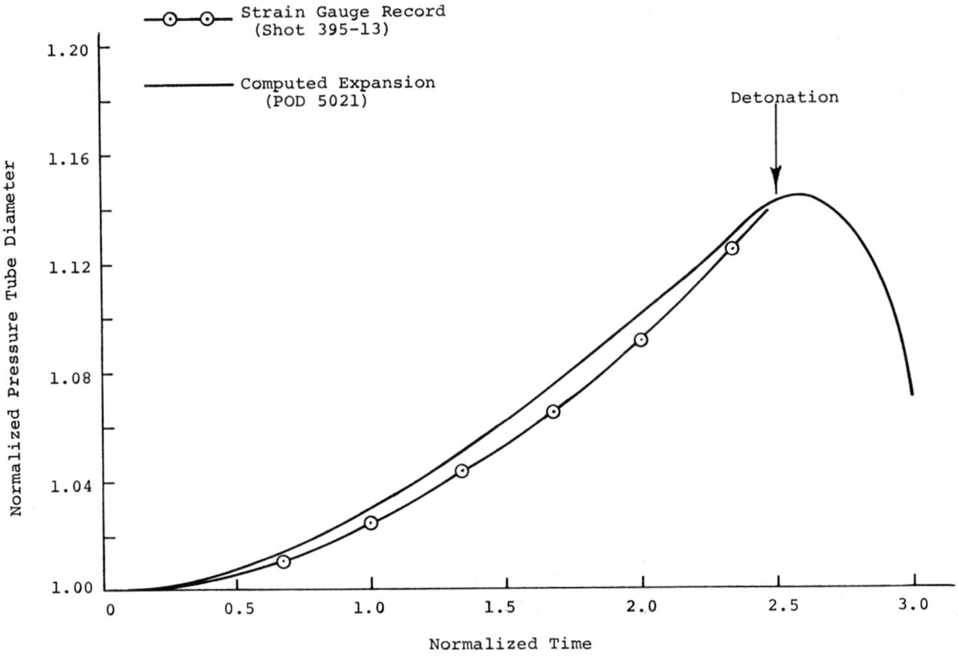

Figure 5
Pressure tube expansion

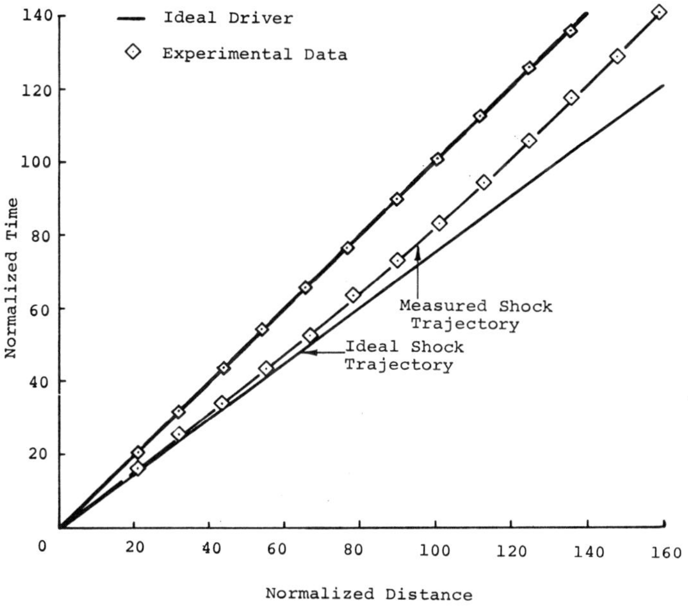

Figure 6
Influence of boundary-layer growth on explosive driver performance

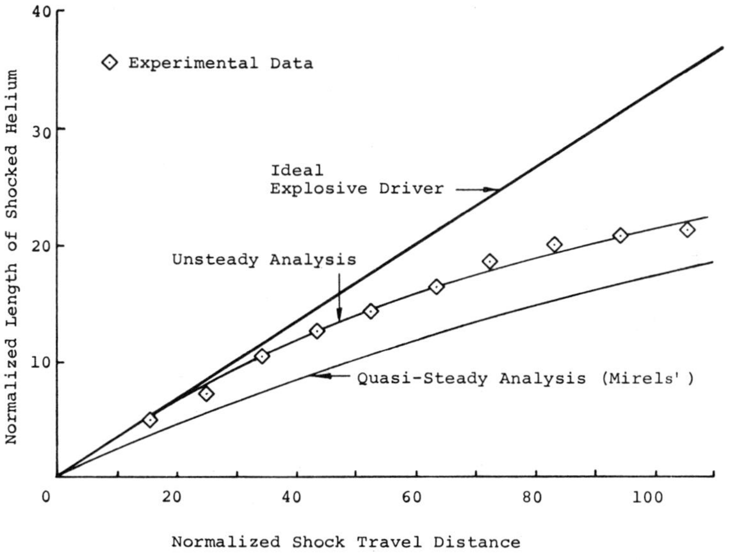

Figure 7
Boundary-layer effects on shocked gas length

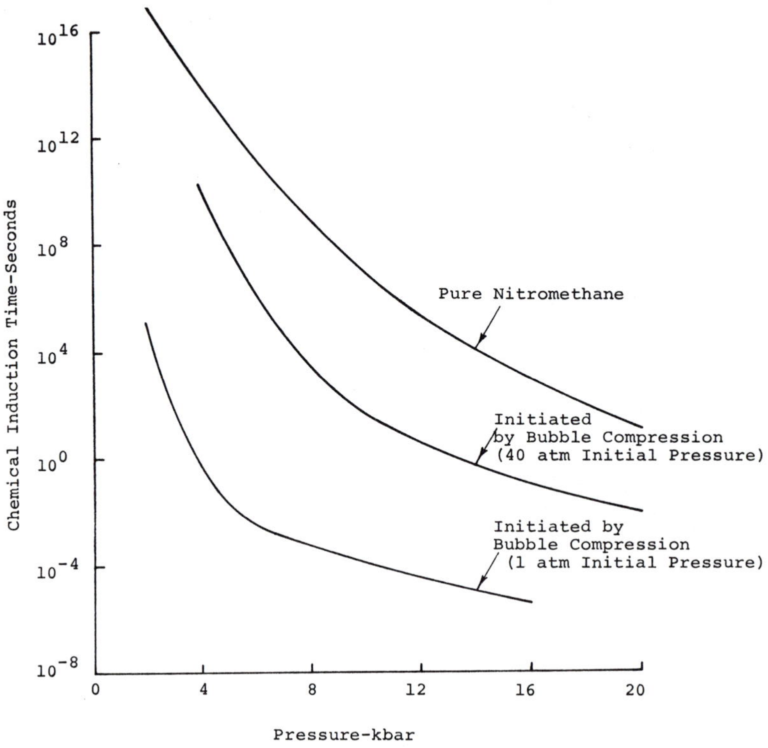

Figure 8
Initiation times in nitromethane as a function of pressure

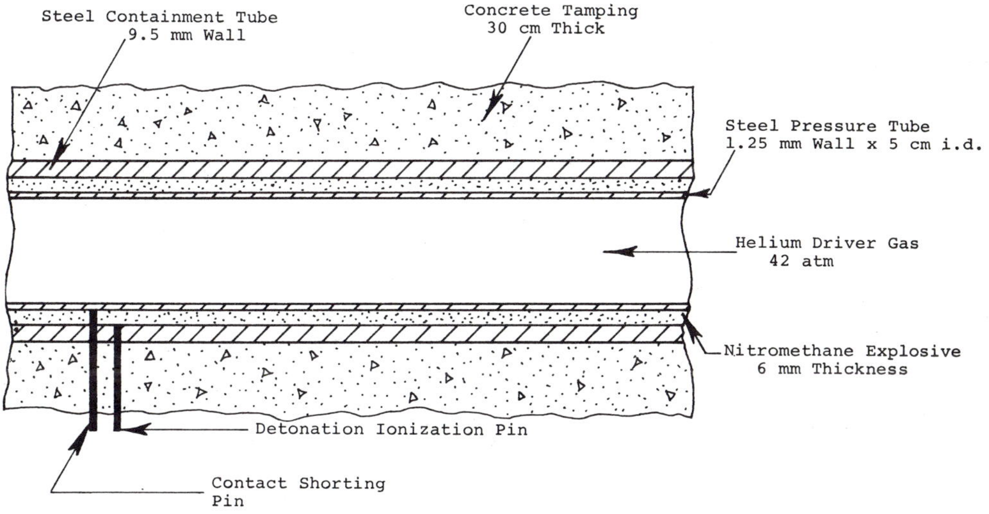

Figure 9
Explosive driver configuration for hypervelocity shock tube experiments

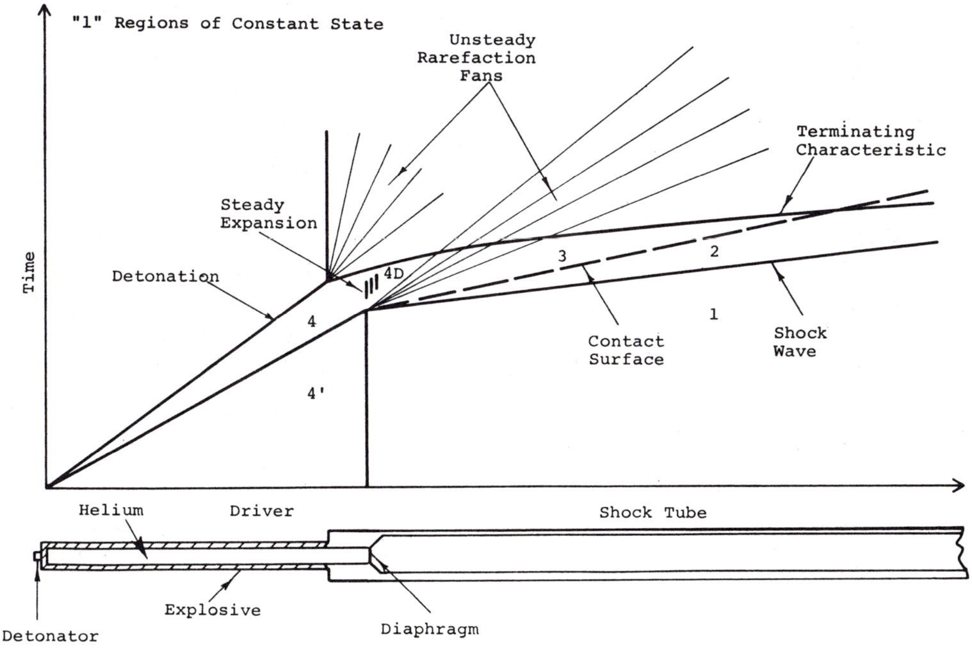

Figure 10
Wave diagram of explosively driven shock tube

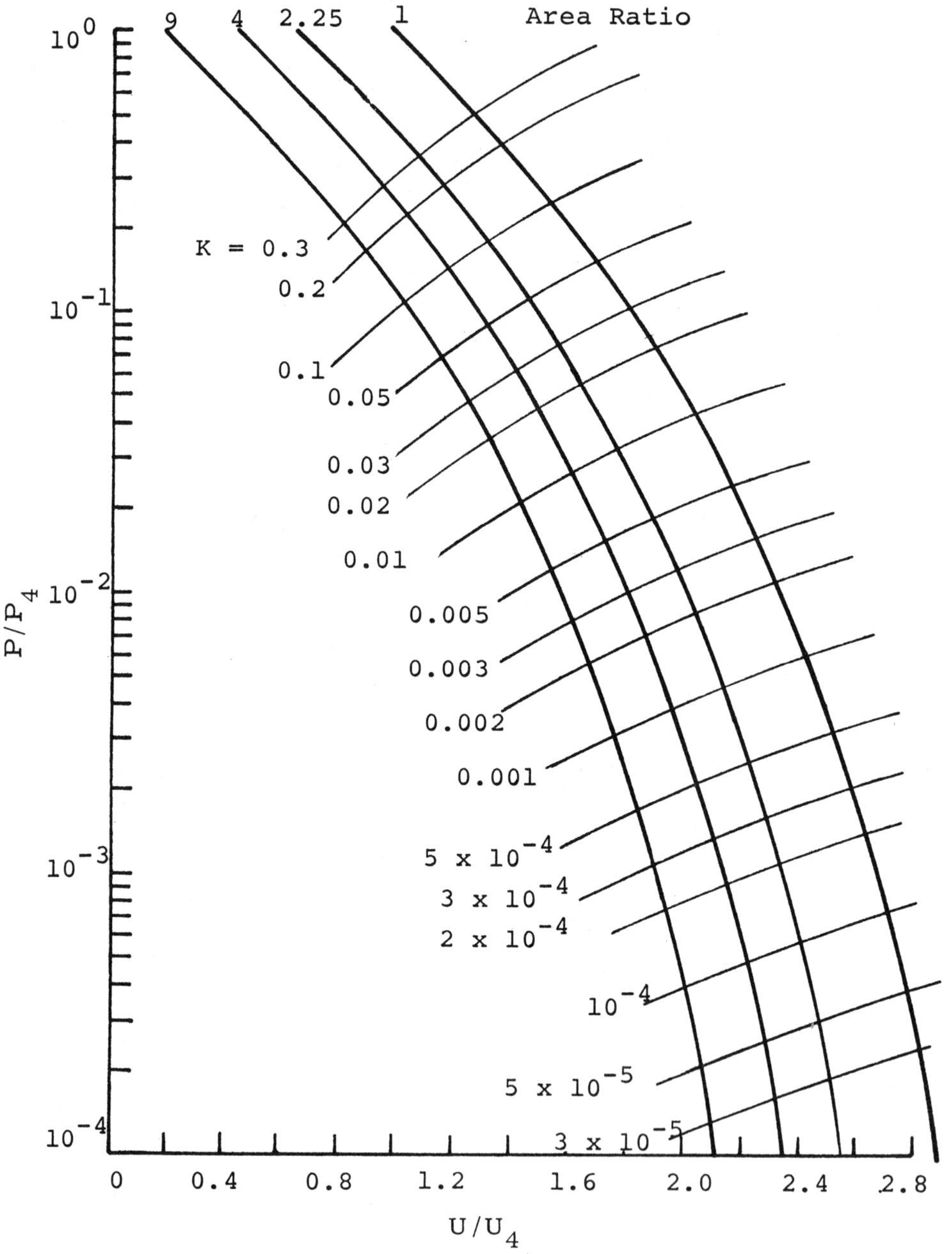

Figure 11
Pressure-velocity relationships in an explosively driven shock tube

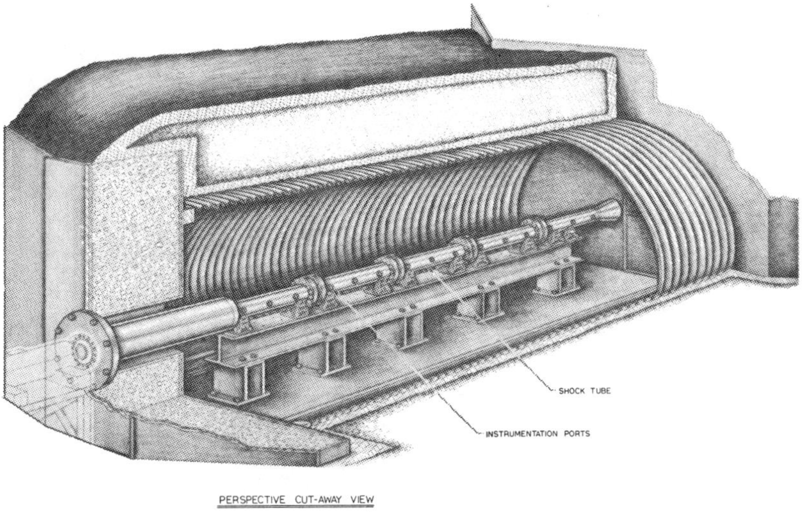

Figure 12
Explosively driven shock tube laboratory
at physics international tracy test site

Figure 14
Recording instruments in instrumentation and
control bunker

Figure 13
Shock tube instrument setup

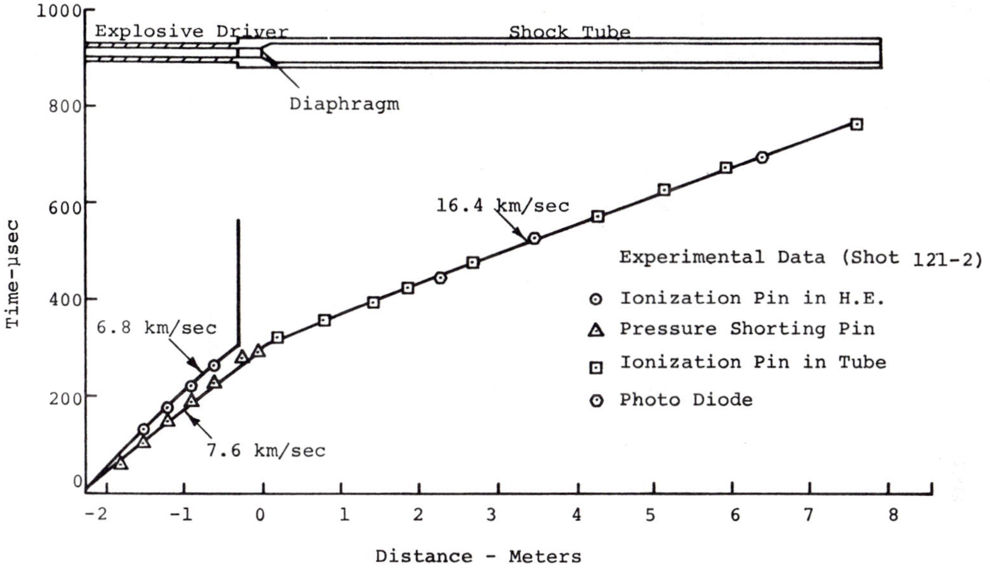

Figure 15
Experimental wave diagram of explosively driven shock tube

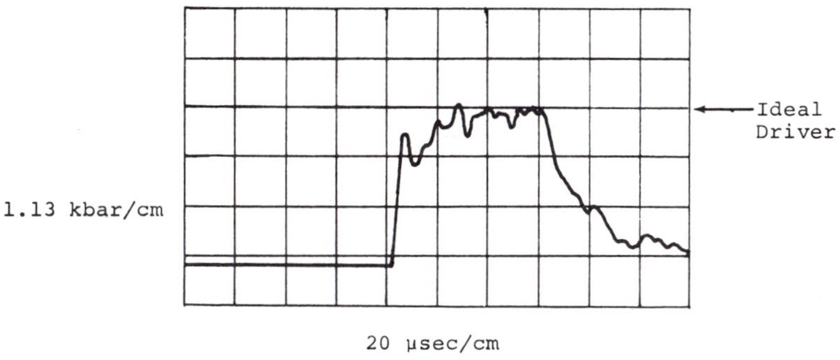

Figure 16
Explosive driver pressure record

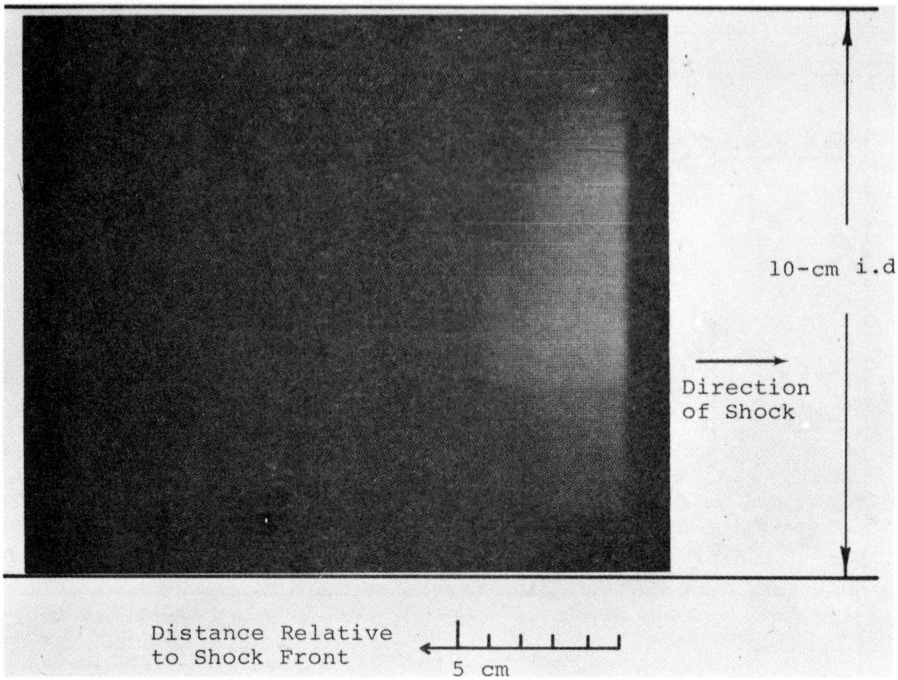

Figure 17
Streaking camera photograph of shocked air

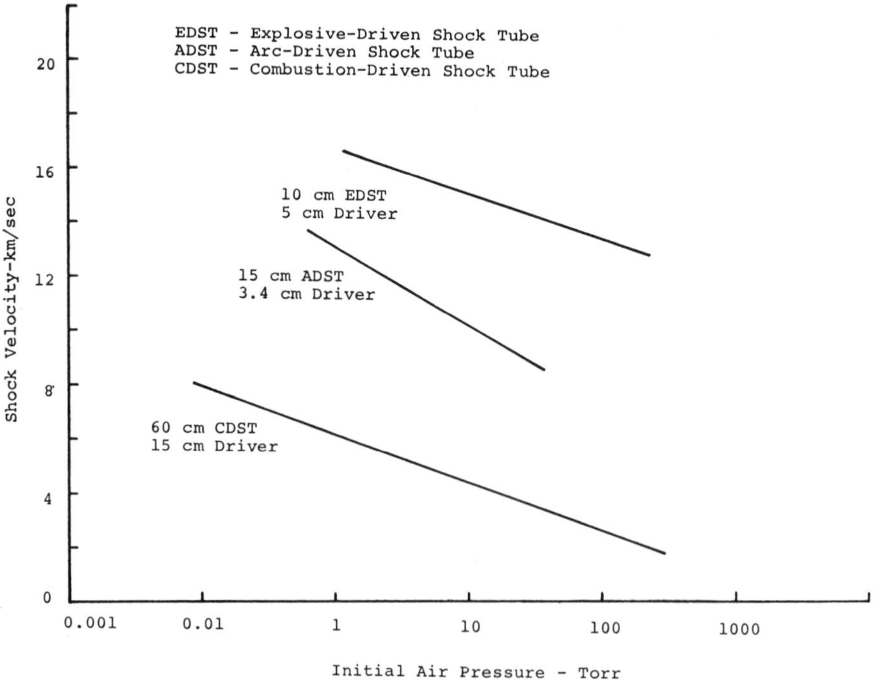

Figure 19
Approximate performance limits of existing shock tubes

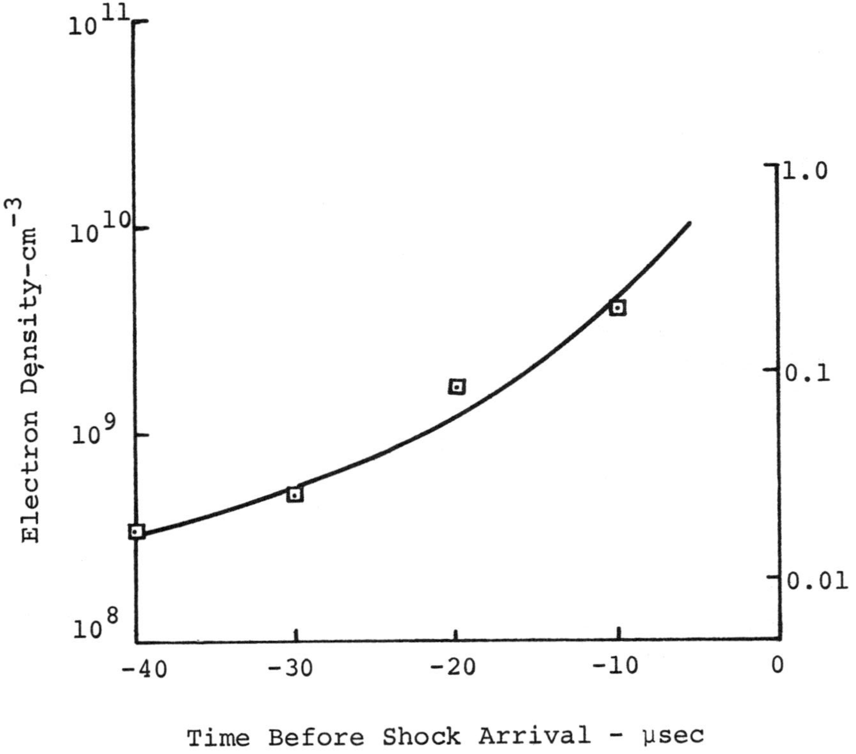

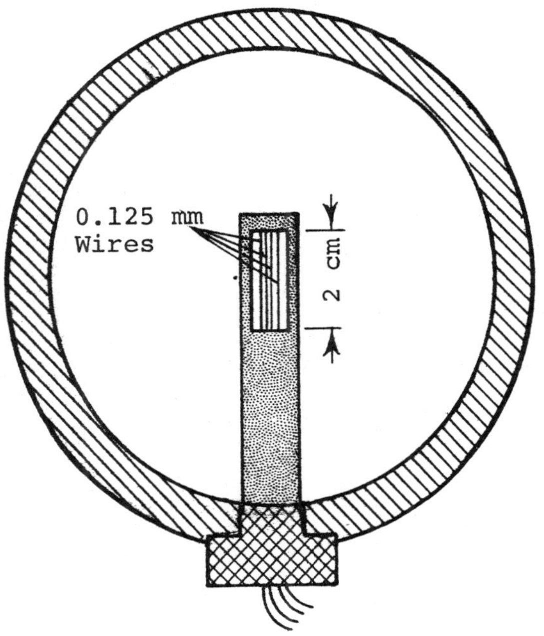

Four Wire Conductivity
Probe Configuration

Figure 18
Precursor conductivity measurements

DESCRIPTION AND PERFORMANCE OF A CONICAL SHOCK TUBE NUCLEAR AIR BLAST SIMULATOR

D.W. CULBERTSON
U.S. Naval Weapons Laboratory
Dahlgren, Virginia

SUMMARY

A large-scale, explosive-driven, conical shock tube nuclear air blast simulator has been developed at the Naval Weapons Laboratory under the sponsorship of the Defense Atomic Support Agency (DASA). Total development effort, including exploratory research, design, construction and final engineering development and evaluation, has encompassed the period 1960 through 1968 as documented by references (a) through (f). This paper describes the major physical characteristics of the completed shock tube and presents performance characteristics as determined from experimental investigations conducted to date.

Results of the experimental performance trials show that the explosive-driven Conical Shock Tube provides excellent simulation of free air blasts in the kiloton yield range. Actual blast amplification effects have been found to exceed design predictions by 10-15% in the larger-diameter terminal portions of the shock tube expansion chamber.

BACKGROUND AND INTRODUCTION

The concept of a high explosive-driven conical (or sector) shock tube as a practical blast simulation research device grew out of experimental studies conducted principally by the Naval Ordnance Laboratory, White Oak, during the late 1950's and early 1960's. Experiments, conducted with relatively small scale laboratory devices, indicated that a small charge of high explosive, when detonated while confined in the apex of a right circular conical tube, produced blast wave parameters which were characteristic of those produced by much larger weights of explosive detonated in free air, references (a) through (d). This observed amplification effect was attributed to the geometry since, as shown by Figure 1, a small solid angle of a full spherical detonation is confined within the conical shell. The ideal blast amplification for such a device was reckoned to be the ratio of the solid angle of the equivalent free-air sphere to the solid angle of the confined conical charge, i.e., ratio of the volume, or weight, of the simulated spherical charge to that of the conical driver charge.

Essentially these initial studies established that, within a conical shock tube, the wave decays in a manner typical of free-air spherical blasts, the actual amplification factor is on the order of 20-30% of the ideal value (depending on cone angle), and once an "efficiency" factor (ratio of actual to ideal amplification factors) is established, the weight scaling law for high explosives is applicable for varying driver charge weights.

These results led to a proposal for constructing a large-scale conical shock tube providing a capability of simulating approximately 20 kiloton blast waves. Eventually this proposal resulted in design of the 2400 ft long x 0.56° Conical Shock Tube at the Naval Weapons Laboratory. This design geometry resulted in an ideal amplification factor of about 160,000 which, assuming an "efficiency" of about 25%, was predicted to provide 20 KT free air blast simulation with 1000 lb TNT driver charges.

This large scale facility has been funded by the Defense Atomic Support Agency and designed and constructed under the cognizance of the Ship Systems Command and the Naval Facilities Engineering Command respectively. Facility construction was initiated in April 1965, was initially scheduled for completion in October 1966, and ultimately accepted from the contractor in August 1967. Preliminary evaluation and calibration experiments were initiated in September 1967 extending through late summer of 1968 and continuing through the present time in conjunction with applied shock tube programs.

DESCRIPTION OF FACILITY

General Description

The principal physical characteristics of the large-scale facility are shown by Figures 2 and 3. The device consists of a 186 ft long detonation (or driver) chamber, which has a constant inside diameter of 16 inches, and a 2268 ft long conical expansion chamber which has a maximum inside diameter of 24 ft. The detonation chamber breech is closed by a "recoil transfer" column, or piston, which transfers the force of the detonation to a reinforced concrete recoil mass. The initial 1255 ft length of the facility may be converted to an altitude chamber by inserting a diaphragm at the 12 ft diameter station of the expansion chamber; the resulting closed volume, approximately 45,000 ft^3, may be evacuated by means of provided vacuum pumps to simulated pressure altitudes of up to about 100k ft. Permanent, "hardened", test areas are located at the 10, 15 and 22 ft diameter stations of the expansion chamber.

Detonation Chamber

The detonation chamber is fabricated from four 16"/45 naval guns. The thinner-wall muzzle-end sections have been removed from the first three guns while the fourth gun is utilized full length, resulting in a total detonation chamber length of 186 ft. The breech and propellant chamber of each gun has been modified with filler inserts to provide a continuous 16 inch inside diameter throughout the chamber length. The gun sections are joined by sliding joints equipped with metal sleeve seals.

The high explosive driver charge, in the form of a long cylindrical "stick" charge, is positioned coaxially within the detonation chamber. Figure 4 illustrates the "flotation" technique used for charge loading: the detonation chamber is blocked on each end and partially filled with water; the charge, supported on rigid foam trays, is floated into the chamber and the water is drained allowing the trays to settle against the chamber walls. The selected support tray width automatically centers the charge within the bore.

Based on both experimental and analytical studies, references (g) and (h), the detonation chamber is predicted to be capable of reliably confining the detonation of 1000 lb TNT charges in the form of a 3.5 inch diameter x 148 ft long cylinder (charge density of 1.6). The experimental studies of reference (g) indicated that undesirable plastic deformation of the chamber wall commences when the charge diameter exceeds about 4.0 inches and further indicated the desirability of restricting the charge diameter to about 3.5 inches.

The driver charge consists of 30 inch long segments of cast explosive placed end-to-end on the rigid foam support trays. Total charge weight is controlled by varying the number (and hence total charge length) of constant diameter segments or by varying the segment diameter. The driver charge is initiated at the breech end via a 50/50 Pentolite booster and an appropriate detonator. Figure 5 illustrates driver charge weight and geometrical characteristics for TNT explosive having a density of 1.6.

Expansion Chamber

The expansion chamber cone increases in diameter at the uniform rate of one foot per 100 feet of axial length, i.e., a plane angle of 0.56°. Thus the theoretical apex of the cone is 2400 ft from the open end, or as shown by Figure 2, within the detonation chamber, 133 ft aft of the cone - detonation chamber junction.

The conical shell is constructed in sections varying in length from 37 ft to 100 ft, in diameter (inside) from 16 inches to 24 ft, and in thickness from 1.875 inches to 0.375 inches. Shell material is HY 80 steel for the initial 731 ft, ASTM A441 carbon-manganese steel for the mid 681 ft and ASTM A36 structural steel for the terminal 855 ft. The sectional construction was adopted so that narrow, relatively easily sealed, temperature expansion joints could be used between sections and to simplify possible future alterations by allowing for removal or modification of discrete sections rather than involving the entire shell.

The axial thrust load resulting from pressurization of the conical shell is reacted at the large diameter end of each structural section via stiffener support/thrust rings attached to either cantilevered buttresses or embedded anchors. The small diameter end of each shell section is supported by ring stiffeners on movable joints. Intermediate stiffener rings have been added to provide necessary buckling stability when the shell is subjected to external pressure loads resulting from the negative phases of the blast wave.

Test Areas

Permanent test areas are provided at the 10, 15 and 22 ft diameter stations for mounting and observing test specimens. Major details of these test areas are shown by Figure 6.

Each test area consists of a reinforced or "hardened" shell section attached via a 3 inch thick steel "mounting plate" and anchor beams to massive reinforced concrete foundations. Each area is equipped with several circumferentially-spaced, 8 inch diameter, flanged-ports, instrumentation access conduit terminating in the mounting plate, a personnel access hatch, and either a removable shell segment type of hatch or a removable shell section to provide for test specimen access.

Altitude Chamber

To form the altitude chamber, an expendable diaphragm is inserted in a short removable section of the expansion cone at the 12 ft diameter station. The expansion joints between the structural shell sections within the altitude chamber and between the detonation chamber sections are equipped with inflatable seals to maintain the vacuum. The closed volume between the diaphragm and the detonation chamber breech, approximately 45,000 ft^3, can be evacuated to a simulated pressure altitude of up to about 100k ft (about 8mm of mercury) by means of three mechanical vacuum pumps, each having a displacement rate of 850 CFM.

Instrumentation

The facility's instrumentation system is diagramed by Figure 7. The system consists of a centralized control/recording center with automated data acquisition and reduction equipment. Initially, the system provides 100 channels of direct FM recording capability which can be concentrated at any of the three fixed test areas or distributed over the entire facility's length. The cabling system, in above ground metal conduit, provides permanent 100 channel patch boards located in weather-proofed enclosures adjacent to each test area. The primary data recording mode is on analog magnetic tape which can be digitized and processed on site by the system's small general purpose automatic data processor. The digital subsystem can also provide some direct on-line data digitizing for lower frequency data acquisition.

FACILITY PERFORMANCE

Description of Experiments

To date, performance of the Conical Shock Tube has been experimentally investigated during conduct of approximately 45 firing trials. The first 22 trials were specifically conducted for preliminary performance investigation purposes while the remaining trials have been conducted for both performance investigation and application purposes. The major objectives of the performance trials have been to determine blast simulator performance of the facility, to determine structural loading and response, to determine design reliability and operability, and to establish safe operational techniques.

These trials have predominantly used cast TNT explosive driver charges ranging in weight from about 25 lbs through 700 lbs. Cyclotol has been used for some of the smaller driver charges because of difficulties experienced in reliably detonating small-diameter (<2.0") cast TNT cylinders. All experiments have been performed with the driver and expansion chambers under ambient atmospheric conditions; investigations of the shock tube's altitude simulation capability have been deferred until spring and summer of 1969 in order to expedite investigations pertinent to current applied research efforts at "sea level" conditions.

The experimental firings have subjected the CST structure to approximately equivalent design working loads. No significant failures or deformations of the main superstructure and foundations were observed. It was found necessary to redesign and replace the joint seals within the detonation chamber and portions of the expansion chamber primarily to reduce possibility of secondary damage by projected seal fragments.

General Performance Results

The shock tube's experimentally determined performance trend is illustrated qualitatively, in comparison with free-air blast characteristics, by Figure 8. This illustrates an observed "split-simulation" performance trend which is attributable to the shock tube geometry: blast amplification effects are lower in the initial portions of the expansion chamber than in the terminal portion because the actual CST geometry does not allow generation of the high initial pressures associated with free-air spherical charge detonations. This trend is shown quantitatively, on a peak static overpressure-distance basis, by Figure 9. Figure 9 also illustrates that the observed blast amplification factor in the terminal 800-1000 ft portion of the CST expansion chamber is on the order of 48,000, about 20% greater than the design predicted value of 40,000.

Experimentally determined positive impulse data for each of the shock tube's test areas are correlated with free-air scaled distance data in Figure 10. Correlation is effected on the basis of simulated spherical charge weight (TNT) determined from P_{so} versus $R/W^{1/3}$ scaled distance curves for experimentally determined values of P_{so}. Agreement is excellent for the 10 ft and 15 ft diameter stations. For the 22 ft diameter station, actual impulse is about 15% less than the free air values because the positive phase duration is shortened here by exit plane reflection effects as discussed below.

Test Area Performance

Shock tube performance at the fixed test areas is summarized, as a function of driver charge weight, in Tables 1 through 3. As noted, the maximum performance level, for the 850 lb driver charge weight, has not been experimentally achieved at this time. However, because of the performance consistency demonstrated by driver charge weights up through 700 lbs, the projected performance values are considered highly reliable. It should be noted that, because of the "split-simulation" trend, the projected 850 lb driver charge will result in a maximum free-air simulation capability of about 14 KT in the initial portion of the CST expansion chamber. Use of larger driver charges, up to perhaps the estimated 1000 lb safe driver chamber capacity, may be a possible means of increasing the initial portion's simulation capability to about 15 KT. However, this will result in working stresses greater than design levels for the terminal portion of the expansion cone structure and is therefore not firmly contemplated at this time.

Typical static overpressure-time histories observed in the shock tube are shown by Figure 11. It is seen here that the pressure decay is near "classical" in form, especially at the 10 ft and 15 ft diameter stations. The disturbances visible near the end of the p-t record for the 22 ft diameter area are due to end reflections which shorten the positive phase duration at this station, resulting in the decreasing trend of the duration and impulse vs distance curves shown qualitatively by Figure 8.

End Reflection Effects

Open-end reflection effects presented a significant problem during final engineering development of the shock tube. Reflected rarefactions were found to produce pronounced negative pulses which adversely affected the free-air blast simulation, especially in the terminal half of the expansion chamber. Accordingly, the reflected rarefaction has been successfully reduced/eliminated through development of an "end-reflector" technique: a reflecting bulkhead was placed normal to the emerging shock, approximately five feet from the shock tube's exit plane; the shock reflected from this bulkhead travels back into the expansion chamber, overtaking and interacting with the foregoing rarefaction. This action is responsible for the low-level secondary shock visible in the pressure-time profile for the 22 ft diameter area, Figure 11. The end reflector used for the experiments covered herein was a temporary massive structure, not readily movable to allow adjustments in the exit plane-reflecting surface stand-off distance. Consequently, reflection "elimination" action has been mainly optimized or "tuned" for approximately the 6 psi overpressure level at the 22 ft diameter test area. For other conditions, the end reflected shock arrives at the 22 ft diameter station either ahead of or behind a reflected rarefaction. An adjustable reflector is currently under construction to provide a convenient means of varying stand-off distance.

Driver Technique

Generally, the explosive driver technique employed by the CST has been satisfactory. However, TNT explosive is not considered to be the "ultimate" choice of a driver charge. This material was initially selected from the standpoint of economics, availability, and relatively desirable "confinability" (low brisance) characteristics. While performance resulting from use of TNT has been fairly consistent, there have been variations attributable to detonation performance; at least one trial has shown pronounced evidence of afterburning occurring in the CST expansion chamber. Additionally, the unbalanced nature of the TNT reaction, further aggravated by presence of polystyrene charge support material, results in collection of relatively large quantities of residual carbon within the initial portions of the expansion chamber.

Consequently, efforts are currently underway to supplant the TNT driver with a more desirable material. Major selection criteria include a rigid cast explosive, preferably of low brisance, and having a balanced reaction. However, alternate techniques, including propellants and detonable gases, are also under consideration.

REFERENCES

(a) Filler, W.S., Phys. Fluids 3, 444 (1960).
(b) Filler, W.S., Design Characteristics of a Conical Shock Tube for the Simulation of Very Large Charge Blasts, U.S. Naval Ordnance Laboratory, White Oak, NAVORD Report 6844 (1960).
(c) Filler, Park, Pages, A Conical Shock Tube Nuclear Blast Simulator, Proceedings of the Fifth International Shock Tube Symposium, published by the U.S. Naval Ordnance Laboratory, White Oak, Silver Spring, Maryland in May 1966.
(d) Filler, W.S., NOL Model Nuclear Blast Simulator, Proceedings of the Fourth Shock Tube Symposium, 18-20 April 1961.
(e) Moulton, J.F., Jr., Final Report of the Conical Shock Tube Advisory Committee, U.S. Naval Ordnance Laboratory, White Oak, NOLTR 61-169 (1961).
(f) Culbertson, D.W., Conical Shock Tube Blast Simulator Facility, Description and Capabilities, U.S. Naval Weapons Laboratory, Dahlgren, Virginia, Technical Memorandum No. T-15/67 (October 1967).
(g) Culbertson, D.W., An Investigation of the Recoil Forces Produced by Detonation of High Explosives in the 16"/45 Cal Gun, U.S. Naval Weapons Laboratory, Dahlgren, Virginia; Technical Memo No. T-3/62, (1962).
(h) Balcerzak, M.J., Thermochemical and Thermodynamic Calculations for the Conical Shock Tube, American Machine and Foundary Company Technical Note MR 1208, August 1962.

Table 1
CST performance summary, 10 ft dia test area

Driver charge wt, lbs[(1)]	P_{so} psi	t^+ msec	I_s psi-msec	Shock Mach	Simulated yield KT(TNT)[(2)]
100	20	250	1800	1.47	2.5
300	42	200	2900	1.85	7.5
500	50	190	3500	1.97	10.0
700	79	175	4000	2.36	11.0
850[(3)]	85	170	4400	2.40	13.5

(1) Nominal weights, TNT (2) Based on peak static overpressure
(3) Projected performance

Table 2
CST performance summary, 15 ft dia test area

Driver charge wt, lbs[(1)]	P_{so} psi	t^+ msec	I_s psi-msec	Shock Mach	Simulated yield KT(TNT)[(2)]
100	9	400	1300	1.23	2.5
300	17	430	2200	1.44	7.5
500	26	450	3200	1.58	12.5
700	31	496	4000	1.67	17.0
850[(3)]	40	500	4200	1.82	20.0

(1) Nominal weights, TNT (2) Based on peak static overpressure
(3) Projected performance

Table 3
CST performance summary, 22 ft dia test area

Driver charge wt, lbs[(1)]	P_{so} psi	t^+ msec	I_s psi-msec	Shock Mach	Simulated yield KT(TNT)[(2)]
100	5	385	900	1.14	2.5
300	8	400	1600	1.21	7.5
500	12	425	2000	1.31	12.5
700	14	450	2600	1.35	17.0
850[(3)]	16	475	3000	1.40	20.0

(1) Nominal weights, TNT (2) Based on peak static overpressure
(3) Projected performance

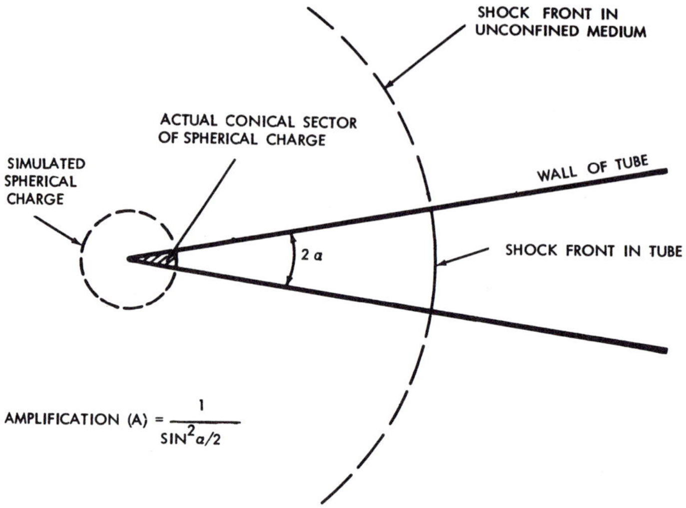

Figure 1
Explosive amplification effect in a conical shock tube

Figure 2
Aerial view of 2400 ft long conical shock tube facility

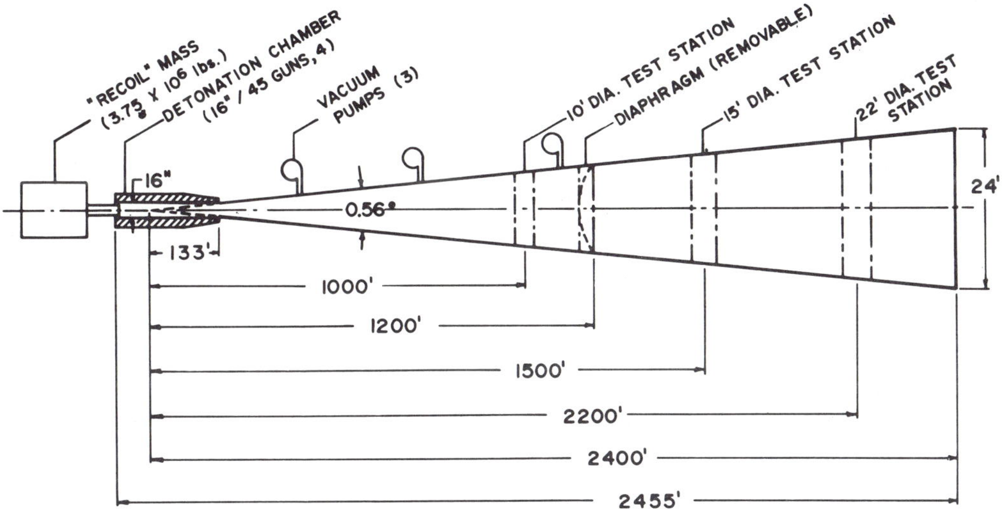

Figure 3
Major dimensional characteristics of the conical shock tube

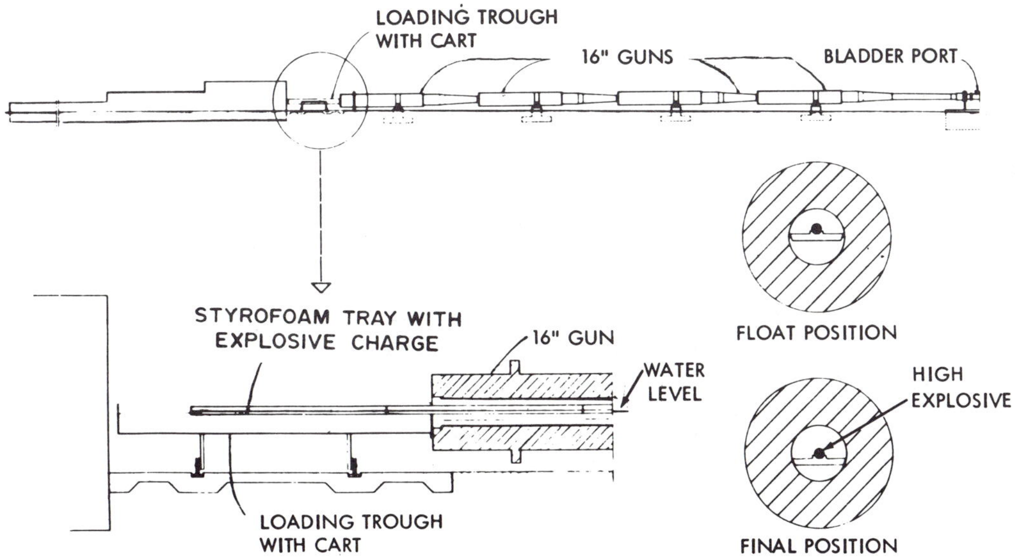

Figure 4
Driver charge loading technique for the conical shock tube

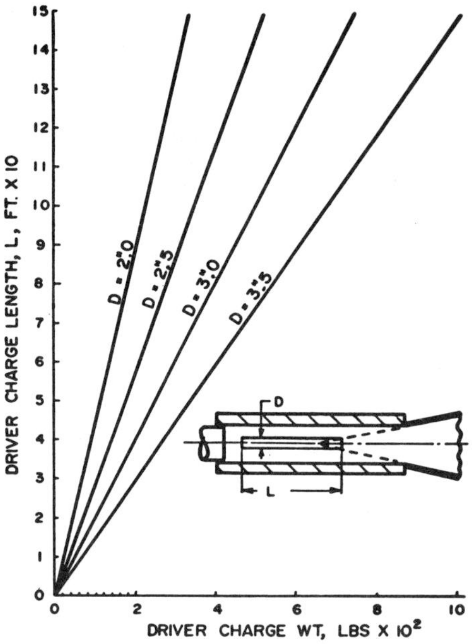

Figure 5
Characteristics of TNT driver charges for the conical shock tube, explosive density = 1.6

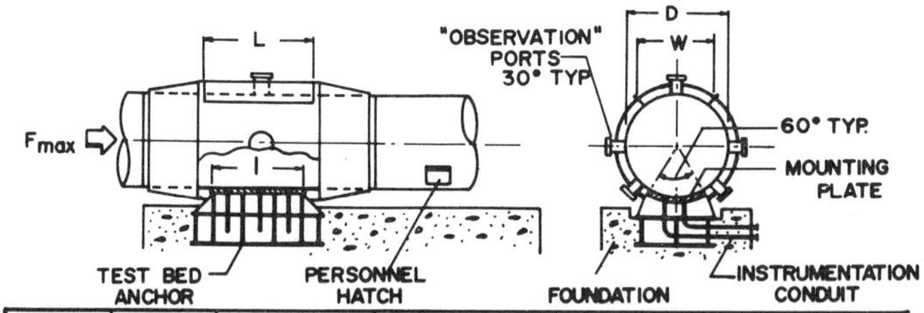

TEST STATION NO.	INSIDE DIAMETER D,ft.	TEST BED LENGTH l,ft.	HATCH DIMENSIONS		FOUNDATION WEIGHT lbs X 10⁶	DESIGN FORCE Fmax,K-LBS.
			W,ft.	L,ft.		
1	10	7.5	NO HATCH REMOVABLE SECTION FWD OF TEST AREA		4.08	5660
2	15	9.0	13	12	1.84	2500
3	22	12.0	18	22	1.82	1824

Figure 6
Conical shock tube test station details

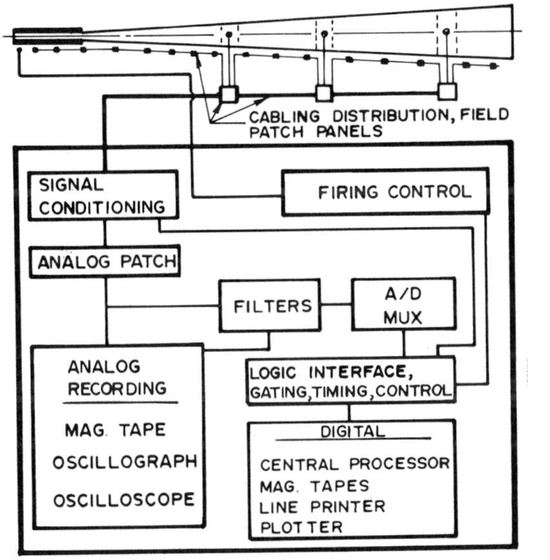

Figure 7
Diagram of the conical shock tube instrumentation system

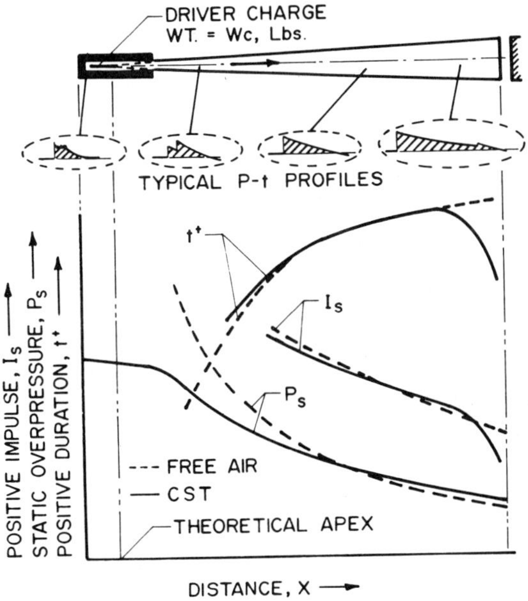

Figure 8
Conical shock tube's qualitative performance trend

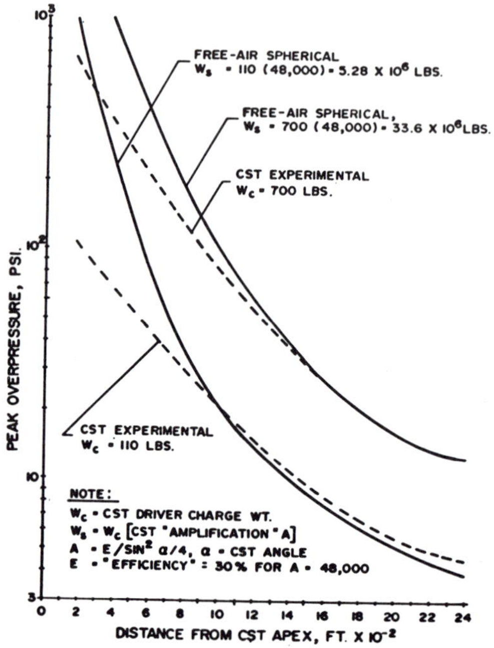

Figure 9
Conical shock tube static overpressure-distance data compared with that of free-air spherical (TNT) charges

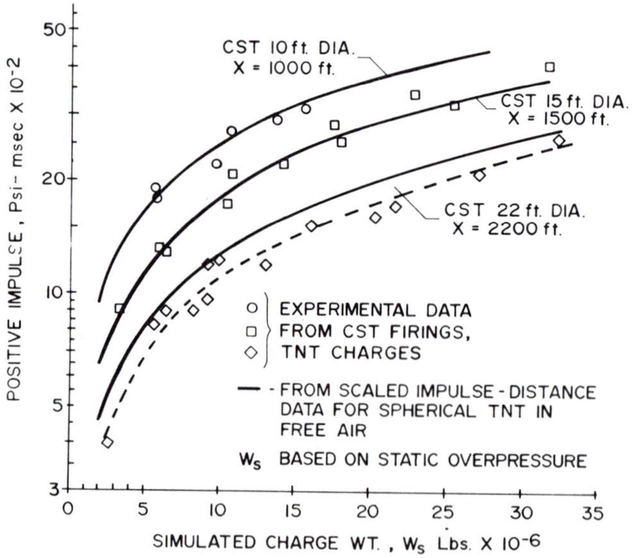

Figure 10
Correlation of conical shock tube positive impulse data

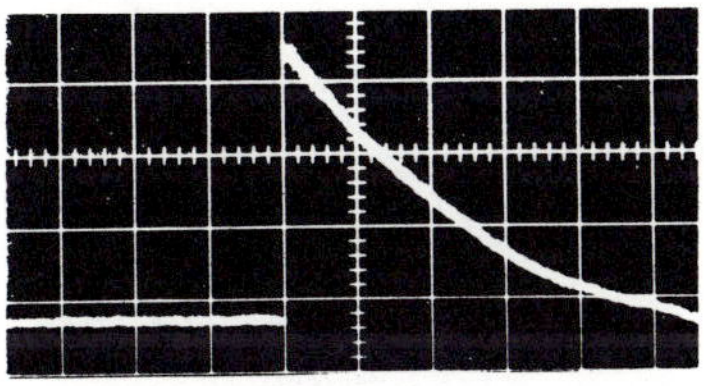

CST 10 ft diameter
Peak P_{so} = 7.5 psi
Positive duration = 220 msec

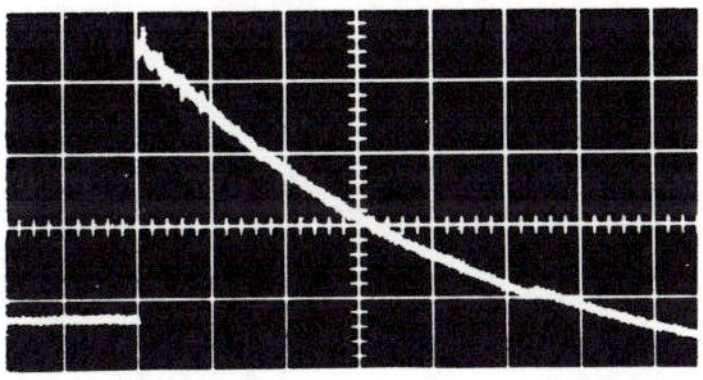

CST 15 ft diameter
Peak P_{so} = 7.5 psi
Positive duration = 325 msec

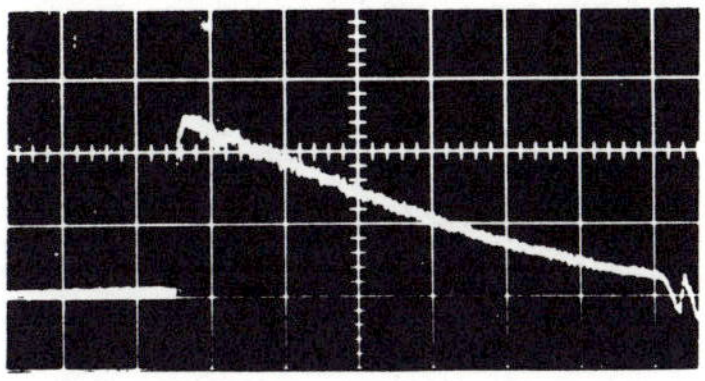

CST 22 ft diameter
Peak P_{so} = 2.0 psi
Positive duration = 335 msec

Figure 11
Typical static overpressure-time histories observed at the conical shock tube's test areas

A COLLISIONAL TREATMENT OF THE REFLECTION OF AN IMPLODING HEMISPHERICAL SHOCK WAVE

A. K. MACPHERSON
Institute for Aerospace Studies
University of Toronto
Toronto, Canada

SUMMARY

The reflection of a hemispherical imploding shock wave was studied in the region very close to the centre. Two aspects which were of particular interest were the effects of the presence of a cold wall and the effects of performing the experiment in a very dense gas. A Monte Carlo technique substantially as proposed by Bird[1] was used to examine the difference between a specularly reflecting wall and a cold wall. The implosion in a dense gas was simulated numerically by considering the first terms of a Taylor series expansion of the position and velocity vectors. The interparticle potential law was a combination of the Thomas-Fermi-Dirac and Lennard-Jones laws as proposed by Redhead, Hobson and Kornelsen[2]. However, as this latter technique is still under development, the results can only be proposed tentatively.

INTRODUCTION

In the analytic study of the performance of the UTIAS implosion-driven hypervelocity launcher[3], the assumption of spherical symmetry and the artificial viscosity techniques have been successful in studying the overall implosion phase. However, the assumptions are suspect when the collapse radius is reduced to a few mean free paths. Further, the difficulty arises of obtaining a realistic equation of state at the extreme densities expected at the centre. It was thus considered that a solution on a molecular level may provide a better insight into the physical processes involved.

As the solution of unsteady problems using the Boltzmann equation are relatively complex, Chu[4], the present study was undertaken using Monte Carlo techniques. These methods have the limitation generally of only allowing a study of small regions of flow, but in the present case this restriction was not very important. The related problems of the collision of two planar shock waves and the reflection of a planar shock wave from a cold wall were first solved[5]. It was found that the results agreed with results from continuum theory. The assumptions of hard sphere molecules and binary collisions were made in these studies.

The effects of the cold wall upon a hemispherical implosion were studied using the technique developed by Bird[1]. In this analysis the assumptions of hard sphere molecules and binary collisions were made. The shock was generated by a specularly reflecting piston and the wall was considered to be diffuse reflective. As the binary collision assumption probably breaks down near the centre, a hemisphere of radius six mean free paths, in the undisturbed gas, was assumed to surround the centre. The hemisphere was considered to be specularly reflective. A solution of the same problem assuming spherical symmetry was produced by assuming all surfaces to be specularly reflective. The region of flow very close to the centre was studied by a method which is still being developed. Thus the results at this stage are only presented tentatively. The gas was taken to be argon which was assumed not to ionize. With this technique it was possible to allow n-body collisions and to calculate the interparticle potential pressure.

BIRD MONTE-CARLO TECHNIQUE

Although Bird[1] has described his technique a number of times, it appears worthwhile to give a brief account from a different viewpoint. Also the present author considers that a modification to the time advance parameter is necessary to produce a logically correct technique. The effect upon the overall result will probably be small except for some special problems.

An alternative view of the Bird Monte-Carlo technique

Consider the gasdynamic system to be made up of a series of small macroscopic cells. These cells are composed of a great number of particles with random locations in configuration space and following the Maxwell-Boltzmann distribution law in velocity space. Choose a number of particles in each cell which are typical of those in the cell and store their positions and velocities in a computer.

Information can only be transferred from one cell to the next by the particles at the surface of the cell. Consider the typical cell particles to be also representative of the information particles. In the role of transferring information the cell particles can be considered to be grouped at one place and their positions in configuration space ignored. Allow the particles to collide for a small length of time ΔT_m in their role of transferring information and then starting from their positions in configuration space, allow them to move without colliding with other information transferring particles for a time ΔT_m. During this period they will collide with other particles but at the end of their motion the condition of the gas will not have changed so they will still be representative of the region they are in. New cells are then formed and the cycle repeated.

One of the problems is in deciding how to advance time in a cell according to the type and number of collisions. In the present work the value of ΔT_m was chosen so that the fastest particle in the system would not cross more than two cells during ΔT_m. This was to prevent the transfer of information by direct diffusion rather than by collisions. This assumption placed a restriction upon the minimum cell width as otherwise ΔT_m could become ridiculously small. The time ΔT_i required for the collision of two particles with relative velocity V_{Ri} is chosen by Bird[1] to be

$$\Delta T_i = \frac{1}{ANV_{Ri}\ N_c/2} \tag{1}$$

where A = collision cross section
$N_c/2$ = number of colliding pairs of particles
N = local number density

The probability of collision of any two particles is chosen proportional to V_{Ri}. Collisions continue in a given cell till

$$\sum_{i=1}^{k} \Delta T_i = \Delta T_m = \sum_{i=1}^{k} \frac{1}{ANV_{Ri}\ N_c/2} \tag{2}$$

This intuitively will give the correct mean time between collisions. On the other hand it can be shown[6]

$$\Delta T_p = -\frac{\ell n\ \xi_i\ .\ N_c}{NA \sum_{j=1}^{N_c} |V_p - V_j|} \tag{3}$$

where ΔT_p is time of flight of the pth particle before a collision

ξ_i is a random number between 0 and 1
V_j is the velocity of the ith particle.

By choosing ΔT_p to be a minimum the first particle to collide will be found. Thus it appears that over an extended period of time, the Bird time parameter and particle collisions will be correct. However, over the short time, the result could be suspect. This is supported by the fact that, in choosing a time parameter incorrectly with respect to the cell width, Bird[7] obtained the wrong values for shock thickness. Further, from equation (3) it can be seen that ΔT_p can be zero for any V_R if $\xi_i = 1$; however from equation (1), ΔT_i can only be zero if V_{Ri} = D

Modified time advance parameter

Consider equation (2) and let $\Delta T_m = 2.0$, $\Delta T_1 = 1.3$ and $\Delta T_2 = 1.1$. The question arises: should k = 1 or 2? If k = 1 the particles in the cell will only move for $\Delta T = 1.3$ and hence not move far enough with respect to particles in adjacent cells. If k = 2, $\Delta T = 2.4$ for the particles and they will move too far. The solution proposed is to calculate the probability of any collision occurring during time δt. An estimate of the average collision time ΔT_A can be obtained for the present problem by substituting for V_R twice the average absolute value of the peculiar velocities in the cell, in equation (1). This factor of 2 was determined "experimentally" from the program. It can not be obtained theoretically as the shock structure does not obey the Maxwellian distribution. As the value of the collisional velocity is not used in the actual time advance, small deviations from δt should not be important in the overall answer. Further, the cell widths and values of ΔT_m were always chosen so that this averaging technique did not dominate the system. That is ΔT_m was not reduced so that only 1 collision occurred in each cell both upstream and downstream of the shock.

DENSE GAS DIRECT SIMULATION TECHNIQUE

Consider a small sector of gas such as shown in Figure 2. This three dimensional sector is filled with argon atoms the correct distance apart for 110 atmospheres. Although the positions are chosen at random, when inserting each new particle, the location of already existing particles must be examined to ensure that two atoms are not placed too close together. Otherwise the temperature will rise locally and atomic velocities will increase unrealistically. The periodic boundary condition is invoked to satisfy boundary conditions. This consists of assuming that the entire sphere is constructed of many sectors with atoms in exactly the same position in each sector. The inner and outer boundaries are dense argon gas surfaces. The positions of the inner atoms do not change during the experiment and are 1.5 Å apart. The positions of the piston atoms move along radial paths, again initially 1.5 Å apart. When the piston moves, the surface atoms must come closer together to prevent a circumferential flow developing. The atoms are initially given a Maxwellian velocity distribution appropriate to 300°K.

Let the ith particle of mass m have position vector $\hat{r}_i$ at time t_1 and velocity vector $\hat{v}_i$ at t_1. The position at time t is given by

$$\hat{r}_i(t) = \hat{r}_i(t_1) + \dot{\hat{r}}_i(t_1)(t-t_1) + \ddot{\hat{r}}_i(t_1)\frac{(t-t_1)^2}{2!}$$

$$+ \dddot{r}_i(t_1)\frac{(t-t_1)^3}{3!} + \ldots \qquad (4)^*$$

$$\ddot{\hat{r}} = \frac{1}{m}\frac{\partial\phi}{\partial r_i} \text{ - where } \phi \text{ is the potential function at } r_i \qquad (5)$$

In the present case using the Redhead et al[2] potential curve, ϕ is generated by the surrounding atoms at r_j so that

$$\ddot{\hat{r}}_i = \sum_{\substack{j \\ i\neq j}} \frac{1}{m}\frac{\partial\phi_{ij}}{\partial \hat{r}_{ij}} \cdot \frac{\partial r_{ij}}{\partial r_i} \qquad (6)$$

$$= \sum_{\substack{j \\ i\neq j}} \frac{1}{m}\cdot\frac{\partial\phi_{ij}}{\partial r_{ij}}\cdot\frac{r_i - r_j}{r_{ij}} \qquad (7)$$

Similarly $\dddot{r}_i$ may be derived.

$$\text{Now } \dot{\hat{v}}_i = \hat{v}_i + \ddot{\hat{r}}_i(t-t_1) + \dddot{r}_i\frac{(t-t_1)^2}{2!} + \ldots \qquad (8)$$

*I am indebted to Dr. P.C. Hughes for suggesting and formulating the high order terms in this and subsequent equations.

Thus, by substituting and allowing $(t-t_i)$ to be sufficiently small, the motion of all particles may be determined. In the present case only the first order terms were considered. Under these circumstances greater accuracy may be obtained by considering the energy to be conserved and hence,

$$\Delta\phi = \Sigma \frac{\partial \phi_{ij}}{\partial r_i} (t-t_1)\ v_{ij}(t_1) \qquad (9)$$

$$v_i(t) = SQRT(v_i(t_1)^2 - 2\ \Delta\phi . v_i(t_1)(t-t_i)/m) \qquad (10)$$

where m is the atomic mass of argon.

Before the piston was set in motion the atoms were allowed to collide for a short length of time to establish equilibrium. The interparticle potential energy was converted to an equivalent kinetic pressure ratio by dividing by the product of the atomic mass and the most probable velocity squared. This value agreed well with the van der Waals' equation for the undisturbed gas.

It was found that high energy gas atoms could overcome the surface potential of the piston and penetrate the monolayer. Under these circumstances specular reflection was assumed to occur at the centre line of surface atoms. The time interval $(t-t_1)$ was chosen so that the fastest atom in the system did not move more than one quarter of the zero potential radius of the argon atoms. If longer times were considered, it was found that instabilities in the undisturbed gas were generated by atoms coming too close together. When higher order terms are considered, it should be possible to use longer time steps as a given atom has earlier warning of the approach of a colliding atom. However, even with careful programming it required 9 hours on a 360/65 to produce the results in this report. It is felt that this can be reduced by the use of high order terms.

RESULTS

Physical details of simulated gaseous regions

The initial piston radius was approximately 45 mean free paths (λ) of the undisturbed gas for the cold wall study as shown in Figure 1. The cells were produced from j hemispherical shells Z_{ij}, initially 1.2 λ in width reducing as the piston moved in to 0.5λ. After this the number of shells were reduced to keep the width approximately constant in width. In shells where sufficient atoms were present to allow 10 atoms in the smallest cell, four cells Z_{ij} (i=1,-4) were formed. The cell Z_{ij} was formed by the volume between the azimuthal angles (measured from the x axis) 0 and 4.5^o, Z_{2j} between 4.5^o and 9°, Z_{3j} between 9° and 13.5^o, and Z_{4j} between 13.5^o and 90° of the radius vector when swept through all polar angles. On all other occasions only 3 zones were formed where the azimuthal angle moved in steps of 30°. The centre of the sphere was surrounded by a sphere of radius 6λ. Initially the atoms were distributed throughout the whole volume with a Maxwellian velocity distribution. While in the information transferring mode, an assumption of spherical symmetry was made with each cell so that the azimuthal (θ) and polar (ϕ) angles could be ignored. Thus each particle's velocity was transformed to a set of axes with the radius vector (r) as the x_1 axis, the x_2 axis as $r\theta$, and the x_3 axis as $SQRT(x^2 + y^2).\phi$. After the collisional process was complete for time ΔT_m, the velocities were transformed back to the original axes for the purpose of moving the particles. A total of 4,000 particles were used to simulate the process, which had an initial number density of 2×10^{18}.

As mentioned previously, in the dense gas study, the initial pressure was 110 atmospheres, temperature 300°K, and the gas argon, which obeyed a potential law as in Redhead et al[2]. The resultant number density was 2.7×10^{21}, the most probable velocity V_m was 4.33×10^4 cms/sec. and the mean free path λ was 8.83Å. The value of λ was calculated assuming an atomic radius of 0.9 of the zero potential radius r_o, where $r_o = 3.42$ Å. For the purpose of calculating pressure, temperature and density ratios, zones were constructed as in Figure 2. There were 30 zones set up at all times. A sphere of radius 2λ was taken to surround the centre as measurements of temperature and pressure become meaningless when the number of particles become very small. The initial piston radius was 110λ, and in both studies the piston velocity was kept constant at 6.778 V_m. This is the velocity to generate a Mach number 10.0 shock wave with a plane piston.

Effects of a cold wall on an imploding shock wave

The origin of coordinates has been taken to be the surface of the sphere surrounding the origin, that is in the present case, 6λ from the hemisphere's centre. The variation of temperature, pressure and density through the shock wave with a specularly reflecting wall is shown in figure 3 at a time $0.002762/V_m$ sec. after the piston started moving. This is the same as the spherically symmetric case which will be used for comparison. A pronounced peak is formed in the temperature profile at the shock front. The reciprocal slope shock thickness, when compared with the planar case[5], shows that the shock waves here are approximately twice as thick as in the planar case. The corresponding case with a cold wall for the region where only 3 cells could be formed in each shell is shown in Figure 4. This result is for the second cell at Z_{2j}, Figure 1. Only the front foot of the shock is found here and by comparison with Figure 3, it can be seen that the shock is in the same position as for the spherically symmetric case. Where four cells per shell were formed the situation is vastly different, Figure 5. Zone 1 are the cells closest to the wall at Z_{1j}, Figure 1 and Zone 4 are the cells Z_{4j}. The pressure and density profiles in zone 1 show that the pressure and density are much higher close to the piston than in the symmetric case. The temperature is lower but it can be seen that the most significant feature is that the shock front is retarded. This retardation is seen in the temperature and pressure profiles in zone 2. However, the pressure and density in zones 2, 3 and 4 are approximately those in the symmetric case. The retardation of the temperature is not seen in zone 3 but the value is lower than for the symmetric case.

Although it was felt that insufficient data were available to be confident that statistical fluctuations had been eliminated, the temperature profile in zone 4 appears to have a similar peak as in Figure 3. The value at the top of the peak and at the piston also correspond. Thus the picture which evolves is of a shock which has been retarded by the cold wall and is slightly elliptical. However the values of density or pressure a short distance from the wall are not greatly affected.

Generally the most interesting region is the part of the flow around the origin during the reflection process. The pressure and temperature profiles for the simulated spherically symmetric case are shown in Figure 6. A steady increase in both temperature and pressure are found except for an overshoot in the temperature for a short time. The same case with a cold wall is shown in Figures 7 and 8 showing pressure and temperature profiles at approximately the same times as in Figure 6. Zone 1 is the cells closest to the wall, Z_{1j} in Figure 1 and zone 3 is Z_{3j} where the conditions were such that only 3 cells per shell could be formed. Zone 3 behaved very much like the spherically symmetric case although in this case the overshoot was in pressure rather than temperature. However, the final results were very similar. In zone 1 the shock was initially slightly retarded and both pressure and temperature were a little less than in zone 3. The significant feature was the presence of a thermal boundary layer near the centre which also reduced the pressure. This was most marked at times T_4, T_5 and T_6. It appeared that as the density built up at the centre, that this vanished. Neither the temperature nor the pressure in zone 1 reached those in zone 3. However, the reflected shock front did not lag behind in zone 1 at time T_7 significantly.

The final results produced were for spherically symmetric case at time $0.004639/V_m$ sec. after the motion started. Fairly smooth curves were obtained for pressure and velocity profiles. A kink is shown in the temperature profile as the results indicated one. This probably is not meaningful and may be due to statistical scatter. The shock front position is well defined by the density ratio. In this case the density has started falling at the origin due to expansion waves generated by the outward motion of the shock. The pressure reciprocal shock thickness shows that the shock is about three times as thick as in the planar case. The temperature profile is not so well defined, but is at least as much as three times thicker than the planar case. The results for the corresponding cold wall case are shown in Figures 10 and 11. Figure 10 represents those cells closest to the wall Z_{1j} and Figure 11 shows cells Z_{3j} under conditions where only 3 cells per shell were formed. The profiles for zone Z_{3j} are very similar to those for the spherical case. The cells close to the wall show a weaker reflected shock as the peak values of pressure, temperature and density are lower in Figure 11 than in Figure 10. In addition, expansion waves have travelled faster here than in the centre zone and the profiles of temperature, pressure and density all show a decrease at the origin. However, the velocity of the shock front is the same in both cases.

Thus the overall picture here is both the incoming and reflected waves will move at the same speed and hence retain a hemispherical shape. The region near the wall will, however, have a weaker shock in terms of temperature, density and pressure.

Imploding spherical shock in dense gas

The shock structure after an imploding piston had moved for 2.56 x 10^{-12} sec. is shown in Figure 12. The pressure ratio is approximately that predicted by ideal gas continuum mechanics. On the other hand the density is less than half the ideal gas value and the temperature more than double. Further, the shock thickness is five or six times greater than for the planar shock, hard sphere molecule theory. The reason appears to be that information is transmitted very rapidly by the dense gas. Thus the velocity of many molecules can be increased without the density undergoing a great increase. A spike is shown in the temperature and pressure profiles in front of the shock position. This is due to the presence of one high speed atom in a zone which has a small net velocity. Such an event is physically realistic because the normal assumption that in a small region of gas there is almost an infinite number of atoms, breaks down. Here the number is definitely finite and an atom with an order of magnitude velocity greater than V_m can produce large changes. As the piston moved further towards the centre and more atoms were sent ahead of the shock the number of peaks increased till the shock was no longer well defined.

The variation of pressure at the origin is shown in Figure 13. The sharp peaks and valleys in the early part are due to the arrival and departure of a few high speed molecules. However, once the density also increases the peaks no longer occur and a fairly steady value is obtained. This pressure ratio of about 10^5 is approximately that which would be expected from artificial viscosity calculations if a sufficiently strong shock was present.

The pressure, temperature and density profiles are shown in Figure 14, at a time 9.83 x 10^{-12} sec. after the piston started. The presence of a reflected shock is clear without sharp peaks due to abnormal regions. This development is most advanced in the temperature profile. The shock thickness of the temperature and pressure profiles, using maximum slope thickness, is now about 25 times that of a planar Mach 10.0 shock in hard sphere molecules. Also shown is the net interparticle potential energy in each zone. This has been converted to an equivalent kinetic pressure ratio by dividing by 1.5 V_m^2 x atomic mass. In only two zones was the value negative and these could not be shown. It can be seen that even though there is considerable fluctuation there is a general decrease through the reflected shock. The peak value is 40 at the origin but the average after a steady pressure was reached was about 10. Thus the atoms are able to move even under these conditions of extreme pressure to effectively minimize the potential forces.

CONCLUSION

The effect of a cold wall upon an imploding hemispherical shock has been found to decrease the pressure, temperature and density in the region near the wall. However, this is a localized effect as the velocity of the reflected shock, including that within the wall region, is virtually unaltered. It would be expected that some distance from the centre the effect would be negligible. Viscous forces and interparticle potential forces do not appear to significantly alter the pressures predicted by artificial viscosity calculations, when using a reasonable equation of state, for the implosion of a hemispherical shock. This is due to the ability of gaseous atoms to arrange themselves very rapidly to minimize potential forces.

ACKNOWLEDGEMENTS

I would like to express my appreciation of the interest and encouragement I have received from Dr. G. N. Patterson and Dr. I. I. Glass. I would also like to acknowledge the helpful discussions I have had with Dr. J. H. deLeeuw, Dr. N. A. Derzko and Dr. D. E. Roberts in the course of this work.

The financial support received from the Canadian National Research Council and the Aerospace Research Laboratory of the United States Air Force under Contract No. AF33(615)-5313 is acknowledged with thanks.

REFERENCES

1. Bird, G.A. Direct Simulation Monte Carlo Method - Current Status and Prospects. Sixth International Symposium on Rarefied Gasdynamics (In Press).
2. Readhead, P.A., Hobson, J.P., and Kornelsen, E.V. The Physical Basis of Ultrahigh Vacuum. p. 15, Chapman and Hall, 1968.
3. Poinssot, J.C. A Preliminary Investigation of a UTIAS Implosion-Driven Shock Tube. UTIAS T.N. 136.
4. Chu, C.K. Kinetic-Theoretic Description of the Formation of a Shock Wave. Phys. Fluids, 8, 12, 1965.
5. Macpherson, A.K. Shock Collision at a Molecular Level. (To be published).
6, Denisik, S.A., Macama, Yu G., Polak, L.S. and Rezvanov, R.A. Application of the Monte-Carlo Method in the Solution of Some Problems of Physical and Chemical Kinetics. Teplofizika Vysokikh Temperatur, 5, 1011, 1967.
7. Bird, G.A. The Velocity Distribution Function Within a Shock Wave, J.F.M. 30, 479, 1967.
8. Wood, W.W. and Parker, F.R. Monte Carlo Equation of State of Molecules Interacting with the Lennard-Jones potential. J. Chem. Phys. 27, 3, 720, 1957.

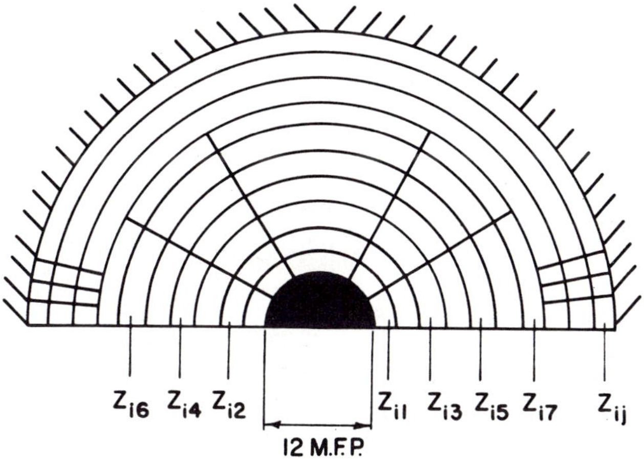

Figure 1.
Cell arrangements of hemispherical imploding piston adjacent to cold wall.

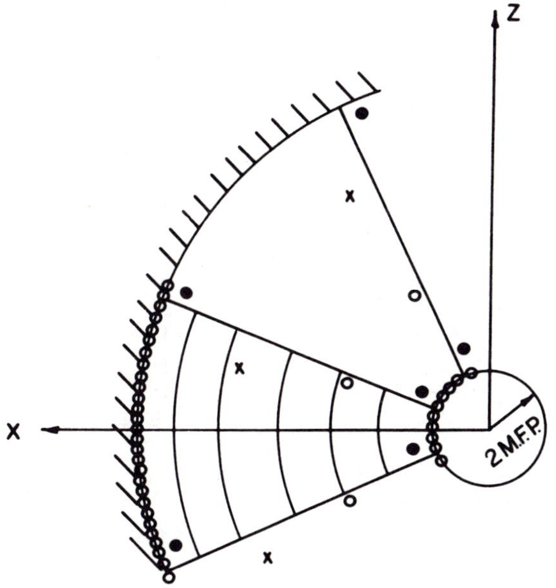

Figure 2.
Periodic boundary condition and cell arrangement for implosion in dense gas.

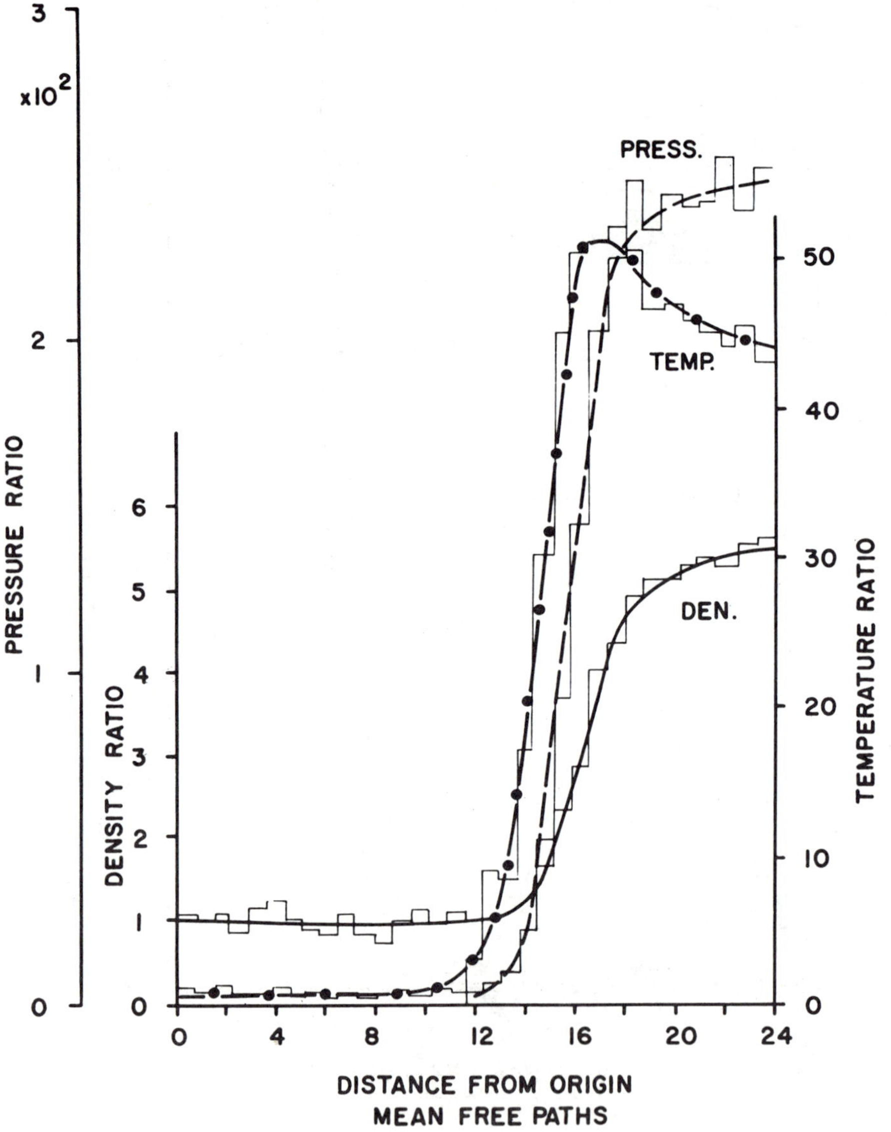

Figure 3.
Density, pressure and temperature profiles of gas driven by an imploding hemispherical piston at time $0.002762/V_m$ sec. after piston commenced moving. Specularly reflecting wall. Origin of coordinates 6 mean free paths from centre.

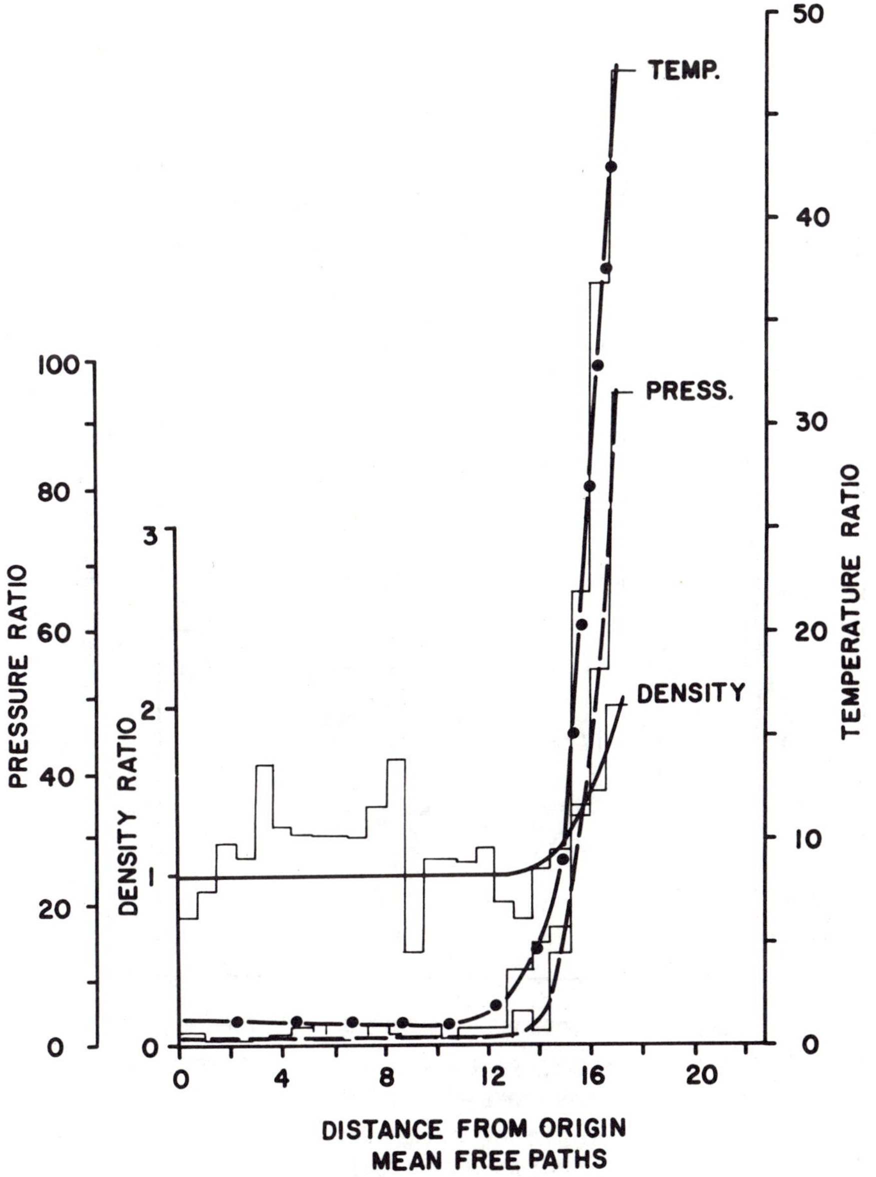

Figure 4
Density, pressure and temperature profiles of imploding shock wave at time $0.002726/V_m$ sec after piston commenced moving. Diffuse reflective wall. Regions where 3 cells per shell formed. Origin of coordinates 6 mean free paths from centre.

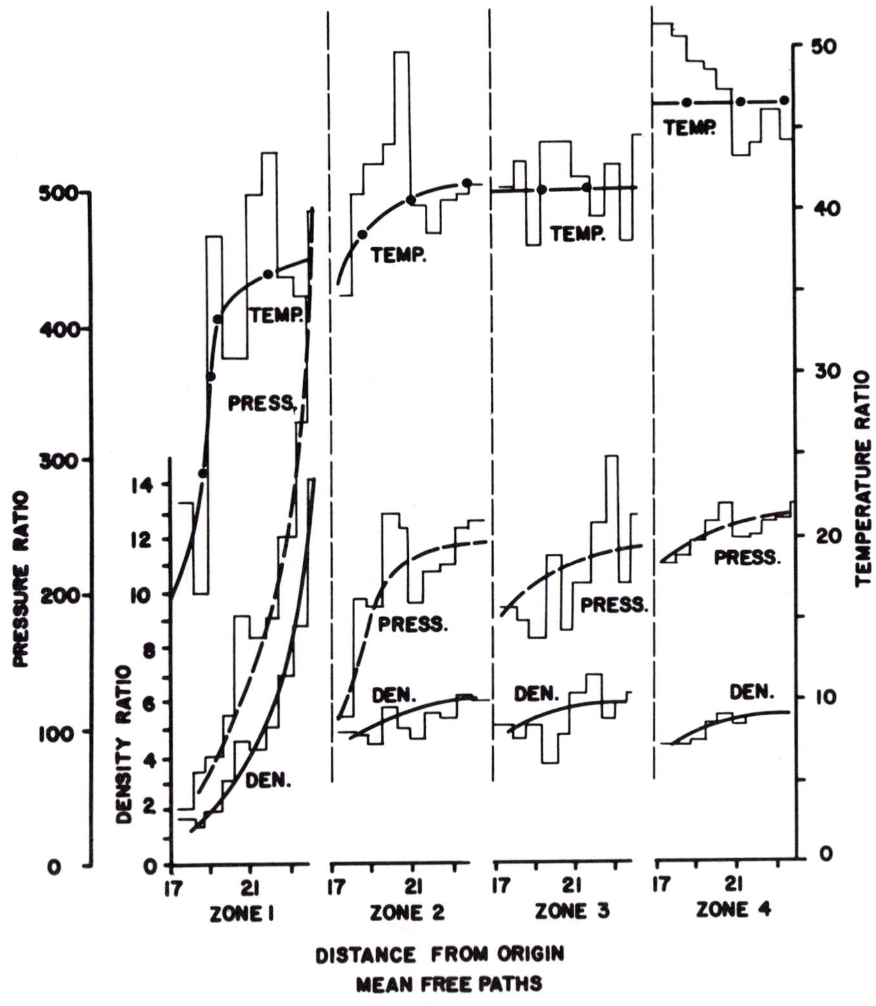

Figure 5
Density, pressure and temperature profiles of imploding shock wave at time $0.002726/V_m$ sec. after piston commenced moving. Diffuse reflective wall. Regions where 4 cells per shell formed. Zone 1 are cells closest to wall and Zone 4 is furthest from wall. See Figure 1. Origin of coordinates 6 mean free paths from centre.

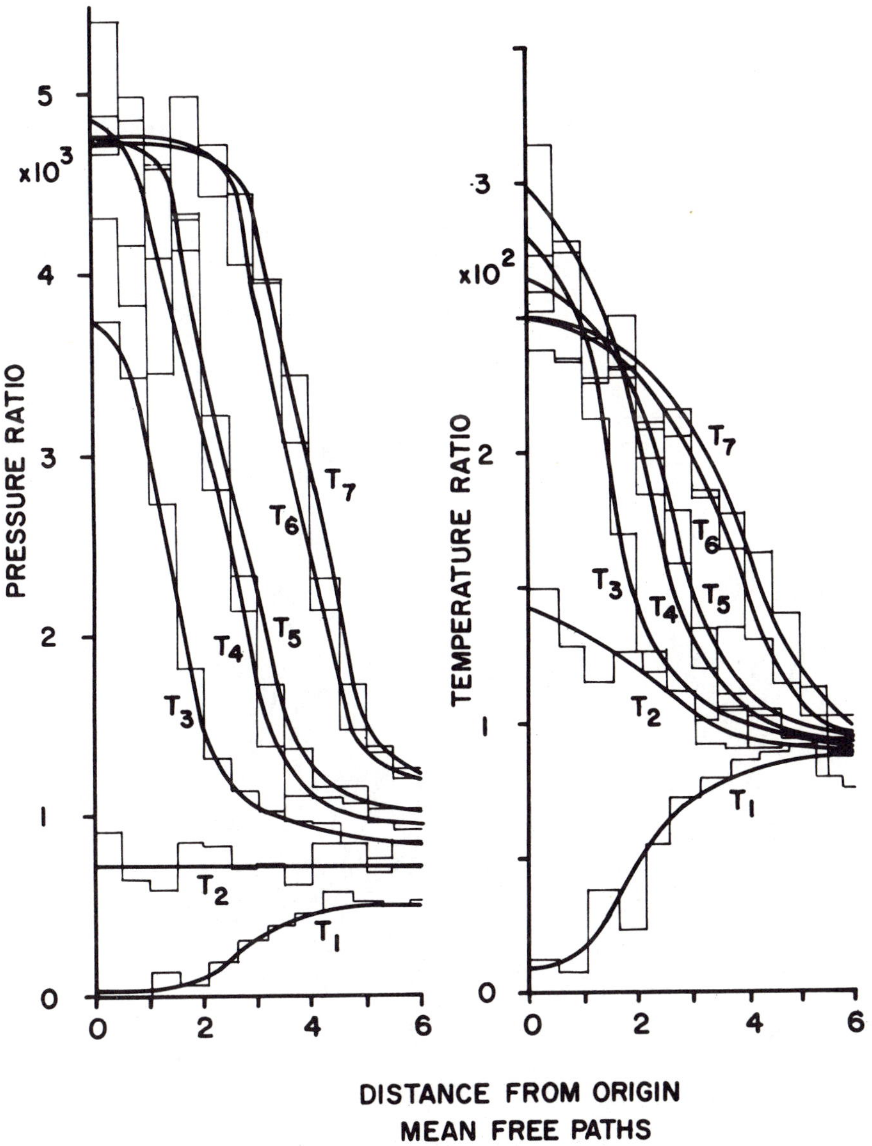

Figure 6
Pressure and temperature profiles during the reflection of a hemispherical imploding shock wave at various times T. Specularly reflecting wall. Origin of coordinates 6 mean free mean paths from centre. $T_1 = 0.003801/V_m$ sec., $T_2 = 0.003946/V_m$ sec., $T_3 = 0.004087/V_m$ sec., $T_4 = 0.004207/V_m$ sec., $T_5 = 0.004291/V_m$ sec., $T_6 = 0.004404/V_m$ sec., $T_7 = 0.004486/V_m$ sec.

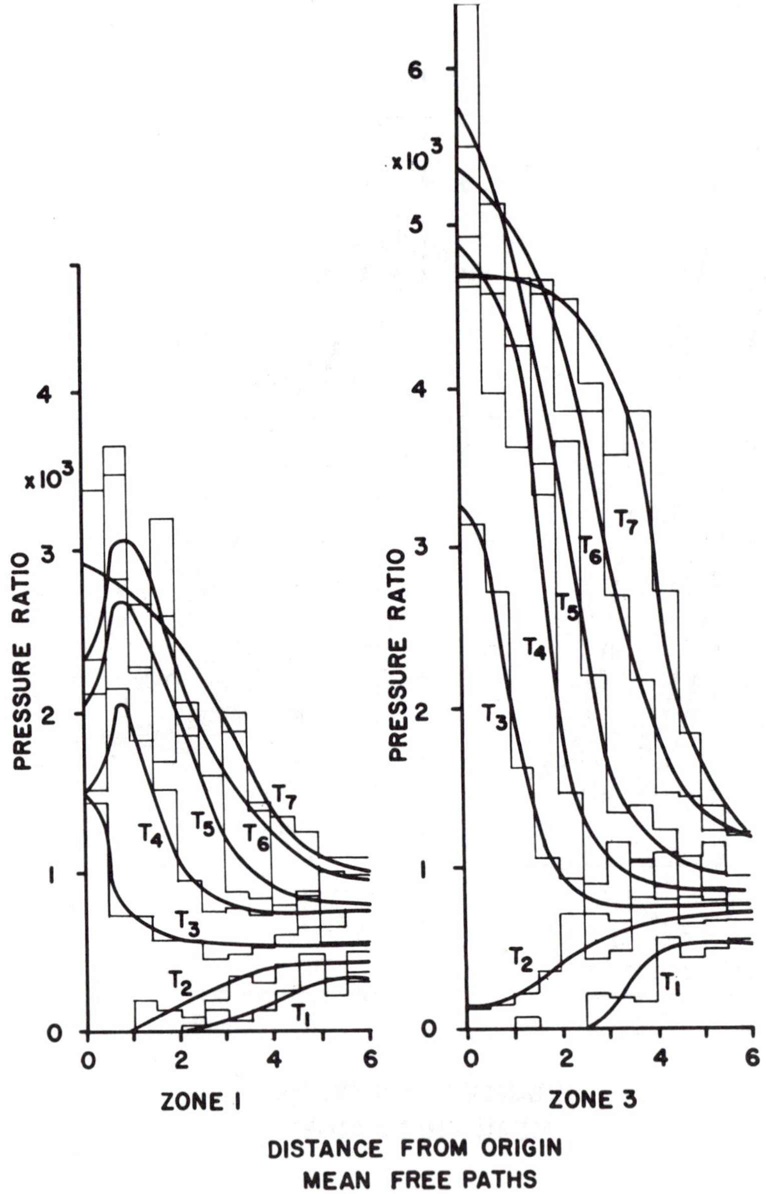

Figure 7
Pressure profiles during the reflection of a hemispherical imploding shock wave at various times T. Diffuse reflecting wall. Origin of coordinates 6 mean free paths from centre. $T_1 = 0.003795/V_m$ sec., $T_2 = 0.003936/V_m$ sec., $T_3 = 0.004081/V_m$ sec., $T_4 = 0.004204/V_m$ sec., $T_5 = 0.004298/V_m$ sec., $T_6 = 0.0044/V_m$ sec., $T_7 = 0.004501/V_m$ sec.

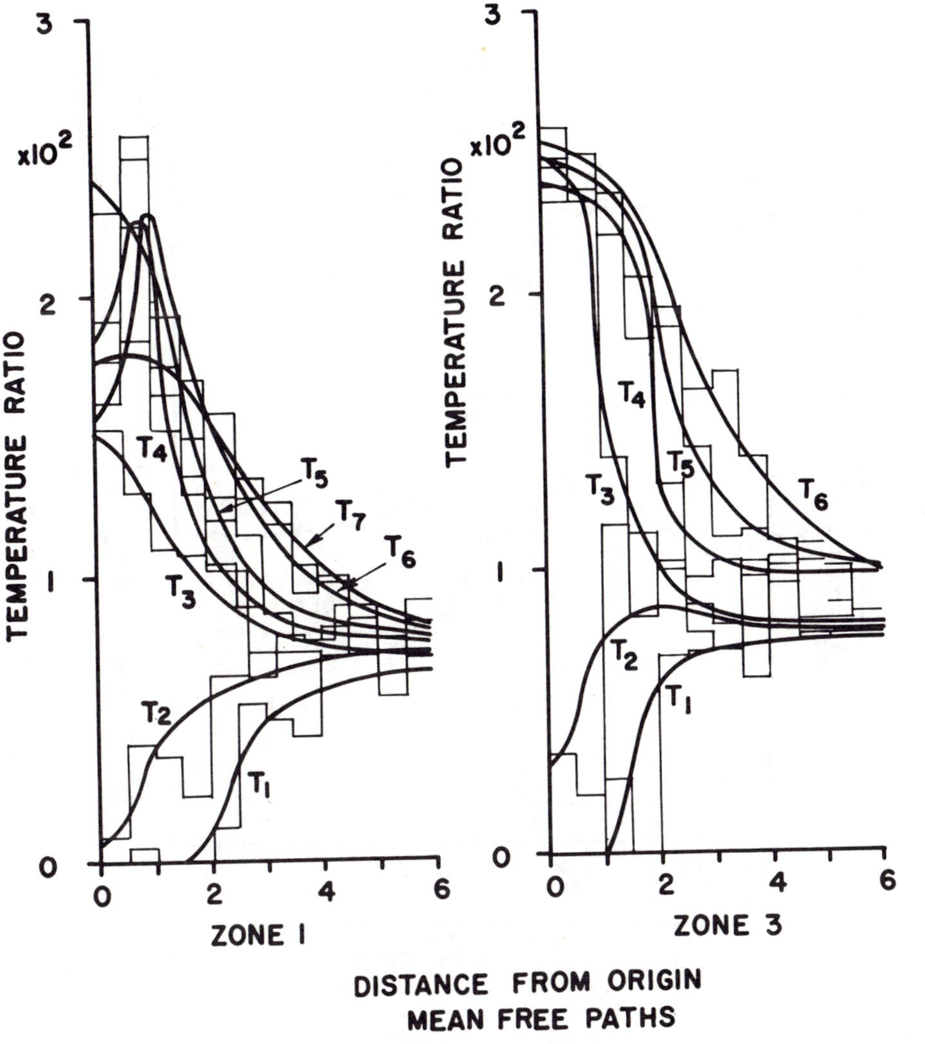

Figure 8
Temperature profiles during the reflection of a hemispherical imploding shock wave at various times T. Diffuse reflecting wall. See Figure 7 for details.

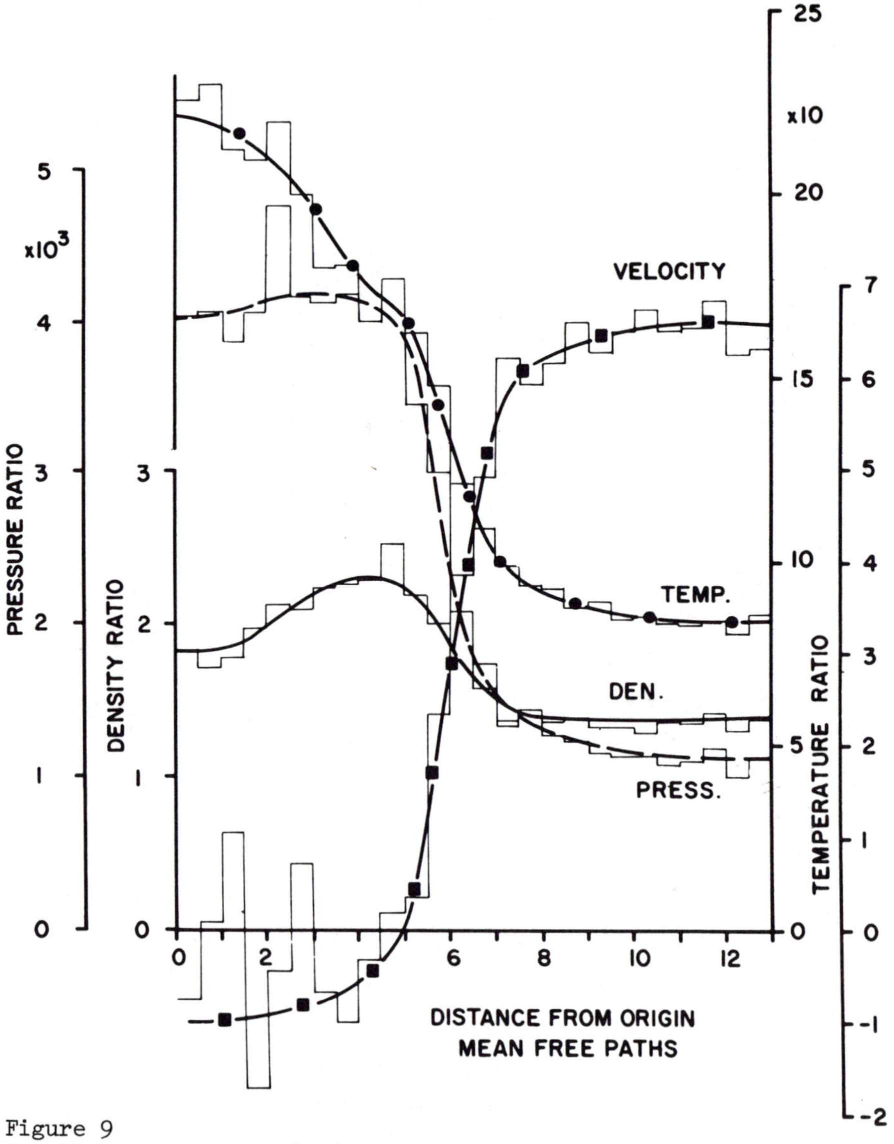

Figure 9
Density, pressure, velocity and temperature profiles as in Figure 3 at time 0.004639/ V_m sec.

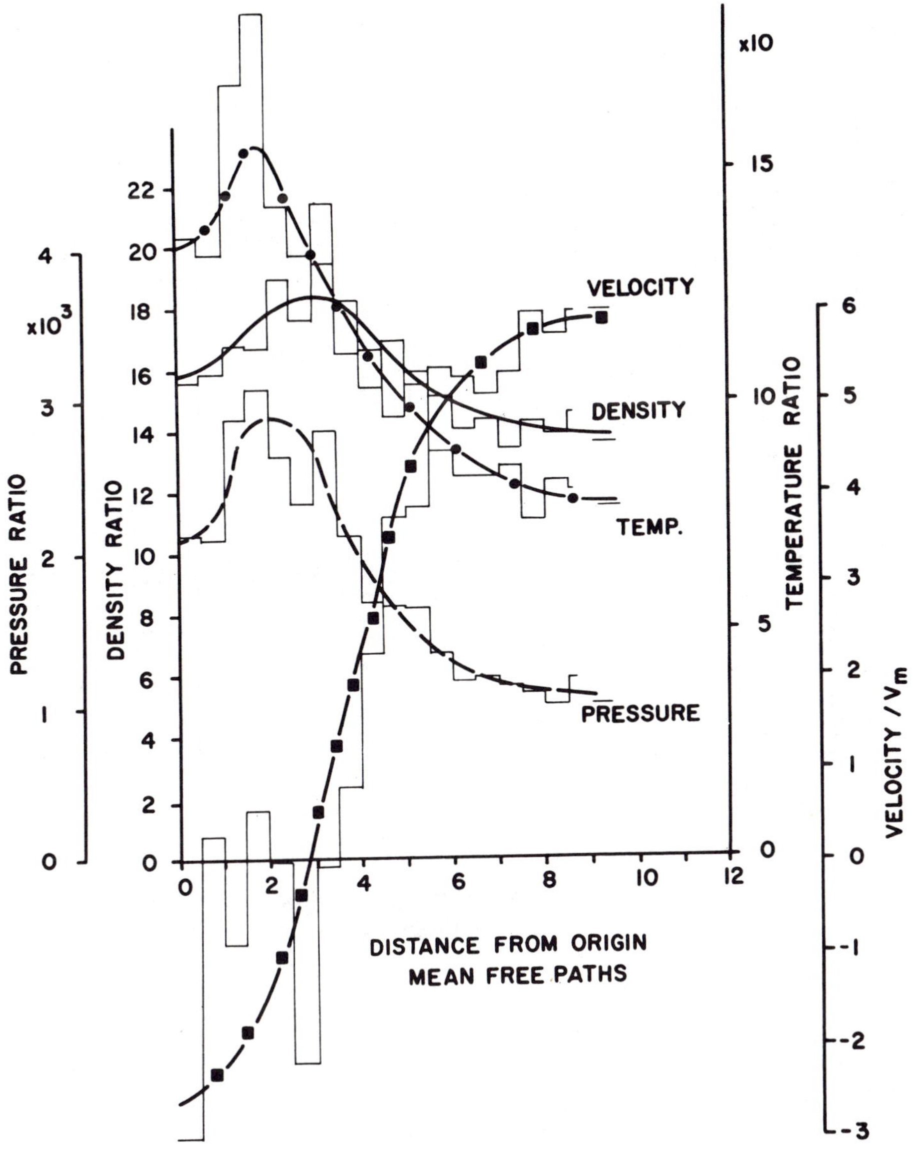

Figure 10
Density, pressure, temperature and velocity profiles of imploding shock wave at time $0.004624/V_m$ sec. after piston commenced moving. Diffuse reflective wall. Origin of coordinates 6 mean free paths from centre. Region where only 3 cells per shell were formed Z_{1j} Figure 1.

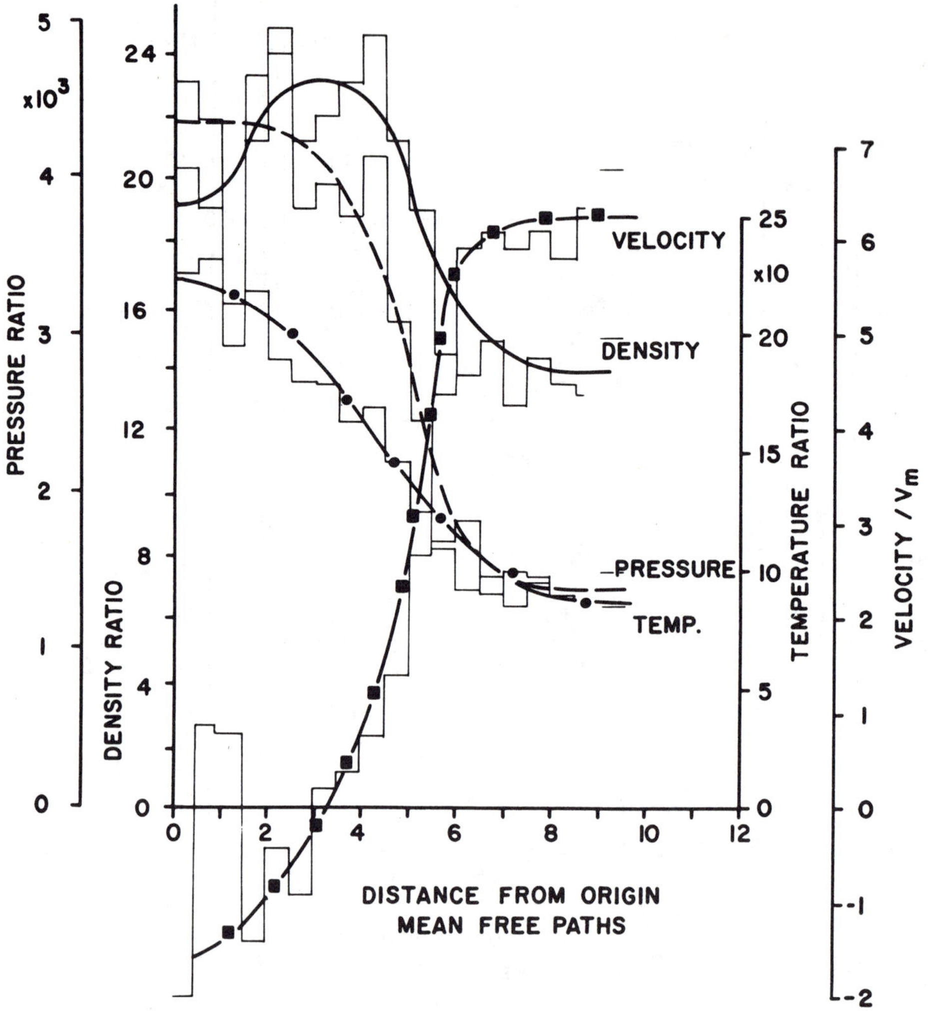

Figure 11
Density, pressure, temperature and velocity profiles as in Figure 10 for cell Z_{3j}.

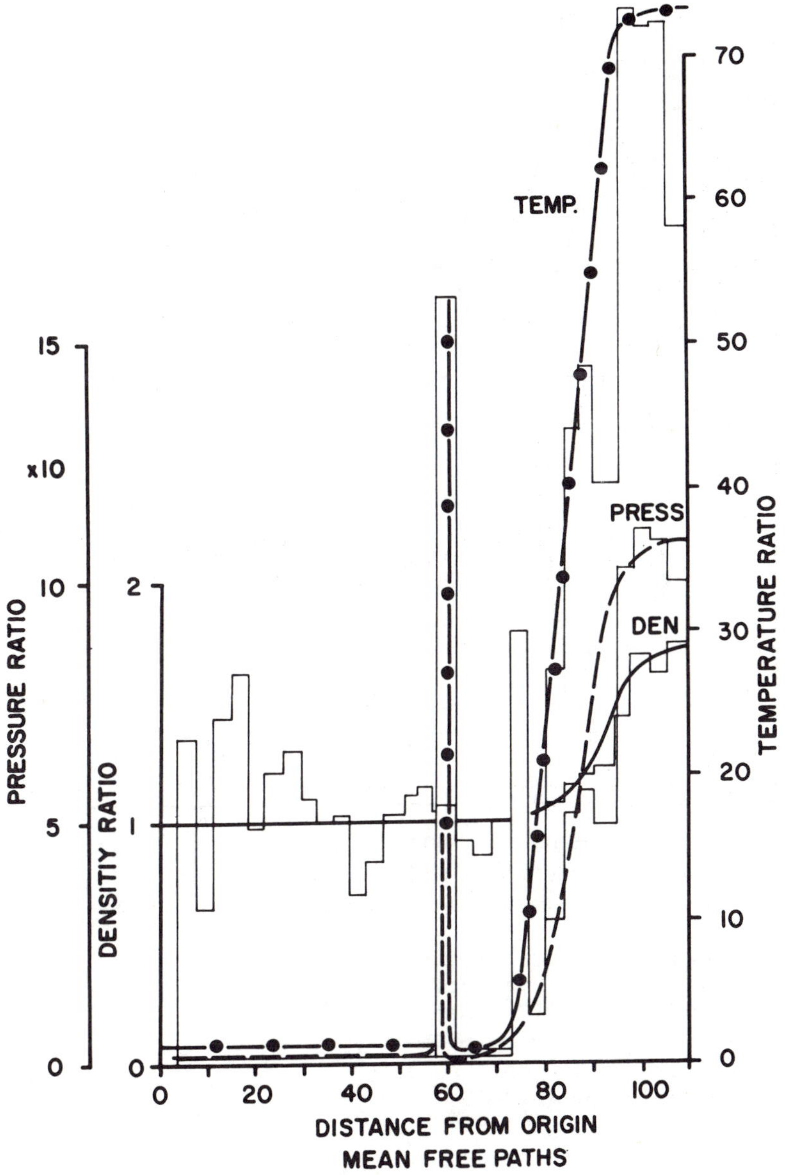

Figure 12
Density, pressure and temperature profiles of imploding shock wave at time 2.56 x 10^{-12} sec. after piston commenced moving into dense argon, initial number density 2.7 x 10^{21}. Origin of coordinates 2 mean free paths from centre.

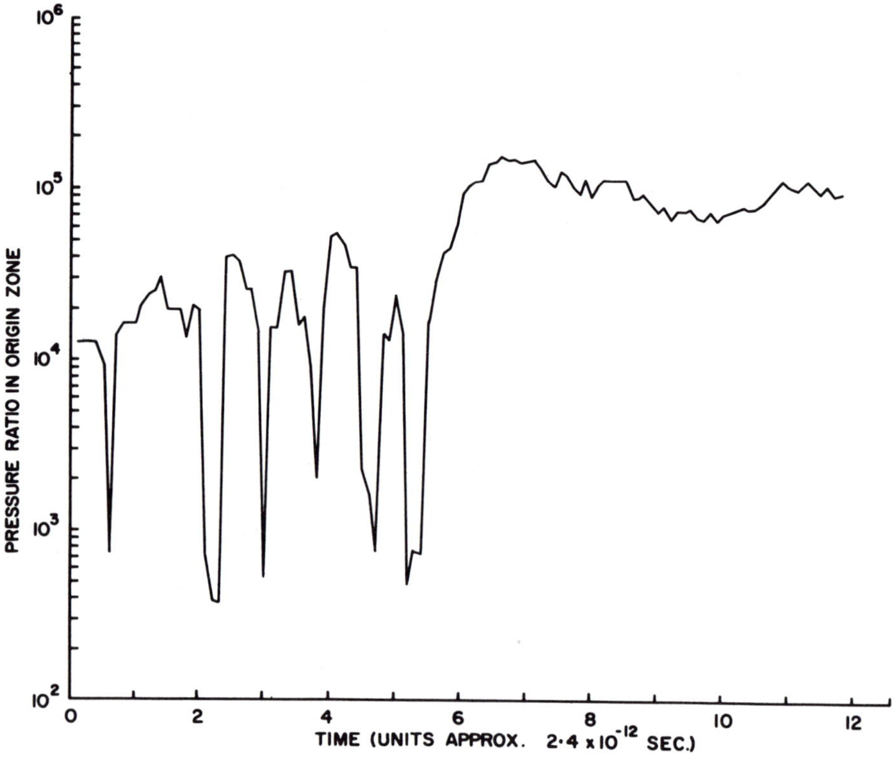

Figure 13
Pressure variation with time in the zone closest to the origin. See Figure 2. Time of commencement is approximately 6.98×10^{-12} sec. after piston started moving.

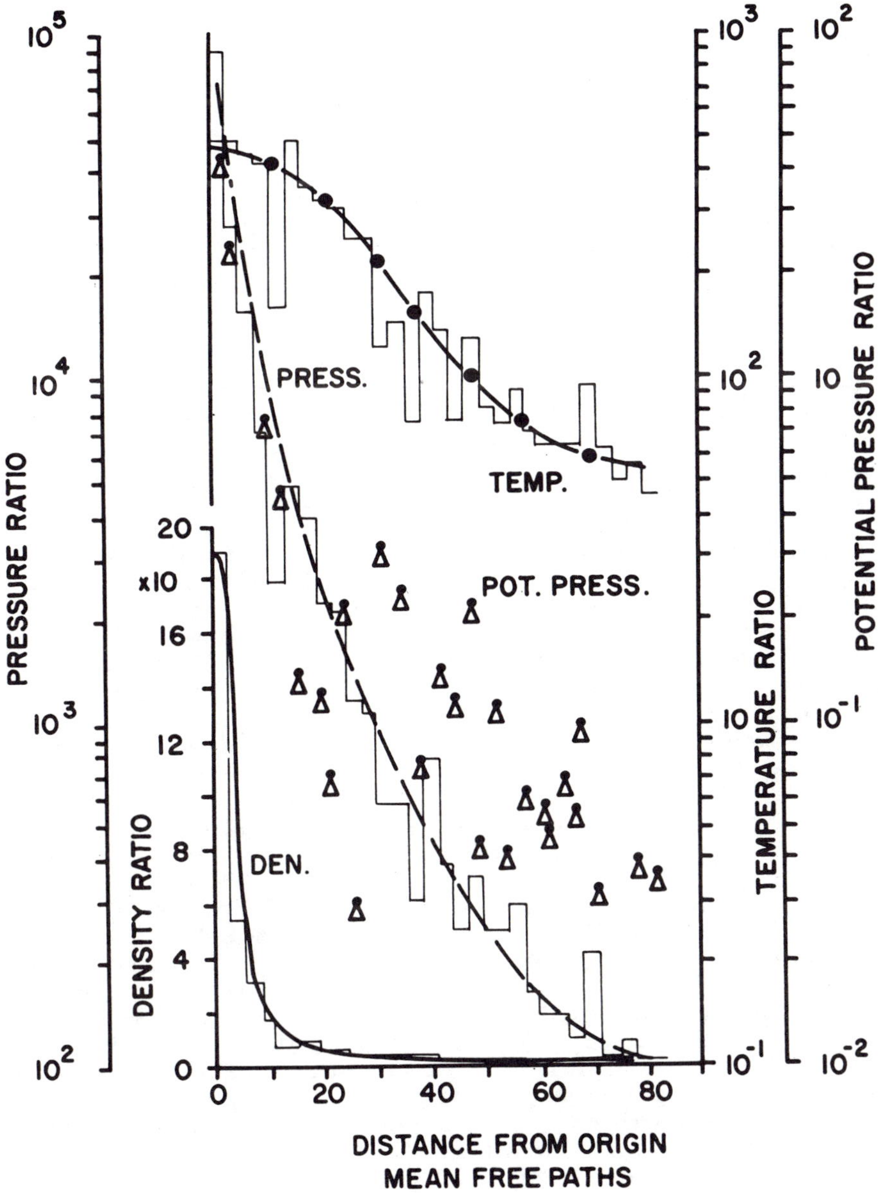

Figure 14
Density, pressure and temperature profiles at a time 9.83 x 10^{-12} sec. after piston commenced motion. Inter-particle potential energy expressed as equivalent kinetic pressure is shown by Δ. Conditions as in Figure 12.

STRONGLY IONIZING HIGH EXPLOSIVE SHOCKS*

M. COWAN and D. A. FREIWALD
Sandia Laboratories
Albuquerque, New Mexico

INTRODUCTION

This report describes a technique which utilizes high explosive for generating ionizing shock waves and high speed gas flow. Previously reported techniques differ in important ways from the one described here. Koski, et al., produced gaseous metallic jets with velocities as high as 90 km/sec by collapsing metal cylinders with cylindrical implosion systems.[1] Although the velocities were impressively high, the associated flow energy was relatively low. That is, an extremely small fraction of the high explosive energy was carried by the high speed flow. The "shock tubes" were ~½ inch in diameter and had to be well evacuated to avoid large deceleration of the flow.

Recent Russian work has developed a technique by which a metal plate is explosively driven with a plane detonation wave into a metal hemisphere or dome which contains the driver gas.[2,3] An expansion tube is attached to the top of the dome. Shock velocities as high as 90 km/sec have been produced, and a higher, though still small, fraction of the high explosive energy is associated with the flow than was the case with the Koski technique. Again, the high velocities reported were obtained only in small diameter tubes, ~½ inch in diameter.

The technique reported here compresses the driven gas directly with the gaseous products of the high explosive detonation. The explosive has a geometry which produces high velocity detonation products (18 km/sec) which carry more than half of the detonation energy. Combining this shaped explosive with converging geometry and double chamber operation, high velocity shock waves are established in relatively large diameter tubes more suitable

---

*This work was supported by the U. S. Atomic Energy Commission.

for experimental use of the high-temperature, high-speed flow. When hydrogen gas is used, the gaseous products of a high explosive detonation can be made to interact strongly with a magnetic field through the strong interaction which can be caused between the shocked hydrogen and the field.

PRINCIPLES OF OPERATION

Figure 1 shows schematically one shock tube geometry which has been tested. The main features of the device are: (1) the high explosive which generates the original shock wave, (2) the shock chamber with geometric convergence produced by a truncated conical section, (3) a very thin mylar diaphragm, and (4) the cylindrical expansion chamber.

Operation of this shock tube is similar to that of an ordinary pressure driven shock tube operated in the so-called expansion tube mode.[4] Figure 2 shows an x, t diagram of the flow. The shock chamber is filled with the driven gas at pressure P_6 while the expansion tube is filled with the accelerating gas to the much lower pressure P_1.

The explosive arrangement used consists of a plane wave generator fitted with a specially shaped one-inch thick pad of plastic bonded explosive (PBX-9404). The PBX pad has a matrix of cylindrical cavities arranged so that their axes are perpendicular to the detonation front. When the cavities collapse as a result of the detonation, jets are formed by the Munroe effect. The approximate theory (for incompressible flow) of jets formed by collapse of such cavities, predicts jet velocity u_5 (see Figure 2 for the meaning of numerical subscripts) to be

$$u_5 = 2u_d \qquad (1)$$

where u_d is detonation velocity.[5] The escape velocity of explosive detonation gases from a smooth plane explosive surface has a maximum value $\cong u_d$. Observations confirm that the maximum velocity of detonation gases is effectively doubled by the cavities.

Assuming an ideal polytropic gas in the shock chamber, the velocity increase due to convergence in the conical section of the chamber may be estimated following Chisnell.[6] Using his simple approximate expressions, one finds for strong shocks that shock strength Z scales with surface area of the shock A as

$$\frac{Z_4}{Z_5} \doteq \left(\frac{A_5}{A_4}\right)^k \qquad (2)$$

where A_4 is the cross-sectional area of the expansion chamber, A_5 is the surface area of the shaped explosive,

$$\frac{1}{k} \doteq \frac{1}{2} + \frac{1}{\gamma} + \left\{\frac{\gamma}{2(\gamma-1)}\right\}^{\frac{1}{2}} \qquad (3)$$

and γ is the ratio of specific heats. Since for strong shocks (and constant γ) $Z \sim u^2$ where u is flow velocity

$$\frac{u_4}{u_5} = \left(\frac{A_5}{A_4}\right)^{k/2} \qquad (4)$$

Thus, the flow speed of the driven gas into the diaphragm u_4 from Eqs. (1) and (4) is given by

$$u_4 \doteq 2u_d\left(\frac{A_5}{A_4}\right)^{k/2} \qquad (5)$$

After the diaphragm bursts, the driven gas undergoes unsteady expansion into the expansion chamber, producing a shock in the accelerating gas. Since the density of the shocked driven gas is very large compared to that of the undisturbed accelerating gas, the velocity of its expansion into the expansion chamber u_2 is given approximately by

$$u_2 \cong u_4 + u_{ex} \qquad (6)$$

where u_{ex} is the velocity of expansion into vacuum for the shocked driven gas, and

$$u_{ex} = \frac{2}{\gamma-1} a_4 \qquad (7)$$

where a_4 is the speed of sound of the shocked driven gas. Using the following relation for a strong shock wave in an ideal gas,

$$RT_4 = \frac{u_4^{\,2}}{2} (\gamma-1) \qquad (8)$$

and expressing a_4 as

$$a_4 = \sqrt{\gamma RT_4} \qquad (9)$$

where R and T are the specific gas constant and the temperature, it follows from Eqs. (6), (7), (8), and (9) that

$$u_2 = u_4\left\{1 + \left(\frac{2\gamma}{\gamma-1}\right)^{\frac{1}{2}}\right\} \qquad (10)$$

And using Eq. (4) one obtains

$$u_2 = 2u_d\left(\frac{A_5}{A_4}\right)^{k/2}\left\{1 + \left(\frac{2\gamma}{\gamma-1}\right)^{\frac{1}{2}}\right\} \tag{11}$$

Equation (11) expresses the maximum flow velocity to be expected for a fully evacuated expansion tube when the driven gas has constant γ and the effect of the diaphragm is negligible.

Detonation velocity u_d may be as high as 9 km/sec, and if $\gamma = 5/3$, one obtains from Eqs. (3) and (11)

$$u_2 = 59\left(\frac{A_5}{A_4}\right)^{0.225} \quad \text{km/sec} \tag{12}$$

From Eq. (12) one may anticipate flow velocities high enough to strongly ionize hydrogen since ionization for this gas begins at a shock velocity of ~30 km/sec and is virtually complete at ~70 km/sec.[7] However, a real gas subject to dissociation and/or ionization must be used in the shock chamber. Also, the diaphragm and expansion tube walls may produce important effects. Thus, Eq. (12) would be expected to overestimate experimental values of u_2.

EXPERIMENTAL APPARATUS

The explosive unit is a plane wave generator fitted with a perforated pad of plastic bonded explosive (PBX-9404) as shown in Fig. 3. The PBX pads are one-inch thick pressings with a square array of ¼-inch diameter holes having 3/8-inch center-to-center spacing. The diameter of the plane wave generator shown in Fig. 3 is four inches; however, 8 and 12-inch diameter plane wave generators have also been used in this type of shock generator. The converging section in the shock chamber has an included angle of 60°, and the distance between the exposed surface of the explosive and the diaphragm is 20 inches.

Suitable steps are taken in constructing the shock generators for evacuating and backfilling both the shock and expansion chambers with the desired gases. The diaphragm separating the shock and expansion chambers is made of type C mylar with a thickness no greater than that required to hold the necessary pressure P_6. When atmospheric pressure (~12.5 psi) is used in the shock chamber, 0.5 mil mylar is adequate for a 2-inch I.D. chamber.

Single chamber constant diameter shock tubes have also been used to investigate the kinetic energy-velocity distribution produced in the detonation products by the cavities in the explosive pads. Also, hemispherical arrangements for the explosive units were used to see if results were sensitive to the means of achieving convergence.

Shock generator performance was determined by high speed photography utilizing both streak and framing cameras.

EXPERIMENTAL RESULTS

All double chamber shock tube geometries have been tested at least once with one atmosphere (~12.5 psi) of hydrogen in the shock chamber and ~50μ of hydrogen in a 2-inch inner diameter expansion chamber. Results from these tests which are summarized in Table I allow comparison of the performance of the various configurations. The first three entries of Table I refer to the type of shock generator shown in Fig. 1 with the indicated inner diameter for the cylindrical part of the shock chamber. The last entry is for a hemispherical arrangement of ten 4-inch diameter explosive units like the one shown in Fig. 3. For such a low expansion tube pressure, one would not expect an important difference between shock speed and contact surface speed u_2; thus, experimental values listed for u_2 are the speeds observed for the luminous front.[8] Measured values are compared with those calculated from Eq. (12). This comparison is made graphically in Fig. 4. Calculated expansion speeds are consistently a factor of ~2 higher than measured speeds.

It is not surprising that Eq. (12) would overestimate speeds since no accounting was made for: (1) the effects of the diaphragm, (2) the effects of the expansion tube walls, (3) real gas effects (dissociation and ionization), and (4) the fact that expansion was not into perfect vacuum. However, it appears that shock tube performance scales reasonably well with $(A_5/A_4)^{0.225}$ as indicated by the equation.

Expansion chamber tubes were 0.5 to 1 m long, but streak camera records show no evidence of flow attenuation.

By testing the same shock generator geometry with and without a diaphragm, it was possible to establish that most of the discrepancy between the mathematical model and actual devices has to do with the role played by the diaphragm. Several tests indicated that the model is relatively good when no diaphragm is used, especially if real gas effects are considered when calculating k for use in Eq. (5). Still, there is a large speed increase across the diaphragm which allows the driven gas to run away from the gaseous products of detonation, producing a relatively long column of "clean" hydrogen which may be utilized in experiments.

Shown in Fig. 5 is measured shock speed as a function of ambient gas density in a two-inch diameter expansion chamber for the 12-inch linear shock generator. The streak camera record for point 4 did show a measurable slowing down of the flow, and the value listed is an average for the first 50 cm.

Figure 6 shows a plot of shock velocity vs. distance for a linear shock tube of constant 4-inch inner diameter filled with argon at atmospheric pressure ($\rho = 1.4\ kg/m^3$). This relatively high density gas loads the expanding products of detonation enough to noticeably decelerate them. Use of these data will be made in the final section to estimate the kinetic energy associated with the flow.

The fact that flow produced by these explosive devices can strongly interact with a magnetic field is illustrated in Fig. 7. For this experiment, which will not be described in detail here, a single-turn coil with one-half-inch cross section and 2.3-inch I.D. was positioned 50 cm downstream around the expansion chamber tube of a spherical shock generator (operated like the last entry of Table I). At the time the flow approached the coil, the magnetic field was about 0.6 megagauss at the center of the coil. The leading portion which is almost invisible at first eventually becomes very bright as induced electric fields produce currents which heat the plasma and make the interaction stronger still. In its own reference frame, the plasma in this experiment experiences effects much like those of a conventional theta pinch. However, the energy for this pinch must be supplied by the kinetic energy of the flow.

When a specific experiment was repeated, measured velocities showed a variation which did not exceed ±5%. This serves as an estimate of device performance reproducibility and experimental accuracy combined.

DISCUSSION AND SUMMARY

Using the shaped explosive described here, it is possible effectively to double the maximum escape velocity of detonation products from the explosive "surface." That is, the upper limit for shock speeds which may be produced in plane wave geometry by shaped explosive is twice that for unshaped explosive. In addition, the fraction of the total kinetic energy of detonation products associated with a velocity range near the maximum escape velocity is much greater for the shaped than for the unshaped explosive.

This second advantage of shaped explosive is predicted by the simple theory of jets from metal-lined cavities.[5] For a detonation running parallel to the axis of a cylindrical cavity, if a jet is produced, jet velocity is twice detonation velocity $2u_d$, regardless of the velocity of the liner collapse. Thus, even if the liner has a distribution of velocities, only one jet velocity $2u_d$ is established by the collapse of the cavity. Also, a very large fraction ~ 1 of liner kinetic energy is predicted to appear in the jet.

The unlined cavity may be expected to perform much like the lined one with the exception that the resulting jet is always gaseous for the former. Of course, the compressibility of the detonation products must meet requirements of jet formation for the geometry and conditions involved.[9]

Measurements have confirmed that a large fraction of the energy of shaped explosive is associated with gaseous detonation products having velocity $\sim 2u_d$. This may be illustrated with the data shown in Fig. 6. Here the detonation products from a one-inch thick pad of shaped explosive were directed into atmospheric argon in a constant cross-section four-inch diameter tube.

The deceleration of the shock provides information from which estimates of flow energy may be made. We assume conservation of momentum by the equation

$$2Mu_d = (M + \rho A\ell)u \qquad (13)$$

where M is the mass of detonation products which have velocity $2u_d$ when directed into vacuum and velocity u after "snow plowing" a distance ℓ through a tube with cross section A which is filled with gas having density ρ. Then the kinetic energy associated with the snow plow is

$$E = \frac{A\rho\ell u^3}{2(2u_d - u)} \qquad (14)$$

If this equation is evaluated using the data of Fig. 7 for $\ell = 0.13$ m, u = 15 km/sec, $A = 78 \times 10^{-4}$ m^2, and $\rho = 1.4$ kg/m^3, the kinetic energy of the snow plow is E = 0.8 Mjoules. The mass of the shaped PBX-9404 explosive pad was 0.26 kg, and the heat of detonation for this explosive is 5.9×10^6 joules/kg. Thus, this calculation indicates that about 50% of the energy released by the explosion was associated with the snow plow at $\ell = 0.13$ m. One would expect the kinetic energy of the snow plow to decrease as it moves down the tube. Thus, the original kinetic energy of the jets must have accounted for more than half of the energy released, giving qualitative agreement with simple jet theory. Also, from simple theory, one would expect the length of jets formed by detonation products to be approximately equal to the original thickness of the shaped explosives. From this and the preceding estimates for energy in the flow, we estimate the dynamic pressure exerted by the detonation products to be in the range 4-8 $\times 10^3$ newtons/m^2 (40-80 kbars) and the density of the detonation gases to be in the range 25-50 kg/m^3.

Shock velocities > $2u_d$ are produced in the devices described here by employing convergence and double-chamber operation. The maximum shock velocity possible is reasonably well predicted by the semi-empirical formula

$$u_2 \approx 3(A_5/A_4)^{0.255} \times 10^4 \text{ m/sec}$$

where A_5 is the area of the original shock front (the area of the shaped explosive), and A_4 is the area of the final shock front (the cross-sectional area of the expansion chamber). Results are not sensitive to the geometry used to achieve convergence. A simple conical section in the shock chamber works as well as a hemispherical arrangement for the explosive.

Applications for explosive shock generators which are of current interest call for strong interaction between the plasma which they produce and magnetic fields. A convenient measure of the ability of a plasma to produce a strong interaction is its magnetic Reynolds number $R_m = \mu_0 \sigma v \ell$, where σ, v, and ℓ are conductivity, velocity, and a characterictic length. If $R_m \geq 1$, a "positive feedback" is possible. That is, the magnetic field into which the plasma flows may be increased by induced currents in the plasma, making the interaction stronger. Furthermore, ohmic heating resulting from the eddy currents may increase σ causing the value of R_m to increase and so on.

Using the calculations of Freeman[7] and Brezing,[10] R_m vs. shock speed was calculated for hydrogen and deuterium, and results are shown in Fig. 8. A characteristic length of 0.1 m was chosen for this calculation since it is representative of experimentally practical sizes. Absolute shock speed was used to emphasize the important differences between the two gases with regard to high explosive-plasma experiments which might be done. For example, at a shock speed of 18 km/sec, which is achieved in plane wave geometry with shaped explosive, the value of R_m for deuterium is 0.2, but it is several orders of magnitude lower for hydrogen. At higher speeds deuterium is still significantly better.

REFERENCES

1. W. S. Koski, F. A. Lucy, R. G. Shreffler, and F. J. Willig, "Fast Jets from Collapsing Liners," J. Appl. Phys., 23, No. 12 (December 1952).
2. A. E. Voitenko, "Generation of High Speed Gas Jets," Soviet Physics--Doklady, 9, No. 10, p. 860 (April 1965).
3. A. E. Voitenko, "Strong Shock Waves in Air," Soviet Physics--Technical Physics, 11, No. 1, p. 128 (July 1966).
4. R. L. Trimpi, "A Preliminary Theoretical Study of the Expansion Tube, A New Device for Producing High-Enthalpy Short-Duration Hypersonic Flows," NASA TR R-133 (1962).
5. G. Birkhoff, D. P. MacDougall, et al., "Explosives with Lined Cavities," J. Appl. Phys., 19, p. 563 (June 1948).
6. R. F. Chisnell, "The Motion of a Shock Wave in a Channel, with Applications to Cylindrical and Spherical Shock Waves," J. Fluid Mech., 2, p. 286 (1957).
7. J. R. Freeman, "Equilibrium Hydrodynamic Variables Behind Incident and Reflected Shock Waves in Hydrogen," Sandia Laboratories Research Report, SC-RR-68-687 (October 1968).
8. R. E. Duff, "Shock Tube Performance at Low Initial Pressure," Phys. Fluids, 2, No. 2, p. 207 (1959).
9. J. M. Walsh, R. G. Shreffler, and F. J. Willig, "Limiting Conditions for Jet Formation in High Velocity Collisions," J. Appl. Phys., 24, No. 3 (March 1953).
10. Dieter Brezing, "Transport Properties of Hydrogen," AIAA, 3, No. 8 (August 1965).

Table I

Shock generator performance for relatively low H_2 pressures in the expansion chamber (~50μ) and one atm H_2 in the shock chamber. All results for 2-inch inner diameter expansion chamber.

Type of shock chamber	Measured expansion speed u_2 (10^4 km/sec)	Expansion speed from Eq. (12) (10^4 km/sec)
Linear (4-inch)	43	81
Linear (8-inch)	55	110
Linear (12-inch)	61	125*
Hemispherical	70	135

* The value of A_5/A_4 used for calculation in this case was 28 rather than 36 since seven 4-inch pads were used rather than one 12-inch pad.

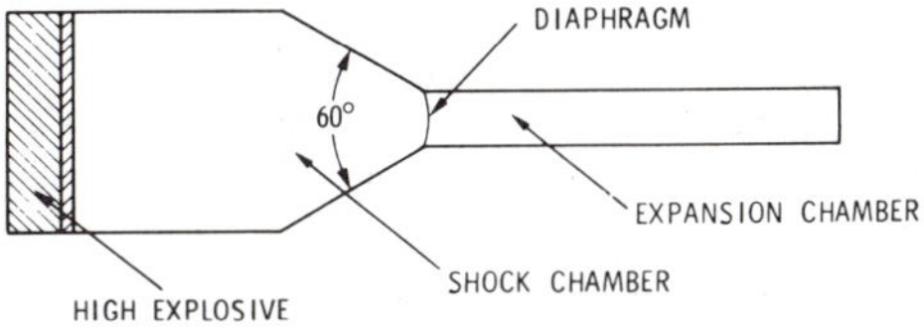

Figure 1
Schematic diagram of the linear shock generator

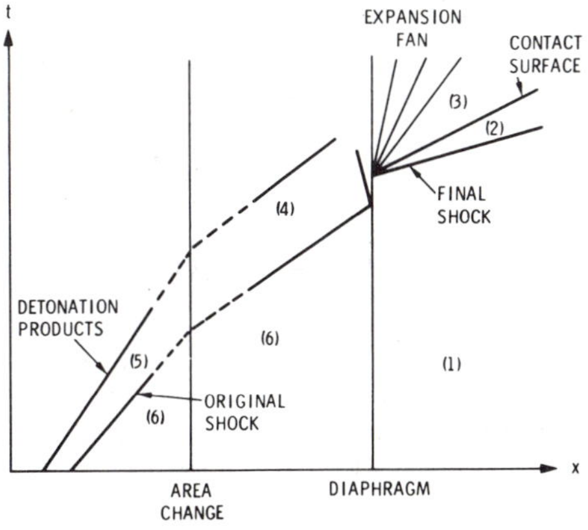

Figure 2
Shock generator x-t diagram

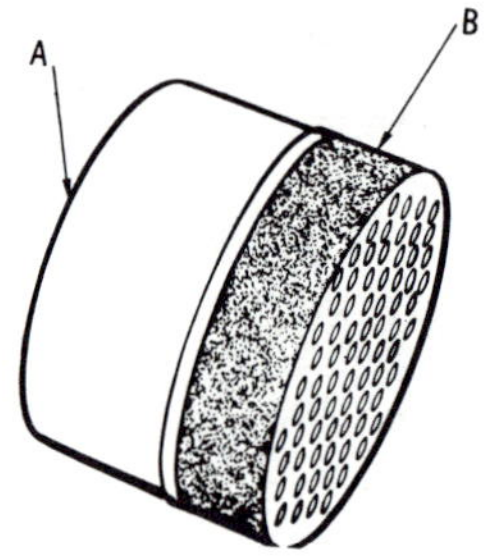

Figure 3
Line drawing of the explosive unit. A - plane wave generator. B - one-inch thick pad of shaped explosive

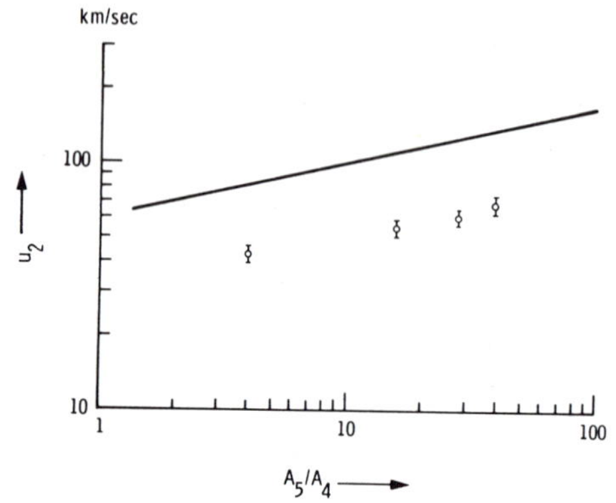

Figure 4
Expansion velocity u_2 vs. ratio A_5/A_4 where A_5 is the surface area of the shaped explosive and A_4 is the cross-sectional area of the expansion chamber. Solid curve computed from Eq. (12)

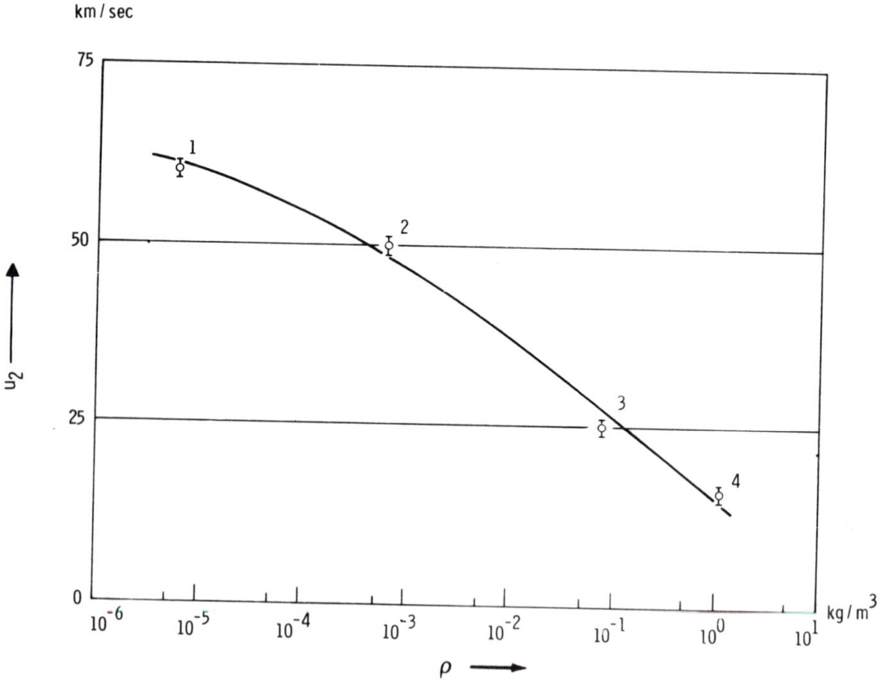

Figure 5
Expansion chamber velocity vs. density of ambient gas in expansion chamber ρ. Points 1, 2, and 3 for hydrogen, and point 4 for air in expansion chamber. Curve drawn by inspection

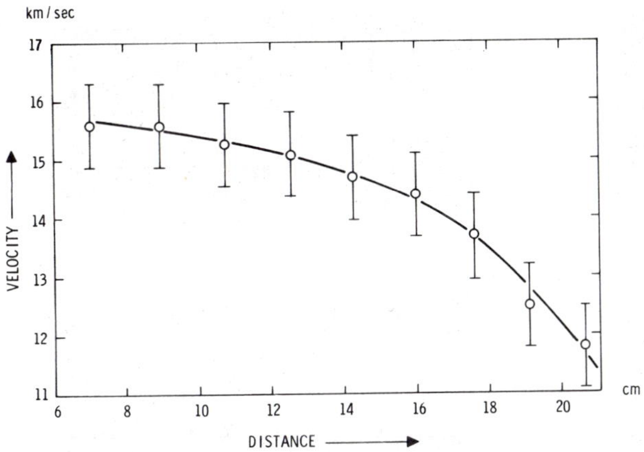

Figure 6
Shock velocity vs. distance in 1.4 kg/m^3 argon. A one-inch thick pad of shaped explosive was used in a 4-inch diameter shock chamber. Solid curve drawn by inspection

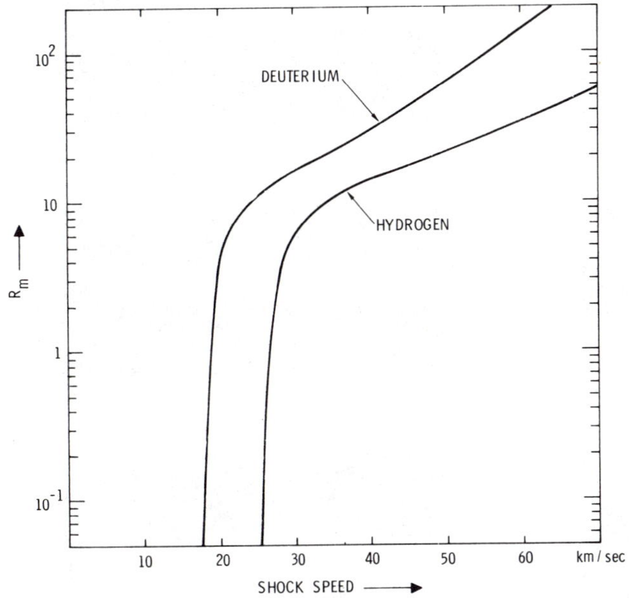

Figure 8
Magnetic Reynolds number R_m vs. shock speed for deuterium and hydrogen. A characteristic length of 0.1 m was used for the calculation.

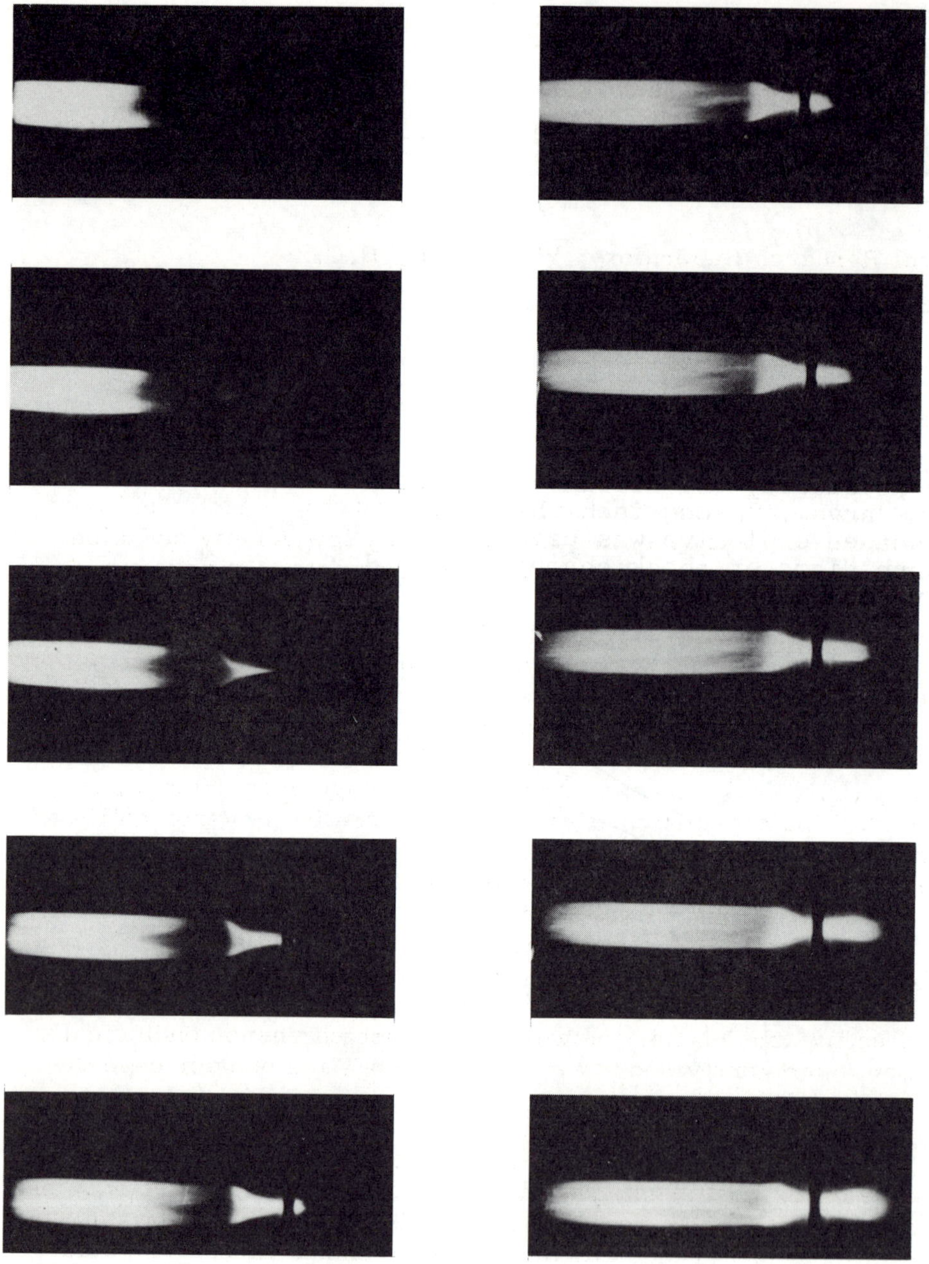

Figure 7
Framing camera pictures (2 frames/ μsec) of 70 km/sec flow through axially symmetric magnetic field of short (½-inch) θ coil. Coil shows as dark vertical band

APPLICATION OF ELECTROMAGNETIC SHOCK TUBES IN THE STUDY OF PLASMAS

A. C. KOLB
Naval Research Laboratory, Washington, D.C.

SUMMARY

During the past few years there has been a substantial effort in many laboratories to produce and measure the properties of large amplitude waves propagating in a relatively low density collision-free plasma. In these experiments one attempts to generate a quasistationary shock which is separated from the magnetic piston. Because binary collisions are not frequent enough to provide dissipation and a change of entropy, the understanding of the formation and structure of such waves involves a basic understanding of the instability mechanisms and resulting turbulence which provide the collective, randomizing wave particle interactions. Perpendicular shocks ($B \perp k$) at low Mach numbers ($M \lesssim 3$) have been extensively studied. The dissipation is provided by effective electron collisions which lead to an anomalous resistivity and so called resistive shocks where the heating is provided by the conversion of magnetic energy into random particle energy. At higher Mach numbers the energy reservoir in the compressed magnetic field is not sufficient to bring the plasma up to the steady state Rankine-Hugoniot temperature. It is suspected that at higher Mach numbers there is an effective ion-ion friction which is connected phenomenologically with an anomalous viscosity. For the high Mach number case the experimental data is very meagre and there is no general agreement on the instability mechanisms, nor is there anything resembling a complete theory to back up various conjectures. In the talk the experimental situation will be reviewed and, where possible, correlated with theoretical models. This conference follows an international conference on collision-free shock waves to be held at Frascati, Italy. The current status in this field will emerge at that time and I will also attempt to summarize that conference and present any new information.

SPECTROSCOPIC STUDIES OF HELIUM-HYDROGEN PLASMAS IN THE FIRST REFLECTED SHOCK REGION IN AN ELECTROMAGNETIC T-TUBE

K. FUKUDA, R. OKASAKA, and T. FUJIMOTO
Kyoto University
Kyoto, Japan

SUMMARY

In order to examine the effects of hydrogen on the ionization relaxation of a helium plasma, helium-hydrogen mixtures containing molecular hydrogen in the concentration range from 0.005 to 1.5 mole percent have been heated in an electromagnetic T-tube, thereby helium plasmas containing atomic hydrogen in varied concentrations have been produced in the first reflected shock region.

For the helium-hydrogen plasmas the change in electron density behind the front of the first reflected shock has been measured with a Gerardo-Verdeyen type interferometer with He-Ne laser light of λ6328A. Further, the changes in the emission intensities of some hydrogen and helium lines have been observed.

The results of the experiment will be discussed in comparison with calculations of thermodynamic states and ionization relaxation in the first reflected shock region.

INTRODUCTION

The relaxation process of ionization behind the shock front has been studied extensively by many investigators for argon, xenon, and krypton,[1-10] especially relative to the atomic process in gas under non-equilibrium conditions, but it is not yet fully understood to what extent the ionization behind shock front is affected by the presence of a precursor[11-15] ahead of the shock front or by the mixing of an impurity of low ionization potential in the test gas.[1,6,7]

The present paper is concerned with the spectroscopic observation and analysis of the relaxation process for the ionization of pure helium and helium-hydrogen mixtures containing molecular hydrogen in the concentration range 0.005 to 1.5 mole percent.

The generation of a clean shock wave was accomplished successfully in an electromagnetic T-tube with L-shaped electrodes, as reported in our previous paper.[16] The Mach number of the primary shock wave was about 10, and time-resolved spectroscopic observations were performed to observe the changes in the emission intensities from some hydrogen and helium lines in the first reflected shock region. The change in electron density was also measured with a Gerardo-Verdeyen interferometer.[17,18]

Calculations of thermodynamic states and ionization relaxation under the conditions of the present experiment indicate that in the primary shock region molecular hydrogen is completely dissociated but no ionization occurs in both hydrogen and helium. In the first reflected shock region the mixture of helium and atomic hydrogen is heated to about 2.3×10^4 °K immediately behind the reflected shock front and the ionization proceeds along the paths of regimes I, II, and III, as described by Wong and Bershader.[5]

For the mixtures containing 1.5 mole percent hydrogen the maximum electron density of $1.5 \times 10^{17} cm^{-3}$ appeared about 2 μsec after the arrival of the first reflected shock, but as the mixing ratio was lowered to 0.005 mole percent, the maximum density became $0.5 \times 10^{17} cm^{-3}$ at 8 μsec. The corresponding temperature change has been determined by assuming Saha-Boltzmann equilibrium for the values of electron density. The results will be discussed in comparison with calculations of thermodynamic states and relaxation time.

For mixtures containing less than about 0.05 mole percent hydrogen the behavior of the emission intensities of H_α and H_β lines in the first reflected shock region is qualitatively explained by the quasi-steady state solution[19,20] of a helium-hydrogen plasma.

THE SHOCK WAVE AND ITS THERMODYNAMIC STATES

Generation of the Shock Wave

In the present experiment pure helium and helium-hydrogen mixtures were used as the test gas. The mixing ratio of hydrogen molecules in the latter ranged from 0.005 to 1.5 mole percent, and the hydrogen content in the former was less than the lowest mixing ratio.

A quartz plate reflector adjustable in position was set 50 cm downstream from the ends of the arms of the L-shaped electrodes and the tube was operated at the initial gas pressure of 11 Torr with a circuit of V=12 kV, C=40 μF, and ringing frequency 20 kHz. Then the front of the driving discharge plasma almost stopped after colliding with the first reflected shock so that the first reflected shock region could be kept almost free from rarefaction or compression waves in about 30 μsec.

An example of a streak picture and the corresponding x-t diagram of the shock wave trajectories are shown in Figure 1. The Mach number of the primary shock wave is estimated to be about 10 from the observed average speed of the driving plasma of 0.77 cm/μsec at a distance 10 cm in front of the reflector. The spectroscopic and interferometric observations were carried out at a position 10 mm from the reflector surface as shown in Figure 1(b).

Thermodynamic States in The Primary and First Reflected Regions

The Temperature T, particle density N, and electron density n_e have been calculated for Mp=10 by assuming various states of partial and full equilibrium in both primary and first reflected shock regions. The calculations are based on Rankine-Hugoniot relations and Saha equations applied to the helium-hydrogen mixtures, and the calculated quantities are given as a function of the mixing ratio of atomic hydrogen, n_H/n_{He}, in Figure 2. Here the numbers 2 and 5 on the curves correspond to the primary and first reflected shock regions, respectively, as shown in Figure 1(b).

Region 2 indicates the state in the primary shock region, when hydrogen molecules are completely dissociated and the ionization of resulting atomic hydrogen is in equilibrium. Further, here it is assumed that there is no ionization of helium, i.e. the ionization ratio of helium α_{He} is zero.

Region 5-1 corresponds to the state immediately behind the front of the first reflected shock, in which the gas in region 2 has been compressed by the reflected shock but no further ionization takes place. In this state the ionization ratio of hydrogen α_H is the same as that in region 2 and $\alpha_{He}=0$.

Region 5-2 indicates the state in which ionization of hydrogen is equilibrated but no ionization of helium occurs ($\alpha_{He}=0$).

Region 5-3 corresponds to the state in which ionization of helium is equilibrated but no ionization of hydrogen takes place ($\alpha_H=0$).

Region 5-Exact is the state of full equilibrium, in which ionization of both hydrogen and helium is equilibrated. In this case the Saha equations for helium and hydrogen must be connected by a common value of electron density:

$$S_{He}(T) = \frac{n^+_{He} n_e}{n^o_{He}} = (2Z^+_{He}/Z^o_{He})(2\pi mkT/h^2)^{3/2}\exp(-I_{He}/kT) \quad (1)$$

and

$$S_H(T) = \frac{n_H^+ n_e}{n_H^o} = (2Z_H^+/Z_H^o)(2\pi mkT/h^2)^{3/2}\exp(-I_H/kT), \quad (2)$$

where Z is the electronic partition function and I is the reduced ionization energy of the species indicated by the subscript. n^o and n^+ denote the number density of neutral and singly ionized atoms, respectively, and $n=n^o+n^+$ for each species. Here, of course, the condition of macroscopic neutrality

$$n_e = n_H^+ + n_{He}^+$$

is satisfied identically.

The values of T and N in regions 2 and 5-1 change very little even if no ionization of atomic hydrogen is assumed ($\alpha_H=0$).

CONSIDERATIONS OF IONIZATION RELAXATION

Relaxation in the Primary Shock Region

The above results allow one to consider first the ionization of the helium-atomic hydrogen mixture in the primary shock region since molecular hydrogen is completely dissociated in a very short time[21] at $T=1x10^4$ °K and $N=1.4x10^{18}cm^{-3}$ (see the values of Region 2).

The ionization process, in general, proceeds in distinct stages, described as regimes I, II, and III.[5] In regime I, in which atom-atom inelastic collisions are responsible for ionization, the rate of electron production from the mixture is given approximately by[6-9]

$$dn_e/dt = k(H^*\text{-He})\, n_H n_{He} + k(H^*\text{-H}) n_H^2$$
$$+ k(He^*\text{-He}) n_{He}^2 + k(He^8\text{-H}) n_{He} n_H \quad (4)$$

where, for example, $k(X^*\text{-}Y)$ is the rate constant for the excitation of X-atoms due to XY-binary collisions. Since the mixing ratio of atomic hydrogen of interest, n_H/n_{He}, is less than $3x10^{-2}$, Eq.(4) is further simplified as follows:

$$dn_e/dt = k_H{}^* n_H n_{He} + k_{He}{}^* n_{He}{}^2, \quad (5)$$

where

$$k_H{}^* = k(H^*\text{-He}) = 2\sqrt{2}S_H{}^*(\pi\mu_{H\text{-}He})^{-1/2}(kT)^{3/2}\left(\frac{T_{H^*}}{2T}+1\right)\exp(-T_{H^*}/T) \quad (6)$$

and

$$k_{He^*} = k(He^* - He) = 4S_{He}^*(\pi m_{He})^{-1/2}(kT)^{3/2}\left(\frac{T_{He}^*}{2T}+1\right)\exp(-T_{He}^*/T). \quad (7)$$

Here T* corresponds to the excitation energy $E^*=kT^*$ from the ground to the lowest excited state for the species noted in the subscript.[10] m_{He} is the mass of the helium atom and μ_H-He is the reduced mass of the helium and hydrogen atoms. $S_H^*=S(H^*-He)$ and $S_{He}^*=S(He^*-He)$ are the cross-section constants for the binary collisions of interest.

No definite data on S_H^* and S_{He}^* are available, and it is assumed here that $S_H^*=S_{He}^*=1\times10^{-20}$ by referring to the data on cross-sections for He-He[22,23] and He-H_2[24] collisions and also to S^*-values of rare gas atoms.[6-9] Then we obtain

$$k_H^*=7.5\times10^{-19}cm^3sec^{-1} \text{ and } k_{He}^*=1.1\times10^{-23}cm^3sec^{-1} \quad (8)$$

at $T=1\times10^4$ °K. From Eq. (5) with these values it is seen that in regime I electrons are produced mainly from hydrogen even for the mixture of lowest hydrogen concentration, $n_H/n_{He}=1\times10^{-4}$.

In regime II electron-atom collisions predominate and the rate of electron production is described approximately by

$$dn_e/dt = q_H^*n_en_H + q_{He}^*n_en_{He} \quad (9)$$

where q_H^* and q_{He}^* are the rate constants of the excitation due to electron-atom collisions for hydrogen and helium, respectively. The q* values have been calculated from the excitation cross-sections considered to be most reliable on the basis of a comparison of recent data[25-36] (see Table 1B for helium):

$$q_H^*=2.6\times10^{-13}cm^3\ sec^{-1} \text{ and } q_{He}^*=4.6\times10^{-19}cm^3sec^{-1}. \quad (10)$$

The cross-section[37,38] used by McWhirter and others[19,20] in their quasi-steady-state solutions of plasma leads to a value of q_H^* about five times larger than that calculated above, but the former seems to be less reliable. For the values of Eq. (10) the second term can be neglected in Eq.(9) even for the mixture of lowest hydrogen concentration. Therefore, in regime II also, electrons are supplied by the ionization of hydrogen atoms.

When the electron density increases in regime I and reaches some critical value, the ionization process turns from regime I into regime II. Here the critical value of the electron density n_e^c is defined as the density for which the rates of Eqs.(5) and (9) become equal. Further, the relaxation time τ_I of regime I is defined as the time in which the electron density reaches the value of n_e^c.

From Eqs. (5) and (9) we obtain

$$n_e^c = k_H^* n_{He}/q_H^* = 4 \times 10^{13} cm^{-3} \quad (11)$$

for $n_{He}=1.4 \times 10^{18} cm^{-3}$ in region 2.

From Eq. (5) we obtain $n_e=k_H^* n_H n_{He} t$ and hence $\tau_I=1/q_H^* n_H$= 30 and 0.3 msec for $n_H/n_{He}=10^{-4}$ and 10^{-2}, respectively. These values of τ_I are very long in comparison with the time during which particles stay in the primary shock region for the observations of the present experiment (See Figure 1(b)). In the above calculations n_H and n_{He} are assumed to be constant, and this assumption is approximately valid since n_e^c is smaller than n_H even for $n_H/n_{He}=10^{-4}$.

The above considerations allow one to consider that the ionization of helium-atomic hydrogen mixtures does not proceed at all in the primary shock region but only successively along the paths of regimes I, II, and III in the first reflected shock region.

Relaxation in the First Reflected Shock Region

In the first reflected region the ionization of helium-atomic hydrogen mixtures starts from the state of region 5-1, where T=2.3~2.2x10^4 °K and n=3.4~3.6x10^{18} cm^{-3} for $n_H/n_{He}=1 \times 10^{-4}$ to 3×10^{-2} as shown in Figure 2. Here, of course, no ionization is assumed for both hydrogen and helium ($\alpha_H=\alpha_{He}=0$).

The critical value n_e^c of the electron density is determined in the same way as described above, but in this case it must be taken into account that the electron temperature T_e becomes somewhat lower than the gas temperature T.

The value of T_e has been estimated to be 2.2x10^4 °K for $n_H/n_{He} \lesssim 10^{-2}$ and $n_e \lesssim 1 \times 10^{16}$ cm^{-3} with the energy balance equation[1] between elastic electron-atom and electron-ion collisions and inelastic electron-atom collisions for the mixtures. In this calculation the rate constants for the latter have been calculated as a function of electron temperature using recent data on cross-sections.[25-36]

In regime I of the first reflected shock region the rate constants in Eq. (5) are calculated with Eqs. (6) and (7):

$$k_H^* = 1.5 \times 10^{-15} cm^3 sec^{-1} \text{ and } k_{He}^* = 1.3 \times 10^{17} cm^3 sec^{-1} \quad (12)$$

at T=2.3x10^4 °K.

In regime II the rate constants in Eq. (9) are given by

$$q_H=1.4\times10^{-10}cm^3sec^{-1} \text{ and } q_{He}=1.1\times10^{-13}cm^3sec^{-1} \qquad (13)$$

at $T_e=2.2\times10^4$ ^oK for $n_H/n_{He} \lesssim 10^{-2}$ and $n_e \lesssim 10^{-16}cm^{-3}$ (see Table 1).

From Eqs. (5) and (9) we get for the critical value of the electron density

$$n_e^c = n_{He}\left\{k_H^*\left(\frac{n_H}{n_{He}}\right)+k_{He}^*\right\}\Big/\left\{q_H\left(\frac{n_H}{n_{He}}\right)+q_{He}\right\} \qquad (14)$$

and for the relaxation time of regime I τ_I

$$\tau_I= 1/\left\{q_H\left(\frac{n_H}{n_{He}}\right)+q_{He}\right\}n_{He}. \qquad (15)$$

The values of n_e^c and τ_I have been calculated for various hydrogen concentrations, n_H/n_{He}, and are shown in Table 2. For lower mixing ratios of $n_H/n_{He} < 1\times10^{-2}$, electrons are produced mainly from helium in regime I and n_H may be considered to be constant. The electron density $n_e^c(H)$ produced from hydrogen at $t=\tau_I$ is given in Table 2 together with the values of n_H.

In regime II electron-atom collisions dominate the ionization process, and in regime III three-body recombinations of electrons and ions play an important role in equilibrating the plasma. The whole ionization process must be explained by computing the rate equations, which include possible forward and backward moving processes for the binary system of helium and hydrogen, but the rate constants concerned are not known exactly. The approximate values of the relaxation time for ionization will be estimated below.

We use the following notations below: $x_o=n_H$; $y_o=n_{He}$; and x_1 and x_j are the number densities of hydrogen atoms in the ground and jth level, respectively, and y_1 and y_j have the same meaning for helium.

The excitation processes in regime II are described by the following rate equations:

$$dx_1/dt = -q_Hx_1f', \qquad (16)$$

$$dx_j/dt = \{q_H^*(j-1,j)x_{j-1} - q_H^*(j,j+1)x_j\}f', \qquad (17)$$

and

$$x^+ \simeq x_o - x_1 \qquad (18)$$

with

$$f(t) = \int_o^t n_e dt. \qquad (19)$$

Here $q_H^*(j,j+1)$ is the rate constant from the jth to the (j+1)th level and the lower limit of the integral in Eq. (19) corresponds to the time $t=\tau_I$. The solutions are as follows:

$$x_1 = \{x_o - n_e^c(H)\} \exp(-q_H f), \tag{20}$$

$$x_j = \{x_o - n_e^c(H)\}\{q_H/q_H^*(j,j+1)\}$$
$$\times\left[\exp(-q_H f) - \exp\{-q_H^*(j,j+1)f\}\right] \tag{21}$$

and

$$x^+ = x_o\left[1 - \{x_o - n_e^c(H)\}/x_o \cdot \exp(q_H f)\right]. \tag{22}$$

These solutions are approximately valid when the following conditions are satsified for any j-value: $q_H^*(j,j+1) > q_H^*(j,j+1)$ for i=2,3,...and...$> q_H^*(j,j+1) > q_H^*(j-1,j) > ...$

For the pure atomic hydrogen system we have $x^+ = n_e = df/dt$ and obtain

$$f_H = (1/q_H)\log F_H \tag{23}$$

and

$$F_H = \{n_e^c(H)/x_o\}\exp(q_H x_o t) + \{x_o - n_e^c(H)\}/x_o. \tag{24}$$

Then we obtain

$$x_1 = \{x_o - n_e^c(H)\} F_H^{-1} \tag{25}$$

and

$$n_e(H) = x_o\left[1 + \{x_o - n_e^c(H)\}/n_e^c(H)\exp(q_H x_o t)\right]^{-1}. \tag{26}$$

For the helium system the same treatment gives solutions similar to Eqs. (20)-(22) and for pure helium the following expressions are obtained for $y^+ = n_e$:

$$f_{He} = (1/q_{He})\log F_{He} \tag{27}$$

and

$$F_{He} = \{n_e^c(He)/y_o\}\exp(q_{He} y_o t) + \{y_o - n_e^c(He)\}/y_o. \tag{28}$$

Then the electron production from pure helium is given by

$$n_e(He) = y_o\left[1 + \{y_o - n_e^c(He)\}/n_e^c(He)\exp(q_{He} y_o t)\right]^{-1}. \tag{29}$$

When the equilibrium value of the electron density $n_e^E(He)$ is much smaller than $y_o - n_e^c(He) \simeq y_o$, Eq. (29) can be written as

$$n_e(He) = n_e^c(He)\exp(q_{He} y_o t) \tag{30}$$

since the last term in the denominator of Eq. (29) is larger than unity. The relaxation time $\tau_{II}(He)$ is given by

$$\tau_{II}(He) = (1/q_{He} y_o)\log(n_e^E(He)/n_e^c(He)). \tag{31}$$

The value of τ_{II}(He) for n_H/n_{He} in Table 2 is calculated from Eq. (31) for $n_e^E(He) \approx n_e^E$ and $n_e^C(He) = n_e^C$. The approximate value of the relaxation time is given by $\tau = \tau_{I+II}(He)$ for $n_H/n_{He} = 0$ and $1x10^{-4}$.

For the mixture of low hydrogen concentration, that is, $n_H/n_{He} < 1x10^{-3}$, the value of τ_{II}(He) changes very little. In this case for hydrogen the population density in the ground and jth levels must be expressed Eqs. (20) and (21) with $f = f_{He}$ of Eq. (27). The last term of Eq. (21) gives a very rapid rise of population in the jth level and, therefore, this term may be neglected:

$$x_j = \left\{ q_H / q_H^*(j,j+1) \right\} x_1 \text{ and } x_1 = \left\{ x_o - n_e^C(H) \right\} (F_{He})^{-q_H/q_{He}}. \tag{32}$$

The time of τ_{II}(H) for $n_H/n_{He} = 1x10^{-4}$ in Table 2 is the time in which the population density of x_1 or x_j decreases to one-tenth of its initial value; the value of τ_{II}(H) is calculated from $(F_{He})^{q_H/q_{He}} = 10$ in Eq. (32).

For the mixtures of higher hydrogen concentration, that is, $n_H/n_{He} \geq 1x10^{-3}$, the ionization process is complicated since the rate of production of electrons from hydrogen becomes equal to or larger than that from helium. However, it is assumed that for $n_H n_{He} \geq 3x10^{-3}$ atomic hydrogen is first ionized and reaches 90 percent ionization ($\alpha_H^E \simeq 0.9$) and then the ionization from helium starts at the electron density of 90 percent of n_H. The time τ_{II}(H) for 90 percent ionization of hydrogen, or that for one-tenth decay of x_1 or x_j, is calculated from Eq. (26) or from Eq. (25), respectively, and the relaxation time τ_{II}(He) of helium ionization is determined from Eq. (31) with n_e^C(He) replaced by $0.9xn_H$. The values are listed in Table 2 together with the values of $\tau = \tau_I + \tau_{II}(H) + \tau_{II}(He)$ for $n_H/n_{He} = 0.3$ to $3x10^{-2}$.

The above process of ionization corresponds to the path from region 5-1 to 5-Exact through region 5-2 in Figure 2(a) and (b), in which $\alpha_{He} = 0$ and hydrogen is completely ionized ($\alpha_H = 1$). In regime III, where the plasma approaches equilibrium, the gas temperature T decreases and the electron temperature T_e also decreases slightly.[3] Hence the above-calculated relaxation time will be somewhat shorter.

EXPERIMENT AND RESULTS

The electron density behind the front of the first reflected shock was measured with a Gerardo-Verdeyen type interferometer[17,18] with He-Ne laser light of λ6328A. The sensitivity factor was 4(=s), which gave one fringe jump for a change in helium particle density of

$2.2 \times 10^{18} cm^{-3}$ ($\Delta N = \Delta n_{He}$) and in electron density of $1.6 \times 10^{16} cm^{-3}$ ($= \Delta n_e$).

The T-tube was set in the reference cavity and the measurement was performed at a position 10 mm in front of the reflector surface by passing laser beams through glass windows on the tube wall. The bundle of laser beams was 3 mm in width, from which the time-resolution of the measurement is estimated to be 0.3 and 0.6 μsec for the speeds of incident and first reflected shocks of about 10 and 5 mm/μsec.

Figure 3 shows an example of fringe jumps for the mixture with $n_H/n_{He} = 1 \times 10^{-4}$. The time between the arrivals of incident and first reflected shocks shows a good correspondence with the shock wave trajectories on the streak picture in Figure 1(a). A half-fringe jump on the arrival of the incident shock corresponds to a particle density change of $1.1 \times 10^{18} cm^{-3}$, which is approximately equal to the particle density $1.4 \times 10^{18} cm^{-3}$ of region 2 of Figure 2(b). However, one fringe jump expected from the density change of $2.0 \times 10^{18} cm^{-3}$ between the incident and reflected shock is not clearly seen in Figure 3. From the inspection of many data it is considered that the second half of one fringe jump is canceled by the initial half of the following fringe shift in the reverse direction, which is caused by the gradual increase in the electron density behind the reflected shock front.

The observed change in the electron density behind the first reflected shock is shown in Figure 4 for pure helium and the mixtures over the whole range of hydrogen concentration. The temperature behind the reflected shock front is calculated from the observed electron density by assuming Saha-Boltzmann equilibrium of Eqs. (1) and (2); this result is also shown in Figure 4.

The change in the emission intensities from H_α and H_β and some He lines was observed also at the position 10mm from the reflector surface with a grating spectrometer with two channels of photomultipliers. The dispersion was 16A/mm and 8A/mm for the gratings with 600 and 1200 lines/mm, respectively. Figure 5(a) and (b) show the changes in the intensities of H_β and He λ5876A lines, respectively.

DISCUSSION

As shown in Figure 4 the observed electron density and the corresponding Saha temperature reach their maxima about 2, 4, and 6 μsec after the arrival of the first reflected shock for the mixtures containing 3, 1, and 0.3 atomic percent hydrogen respectively. These times for maximum electron density or temperature do not greatly disagree with the calculated relaxation times in Table 2, and the electron densities at these times (1.5, 1.2, and 1.1×10^{17} cm^{-3}) are in very good agreement with the calculated n_e^E of region 5-Exact in Table 2. Further, the corresponding Saha temperatures (1.74, 1.74 and 1.77×10^4 °K) are also in agreement with those of region 5-Exact (1.73, 178, and 1.8×10^4 °K). The Saha temperature, however, becomes lower with the lapse of time: $T = 1.6 - 1.5 \times 10^4$ °K for $t \approx 12$ μsec. This fact demonstrates the existence of cooling in the first reflected shock region. The emission intensities from H_β and He λ5876A lines (Figure 5) are approximately consistent with the above Saha temperature for the mixing ratio concerned under the assumption of Saha-Boltzmann equilibrium.

For the mixtures containing less than 0.3 atomic percent hydrogen, the maximum electron density and Saha temperature ($n_e = 0.8$ and 0.5×10^{17} cm^{-3}, $T = 1.7$ and 1.6×10^4 °K for $n_H/n_{He} = 1 \times 10^{-3}$ and 3×10^{-4} to 0, respectively) are considerably lower than those of the thermodynamic states ($n_e = 1.1 \times 10^{17}$ cm^{-3} and $T = 1.8 \times 10^4$ °K for $n_H/n_{He} = 1 \times 10^{-3}$ to 0). Further, the time of about 8 μsec to the maximum electron density is much shorter than the time τ of about 17.5 μsec in Table 2. This fact seems to be also due to the cooling mentioned above, which prevents the plasma from reaching the equilibrium expected from the thermodynamic calculations.

In this range of hydrogen mixing ratio (<0.3) the emission intensity from the He λ5876A line about 5 μsec or more after the arrival of the reflected shock increases with increasing hydrogen mixing ratio, whereas the intensity shows saturation for a mixing ratio higher than about 0.1 atomic percent. This fact is complicated to explain, but it is possible to consider that a small amount of hydrogen scarcely changes the thermodynamic state of the helium plasma.

For a mixing ratio of the order of 0.1 atomic percent, the H_β (or H_α) emission shows a sharp spike immediately behind the reflected shock front and then a following slow rise. The height of the spike and its decay are proportional to the initial number density of hydrogen in the ground state and its depopulation due to ionization, respectively, as shown in Eq. (32). At the minimum emission intensity the rate equations including forward and backward moving processes are balanced relative to the the jth level (j=4 for the H_β line) and then the population increases as more electrons are supplied from helium.

From Eq. (32) and Table 1A the initial population density of $x_4 = (q_H/q_H^*(4-5))\, n_H = 10^{-5} n_H$ is obtained, which is much smaller than the x_4^E obtained under Saha-Boltzmann equilibrium at $T = 1.6 \times 10^4$ °K (the temperature about 10 μsec after the arrival of the reflected shock). In a high-density plasma populations in excited states are very rapidly equilibrated in the Boltzmann distribution through atom-atom collisions and, therefore, the initial population density may be given approximately by $x_4 = (q_H/q_H^*(2-3))n_H = 10^{-3}n_H$, which is nearly equal to the above value of x_4^E. Further, Eq. (32) leads to a time of one-tenth decay of $\tau_{II}(H)=5.5$ μsec for hydrogen atoms in the ground state with $n_H/n_{He}=1\times10^{-4}$ (see Table 2). The above considerations give a qualitative explanation for the behavior of the H_β emission, but not a quantitative one; the observed height of the initial emission spike is very high and its decay time is very short in comparison with the results predicted above (see Figure 5(a)).

The observation of the H_β emission was performed at a position 5, 10, and 15 mm in front of the reflector surface. The height of the emission spike decreases as the distance from the reflector increases. As the mixing ratio of hydrogen was increased, the height of the initial spike did not increase linearly, which indicates that a considerable amount of the hydrogen is ionized before the arrival of the first reflected shock. In fact, weak luminosity as well as a fringe shift of less than a half-fringe due to electrons was observed near the contact surface in the incident shock region for $n_H/n_{He} \simeq 1\times10^{-2}$.

These observations indicate that hydrogen atoms are ionized in the incident and first reflected shock regions by the radiation from the driving discharge plasma. This fact makes it difficult to analyze the hydrogen emission quantitatively.

REFERENCES

1. Petschek, H. and Bayron, S. Ann. Phys. 1, 270, 1957.
2. Bond, J. W. Phys. Rev. 105, 1683, 1957.
3. Biberman, L.M. and Yakubov, I.T. Soviet Phys. Tech. Phys. 8, 1001, 1964.
4. Yakubov, I.T. Soviet Phys.Tech. Phys. 9, 676, 1964: Proc. Int. Conf. Phenomena of Ionized Gas, Belgrade,II, 797, 1965.
5. Wong, H. and Bershader, D. J. Fluid Mech. 26, 456, 1966.
6. Harwell, K.E. and Jahn, R.G. Phys. Fluids 7, 214, 1964.
7. Morgan, E.J. and Morrison, R. D. Phys. Fluids 8, 1608, 1965.
8. Kelly, A.J. J.Chem.Phys. 45, 1724, 1966.
9. Mclaren, T.L. and Hobson, R.M. Phys. Fluids 11, 2165, 1968.
10. Johnston, H.S., and Kornegay, W. Trans Faraday Soc. 57, 1563, 1961.
11. Mclean, E.A., Faneuff, C.E. and Kolb, A.C. Phys. Fluids 3, 843, 1960.
12. Gerardo, J.B., Hendricks, C.E. and Goldsten, L. Phys. Fluids 6, 1963.

13. Lagar Kov, A.N. and Yakubov, I.T., Opt. i. Spekty. 14, 199, 1963.
14. Chu, C.K. Phys.Fluids 10, 1, 1964.
15. Lubin, J.M. and Resler, E.L. Phys.Fluids 10, 199, 1967.
16. Fukuda, K., Okasaka, R. and Fujimoto, T. Japanese J. Appl. Phys. 6, 995, 1967.
17. Gerardo, J.B. and Verdeyen, J.T. Proc.IEEE 52, 690, 1964.
18. Gerardo, J.B., Verdeyen, J.T. and Gusinow, M.A. J.Appl. Phys. 36, 2146, 1965.
19. McWhirter, R.W.P. and Hearn, A.G. Proc.Phys.Soc. 82, 641, 1963.
20. Bates, D. R. and Kingston, A. E. Planetary and Space Sci. 11, 1, 1963.
21. Sutton, E.A. J.Chem.Phys. 36, 2923, 1966.
22. Hyden, H.C. and Amme, R.C. Phys.Rev.141, 30, 1966.
23. Soloriev, E.S., Ilin, R.N., Oparin,V.A. and Fedorenko, N.V. Atomic Collision Processed. M.R.C. McDowell North-Holland Pub.Co., 1964, p.692.
24. Hayden, H.C. and Utterback, N.G. Phys.Rev. 135, 1575, 1964.
25. Burke, P.G., Taylor, A.J. and Ormonde, S. Proc.Phys. Soc. 92, 345, 1967.
26. Fite, W.L., Stebbings,R.F. and Brackman, R.T. Phys. Rev. 116, 356, 1959.
27. Chamberlain, G.E., Smith, S.J. and Heddle, D.W.O. Phys. Rev.Letters 12, 647, 1964.
28. Morrison, D.R.T. and Rudge, M.R.H. Proc.Phys.Soc. 89, 45, 1966.
29. Drawin, H.W. EUR-CEA-FC-236 Report 1963; 383 Report 1966.
30. Burke, P.G., Cooper, J.W. and Ormonde, S.Phys.Rev. Letters 17, 345, 1966.
31. Holt, H.K. and Krotkov, R. Phys.Rev.144, 82, 1966.
32. Jobe, J.D. and John, R.M.St. Phys.Rev.164, 117, 1967.
33. Schulz, G.J. and Fox, R.E. Phys. Rev.106, 1179, 1957.
34. Chamberlain, G.E. and Heideman, H.G.M. Phys.Rev. Letters 15, 337, 1965.
35. Vriens, L. Physica 31, 1081, 1965.
36. Druyvesteyn, M.J. and Penning, F.M. Rev.Mod.Phys. 12, 385, 1940.
37. Gryzinski, M. Phys.Rev. 115, 374, 1959.
38. McCarroll, R.M. Proc.Phys.Soc. A70, 460, 1957.

Table 1
Excitation and ionization rate constants of electron-atom collisions.

A. Hydrogen; $T=2.2\times10^{4}$ °K.

	$q_H = \sum q^*_H(1-n)$	$q^*_H(1-n)$ 1-2 2s+2p	1-3	1-∞	$q^*_H(n-n')$ 2-3	3-4	4-5
a.	0.8(-10)	0.7(-10)	0.4(-11)	0.1(-10)	1.6(-7)	2.8(-6)	1.4(-5)
b.	1.1(-10)	1.0(-10)					
c.	1.4(-10)	1.3(-10)					

References. a:25, b:22-24, c:21.

B. Helium.

T °K	q^*_{He} from the ground state 1^1S to 2^1S(d)	2^1P(e)	2^3S(f)	2^3P(e)	n=∞(g)	$q_{He} = \sum q^*_{He}$
1×10^{4}	1.2(-19)	7.8(-21)	2.6(-19)	7.5(-20)	8.7(-22)	4.6(-19)
2.2×10^{4}	3.8(-14)	8.5(-15)	2.9(-14)	3.3(-14)	3.2(-15)	1.1(-13)

References. d:26,27, e:28, f:29,30, g:31.

The numbers in parenthesses indicate the power of 10 by which the entries must be multiplied.

Table 2
Critical and equilibrium electron densities and relaxation times of helium-hydrogen plasmas.

n_H/n_{He}	n_H	n_e^c	n_e^c(H)	n_e^E	τ_I (H)	τ_{II} (H)	τ_{II} (He)	τ (He)
		$cm^{-3}sec^{-1}$			μ sec			
0	0	4.0(14)	0	1.1(17)	2.6		15	17.6
1(-4)	3.4(14)	3.6(14)	4.1(12)	1.1(17)	2.4	5.5	15	17.4
1(-3)	3.4(15)	2.0(14)	2.1(13)	1.1(17)	1.2			
3(-3)	1.0(16)	1.1(14)	2.6(13)	1.1(17)	0.5	5.7	6.3	12.5
1(-2)	3.5(16)	0.7(14)	3.5(13)	1.2(17)	0.2	1.8	4.7	4.7
3(-2)	1.1(17)	0.4(14)	3.6(13)	1.5(17)	0.07	0.7	0.8	1.6

The numbers in parentheses are the power of 10.

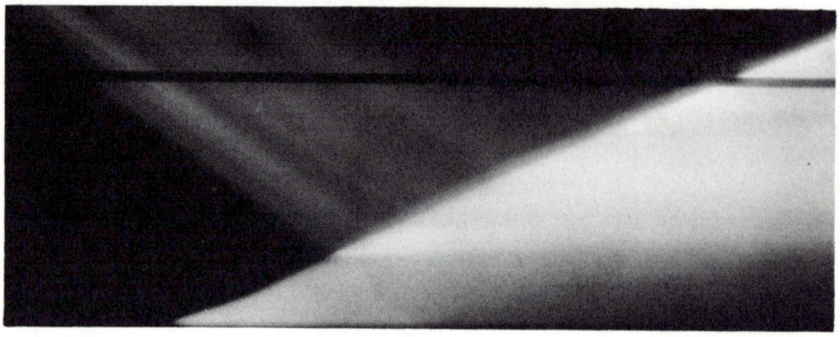

(a)

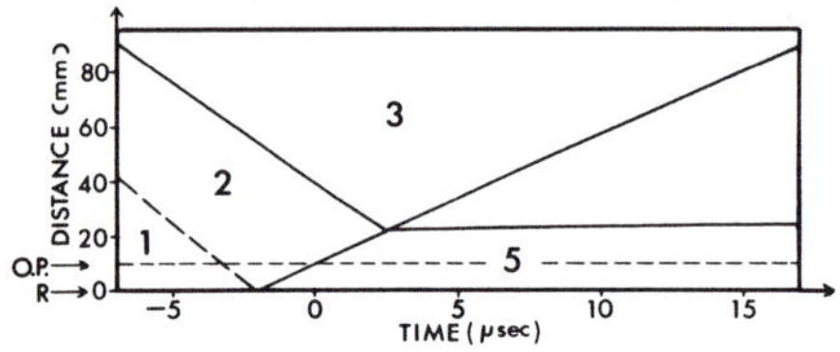

(b)

Figure 1
(a) A streak picture of shock waves for a helium-hydrogen mixture containing 1.5 mole percent hydrogen.
(b) The corresponding x-t diagram of the shock wave trajectories
O.P.: observation position, R:reflector

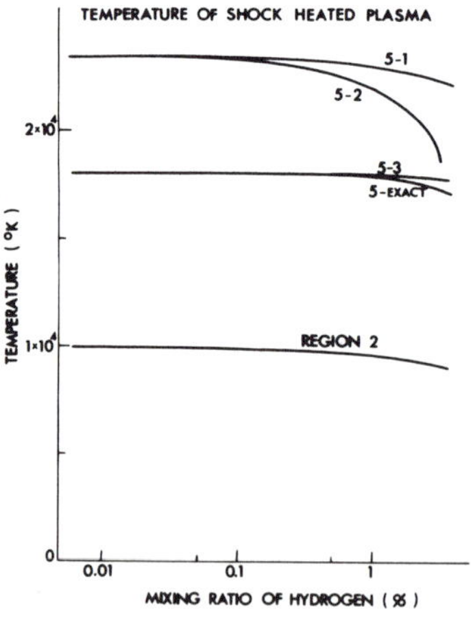

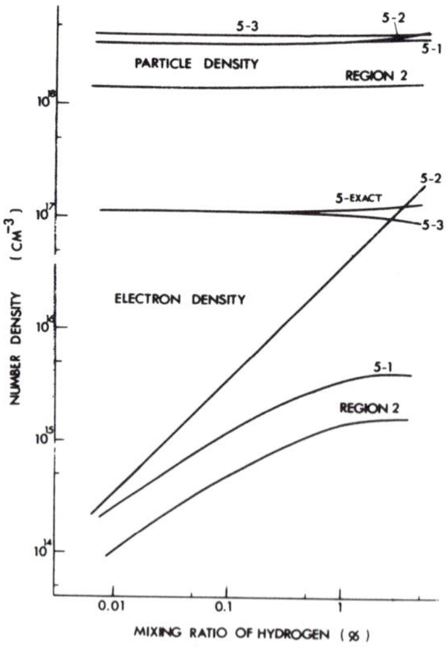

(a) (b)

Figure 2
Temperature (a) and particle and electron densities (b) as a function of the hydrogen mixing ratio (in atomic percent) in various states of the incident and first reflected shock regions

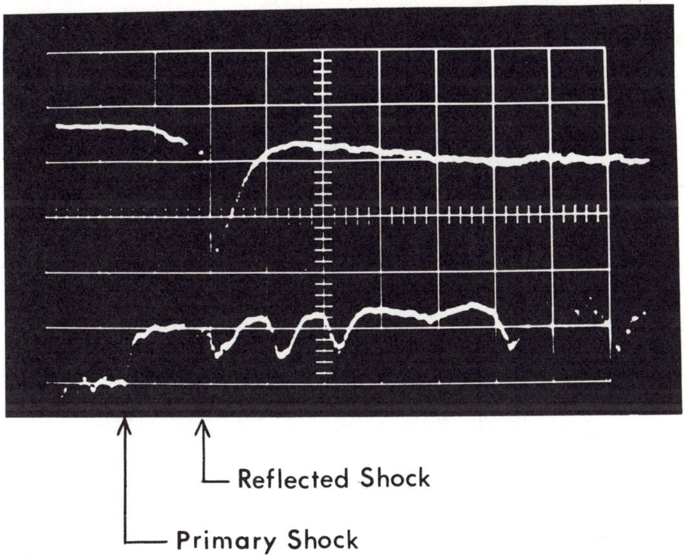

Figure 3
Interference fringe jumps at the incident shock front and behind the first reflected shock for the mixture containing 0.01 atomic percent hydrogen. Sweep rate: 2 μsec/div

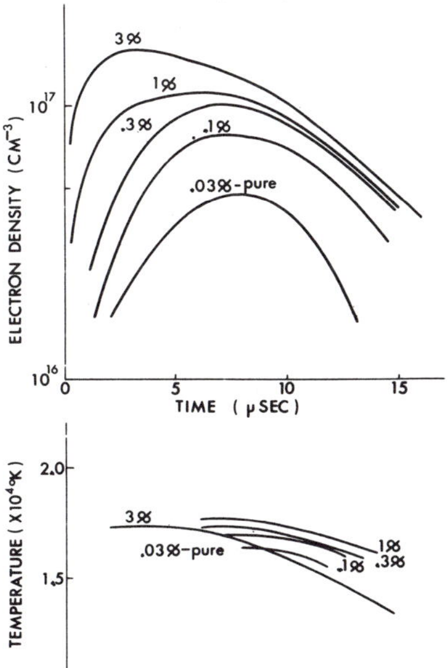

Figure 4
Observed electron density (top) and the corresponding temperature (bottom) behind the first reflected shock

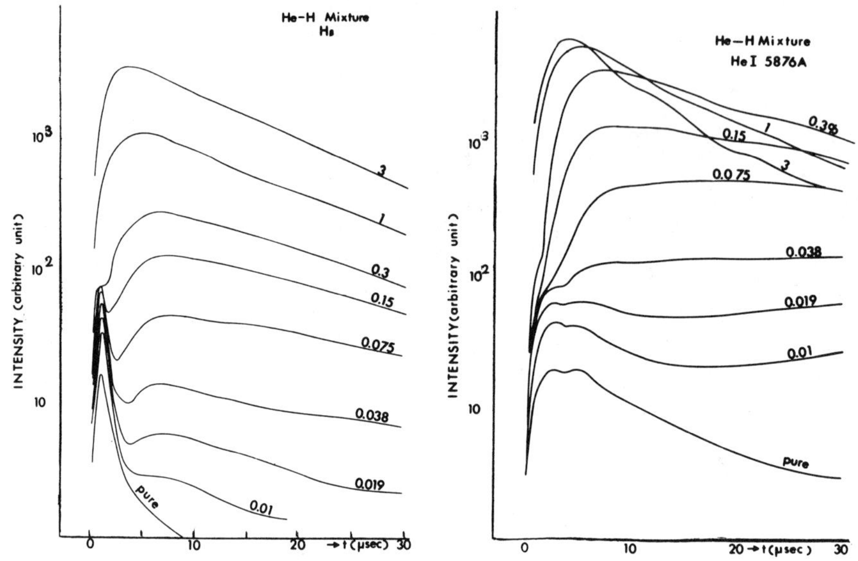

(a) (b)

Figure 5
The emission intensities from (a) the H_β line and (b) He λ5876A in the first reflected region. The number attached to each curve indicates the hydrogen mixing ratio in atomic percent

EXPERIMENTAL STUDY OF HIGH MACH NUMBER COLLISIONLESS SHOCK WAVES

L. S. LEVINE, I.M. VITKOVITSKY, and A. C. KOLB
Naval Research Laboratory
Washington, D.C.

SUMMARY

Experimental observations of high Alfven Mach number collisionless shock waves are reported. The shocks were produced in a low density argon plasma by an extremely fast theta pinch discharge. We have measured rapid heating of the electrons to energies comparable to the ion flow kinetic energies. The measured shock thickness is of the order of the nonthermal ion Debye length (u/ω_{pi}). Under comparable conditions in a hydrogen plasma, no shock compression was observed.

INTRODUCTION

There are now many experiments which demonstrate that a large amplitude wave launched into a magnetized plasma can steepen to form a shock wave under circumstances where the binary collision times are long compared to the relevant observational time; i.e. the flow time of gas through the shock transition layer[1]. Qualitatively, the dissipation processes which are necessary to change the entropy arise from instabilities which produce turbulence. The ions and electrons in the plasma are randomized (thermalized) by their interaction with these turbulent waves. There is now considerable experimental and theoretical evidence that these interactions can be adequately treated phenomenologically as "effective" transport coefficients in a fluid-like model of collisionless shock waves.

Below a calculable critical Alfven Mach number $(M \approx 3)$, the final state energies can reach their Rankine-Hugoniot values by the resistive dissipation of magnetic energy. The resistive shock thickness δ_r is approximately

$$\delta_r \simeq \left(\frac{c^2}{4\pi}\right) \frac{\eta_{eff}}{u(M-1)} = \left(\frac{c}{\omega_{pe}}\right)^2 \frac{\nu_{eff}}{u(M-1)} \quad ,$$

where η_{eff} and ν_{eff} are the effective resistivity and associated collision frequency, M is the Alfven Mach number and u is the flow speed. Experiments show that ν_{eff} is typically $(0.2\text{-}0.5)\omega_{pi}$, where ω_{pi} is the ion plasma frequency, and the shock thickness is

$\delta \simeq 10(c/\omega_{pe})$.[2] A typical (schematic) structure of a low Alfven Mach number shock is shown in Fig. 1.

At high Alfven Mach numbers ($M \gtrsim 3$) resistive dissipation can no longer provide the energy required by conservation laws for a steady state shock. Another dissipative mechanism analogous to an effective ion-ion friction must exist in such shock waves. The classical analogy is the friction which produces viscous stresses in an ordinary fluid.

There have been relatively few experiments in this high Mach number regime. Under conditions in which the Mach number is only slightly greater than the critical value, structures have been observed with thicknesses of $\delta \simeq 0.3$ to $2\ (c/\omega_{pi})$.[3-6] There have also been satellite observations of the earth's bow shock which show comparable widths. The increase of the characteristic length from $\delta \approx 10(c/\omega_{pe})$ at low Alfven Mach numbers is not surprising since $M(c/\omega_{pi})$ is of the order of the cyclotron radius of ions at the Rankine-Hugoniot temperature, and ion motions should dominate the structure of high Alfven Mach number shocks. On the other hand, there have been observations of the bow shock in which the electron heating occurs over distances two orders of magnitude less that (c/ω_{pi}).[7]

Ion heating in high Alfven Mach number shocks was originally[8] attributed to an ion streaming instability produced by the interaction of incident ions and ions reflected from the shock transition region. This instability leads to shock thicknesses of the order of the ion Larmor radius or $M(c/\omega_{pi})$. Recently, Tidman has shown[9] that in a hot electron plasma, electrostatic ion heating could be produced over distances which are only of the order of the nonthermal ion Debye length

$$\delta \simeq \lambda_D = \sqrt{\frac{W_i}{4\pi ne^2}} \simeq (u/\omega_{pi}).$$

Such electrostatic shocks could be formed in times of the order of $(\omega_{pi})^{-1}$, which can be much shorter than the ion gyroperiod required for the ion-streaming instability to develop. This difference in formation time may be critical in interpreting the wide variations in observations since in laboratory experiments studying very high Alfven Mach number shocks the experimental times may be much less than an ion gyroperiod.

In the present experiments an attempt was made to produce a shock wave in a time much shorter than an ion gyroperiod ($\tau_{ci} \simeq 500\ t_{exp}$). Large amplitude magnetic compression waves were observed and they were found to have a thickness of the order of the nonthermal ion Debye length.

EXPERIMENTAL APPARATUS AND DIAGNOSTICS

The collisionless shock waves are produced by a small theta pinch coil, 50 cm long and 12 cm inner diameter. The pinch is driven by a fast rising ($\tau \approx 20$ nsec), low impedance ($Z_0 = 1.4\ \Omega$) energy storage system. This consists of a mylar transmission line designed in a folded Blumlein configuration with the pinch coil as the load. The charged line delivers its energy (560 Joules) to an empty coil in 50 nsec, with a peak voltage of 250 kV and a peak current of 240kA, corresponding to a piston field of 6 kG.

The line is initially pulse charged in 700 nsec by the triggered discharge of two ordinary high voltage capacitors through a 5:1 step-up transformer. The Blumlein is discharged by the over-volting of a multiple-stab dielectric switch placed between the inner and top plates. The electrical system was designed to perform as an ideal pulse line matched to the effective resistance, dL/dt, produced by the imploding shock.

The applied bias magnetic field is produced by a 5 kJ capacitor bank connected across the outer plates of the Blumlein. The bank is discharged through triggered air gap switches and after one quarter period, the discharge is crowbarred through a set of evacuated triggered switches. This produces a slowly decaying ($\tau \simeq 100\mu$sec) current in the coil, and the axial magnetic field falls from a peak value of 3300 G. It can be chosen to be of either polarity with respect to the piston magnetic field.

The initial plasma is produced by a conical theta pinch gun placed along the axis of the main pinch coil and external to it. The plasma flows down the axis of a 120 cm long pyrex tube (10 cm inner diameter) into the theta pinch region. The tube is initially evacuated and a puff of gas (usually hydrogen or argon) is injected by a solenoid operated valve into the region of the conical gun. At a fixed delay time after gas injection, the capacitor driving the conical theta pinch gun is fired. The ringing discharge ionizes the puff of gas during the first half cycle and then pinches and ejects it down the tube on the second half cycle. The plasma properties can be controlled by varying the energy in the gas valve solenoid or the conical gun, the gas pressure behind the valve, and the delay time between injection and the firing of the gun. A schematic of the experiment is shown in Fig. 2.

The gun plasma moves down the tube at about 2.5 cm/μsec. Initially the peak density is of the order of 10^{16} cm^{-3} and at a temperature of several eV but it expands and cools as it propagates. The low density plasma into which the collisionless shocks are launched is estimated to be only 50% ionized but, because of the extremely short times of shock propagation, the presence of such neutral densities is tolerable. ‡

A double Langmuir probe of the "hairpin" variety[10] was developed to

‡ Typical ionization cross sections are $\sim 10^{-17}$cm^2 so, if $n_e \simeq 10^{13}$cm^{-3} and $u \simeq 10^8$cm/sec, typical ionization times are 10^{-4}sec. Similarly, ion-neutral charge exchange interactions can be ignored if the neutral density is less than 10^{14}cm^{-3}.

measure electron density and temperature in the initial plasma as a function of time and position within the pinch coil. Probe characteristics were determined on a shot-to-shot basis with an estimated plasma reproducibility of ± 10%. Typical plasma conditions were $n_e = 5 \times 10^{11}$ to 5×10^{13} cm^{-3}, and $T_e = 0.3$ to 1 eV. The density measurement was compared to that obtained from an 8 mm microwave interferometer and agreement was found (with no applied magnetic field) to within 35%.

Density profiles within the theta pinch coil were measured with the double Langmuir probe. The radial and axial profiles are shown for the case of applied magnetic bias field in Fig. 3a. A comparison of radial profiles with and without bias field is shown in Fig. 3b. The radial density dependence in the absence of applied field is typical of decaying plasma and is due to wall effects. On the other hand, the sharply contracted density profile in the presence of bias field is due to channelling of the gun plasma toward the axis by the fringing magnetic fields at the entrance to the theta pinch coil.

At a time corresponding to a predetermined value of initial plasma conditions and bias magnetic field, the fast energy storage system was discharged, and a magnetic piston formed which propagated into the plasma. Propagation velocities in this high voltage experiment were extremely fast, ranging from 50 to 250 cm/μsec. They were measured from time correlated signals from three small magnetic probes placed at r = 1, 2 and 3.7 cm in the midplane of the tube. The probes consist of single loops of 1 mm diameter of bare wire, each connected to a miniature 50 Ω coaxial cable all of which pass out of the tube through a copper tube. The output of the probes is a direct measure of the time derivative of the axial magnetic field. The signals were terminated and displayed on the time correlated (to within 2 nsec) beams of two Tektronix 556 oscilloscopes. The average speeds were determined, with a precision of ~ 20%, from the intercepts on the axis of the rising piston portions of the probe signals. To observe high frequency oscillations in the magnetic signals, the probe outputs were also displayed on a travelling wave oscilloscope ($\tau \simeq 0.1$ nsec). Graphical integration of the probe signals has been used to obtain the magnetic profile of the shock and piston as a function of time.

In addition to magnetic probes, the principal diagnostic was detection of the x-ray bremsstrahlung spectrum emitted by the plasma. A measurement of the bremsstrahlung spectrum transmitted through pairs of metallic foils provides a good estimate (to within ± 20%) of the average temperature of the hot electrons in the shock or piston.[11] In the present work, thin aluminum and gold foils were used, the aluminum foils being useful at relatively low temperatures (1 to 15 keV) and the gold foils at higher temperatures. The foils were mounted in a two or three channel detector. Each channel consisted of a foil, a length of scintillator long enough to absorb nearly all of the x-ray photons passing through the foil, and a high gain photomultiplier. The relative amplitude of the x-ray signals was then determined from the amplitudes of the photomultiplier signals. Calibration of the channels was accomplished in situ by inserting identical foils into each channel. To check that the x-rays were due to bremsstrahlung and not line radiation, the temperature was measured under similar plasma conditions with aluminum and with gold foils. The two deduced temperatures agreed to within their experimental errors and the estimated reproducibility of the shocks. Since the two metals have different absorption behavior, this demonstrates that line radiation could not be very important in the measured x-ray signals.

EXPERIMENTAL RESULTS

The experiments were conducted with hydrogen and argon plasmas. Observations were made over a wide range of densities ($n_e = 5 \times 10^{11}$ to 5×10^{13} cm^{-3}) and over a wide range of bias magnetic field values ($B_0 = 75$ to 1200 G). The applied magnetic field was produced both parallel and anti-parallel to the piston magnetic field.

In all cases the magnetic piston was observed to propagate at velocities that were consistent with the field strength and gas density. For example, in Fig. 4, the observed piston velocity in hydrogen is plotted as a function of the mean particle density in the tube. The behavior estimated by the snowplow model, using measured values of density and piston field, is also shown for comparison. The agreement is within 15% except at the lowest densities, where the predicted speeds are ~50% greater than those measured. The experimental precision ranges from 15% to 30% at the higher speeds.

When hydrogen was used as the working gas, under no conditions was a shock wave observed to develop ahead of the piston. Careful attempts were made to detect a shock wave with the magnetic probes but no compression wave separated from the piston was seen.

Even in the absence of shock formation, pulses of x-ray emission were observed and the electron heating was measured from the x-ray bremsstrahlung. In Fig. 5, the measured electron energies at the time of the peak x-ray signal are plotted as a function of piston velocity for the hydrogen case. The maximum in the emission occurs when the plasma cumulates on the axis, but kilovolt electron energies were also observed during the implosion. For reference, we have also plotted in Fig. 5, the calculated electron temperature behind a shock moving at the observed piston speed, assuming $T_e = T_i$. The heating here is clearly due to energy dissipation in the piston due to some strong instability. Finally, estimates of the absolute intensity of the x-ray bremsstrahlung suggest that essentially all of the electrons are being heated to these high energies.

Argon was then substituted for hydrogen as the working gas. The mass density and flow speed were kept in the same range as in the hydrogen experiment by working at much lower number densities of argon. For conditions comparable to those used in the hydrogen cases, well-defined compression waves were formed in argon. The evolution of such a shock wave is shown in Fig. 6. The development of the anti-parallel magnetic field compression in front of the advancing piston can be clearly seen. Also shown is the time-correlated signal from one of the x-ray detectors.

The amplitude of the shock compressed magnetic field B_1, was measured from a series of five graphically integrated signals from the middle probe. While the front speeds varied by more than a factor of two, the magnetic field jump remained fairly constant. Its mean value was $\Delta B \equiv B_1 - B_0 = -(220 \pm 10)$G, where the error represents the standard deviation of the measurements. If one assumes steady shock jump conditions with a compression ratio of 4, the measured field jump implies an initial field $B_0 \approx -75$ G. This field originates from the field trapped in the plasma when it was initially made by the theta pinch gun, and its magnitude agrees well with that estimated from the resistive decay of the gun field. No shocks were observed when an additional field was produced by the bias bank. This is presumably because of the highly nonuniform radial density distribution produced by the bias bank (see Fig. 3b).

An upstream field of the order of -75 G implies that the Alfven Mach numbers of these shocks ranged from 20 to 30, which is far above the critical Mach number.

The electron heating was also measured for the case of the argon shock waves. Careful time correlation of the x-ray signals with the magnetic probe signals showed that the earlier, smaller x-ray peak such as can be seen in Fig. 6, must be associated with the implosion phase, while the main peak appears after the shock heated plasma has all reached the axis. The measured temperatures from the x-rays are plotted in Fig. 7 as a function of piston speed. As in the hydrogen case, the bulk of the electrons are probably heated in the piston rather than the shock.

At these high Alfven Mach numbers there is evidence of strong magnetic turbulence. A typical trace from the fast travelling wave oscilloscope is shown in Fig. 8. Evident in the trace are magnetic field fluctuations through the piston. They are irregular and of quite high frequency ($\omega = 3.6 \times 10^9$ sec^{-1}). This frequency is well above the ion plasma frequency or any cyclotron frequency, but is less than a typical electron plasma frequency. The fluctuations in the piston are most likely related to the heating observed but the precise nature of the interaction remains to be determined.

The measured thickness of the magnetic compression layer is $\delta = (1 \pm 0.3)(u/\omega_{pi})$, where the error represents the standard deviation of the measurements. The thickness would be a few times the ion Debye length, if the compressed density were used in computing ω_{pi}, rather than the initial density. In contrast, the thickness is more than two orders of magnitude less than c/ω_{pi} and 0.1% of the cyclotron radius of an ion moving with the flow velocity. Finally, we should note that the electrons have sufficient time to respond to the observed magnetic pulse since the thickness (~1 cm) is also comparable to the inertial skin depth c/ω_{pe}.

The question still remains why no shock wave is formed in the hydrogen case. One possible explanation of this is found in the following model of a leaky piston. In the frame of reference of the piston, ions are flowing into the piston at a velocity u_p. For a shock to form, the piston must stop these ions by establishing an electric potential such that the potential energy is comparable to the ion kinetic energy. This electric field is normally established by the separation of charge from electrons bound to the magnetic field lines within the sheath. The characteristic length for ion acceleration when electrons are tightly bound to the field is $\lambda_D \approx (u/\omega_{pi})$, the nonthermal ion Debye length.

Now suppose that the electron cyclotron radius is comparable to or greater than λ_D. Then one might expect the acceleration time to increase and the piston to become leaky. We postulate that the ratio of these two lengths, r_{ce}/λ_D, is an important parameter in the behavior of the piston, and the present experiments provide some support for this model. Thus consider the magnitude of the quantity

$$\frac{r_{ce}}{\lambda_D} = \frac{cv_e}{u^2} \frac{m_e}{m_i} M ,$$

For the high Alfven Mach number hydrogen experiment

$$\frac{c}{u} \frac{m_e}{m_i} M = 1.5 \text{ to } 2.5$$

Taking v_e to be either the flow speed or the electron thermal speed leads to a ratio (r_{ce}/λ_D) of order unity or greater. In contrast, for the argon experiments, the ratio is reduced to a value much less than unity (0.03 to 0.1).

CONCLUSION

Attempts to produce a large amplitude magnetic compression pulse with a fast magnetic piston moving in a low density magnetized plasma were successful when argon was substituted for hydrogen. Conditions were chosen such that the ion cyclotron period was several orders of magnitude greater than the experimental time.

The experiment was motivated by the question of whether the compression was influenced by reducing the ratio of the electron cyclotron radius to the nonthermal ion Debye length. The latter is the characteristic length for ion acceleration in space charge electric fields and is also the characteristic length in Tidman's theory of high Mach number electrostatic shocks. The observed thickness was of this order of magnitude.

Another interpretation of the results is that a shock forms when the ion Debye length is increased to a value comparable to the response distance of the electrons; namely, the inertial skin depth c/ω_{pe}.

These results, as well as observations of the scale length for heating in the earth's bow shock, are suggestive that there is strong electrostatic dissipation on the scale of the ion Debye length.

Better control of the uniformity of the initial magnetic field and measurements of ion energies with gases of different atomic weight are needed to remove the ambiguities in the detailed interpretation.

REFERENCES

1. E. Hintz, in "Methods of Experimental Physics: Plasma Physics" (H.R. Griem and R.H. Lovberg, Eds.), Academic Press, New York, to be published.
2. N.A. Krall, in Proc. E.S.R.O. Conf. on Collisionless Shock Waves (Frascati, Italy, 1969). To be published.
3. J.W.M. Paul, et al, in Proc. Seventh Int. Conf. on Phenom. in Ionized Gases (Gradevinska Knjiga, Beograd, Yugoslavia, 1966), Vol. II, p. 819.
4. S.G. Alikhanov, et al, in Proc. Third IAEA Conf. on Plasma Physics and Controlled Nuclear Fusion Research (Novosibirsk, 1968), Paper CN-24/A-1. To be published.
5. E. Hintz, in Proc. Third IAEA Conf. on Plasma Physics and Controlled Nuclear Fusion Research (Novosibirsk, 1968), Paper CN-24/A-2. To be published.
6. R. Chodura, et al, in Proc. Third IAEA Conf. on Plasma Physics and Controlled Nuclear Fusion Research (Novosibirsk, 1968), Paper CN-24/A-3. To be published.
7. M. Montgomery, in Proc. E.S.R.O. Conf. on Collisionless Shock Waves (Frascati, Italy, 1969). To be published.
8. R.Z. Sagdeev, Sov. Phys.-Tech Phys. 6, 867 (1962).
9. D.A. Tidman, Phys. Fluids 10, 547 (1967).
10. T.I. McLaren, J.N. Fox, and R.M. Hobson, Nature 198, 1264 (1963).
11. R.C. Elton, "Determination of Electron Temperatures between 50 eV and 100 keV from X-ray Continuum Radiation in Plasmas", Naval Research Lab. Report No. 6738 (1968).

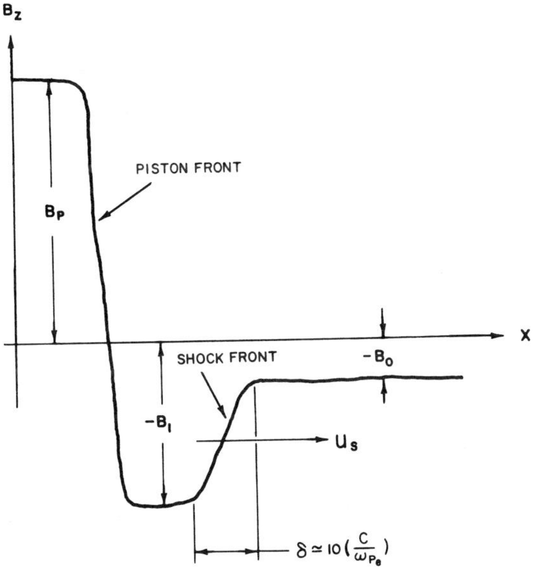

Figure 1
Schematic profile of a subcritical collisionless transverse shock

1. THETA PINCH COIL.
2. FOLDED MYLAR BLUMLEIN.
3. CONICAL THETA PINCH GUN WITH CAPACITOR BANK AND SWITCH.
4. DIELECTRIC STAB SWITCH (BETWEEN TOP AND MIDDLE PLATES).
5. INITIAL ENERGY STORAGE SYSTEM WITH SWITCH AND 5:1 STEP-UP TRANSFORMER.
6. BIAS MAGNETIC FIELD BANK WITH CROWBAR SWITCH.

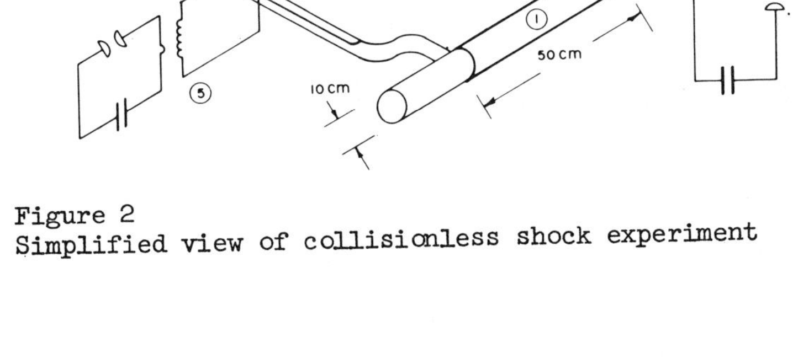

Figure 2
Simplified view of collisionless shock experiment

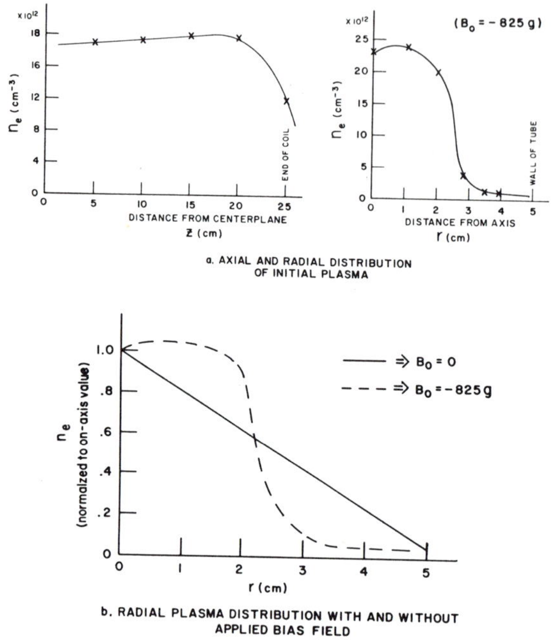

Figure 3
Density distribution of initial plasma

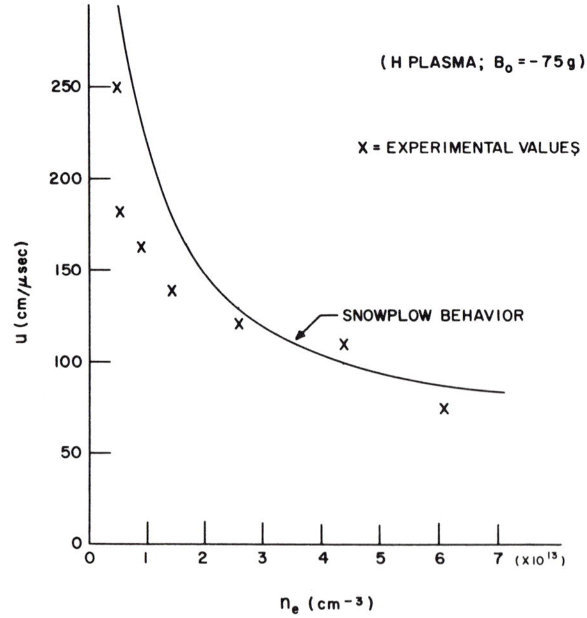

Figure 4
Wave speed as a function of initial plasma density

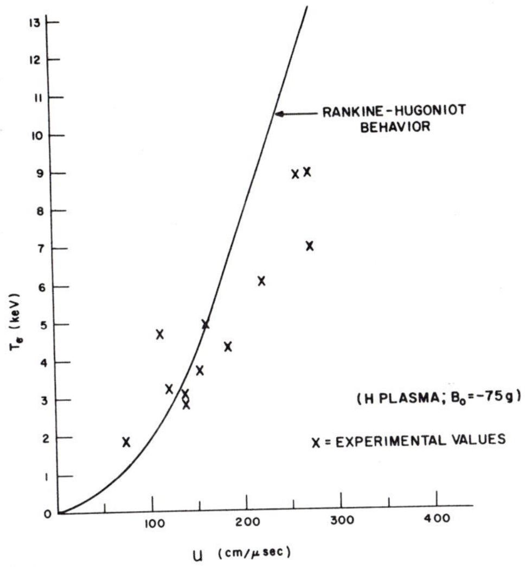

Figure 5
Electron energy as a function of wave speed in a hydrogen plasma

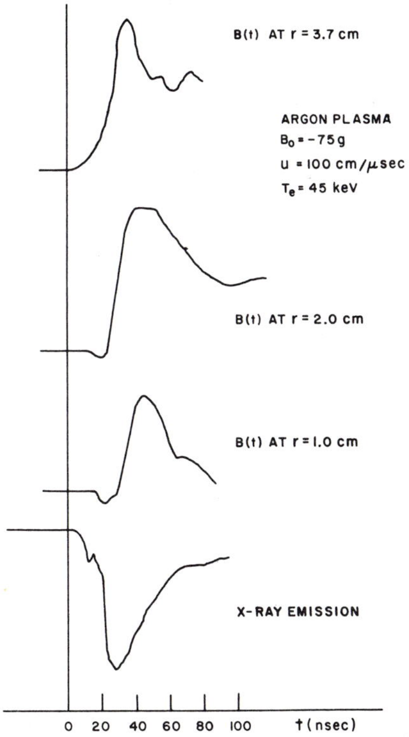

Figure 6
Time correlated magnetic probe and x-ray signals

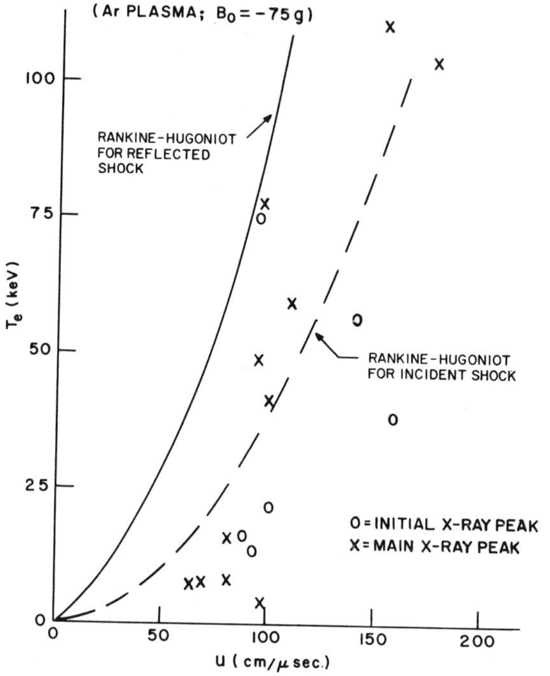

Figure 7
Electron energy as a function of wave speed in an argon plasma

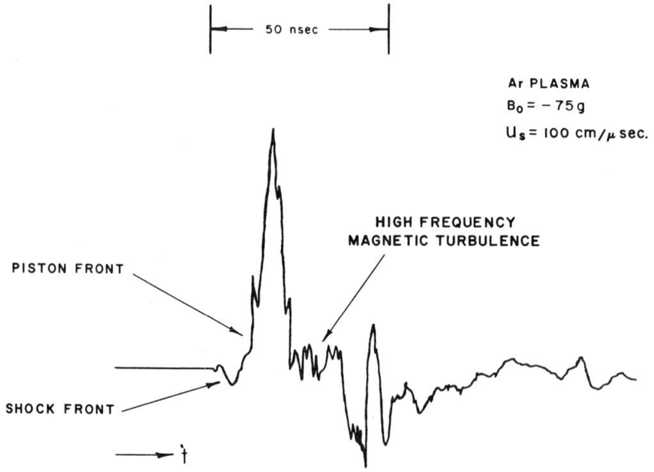

Figure 8
Magnetic probe signal at r = 2 cm

SHOCK TUBE FLOW INTERACTION WITH AN ELECTROMAGNETIC FIELD*

J. ROSCISZEWSKI**
Air Vehicle Corporation
San Diego, California

W. GALLAHER
General Dynamics/Convair
San Diego

SUMMARY

In the present paper, the problem of strong interaction of a shock tube flow with a linear segmented-electrode MHD generator is discussed. Solution of the problem was obtained using a modified two-step Lax-Wendroff finite difference method of solving a system of partial differential equations. A particular example of shock tube flow with xenon as driven gas and initial shock Mach number of 11.2 has been calculated for different generator loads to match experimental conditions of B. Zauderer (the following paper at this Symposium). Computer calculations have been performed assuming one-dimensional non-steady flow with equilibrium ionization.

INTRODUCTION

Interaction of a shock wave and a magnetic field has been investigated experimentally by many investigators (J. H. deLeeuw, 1958; K. Dolder and R. Hide, 1958; R. M. Patrick and T. R. Brogan, 1959; H. J. Pain and P. R. Smy, 1960-61). Theoretical analysis of the interaction of a semi-infinite slug of gas behind the shock wave with a semi-infinite segmented MHD generator was discussed previously by one of the authors of this present paper (J. Rosciszewski, A. K. Oppenheim, 1963). In that paper, a closed form solution was obtained for a variable shock wave speed and the whole flow field was calculated only in a linearized approximation. Numerical calculations of the shock wave-induced flow with a linear MHD generator section using the method of characteristics have been discussed by Johnson (M. R. Johnson, 1966).

---

* This paper has been supported by the Air Force Office of Scientific Research under Contract No. F44620-68-C-0052.

** Consultant.

The method of solution is the modified Lax-Wendroff technique (P. Lax, B. Wendroff, 1960). The Lax-Wendroff technique is applied to the conservation form of inviscid flow equations. These equations do not properly describe shock structure, but satisfy matching conditions across the shock wave. The shock is smeared over a few spatial meshes instead of being a sharp discontinuity. Therefore, shocks are obtained from calculations as locations of steep gradients without a need of a priori assuming of existence of shocks and using separate matching conditions as is the case in the method of characteristics. This is especially convenient in the case of shock formation and growth of a shock, where the method of characteristics can lead to problems.

For the case of strong shock waves, the Lax-Wendroff procedure leads to an oscillatory pressure overshoot on the downstream side of the shock. This overshoot can be suppressed by introducing a damping factor, some sort of artificial viscosity (H. U. Thommen, L. D'Attorre 1965). Application of this technique to prediction of pressure pulses due to rocket engine starting was discussed in the work of Magnus (R. J. Magnus, 1966).

In the present paper, the modified Lax-Wendroff technique was applied to the interaction of a shock tube flow with the linear MHD generator. Ionization energy, calculated on the basis of equilibrium ionization, has been included in the energy equation as a source term. Inclusion of this term under differential operators was essential and showed a large discrepancy from constant, "mean" approximation or from the case where calculation of ionization in the source term is calculated 1/2 time-step behind. However, introduction of the ionization term requires the solution of a transcendental equation for internal energy or degree of ionization at each time-step. The degree of ionization is based on the Saha equation.

Calculations presented herein were obtained by neglecting induced magnetic field (justified by the small magnetic Reynolds number) and taking into account Lorentz force, Joule heating, and radiation loss of energy. Actual generator load including electrode loss was taken from the experimental data of B. Zauderer (B. Zauderer, Tate 1969). Calculations including the ionization energy under the differential operators lead to some problems when the interface reaches the MHD section, namely a sudden pressure peak appears with a speed of propagation close to that of the interface. This could not be interpreted as a real physical phenomenon, but represents some sort of a numerical instability. Moreover, a constant γ solution does not give any such pressure peak.

SYSTEM OF EQUATIONS AND THE NUMERICAL TECHNIQUE

Conservation of mass

$$\frac{\partial \rho}{\partial t} + \frac{\partial \rho u}{\partial x} = 0 \tag{1}$$

Conservation of momentum

$$\frac{\partial \rho u}{\partial t} + \frac{\partial(\rho u^2 + p)}{\partial x} = -jB \tag{2}$$

Conservation of energy

$$\frac{\partial}{\partial t}\left\{\rho\left(\frac{u^2}{2}+e+\frac{V_i}{\rho}\alpha\right)\right\}+\frac{\partial}{\partial x}\left\{\rho u\left(\frac{u^2}{2}+h+\frac{V_i}{\rho}\alpha\right)\right\}=Ej-Q_{RAD} \quad (3)$$

where ρ denotes density, u velocity, p pressure, j current density, B magnetic field, $e = \frac{1}{\gamma - 1}\frac{p}{\rho}$ internal energy, V_i ionization energy, $\alpha = \frac{n_e}{n}$ degree of ionization, E electric field, Q_{rad} radiated energy.

Equations (1 - 3) can be written in an abbreviated form

$$W_t - F_x = S \quad (4)$$

where the subscripts denote partial differentiation and W, F, and S are following column vectors:

$$W = \begin{pmatrix} \rho \\ \rho u \\ \rho\left(\frac{u^2}{2}+e+\frac{V_i}{\rho}\alpha\right) \end{pmatrix} \qquad F = \begin{pmatrix} \rho u \\ \rho u^2 + p \\ \rho u\left(\frac{u^2}{2}+h+\frac{V_i}{\rho}\alpha\right) \end{pmatrix} \quad (5)$$

$$S = \begin{pmatrix} 0 \\ -jB \\ Ej - Q_{rad} \end{pmatrix}$$

In the system of equations, three quantities W_1, W_2, W_3 are treated as independent variables. Physical variables ρ, u, p, T, and vectors F and S can be expressed through W_1, W_2, W_3. For example,

$$\rho = W_1, \qquad u = \frac{W_2}{W_1}, \qquad F_1 = W_2, \qquad \text{etc.}$$

In order to find temperature, pressure, or α, a transcendental equation for e or p is solved using the relationship $\alpha = (T, p)$ given by the Saha equation.

Electric current density was calculated from Ohm's law

$$j = \sigma(E - uB) \quad (6)$$

where $\sigma(T, n_e)$ is the electric conductivity. For the case of the present generator under investigation, $uB > E$. Following the notation of Zauderer, one can introduce load parameter K related to electric field E as follows:

$$E = K V_o B \tag{7}$$

where $K(x,t)$ is the load parameter and V_o denotes initial velocity behind the shock wave before interaction occurs. If one denotes R the resistant connecting two electrodes (Fig. 1) and the ratio of external resistant to internal is $\lambda = \frac{\sigma R A}{\ell}$ (as defined in the work of Rosciszewski and Oppenheim), A, denoting electrode area, ℓ channel width, neglecting electrode potential drop, relationship between K and λ is

$$K = \frac{u B \lambda}{(1+\lambda) V_o} \tag{8}$$

The numerical solution of Eqs. (4) is based on the modified two-step Lax-Wendroff difference scheme. Knowing values of W, F, S at $t = t_o$, one calculates corresponding values at $t = t_o + \Delta t$. The half-step approximation reads

$$W^{1/2}_{\pm 1/2} = \frac{1}{2}(W^0_0 + W^0_{\pm 1}) + \frac{\Delta t}{2\Delta x}(\pm F^0_0 \mp F_{\pm 1}) + \frac{\Delta t}{2}(S^0_0 + S^0_{\pm 1}) \tag{9}$$

where

$$W^{\mu}_{\nu} = W(t_o + \mu \Delta t, x_o + \nu \Delta x)$$

The full-step value is

$$W'_0 = W^0_0 + \frac{\Delta t}{\Delta x}(-F^{1/2}_{1/2} + F^{1/2}_{-1/2}) + \frac{\Delta t}{2}(S^{1/2}_{1/2} + S^{1/2}_{-1/2}) \tag{10}$$

Quantities $F^{1/2}_{1/2}$, etc., were expressed by quantities $W^{1/2}_{1/2}$.
In order to suppress oscillation behind the strong shock, a second approximation of the value of W'_o has been introduced, which uses an artificial viscosity term denoted as D:

$$(W'_0)_2 = \frac{1}{2}(1-D)(W^{1/2}_{1/2} + W^{1/2}_{-1/2} - W^0_0) + \frac{1}{2}(1+D)W'_0 \tag{11}$$

for $D = 1$, $(W'_o)_2 = W'_o$.

A value of $D = 0.7$ was used, giving good suppression of oscillations (Thommen, D'Attorre, 1965).

One large problem in applying the Lax-Wendroff technique, to Eulerian system of equations, is associated with contact discontinuities where there is a sharp change of temperature and density, and dependent variables W_1, W_2, W_3. Because the contact surface crosses mesh points in an unknown position, to resolve any discontinuity in pressure and velocity, one assumes small disturbance waves propagating from the contact surface in both sides so that pressure and velocities are matched at the contact surface (Magnus, 1966).

Stability analysis of the Lax-Wendroff scheme is a very difficult problem and has been performed only by using a linearized form (Richtmeyer, Thommen-D'Attorre). In the latter paper, using constant coefficients, the eigenvalues of amplification matrix were found and the von Neumann stability criterion requiring that the maximum value of the eigenvalue of the amplification matrix should not be greater than one has been checked, for assumed values of

$$\lambda = \frac{c\Delta t}{\Delta x} \quad \text{(c being speed of sound).}$$

This leads to values of

$\lambda = \frac{c}{|u_I| + c}$. The condition of $\lambda \leq 1$ is well-known Courant-Friedrichs-Lewy condition for hyperbolic equations, which says that finite difference mesh is contained in a characteristic triangle. The present damping coefficient affects the stability and results in a smaller value denoted as $\lambda' = \lambda\sqrt{\frac{2D}{1+D}}$ for a linearized analysis (D'Attorre, Gallaher, 1967).

RESULTS AND DISCUSSION

Calculation of the shock tube flow with a finite slug of hot gas has been performed assuming constant mean value of $\gamma = 1.47$. This value corresponds to a matching condition across the shock before it enters the MHD section. Two different initial values of K based on the velocity V_o behind the oncoming shock have been assumed. Resistance R of the outside circuit has been taken constant. The larger value of K corresponds to a weaker interaction (larger resistance R) and the smaller value of K to a stronger interaction (smaller resistant R).

In the case of stronger interaction (K = 0.735), the pressure distribution (Fig. 2) indicates formation of a standing upstream-faced shock wave of growing intensity even after contact surface has already passed MHD section. In the case of a weaker interaction (K = 0.9), a weaker pressure front is being swept downstream as the contact surface passed MHD section. Temperature distribution (Fig. 3) indicates a significant temperature rise with an abrupt drop at the contact surface.

Temperature distribution for the case where ionization energy is taken into account in the energy equation shows (Fig. 7) a much smaller temperature increase, indicating a strong ionization "cooling" effect.

In Figure 4 the t-x diagram of shock and contact surface motion is presented, indicating small shock wave attenuation and large velocity change of the contact surface.

Pressure and velocity distribution presented in Figs. 5 and 6 which corresponds to the case of ionization energy being taken into account shows the same, but even stronger, secondary standing shock as in the case of constant γ (Fig. 2).

Electric current density (Fig. 9) indicates a larger value of current extracted at the front electrodes than at the middle and downstream electrodes, despite a lower conductivity at the front (Fig. 10). There is also a noticeable current decrease as time increases. After the shock passes, the electric conductivity indicates small variation with time (Fig. 10).

Comparison of these results with experimental data is presented in the following paper.

Concluding, we are confident that the present numerical technique could give useful information of complicated interaction process of the shock tube flow. However, further refinement is necessary of the numerical procedure dealing with the contact surface interaction in ionizing gas.

ACKNOWLEDGMENT

The authors wish to express their gratitude to Dr. R. Magnus for fruitful discussions and to Mr. M. Schoonover for help in programming and running the computer calculations.

REFERENCES

1. D'Attorre, L.D., Gallaher, W. A Difference Scheme of Improved Damping. General Dynamics/Convair Rept. GDE-DBE-67-015.
2. deLeeuw, J.H. The Interaction of a Plane Strong Shock Wave with a Steady Magnetic Field. University of Toronto, Institute of Aerophysics, Rept. No. 49, 1958.
3. Dolder, K., and Hide, R. An Experiment on the Interaction Between a Plane Shock and a Magnetic Field. Nature, 181 1116, 1958.
4. Johnson, M.R. A Gasdynamic Analysis of the One-Dimensional Interaction of Shock Induced Flow with Transverse Magnetic Fields of Finite Extent. Northwestern University, Ph.D. thesis, 1966.
5. Lax, P.D. Wendroff, B. Systems of Conservation Laws. Com. Pure Appl. Math., Vol XIII, 217-237, 1960.
6. Magnus, R.J. Analytic Prediction of Pressure Pulses Due to Rocket Engine Starting. General Dynamics/Convair Rept. GDC-ERR-AN-984.
7. Pain, H.J., and Smy, P.R. Experiments on Power Generation from a Moving Plasma. Journal of Fluid Mechanics, 10, 51, 1961.
8. Pain, H.J., and Smy, P.R. Magnetic Field Interactions with Shock Ionized Argon. Proceedings of the Physical Society, 76, 849, 1960.

9. Rosciszewski, J., and Oppenheim, A.K. Shock Interaction with an Electromagnetic Field. The Physics of Fluids, 6, 5, 689, May 1963.
10. Thommen, H.U., and D'Attorre, L.D. Calculation of Steady, Three-Dimensional Supersonic Flow Field by a Finite Difference Method. General Dynamics/Astronautics Rept. GDC-ERR-AN-718.
11. Zauderer, B. and Tate, Interaction of an Incident Shock-Tube Flow with an Electromagnetic Field: Part II Experiment (This volume).

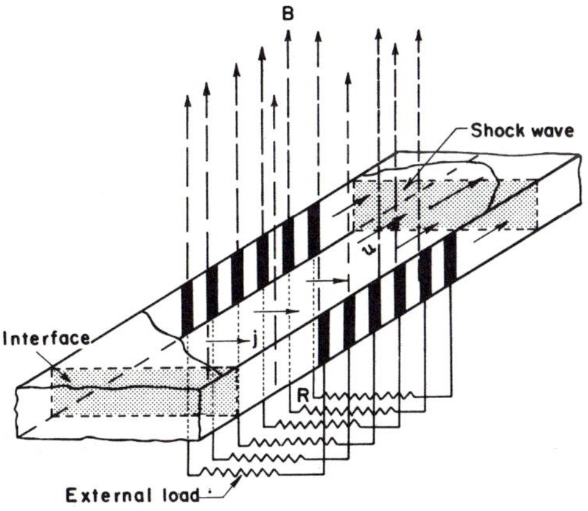

Figure 1
Shock wave propagating through a section of a linear MHD generator.

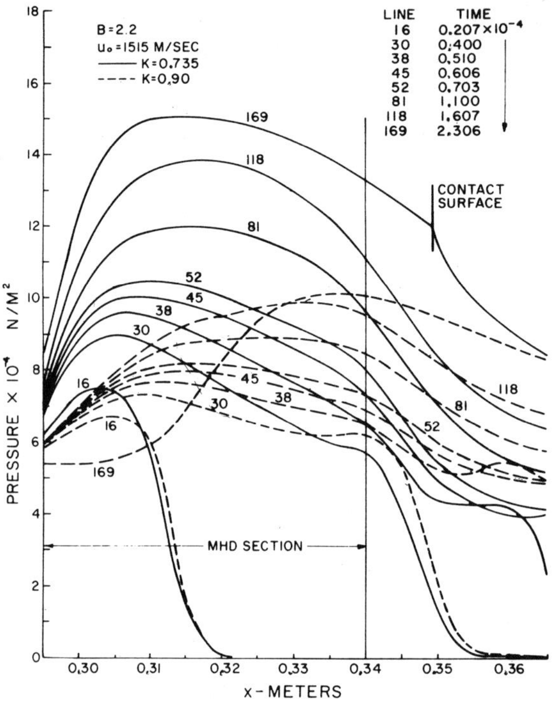

Figure 2

Pressure distribution resulting from interaction of a shock tube flow with a section of a linear MHD generator. Case of $\gamma = \gamma_m = 1.47$ = const with two different load parameters with constant resistance R.

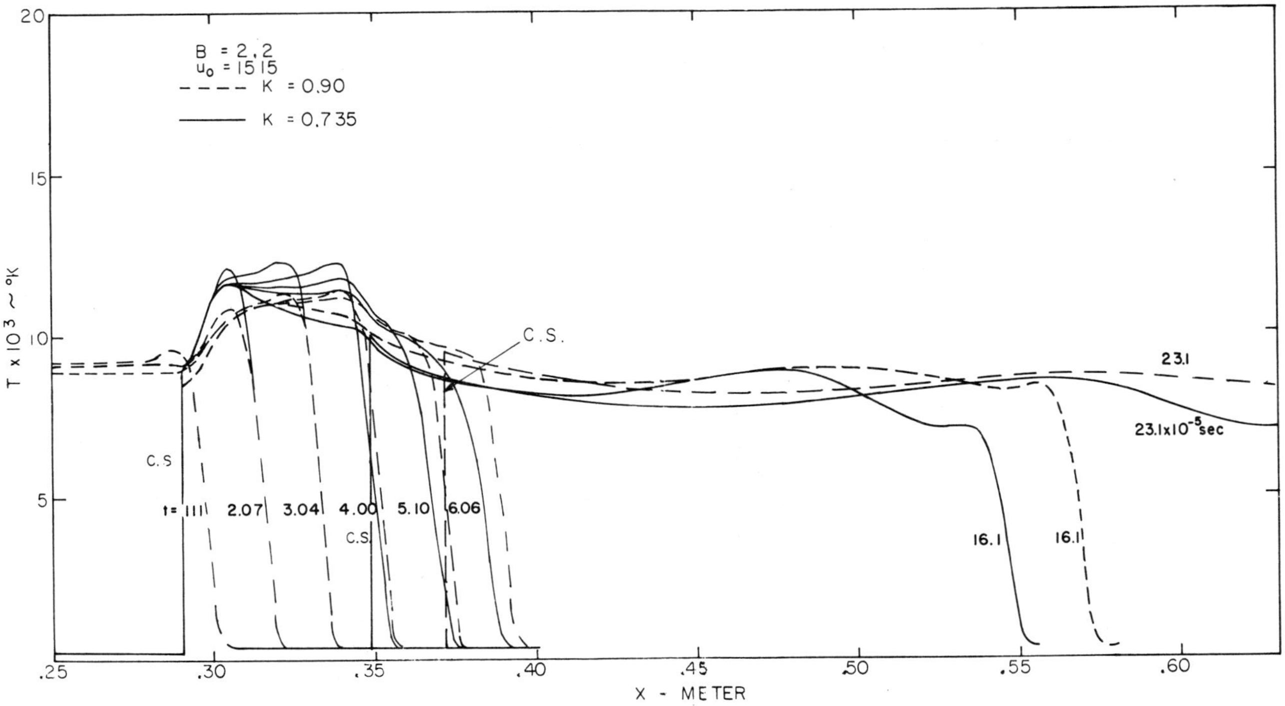

Figure 3

Temperature distribution resulting from the interaction of a shock tube flow with a section of a linear MHD generator. Case of γ = const with two different loads K and the constant resistance R.

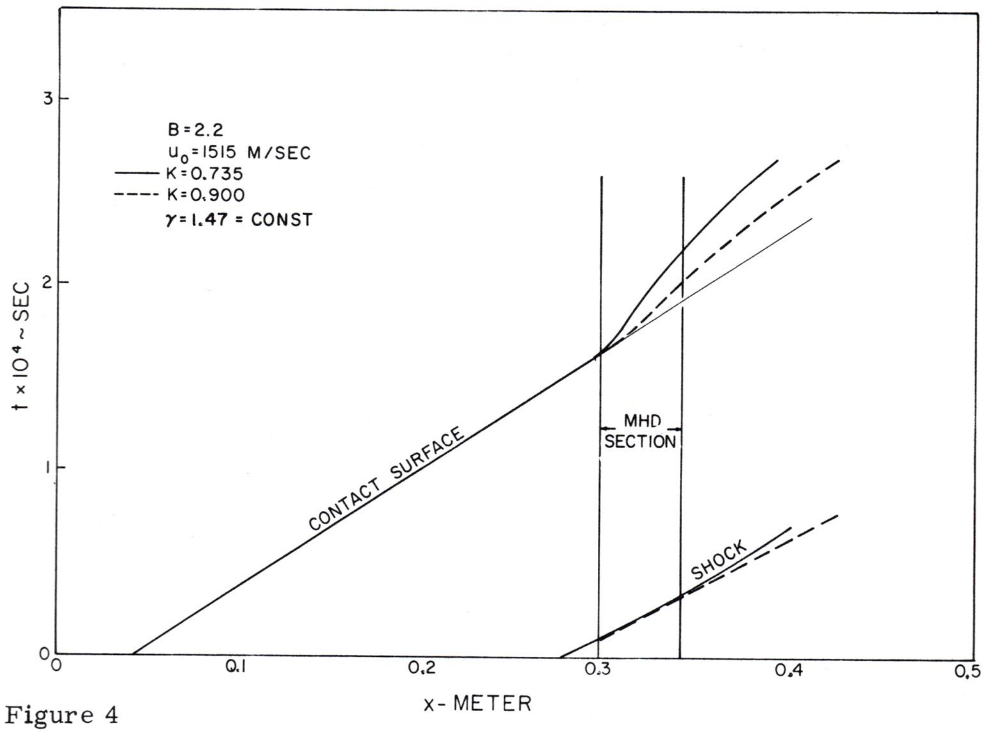

Figure 4

x - t diagram of the interaction process. Case $\gamma = 1.47$ = const.

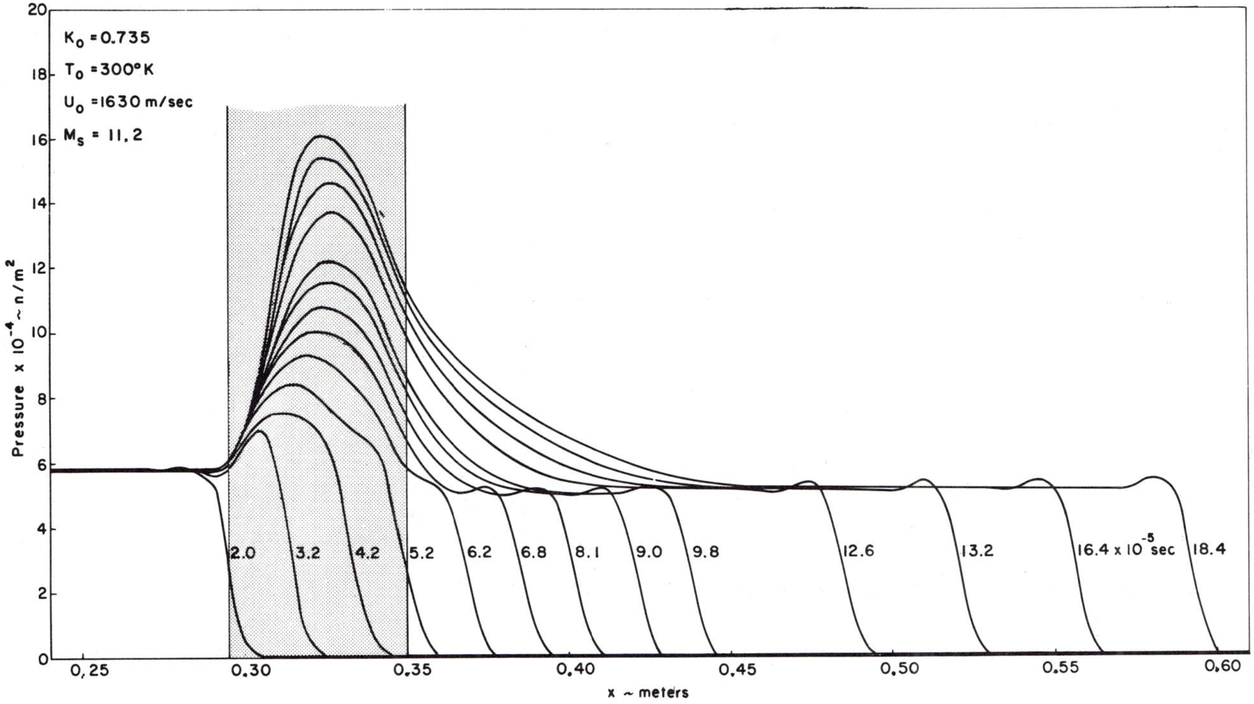

Figure 5
Pressure distribution resulting from interaction of the flow behind the shock wave with a linear MHD generator. Ionization energy included.

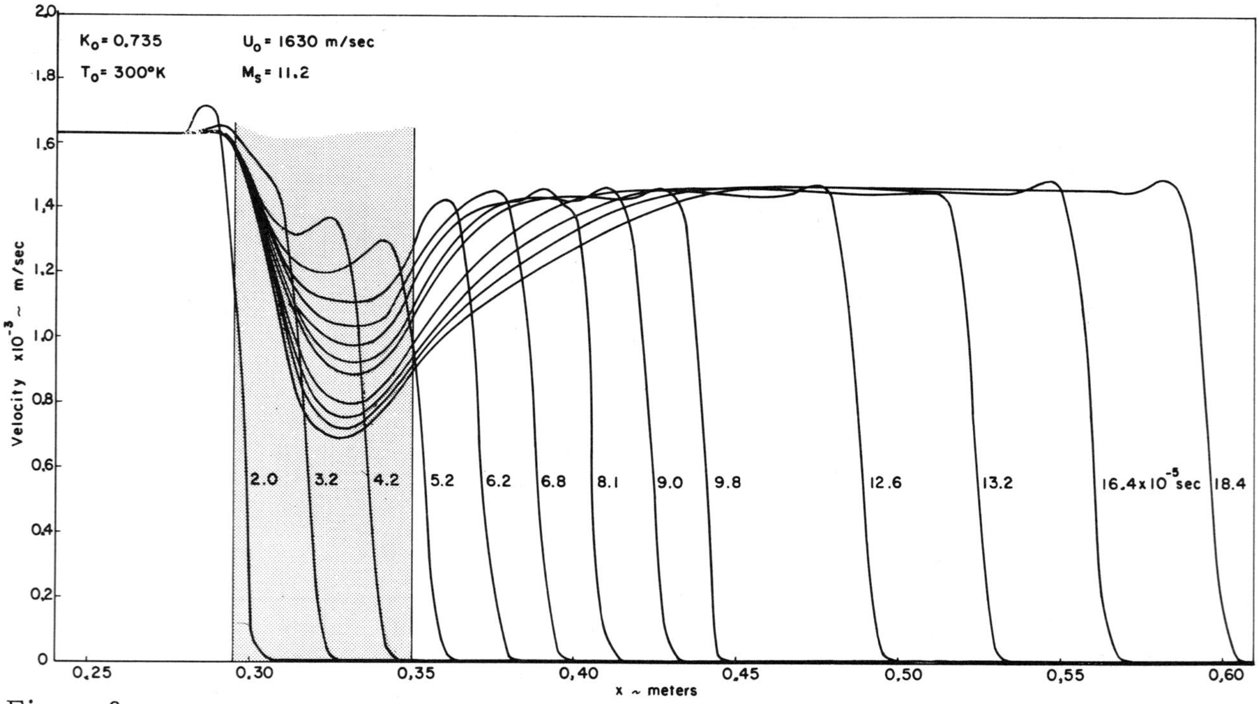

Figure 6

Velocity distribution resulting from interaction of the flow behind the shock wave with a linear MHD generator. Ionization energy included.

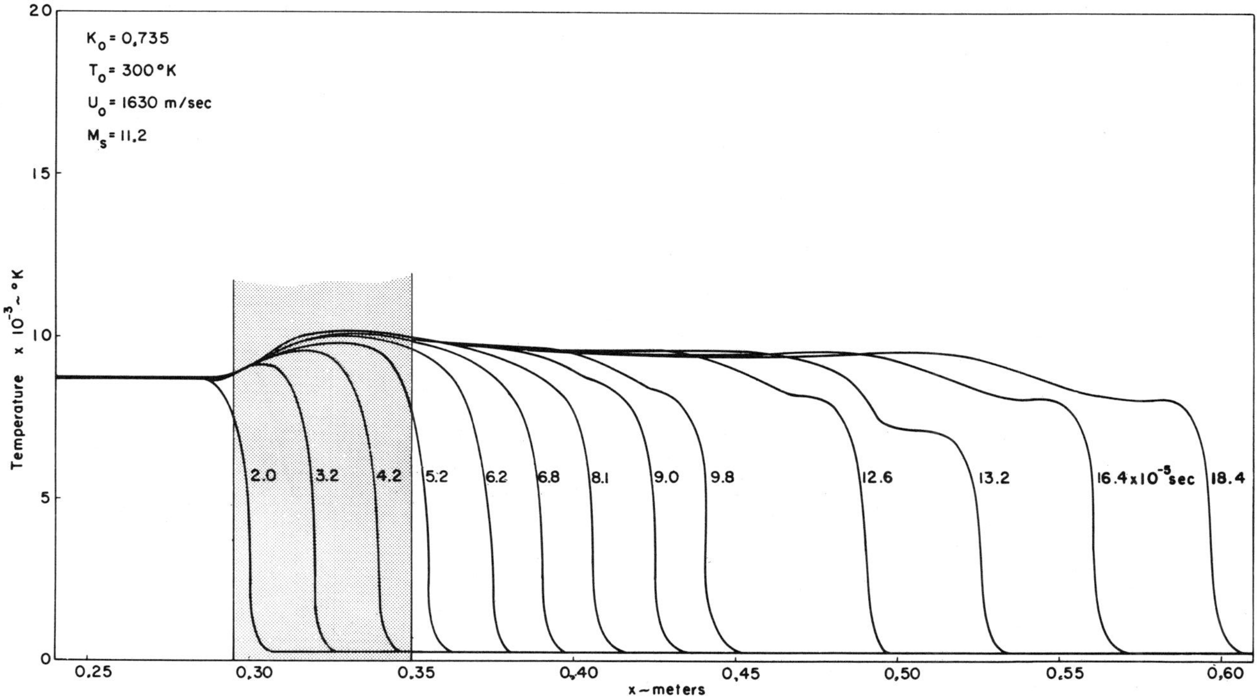

Figure 7
Temperature distribution resulting from interaction of the flow behind the shock wave with a linear MHD generator. Ionization energy included.

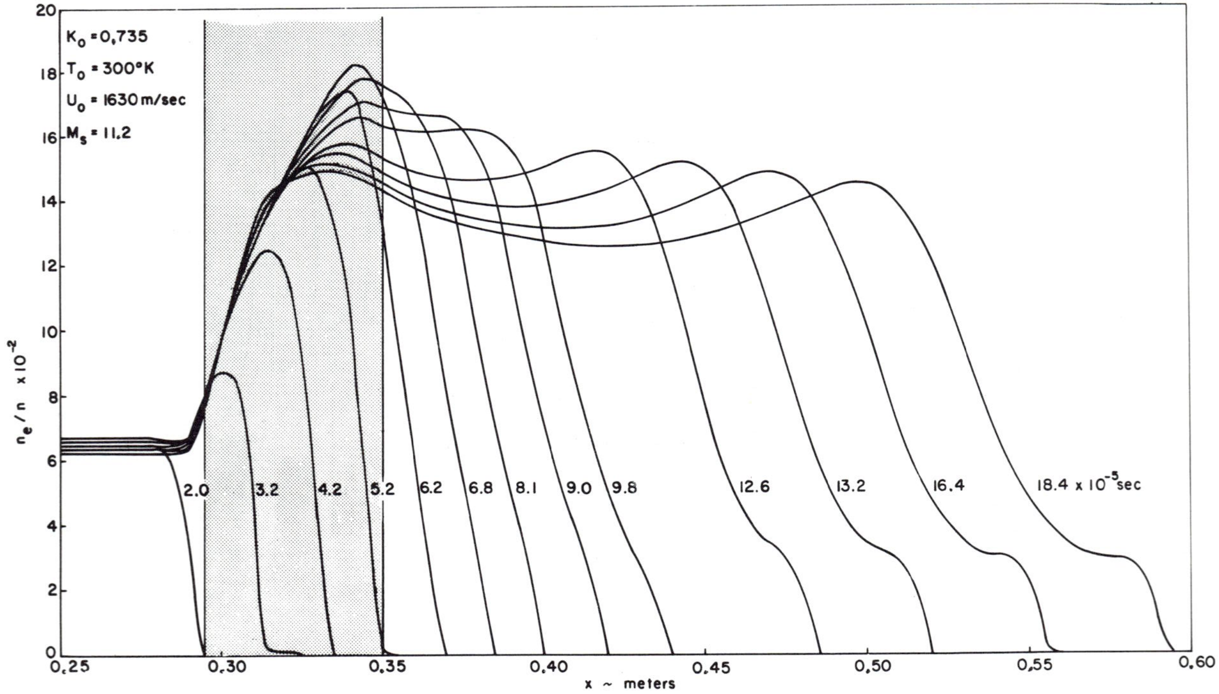

Figure 8

Degree of ionization resulting from interaction of the flow behind the shock wave with a linear MHD generator. Ionization energy included.

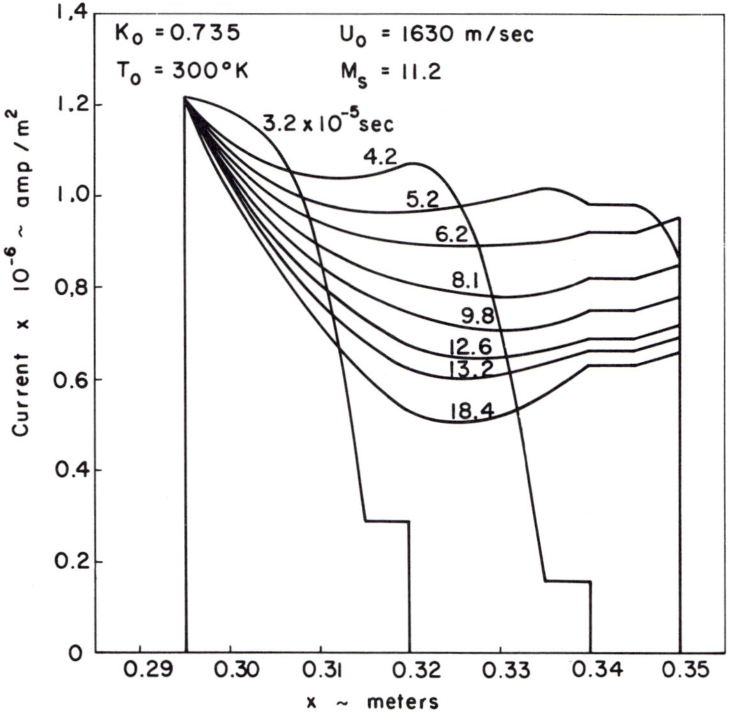

Figure 9 Electric current distribution resulting from interaction of the flow behind the shock wave with a linear MHD generator. Ionization energy included.

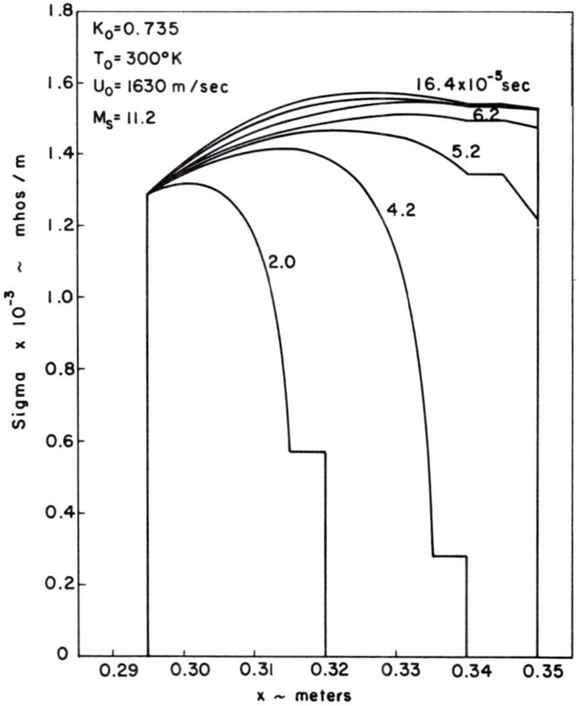

Figure 10 Distribution of the electric conductivity resulting from interaction of the flow behind the shock wave with a linear MHD generator. Ionization energy included.

INTERACTION OF AN INCIDENT SHOCK-TUBE FLOW WITH AN ELECTROMAGNETIC FIELD: PART II - EXPERIMENT

B. ZAUDERER and E. TATE
Space Sciences Laboratory
General Electric Company
King of Prussia, Pennsylvania

An experimental study of the interaction of a high conductivity shock tube flow with a one-dimensional magnetohydrodynamic channel was performed. The aim of the study was to obtain sufficient diagnostic measurements of the gas dynamic, plasma and electrical properties of the interaction process to allow a detailed comparison to be made with a time dependent theory. This theory, which is based on the Lax-Wendroff method, includes the incident shock front and contact surface trajectories and it can predict the formation of secondary shocks in the interaction region. For simplicity the experiments were performed in xenon under thermodynamic equilibrium conditions and at Hall parameters less than one in the interaction region. Reasonably good agreement was obtained between theory and experiment. Exact agreement could not be obtained due to difficulties in theoretically reproducing the initial experimental plasma conditions. Both the theory and the experiments showed that a secondary shock wave is formed as the contact surface enters the interaction region. It was also shown that although the gas slug length was five times the magnetohydrodynamic channel length, the flow was non-steady in the interaction zone.

I. INTRODUCTION

Due to its ability to produce high electrical conductivity and high velocity gas flows, the shock tube has been used extensively to study the phenomena occurring in a magnetohydrodynamic channel flow. However, the test time, which is of the order of hundreds of microseconds, in pressure driven shock tubes, is relatively short. Thus in using the shock tube one must first ascertain whether the hot gas flow behind the shock is of sufficient length to achieve quasi-steady-state conditions in the magnetohydrodynamic channel. The study of the magnetohydrodynamic channel can be divided into three major subject areas. The first one concerns the plasma properties in the channel; the second, the electrical characteristics of the channel; and the third, the interaction of the gas dynamic forces with the electromagnetic (Lorentz) forces in the channel. The plasma properties and the electrical characteristics depend primarily on the ionization relaxation time which is of the order of tens of microseconds. Thus in these two cases, quasi-steady-state conditions can be attained in a shock tube.[1,2] However, the interaction of the Lorentz and gas dynamic forces has a relaxation time of the order of the acoustic gas velocity, which is of the order of the test time. Thus, the interaction process in the shock tube will be generally non-steady.

This paper presents the results of a study of the performance of a magnetohydrodynamic channel under conditions where the Lorentz force was of the order of the gas dynamic forces. This problem has been the subject of a number of previous experimental and theoretical investigations. The earliest experimental investigation was by Patrick and Brogan[3] who used a circumferential and a disc magnetohydrodynamic channel configuration. Pain and Smy[4] used the linear channel configuration in their shock tube experiment (Figure 1). These studies[3,4] were performed under conditions where the Lorentz force was of sufficient strength to produce a secondary shock inside the magnetohydrodynamic channel. The diagnostics were limited to streak camera measurements of the wave processes and to electrode current measurements. Based on the observation that the secondary shock in the channel was stationary, the authors[3] concluded that the flow in the channel was quasi-steady and they used a steady-state one-dimensional theory to analyze the interaction. In another shock tube study, McChesney and Abdel-Ghany showed from static pressure measurements in the magnetohydrodynamic channel that the shock tube flow was non-steady. Nevertheless, McChesney et al also used a steady-state theory to analyze the interaction process. Klingenberg[6] studied primarily the electrode conduction mechanism and the shock wave trajectories in the magnetohydrodynamic interaction zone. Zaitsev et al[7] performed a detailed investigation of the plasma state and the wave trajectories in the interaction zone. However, they did not measure the electric field and current distribution. Recognizing the unsteadiness of the interaction process, other authors[8,9] have used the method of characteristics to analyze the shock wave flow magnetohydrodynamic channel interactions. This latter method has two major drawbacks. To simplify the analyses, it is assumed that γ, the ratio of specific heats, is constant in the channel. This assumption is generally incorrect for moderate to strong interactions. In addition, in a decelerating magnetohydrodynamic channel flow the compression waves in the channel coalesce into a shock wave at relatively moderate interaction levels. Under these conditions, the method of characteristics fails.

The aim of the present experimental study was to obtain sufficient diagnostic measurements of the electromagnetic-gas dynamic interaction in the channel to allow a detailed comparison to be made with a rigorous, time-dependent, one-dimensional theory. The details of this theory, which is based on the Lax-Wendroff technique, are given in a companion paper[10] and they will not be repeated here. One of the objects of both studies was to determine under what conditions a steady state would be achieved in the magnetohydrodynamic channel. Since the primary interest centered on the gas dynamic-Lorentz force interaction, the experiments were performed under thermodynamic equilibrium plasma conditions and at Hall parameters of less than one in the magnetohydrodynamic channel. The presence of large Hall parameters and non-equilibrium plasma conditions produces effects in the channel which complicate the interpretation of the Lorentz force effects[1,2]. In Part II, the experimental method and the diagnostic techniques are given. The choice of the test gas conditions and the experimental results will be presented in Part III of this paper. Finally, the results will be compared to numerical computations obtained by using the theory of reference 10.

II. EXPERIMENTAL DIAGNOSTICS

The study was performed in the previously described[1,2] 4.85 cm square, pressure driven shock tube. The magnetohydrodynamic channel (Figure 1) had 30 flush-mounted copper electrode pairs. The axial electrode pitch divided by the channel width was 0.2. The ratio of conductor width to insulator width on the electrode wall was 0.33. The effective electrode area was 4.5 cm^2. The channel was operated in the generator mode with five of the 30 electrode pairs connected to electrically isolated load resistors. This short generator length of 5 cm was dictated by the relatively short (25 cm) hot gas slug length in this series of experiments. It was desirable to maximize the ratio of the gas slug length to the generator channel length in an attempt to attain steady state gas dynamic conditions inside the generator channel. Any further reduction in the generator length below five electrode pairs (5 cm.) would have resulted in large three-dimensional electrical end-effects. With large end-effects the assumption of one-dimensionality made in the theoretical analyses[10] would be questionable. All the experiments were performed at a magnetic field strength of 2.15 Webers/m^2. In each experiment, identical load resistors, R_L, were placed at each of the five electrode pairs. Experiments were performed with R_L values of .1 Ω, .23 Ω, .43 Ω, 1.6 Ω and open circuit.

The following diagnostic methods were used: A continuous writing, streak camera was used to record the incident shock wave and contact surface trajectory and to detect the possible existence of secondary shock waves inside the generator region. The current at each of the five electrode pairs was measured. Static pressure measurements were obtained with piezoelectric pressure transducers, flush mounted on one of the two channel insulator walls at the 10th and 20th electrode position and at locations 40 cm upstream and 40 cm downstream of the electrode region. Continuum radiation measurements were made at the 7th and 21st electrode position. This radiation is proportional[1] to $n_e^2 T_e^{-.5}$, where n_e is the electron density and T_e is the electron temperature. Measurements of the induced electric field, E_y, which is mutually perpendicular to the axial gas velocity, U_x, and the transverse magnetic field, B_z, were obtained with pairs of metal probes which protruded from one of the insulator walls at the 3rd, 14th and 25th electrode positions. To obtain a complete axial distribution of the above mentioned measurements it was necessary to repeat an experiment at a specific set of conditions a number of times. During each repetition a different set of five electrode pairs was connected to the load resistors. This method of obtaining the axial distribution of the various measured parameters is the primary source of data scatter in the experiments, since it was not possible to exactly reproduce the initial shock Mach number of 11.2 to better than 2% from experiment to experiment. The electrical conductivity of the hot gas slug was measured by the Lin method[11].

III. EXPERIMENTAL RESULTS AND DISCUSSION

A. Initial Plasma Properties

All the experiments were performed at one set of initial gas conditions. The test gas was xenon at an initial static pressure of 2.73 torr. The incident shock Mach number, Ms, was 11.2 in the region immediately upstream of the magnetohydrodynamic channel. There was considerable attentuation of the incident shock wave. 1.5 meters upstream of the electrode region Ms was 12. The value of Ms of 11.2 represents the average Mach number in a one meter section of the shock tube having the magnetohydrodynamic channel at its center. Due to the short gas slug length ($\approx$25 cm) Ms of 11.2 was used to compute the gas properties immediately behind the shock. For Ms = 11.2 and P_1 = 2.73 torr, the computed equilibrium gas properties are: gas pressure = .62 atm; gas temperature = 8760°K; gas velocity = 1630 m/s; electron density, $n_e = 2.76 \times 10^{16}$ cm^{-3}, local Mach number $\approx$1.9. The computed electrical conductivity obtained with the Shkarofsky method[12] is 2200 Mhos /m. The measured σ reached a peak value of 1500 Mhos /m immediately behind the shock front.

It subsequently decreased linearly to 900 M hos/m at the driver-driven gas interface. The decrease in σ is caused in part by the effect of radiation cooling of the gas. This can be seen in Figure 2, which shows the measured change in n_e as a function of time behind the arrival of the shock front at the measuring station. Following a 15 μs ionization relaxation period in which n_e rises from zero to a maximum, n_e decreases continuously with increasing distance behind the shock. The upper theoretical curve in Figure 2 was obtained by using the Kramer-Unsold theory[13]. The lower curve, which approximately fits the experimental points, is empirical. It was obtained by multiplying the constant in the Kramers-Unsold theory by 2.5. The discrepancy between the theory and the experiments can be attributed to the effect of shock attenuation which was not considered in computing the radiation cooling of the plasma. Assuming that the maximum n_e in Figure 2 corresponds to the value at Ms = 11.2 ($2.76 \times 10^{16} cm^{-3}$), one obtains from Figure 2 an experimental n_e at the end of the test time of $8.60 \times 10^{15} cm^{-3}$. Assuming thermodynamic equilibrium during the gas cooling this final n_e corresponds to a gas temperature of about 7600°K and a computed σ of 1330 Mhos/m. The measured electrical conductivities at the beginning and end of the test time are both 68% of the theoretical values. Considering the approximations used in the Shkarofsky theory and the possible errors occurring in the calibration and data analyses of the conductivity probes, the above agreement is good. In the data reduction, the experimental σ values were used to compute the generator performance. The above plasma properties satisfy the two conditions for this study which were mentioned in the Introduction. They are: For all current levels in the generator the electron temperature is approximately equal to the gas temperature and the Hall parameter is less than one.

To obtain a quantitative comparison between the theory[10] and the experiments in the magnetohydrodynamic interaction region it was necessary to include in the theoretical analyses the initial gas conditions upstream of interaction region. Accordingly in the theoretical analyses an empirical constant was used for the radiation cooling term which fit the experimental n_e variation in Figure 2. The computations showed that for Ms = 11.2 and P_1 = 2.73 torr, the degree of ionization, α, decreased from 0.057 behind the shock to approximately 0.04 5 cm downstream of the shock. α then increased linearly to about 0.06 at the contact surface. An explanation for this behavior, which is sharply at variance with the experiments, is not known. It is felt that it arises from the imposed shock front and contact surface trajectories in the analyses which may differ from the experiment trajectories. Due to this behavior of the upstream electron density an absolute comparison of the n_e increase in the generator and of the current variation in the generator was not made.

B. Magnetohydrodynamic Channel Performance

1. Electrical Characteristics. Figure 3 shows a series of oscilloscope traces of electrode current, I_y, measurements as a function of time. Figure 3A shows current traces obtained when the electrode load resistance, R_L, was equal to or greater than 0.23 Ω. The non-horizontal baseline in the oscilloscope trace is caused by electromagnetic pickup in the measuring circuit from the pulsed magnetic field. Figure 3A shows that for $R_L \geq .23\ \Omega$, the current is approximately independent of time. Figure 3B shows four electrode currents for the experiments in which R_L was equal to 0.1 Ω. In this case, I_y decreases with time, i.e., with increasing distance behind the shock front. In Figure 4 the measured electrode currents are plotted as a function of the electrode position. The current was approximately equal at all the electrodes for $R_L \geq .23\ \Omega$. For $R_L = .1\ \Omega$, I_y decreased with increasing x and with increasing time.

Figure 5 shows the axial distribution of the average transverse electric field, E_y, for the various current levels in the generator. In general, there was a monotonic variation in E_y, which ranged from zero to 10% during the test time. In the E_y measurements where the variation exceeded 10%, an error flag has been shown at that point in Figure 6. One notes from Figure 6 that at open circuit (i.e., $I_y = 0$), the measured field is within 10% of the theoretical $U_x B_z$ field of 3500 V/m (U_x = 1630 M/s, B_z = 2.15 W/m^2). For $I_y \leq$ 420 A, (i.e., $R_L >$ .23 Ω), E_y is approximately constant in the current carrying electrode region, whereas for $I_y >$ 420 A, there is some variation in E_y in this region. There exists some current fringing at the ends of the five electrodes as can be seen from the upstream E_y measurements. This end effect increases from one electrode width (1 cm.) at I_y = 75 A to three electrode widths at I_y > 420 A. One also notes that far downstream of the current carrying electrode region, where current fringing effects should be very small, E_y does not return to its initial open circuit value. The downstream E_y measurements are equal to the local $U_x B_z$ and they can be used to obtain the gas velocity distribution in this region. As was noted in Section II, due to the limited number of E_y probes the measurements were obtained in a repetitive series of runs. Since exact reproduction of the initial conditions was not possible from run to run the E_y data show some scatter in the region downstream and upstream of the current electrodes. Nevertheless, the data clearly show that with increasing Lorentz force (i.e., increasing I_y), the downstream gas velocity is reduced. Finally, it should be noted that E_y exceeds the value of $I_y \cdot R_L/D$, where D is the channel width. The difference, which equals 500 to 900 V/m, is due to the equivalent electric field arising from the electrode voltage losses. This electrode voltage loss, for fixed B_z, depends on I_y.

In comparing the present experiments to the theory of reference 10 the measured E_y values for $0 < x < 10$ cm. in Figure 5 were used as input conditions in the analyses. In the $R_L = .1\Omega$ experiment, E_y ranges from 1800 to 2100 V/m in the current carrying electrode region. The corresponding electrode current decreases from an average of 600 A. at the first electrode to an average value of about 450 A. at the fifth electrode (Figure 4). The current also decreases with time at any electrode position (Figures 4 and 5). The theoretical current is in approximate agreement with this result in that it decreases from an average of 600 A. at the first electrode to 360 A. at the fifth electrode. On the other hand in the $R_L = .23\Omega$ experiment the measured current is constant at about 420 A. at all five electrodes. The theory for this case shows that I_y decreases from 300 A. to 180 A. between the first and fifth electrode. The theory shows[10] that the above behavior of I_y is a consequence of Lorentz force effect on Ohm's law. For the present conditions Ohm's law is given by:

$$J_y = \sigma (E_y - U_x B_z)$$

For the above experiments the Lorentz-force-induced flow deceleration exceeds the compressional gas heating and consequent σ rise. Hence, the theoretical J_y decreases in both experiments. The fact that the experimental J_y is constant in the $R_L = .23\Omega$ experiment indicates that σ and U_x compensate each other. The discrepancy between theory and experiment in the $R_L = .23\Omega$ case is due to the higher initial n_e distribution used in the theory. (See Section III A). Thus, the initial σ and hence the $J_y B_z$ force is greater in the theory than in the experiment. In the $R_L = .1\Omega$ experiment the Lorentz force is sufficiently large that the initial n_e distribution has a lesser effect on the current distribution.

2. Plasma Properties. Figure 6 shows the electron density deduced from the radiation measurements as a function of time behind the incident shock for the $I_y = 420$ A. (i.e., $R_L = .23\Omega$) experiment. The initial electron density upstream of the five current carrying electrode region, as given in Figure 2, was normalized to 1 in plotting Figure 6. The relative n_e increase at the first, second and fourth of the five electrode pairs is shown. These measurements were performed with both monochrometers and the error flags in the figure show the range of measurements obtained with them. Most of the scatter is attributable to the difficulty in exactly determining (within .5 cm.) the point in the channel on which the monochrometer focused. The theoretical analyses of this experimental run show qualitatively the same order of magnitude electron density increase as the experiment. The discrepancy between theory and experiment is again attributed to the difference in the initial n_e distribution in the two cases. (See Section III A.)

Only isolated measurements of the electron density for the $R_L = .1\Omega$ experiments were taken. However, it is estimated the relative increase in n_e is the central region of the generator ranged from a factor of 1.5 at the shock front to a factor of 5 in front of the contact surface. The theoretical results for this case show that n_e increased from a factor of 2 to 3 inside the generator.

3. Gas Dynamic Characteristics. This section discusses the influence of the magnetohydrodynamic interaction on the gas dynamic properties of the flow. Figure 7 gives the shock Mach number downstream of the generator section as a percentage of its upstream value, versus the interaction parameter, $J_y B_z \cdot L/\rho U_x^2$. L is the generator length and ρ is the gas density. The measurements were obtained from the streak camera photographs and from shock velocity measurements with piezoelectric transducers located upstream and downstream of the electrode region. Both methods are in agreement. The results show that although the interaction parameter is fairly large, the incident shock velocity is reduced by less than 18%. No secondary shocks were detected on the film inside the generator interaction zone. However, it was observed that when the contact surface reached the interaction region it divided into two wave fronts (see Figure 8) which propagated downstream. The pressure measurements, to be given below, indicate that the front wave is a shock. The theory[10] also predicts shock formation at the contact surface for the $R_L = .23\Omega$ and $.1\Omega$ experiments. The theoretical analyses for these two cases show that the incident shock is not attenuated in traversing the generator, whereas Figure 7 shows that the measured attenuation is 10% and 17% respectively. The theoretical gas properties behind the transmitted shock indicate that the shock attenuates. A possible explanation for the zero theoretical attenuation is that the shock velocity is obtained by graphical construction and it is thus more subject to error. For the $R_L = .23\Omega$ experiment, where $J_y B_z \cdot L/\rho U^2 = .35$, one obtains from Figure 7 a transmitted Mach number of 11.2 x .89 which equals 10. The theoretical gas pressure and gas velocity behind a Mach 10 shock are, for $P_1 = 2.73$ torr, .48 atm. and 1410 M/sec. The measured pressures ranged from .32 to .48 atm. The local values of U_x deduced from the downstream Ey measurements in Figure 5 ranged from 1430 to 1520 M/sec. The theoretical value for this case obtained from Reference 10 was 1460 M/sec. for the velocity and .55 atm. for the pressure. In the $R_L = .1\Omega$ experiments the theoretical U_x, based on the transmitted shock Mach number of 9.3, was 1270 M/sec. The U_x values deduced from the E_y measurements ranged from 1320 to 1220 M/sec., and the measured static pressure behind the transmitted shock was .48 atm. For this case the theory[10] gives a U_x value of about 1350 M/sec, and

a pressure of .48 atm. Thus the shock and gas velocity and gas pressure measurements far downstream of the interaction zone are consistent. They are also in reasonable agreement with the theoretical values.

The gas velocity inside the interaction region, i.e., inside the generator, can be deduced from Ohm's law, since J_y, E_y and B_z are measured functions of x and time. σ can be determined from the spatial and temporal n_e measurements. One could, therefore, construct the spatial and temporal distribution of U_x inside the generator. However, due to the various sources of error which were previously enumerated in connection with the J_y, E_y, σ, n_e, and B_z measurements no attempt was made to obtain such a description of U_x. It was found with the use of Ohm's law that U_x ranged from 1400 to 1500 m/sec. in the $R_L = .23\,\Omega$ experiments and from 1100 to 1200 m/sec. in the $R_L = .1\,\Omega$ experiments. The theoretical values[10] corresponding to the $R_L = .23$. ranged from 1200 to 1300 m/sec. and in the $R_L = .1\,\Omega$ case they ranged from 850 m/sec. to 1000 m/sec. The theoretical values are about 15% below the experimental values. The lower theoretical U_x values account for the reduction in the theoretical current below the experimental values, as noted in Section III-B-1.

Figure 8 shows three typical oscilloscope traces of the static pressure in the channel as a function of time for $R_L = \infty$ (Figure 8 A), $R_L = 8.6\,\Omega$ (Figure 8 B) and $R_L = .1\,\Omega$ (Figure 8 C). At open circuit ($R_L = \infty$), the pressure rises only slightly with increasing distance behind the shock front. For weak interactions, where $J_y B_z / \rho V^2 << 1$ (Figure 8 B), a monotonic, moderate increase in pressure occurs with increasing time behind the shock front. For strong interactions, where $J_y B_z \cdot L/\rho U^2 \approx .5$ (Figure 8 C) the pressure changes considerably during the test time. Prior to the arrival of the contact surface a further pressure peak occurs which falls off rapidly as the contact surface enters the generator section. This pressure peak is predicted by the theory[10]. It is caused by the formation of a secondary shock wave as contact surface enters the generator section. Figure 9 is a cross plot of the pressure measurements as a function of axial distance in the test section, for the $R_L = .23\,\Omega$ experiment. The data points show the pressure at any position X in the test section at zero, 50 μs., 100 μs., and 150 μs., after the passage of the shock front through the position X. The following features in these figures are of interest: One, the pressure in the channel is clearly unsteady. Thus, the assumption of steady state gas dynamic conditions made in previous studies is not justified, Two, the maximum pressure at any position reaches a maximum at the contact surface. For t > 150 μsec. the pressure decreases with increasing time (See Figure 8 C). This pressure overshoot is due to the previously mentioned secondary shock which forms as the contact surface enters the interaction region. (See Reference 10.) The theoretical pressure for t = 150 μsec.

behind the shock is not shown in Figure 9 because numerical difficulties occurred in this region. (See Reference 10). Three, one notes that as the shock front enters the generator the pressure behind the shock (points labelled "X" in Figure 9) decreases below its upstream value. Only for $t > 50$ μsec. does the pressure in the interaction zone exceed its upstream value. This reduction in pressure is caused by the attenuation of the shock front. The experimental points for $t = 0$ μs. lie below the theoretical curve for $t = 0$ μs. This is probably caused by the neglect in the theory of the 15 μs. ionization relaxation region. The gas in this region is not affected by the electromagnetic field. Hence, its pressure drops below the theoretical values due to the expansion waves which propagate downstream from the conducting gas slug. Four, on comparing the theoretical curve in the central region of the gas slug (i.e., 50 μs. $< t <$ 100 μs.), one notes that the agreement is reasonably good inside the interaction region. Downstream of the interaction region the experimental pressure is higher. It is felt that this discrepancy is also related to the above-mentioned numerical difficulties near the contact surface. The pressure data for the $R_L = .1\ \Omega$ experiment exhibited the same features as in Figure 9 and they will not be reproduced.

CONCLUSIONS

The present experimental study has shown that to obtain a complete description of the interaction process in a one dimensional magnetohydrodynamic channel, diagnostic measurements of the gas pressure, gas velocity, shock velocity, electron density, electric fields and generated currents are necessary. It was shown that measurements of the shock velocity in the test section are insensitive indicators of the details of the interaction process.

Reasonably good agreement was obtained with a one dimensional time dependent theory[10] which included the shock front and contact surface trajectory in the analyses. It is believed that the discrepancies between this theory and the experiments are due to the difficulty in prescribing a contact surface trajectory which is in accord with the experimental one. Accordingly, the initial upstream electron density used in the theoretical analyses was different than the experimental value. This discrepancy accounts for much of the deviation between the theory and experiments. As predicted by the theory, it was observed that a secondary shock wave was formed as the contact surface entered the interaction region.

Both the theory and the experiments show that, although the hot gas slug length is five times the generator length, the gas dynamic properties of the flow are non-steady. Thus to simulate a steady state magnetohydrodynamic device in a shock tube longer test times are required.

ACKNOWLEDGEMENTS

This work was supported in part by the Office of Naval Research, Power Branch.

REFERENCES

1. Zauderer, B., "Experimental Study of Non-Equilibrium Ionization in a Linear MHD Generator", AIAA Journal 6 701 (1968).
2. Zauderer, B. and Tate, E., "Electrical Characteristics of a a Linear, Non-Equilibrium, MHD Generator", AIAA Journal 6 1685 (1968).
3. Patrick, R. M. and Brogan, T. R., "One Dimensional Flow of an Ionized Gas in a Magnetic Field", J. Fluid Mech. 5 289 (1959).
4. Pain, H. T. and Smy, P. R., "Experiments on Reflected Shock Waves in a Plasma Flow-Magnetic Field Interactions", Brit. J. Appl. Phys. 17 1585 (1966).
5. McChesney, M. and Abdel-Ghany, S. M., "Experiments in a Shock Tube Magneto-Fanno Process", in International Symposium on Prop. and Appl. of Low Temperature Plasmas, Moscow, U.S.S.R., July 1965.
6. Klingenberg, H., "Arc Phenomena and Gas Dynamic Effects Due to Interaction of Shock Waves with Magnetic Fields", Z. Naturforschg. 23a 1929 (1968).
7. Zaitsev, S. G., Lazareva, E. V., Chebotareva, E. I., and Chekalin, E. K., "Structure of a Supersonic Flow of Conducting Gas in a Transverse Magnetic Field", in Electricity from MHD (International Atomic Energy Agency, Vienna, 1968) Vol. II, p. 751. In Russian.
8. Rosciszewski, J. and Oppenheim, A. K., "Shock Interaction with an Electromagnetic Field", Phys. Fluids 6 689 (1963).
9. Johnson, M. R., "Shock-Induced Flow Interactions with Transverse Magnetic Fields", Phys. Fluids 10 539 (1967).
10. Rosciszewski, J. and Gallaher, W., "Shock-Tube Flow Interaction with Electromagnetic Field: Part I - Theory" in Proceedings 7th International Shock Tube Symposium (University of Toronto, Toronto, Canada, 1969).
11. Lin, S. C., Resler, E. L., and Kantrowitz, A., "Electrical Conductivity of Highly Ionized Argon Produced by Shock Waves", J. Appl. Phys. 26 95 (1955).
12. Zauderer, B., "Measurement of Electrical Transport Properites in a Shock Tube" in Proceedings of the Seventh International Conference on Phenomena in Ionized Gases (Gradevinska Knyiga Publishing Co., Belgrade, Yugoslavia, 1966), Vol. II, p. 99.

13. McChesney, M. and Al-Attar, Z., "Continuum Radiation Losses in Shock Heated Argon", J. Quant. Spectros. Radiat. Transfer. 5 553 (1965).

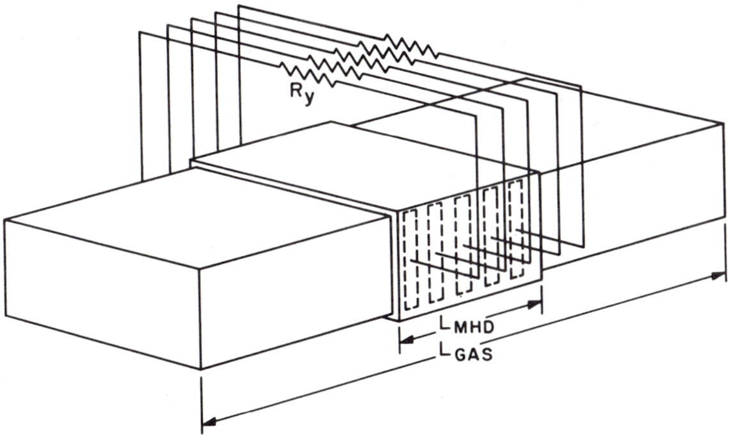

Figure 1
Schematic of the linear magnetohydrodynamic generator

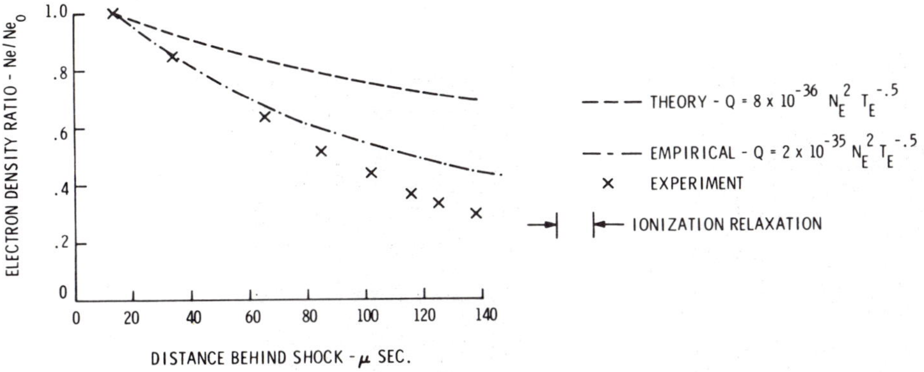

Figure 2
Theoretical and experimental radiation cooling behind a $M_s = 11.2$, $p_1 = 2.73$ torr shock wave in xenon. The empirical curve is obtained by multipyling the constant in the Kramer-Unsold theory by 2.5. The units of the radiation cooling term, Q, are joules $m^{-3}sec^{-1}$.

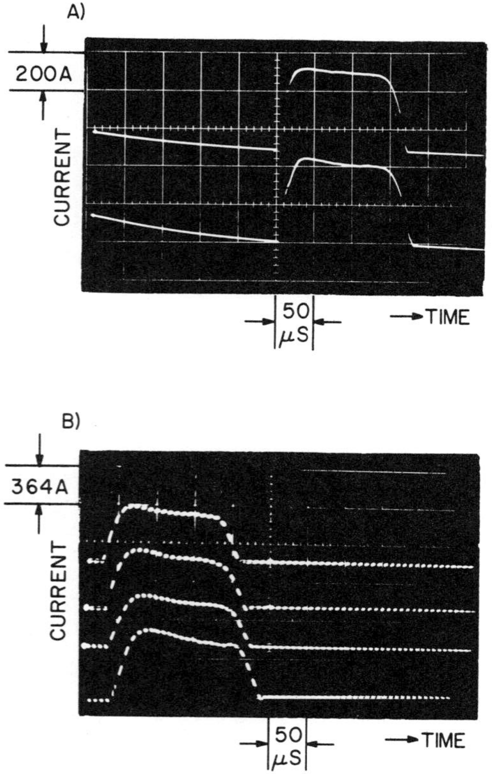

Figure 3
Oscilloscope traces of the electrode currents. In A, R_L was 0.23 Ω, and in B, R_L was 0.1Ω

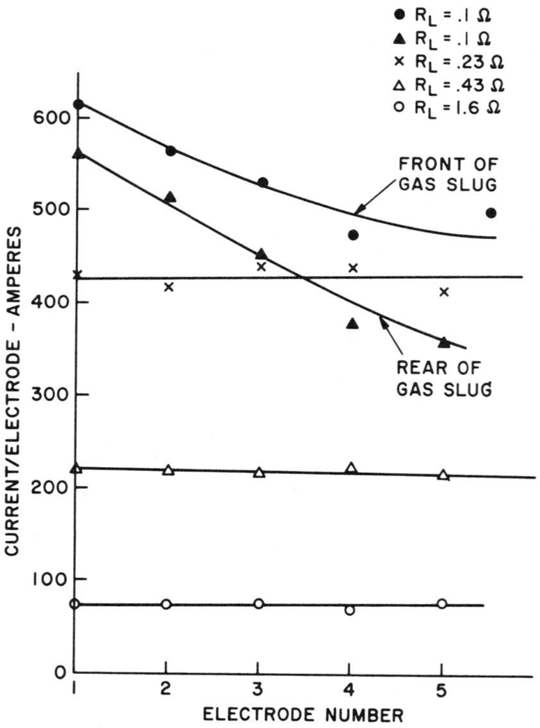

Figure 4
Electrode current, I_y, at each of the five electrodes. For $R_L \geq .23\Omega$, I_y was independent of time. For $R_L = .1\Omega$, the current at each electrode decreased from a maximum at the shock front to a minimum at the contact surface

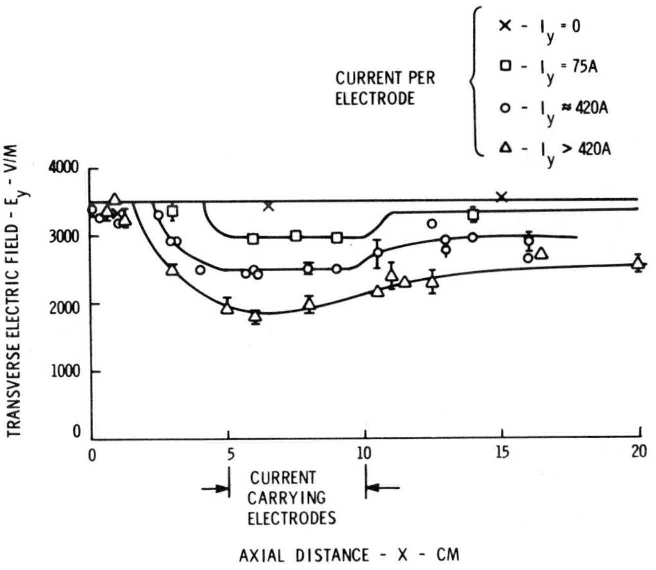

Figure 5
Transverse electric field, E_y, as a function of distance in the channel for various load resistances

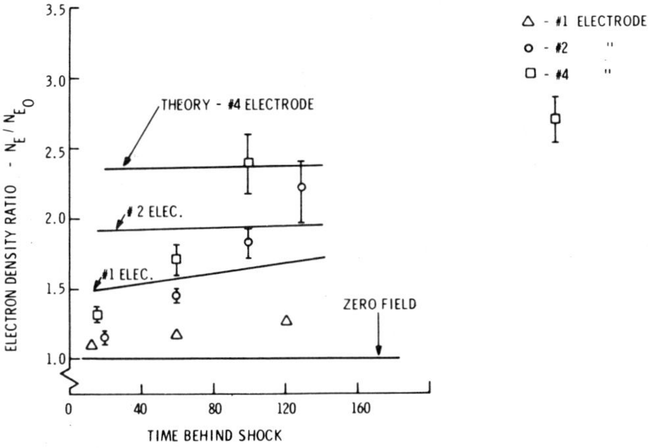

Figure 6
Theoretical and experimental electron density in the generator at various electrode positions, as a function of time behind the shock. The upstream electron density has been normalized to one

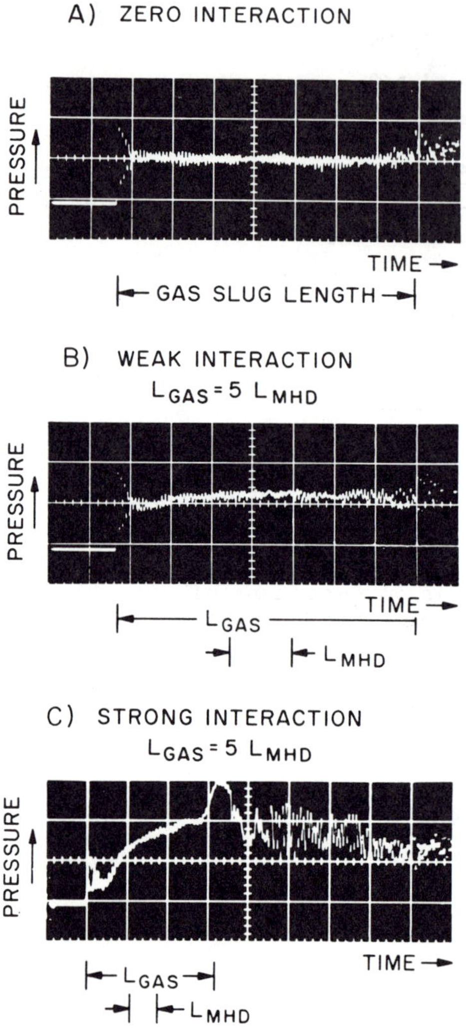

Figure 8
Static pressure at one point in the interaction zone as a function of time after the passage of the shock. In A, B_z was zero, in B, R_L was 1.6Ω, and in C, R_L was 0.1Ω. The gas slug length and the interaction (generator) length are shown on the time axis

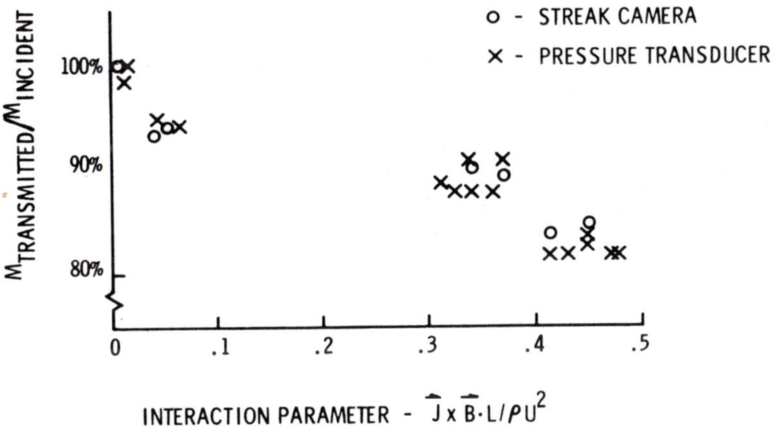

Figure 7
Incident shock front attenuation in traversing the magnetohydrodynamic generator section, as a function of the Lorentz force interaction parameter

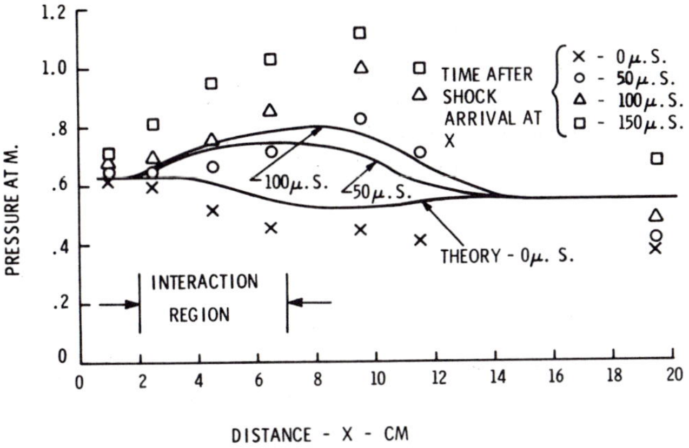

Figure 9
Static pressure distribution in the test section for the R_L = .23Ω experiment, showing the theoretical and experimental pressure at various times after the passage of the shock front at position "X"

VALIDITY OF THE LOCAL THERMAL EQUILIBRIUM ASSUMPTION IN ELECTROMAGNETIC SHOCK TUBES

T.N. LIE, M.J. RHEE
The Catholic University of America
Washington, D.C.

E.A. MCLEAN
Naval Research Laboratory
Washington, D.C.

SUMMARY

The electron temperature has been determined using two techniques simultaneously, one which is strongly dependent on the local thermal equilibrium assumption, namely, the ratio of two spectral line intensities He II 4684 Å and He I 5876 Å, and one which is independent of the local thermal equilibrium assumption, namely, the light scattered by the plasma from a ruby laser beam. These measurements were made on a reflected and incident shock wave produced in an electromagnetic T-tube which had been prefilled with helium or hydrogen to a pressure of 0.5 to 1 torr. Within the experimental uncertainty of ± 15 percent, the two electron temperatures agreed, indicating that the assumption of local thermal equilibrium used in the spectroscopic temperature determination is a reasonable approximation. Validity criteria for local thermal equilibrium given previously are discussed and a justification for the experimental result is made.

I. INTRODUCTION

Spectroscopic techniques have been used for some time to measure the temperature and particle densities in shock tubes[1-4]. These methods provide a convenient and accurate measurement of plasma parameters without disturbing the plasma. However, to use many of these methods, such as absolute line intensities, ratios of line intensities, continuum intensities, line-to-continuum ratios, and various others, one needs to assume that the velocity distribution function of the electrons is Maxwellian, the populations of the excited states are governed by a Boltzmann distribution, and the distribution of atoms, ions and electrons obeys the Saha equation, i.e., a condition of local thermodynamic (thermal) equilibrium (LTE) exists.

This assumption of LTE which is so essential to spectroscopic diagnostics has been questioned by a number of authors in applications where the plasma is very transitory in nature, as is the case in the electromagnetic shock tube.

For example, Eckerle and McWhirter[5] reported that the LTE condition does not exist in the plasma ($N_e \sim 10^{17}$ cm^{-3}) created by a reflected shock wave in a T-tube containing a 99 percent H_2 and 1 percent He mixture at 0.5 torr. In this experiment, temperatures measured using the helium ion to neutral line ratio (a technique which depends on the LTE assumption) disagreed by a factor of two with a measurement of the hydrogen Balmer discontinuity, a method which had little dependence on the LTE assumption. They concluded that the reason for this is the departure from LTE which is caused by the fact that the recombination rate is too small for the plasma to follow rapid changes in temperature.

Also Isler and Kerr[6] using a linear electric shock tube (a modified conical discharge tube) spectroscopically observed incident shock waves in helium at 1.0 torr ($N_e \sim 10^{17}$ cm^{-3}, $T_e \sim 3$eV) and found the ratio of doubly-charged to singly-charged ions to always be greater (up to 10^5 times) than the value expected for ionization equilibrium (or LTE). They suggested that the plasma is initially heated by the discharge and that this plasma decays or recombines as it propagates down the shock tube.

On the other hand, there have been a number of papers[1,3,4] in the literature which used spectroscopic techniques to determine plasma temperature and which showed very good agreement between various methods. This would seemingly imply that the local thermal equilibrium assumption is valid; however, it is obvious that the ultimate proof of the validity of LTE has to be made by an experiment which measures the electron temperature using both a technique which depends on the local thermal equilibrium assumption and one which does not depend on this assumption.

It is the purpose of this experiment to compare the measured electron temperature of a helium plasma using the line intensity ratio of He II 4686 Å to He I 5876 Å, which has a sensitive dependence on LTE, and the electron temperature measured using a laser scattering technique, which is independent of LTE. The plasma is produced in a T-type electromagnetic shock tube. The light scattering technique only depends on the assumption of the Maxwellian velocity distribution of the electrons, which is verified in this experiment by the measured spectral profile of the scattered laser line.

Also, to better understand the plasma used for this comparison, additional spectroscopic measurements of the electron temperature and density were made at positions closer to the arc. Finally to see if one could detect the high T_e in hydrogen as described in reference (5), a laser scattering measurement of the temperature was made in hydrogen.

II. VALIDITY CRITERIA FOR LTE

Griem[7] and others[8,9] have derived useful validity formulas which can be applied to plasmas to determine when the local thermal equilibrium assumption can be used, provided approximate electron densities and temperatures of the plasmas are known. First, we consider the validity criterion for LTE in a homogeneous, transient plasma. It is necessary to ascertain if the relaxation for the establishment of LTE is smaller than the time variations of the densities and the temperature. The plasma that is being observed is relatively homogeneous perpendicular to the axis of the tube, i.e., along the line of sight (except the thin boundary layer) and has only gradual variations along the tube axis. Since collisional ionization occurs mainly via intermediate excited states (stepwise ionization), the ionization relaxation time is determined by the slowest process in the excitation chain, i.e., the excitation from the ground state to the upper state of the resonance line. This relaxation time, including a factor for the equilibrium fraction of atoms or ions in the next stage of ionization, is given by Griem[7] as

$$\tau_1^{z-1} \approx \frac{1.1 \times 10^7 z^3 N^z}{f_{21} N_e (N^z + N^{z-1})} \frac{E_2^{z-1}}{z^2 E_H} \left(\frac{kT}{z^2 E_H} \right)^{1/2} \exp\left(\frac{E_2^{z-1}}{kT} \right) \text{ sec.} \qquad (1)$$

Here, z is the number of effective charges acting on the radiating electron, (z = 1 for an atom, z = 2 for a singly charged ion, etc.), E_H is the ionization energy of hydrogen, f_{21} is the absorption oscillator strength of the resonance line, N_e is the electron density, E_2^{z-1} is the excitation energy of the resonance line, and N^z, N^{z-1} are the particle densities of the indicated degree of ionization.

A calculation of the relaxation time to establish first ionization yields times (10^{-8} - 10^{-7} sec) usually much shorter than the plasma lifetime (approximately 1 μsec). Also, collisional-radiative calculations[10,] show that the bound states of He I are in local thermal equilibrium; however, the question is whether population of excited states of He II and the number of He III ions is in LTE relative to the ground-state population of He II. A calculation of the relaxation time for second ionization yields values ranging from 0.3 μsec for $N_e \gtrsim 10^{17}$ to several microseconds for the lower values of N_e. However, the plasma lifetime may be lengthened because the excited particles are carried along with the plasma. Generally, then it must be said that the Griem LTE criterion is only marginally met in the higher density cases, and is not fulfilled in the lower density cases.

In addition to the first validity criterion for LTE described above, it is necessary to fulfill the criterion for LTE in a homogeneous and time-independent plasma. If the resonance line is optically thick, as is usually the case for such plasmas, this criterion[7] for the electron density is

$$N_e \gtrsim 9 \times 10^{16} \left(\frac{E_2^{z-1}}{E_H} \right)^3 \left(\frac{kT}{E_H} \right)^{1/2} \text{cm}^{-3} , \qquad (2)$$

with the same meaning for the symbols as in the previous equation. If this criterion is fulfilled, the population of the ground state is in LTE with respect to higher levels to about 10 percent. A typical example of the electron densities necessary for complete LTE to within a factor of 2 or 3 (rather than the 10 percent) for ionized helium with optically thick resonance lines at kT = 4 eV would be $N_e \simeq 10^{17} \text{cm}^{-3}$. Our experimental electron densities are approximately within this range. A similar result has also been obtained by Mewe[11].

III. EXPERIMENT

A schematic diagram of the experimental setup is shown in Fig. 1. A quartz T-tube is shown with a side arm extending about 12 cm from the electrode. The electrodes are connected to a 2.2 μF capacitor through a spark-gap switch. The charging voltage was fixed at 15 kV. All of the observations of the laser scattered light were made at a position 3 mm beyond the end of the side arm, and for the reflected shock measurements the reflector was located 6 mm beyond the end of the side arm. The gas used is helium or hydrogen at a filling pressure of between 1/2 and 1 torr.

To make the 90-degree laser scattering measurements, a Q-switched ruby laser (~100 MW) was directed perpendicular to the T-tube side arm. The laser beam was focused to a spot 1 mm in diameter at the center of the side arm. The synchronization of the laser pulse has been made using a light guide and a delay generator. The spectral observations of the scattered laser light were made perpendicular to the plane of the diagram (Fig. 1) using a mirror to deflect the scattered light into a monochromator. To check reproducibility the laser output was monitored shot-to-shot with a photodiode which was calibrated previously using a calorimeter. The duration of the laser pulse was about 30 nanoseconds and the total energy in a laser pulse was 2-3 J. The stray laser light scattered from the walls of the system was practically eliminated by use of a number of diaphragm baffles and a beam dump composed of a Brewster-angle window with an absorber as shown in the diagram.

Simultaneously with the laser scattering measurements, observations of the spectral line intensities of He II 4686 Å and He I 5876 Å were made at the same position of the T-tube using a polychromator. Also the spectral line intensities and line profiles were determined at positions closer to the arc. The polychromator was calibrated using a standard tungsten ribbon-filament lamp.

IV. RESULTS AND DISCUSSION

The laser scattering data was analyzed using the standard theory[12,13]. However, the analysis was greatly facilitated by making use of a report by W.H. Kegel[14] which gives a calculated set of graphs of the spectral distribution of light scattered by a plasma. The spectra were calculated with the following data: Wavelength $\lambda = 6943$ Å (ruby laser light) and scattering angle $\theta = 90°$. The variable parameter was $\alpha = \lambda/(4\pi D_e \sin\theta/2)$, where D_e is the electron Debye length. Both the electron temperature and density are then determined by comparing the experimentally obtained spectral profile with these calculated profiles. (By using this method of analysis, one does not have to make an absolute calibration of the intensity of the scattered spectral line to determine the electron density.)

A typical result of the scattering measurement of a reflected shock wave in helium at 0.5 torr is given in Fig. 2. The experimental points are obtained from a shot-to-shot scan of the laser line profile, each point being the average of about 5 shots. The solid curve is a theoretical curve taken from the Kegel[15] report which best fits the experimental data. The temperature to which the line profile corresponds is 4.5 eV and the electron density is 1.4×10^{17} cm^{-3}. Also shown on the same plot is the profile which would be given by a plasma with the same density but a temperature of 8.6 eV, or about 2 times the measured temperature. One notes that the 8.6 eV curve falls well outside of the error brackets of the experimental data.

Figure 3 gives a summary of the values of temperature and density measured using these two techniques at various times behind the reflected shock plasma. The time refers to the time after the peak of the continuum signal. One sees that the spectroscopic temperatures determined from the ratio of He II 4686 Å to He I 5876 Å are consistently lower than the values obtained from laser scattering; however, they do agree within the overlapping error brackets.

Since one of the cited papers[5] used a mixture of 99 percent hydrogen and 1 percent helium in their study, an experiment was done to observe reflected shock waves in hydrogen at a fill pressure of 0.5 torr. Special care was taken to measure the temperature very soon (0.1 μsec) after the density peak. The scattering profile is shown in Fig. 4, yielding an electron temperature of 4.6 eV $\pm$ 10 percent and an electron density of 1.7×10^{17} cm^{-3}. It is noted that the temperature is more than a factor of two less than that reported previously[5].

By removing the reflector from the end of the T-tube, observations could be made on the incident shock wave in a similar manner to those for reflected shock waves. Using helium at a fill pressure of 1 torr, other conditions remaining the same, the electron temperature and density were measured behind the shock front in a similar fashion to that used for reflected shock waves in helium. Figure 5 is the profile of the scattered laser line showing the experimental points plotted on the theoretical (solid) curve that best fits the data and the dashed curve represents the theoretical profile if the density is the same and the temperature is 9 eV, or approximately 2 times greater. The electron temperature is found to be 3.9 eV and again one can see that the 9 eV curve falls well outside of the error brackets of the experimental data.

Figure 6 gives the comparison between the laser scattering temperature and the spectroscopic temperature for the helium incident shock waves, and also the laser scattering electron density at various times after the peak of the continuum intensity signal. One notes that the spectroscopic temperatures are consistently lower than the values obtained from laser scattering; however, again the error brackets for the two measurements overlap.

Although the difference between the laser scattering and the spectroscopic temperature is about the same as the estimated error, the deviation between the two could be due to (1) the departure from LTE in the plasma, and (2) the cold boundary layer at the tube wall[15]. The former explanation would make the spectroscopic temperature lower than the scattering temperature as obtained in this experiment and so would the latter explanation, because the value obtained from scattering is essentially the temperature of a small portion of plasma (0.1 mm x 3 mm) at the tube axis whereas the spectroscopic temperature is the average value along the entire tube diameter (2.8 cm). In order to observe this contribution to the temperature difference, the electron temperature near the tube wall was measured by the laser scattering method for a reflected shock wave in helium. The result is shown in Fig. 7. One does see, as expected, a tendency of a lower temperature near the tube wall although the difference is only a few percent and is within the estimated error. The remaining deviation between the two measured temperatures could be due to the departure from LTE.

To ascertain if this plasma is a recombining plasma in which the heating has been done solely in the arc region, spectroscopic measurements were made of T_e and N_e at various distances from the discharge. The temperature was measured from the ratio of He II 4686 and He I 5876 line intensities and the densities determined from the Stark profiles of these two lines using the theoretical calculations of Griem, et al[16]. Table I gives the distance z from the arc, the optical layer length l, and the values of T_e and N_e. The value of N^+ was calculated from the Saha and quasi-neutrality equations using measured values of N_e and T_e. With this information one can calculate the intensity of the He II 4686 line and compare the intensity with the measured absolute intensity of the same line. This was done and is tabulated. It is seen that the intensity at 2 cm from the arc, where there is still a distinct filament, gives a good comparison of measured and calculated values; whereas, at 6 cm and 12 cm the calculated value differs from the measured value by a factor of 2 - 3. This indicates that there could, indeed, be a deviation from LTE populations further down the tube of maybe a factor of 3; however, this would lead to an error in the temperature measurement of only $\sim$ 15 percent.

It is interesting to note that the peak temperature near the arc is only slightly higher[17] (4.1 eV) than the peak temperature 12 cm from the arc (3.7 eV). Isler and Kerr[6], and Cloupeau[18] maintain that the radiation from the luminosity fronts comes from gas excited primarily in the discharge rather than gas which had been heated by a shock wave. Because of this, Isler and Kerr feel the temperature is decreasing too rapidly to permit the ion concentrations in the plasma to relax to their steady-state values. The data presented here corresponds to a lower temperature in the arc region than originally expected[1] and the decay in temperature is not as fast as one would expect even for a simple adiabatic expansion. Thus, it appears that the plasma here is not only heated by the arc, but is also continually heated by the shock wave that forms at the leading edge of the plasma, although for these conditions the shock wave does not separate from the driver gas[19].

V. CONCLUSIONS

The data presented here indicate that the population densities of the excited states of the singly-charged helium ion deviates from LTE populations by perhaps a factor of 3; however, the temperature as measured from the ratio of He II 4686 to the He I 5876 line intensities deviates by only about 15 percent if a few percent correction is made for the temperature gradient near the walls of the tube. This is well below the factor of two discrepancy claimed by others[5]. The calculated relaxation times for excitation and ionization of the singly charged helium ion in certain cases are somewhat longer than the lifetimes of the observed plasmas, which would corroborate the validity criterion for LTE[8]. Using the same method for comparison as Isler and Kerr[6], the ratio of doubly-charged to singly-charged ions was found to be the same as the LTE ratio within a factor of two, well below the several orders of magnitude which they claim.

In conclusion, then it can be said that the helium plasmas produced in the T-tube ($N_e \sim 10^{17}$ cm^{-3}, $T_e \sim 4$ eV) are in approximate LTE, so that spectroscopic measurements (spectral line ratios) of the electron temperature should be accurate to $\sim$15 percent. However, measurements of absolute transition probabilities, etc. which depend very sensitively on population densities of the excited states should be made with caution, preferably doing the experiment at higher electron densities than those reported here, or else making systematic correction for LTE.

ACKNOWLEDGMENT

We would like to thank Professor H.R. Griem for his interest and many stimulating discussions on this subject. His suggestions greatly aided the preparation of this paper. Also, we would like to thank Dr. R.W.P. McWhirter for his active interest and useful criticism of this work and Dr. A.W. Ali for number of interesting discussions. We are indebted to Professor C.C. Chang and Dr. A.C. Kolb for their support of this study.

This research was supported by the National Aeronautics and Space Administration under Research Grant NGR-09-005-025.

REFERENCES

1. McLean, E.A., Faneuff, C.E., Kolb, A.C., and Griem, H.R., Phys. Fluids 3, 843 (1960).
2. Wiese, W., Berg, H.F., and Griem, H.R., Phys. Rev. 120, 1079 (1960).
3. Lie, T.N., Ali, A.W., McLean, E.A., and Kolb, A.C., Phys. Fluids 10, 1545 (1967).
4. Costa, S. and Tondello, G., Ricerca Sci. 35, 949 (1965).
5. Eckerle, K.K. and McWhirter, R.W.P., Phys. Fluids 9, 81 (1966).
6. Isler, R.C., and Kerr, D.E., Phys. Fluids 8, 1176 (1965).
7. Griem, H.R., Plasma Spectroscopy (McGraw-Hill, Inc., New York, 1964), Chap. 6.
8. Wilson, R., J. Quant. Spectrosc. Radiat. Transfer 2, 477 (1962).
9. McWhirter, R.W.P., a paper included in A Survey of Phenomena in Ionized Gases (Invited Papers) (International Atomic Energy Agency, Vienna, (1969)) p. 667.

10. Bates, D.R., Kingston, A.E., and McWhirter, R.W.P., Proc. Roy. Soc. (London) A 267, 297 (1962).
11. Mewe, R., Brit. J. Appl. Phys. 18, 107 (1967).
12. Saltpeter, E.E., Phys. Rev. 120, 1528 (1960).
13. Kunze, H.J., Chapt. 9 in Plasma Diagnostics (John Wiley and Sons, Inc., New York, 1968) edited by W. Lochte-Holtgreven.
14. Kegel, W.H., Institut fur Plasmaphysik, Garching bei Munchen, Report No. IPP 6/34 (March 1965).
15. McLean, E.A. and Ramsden, S.A., Phys. Rev. 140, A1122 (1965).
16. Griem, H.R. and Shen, K.Y., Phys. Rev. 122, 1490 (1961); Griem, H.R. Baranger, M., Kolb, A.C., and Oertel, G., Phys. Rev. 125, 177 (1962).
17. Greig, J.R., Phys. Fluids in press.
18. Cloupeau, M., Phys. Fluids 6, 679 (1963).
19. Ferguson, J.B. and Barach, J.P., Phys. Fluids 12, 764 (1969).

Table I
Comparison of total line intensity of He II 4686 measured and calculated under the assumption of LTE

z	l	T_e	N_e	N^+	He II 4686 measured	He II 4686' calculated
cm	cm	eV	cm^{-3}	cm^{-3}	watts sr^{-1} cm^{-2}	watts sr^{-1} cm^{-2}
2	0.9	4.1	3.5×10^{17}	2.7×10^{17}	82	79
6	2.8	3.7	6.8×10^{16}	5.3×10^{16}	29	13
12	2.8	3.7	5.4×10^{16}	4×10^{16}	30	9

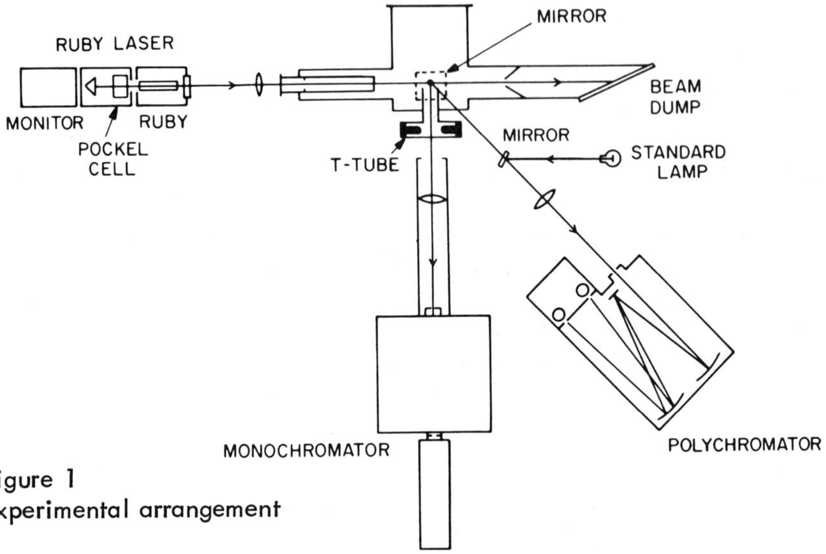

Figure 1
Experimental arrangement

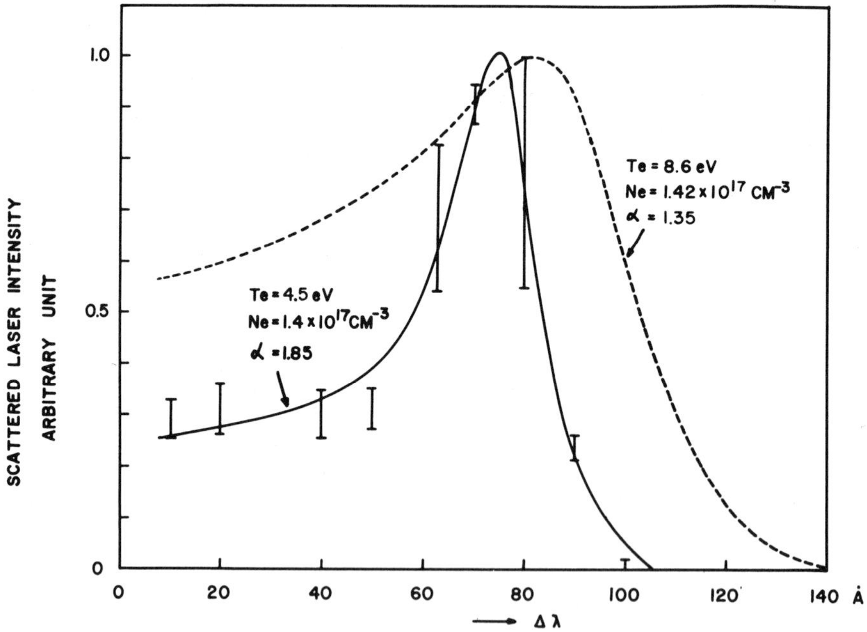

Figure 2
Spectral profile of scattered laser radiation behind the reflected shock in He at 0.5 torr. Dotted curve indicates the profile which would be given by a plasma of T_e = 8.6 eV. and N_e = 1.42 x 10^{17} cm^{-3}.

TIME	LASER SCATTERING T_e	N_e	SPECTROSCOPIC T_e
0.2 μsec	4.5 eV ± 10 %	$1.4 \times 10^{17} cm^{-3}$ ± 10 %	3.7 eV ± 15 %
0.6 μsec	4.1 eV ± 10 %	$8.5 \times 10^{16} cm^{-3}$ ± 10 %	3.5 eV ± 15 %
1.1 μsec	4 eV ± 10 %	$7 \times 10^{16} cm^{-3}$ ± 10 %	3.4 eV ± 15 %

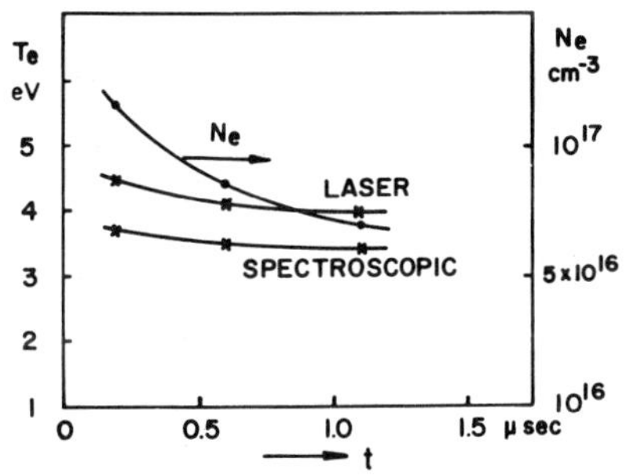

Figure 3
Electron temperature and density behind the reflected shock in He at 0.5 torr.

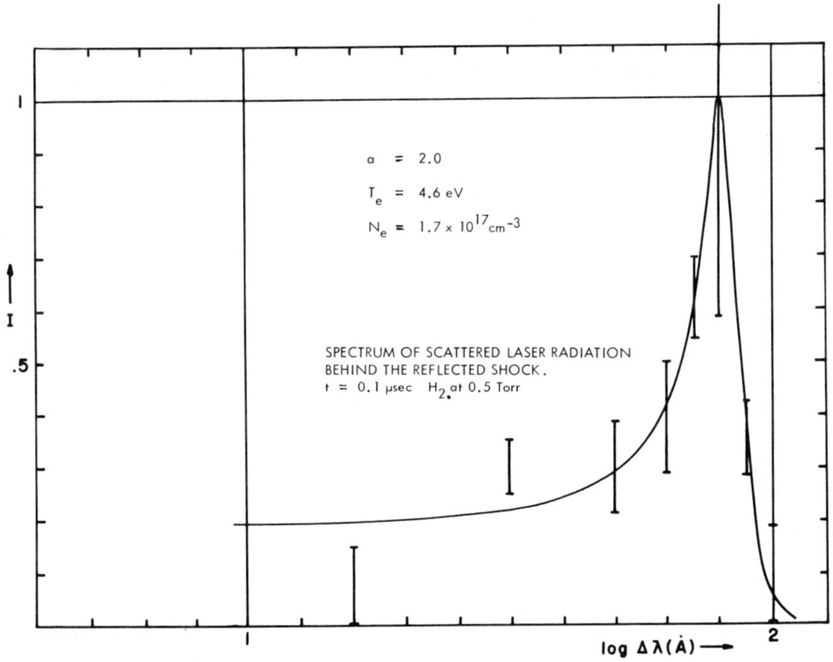

Figure 4
Spectral profile of scattered laser radiation behind the reflected shock in H_2 at 0.5 torr.

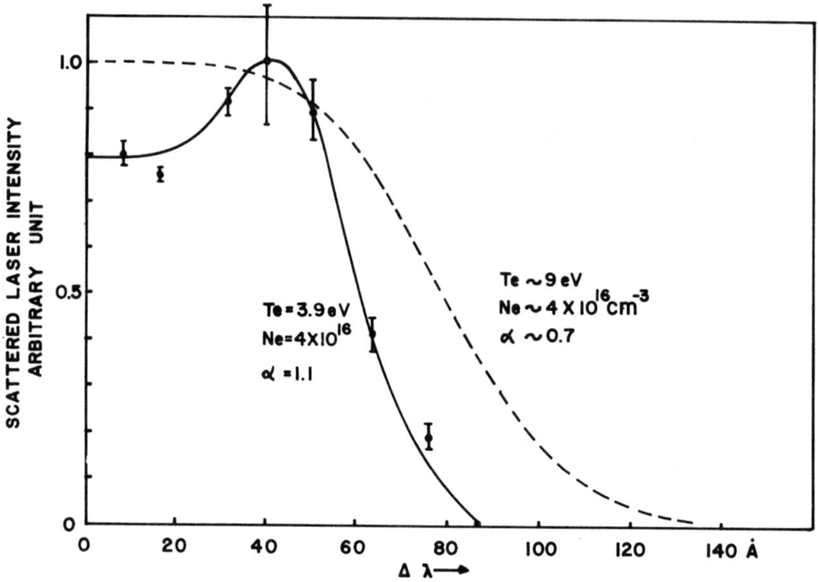

Figure 5
Spectral profile of scattered laser radiation behind the incident shock in H_e at 1 torr. Dotted curve indicates the profile which would be given by a plasma of Te = 9 eV, N_e = 4×10^{16} cm^{-3}.

TIME	LASER SCATTERING T_e	N_e	SPECTROSCOPIC T_e
0.3 μsec	4.0 eV ± 10%	$3.2 \times 10^{16} cm^{-3}$ ± 10%	3.5 eV ± 15 %
0.8 μsec	3.9 eV ± 10%	$4.0 \times 10^{16} cm^{-3}$ ± 10%	3.3 eV ± 15 %
1.3 μsec	3.7 eV ± 10%	$2.4 \times 10^{16} cm^{-3}$ ± 10%	3.2 eV ± 15 %
1.8 μsec	3.4 eV ± 10%	$2.2 \times 10^{16} cm^{-3}$ ± 10%	

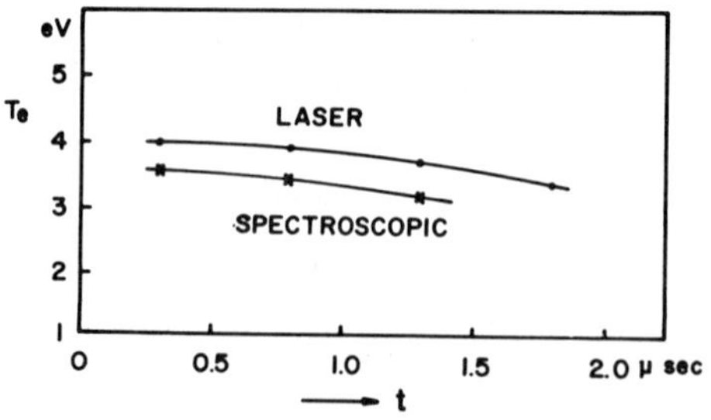

Figure 6
Electron temperature and density behind the incident shock in H_e at 1 torr.

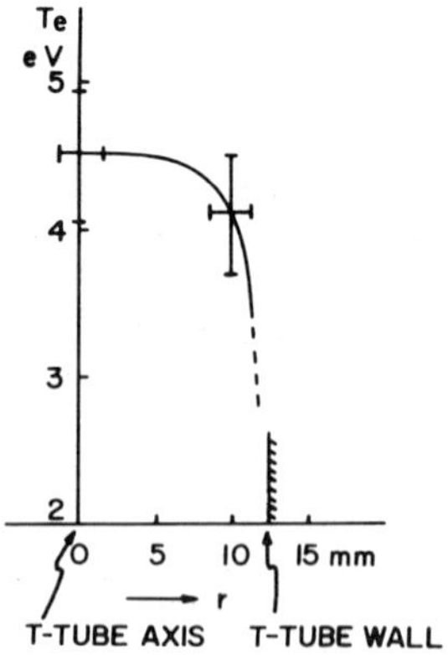

Figure 7
Temperature profile along the tube radius.

A PARAMETRIC STUDY OF THE CURRENT-SHEET SPEED IN A MAGNETICALLY DRIVEN SHOCK TUBE

C.T. CHANG and M. POPOVIĆ
Danish AEC Research Establishment
Risö, Roskilde, Denmark

U. KORSBECH
Technical University
of Denmark,
Lyngby, Denmark

SUMMARY

By varying the discharge voltage, pressure and test gas, a parametric study is made of the variation of the current-sheet speed in a shock tube with parallel-plate electrodes and a constant-voltage energy source. The experimental results are compared with theoretical predictions derived from the snow-plough model. The discrepancies are discussed, and temporary explanations are put forward.

INTRODUCTION

In an electromagnetically driven shock tube, the shock is generated by a moving magnetic piston or a current-sheet. The moving current-sheet introduces a variable impedance into the electric circuit and has a feedback effect on the driving current.

Among other investigators (Wright and Black, 1959; Dattner, 1960; Moore, 1962) Chang and Kofoed-Hansen (1968) analysed this interaction problem for a shock tube with parallel-plate electrodes driven by a constant-voltage energy source. Their analysis, based mainly on the snow-plough model (Rosenbluth, 1957), showed that the current-sheet accelerates quickly along the electrodes to a steady state. The immediate implication is that, on theoretical ground, shocks of constant speed can be formed in such a device. It is also shown that for various combinations of discharge voltage, pressure and test gas, the total current supplied to the electrodes as well as the speed of the current-sheet depend on a single dimensionless parameter only,

$$\epsilon = (R_b/R_o)^2, \tag{1}$$

i.e. the ratio between the "running impedance" R_b of the shock tube and the characteristic impedance R_o of the energy source. The "running impedance" *) of the shock tube R_b is defined as

*) The assembly of electrodes viewed as a transmission line has its own characteristic impedance, which depends on the geometry, but not the operating conditions.

$$R_b^2 = \left(\frac{\mu}{2\rho_1}\right)^{\frac{1}{2}}\left(\frac{L_1}{b}\right) V_o \,, \tag{2}$$

where μ is the magnetic permeability, ρ_1 the density of the filling gas, L_1 and b the inductance per unit length and the width of the electrodes respectively, and V_o the running voltage of the battery. At the steady state the speed of the current-sheet $\hat{u}$ is related to the driving current $\hat{I}$ and the density of the filling gas through the simple expression

$$\hat{u} = \left(\frac{\mu}{2\rho_1}\right)^{\frac{1}{2}} \frac{\hat{I}}{b} \,. \tag{3}$$

This expression is obviously just the restatement of the snow-plough assumption at the steady state.

The main assumptions used in the analysis by Chang and Kofoed-Hansen are:

(I) The total current is confined within a thin sheet.
(II) The magnetic pressure, taken as the only driving mechanism, is estimated by neglecting the end effect of the plates; thus the lines of force are considered straight and parallel, and the field is uniform behind the sheet.
(III) The gas in the volume swept by the current-sheet is completely picked up by the sheet ("snow-plough").
(IV) The impedance of the plasma is negligible.
(V) Wall friction plays no significant role.
(VI) The effect of impurity is negligible.
(VII) The processes of excitation and ionization are irrelevant.

In fact, their analysis is essentially aimed at studying the interaction between the driving current and the moving current-sheet, and is much too simple to cope with detailed phenomena actually observed. Departure from the theoretical predictions would be expected.

In the present paper we report some results obtained from an experimental set-up closely resembling the one used in the analysis cited. First a comparison between the observed data and the theoretical predictions is presented. The discrepancies are then discussed, and temporary explanations are given whenever feasible.

We would mention here that experiments with a similar arrangement of electrode configuration and energy source have been performed previously by Liebing (1963). However, his experiments were conducted in argon under fixed discharge conditions as to voltage and filling pressure; the time history and the speed of the current-sheet under various discharge conditions (as described in the present paper) were not investigated systematically. On the other hand, a similar parametric study of the current-sheet speed has been made by Keck (1964) with a shock tube having a coaxial geometry and a large radius ratio. A comparison between his result, obtained from a self-field strongly dependent on the radius, and ours, with a rather uniform field, should be of some interest.

EXPERIMENTAL CONDITIONS

The shock tube consists of two stainless-steel plane-electrodes, 40 cm long, 8 cm wide, set parallel to each other 2 cm apart.

The driving source is made up of six parallel-connected transmission lines, each line composed of thirty-three 0.5 μF/25 kV - KB Bosch Condensers; the bank has a characteristic impedance of 0.097 Ω and a running time of 18 μs. Operating at 15 kV, it reaches a peak current of 150 kA around 0.5 μs. A general view of the shock tube and the energy bank is shown in Fig. 1.

The test gases used were helium, argon and nitrogen. Measurements were taken over a range of discharge voltages V_o from 7 to 18 kV and initial pressures p_1 from 0.1 to 10 Torr. Most of the observations reported here were carried out with a magnetic probe of one mm diameter with 10 windings and a linear response up to 10 MHz. The total current supplied to the electrodes was recorded by a Rogowski coil placed at the connection between the collecting plates and the electrodes.

COMPARISON WITH EXPERIMENTAL RESULTS

(a) Total Current at the Steady State

Fig. 2 shows two oscillograms of the total current supplied to the electrodes, taken in He at a discharge voltage of 13 kV. The upper trace was taken at a discharge pressure p_1 = 10 Torr, the lower one at p_1 = 0.5 Torr. One notices that only the amplitude, not the shape, of the current trace is affected by the discharge pressure. According to the theory, the steady-state current $\hat{I}$ varies with the pressure p_1 through the parameter ϵ (eqs. (1) and (2)) as follows:

$$\hat{I} = \left(\frac{V_o}{R_o}\right) \frac{(1 + 4\epsilon)^{\frac{1}{2}} - 1}{2\epsilon} \quad . \qquad (4)$$

A quantitative comparison between the experimental result and the theoretical prediction is given in Fig. 3. One observes that the experimental data follow a curve lying above the theoretical one.

(b) The History of the Current-Sheet Front

To record the history of the current-sheet, we first check the reproducibility of the test runs; the x - t trace of the current-sheet is then constructed from the onset time of the magnetic probe signal obtained at various axial locations along the tube. Fig. 4 shows some typical curves obtained in this way in helium. Similar curves are obtained in nitrogen and argon.

(c) The Variation of the Current-Sheet Speed along the Tube

From the slope of the x - t trace of the current-sheet, its speed (at different locations) is deduced. As shown in Fig. 4 for values of ϵ covered in the experiment, the current-sheet always reaches a steady state after travelling a distance around 20 cm from the driving end of the tube. According to the theory, the current-sheet reaches the steady state only asymptotically. However, as shown in Fig. 5, numerical computation indicated that for values of $\epsilon < 0.5$ the current-sheet speed $\dot{\eta}$ at $\eta = 1.5$ differs no more than 2% from its value at $\eta = \infty$. Since η is related to the space coordinate x by $x = \eta\, L_o/L_1$, for our experimental set-up (L_o = 22.8 nH, L_1 = 2nH/cm), one would expect the current-sheet to reach its "steady-state" theoretically after traveled 17 cm along the electrodes. Comparing this with the experimental value of x = 20 cm, we see that it is about 18% longer than expected. A careful examination of the curves shown in Fig. 4 indicates that the approach to the steady state is somewhat faster for smaller values of ϵ, which is in agreement with the theoretical prediction.

At lower values of ϵ the current-sheet seems to accelerate quickly to its maximum speed of the steady state. However, entirely unexpectedly, at higher values of ϵ (for He, $\epsilon > 0.2$ approximately), the current-sheet behaves like a "blast Wave" (Kolb, 1957; Hess et al., 1961; Muntenbruch, 1967) within the first 20 cm and then proceeds onward at a constant speed. The most interesting result is that for intermediate values of ϵ the current-sheet reaches its maximum speed before the steady state, usually around 5 to 7 cm from the end-plate.

The same trend is observed in argon and nitrogen.

(d) Variation of the Current-Sheet Speed with Respect to the Parameter ϵ

For various combinations of the discharge voltage, pressure and test gas, the theory predicts that the speed of the current-sheet should depend only on the parameter ϵ, that is, at the steady state:

$$\hat{u} = (1/2)\left(\frac{R_o}{L_1}\right)\left[(1 + 4\epsilon)^{\frac{1}{2}} - 1\right]. \qquad (5)$$

Fig. 6 shows that for a given gas, within experimental scattering, the experimental points follow reasonably well a single curve which, however, is much below the theoretical one. Besides, the three gases tested, instead of all following a single curve, seem to have three separate curves, with the heavier gas Ar closer to the theoretical one than the lighter gas He.

(e) Validity of the Snow-Plough Assumption

Taking a room temperature of 300°K and using the equation of state, we can replace the gas density ρ_1 in eq. (3) by its corresponding pressure p_1. For the electrodes used in our case, the relationship between the current and the current-sheet speed at the steady state can be rewritten as

$$\hat{u} = \frac{4.29\ \hat{I}}{(p_1 M)^{\frac{1}{2}}} \qquad (3a)$$

where $\hat{u}$ is measured in cm/μs, $\hat{I}$ in kA and p_1 in microns, and M is the molecular weight of the test gas. A comparison of the experimental data with the theory is shown in Fig. 7 for both He and Ar. One observes that at lower values of $\hat{I}/(p_1 M)^{\frac{1}{2}}$ the experimental data fit reasonably well into a straight line, but with a slope much less steep than the theoretical one; at higher values of $\hat{I}/(p_1 M)^{\frac{1}{2}}$, the departure from linearity is evident.

DISCUSSION

In summary we note that, generally speaking, the experimental results agree rather well qualitatively with theoretical predictions. However, departures are observed; noticeably, the total current is much higher and the speed of the current-sheet much lower than expected (Fig. 3 and 6).

At the steady state the effective impedance of the electrodes is purely resistive, namely $L_1\hat{u}$. For some reason, if the current-sheet moves more slowly than expected, the corresponding resistance of the electrodes is smaller, and consequently the current will be higher.

In estimating the magnetic pressure in the region behind the current-sheet we have assumed that the lines of force are straight and parallel. In reality, because of the finite width of the electrodes, the lines are curved. Not only is the field strength reduced, but since the component of the Lorentz force normal to the sheet is responsible for its axial acceleration, the net result is equivalent to a reduction of the driving current and consequently a lowering of the parameter ϵ (see eq. (3), Chang and Kofoed-Hansen, 1968). The presence of the plasma resistance may in effect be considered as an increase of the battery resistance R_o, which reduces the value of ϵ. If, during the acceleration, impurities due to electrode erosion or other causes are continuously fed into the gas *), the effective value of ϵ could also be reduced through an increase of the apparent gas density. This might explain why the three gases tested do not follow a single curve, and why the experimental data of the heavier gas, argon, as shown in Fig. 6, agree better with the theoretical prediction.

---

*) According to the investigation of Chang and Mondrup (1968), the "steady-state" current-sheet speed $\hat{u}$ will not be reduced if a constant amount of impurity is introduced at the beginning of the acceleration process.

When considering the acceleration of the current-sheet, one usually assumes that all the current supplied to the electrodes participates in the acceleration process. On the other hand, our experimental results shown in Fig. 7 might indicate that only a fraction of the total current is effective. Thus for He at values of $\epsilon < 0.1$ this might amount to 65% of the total current; at higher values of ϵ the fraction of the total current that is used for the acceleration process might be even smaller. In this connection it is interesting to note that many investigators - on the bases of experiments with shock tubes of the coaxial type - have either suspected or actually observed that a portion of the total current supplied by the energy bank flows near the insulator end of the electrodes (Dattner, 1964; Keck, 1964; Lindberg and Jacobsen, 1964).

Although the above mechanism might be the main cause of the lowering of the current-sheet speed, other losses such as those caused by the friction (Thom et al., 1964) and the thermal conduction at the surfaces of the electrodes (Jensen et al., 1961), by the erosion of the electrodes and the back insulator-plate (Keck, 1964; Cormack, 1963) might also be of some importance.

The most interesting phenomenon observed in our present experiments is that under certain discharge conditions the current-sheet attains its maximum speed before reaching the steady state. The occurrence of a maximum speed of the current-sheet* near the driving end of the shock tube has been reported previously by many other investigators (Kolb, 1957; Hess et al., 1961; Matsuzaki, 1967). However, the discharge currents are oscillatory in all these experiments; the acceleration process is most likely limited to the first half cycle. The observed maximum speed of the current-sheet can be related to the maximum of the discharging current at its quarter cycle (Hess et al., 1961; Matsuzaki, 1967). The shock produced behaves rather like a blast wave, and never reaches a steady state. In comparison, the current-sheet observed in our experiments, after passing its maximum speed is first attenuated, but finally moves at a constant speed. Furthermore, at a given discharge voltage, the appearance of the maximum current-sheet speed prior to its steady state usually occurs at lower discharge pressures (see Fig. 4); oscillograms of the current probe usually indicate a splitting of the total current. Whether there is a definite correlation between these two phenomena, or it is a mere coincidence, remains to be investigated. At the moment, the main cause of these phenomena is still not clear. The presence of a large amount of impurities from the erosion of the electrode and insulator-plate materials, as suggested by Cormack (1963), might be a possible cause; inefficient mass accumulation of the magnetic piston, e.g. due to ionization relaxation etc., during the earlier phase of the acceleration process is another possibility. On the other hand, under some discharge conditions in our experiments it is not impossible that the

---

* More precisely, the luminous front in the examples cited.

driving mechanism may have been a hybrid of the magnetic pressure and the electron gas pressure (Fowler et al., 1961; Fowler and Fried, 1961). The occurrence of the maximum current-sheet speed could then be related to the breakdown conditions of the discharge.

REFERENCES

1. Chang, C.T. and Kofoed-Hansen, O. (1968), Plasma Physics 10, 137.
2. Chang, C.T. and Mondrup, K. (1968), Plasma Physics 10, 1106.
3. Cormack, G.D. (1963), Can. J. Phys. 41, 1591.
4. Dattner, A. (1960), Proc. 4th Int. Conf. on Ionization Phenomena in Gases, Uppsala 1959, vol. II, p. 1151.
5. Dattner, A. and Eninger, J. (1964) Physics Fluids Suppl 7, S 41.
6. Fowler, R.G., Paxton, G.W. and Hughes, H.G. (1961) Physics Fluids 4, 234.
7. Fowler, R.G. and Fried, B.D. (1961), Physics Fluids 4, 767.
8. Hess, H., Kischel, K., Morgenroth, H., Radmacher, K., and Seliger, K. (1961), Ann. Physik 8, 175 (Theory) and 189 (Experiment).
9. Jensen, V.O., Sillesen, A.H., Heikel Vinther, F., and Wandel, C.F. (1961), Risö Report No. 28.
10. Keck, J. (1964), Physics Fluids Suppl. 7, S16.
11. Kolb, A.C. (1957), Phys. Rev. 107, 345; also in "Magnetohydrodynamics", edited by Landshoff, Stanford Univ. Press, p. 76.
12. Liebing, L. (1963), Institut Für Plasmaphysik, Garching b. München, Report IPP 3/16; Physics Fluids 6, 1035.
13. Lindberg, L. and Jacobsen C.T. (1964), Physics Fluids Suppl. 7, S44.
14. Matsuzaki, Ri'ichi (1967), Japanese J. Appl. Phys. 6, 430.
15. Moore, T.F. (1962), Physics Fluids 5, 596.
16. Muntenbruch, H. (1968), Proc. 6 International Shock Tube Symposium, Freiburg i.B. 1967. Ernst-Mach-Institut Wissenschaftlicher Bericht Nr. 4/68, p. 327.
17. Rosenbluth, M. (1957), in: "Magnetohydrodynamics" edited by Landshoff, Stanford Univ. Press, p. 57.
18. Thom, K., Norwood, J., and Jalufka, N. (1964), Physics Fluids Suppl. 7, S 67.
19. Wright, J.K. and Black, M.C. (1959), J. Fluid Mech. 6, 219.

Figure 1
General·view of the shock tube and the energy bank.

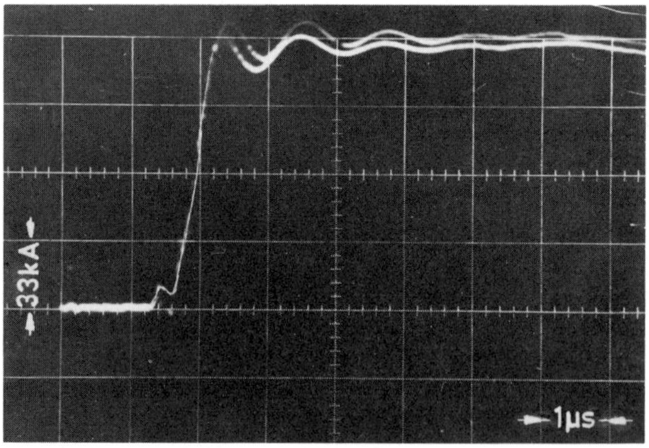

Figure 2
Oscillograms of the total current supplied to the electrodes. The upper trace was taken at a discharge voltage V_o = 13 kV and a discharge pressure p_1 = 10 torr, the lower one at V_o = 13 kV and p_1 = 0.5 torr.

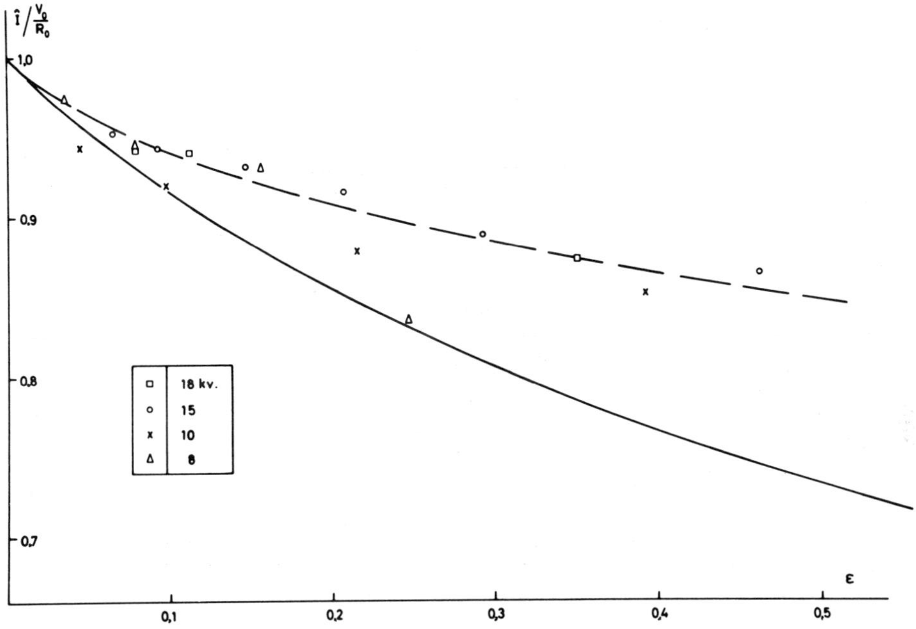

Figure 3
Variation of the total current $\hat{I}$ at the steady state vs, the parameter ϵ. The dashed curve is fitted over the experimental data, the solid curve is drawn according to eq. (4).

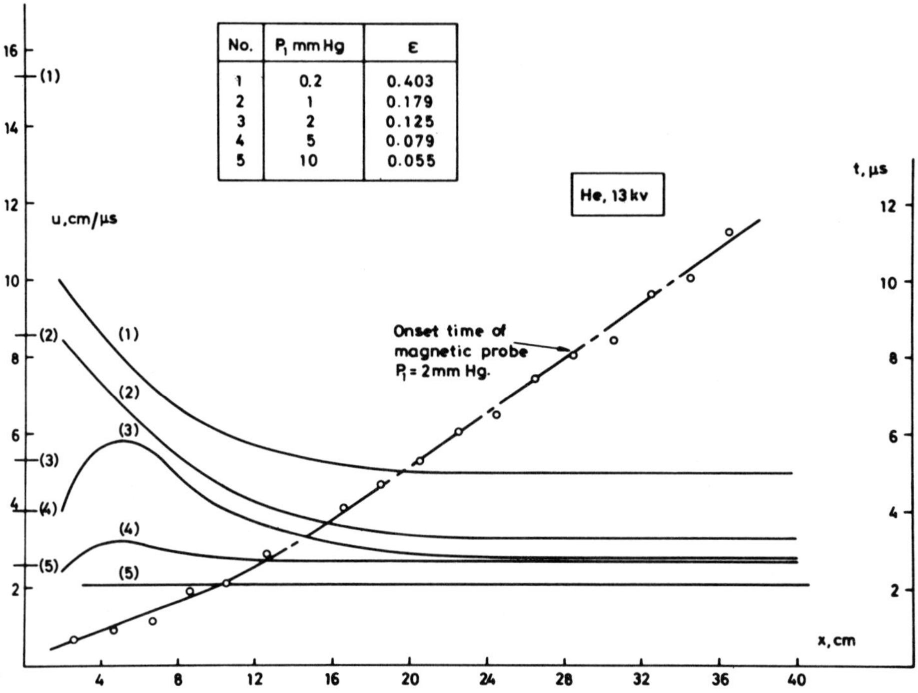

Figure 4
x-t trace of the current-sheet front and the variation of its speed along the electrodes with ϵ as the parameter. The horizontal bars on the u-axis are the theoretical steady-state current-sheet speed.

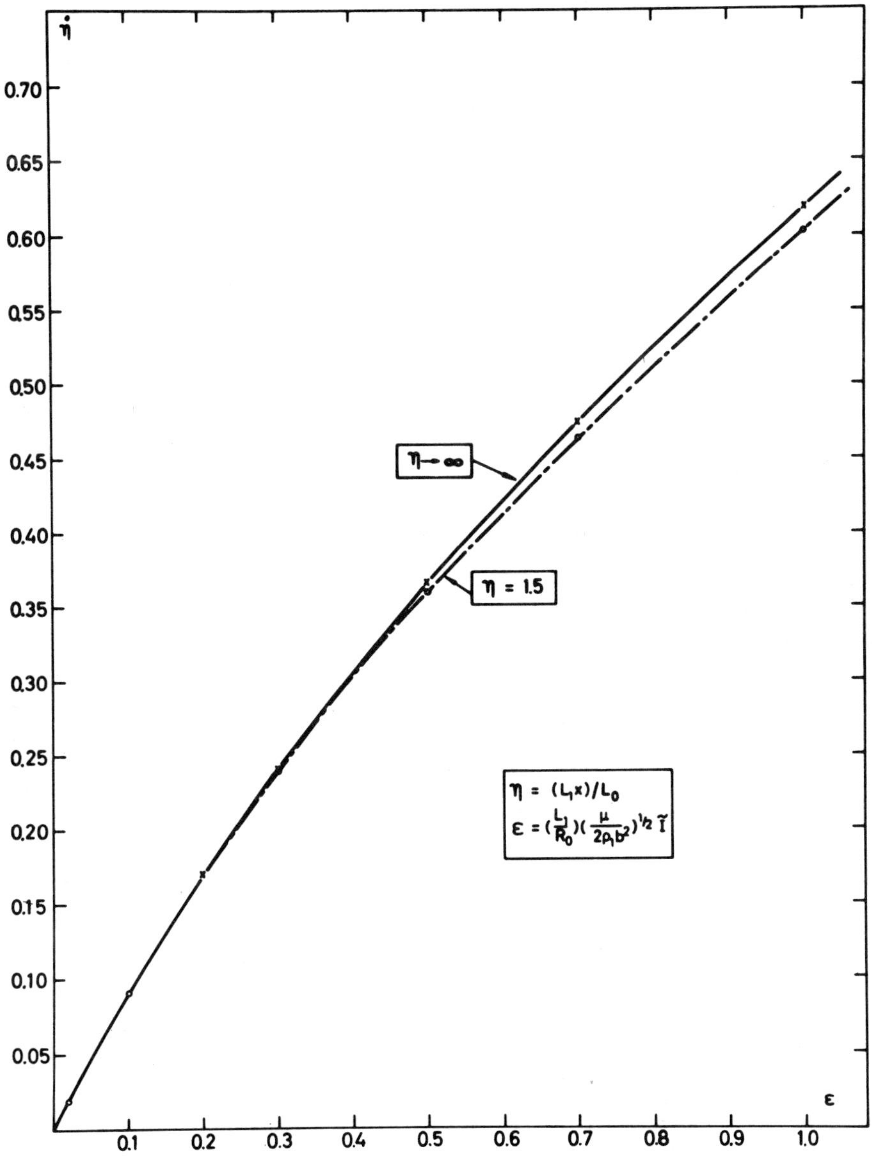

Figure 5
Theoretical variation of the dimensionless current-sheet speed $\dot{\eta}$ vs. the parameter ϵ. The solid curve is the value of $\dot{\eta}$ at the steady state, the dashed curve its value at $\eta = 1.5$.

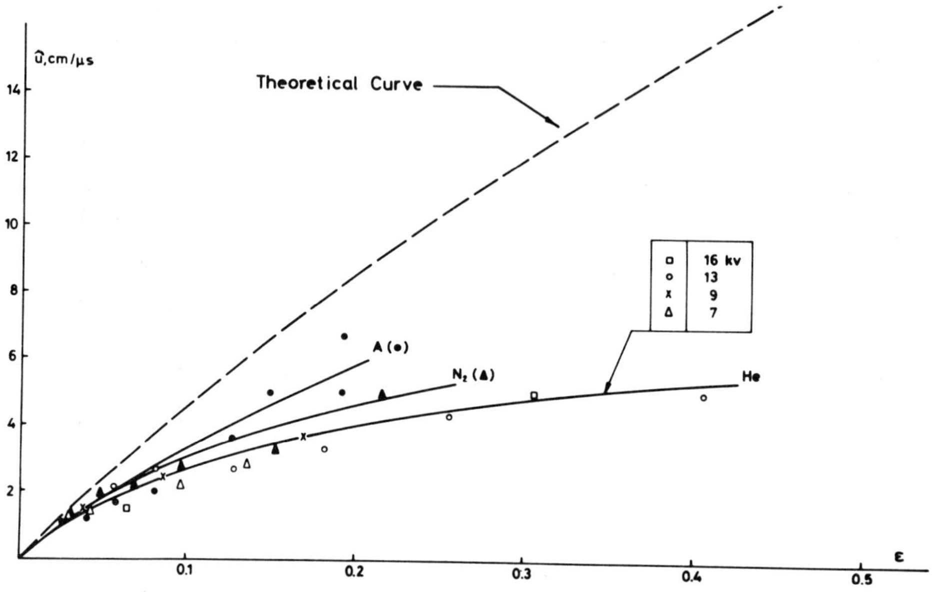

Figure 6
Variation of the steady-state current-sheet speed $\hat{u}$ vs. the parameter ϵ. The dashed curve is drawn according to eq. (5).

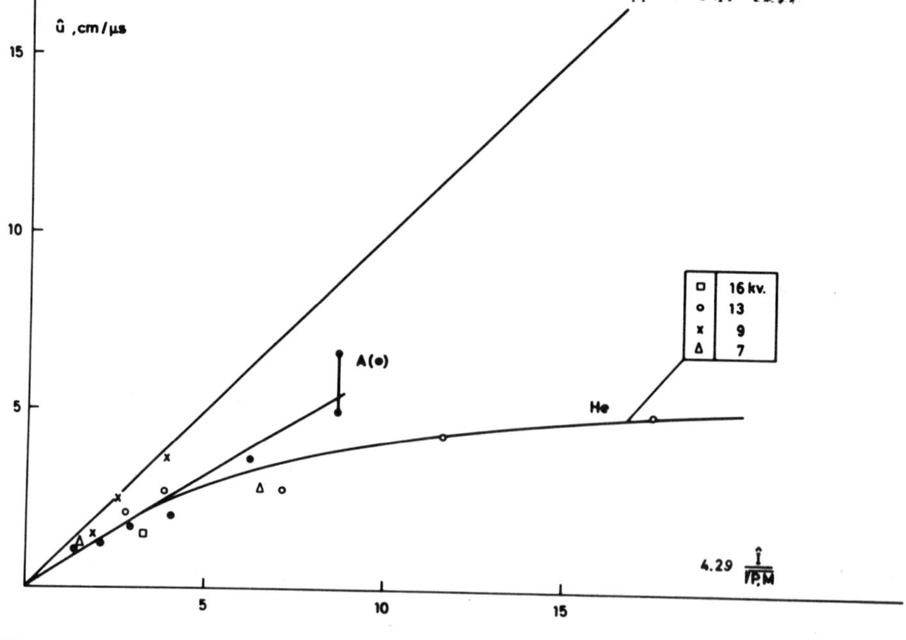

Figure 7
Steady-state current-sheet speed $\hat{u}$ vs. the parameter $\hat{I}/(p_1 M)^{\frac{1}{2}}$. The straight line is drawn according to the snow-plough model, eq. (3a).

MAGNETO-FLUID DYNAMIC STUDIES IN A CESIUM SHOCK TUBE

HARLOW G. AHLSTROM AND PHILIP A. PINCOSY
University of Washington

SUMMARY

Theoretical efforts in compressible magneto-fluid dynamics MFD, has largely been for the limiting cases of either magnetic Reynolds number, $Rm \rightarrow 0$ or $Rm \rightarrow \infty$ or for the development of fundamental solutions to the general cases. General flows with arbitrary parameters are still largely unexplored. A shock tube using cesium vapour as the driven gas has been built to perform simple but definitive experiments in compressible MFD. This shock tube incorporates a number of new features in order to produce interesting test conditions. The driven gas is pure cesium vapour at approximately one Torr. Cesium was chosen because of its low ionization potential; it is fully ionized when $M_s = 16$. The test gas; 2 ft. long, 6 in. diameter and $16,000^{\circ}K$ will provide a unique plasma for studying magneto-fluid dynamics. The aligned fields condition is obtained by placing a solenoid over the full length of the driven tube. The flow parameters will be Mach number, $M > 1$; Alfvén Mach number, $A \lesssim 1$; magnetic Reynolds number, $Rm = O(1)$; and interaction parameter. $Rm/A^2 = O(1)$. The MFD experiments and the constructional features of the shock tube are discussed.

INTRODUCTION

A large amount of theoretical work has been done in MFD:[1,2,3,4,5] however, most of the work has been done for limiting cases which remove many of the complexities or interesting flow features. Other more general treatments study fundamental solutions of the linearized continuum equations, but they have not attempted to construct observable flows.

A number of illuminating experiments have been done with conducting liquids.[6,7,8]. The experimental problems were significant, but new and interesting flow features were discovered and explained for incompressible flows. However, the experimental difficulties are even more significant for compressible flow studies. Many ingenuous methods have been devised to produce highly conducting gases for MFD studies, but in most cases the research has been sidetracked into studying the plasma production device rather than using it as a facility to study MFD flows.

The purpose of the program at the University of Washington was to build a facility which would produce a uniform slug of high conductivity plasma which could be used for fundamental studies of compressible MFD flows. Since a considerable amount of theoretical work has been done on the "aligned fields" case:, i.e., the flow and magnetic field aligned at infinity, this area was chosen as the initial area of concentration. The shock tube is a logical choice for the production of a high enthalpy, uniform slug of gas. However, because of the high ionization potential of most gases, > 10 ev, it is very difficult to achieve conditions which would be interesting for MFD studies. There are two approaches to the solution of this problem. The first is to use an explosive driver or arc heated driver; the second is to use an alkali metal vapour for the driven gas so that the ionization potential is low, e.g., $I_p(Cs) = 3.98$ eV. There are difficulties associated with either approach; however, the initial design analysis indicated that the research objectives could best be achieved with a shock tube using cesium. Previously A. Haught of Princeton[9] and J. Louis of Avco-Everett[10] have used cesium in shock tubes. Their apparatus and experience was very valuable, but the requirements for this program were sufficiently different that many additional problems had to be solved.

In order to achieve the aligned fields condition, a solenoid is wound concentric with the shock tube the full length of the driven section. In this way the plasma is produced as the shock wave is moving along the magnetic field lines and is not perturbed until it reaches the test section. This design makes the solenoid and power supply a major portion of the required equipment; however the injection of a plasma into a magnetic field is a problem in itself and the nonuniformities induced in the plasma could obscure the effects to be observed.

Shock Tube Design

The critical design requirements that had to be met in order to achieve the research objectives were:

1. Operating temperature $T_1 < 600^\circ F$
2. Degree of ionization $\alpha = 100\%$
3. Test length $\doteq$ 2.0 ft.
4. Magnetic Reynolds No. $Rm \geq O(1)$
5. Magnetic Interaction Parameter $Rm/A^2 \geq O(1)$

6. Capacitor energy storage for magnetic field < 360,000 Joules.

The operating temperature is most important because through it the initial pressure, p_1, is determined. A maximum $T_1 = 600^\circ F$ was set so that simple non-metallic materials could be used in contact with the shock tube. It was very fortunate indeed that for cesium $p_1 = 1$ Torr corresponds to $538^\circ F$. Also for $p_1 = 1$ Torr, a reasonable shock tube diameter is 6 in.

In order to achieve significant conductivity and test length, the plasma should be completely singly ionized. The difference of the first and second ionization potentials should also be large; for Cs $(Ip_I) = 3.98$ eV and $Ip_{II} = 25.1$ eV, so that as $\alpha \to 100\%$ the excess enthalpy is used to heat the electrons. Figure 1 shows that for $M_s = 16$, $\alpha \doteq 100\%$ and figure 2 shows that with a H_2 driver gas at 185 psi and Cs at 1 Torr that this condition is achieved. Helium will also be used as a driver gas but for the same driver and driven gas pressures only 54% ionization will be achieved.

It is expected in the MFD experiments that both elliptic and hyperbolic flow regions will be studied. Therefore if the plasma is to be used to study steady or quasi-steady elliptic flows, the test length must be large compared to the perturbation source. Based on non-MFD flows, a length factor of approximately 20 is necessary. The ideal test length for a 30 ft. long tube is 30 in. for $M_s = 16$ and $p_1 = 1$ Torr. Allowing for the tube boundary layer and a finite contact region, the test length would be reduced to 20 in. for a 6 in. tube diameter.

The Rm and Rm/A^2 should both be at least of order one so that both the magnetic field and velocity field perturbations are large. Figure 3 shows the predicted operating parameters for the shock tube.

Figure 3 also shows that a magnetic field of approximately 1 $Weber/m^2$ is needed. A DC power supply for this requirement is beyond reason for a university laboratory, so the obvious solution is a capacitor bank. A 312 turn/meter solenoid was wound with copper wire which with 225 kJ produces 0.90 W/m^2. Additional capacitors will be added to bring the energy up to 360 kJ so the maximum field will be 1.1 W/m^2.

The shock tube is shown schematically in figure 4. Some important features are listed below:

1. The tube is 36 ft. long and made of stainless steel tubing 6 in. OD and 5 3/4 in. ID.
2. All seals are Varian con flat flanges with copper gaskets. The valves are stainless steel copper seat Varian bakeable valves. The use of the Varian copper gasketed flanges and bakeable valves is critical to the feasibility of this shock tube. Several years ago this high vacuum, high temperature technology was not available. Both of these features are absolutely necessary to the success of a cesium shock tube.

3. The high vacuum pumping is done with a 140 liter/sec. vac ion pump so that the base pressure at 551°K is 4×10^{-6} Torr. before the cesium is introduced into the driven section.
4. The flow is observed with a laser schleiren in the test region through four sapphire windows. The windows were another critical problem. It has been shown that all standard glass to metal seals are destroyed by cesium. (11) The metal oxide in these seals is attacked by the cesium. However since sapphire can be metalized, the seal is made by welding to the metalized portion of the sapphire.

Preliminary Experiments

The shock tube has been run with He driver gas and Argon driven gas to check the vacuum, pressure, heating and instrumentation systems. Four runs have been made with He driver gas and Cs driven gas. Three of these runs were successful. The results are summarized in table I.

Table I
Experimental Results with Cesium

M_s	p_1 psi	p_2 psi	p_4 psi	Test Length ft	(T_1)ave °K
12.5	1.51×10^{-2}	3.6	187	1.5	559
11.0	1.76×10^{-2}	3.3	164	---	557
11.5	1.95×10^{-2}	4.0	187	1.1	553

The shock speed and p_2 are measured using five Kistler model 603A-107 pressure transducers spaced along the tube. An estimate of the test time has been obtained using a photomultiplier to observe the passage of the luminous plasma. The calculated test length was approximately 1.5 ft. Using the measured values of p_2 and the shock Mach No., p_1 was calculated. The calculated values of p_4/p_1 are shown with the experimental points in figure 5. The experimental values indicate that less pressure ratio is required to achieve a value of M_s than the calculations based on a Mollier diagram. However the Kistler gages used to measure p_2 are being used at their maximum operating temperature so that their calibration could have changed. Figure 6 shows the typical temperature variation in the shock tube and the corresponding vapour pressure variation. As the instrumentation and operating techniques are improved, more reliable data will be obtained.

The most significant result of these preliminary experiments is that the shock tube does operate with cesium with no noticeable degradation of the system. The cesium is effectively removed from the shock tube by the pumping system after each run. There is a slight reduction in optical transmission thru the test section due to scattering from the cesium or a small amount of cesium on the windows. Several milliseconds after the passage of the test gas the cesium begins to condense on the test section windows. At the conclusion of a shot the windows have become good mirrors. However when the tube is pumped out with continued bake out the cesium vapourizes and the windows are again clear.

The solenoid for the shock tube has been wound and tested but not calibrated. The solenoid will be installed after the instrumentation and operation with cesium have been made reliable.

MFD Research

The initial MFD experiments will be to observe the flow over a cone. The Mach number of the flow behind the shock wave is 2.0. Therefore when B is small and $A >> 1$, then regardless of the value of Rm the flow will be hyperbolic with a gas dynamic conical shock wave. Figure 7 illustrates the flow regimes[12] which are generated for $Rm = \infty$ and the conditions of Alfven Mach No. and Mach No. that will be achieved in the cesium shock tube. It is interesting to note that for $A < 1$ and $M > 1$ the flow is elliptic for $Rm = \infty$, however it is unclear what the nature of the flow will be for $Rm = O(1)$. This is the first of many real MFD flows that will be examined in this shock tube.

This work was supported by the National Science Foundation under grant numbers GK 996, GK 1001, and GK 2694.

REFERENCES

1. McCune, J. E. and Resler, E. L. Jr. J. Aerospace Sci. 27 (1960).
2. Clauser, F. H. Phys. of Fluids, 6 (2; Feb. 1963).
3. Salathe, E. P. and Sirovich, L. Phys. Fluids, 10 (7; July 1967).
4. Salathe, E. P. and Sirovich, L. J. Fluid Mechanics, 33, Part 2 (1968).
5. Gratzer, L. B. Ph.D. Thesis, University of Washington (1968).
6. Ahlstrom, H. G. J. Fluid Mechanics, 15, Part 2 (1963).
7. Maxworthy, T. J. Fluid Mechanics, 31, Part 4 (1968).
8. Yonas, G. J. Fluid Mechanics, 30, Part 4 (1967).
9. Haught, A. F. Phys. Fluids, 5 (11; Nov. 1962).
10. Louis, J. F. Proc. 5th International Shock Tube Symposium (April 1965).
11. Haught, A. F. United Aircraft Research Laboratories, Private Communication.
12. Shercliff, J. A. A Textbook of Magnetohydrodynamics (Pergamon Press, 1965).

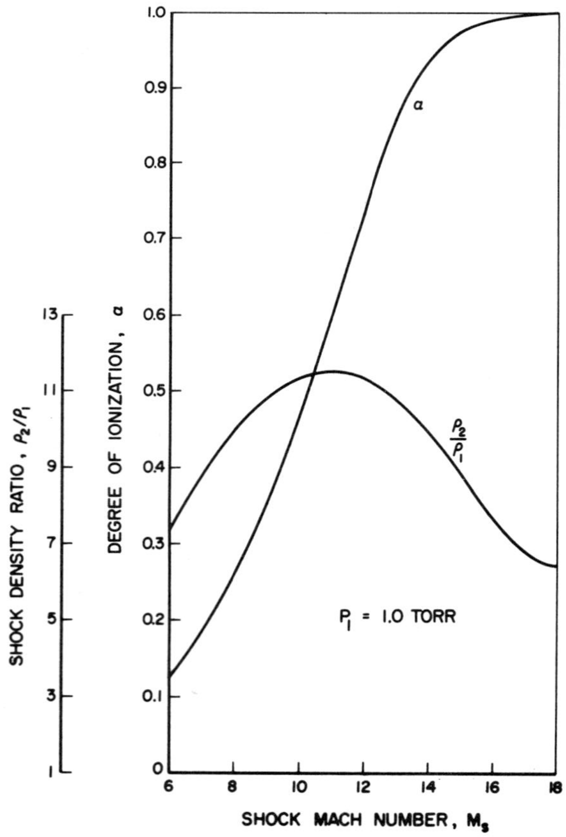

Figure 1
Shock tube operating parameters.

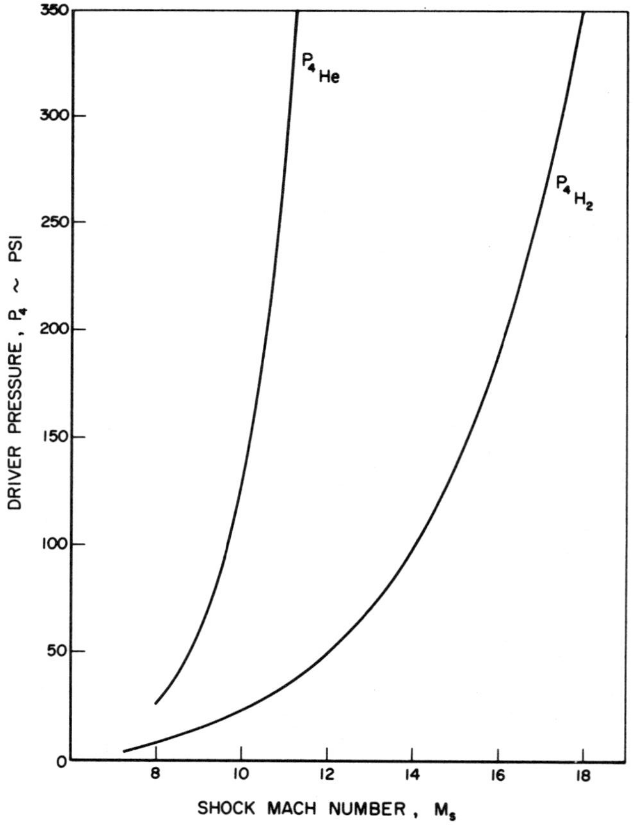

Figure 2
Diaphragm pressure ratio

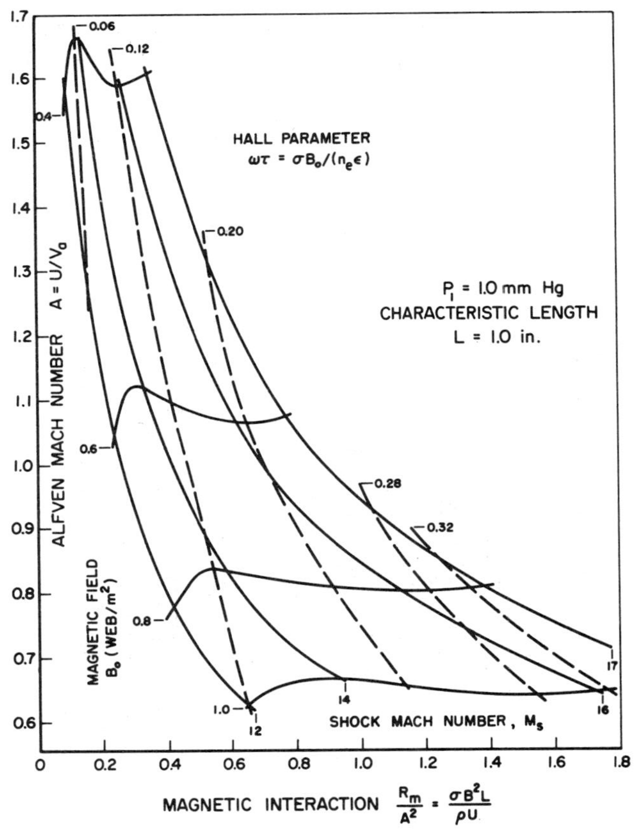

Figure 3
Plasma parameters

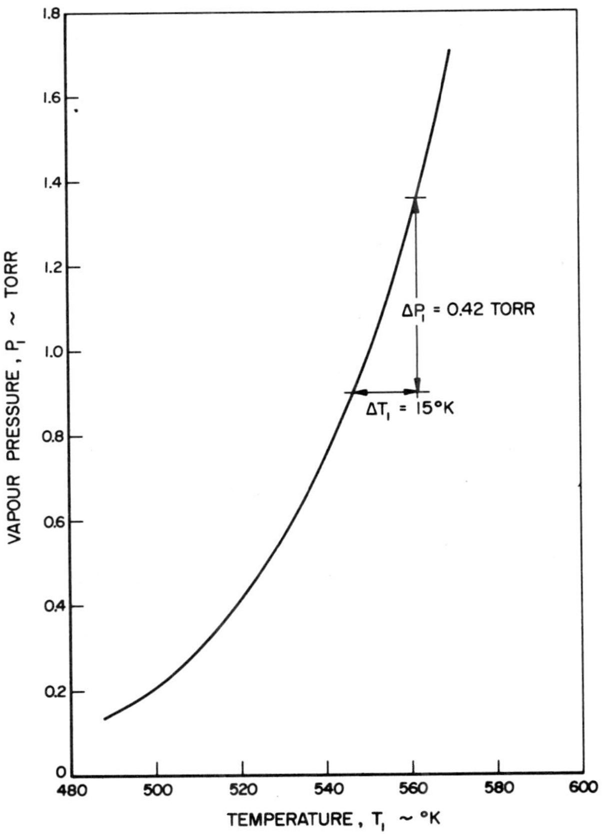

Figure 5
Calculated driven section pressure
as a function of shock tube temperature

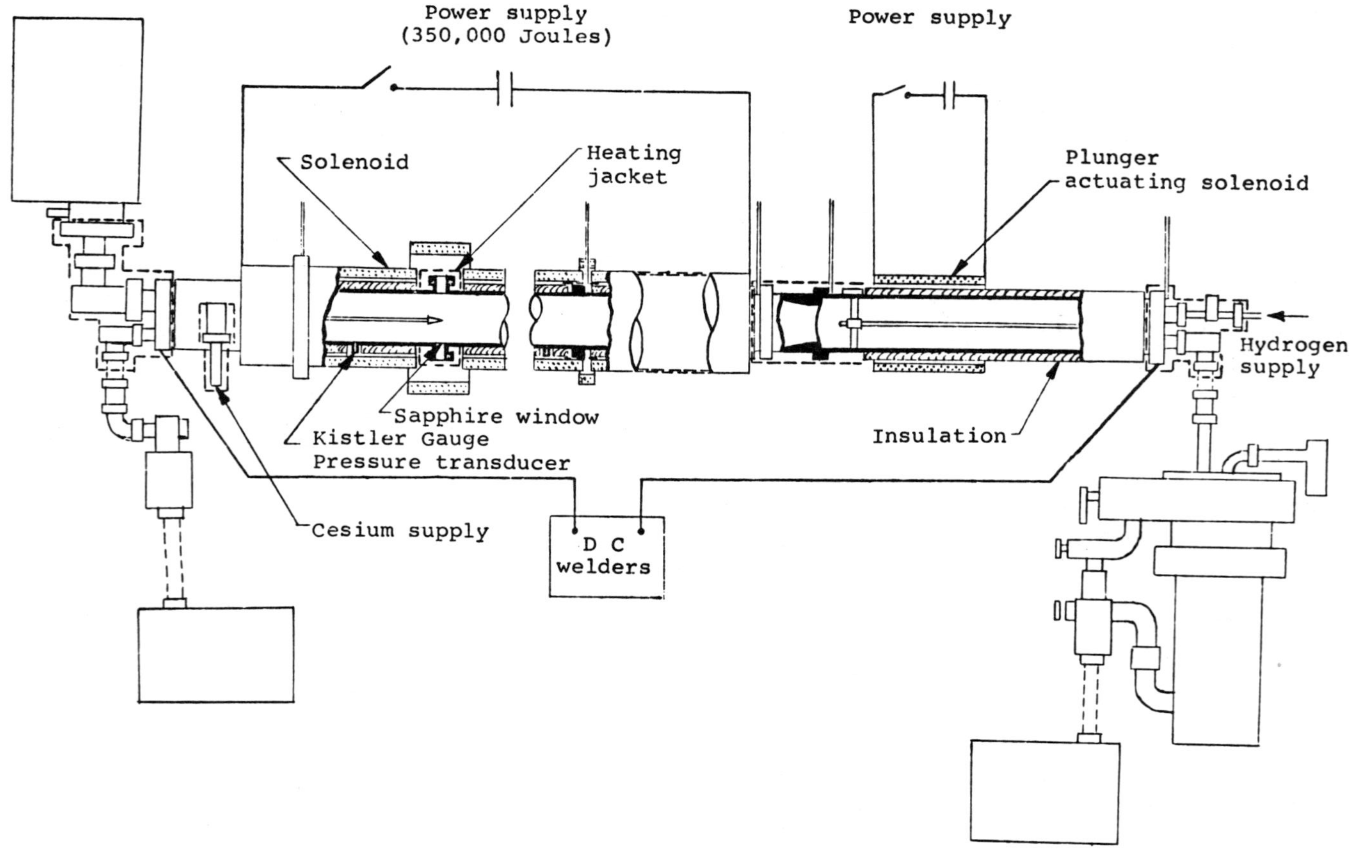

Figure 4
Cesium Shock Tube Assembly

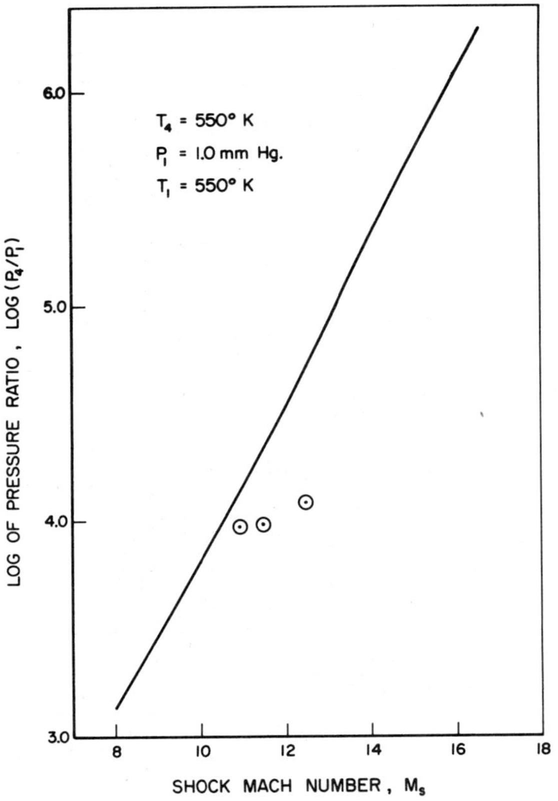

Figure 6
Comparison of the calculated and measured shock speeds for a He driver

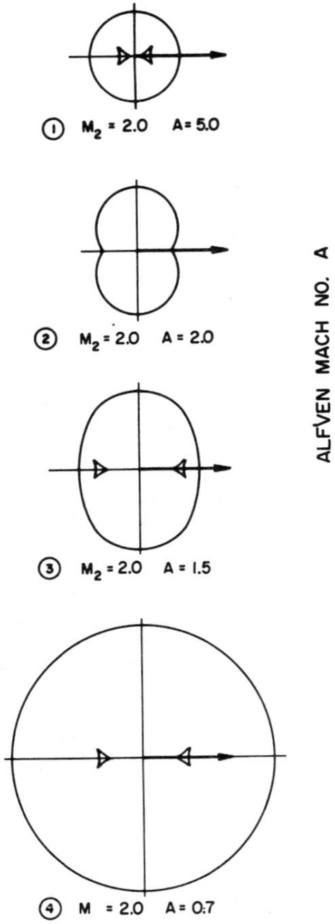

WAVE DIAGRAMS FOR DISTURBANCE

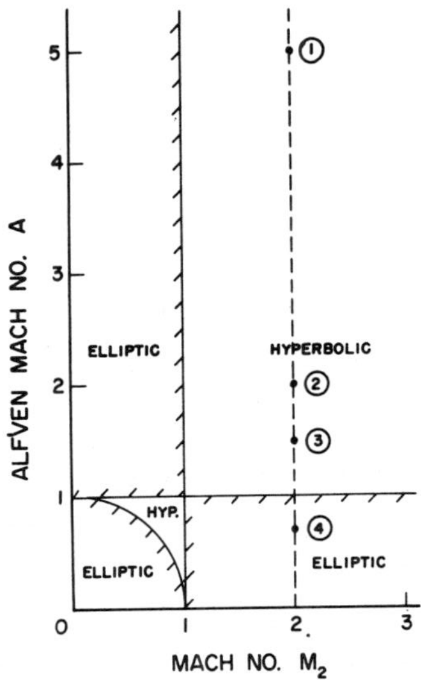

FLOW REGIMES FOR ALIGNED FIELDS

FIGURE 7

STUDY OF IONIZING FLOW THROUGH A TRANSVERSE MAGNETIC FIELD

S.G. ZAYTSEV
E.V. LAZAREVA
I.K. FAVORSKAYA
E.I. CHEBOTAREVA
M.B. BORISOV
G. M. Krzizhanovsky Power Institute, Moscow

This paper presents the results of an experimental and theoretical investigation of the propagation of an ionizing shock wave through a magnetohydrodynamic (MHD) channel. The experimental investigation was carried out in a shock tube. The motion of the ionizing shock wave was determined in the MHD channel and beyond. Measurements were made of the induction emf current as well as the density and electron concentration in the channel by means of interferometry. The generation and development of the shock wave was studied within the motion of the supersonic flow through the MHD channel. The propagation of the ionizing shock wave through the MHD channel was calculated, the results were in agreement with the experiment.

INTRODUCTION

As is known, the propagation of an ionization shock wave through a transverse magnetic and electric field is accompanied by considerable disturbance of the flow (interaction parameter $S \geq 1$). Under certain conditions a stagnation shock appears in this flow. A most interesting question is that of the time required for the shock to form in the channel and the flow parameters following it, in particular for the production of steady-state flows in the MHD channel.

The mechanism of the electric current in the flow greatly influences the process of flow development in the MHD channel. Results available from studies of this process show that the lines of the induced current appearing during the flow of the ionized gas through the MHD channel differ considerably from those connecting the amode and cathode. This circumstance probably causes a deviation from one-dimensional flow, which is of special importance at the initial moment of the stagnation shock development. An analysis of the transient flow in the channel of the shock tube within the limitations of linearized theory is contained in Ref. 1. In a further consideration of the problem the characteristic method was used.[2] In the present article a numerical method was applied to solve the system of magnetogasdynamic equations,[3] artificial viscosity being introduced to make it possible to carry out calculation during the production of the shock front.

EXPERIMENTAL SETUP

Experiments were conducted in a shock tube having a low pressure chamber 5.5 m long, and a channel cross-section of 7.2x7.2 cm^2. A magnetogasdynamic channel made of dielectric material was placed at the end of the low-pressure chamber and a pulsed magnet and optical windows were mounted in the side walls of the channel to enable optical measurements to be made. On the upper and lower walls of the channel a sectional system of electrodes were placed, the electrode area being 6.0x1.0 cm^2, and the distance between adjacent electrodes 0.5 cm. Load resistors made of 0.3x1x50 cm^3 copper strip were connected to each pair of electrodes. The load resistors in the experimental condition were inductive, with a resistance of about $10^{-2}\,\Omega$ (for a frequency of $\sim 10^4$ Hz). The change in magnetic field strength H was $\pm$ 25% and 5% along the x and y axes respectively. The pulsed magnetic field strength was in the centre of the MHD channel, and was equal to 1.5 teslas in all experiments.

The change in H during the passage through the MHD channel of the ionizing flow was 5%.

The ionization gas flow was produced by a shock wave propagating at Mach number M_S = 10-14 in argon of initial pressure 5 torr. The ionized gas flow behind the shock was 60 cm long. The length of the MHD channel (L) was determined by the number of electrode pairs connected to individual, electrically isolated load resistors, L varying from 1 to 14.5 cm. A study of the flow development in the channel was carried out with the help of flow field interferometry, current measurements at load resistors, and observations of the ionization shock wave propagation before and after the MHD channel.

EXPERIMENTAL RESULTS

The ionizing shock wave propagated down the MHD channel (Figure 1) with uniform velocity (3%), which tends to decrease considerably on leaving the channel, as recorded by a pressure transducer.

The trajectory of the shock wave through the MHD channel is shown in Figure 1. The Mach number of the shock S before entering the channel $M_S^o = 12.9 \pm 0.2$. The plates drawn about the abscissa in Figure 1 show the electrode position by number.

Figure 1 also shows the coordinate system. Points 1 correspond to the path of the shock S through the channel in which only the fifth pair of electrodes was connected to load resistors. Points 2 correspond to the path of the shock wave S through the MHD channel with all ten pairs of electrodes connected to individual, electrically isolated load resistors. The dash-dot line corresponds to the shock wave path S in the absence of a magnetic field. The development of the stagnation shock T in the MHD channel and its interaction with the contact surfaces was analysed in previous articles.[4] This work attempts to determine the form of the shock T. For this purpose interferometric pictures of the process were continuously registered at a definite cross-section of the channel.

A number of experiments conducted with different positions of interferomatric fringes and the slit through which the interferograms were registered, as well as different directions of film movement, made it possible to reconstruct the time picture of development and deformation of the stagnation shock wave (Figure 2). At the start there occured a disturbance in the flow near the surface of the electrodes with further disturbances moving to the central part of the flow where they united. From that moment the front T separating the undisturbed flow from the disturbed region began to move against the undisturbed flow. Figure 1 gives the path of the central part of front T (cross-section z=0) at a moment t^* corresponding to the development of T. The load resistors were connected to the fifth pair of electrodes ($M_S^o = 12.9 \pm 0.2$). For $t > t^*$ (see Figure 1) the supersonic flow entering the channel propagated upstream. Consequently, front T can be regarded as the shock front. While T was propagating upstream its curvature was decreasing (Figure 2); this is due to the considerably lower velocity of the regions of T adjacent to the walls than at cross-section $z = 0$. In the process of the development of the shock its path on the interferograms became more visible and its velocity approached a constant value V_T. The value V_T for experiments with $M_S^o = 11.6$ was 0.052 cm/μ sec (one pair of electrodes having a load resistor). The time necessary to obtain this constant velocity increased with the length of the MHD channel. The successively increasing length of the MHD channel did not change the general nature of the development of shock T.

The scheme given in Figure 2 corresponds to experiments with $M_S^o = 12.7$, with a load resistor connected to the fifth electrode pair. The shaded area behind T corresponds to the nonequilibrium region of the flow. The current measurements at the load resistor were made with the Rogovski coil and a shunt. In experiments with load resistors connected to a single electrode pair the peak current totalled 12 x 10^3 amps (M_s^o = 13.5). The beginning of the current signal coincided with the moment the flow equilibrium region entered between the electrodes, while the beginning of the e m f corresponded (with an accuracy of ± 5) to the moment when the shock wave S entered the region between the electrodes.

The experiments with an MHD channel 14.5 cm long the distributions of electron concentration and density of atoms and ions were measured. In assessing the measurements, it should be noted that the stagnation shock is not flat and consequently the flow near it is not one-dimensional. The quantities were measured in the region between the electrode pairs 5-8 using two-wavelength interferometry. The optical slit was parallel to z, and the interferometric fringes and this direction of motion of the film were along the x-axis. The absolute errors of the density determination and the electron concentration were 0.3×10^{-5} g/cm^3 and 5×10^{15} cm^{-3}, respectively. The distribution of ρ and α (degree of ionization) are given in Figures 3 and 4, respectively (for cross-section z = 20 cm; M_s^o = 11.2 ± 0.2). At the initial moment of time $t \simeq 30$ μsec the values of ρ and $\bar{\alpha}$ were constant and equal to their values in the flow behind S in the absence of a magnetic field. In subsequent moments ρ and α increased. During the interval of time ~100 μsec, $\rho(x)$ and $\alpha(x)$ were different at different cross-sections z, which testified to the fact that the flow was not one-dimensional. For $t > 100$ μsec the distributions of $\rho(x)$ and $\alpha(x)$ within the accuracy of the measurements were not dependent on z ($|z| < 3.0$ cm). This indicated that the flow in the MHD channel was becoming one-dimensional.

Calculation of shock wave propagation

A one-dimensional approximation was considered. The movement of the medium was directed along the x-axis. The magnetic field H was assumed to be constant and directed along y. The axial current was assumed to be zero; viscosity and thermal conductivity were not taken into account. The electrical conductivity σ was a function of the flow parameters, and the electrical field E was directed along z. With these assumptions the flow equations behind the shock wave S have the form:

$$\frac{\partial \rho}{\partial t} + \rho \frac{\partial u}{\partial x} = 0,$$

$$\frac{du}{dt} = -\frac{1}{\rho}\frac{\partial P}{\partial x} - \frac{1}{c\rho} jH,$$

$$\frac{d}{dt}\left(\epsilon + \frac{u^2}{2}\right) = -\frac{1}{\rho}\frac{\partial}{\partial x}(Pu) + \frac{jE}{\rho}, \qquad (1)$$

$$P = P(\rho, T)$$

$$\epsilon = \epsilon\ (\rho, T),$$

$$j = \sigma\ [E + \frac{1}{c}[u \times H]]\ ,$$

where ϵ, u, P and T are the internal energy, speed, pressure and temperature of the fluid; j is the current and c is the speed of light (3×10^{10} cm/sec).

Initial conditions: At time $t = 0$ the shock wave S is located directly in front of the MHD channel ($x = 0$). The values ρ^o_1, u^o_1 and T^o_1 before S are connected to the Hugoniot values ρ^o_2, u^o_2 and T^o_2 behind S (the equilibrium parameters of the flow behind S). It is assumed that directly behind the front the temperature, velocity, etc. have their equilibrium values. The degree of ionization and conductivity in the nonequilibrium region of the flow are zero. At the end of the nonequilibrium region, by a jump, σ and α reach the values of the corresponding equilibrium flow parameters.

Boundary conditions: The solution of the set of equations (1) is determined in the limited region $0 \leq x \leq \ell$. For $x = 0, \ell$, the velocity of the flow u, or its pressure P, is given as a function of time. Use was made of the finite difference method developed in Reference 3 and the Lagrange variables. The Lagrange mass coordinate m is introduced in the direction of motion $dm = \rho\, dx$. In the right hand sides of the set (1), P is substituted for the function $G = P + w$, where w is the artificial viscosity. The thickness of the shock wave is spread over three to five of the mass intervals through the choice of w. The set (1) in non-dimensional form becomes

$$\frac{\partial m}{\partial X} = \rho,$$

$$\frac{\partial x}{\partial t} = u,$$

$$\frac{\partial u}{\partial t} = - \frac{\partial G}{\partial m} - R_M \frac{jH}{\rho},$$

$$\frac{\partial}{\partial t}\left(\epsilon + \frac{u^2}{2}\right) = - \frac{\partial (Gu)}{\partial m} + R_M \frac{jE}{\rho}, \qquad (2)$$

$$P = \rho\,(1 + \alpha)$$

$$\epsilon = (1 + \alpha)\, C_V + U\alpha n, \qquad j = \frac{\sigma uH}{1+K}, \qquad E = - \frac{K}{K+1}\, uH,$$

$$\omega = -V_o\, \Delta m^2 \rho \left| \frac{\partial u}{\partial m} \right| \left[\frac{\partial u}{\partial m} - \left| \frac{\partial u}{\partial m} \right| \right],$$

where C_V is the specific heat at constant volume, R is the gas constant, U the ionization potential, n the number of particles per unit mass, K the load factor, $R_M = \sigma u LH^2/C^2P$ is a dimensionless parameter, σ is determined by correlation[6] and α is determined by means of the Saha equation.

The set (2) is solved by iteration[5]. Numerical calculations were performed for a shock S moving into argon ($M^o{}_S = 12.0$) with P_o = 10 torr, H = 2 tesla and a relaxation time of 25 μ sec or zero. The results are given in Figs. 5 and 6. The paths of S and T are shown on Fig. 5. The curves on Fig. 6 are those of the dimensionless parameters $u/u^o{}_2$ and $\rho/\rho^o{}_2$ at a fixed time. The shaded areas on Fig. 6 indicate the regions of the shock fronts S and T. The continuous change in the parameters S and T in the shocks are due to the artificial viscosity.

DISCUSSION OF THE CALCULATIONS AND EXPERIMENTAL RESULTS

The experimental results show that on interaction of a shock wave S with a magnetic field, the shock wave S is attenuated, and in the flow a stagnation shock wave T begins to form under the influence of electromagnetic forces. On being formed the wave T begins to move upstream. It is assumed that the moment of the appearance of shock wave T is t*, the time at which the density disturbances in the channel begin to propagate upstream. The velocity of the shock T reaches a constant value V_T comparatively quickly. The results obtained for the above-mentioned regime by calculation show that when the front leaves the MHD channel, the flow becomes stationary. Paths T_o and S_o in Fig. 5 correspond to the solution of the problem $\tau = 0$, i. e, neglecting the nonequilibrium region behind front S. A rough calculation of the relaxation region leads to a satisfactory correlation of the motion of shock wave T for $x > 15$ cm with the experimental data. The points in Fig. 5 correspond to the experiment with $M^o_s = 12.9 \pm 02$, $P_o = 5$ torr. The existence of a relaxation zone practically does not change the value of v_T or the formation time t^* of wave T obtained by calculation. Only the position of its appearance ℓ^*(Fig. 5) changes slightly. The density distribution in the channel (Fig. 6) is in satisfactory agreement with the experimental data. The points in Fig. 6 correspond to measure values of ρ obtained in the experiment at $M^o_s = 12.02$, $P_o = 5$ torr at time $t = 115$ μsec (Fig. 3). Table I shows some results of the calculations (t_1 is the time when the wave leaves the MHD channel).

Table I

Parameters of an ionizing flow through a transverse magnetic field

M^o_s	τ (μsec)	P_o (torr)	H_o tesla	L (cm)	v_T (cm/μsec)	t^* (μsec)	t_1 (μsec)	ℓ^* (cm)
12	0	10	2	14.5	0.065	45	76	2.2
12	25	10	2	14.5	0.065	45	96	3.4

In conclusion it is to be noted that the given calculation scheme satisfactorily reflects some properties of the process of the propagation of a shock wave through the MHD channel. At the same time, the experimental results show a significant deviation from a one-dimensional flow at the initial moments of the appearance of the stagnation shock front.

REFERENCES

1. J. Roscizewski and A. K. Oppenheim. Shock Interaction with Electromagnetic Field. Phys. Fluids 6, 5 (1963)
2. M. R. Johnson. Shock Induced Flow Interactions with Transverse Magnetic Fields. Phys. Fluids 10, 3 (1967)
3. A. A. Samarskii, P. P. Volocevich, M. S. Volchinsky, and S. P. Kurdumov. Numerical Methods of Solution of One-Dimensional Non-Stationary Problems of Magnetohydrodynamics. J. Vycisl. Matimat. Fis. 8, 4 (1968).

4. S. G. Zaytsev, E. I. Chebotareva, E. V. Lasareva, and E. K. Chekalin. The Structure of Supersonic Flow of Conducting Gas in a Transverse Magnetic Field. Electricity from MHD, 1968, Vol. II, SM-107/157. International Atomic Energy Agency, Vienna, 1968.
5. A. N. Tihonov, and A. A. Samarskii. The Equations of Mathematical Physics. Moscow, 1966.
6. G. I. Kozlov. The Question of the Conductivity of Argon and its Coefficients of Recombination. Thermal Physics of High Temperatures, 3, 4 (1965), 510-519.

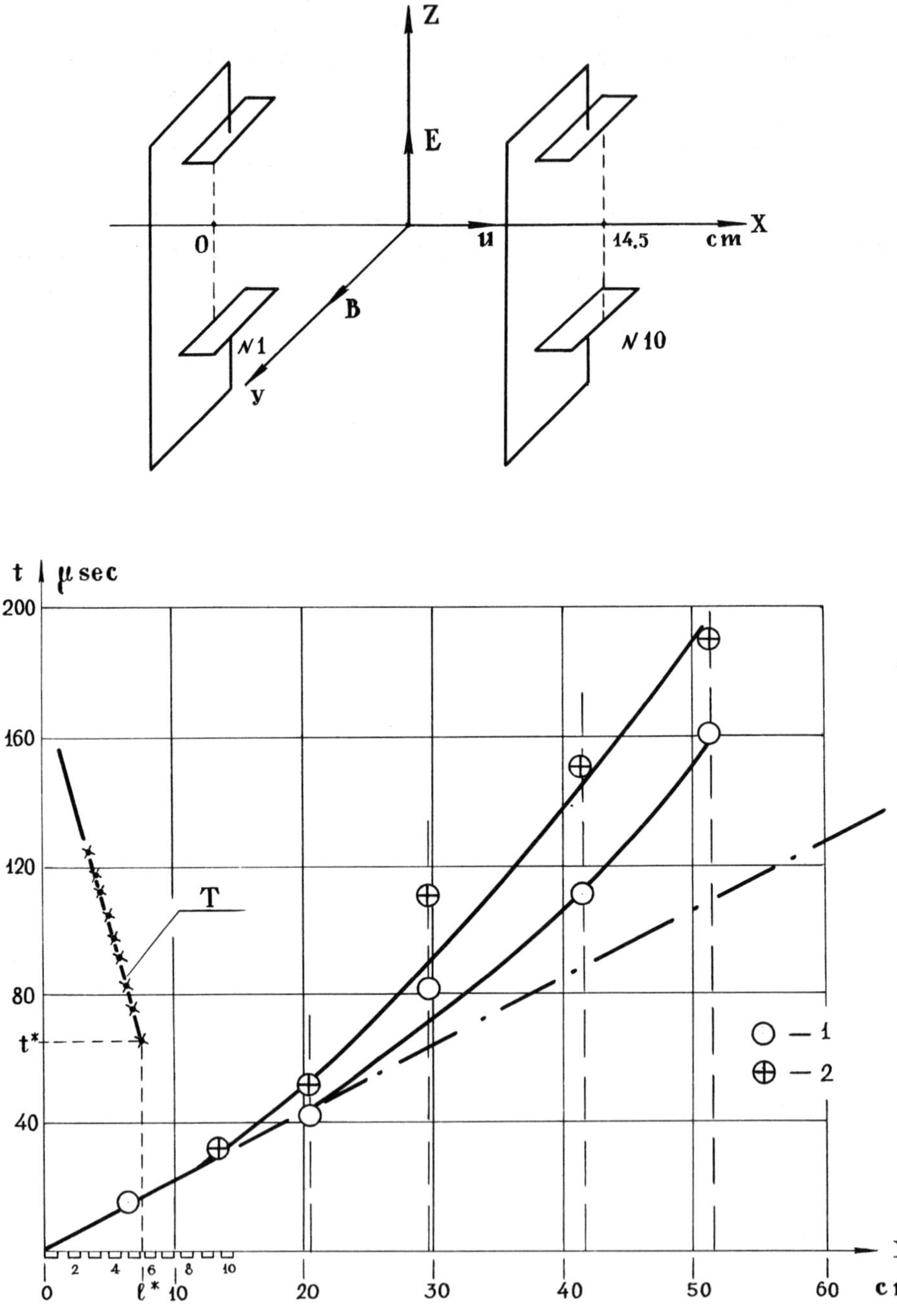

Figure 1
Physical set-up and time-distance diagram of the interaction

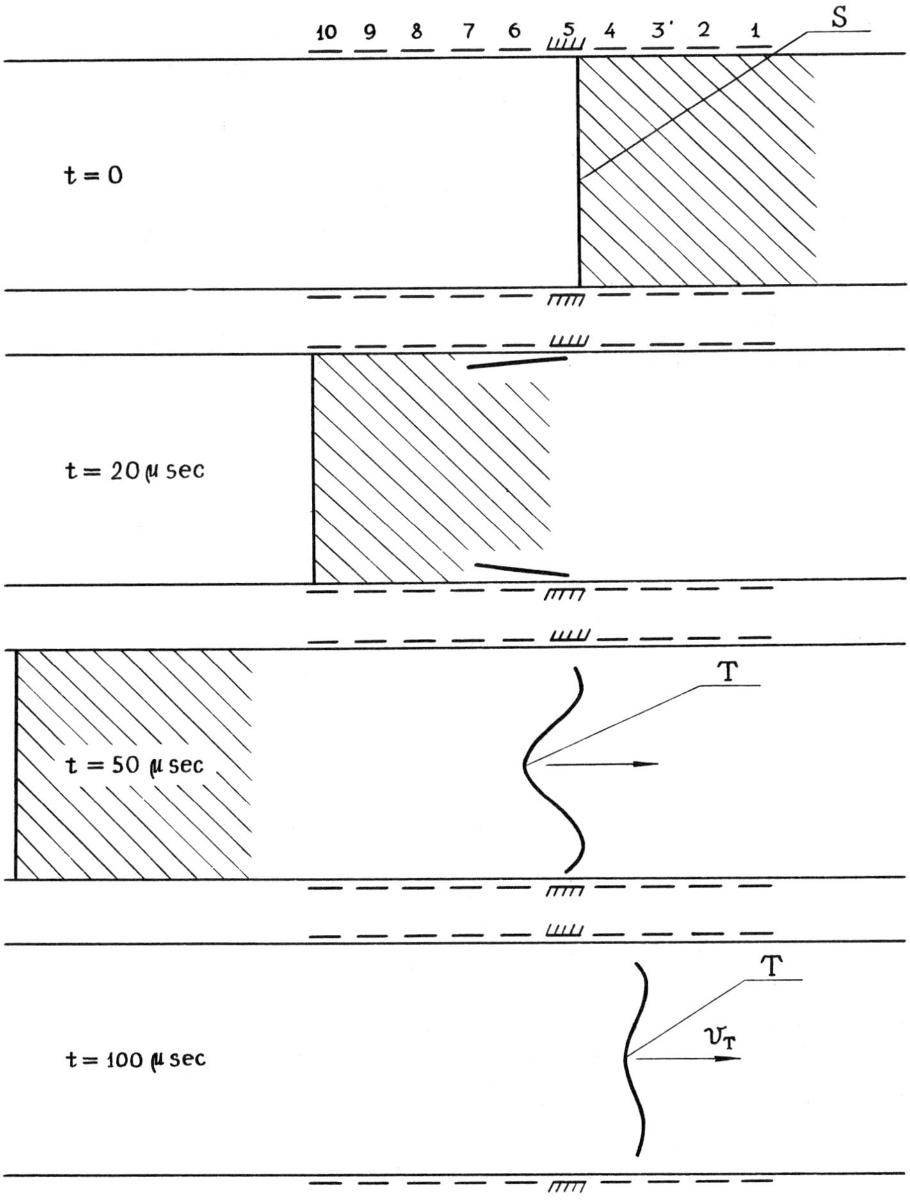

Figure 2
Schematic diagram of the formation of the reflected shock wave T

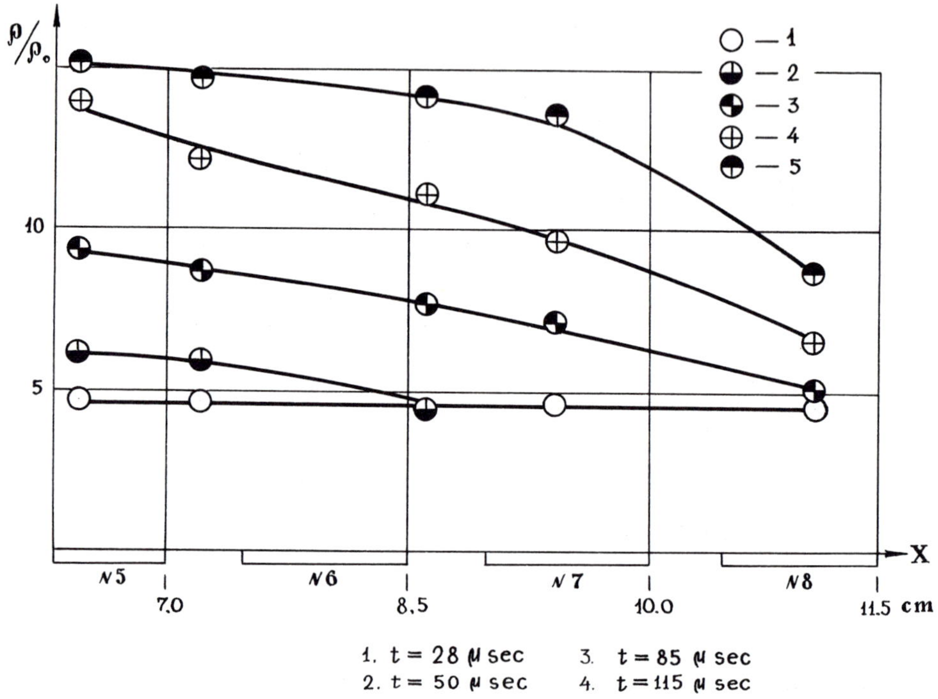

Figure 3
Density distribution in the MHD channel

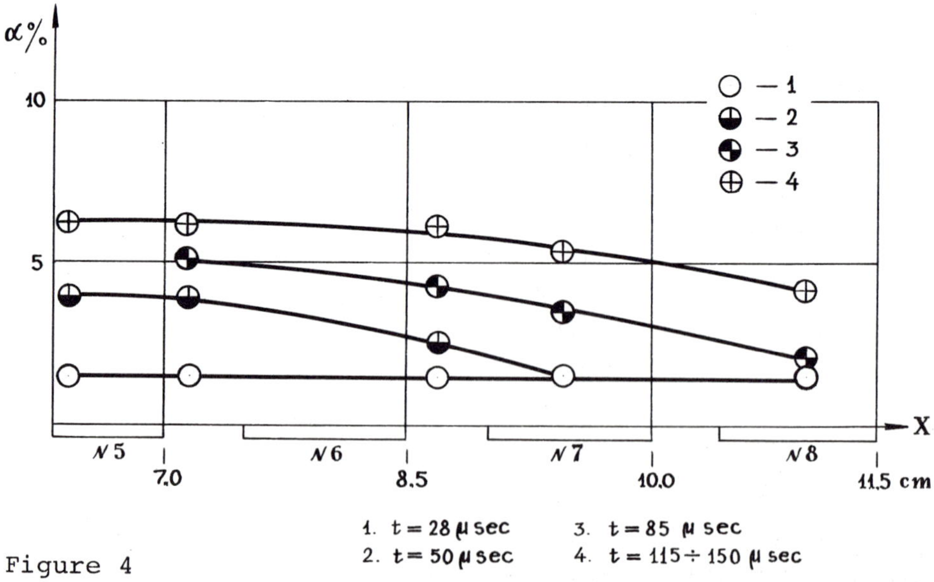

Figure 4
Degree of ionization with distance in the MHD channel

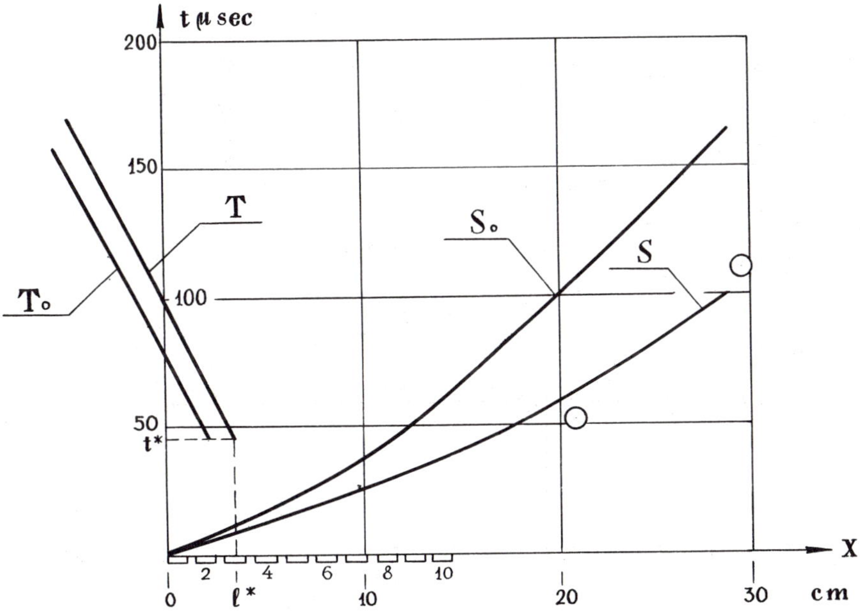

Figure 5
Calculated and experimental time-distance diagram

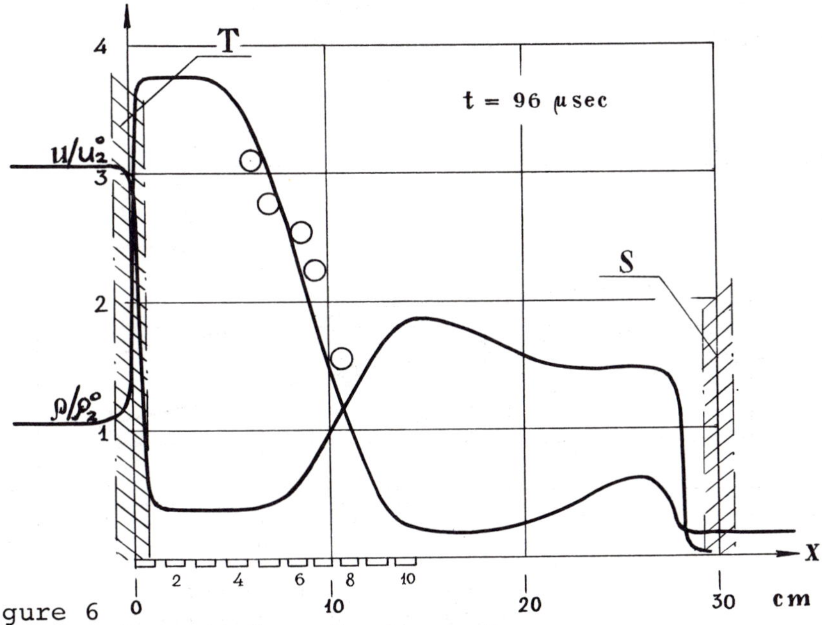

Figure 6
Calculated particle velocity and
density (measured) profiles

SPECTROSCOPIC AND KINETIC STUDIES IN SHOCK TUBES

R.W. NICHOLLS and H.O. PRITCHARD
Centre for Research in Experimental Space Science
York University, Toronto

SUMMARY

The energetic collision processes which take place in shock-heated gases pass through rapidly changing non-equilibrium regimes before local thermodynamic equilibrium is achieved. Light emitted by the excited species is often used diagnostically to follow kinetic processes. It is also used, when L.T.E. has been established, for spectroscopic studies, both to determine energy level arrays and also to determine transition probability parameters. In this review paper, spectroscopic and kinetic studies are reviewed. It is maintained that while the shock tube has become an important and productive tool of spectroscopic research that there are nevertheless serious experimental and theoretical problems involved in its meaningful applications to kinetic studies.

INTRODUCTION

The shock tube provides a simple means for the controlled production of high temperature ($\sim 10^4$°K) gas in which the spectroscopy of light emitted as a result of energetic collision processes of many kinds may readily be studied (Gaydon and Hurle 1963). Spectroscopic methods have also been used as diagnostic tools to follow the kinetics of the rate processes taking place in the hot gases. The shock tube has thus become a very popular research tool in chemical physics, astrophysics, atomic collisions, gas dynamics and other fields. Much experimental and theoretical work has been done over the past few years on the effects of individual elementary energy exchange processes which take place in real hot gases and an authoritative body of knowledge has thereby developed in this field. (Gaydon and Hurle 1963, Bradley 1962, Bond, Watson and Welch, 1965, Vincenti and Kruger 1965, Zel'dovich and Raizer 1966).

In this paper we review recent shock-tube spectroscopic and kinetic research. Much of the spectroscopic work has been done when the shock-heated gas had reached L.T.E. For such work the shock tube has been a very successful spectroscopic light source. On the other hand many of the shock tube kinetic studies have been performed in regimes before equilibrium had been attained and results are much harder to interpret unambiguously.

In what follows the pattern of recent shock tube spectroscopic work (Excitation-Identification-Wavelength measurement and level assignment-Intensity Measurement-Diagnosis of state of shock-heated gas) is first reviewed and this is followed by a critical discussion of recent work on the kinetics of mechanisms of rate processes which take place in the hot gas.

SHOCK TUBE SPECTROSCOPY

One of the principal attractions of the shock tube as a spectroscopic light source, is its ability to sustain for milliseconds (particularly if tailored conditions are used) predetermined high temperature LTE conditions in the shocked gas. A 1000° - 20,000°K temperature range is the most common for shock tube spectroscopy. In this temperature range the average kinetic energy per particle is between 0.1 & 2.0 ev. Energetic atom-atom (or molecule) collisions are thus the primary excitation mechanism (Bauer 1969, Gilmore, Bauer and McGowan 1969), though the auxillary effects of electrons, ions, and radiation (Brabbs 1968), when produced by the primary collisions, cannot be overlooked. Appropriate choice of initial conditions allows one to pre-set temperatures, and thus selectively to excite molecular, atomic or ionic spectra (or mixtures of them). Temperatures comparable to those in stellar envelopes are easily attained in shock tubes which have thus become attractive research tools in laboratory astrophysics.

In common with all spectroscopic research, the reasons for doing shock tube spectroscopy are varied and immediately related to the sequence of operations of the Spectroscopic Method illustrated in Table 1.

As indicated in lines 1 and 2 of Table 1, the shock tube has been used for some years as a convenient source of atomic, molecular and ionic spectra which are only readily excited in a high temperature environment. For example much work has been done on the excitation of metal oxides which are refractory materials at room temperature. The spectroscopy of oxides is of great current interest and many problems of location and assignment of energy levels remain to be solved. The shock tube spectroscopy of powdered solids has been reviewed by Nicholls, Parkinson and Reeves (1963).

Identification of spectral features is the next step (line 3), following the recording of the spectrum. Identification procedures involve the use of wavelength lists and of comparison with atlases. A particularly fine example of such an atlas, and accompanying wavelength lists arising from shock tube spectra is the important work of Weeks and Simpson (1967) on the vacuum ultraviolet spectrum of iron. Definitive identification, though simple in principle is by no means trivial. A set of identification atlases of astrophysically and aeronomically important molecular band systems (and associated compilations of data) is in progress in the authors' laboratory. (Tyte and Nicholls 1964a, Tyte and Nicholls 1965a, Tyte and Nicholls 1965b, Hebert, Innanen and Nicholls, 1967, Tyte, Innanen and Nicholls 1968, Degen, Hebert, Innanen and Nicholls 1969, Hebert, Innanen and Nicholls, 1969, Brocklehurst, Rosevear, Innanen, Hebert and Nicholls 1969).

Some of the spectra in the atlases have been produced by shock tube.
Following identification, much spectroscopic work falls into one of the two categories of wavelength or intensity measurement as inidcated on lines 4, and 5. The wavelength-oriented work aims at the precise location of atomic or molecular energy levels, and the precise determination of constants of atomic or molecular structure. Examples are recent work on Tl in the Harvard Shock tube laboratory, and work on TiO in the authors' laboratory. The intensity-oriented work can be motivated by a desire to provide transition probability or cross-section data for radiative transitions experienced by the shocked gas, or it can use such data to interpret diagnostically the physical conditions existing in the shocked gas. Examples of both types of quantitative spectroscopy are found in many shock tube laboratories. For many years in the authors' laboratory programmes have been running for the provision of transition probability data for many molecular spectra of aeronomical and astronomical interest. Examples are cited below.

Finally, as indicated on line 6, once diagnosis has been performed, conjectures on energy exchange mechanisms which maintain the gas in the measured state may be made. This is the area of kinetic spectroscopy and is closely related to the other discussions of shock-tube kinetics reviewed here.

In what follows, recent shock-tube spectroscopic work in all of these areas is reviewed. Limitations of space prevent this review from being exhaustive.

SHOCK-EXCITATION AND IDENTIFICATION OF SPECTRA

The light associated with the propagation of shock waves in shock-tubes has been known for years, and in the early work was nearly always associated with impurities (eg. Na, Ca, CN, C_2, OH, AlO etc). In fact the ease by which impurity spectra could be excited was exploited by Parkinson and the author in their early work on shock excitation of powdered solids (Nicholls, Parkinson and Reeves 1963). Much of the early work on the absorption and emission spectroscopy of shock-excited gases has been summarised by Gaydon and Hurle (1963).

At relatively low temperatures, the emission radiation from hot gases is dominated by discrete line or band emission features of large oscillator strength, which are easily excited by atomic collisions. At higher temperatures, where the degree of ionization becomes appreciable, plasma spectroscopy effects become evident, (Griem 1964). These include the modification of the discrete features by various line broadening effects, the emission of Kramers continua from free-bound and free-free transitions of electrons in the fields of ions, (Allen, Textoris and Wilson, 1965, McChesney and Al-Attar 1965, Hamberger and Johnson 1965), and the general extension of the wavelength range of interest into the vacuum ultraviolet (Marrone, Wurster and Shatton 1968). As the temperature rises, and the continuous spectra assume greater importance, the spectrum of the hot gas often approaches that of a black-body at the temperature in question (Taylor and Kane 1967).

Intense shock waves have been used for some years as plasma sources, and astrophysically interesting work has been done with them. Particularly good work has been done on the spectra of ions in various degrees of ionizations. Hot shock-produced plasmas are often optically thick (photon mean free path less than geometrical size of gas). In this case the spectral emissivity is determined by the solution of the equation of radiative transfer. For a solid slab of thickness z this is

$$I_{em}(\lambda) = B(\lambda,T)\,(1 - \exp{-k_\lambda' z}) \qquad (1)$$

$$\text{where } k_\lambda' = k_\lambda\,(1 - \exp - hc/\lambda kT) \qquad (2)$$

k is the spectral absorption coefficient or opacity. There have been many experimental and theoretical studies in recent years, of these aspects of shock-heated plasmas (Nelson and Goulard 1968). The emission spectrum of heated air in particular has received much attention (Landshoff and Magee 1969a,b, Armstrong and Nicholls 1967, Buttrey, McChesney and Hooker 1969). Propagation of shock waves through dilute hydrocarbon mixtures have been used to excite such species as C_2 and CN (Fairbairn 1964). Propagation of shock waves through dilute mixtures of various carbonyls (particularly of iron and of chromium) have been used to excite spectra of transition metals. Volatile compounds eg. $VOCl_3$ of other metals have also been similarly employed in the excitation of specific molecular spectra.

The shock tube is a unique instrument for the excitation of lines of high rotation quantum number in molecular spectra, Gibson (1962) and Gibson and Buttrey (1962) have thus studied the excitation of N_2^+ bands to much higher rotational development than had previously been possible.

Such highly developed spectra often have a different gross appearance from that which is common in discharge or flame-excited spectra, and this can lead to difficulties in identification. One particularly interesting recent example of this is identification by Goldberg, Parkinson and Reeves (1965) of bands of the CO Fourth positive system in the solar spectrum between 1500 and 2000 Å. They compared the solar spectrum in this region (Tousey 1964) with a flash absorption spectrum of shock heated CO at about 5000°K. They were thereby able to identify the sharp onset of opacity in the solar spectrum at 1600Å with the very strong overlap in rotational structure of highly developed CO Fourth Positive bands in this region.

SHOCK-TUBE SPECTROSCOPY AND ATOMIC AND MOLECULAR STRUCTURE

Atomic Spectra

The one and two electron spectra of Gp I,II,III elements have been intensively studied in the past two decades particularly with respect to their extension into the vacuum ultraviolet (Garton 1966). Much of this work has been concerned with the definitive establishment of line series, energy levels and series limits through quantum defect plots (see eg. Garton and Codling 1965, 1968) and with the study of broad autoionization features which appear in the one electron ionization continua as a result of mixing of two electron levels with the ionization continua. (see eg. Garton and Wilson 1966, Marr and Heppinstall 1966, Garton, Grasdalen, Parkinson and Reeves 1968, Newsom and Shore 1968). In such work it had been usual to vapourise the absorbing metal in furnaces. In recent years however, the shock-tube and flash absorption techniques have been used at Harvard College observatory to extend the work to materials and spectra which were not susceptible to the study by other techniques. Examples are the measurements of Garton, Parkinson and Reeves (1963, 1966) on Ba, In, Tl and of Newsom (1966) on Ca.

Our understanding of the energy level schemes and of the structure of Gp I-III elements has been greatly increased thereby because of the application of far superior experimental techniques and instrumentation than was used in the first studies of these elements by traditional spectroscopic methods in the first three decades of this century. The transient nature of shock tube experiments has called for the development of reliable reproducible high intensity microsecond flash sources (Garton 1959), microsecond operation pre-slit shutters (Camm, 1960, Clayton 1963, Grasdalen, Haber and Newsom 1968), and time resolving spectrographs for work in the visible (Buttrey 1967) and in the infrared (Camm, Taylor and Lynch 1967).

Molecular Spectra

The shock tube has also been used in recent years for the excitation, identification and analysis of new band systems from which molecular constants have been derived. Examples are found in the work of Tyte (1964) on AlO, Parkinson and Reeves (1963) on TiN, Nicholls and Tyte (1965) on WO, and Linton and Nicholls (1969) on TiO. An important molecular constant is the dissociation energy which has been measured for some molecules using shock tube techniques. For example Seal and Gaydon (1966) made photographic intensity measurements on the NH absorption spectrum using absorption flashes through NH produced in reflected shocks through N_2-H_2-Kr and NH_3-Kr mixtures. The strength of the absorption of the (0,0) band depends upon the number density of NH and upon the oscillator strength of the band. Oscillator strengths are inferred from independent lifetime measurements of Bennet and Dalby (1960) and Fink and Welge (1964). The number density, as determined from thermochemical considerations depends upon the dissociation energy in the exponent of a Boltzman expression. In this way they inferred a dissociation energy of 3.21 ± 0.16 ev.

Tyte (1967) used a direct method of interpretation of the flash absorption spectrum of AlO to place a lower limit of 4.60 ev on the dissociation energy of AlO. A continuum which extends from 2730 Å to shorter wavelengths was interpretated by him as the photodissociation continuum. The location of its long wavelength edge located the dissociation limit.

SHOCK TUBE SPECTROSCOPY AND TRANSITION PROBABILITY MEASUREMENTS

Spectral intensities are inherently controlled by the product of population (or number density) of active species and transition probability parameters. The shock tube is an excellent spectroscopic source for transition probability studies as the existence (to a high degree of approximation) of LTE between the various degrees of freedom accessible to all species (Parkinson and Reeves 1964, Garton, Parkinson and Reeves 1964, 1966) implies that number densities are known and thus transition probabilities can be at once inferred from spectral intensity measurements in emission or in absorption. In the optically thick case the above ideas are applied in such a way as to take account for radiation diffusion, often by using curve of growth techniques. One area of shock-tube spectroscopy in which there has been a great deal of activity is that of the provision of transition probability data of atomic and molecular transitions of importance in astrophysics, aeronomy, combustion etc.

In the discussion of spectral intensities the fundamental transition probability parameter is the transition strength element

$$S_{uL} = \left| \int \Psi_u^* \, M \, \Psi_L \, d\tau \right|^2 \qquad (3)$$

where Ψ_u, Ψ_L are the wave functions of the upper and lower states involved. M is the dipole moment of the transition and $d\tau$ is the element of configuration space. It is well known (Nicholls 1969) that commonly met transition probability parameters are

Emission intensity $I_{UL} \propto S_{UL} \; N_U/\lambda^4_{UL}$ (4a)

Einstein A coefficient $A_{UL} \propto S_{UL}/\lambda^3_{UL}$ (4b)

Einstein B coefficient $B_{UL} \propto S_{UL}$ (4c)

Oscillator strength $f_{UL} \propto S_{UL}/\lambda_{UL}$ (4d)

Absorption coefficients α_{UL}(atomic), k_{UL}(linear), K_{UL}(mass), τ_{UL}(optical depth), Λ_{UL}(photon mean free path) $\propto S_{UL}$ (4e)

The absorption coefficients are related by the Beer law

$$\ln \frac{I_o}{I} = k_{UL}\, x = K_{UL}\rho x = N_L \alpha_{UL}\, x = \frac{x}{\Lambda_{UL}} = \tau_{UL} \qquad (5)$$

where x is the optical path length of the absorber.

Many shock tube spectroscopic determinations of transition probabilities involve the direct interpretation of emission or absorption intensity measurements of shock-heated gases whose thermodynamic state is known. Great care has to be taken in such work to ensure whether optically thin or thick conditions apply and to use the correct interpretation. Care has also to be taken to account for the change in line profiles by collision and other broadening influences.

Another technique for oscillator strength measurements which has been applied with success to shock tube spectra at Harvard College Observatory (Huber 1966), is the application of the interferometric "hook" method by crossing the usual absorption spectral arrangement with a Mach-Zehnder interferometer. The wavelength difference between hooks in the disturbed fringes on either side of an absorption line is a measure of the oscillator strengths. Marlow (1967) has reviewed the method which has been used for many years in static situations on atomic spectra in the Soviet Union, (Meroz 1962). It has been applied in static situations to molecular spectra by Anketell and Pery-Torne (1967).

Atomic Spectra

Many shock tube studies of atomic transition probabilities have been astrophysically motivated, and have concentrated on such elements as Fe, Cr, O etc. (and their ions).

Byard (1967) reports quantitative flash absorption measurements on 43 lines (4603-6180Å) of FeI produced by shock treatment of a dilute mixture of argon and iron carbonyl. He used a curve of growth method to interpret the optically thick spectra and obtained fair agreement with the free burning arc results of Corliss and Bozmann (1962) and Corliss and Warner (1964). Similar and extensive studies were made by Huber and Tobey (1968) on 38 FeI, 5 CrI and 2 CrII lines (3150-3780Å) and by Grasdalen, Huber and Parkinson (1969) on 34 lines of FeI and 12 lines of FeII (for which Van der Waals broadening coefficients were also determined) (3196-3341Å). Very great care was taken to define the experimental conditions, and the results indicate that below 4000Å the results of Corliss and Bozmann and Corliss and Warner are about an order of magnitude (one dex) too high. Above 4000Å the shock tube and our results seem to agree.

CrI and II spectra have been shock excited using chromium carbonyl ($Cr(CO)_6$), and oscillator strengths determined in a similar manner by Shackleford (1965), Byard (1968) and Wolnick, Berthel, Larson, Carnevale and Wares (1968). The lines appear in the (3000-5400Å) region. There appears to be general agreement in the shock tube measurements, which are lower by up to an order of magnitude than those of Corliss and Bozmann (1962) and Corliss and Warner (1964)

Such comparisons emphasise the need, in atomic physics, to make measurements on basic data in different ways in different laboratories. Some confidence may be placed on the shock tube results because of the realistic number density estimates which could be made.

TiI and TiII oscillator strengths were studied by Boni (1968a) using dilute $TiBr_4$ in his test gas. The data for TiI agreed well with earlier work using other techniques. Earlier measurements on TiII lines by Corliss and Bozmann and Corliss and Warner however appeared to be too large.

13 prominent SiI line oscillator strengths have been measured by Miller and Bengtson (1969) using dilute shock excited mixtures of SiH_4 and $SiHCl_3$ in Ne. Relative measures were normalised to a good theoretical estimate from using Coulomb wave functions.

Coates and Gaydon (1966) measured the transition probabilities of 13 argon lines (4159-7635Å).

The very high temperatures which can be shock produced provide sources for ions. Berg, Eckerle, Burris and Wiese (1964) for example used an electrically driven T-tube to produce shock waves in He-O_2 mixtures and thereby to excite 18 OII lines (4649-3271Å) and 18 OIII lines (3760-3722Å). Huber (1966) has used the hook method on shock excited FeII lines to measure oscillator strengths.

Extensive recent measurements have been made at Maryland by Bengston (1968) and Miller (1968) on lines of AlII, CI, ClI, ClII, FI, NeI, OI, BrI, BrII, SiI, SiII, PI, PII, SI, SII.

All of the above work refers to oscillator strength measurements on discrete atomic lines. The shock-produced plasma is also well suited for study of intensities, widths (and thus ionic lifetimes) of broad autoionization features which occur as a result of level mixing, Marr (1967). Garton, Parkinson and Reeves (1964) using flash absorption techniques on the hot gas produced shock-excitation of mixtures of calcium and aluminium sulphides measured widths of broad autoionization features of Al from which lifetimes were inferred. They studied in particular the astrophysically important AlI doublet at 1932 and 1936Å, whose large width corresponds to an autoionization lifetime of 1.1×10^{-13} sec. Newsom (1968) has extended this work to a study of the autoionized Ca lines near 6360Å.

An example of the application of shock tube spectroscopy to the study of photoionization continua (bound-free transitions) and cross sections is found in the work of Rich (1967). He shock-excited dilute trichlorsilane ($SiHCl_3$) and measured by flash absorption techniques the photoionization continua arising from the ground (^3P) and the first excited (^1D) states. The continua correspond to cross-sections of 37 and 33 megabarns respectively at the ionization limit.

Molecular Spectra

During the past decade there has been an increasing demand for firm transition probability data for diatomic molecular spectra which have applications in astrophysics, aeronomy, and the physics of reentry into planetary atmospheres, (Nicholls 1964, 1969b). Many experimental and theoretical methods have been applied to provide these data. For the same reasons as outlined in the preceding section, the shock tube method has been very successfully applied to provide the data. Flash absorption and emission studies have been made. A molecular gas is in general much richer in lines than an atomic one, and hence the effects of opacity and optical thickness are often met particularly if the lowest state of the absorption transition is the ground state. Synthetic spectra methods which take into account lineshapes and the detailed spectral profile have been developed in a number of laboratories (including the authors') to account for this. In such work the specific form of equations (1) and (2) appropriate to the spectra in question are used. It is noticed in these equations that the spectral emissivity is entirely controlled by the absorption coefficient. That is why the spectral absorption coefficient is such important transition probability data in the discussion of radiative transport in hot molecular gases, particularly in cases where the gasdynamic flow is effected by coupling with radiation (as in the case of planetary re-entry processes and in the dynamics of the envelopes of variable stars). Nicholls (1969a) has summarised the basic concepts of molecular spectroscopy assumed in the review which follows.

AlO

AlO is used as a seeded thermometric molecule in aeronomical studies. The synthetic spectrum method (Drake, Tyte and Nicholls 1967) has been applied to the study of intensity distributions in the blue-green system of AlO by Linton and Nicholls (1969a) in which self-absorption effects are important. Relative band strengths for 28 emission bands have been measured. They were shock-excited in powdered Al in a dilute $A-O_2$ mixture. This work corrects a number of earlier arc and exploding wire studies in which the effects of self-absorption have been neglected.

BeO

The BeO blue-green system has aeronomical, re-entry and astrophysical importance. It is strongly self-absorbed in shock-excitation. Drake, Tyte and Nicholls (1967) developed a synthetic spectrum method by which the detailed spectral emissivity is evaluated at a number of points per band-line, for comparison with experimentally excited spectra. The procedure of comparison between experimental and theoretical spectra allows for the determination of band strength and the deliniation of the variation of electronic transition moment with internuclear separation. 19 BeO bands were studied in this work.

CO

The CO fourth positive system is a very important contributor to the solar spectrum (Goldberg, Parkinson and Reeves 1965), and to the atmospheric spectra of Mars and Venus. Rich (1968), by a flash absorption study of shock-excited dilute mixtures of CO and $Cr(CO)_6$ inferred oscillator strengths for 12 strong bands of this system. His results confirm the suggestion of Nicholls (1960) that the electronic transition moment for this system is relatively insensitive to internuclear separation, and that Franck-Condon factors well represent the intensity distribution across the system.

C_2

The C_2 Swan bands play important roles in the atmospheres of stars and in the radiatively coupled flow phenomena of high speed flight in some planetary atmospheres which contain CO_2. They are also very important in many combustion applications. A number of workers have thus supplemented other intensity measurements on this system with shock tubes studies. Sviridov, Sobolev and Sutovski (1965), by interpretation of measurements on five band passes within the $\Delta v = -2,-1,0,+1,+3$ sequences of the shock-excited emission spectrum of C_2 from A-CO mixtures in terms of an optically thin model conclude that the average oscillator strength for the band system is 0.022 ± 0.008.

Harrington, Modica, and Libby (1966) in a similar experiment in which A-CF_4,C_2F_4,C_2ClF_3 mixtures were shocked used the optically thin smeared rotational structure band model of Keck, Camm, Kivel and Wentinck (1959) to assign an oscillator strength at the (0,0) band of 0.028 ± 0.009 to the Swan system and 0.0051 ± 0.0095 to the CF system which they also excited. Modica (1968) later measured the oscillator strengths of the CF_2 system in a similar experiment. Fairbairn (1966) in a broad waveband study of the system shock-excited in mixtures of C_2H_2, C_2N_2, CO, assigned an "average" oscillator strength to the whole system of 0.003 ± 0.012.

Arnold (1968) using a reliable detailed synthetic spectrum method (Whiting, Arnold and Lyle 1967) to interpret intensity measurements on C_2 Swan bands shock-excited in CO_2-A mixtures was able to determine the variation of the electronic transition moment, and assigned an oscillator strength of 0.035 ± 0.005 to the (0,0) band of the system and an average of 0.043 ± 0.006 over the $\Delta v = +1$ sequence. He presents a very useful comparison table of other measurements using shock tube and other techniques.

In many of the above studies attention was centred, photoelectrically, on specific band passes of the $\Delta v = -1, 0, +1$ sequences of bands. Myer (1968) made a relatively high resolution photographic study of the shock-excited C_2 spectrum by the synthetic spectrum method of Drake, Tyte and Nicholls (1967). He studied 12 bands and provided a Deslandres array of relative band strengths which are placed on an absolute basis by comparison with the (0,0) absolute values of Arnold (Myer and Nicholls 1969).

CN

The CN red and violet band systems, like the C_2 Swan system are very important astrophysically (stars and comets) and in hypersonic reentry spectra through CO_2 containing atmospheres. Gippius, Kudryavtsev, Pechenov, Sobolev and Fokeev (1964) have made flash absorption measurements on the CN Red bands (6330-6550A) resulting from shock-excitation of CO-N_2 mixtures. They derived an average electronic transition moment of 0.19 ± 0.09 atomic units. In a more detailed similar study which used image tubes, Dronov, Sobolev, Faizullov and Boiko (1966) showed that the electronic transition moment had a magnitude of 0.11 atomic units.

In a quantitative synthetic spectrum study of the CN radiation from the shock layer of a hypervelocity projectile and using a time-of-flight scanning spectrometer, Reis (1965) measured average oscillator strengths for the $\Delta v = +1$ and -1 sequences of this system as

0.09 ± 0.004 and 0.02 ± 0.004 respectively.

N_2 and N_2^+

Band systems of N_2 and N_2^+ make extremely important contributions to many aspects of atmospheric physics and aeronomy, (Nicholls, 1964, 1969b). Reis (1964) using a similar method to that used in his work on CN, studied N_2 second positive and N_2^+ first negative band emissions for the shock layer around a hypervelocity projectile using a synthetic spectrum method was able to assign oscillator strengths to 0.057 ± 0.01 and 0.053 ± 0.01 and to the $\Delta v = 0$ sequences of the second positive and first negative bands respectively. Wray and Connolly (1965) studied a 66Å band pass at 4272Å of the emission from shock heated N_2 from it inferred an oscillator strength of $0.034 \pm 14\%$ to N_2^+ in that region.

NH

Harrington, Modica and Libby (1966) made quantitative intensity studies on the (0,0), (1,1) and (2,2) bands of NH(A-X) system shock excited in a dilute A-NH_3 mixture and assign oscillator strengths of $(4.17 \pm 0.98) \times 10^{-3}$, $(4.95 \pm 1.2) \times 10^{-3}$ and $(7.04 \pm 2.0) \times 10^{-3}$ to the bands.

NO

There have been many shock tube studies of NO. Recently Cann and Dickerman (1968) have re-examined the shock-excited A-NO spectrum at 32 wavelengths between 2611-5263Å and have redetermined the variation of electronic transition moment with internuclear separation, for the NO beta system. Wray (1969) in a recent study using constricted arc jet air has shown that much of the excess near-infrared (9000-12,000Å) radiation in shock tube studies of N_2 could not be ascribed to N_2 First positive emission. He claims that this emission may well be due to transition between highly excited Rydberg states of NO. He assigned average oscillator strengths to 9 such transitions.

OH

Hydroxyl has also been studied by shock tube spectroscopy for many years. Watson (1964) reports a quantitative absorption study on a 100A pass band near the 3064A (0,0) band from shock excited A-H_2O mixtures. He assigned an average oscillator strength of $3.9 \pm 0.9 \times 10^{-3}$. Watson and Ferguson (1965) investigated the possible effect of the vibration-rotation in the OH molecule, and found a maximum effect of no more than 5%.

SiO

Main, Marswell and Hooker (1968) as part of a continuing study of oxide spectra report on flash absorption studies of shock-excited aerosols of phenolic refrasil and air. Relatively low dispersion spectra were obtained photographically and their intensity distribution was fitted to a synthetic profile. From this work an average oscillator strength for the system of 0.023 ± 0.013 was found.

TiO

Linton and Nicholls (1969b,c) using the synthetic spectrum method employed in the authors' laboratory on AlO and BeO have obtained relative band strengths for the α and γ systems of TiO which are important features of the absorption spectra of the enveloped of M-type stars.

VO

Harrington and Nicholls (1969a,b) have recently made intensity measurements on the red and yellow-green band systems of VO using the synthetic spectrum method to interpret the emission spectrum of VO excited in shock-treated A-$VOCl_3$ mixtures. VO is also an important contributor to the spectrum of cool stars. Band strengths, were determined. The electronic transition moment variation has been studied and chemical kinetics have been followed.

Spectral Absorption Coefficients of Air and other Gases

Much of the work reviewed above relates to the determination of oscillator strength of important diatomic molecular band systems. There has in addition been significant experimental and theoretical effort in the study of absorption coefficients or opacities of gases or gas mixtures with special application to planetary atmospheres, particularly air. The effect of line shapes in the application of synthetic spectrum methods are important here. Whiting, Arnold and Lyle (1967) describe a method for calculation of realistic profiles which does not make an oversimplified assumption of smearing of band structure.

The SACHA (Spectral Absorption Coefficient of Hot Air) computer code has recently been developed at Lockheed Research Laboratories (Churchill, Hagstrom and Landshoff, 1964, Churchill and Meyerott 1965, Churchill, Armstrong, Johnston and Muller 1966). It has been discussed and tabulated in full by Landshoff and Magee (1969a,b). The physical basis of the calculations have been discussed at length by Armstrong and Nicholls (1967).

The SACHA code has been powerfully applied by Buttrey, McChesney and Hooker (1968) to the determination of Stark-broadened molecular line profiles of shock-excited N_2^+ lines in the 9000°-$14,000^\circ K$ temperature region.

Total radiation from shock-heated air has been studied by Gruzezynski and Warren (1967) and by Wood, Hoshizaki, Andrews and Wilson (1969). Near infrared contributions to the radiation from air and N_2 have been measured by Wray and Connolly (1965) and for NO by Wurster and Marrone (1967).

Free-free and free-bound continua in the emission spectra of shock excited N_2, O_2 and air in the temperature regime 8000-14,000°K have been studied by Allen, Texoris and Wilson (1965).

Infrared absorption coefficients of shock-excited CO_2 have been measured by Sulzmann (1964) and similar work on H_2O has been done by Patch (1965). Absorption coefficients of shock-excited Cl_2 have been made by Jacobs and Giedt. Voigt line shape studies on individual shock excited OH band lines near 3090A have been made by Bird and Schott (1965). Intensity measurements on spectra of shock-heated CO_2-N_2 mixtures have been made in a number of laboratories to simulate entry conditions of vehicles into Martian and Venusian atmospheres. Examples are the work of Fairbairn (1964), Thomas and Menard (1966) and Menees and McKenzie (1968). In the first of these estimates of the CN oscillator strengths are also made.

DIAGNOSTICS AND SHOCK-TUBE SPECTRA

As indicated in the introductory remarks, once a firm set of structure and transition probability basic data has been established for atoms and molecules, spectroscopic methods can be employed diagnostically for the determination of physical conditions existing in plasmas. The most commonly sought after diagnostic information is temperature, for when it is defined, LTE conditions exist and it expresses compactly the populations of species in various energy states available to them. Relative populations are the second most sought after data, particularly in the discussion of kinetics. In LTE conditions populations are Boltzmann. Many interesting problems remain however in the discussion of non-equilibrium situations.

The most powerful and most commonly used method for determination of temperatures in shock-heated gases is the pyrometric line reversal method originally adapted to shock-tube studies by Clouston, Gaydon and Glass (1958), as reviewed by Gaydon and Hürle (1963). In such work the gas of interest is "seeded", often by Na-emitting material. The method has been extended by Parkinson and Reeves (1964) and Garton, Parkinson and Reeves (1965,1966) to the thermometric use of lines of species already in the hot gas. The need for seeding is thereby eliminated. They have also established in this work that the physical conditions which occur in many spectroscopic shock tube experiments are very closely LTE for much of the experimental time. Thus the Boltzmann and Saha equations may be used for the specification of populations. Byard and Roll (1965) have also used this method. Blau and Pratt (1968) have suggested an extension of the line reversal method to band spectra. Coates and Gaydon's (1966) work on shock excited organic gases should allow temperatrues in the 6000-15,000A region to be measured by the two line method. A Kirchoff law method in which absorbance and radiance in the infrared of shock heated CO_2 and H_2O has been used by Penzias, Dolin and Kruegle (1966) to measure temperatures in the 2000° range.

SHOCK-TUBE SPECTROSCOPY AND THE KINETIC STUDY OF COLLISION PROCESSES

A vast literature exists on the study of atomic and molecular collision processes which take place within the hot gas of the shock wave (see eg. Bauer 1969, Gordon, Klemperer and Steinfeld 1969). It is not the purpose of this section to review that work particularly as a deeper discussion of chemical kinetics in shock waves is made in the next section. Rather, as an illustration of the final stage of the spectroscopic process (line 6 of Table 1) a few very recent studies of processes (dissociation, ionization etc.) which used spectroscopic methods of measurement will be mentioned.

Reference has also been made to the growing importance of the vacuum ultraviolet region of the spectrum. There is growing interest in this wavelength region in shock tube spectroscopy. For example at the Cornell Aeronautical Laboratory, Marrone, Wurster and Stratton (1968) have developed the technology of a windowless vacuum ultraviolet spectrometer which can be used in a transient manner in conjunction with a shock tube. They have used this in important studies on N^+ and O^+ recombination radiation involved in study of the kinetics of recombination of these ions. Other examples include the measurement (Rothe, 1968) of photoionization cross sections for the 3p $^2P_{\frac{1}{2},3/2}$ state of Na, by monitoring the recombination radiation from the inverse process $e + Na^+ = Na(^2P)$ which occurs in Na seeded shocks; also, the study by Boni (1968b) of the kinetics of ionization of Ti by monitoring TiI and TiII lines emitted in optically thin conditions in shock excited $TiBr_4$-A mixtures.

Dissociation and recombination are very important processes in shock waves and many spectroscopic studies have been made on them. For example, Appleton, Steinberg and Liquornik (1968) have studied the kinetics of nitrogen dissociation caused by heavy particle collisions in shock waves through nitrogen. They monitored the 1176A N_2 absorption Lyman-Birge-Hopfield band from v"=10 as a function of time to measure specific N_2 dissociation from that level. Carbetta and Palmer (1967a,b) have studied dissociation and recombination of chlorine in shock waves, by monitoring the recombination continuum in the visible spectrum. Similarly the kinetics of a number of possible recombination reactions of H and O were studied by Gutman, Hardwidge, Dougherty and Lutz, by monitoring the chemiluminescent O, CO recombination in shock heated H_2-O_2-CO-A mixtures. A study of O-O recombination was made by Myers and Bartle (1968) by monitoring the recombination continuum as a function of time following the shock-treatment of O_3-A mixtures.

The kinetics of excitation processes in shock-heated nitrogen have been worked on in many laboratories. Examples of such studies is the work of Wray (1966) on the excitation of the $B^3\Pi$ state of N_2 and the $B^2\Sigma$ state of N_2^+ in shock-heated N-N_2 mixtures, and similar work by Smekhov and Losev (1968).

The growth of research activities on CO_2 lasers has stimulated much experimental and theoretical work on energy exchange processes in N_2-CO-CO_2 mixtures (Taylor and Bitterman 1969). It has also stimulated shock tube studies of the kinetics of some of the processes. An example of this is the recent work of Russo and Watt (1968) on vibrational excitation and de-excitation of N_2 in the presence of CO.

In all studies of the kinetics of excitation or other processes in which the primary measurements are by photoelectric spectroscopy, extreme care has to be taken in identification of the spectral feature which is monitored. This is particularly difficult when various band passes of a continuum have to be studied and one only has theory to guide one as to the expected intensity distribution of the continuum. When all spectroscopic measurements in hot gases are fraught with problems of optical thickness (which may well change rapidly with time and place in the plasma), extreme care has to be exercised at all times to ensure that the measurements are relevant.

THE SHOCK TUBE AND CHEMICAL KINETICS

The use of shock tubes in chemical-kinetic measurements on neutral species has been reviewed on a number of occasions (Bradley, 1962; Gaydon and Hurle, 1963; Greene and Toennies, 1964; Stupochenko, Losev and Osipov, 1967) and it is not proposed to duplicate any of this material here. When the shock tube first came on the chemical-kinetic scene about 20 years ago, there were hopes that it would solve many of the technical problems that were at that time facing kineticists. In particular the study of reactions at temperatures near 1000°C, mainly by flow techniques, was subject to considerable uncertainties in all the important variables - particularly in the contact time and the mean reaction temperature, and there were always the inevitable heterogeneous side reactions producing soot, tar, etc. Not only did the shock tube seem to offer solutions to these problems - a reasonable time resolution, a well defined temperature and a "wall-free" reaction zone, but it extended the temperature range very much higher than 1000°C to embrace the study of diatomic dissociations up to very high temperatures. However, with the passage of time, it was learned that the retrieval of reproducible kinetic parameters from shock-tube experiments was a difficult task. One of the early casualties was the reflected-shock technique, a method which it had seemed held much promise, and which is now becoming acceptable again. More recently, we have been coming to the realisation that impurities, whether desorbed from the walls before firing, scavenged from the walls during the flow, or originally present in the test gases themselves can play a very important role in the subsequent processes (Windsor, Davidson and Taylor, 1957; Kistiakowsky, 1968; Myerson and Watt, 1968). Thus, while shock-tube technology and instrumentation has improved enormously over the last 20 years, and has yielded most important information on rotational and vibrational relaxation, rates of chemical reactions, rates of ionization etc., one's view of these achievements is coloured by the fact that chemical kinetics is an extremely well-developed area of endeavour. Whilst much of our knowledge of relaxation and ionization processes comes from shock-tube data, chemical kinetics has been with us in quite sophisticated detail since the 1920's (Kassel, 1932; Slater 1959; Bunker 1966), and thus there are many aspects of the theory, or of extrapolation from data known at lower temperatures, which we would like to test. And it is precisely here that the breakdown occurs, since observations of reaction processes which take place in the shock tube are not really consistent with existing notions about chemical reactions. Whether this is simply irreproducibility, systematic experimental errors, or whether there is some deeper underlying difficulty, remains to be seen; or on the other hand, the theories may be quite inadequate for reactions occurring on these short time scales.

Diatomic Dissociation Reactions

It is, therefore, of considerable importance to achieve, for one of the simplest of all chemical reactions, the dissociation of a diatomic molecule, some measure of consistency between the theoretical description and the experimental results. One thing is quite certain (Rush and Pritchard, 1967), that the Arrhenius temperature coefficients of diatomic dissociation rate constants are always less, often considerably less, than the known endothermicities of the reactions. However, the chemist's trusted stand-by, correlation of the abnormalities with other known properties, has failed to materialise, and it seems that the best we can do at the present time is to invoke some non-equilibrium effect.

Non-equilibrium conditions certainly exist in these dissociation reactions. For example, it has been fairly convincingly demonstrated (Wray, 1963) that the vibrational and chemical parts of the relaxation are strongly coupled in the dissociation of oxygen at high temperatures. In other words, the dissociation into atoms commences before the vibrational motion of the molecule has reached equilibrium with the heat bath, and therefore, dissociation is taking place from an assembly of molecules which does not conform to a Boltzmann distribution, at least as far as vibration is concerned. However, no measurement has yet been made of the actual populations during the process, and we do not really know whether the interesting levels are depopulated to the extent of 1%, or 50%, or 90% or what.

From the theoretical point of view, a fairly clear and consistent picture of the dissociation process is emerging from work in progress in our laboratory (McElwain and Pritchard, 1969). These calculations refer to the dissociation of a model diatomic (represented by nuclear-spin-free H_2 at 2000°K) sufficiently diluted in an inert gas so that V-V transitions can be neglected. The principal characteristics which emerge are that in rather less than 1μ sec, most of the vibrational levels of the molecule come into equilibrium with the heat-bath temperature, and since the reaction has not yet advanced very far, this means that these levels are all overpopulated to the same extent. There are, however, a few levels at the top of the potential well (those from which the final dissociation steps take place) which are always underpopulated. As the reaction proceeds, these over-populations of the lower levels, and underpopulations of the upper levels persist, but they shrink monotonically until the final equilibrium distribution is reached. The calculations were carried out with an intuitively reasonable set of T-V transition probabilities amongst the vibrational levels (multiple jumps included), and in order to test the generality of these conclusions, quite gross variations were made in some of the assumed probabilities. It was found that the general pattern of over and under-population was always present, although the severity of the departures from a Boltzmann distribution depend on the actual choice of probabilities. Hence, although we have a good qualitative picture of the process, we are not yet in a position to be able to calculate ab initio rate constants, and thence Arrhenius parameters. Similar calculations have been performed for the recombination process, with analogous conclusions: that the atom and upper-level populations always exceed their equilibrium values, the lower-level populations are always too small, and that all populations approach monotonically their Boltzmann equilibrium values.

One thing is very clear indeed from these calculations, that once the induction period is over, the forward and backward rate constants (k_d and k_r) are strictly related by the equilibrium constant K provided one uses rate constants derived from the slopes of reactant concentrations. If, however, one uses integrated rate constants, the ratio of k_d to k_r can differ from (exceed) the equilibrium constant quite markedly, and depending on the choice of transition probabilities, the extent of the discrepancy could easily be a factor of two or three. This is simply a consequence of the fact that dissociation is assumed to take place from several vibrational levels, and when one has a number of parallel processes taking place at different rates, one cannot integrate the differential form of the rate-constant expression (McElwain and Pritchard, 1969). Also the induction period is very short and, partly because of the assumption of multiple jumps and parallel dissociation paths, the logarithm of the atom concentration is increasing more-or-less linearly with the logarithm of the time, once the initial transient period is over. This period is about 10^{-8} seconds at 1 atmosphere total pressure, and so it would be our conclusion that induction periods of 10^4 microseconds for O_2 near 2000°K and 4mm total pressure (Watt and Myerson, 1969), say 10^{-5} seconds at 1 atmosphere total pressure, are a little on the long side. Care should be taken to distinguish unequivocably between true induction due to vibration-dissociation coupling, and impurity effects such as those observed in the H_2 dissociation (Myerson and Watt, 1968); in the latter experiments, no hydrogen atoms were detected in the early stages of the reaction because they reacted with oxygen impurity in the shocked gas.

Measurements of dissociation reactions cannot yet be made at such high dilutions, although with recent improvements in detection techniques (Myerson and Watt, 1968) there is some hope in this direction. From a theoretical point of view, calculations on undiluted gases are still a long way off, and the best that one can do is to make a mental extrapolation from the results for the dilute case. For example, we know that V-V (and V-R?) processes are much more efficient than the T-V processes which are the basis of the dilute-gas calculation, and it is clear that the departures from equilibrium exhibited in such calculations must, therefore, be overestimates. Thus, it would appear that the results of dilute-gas calculations will bear the same relation to this area of reaction kinetics as does say the Heitler-London calculation to molecular valence theory.

We are, however, still lacking any real theoretical understanding of the low Arrhenius temperature coefficients of these dissociation reactions. What would be feasible in the near future would be to constrain the assumed set of transition probabilities to give the observed k_d at one temperature, and then to see if the value of k_d at another temperature could be reproduced making entirely reasonable changes in the probabilities. However, the long-term solution is to make *a priori* calculations of all the individual transition probabilities involved in the dilute-gas relaxation scheme (if rotational relaxation is assumed to be rapid, one needs about 100 transition probabilities for H_2; if rotational relaxation is included, the requirement increases to about 45,000, and correspondingly more for heavier molecules with more bound rotation-vibration levels). At the present time, it is difficult to judge the correctness or otherwise of these low Arrhenius temperature coefficients. Certainly they seem universal to diatomic dissociations, and also to ionisation processes (McLaren and Hobson, 1968). On the other hand, whether they have anything to do with vibrational disequilibrium (Rush and Pritchard, 1967) has been questioned on the grounds that HCl and DCl dissociate with the same Arrhenius parameters, but they are vibrationally quite different molecules (Jacobs, Cohen and Giedt, 1967). In our view, this observation does not refute the vibrational disequilibrium idea entirely because of our finding that rates of dissociation are rather insensitive to our choice of transition probabilities.

All the same, it should lead us to question whether all these anomolous temperature coefficients, all anomolous in the same sense, could be due to some hidden systematic error in the experiments. One difficulty, which is well known (Hurle, 1967) is that there is a substantial temperature change during the dissociation process, particularly at low dilutions, and unless initial rates are used, there is a problem in defining the temperature at which the rate is being measured. There is, however, no reason to believe that this lies at the root of the anomalies. To carry the re-examination further, there are three basic ingredients in any kinetic measurement, concentration, temperature and time. The measurement of the concentration does not seem to be the basic cause, particularly in view of recent measurements on H_2 and O_2 where the atom concentration was measured directly (Myerson and Watt, 1968; Watt and Myerson, 1969): the observation, however, that the less impurity was present in the H_2, the higher was the Arrhenius temperature coefficient, is a little disturbing although extrapolation to zero concentration of (known) impurity still leaves "E" about 10 kcal/mole below the dissociation energy. It could be, perhaps, that one is systematically overestimating the temperature rise, and that the discrepancy increases with temperature: again, in view of the great amount of work done on temperature measurements (Gaydon and Hurle, 1963) we feel that this is not the cause of the anomaly either. Finally, whilst it is possible to measure the shock speed accurately, one then has to make the transformation from laboratory time to particle time. We have been cautioned recently (Mirels, 1966) that at low densities, and at quite low Mach numbers, the observation of a constant shock velocity does not necessarily imply an ideal shock, and it is not impossible that a thorough re-examination of the ideality or otherwise of these shocks may reveal a small systematic error in the laboratory-particle time transformation.

We suggest, therefore, that the attempt to resolve these discrepancies should proceed on two broad fronts. On the theoretical front, one is currently limited to dilute-gas calculations, but these can be based on more accurate transition probabilities. On the experimental front, recent work (Myerson and Watt, 1968; Olschewski, Troe and Wagner, 1967) has opened the way to working under more dilute conditions: impurities need to be characterised, and their effects eliminated; finally, we need a critical re-evaluation of the validity of the shock-tube equations to ensure that no unsuspected systematic error lies at the root of all this controversy.

We have spent considerable space discussing the diatomic dissociation problem, but in the absence of a thorough understanding of this problem, the extension of the use of the shock tube for other chemical kinetic measurements will always be open to question. However, once a solution is forthcoming, there are many areas of study which will yield information of interest to kineticists, and of which we will present a brief summary, concentrating mainly on the theoretical aspects.

Elementary Reaction Rates

Before so doing, we note that there are very many simple reactions which, in principle, can be studied in a shock tube: an early example (Britton and Davidson, 1955) was the reaction of Br with H_2 using Br_2/H_2 mixtures; a particularly pretty situation occurs when the atoms or radicals can be formed in an incident shock at a relatively low temperature, and their reaction with substrate takes place in the reflected shock after the second heating. Many such reactions are of interest in combustion and in aeronomy or astrophysics, and the advancement of these subjects is often retarded for the lack of definitive knowledge of simple reaction rates - recent compilations (Baulch, Drysdale and Lloyd, 1968) bear witness to the gross uncertainties which still exist for many of these important reactions.

Polyatomic Dissociation Reactions and Unimolecular Rate Theory

Turning from diatomic to polyatomic dissociations, one immediately enters an area, as we have already noted, where the theory of reaction rates, associated with the names of Hinshelwood, Slater and RRKM (Rice, Ramsperger, Kassel and Marcus) is very well developed indeed. The difference between such reactions and the diatomic dissociation lies in the fact that polyatomic molecules can store vibrational energy for a time, characteristic roughly of the size and complexity of the molecule, whereas a diatomic must fall apart within one vibrational period after receiving enough energy to dissociate. This leads (Pritchard, 1969) to the fact that polyatomic dissociations are first order at high pressures and second order at low pressures. Also, the characteristics of unimolecular reactions are such (Slater, 1959) that this transition takes place at moderate pressures for small molecules (say 1 atm. for 3-4 atoms) but at very low pressures for complex molecules (say 10^{-9} atm. for 25-30 atoms). This changeover (fall-off) from 1st to 2nd order behaviour has been studied for many molecules (Bunker, 1966) over a small temperature range in static systems, and is very well characterised. However, the changeover is also temperature-dependent, occurring at higher and higher pressures as the temperature increases (Hole and Mulcahy, 1969). For example, for a molecule like C_2N_2 (Tsang, Bauer and Cowperthwaite, 1962) or N_2H_4 (McHale, Knox and Palmer, 1965) might be at its high-pressure region near 1 atmosphere at room temperature (this means that the recombination will be second order and not third order at this temperature and pressure), but in the 1000-3000°K range where dissociation occurs, the high-pressure region may now be at several hundred atmospheres. Thus, there is a need to study this fall-off characteristic, not as a function of pressure at one temperature, but as a function of temperature at fixed pressure, because of the additional theoretical information which will be provided: such an experiment is only possible because of the enormous range of temperatures accessible by shock-tube techniques. Up till now, many dissociation reactions have been carried out in a pressure region which is intermediate between the 1st and 2nd order regions, and so they do not provide useful theoretical information. Finally, one might ask, is it possible to study a diatomic dissociation in its high-pressure region? This would imply that the pressure was sufficiently high for the time between collisions to be substantially less than one vibrational period, in which case the dissociation would be first order!

There is another area of chemical kinetics in which our knowledge and understanding is rather diffuse, and that is the question as to whether or not the Arrhenius equation

$$k = A \exp(-E/RT) \qquad (6)$$

should strictly relate rate constants at different temperatures for a reaction carried out in the absence of non-equilibrium effects. It is common to describe diatomic dissociation rates by an expression of the form

$$k = BT^n \exp(-E/RT) \qquad (7)$$

where n is chosen either in accordance with some preconceived mechanism, or is found by some best-fit calculation. In the strict Arrhenius form, n = 0, or as the chemist often says, A is independent of T. As we have already noted, there exist non-equilibrium aspects of diatomic dissociation reactions and under such circumstances, equation (7) is to be regarded as an empirical representation, the value of n being characteristic of the non-equilibrium behaviour, amongst other things. Again, in representing recombination reactions, it is usual to fit rate constants to equation (7) with a non-zero value of n, and since the variation in the exponential term is now very small ($E \approx 0$), one would expect definitive conclusions to be reached by a study of such reactions. In fact, recombination rates are known over temperature ranges of several thousand degrees, but unfortunately, such data do not resolve the problem satisfactorily. This is because only a few results, over a very limited range in k_r and T, are obtained by direct observation of recombination processes; the rest are derived from shock-wave dissociation measurements, which provide an acceptable route to k_r, but a relatively inaccurate one for describing recombination temperature coefficients.

Most theories of reaction kinetics result in rate-constant expressions of the form of equation (7) where n is usually small ($\frac{1}{2}$ or 1, say) but because rates are usually dominated by the exponential term, it is difficult to determine n without the use of a very large temperature range. One polyatomic dissociation reaction, the thermal decomposition of di-tert-butyl peroxide (Shaw and Pritchard, 1968) has been studied at its high-pressure limit by conventional methods over nine orders of magnitude in k_d, and the Arrhenius plot, equation (6), shows no signs of curvature. There is no significant hope for extending the temperature range downwards, but the shock tube could provide a considerable extension to higher temperatures, and help to determine whether n is zero or not. This investigation need not be confined to first-order reactions, for example the reaction Cl + H_2 which has a small temperature coefficient ($E \approx 5$kcal/mole) has been studied from 195°K to 1050°K (Westenberg and deHaas, 1968; Steiner and Rideal, 1939), without any marked deviation from Arrhenius behaviour. It should be mentioned that curved Arrhenius plots are not infrequently encountered, but are usually taken as an indication that there are two competing mechanisms of differing activation energy (Billinge and Gowenlock, 1963), perhaps one of them heterogeneous. Even when this is not the case, e.g. the reaction of H + H_2 in the gas phase (LeRoy, Ridley and Quickert, 1967), or the hydrolysis of tert-butyl chloride in solution (Albery and Robinson 1969), where curved Arrhenius plots are well established, no attempt is made to use equation (7), but chemists search for other explanations!

ACKNOWLEDGEMENTS

This work has been supported by research grants from the National Research Council, The Defence Research Board and contracts from The Department of Supply and Services of Canada. The skilful bibliographic assistance to one of us (RWN) by Mrs. Sandra Innanen is also gratefully acknowledged.

REFERENCES

Albery, W.J. and Robinson, B.H., Trans. Faraday Soc. 65, 980, 1969.

Allen, R.A., Textoris, A., and Wilson, J., JQSRT, 5, 95, 1965

Anketell, J., and Pery-Thorne, A., Proc. Roy. Soc. A 301, 343, 1967.

Appleton, J.P., Steinberg, M., Liquornik, D.J., J. Chem. Phys. 48, 599-608, 1968.

Armstrong, B.H., and Nicholls, R.W., Thermal Radiation Phenomena Vol. 2 The Equilibrium Radiative Properties of Air: Theory DASA 1917-2, Lockheed Missiles and Space Co. 1967.
Arnold, J.O., JQSRT, 8, 1781, 1968.
Bauer, E., JQSRT 9, 499, 1969.
Baulch, D.L, Drysdale, D.D., and Lloyd, A.C., High Temperature Reaction Rate Data, Leeds University, 1968.
Bengtson, R.D.: The measurement of transition probabilities and stark widths for CI,FI,NeI,ClI,ClII,BrI,BrII TN:BN 599, Inst. for Fluid Dynamics and Applied Maths, Univ. Md 1968.
Bennett, R.G. and Dalby, F.W., J. Chem. Phys. 32, 1716, 1960.
Berg, H.F., Eckerle, K.L., Burris, R.W. and Wiese, W.L., Ap. J. 139, 751, 1964.
Billinge, B.M.H. and Gowenlock, B.G., Trans. Faraday Soc. 59, 690, 1963.
Bird, P.F., and Schott, G.L., JQSRT, 5, 783, 1965.
Bond, J.W., Watson, K.M. and Welch, J.A.: Atomic Theory of Gas Dynamics, Addison Wesley, 1965.
Boni, A.A., JQSRT 8, 1385, 1968a.
Boni, A.A., J. Chem. Phys. 49, 288, 1968b.
Brabbs, T.A., J. Chem. Phys. 49, 1433, 1968.
Bradley, J.N., Shock Waves in Chemistry and Physics, Methunen, 1962.
Britton, D. and Davidson, N., J. Chem. Phys. 23, 2461, 1955.
Brocklehurst, B., Rosevear, A.H., Innanen, S.H., Hébert, G.R. and Nicholls, R.W. 1969 Identification Atlas of Molecular Spectra, 8 The CN Violet and CN Red Band systems (in preparation).
Bunker, D.L. Theory of elementary gas reaction rates, Pergamon, 1966.
Buttrey, D.E. Applied Optics, 5, 881, 1967.
Buttrey, D.E. and Gibson, J., unpublished work, 1962.
Buttrey, D.E., McChesney, H.R. and Hooker, L.A., JQSRT 8, 717, 1968.
Byard, P.L. and Roll, R.E., JQSRT 5, 715, 1965.
Byard, P.L., JQSRT, 7, 559, 1967.
Byard, P.L. JQSRT, 8, 1543, 1968.
Camm, J.C., Rev. Sci. Inst. 31, 278, 1960
Camm, J.C., Taylor, R.L. and Lynch, R., Applied Optics, 6, 885, 1967.
Cann, M.W.P. and Dickermann, P.J.: Molecular f-numbers from high resolution spectra, AFWL-TR-67-76, Kirtland AFB, 1968.
Carabetta, R.A. and Palmer, H.P., J. Chem. Phys. 46, 1325, 1967a.
Carabetta R.A. and Palmer, H.P., J. Chem. Phys. 46, 1333, 1967b.
Churchill, D.R., Armstrong, B.H., Johnston, R.R. and Muller, K.G., JQSRT, 6, 371, 1966.
Churchill, D.R., Hagstrom, S.A., Landshoff, R.K.M., JQSRT 4, 291, 1964.
Churchill, D.R. and Meyerott, R.E., JQSRT 5, 69, 1965.
Clayton, J.O., Rev. Sci. Inst. 34, 1391, 1963.
Clouston, J.G., Gaydon, A.G., Glass, I.I., Proc. Roy. Soc. A. 248, 429, 1958.

Coates, P.B. and Gaydon, A.G., Proc. Roy. Soc. A 293, 452, 1966.
Corliss, C.H. and Warner, B., Ap. J. Supp 83, 395, 1964.
Corliss, C.H. and Bozman, W.R., Transition Probabilities, NBS Monograph 53, 1962. (US Govt. Printing Office).
Degen, V., Innanen, S.H., Hébert, G.R. and Nicholls, R.W. 1969 Identification Atlas of Molecular Spectra, 6. The O_2 Herzberg I system.
Drake, G.W.F., Tyte, D.C. and Nicholls, R.W., JQSRT 7, 639, 1967.
Dronov, A.P., Sobolev, N.N., Faizullov, F.S. and Boiko, V.A., Opt. Spect. 21, 397, 1966.
Fairbairn, A.R., AIAA Journal 2, 1004, 1964.
Fairbairn, A.R. 1966, JQSRT 6, 325, 1966.
Fink, E. and Welge, K., Zeits. Naturforsch. 19a, 1193, 1964.
Garton, W.R.S., J. Sci. Inst. 36, 11, 1959.
Garton, W.R.S., Spectroscopy in the vacuum ultraviolet, Advances in Atomic and Molecular Physics 2, 93-176, 1966. (Ed. D.R. Bates and I. Estermann, Academic Press).
Garton, W.R.S., and Codling, K. Proc. Phys. Soc. 86, 1067, 1965.
Garton, W.R.S. and Codling, K., J. Phys. B. Ser. 2 1, 106, 1968.
Garton, W.R.S., Grasdalen, G.L., Parkinson, W.H., and Reeves, E.M., J. Phys. B. Ser. 2 1, 114, 1968.
Garton, W.R.S., Parkinson, W.H., and Reeves, E.M., Proc. Phys. Soc. 80, 860, 1963.
Garton, W.R.S., Parkinson, W.H. and Reeves, E.M. Ap. J. 140, 1269, 1964.
Garton, W.R.S., Parkinson, W.H. and Reeves, E.M., Proc. Phys. Soc. 88, 771, 1965.
Garton, W.R.S., Parkinson, W.H., and Reeves, E.M. Can. J. Phys. 44, 1745, 1966a.
Garton, W.R.S., Parkinson, W.H. and Reeves E.M., Proc. Phys. Soc. 88, 771, 1966b.
Garton, W.R.S., and Wilson, M. Proc. Phys. Soc. 87, 841-850, 1966.
Gaydon, A.G. and Hurle, I.R., The shock tube in high temperature chemistry and physics, Chapman and Hall, 1963.
Gibson, J., MS Thesis, Stanford University, 1962.
Gilmore, F.R., Bauer, E., McGowan, J.W., JQSRT 9, 157, 1969
Gippius, E.F., Kudryavtsev, E.M., Pechenov, A.N., Sobolev, N.N. and Fokeev, V.P., High Temp. 2, 159-164, 1964.
Gordon, R.G., Klemperer, W. and Steinfeld, J.I., Vibrational and Rotational Relaxation, Ann. Rev. Phys. Chem. 19, 215, 1968.
Grasdalen, G.L., Huber, M. and Newsom, G.H., Rev. Sci. Inst. 39, 886, 1968.
Grasdalen, G.L., Huber M. and Parkinson, W.H., Ap. J. 156, June 1969 (in press).
Goldberg, L., Parkinson, W.H. and Reeves, E.M., Ap. J. 141, 1293, 1965.
Greene, E.F. and Toennies, J.P., Chemical reactions in shock waves, Academic Press, 1964.
Griem, H., Plasma Spectroscopy, McGraw-Hill, 1964.
Gruszcznski, J. and Warren, R., AIAA Journal 5, 517, 1967.
Gutman, D., Hardwidge, E.A., Dougherty, F.A. and Lutz, R.W., J. Chem. Phys. 47, 4400, 1967.

Hamberger S.M. and Johnson, A.W., JQSRT 5, 683, 1965.
Harrington, J., Modica, A.P. and Libby, D.R., J. Chem. Phys. 44, 3380, 1966.
Harrington, J., Modica, A.P. and Libby, D.R., JQSRT 6, 799, 1966.
Harrington, J. and Nicholls, R.W., Proc. Nat. Com. for Canada of Int. Astron. Union meeting, UBC, May 2,3 1969a.
Harrington, J. and Nicholls, R.W. (1969b) Unpublished work.
Hole, K.J. and Mulcahy, M.F.R., J. Phys. Chem. 73, 177, 1969.
Hurle, I.R., Rep. Prog. Phys. 30, 149, 1967.
Hébert, G.R., Innanen, S.H. and Nicholls, R.W. 1967 Identification Atlas of Molecular Spectra, 4. The Schumann-Runge System of O_2.
Hébert, G.R., Innanen, S.H. and Nicholls, R.W. 1969 Identification Atlas of Molecular Spectra, 7. The CO Fourth positive system (In preparation).
Huber, M., JOSA 56, 1428, 1966.
Huber, M. and Tobey, F.L., Ap J. 152, 609, 1968.
Jacobs, T.A., Cohen, N. and Giedt, R.R., J. Chem. Phys. 46, 1958, 1967.
Jacobs, T.A. and Giedt, R.R., JQSRT 5, 457, 1965.
Kassel, L.S., Kinetics of homogeneous gas reactions, Chemical Catalog Co., 1932.
Keck, J.C., Camm, J.C., Kivel, B. and Wentinck, T., Ann. Phys. 7, 1, 1959.
Kistiakowsky, G.B., lecture delivered at the O.K. Rice symposium on chemical kinetics, Chapel Hill, N.C., November 1968.
Landshoff, R.K.M. and Magee, J.L. Thermal Radiation Phenomena, Vol. 1, Radiative Properties of Air, Plenum Press 1969a.
Landshoff, R.K.M. and Magee J.L., Thermal Radiation Phenomena, Vol. 2, Excitation and Non-Equilibrium Properties of Air, 1969b.
LeRoy, D.J., Ridley, B.A. and Quickert, K.A., Discuss. Faraday Soc. 44, 92 1967.
Linton, C. and Nicholls, R.W., JQSRT 9, 1-11, 1969a.
Linton, C. and Nicholls, R.W., J. Phys. B. Ser. 2, 2, 490, 1969b.
Linton, C. and Nicholls, R.W., Unpublished work 1969c.
Marlow, W.C., Applied Optics 6, 1715, 1967.
Marr, G.V., Photoionisation processes in Gases, Academic Press, 1967.
Marr, G.V. and Heppinstall, H., Proc. Phys Soc. 87, 293, 547, 1966a.
Marr, G.V. and Heppinstall, H. Proc. Phys. Soc. 87, 547, 1966b.
Marrone, P.V., Wurster, W.H. and Stratton, J.E. 1968 Shock tube studies of N^+ and O^+ recombination radiation in the vacuum ultraviolet. CAL AG 1729-A-7 (Cornell Aeronautical Lab, Buffalo, NY).
McChesney, M. and Al-Attar, Z., JQSRT 5, 553, 1965.
McElwain, D.L.S. and Pritchard, H.O., J. Am. Chem. Soc. submitted for publication, 1969.
McHale, E.T., Knox, B.E. and Palmer, H.B., Tenth Symposium (International) on Combustion, 353, 1965.
McLaren, T.I. and Hobson, R.M., Physics of Fluids, 11, 2161, 1968.
Menees, G.P. and McKeznie, R.L., AIAA Journal 6, 544, 1968.
Meroz, F.: Optical Transition Probabilities (translation), National, Science Foundation, 1962.

Miller, M.H., Thermally insensitive determination of transition probabilities for CI,OI,NeI,AlII,SiI,SiII,PI,PII,SI,SII,ClI Tech. Note BN 550 Institute for Fluid Dynamics and Applied Mathematics, Univ. Md. May 1968.
Miller, M.H. and Bengtson, R.D., Ap. J. 156, 393, 1969.
Mirels, H., Physics of Fluids, 9, 1907, 1966.
Modica, A.P., J. Phys. Chem. 72, 4594, 1968.
Myer, J., Cometary Astrophysics, Ph.D. Thesis, York University, Toronto, 1968.
Myer, J. and Nicholls, R.W., Unpublished work 1969.
Myers, B.F. and Bartle, E.R., J. Chem. Phys. 48, 3935, 1968.
Myerson, A.L. and Watt, W.S., J. Chem. Phys. 49, 425, 1968.
Nelson, H.F. and Goulard, R., JQSRT 8, 1351, 1968.
Newsom, G.H., Proc. Phys. Soc. 87, 975-82, 1966.
Newsom, G.H. and Shore, B., J. Phys. B. Ser. 2 1, 742, 1968.
Newsom, G.H., Ap. J. 154, 1095, 1968.
Nicholls, R.W., Nature 186, 958, 1960.
Nicholls, R.W., Ann. de Geophys. 20, 144, 1964.
Nicholls, R.W. 1969a Electronic spectra of diatomic molecules, Chapter 6, Electronic structure of atoms and molecules, Vol. III, Physical Chemistry an advanced treatise (Ed. H. Eyring, D. Henderson and W. Jost) Academic Press.
Nicholls, R.W., Can. J. Chem. 47, 1847, 1969b.
Nicholls, R.W., Parkinson, W.H. and Reeves, E.M., Applied Optics 2, 919, 1963.
Nicholls, R.W. and Tyte, D.C., Proc. 5th International Shock tube symposium 111-121, 1965, US Naval Ordanance Lab. Silver Spring, Md.
Olschewski, H.A., Troe, J. and Wagner, H. Gg., Eleventh Symposium (International) on Combustion, 155, 1967.
Parkinson, W.H. and Reeves, E.M., Can. J. Phys. 41, 702, 1963.
Parkinson, W.H. and Reeves, E.M., Proc. Roy. Soc. A 282, 265, 1964.
Patch, R.W., JQSRT 5, 137, 1965.
Penzias, G.J., Dubin, S.A. and Kruegle, H.A., Applied Optics 5, 225, 1966.
Pritchard, H.O., in Storage and transfer of energy by molecules, John Wiley, London, Vol. 2, 368, 1969.
Reis, V. JQSRT 4, 783, 1964.
Reis, V., JQSRT 5, 585, 1965.
Rich, J.C. Ap. J. 148, 275, 1967.
Rothe, D.E., JQSRT 9, 49, 1969.
Rush, D.G. and Pritchard, H.O., Eleventh Symposium (International) on Combustion, 13, 1967.
Russo, A.L. and Watt, W.S., Infra Red measurement of vibrational excitation and de-excitation rates of N_2 using CO additive Cal AF-2567-A-1, 1968. Cornell Aeronautical Laboratory, Buffalo, NY.
Seal, K.E. and Gaydon, A.G., Proc. Phys. Soc. 89, 459, 1966.
Shackleford, W.L., JQSRT 5, 303, 1965.
Shaw, D.H. and Pritchard, H.O., Can. J. Chem. 46, 2721, 1968.
Slater, N.B., Theory of unimolecular reactions, Methuen, 1959.
Smekhov, G.D. and Losev, S.A., High. Temp. 6, 369, 1968.

Steiner, H. and Rideal, E.K., Proc. Roy. Soc. A173, 503, 1939.
Stupochenko, Y.V., Losev, S.A. and Osipov, A.I., Relaxation in shock waves, Springer Verlag, New York, 1967.
Sulzmann, K.G., JQSRT, 375, 1964.
Sviridov, A.G., Sobolev, N.N. and Sulowski, E.M., JQSRT 5, 525, 1965.
Taylor, R.L. and Bitterman, S., Rev. Mod. Phys. 41, 26-47, 1969.
Taylor, W.H. and Kane, J.W., Applied Optics 6, 1493, 1967.
Thomas, G.M. and Menard, W.A., AIAA Journal 4, 227, 1966.
Tousey, R., Quart. J. Roy Astron. Soc. 5, 123, 1964.
Tsang, W., Bauer, S.H. and Cowperthwaite, M., J. Chem. Phys. 36, 1768, 1962.
Tyte, D.C., Nature 202, 383, 1964.
Tyte, D.C. Proc. Phys. Soc. 92, 1134, 1967.
Tyte, D.C. and Nicholls, R.W., Identification Atlas of Molecular Spectra 1. Blue Green system of AlO, 1964a.
Tyte, D.C. and Nicholls, R.W., 1964b, Identification Atlas of Molecular Spectra 2. The first negative system of N_2^+.
Tyte, D.C. and Nicholls, R.W. 1965, Identification Atlas of Molecular Spectra 3. The second positive system of N_2.
Tyte, D.C., Innanen, S.H. and Nicholls, R.W. 1967, Identification Atlas of Molecular Spectra 5. The C_2 Swan system.
Vincenti, W. and Kruger, C.H., Introduction to Physical Gas Dynamics, Wiley, 1965.
Watson, R., JQSRT 4, 1, 1964.
Watson, R. and Ferguson, W.R., JQSRT 5, 595, 1965.
Watt, W.S. and Myerson, A.L., C.A.L. No. AD-1689-A9. February, 1969. (Cornell Aeronautical Lab. Buffalo, NY.)
Weeks, Dorothy W. and Simpson, Edwina A., Absorption spectrum of iron in the vacuum ultraviolet 2950-1588A, Scientific Report No. 19, Shock tube laboratory, Harvard College Observatory, 1967.
Westenberg, A.A. and de Haas, N., J. Chem. Phys. 48, 4405, 1968.
Whiting, E.E., Arnold, J.O. and Lyle, G.C., JQSRT 7, 725, 1967.
Windsor, M.W., Davidson, N. and Taylor, R.L., J. Chem. Phys. 27, 315, 1957.
Wolnick, S.J., Berthel, R.O., Larson, G.S., Cornevale, E.H., and Wares, G.W., Phys. Fluids 11, 1002, 1968.
Wood, A.D., Hoshizaki, H., Andrews, J.C., Wilson, K.H., AIAA Journal 7, 130, 1968.
Wray, K.L., J. Chem. Phys. 38, 1518, 1963.
Wray, K.L., J. Chem. Phys. 44, 623, 1966.
Wray, K.L., JQSRT 9, 255, 1969.
Wray, K.L. and Connolly, T.J., JQSRT 5, 111, 1965.
Wurster, W.H. and Marrone, P.V., JQSRT 7, 591, 1967.
Zel'dovich, Y.B. and Raizer, Y.P., Physics of shock waves and high temperature hydrodynamic phenomena, Vols. 1 and 2, Academic Press, New York 1966.

Table 1
The Spectroscopic Method

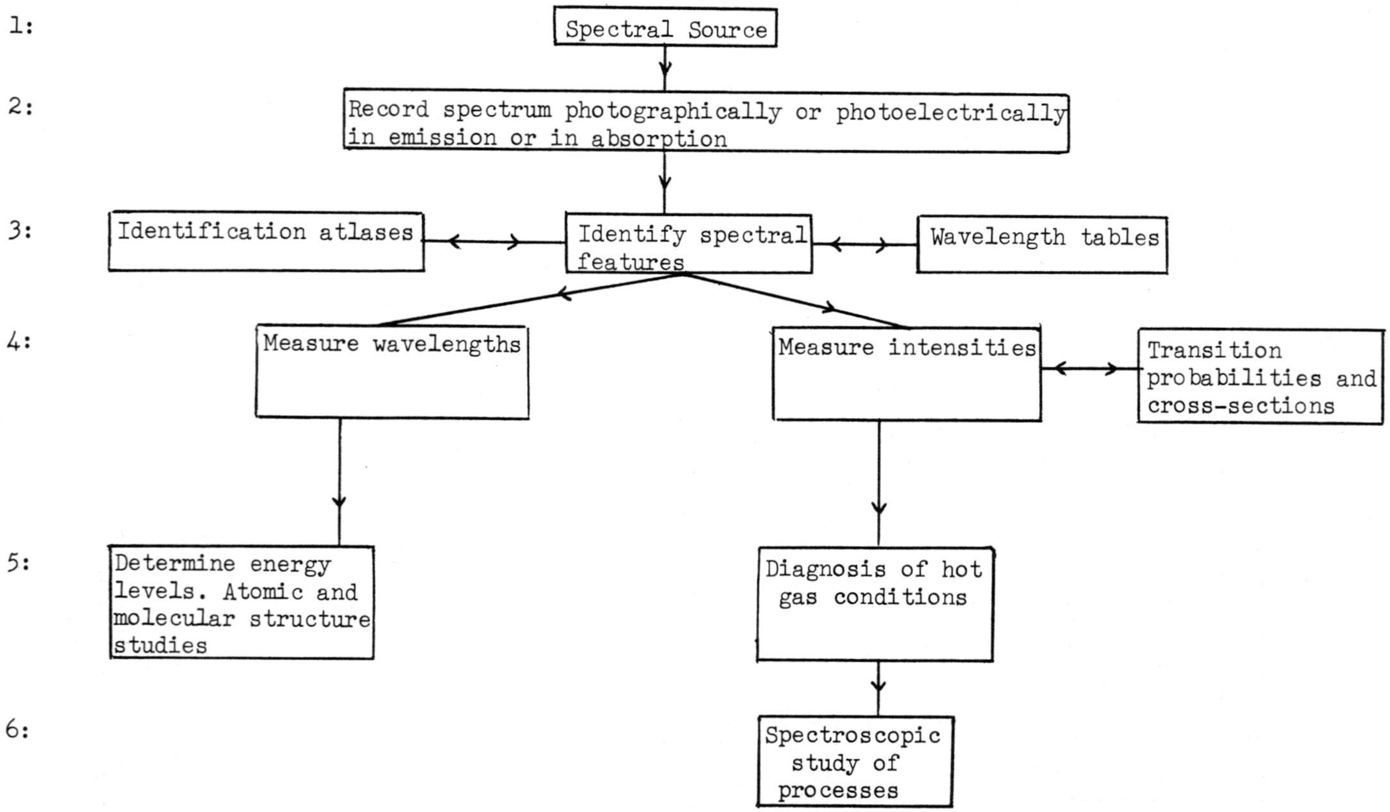

VIBRATIONAL ENERGY TRANSFER IN THE CO_2-N_2-H_2O MOLECULAR SYSTEM

RAYMOND L. TAYLOR and STEVEN BITTERMAN
Avco Everett Research Lab.
2385 Revere Beach Parkway
Everett, Massachusetts 02149

Rates of vibrational energy exchange between various modes of the CO_2-N_2-H_2O molecular system have been determined from simultaneous measurement of the infrared emission from modes ν_2 and ν_3 of CO_2 in shock heated CO_2-N_2 and CO_2-N_2-H_2O gas mixtures. The data have been analyzed by assuming a kinetic model for the collisional transfer of vibrational energy into and between the CO_2, N_2, and H_2O molecules and by using the vibrational energy relaxation equations derived from this mechanism. By using the data previously obtained in CO_2-N_2 mixtures for the rates of the vibration-vibration (V-V) exchange between mode ν_3 of CO_2 and N_2 and between modes ν_2 and ν_3 of CO_2, the rate of the V-V process between mode ν_2 of H_2O and N_2 has been obtained from data in CO_2-N_2-H_2O mixtures. These data cover the temperature range of 700-2000°K and agree well with other measurements and theoretical calculations.

I. INTRODUCTION

The rates and mechanisms of energy exchange between the various internal degrees of freedom of a gas are of considerable practical importance in many high speed flows, e.g. wind tunnels and rocket nozzles. The rates of energy transfer between the translational and vibrational modes of a gas (T-V exchange) have been extensively studied and a large body of experimental data exists on T-V rates together with a good theoretical understanding of such collisional processes. More recently, several experimental techniques have been developed to study the exchange of energy between vibrational modes (V-V exchange).

In this paper we will discuss some recent shock tube measurements of V-V energy transfer in the CO_2-N_2-H_2O molecular system. This molecular system is of interest because of the relevance of the kinetic information to problems involving the CO_2-N_2 laser, energy transfer in the upper atmosphere, and hypersonic flight. This system is also of considerable fundamental interest because it contains examples of various types of V-V exchange, resonant and near-resonant and inter- and intramolecular processes.

Figure 1 is a simplified vibrational energy diagram for the lowest levels of the various modes of CO_2, N_2, and mode ν_2 of H_2O. The arrows indicate the assumed paths for vibrational energy transfer in this system. This mechanism is based in large part on our general knowledge of vibrational energy transfer and theoretical calculations for the CO_2-N_2 system by Herzfeld[1]. This mechanism is used in the data analysis to be discussed subsequently.

Reactions (1), (2), and (3) are examples of T-V processes and some information concerning these rates is available[2]. Reaction (4) is an example of a nearly resonant, intermolecular V-V exchange. (5) is another V-V process, but not nearly as energy resonant as (4). (6) is an example of an intramolecular V-V process between modes ν_3 and ν_2 of CO_2. To be close to energy resonance this process is assumed to be between v = 1 of ν_3 and v = 3 of ν_2, but other possible paths are also possible[2]. Reaction (7) is also an intramolecular V-V process between modes ν_1 and ν_2 of CO_2. Because of Fermi resonance between these modes reaction (7) is assumed to be very fast[1].

II. EXPERIMENTAL

The experiments were performed in a conventional, pressure driven, stainless steel shock tube of 1.5-inch diameter. The experimental arrangement is shown in Fig. 2. Simultaneous measurements were made of the infrared emission from the vibration-rotation bands of modes ν_2 and ν_3 of CO_2 at 15 and 4.3 μ, respectively, behind the incident shock. The measurement system utilized reflective optics, filters to isolate the emission from these bands, and fast response infrared detectors.

Shocks were made into both CO_2-N_2 mixtures and CO_2-N_2-H_2O mixtures. In the latter case the H_2O was introduced into the shock tube by a flowing gas system and its absolute partial pressure determined by means of a calibrated capacitance hygrometer (Panametrics Model 1000). The H_2O flow system is shown in Fig. 3. The temperature of the measurements ranged from approximately 500-2500°K.

A typical oscillogram of the data is shown in Fig. 4 for a shock into 5% CO_2 - 95% N_2. The emission from ν_2 and ν_3 show the following behavior: there is an initial rapid rise in the intensity to some fraction of the final equilibrium level followed by a slower approach to equilibrium. Thus, the data from ν_3 yield three experimental parameters: τ_i, τ_f, and α, the latter defined as the fraction of the final equilibrium intensity (energy) in ν_3 when the transition from τ_i to τ_f takes place. This is illustrated in the lower half of the figure. Similar behavior is found for the ν_2 intensity where β is defined in the same manner as α. Experiments carried out in CO_2-N_2-CO mixtures where the CO was used as a tracer to follow the N_2 show that the N_2 relaxes with one time constant - τ_f.[3] It is observed that τ_f for both modes (and N_2) are identical.

The analysis of the oscillogram shown in Fig. 4 is presented in Fig. 5 where $\ell n\,(\frac{\bar{I}-I}{\bar{I}})$ has been plotted against lab time behind the shock for both modes ν_2 and ν_3. Solid lines have been drawn through the individual data points to determine the initial and final slopes corresponding to τ_i and τ_f for each mode. As can be seen, τ_f for both ν_2 and ν_3 are identical within the scatter of the data. The dashed line represents the relaxation of the N_2, which is not directly measured in this experiment, but has been assumed to be equal to τ_f for modes ν_2 and ν_3. α and β are determined by extrapolation of the τ_f slopes back to t = 0.

When H_2O is added to the CO_2-N_2 mixtures, the relaxation time τ_f for both modes decreases considerably. This is illustrated in Fig. 6 for two runs at about the same shock velocity - the left oscillogram from a 10% CO_2 - 90% N_2 mixture, the right from a 9.3% CO_2 - 83.7% N_2 - 7.0% H_2O mixture. It can be seen that the relaxation time of both modes is shortened considerably in the presence of H_2O. On the basis of the assumed mechanism H_2O can increase the rate at which vibrational energy is transferred to mode ν_2 via reaction (1) with M = H_2O and to mode ν_3 via reaction (6) with M = H_2O or via process (5) to N_2 and then via (4) to ν_3.

τ_f data for modes ν_2 and ν_3 are shown on Fig. 7 for 2 and 10% CO_2-N_2 mixtures. The data are presented as $\tau_f p$ in μsec-torr versus $\overline{T}$ where $\overline{T}$ is defined as the average temperature between conditions where modes ν_2 and ν_1 are relaxed and final equilibrium where all vibrational modes have relaxed. The data for α and β are shown in Fig. 8. The data for β shows considerable scatter; since β is large, i.e. close to unity, it is difficult to determine with precision. Within the scatter β is assumed to be independent of temperature and gas composition.

When H_2O is added to CO_2-N_2 mixtures $\tau_f p$ decreases and α increases as shown in Figs. 9 and 10 respectively. The solid lines on these figures have been fit to 2% CO_2 - 98% N_2 data without H_2O similar to that shown in Figs. 7 and 8. During the course of the H_2O mixture runs, occasional experiments were also made without H_2O as internal checks. These data are shown on Figs. 8 and 9 as the solid symbols. The H_2O concentration on each run varies with changes in flow concentration, however the actual concentration is measured with a calibrated hygrometer. For ease of presentation on Figs. 9 and 10 the various H_2O concentrations have been grouped into three ranges. Runs with H_2O concentrations as high as 7% have also been made.

III. ANALYSIS

The time rate of change of infrared emission from a fundamental vibration-rotation band is proportional to the time rate of change of the vibrational energy of that mode[4]. Hence the analysis of the data utilized the vibrational energy relaxation equations derived from the mechanisms shown in Fig. 1. The vibrational energy relaxation equations were derived with the following two assumptions: (1) The vibrational modes are approximated as harmonic oscillators. This is a reasonable assumption in this study, since, at the temperatures of interest, only the lowest few levels of each mode are appreciably populated; (2) The derivation used the Landau-Teller[5] selection rules for vibrational transitions.

Utilizing the kinetic scheme shown in Fig. 1, the vibration energy relaxation equations for modes ν_1, ν_2, ν_3 of CO_2 and N_2 can be derived[6-8]. The final results are:

$$-\left(\frac{dX_{C2}}{dt} + \frac{2\,\omega_{C2}}{\omega_{C1}}\,\frac{\bar{E}_{C1}}{\bar{E}_{C2}}\,\frac{dX_{C1}}{dt}\right) = \left[\psi_N Z_{CN} P^{10}(C2, N) + \psi_C Z_{CC} P^{10}(C2, C)\right]\theta_{C2} X_{C2}$$

$$-\,2\left[\psi_N Z_{CN} P^{10}_{03}(C3, C2, N) + \psi_C Z_{CC} P^{10}_{03}(C3, C2, C)\right]\theta_{C3}\left\{\left(\frac{1-X_{C2}}{\theta_{C2}}\right)^{3}\left[\frac{1-(1-\theta_{C3})X_{C3}}{\theta_{C3}}\right] - \left(\frac{1-X_{C3}}{\theta_{C3}}\right)\left[\frac{1-(1-\theta_{C2})X_{C2}}{\theta_{C2}}\right]^{3}\right\} \quad (1)$$

$$-\frac{dX_{C3}}{dt} = 2\left[\psi_N Z_{CN} P^{10}_{03}(C3, C2, N) + \psi_C Z_{CC} P^{10}_{03}(C3, C2, C)\right]\theta_{C3}$$

$$\cdot\left\{\left(\frac{1-X_{C2}}{\theta_{C2}}\right)^{3}\left[\frac{1-(1-\theta_{C3})X_{C3}}{\theta_{C3}}\right] - \left(\frac{1-X_{C3}}{\theta_{C3}}\right)\left[\frac{1-(1-\theta_{C2})X_{C2}}{\theta_{C2}}\right]^{3}\right\} -$$

$$\psi_N Z_{CN} P^{10}_{01}(C3, N)\left[X_{C3} - X_N\right] \quad (2)$$

$$-\frac{dX_N}{dt} = \left[\psi_N Z_{NN} P^{10}(N, N) + \psi_C Z_{CN} P^{10}(N, C)\right]\theta_N X_N - \psi_C Z_{CN} P^{10}_{01}(C3, N)\left[X_{C3} - X_N\right] \quad (3)$$

where

$$X = \frac{\overline{E} - E}{\overline{E}} \tag{4}$$

and

$$\Theta = 1 - e^{-hc\,\omega/kT} \tag{5}$$

$\overline{E}$ is the vibrational energy at the translational temperature T, E is the vibrational energy at the local vibrational temperature T_V, ω is the vibrational frequency of the particular mode, and ψ is the mole fraction of a component in the mixture. The subscripts indicate the modes and collision partners: $2 \equiv \nu_2$, etc., $C \equiv CO_2$, $N \equiv N_2$. Z is the total number of binary gas collisions per second. The P's are the probabilities per collision of the various energy transfer processes defined by reactions (1), (2), (4) and (6) of Fig. 1. For example, $P\,(\overset{10}{C3}, \underset{03}{C2}, N)$ is the probability per collision with nitrogen (N) that mode ν_3 of CO_2 (C3) is deactivated from level v = 1 to v = 0 (1 → 0) while mode ν_2 (C2) is activated from level v = 0 to v = 3 (0 → 3). This is the probability for reaction (6) and is the path assumed for energy transfer between modes ν_2 and ν_3 in the derivation of the energy relaxation equations (1) - (3) and the subsequent data analysis. In the derivation of equations (1) - (3) use has been made of the assumption that modes ν_1 and ν_2 are in local equilibrium via the fast reaction (7) to combine $\frac{dX_1}{dt}$ into equation (1). Also, it has been assumed that $\overline{E}_N = \overline{E}_{C3}$ and $\Theta_N = \Theta_{C3}$ to simplify the last terms of equations (2) and (3).

For simplicity equations (1) - (3) have been derived without considering H_2O in the mechanism. Inclusion of H_2O necessitates another T-V term for reaction (1). $_{10}P\,(\overset{10}{C2}, H)$ where $H \equiv H_2O$ in equation (1) and for reaction (2), $P(N, \underset{01}{H})$ in equation (3) with subsequent coupling to ν_3 via reaction (4).

It is obvious that the vibrational energy relaxation equations (1) - (3) are highly non-linear, and it has not been possible to solve them in any closed form. The equations have been linearized for $t > \tau_i$ by using the experimentally observed forms of the vibrational energy time histories, i.e.,

$$X_N = e^{-t/\tau_f}, \quad X_{C3} = (1 - \alpha)\, e^{-t/\tau_f}, \quad X_{C2} = (1 - \beta)\, e^{-t/\tau_f} \tag{6}$$

The following further assumptions have also been employed in the data analysis:

(i) The direct T-V excitation of N_2 via reaction (2) with N_2 and CO_2 as collision partners has been neglected. This allows eqn. (3) to be solved for $P\,(\overset{10}{C3}, \underset{01}{N})$ using data obtained on ν_3 in CO_2-N_2

mixtures. These results have been previously published.[9]

(ii) Having obtained $P^{10}_{01}(C3, N)$, equation (2) can now be solved for the probability for reaction (6) using ν_2 and ν_3 data such as shown in Figs. 7 and 8.

(iii) The effect of H_2O on mode ν_3 of CO_2 can be assumed to be due to reaction (6) with M = H_2O or via V-V coupling to N_2 via reaction (5). A priori one does not know which process, or possibly both, might be important. However, there are results of laser fluorescence experiments[10] in CO_2-H_2O mixtures which have been analyzed for the rate of reaction (6) over the temperature range of 300-1000°K. Using these latter results extrapolated to the shock tube conditions, it is determined that reaction (6) with M = H_2O can not explain the present experimental data obtained in CO_2-N_2-H_2O mixtures. Hence the H_2O data have been analyzed for the rate of V-V reaction (5) using equation (3).

IV. RESULTS

The probability of the intramolecular V-V process between mode ν_3 and ν_2 is shown on Fig. 11. In principle it should be possible to separate the importance of N_2 and CO_2 collisions on the probability for reaction (6), $P^{10}(C3, C2, M)$, by varying the concentration. Moore et al[11] have found $P^{03}_{10}(C3, C2, C) \approx 3\, P^{10}_{03}(C3, C2, N)$ at T = 300°K. However, the present data obtained in 2 and 10% CO_2 concentrations do not show any effect that can be attributed to CO_2-CO_2 collisions, hence the data have been analyzed for $P^{10}_{03}(C3, C2, N)$. The shock tube data are shown as the open squares. The circles are the rate of process (6) obtained by the laser fluorescent experiment of Rosser et al[12]. In their region of overlap the data from these two experiments are in general agreement. Also shown at room temperature are measurements from a number of other techniques. The solid line is a calculation of Herzfeld's[1] multiplied by 10. This calculation sums all the possible V-V rates out of mode ν_3 into various levels of ν_1 and/or ν_2. Clearly this overall probability underestimates the experimental data, although the general temperature dependence of the data is reasonably well reproduced.

Also shown on Fig. 11 are data of Rosser's[10] for the probability of reaction (6) with M = H_2O (triangles). At low temperature this probability is about two orders of magnitude larger than for N_2 as a collision partner, and it also shows a temperature dependence considerably different than that for N_2. It is these results for $P^{10}_{03}(C3, C2, H)$ which have been used with the data of Fig. 8 to determine the proper mechanism for the effect of H_2O on mode ν_3 in the N_2-CO_2-H_2O runs.

The probability for the V-V process between mode ν_2 of H_2O and N_2 (reaction 5) is shown in Fig. 12. These results were obtained from data such as shown in Figs. 8 and 9, assuming that process(5) is the dominant mechanism as explained above. The shock tube data are given by the open squares. The only other data for this process are impact tube measurements of Huber and Kantrowitz[13] shown as the circles. These latter results have been reinterpreted as $P_{01}^{10}(N, H_2)$.[2] The line is a calculation of the probability of this process by means of Herzfeld's theory. As is evident there is excellent agreement between the data and the theory, although this agreement may be considered fortuitous in view of the various assumptions made in analyzing the data and calculating the theoretical line.

V. SUMMARY

Shock tube data in CO_2-N_2 and CO_2-N_2-H_2O mixtures have been used to obtain the probability per collision for various V-V processes in this molecular system. The rates were obtained by assuming a mechanism for this kinetic system and deriving the vibrational energy relaxation equations. These relaxation equations were then solved by using the observed temporal behavior of the infrared emission behind the incident shock from modes ν_2 and ν_3 of CO_2. The experimental results compare well with data obtained by other techniques, particularly the laser fluorescent technique. The data have been compared to theoretical calculations and certain weaknesses of the short range interaction theory for V-V processes are evident.

REFERENCES

1. K. Herzfeld, J. Chem. Phys. 47, 748 (1967).
2. R. L. Taylor and S. Bitterman, Rev. Modern Phys. 41, 26 (1969).
3. R. L. Taylor, M. Camac, and R. M. Feinberg, "Measurement of Vibration-Vibration Coupling in Gas Mixtures", Proceedings of the Eleventh Symposium (International) on Combustion, 14-20 August 1966 (Combustion Institute, Pittsburg, Pa., 1967), pp. 49-65.
4. J. C. Decius, J. Chem. Phys. 32, 1262 (1960).
5. L. Landau and E. Teller, Physik Z. Sowjetunion 10, 34 (1936).
6. R. N. Schwartz, Z. I. Slawsky, and K. Herzfeld, J. Chem. Phys. 20, 1591 (1952).
7. F. I. Tanczos, J. Chem. Phys. 25, 439 (1956).
8. A. Wood and J. D. Teare of the Avco Everett Research Laboratory have derived the energy relaxation equations for a general V-V process.
9. R. L. Taylor and S. Bitterman, J. Chem. Phys. 50, 1720 (1969).
10. W. A. Rosser, Jr. and E. T. Gerry, "The Deexcitation of Vibrationally Excited CO_2 (ν_3) by Collisions with He, O_2, and H_2O", accepted for publication in J. Chem. Phys. (1969).
11. C. B. Moore, R. E. Wood, B. Hu, and J. T. Yardley, J. Chem. Phys. 46, 4222 (1967).
12. W. A. Rosser, A. D. Wood, and E. T. Gerry, "The De-excitation of Vibrationally Excited CO_2 (ν_3) by collisions with either CO_2 or with N_2", in press J. Chem. Phys. (1969).
13. P. W. Huber and A. Kantrowitz, J. Chem. Phys. 15, 275 (1943).

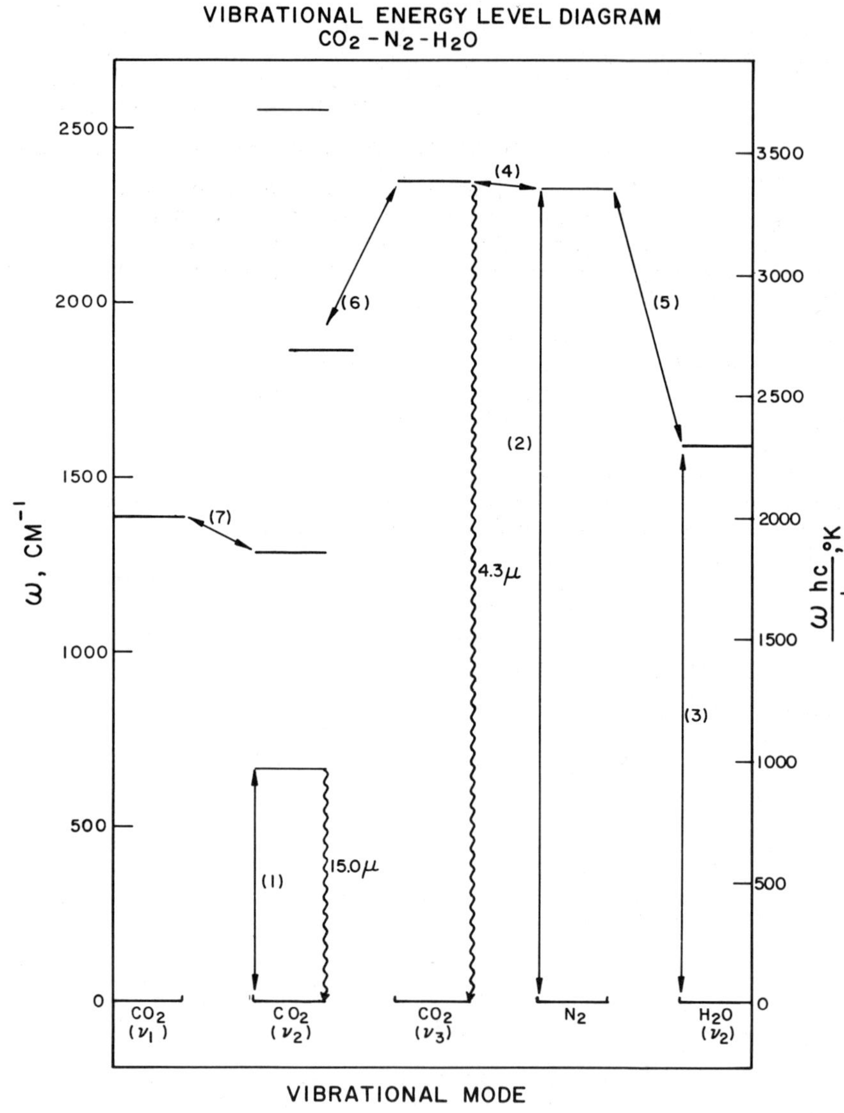

Figure 1
Vibrational energy level diagram for the CO_2-N_2-H_2O molecular system. Shown are the first or the few lowest vibrational levels of the ν_1, ν_2, and ν_3 modes of CO_2, of N_2, and of the ν_2 mode of H_2O. The arrows indicate the assumed paths for vibrational energy transfer within this system of molecules. (The arrows indicate the assumed paths for vibrational energy transfer within this system of molecules.) The optical transitions of the fundamental vibration-rotation bands CO_2, ν_2 and ν_3, are indicated by the wavy lines.

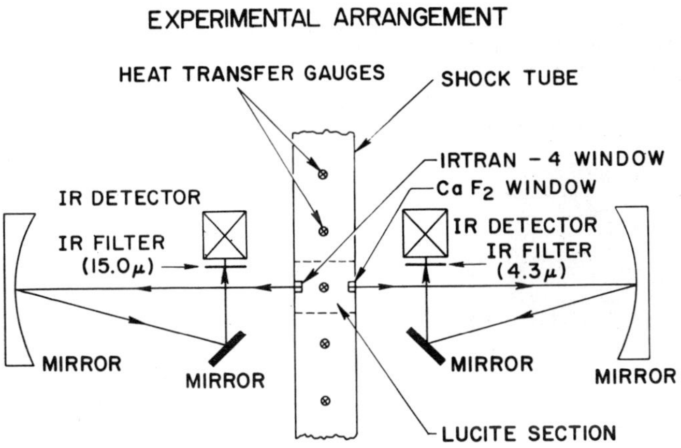

Figure 2
Schematic of the test section on the 1.5-inch-diameter shock showing the dual infrared optical system for simultaneously observing emission from ν_2 and ν_3 of CO_2.

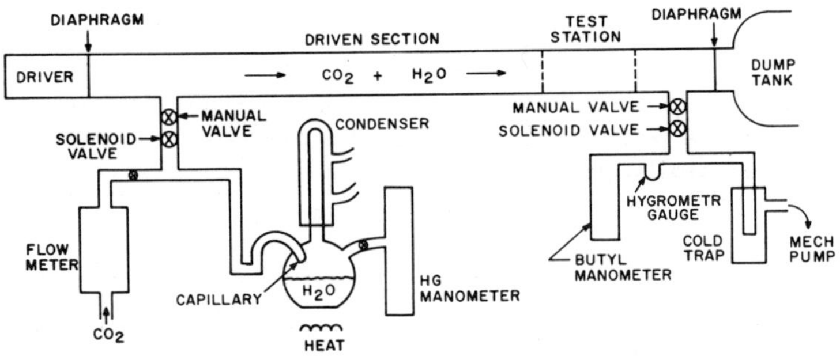

Figure 3
Schematic of shock tube showing principle features of flow system for H_2O measurements. H_2O vapor is introduced into the system through a capillary. The CO_2-N_2 mixture is introduced through an independent system. The gases mix as they flow through the shock tube. The absolute partial pressure of the H_2O vapor in the flow is measured with a capacitance hygrometer. The solenoid valves are used to stop the flow just prior to the run.

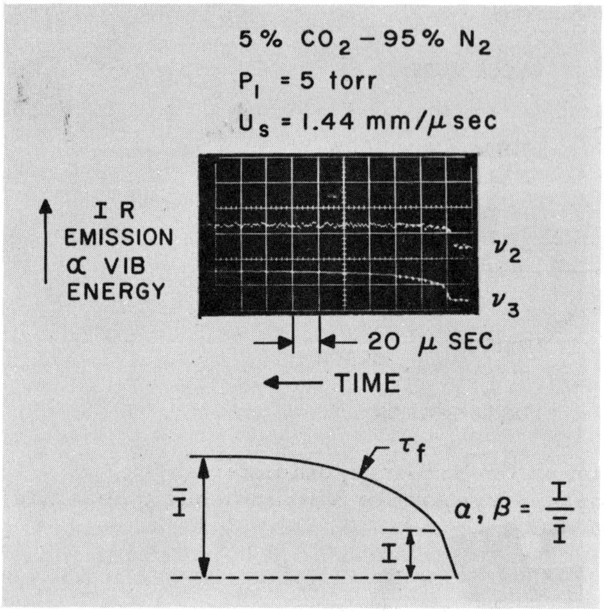

Figure 4
Oscillogram of a shock tube run into 5.0 torr of a 5% CO_2 - 95% N_2 mixture. The emission from both ν_2 and ν_3 is simultaneously displayed. The diagram at the bottom illustrates the temporal features of the emission and the definitions of τ_f, α, and β.

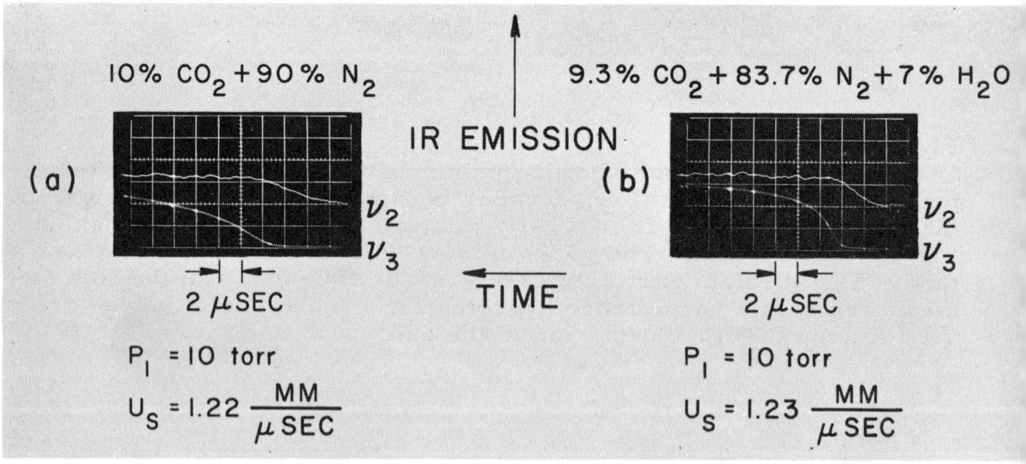

Figure 6
Reduction of the oscillogram of Fig. 4. $\ell n\,(\frac{\bar{I}-I}{\bar{I}})$ for modes ν_2 and ν_3 is plotted versus time behind the shock front, t. The solid lines define τ_i (ν_2), τ_i(ν_3), and τ_f (ν_2) = τ_f (ν_3). The dashed line represents the inferred N_2 relaxation which is not directly measured, α and β are determined by extrapolating the τ_f portion of the curves back to the ordinate.

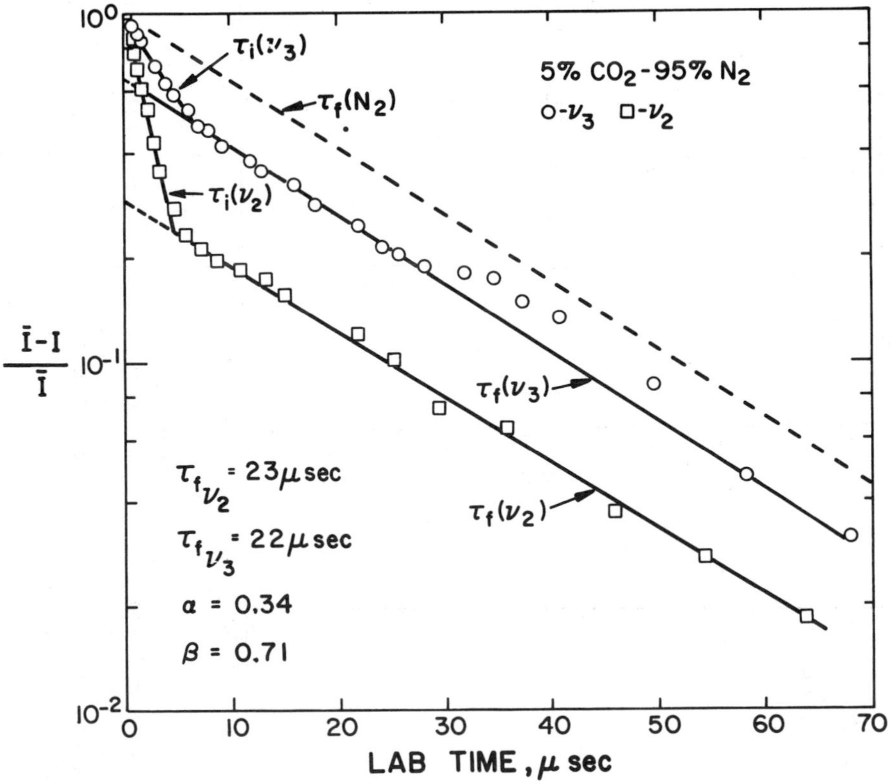

Figure 5
Oscillograms of shock tube runs showing effect of addition of H_2O vapor to a CO_2-N_2 mixture. The left hand oscillogram (a) is obtained in 10 torr of a 10% CO_2 -90% N_2 mixture at an incident shock velocity of 1.22 mm/μsec. (b) shows a run into 10 torr of a 9.3% CO_2 - 83.7% N_2 - 7.0% H_2O mixture at a shock velocity of 1.23 mm/μsec. As can be seen the addition of H_2O increases the rate of equilibrium of both modes ν_2 and ν_3 of CO_2.

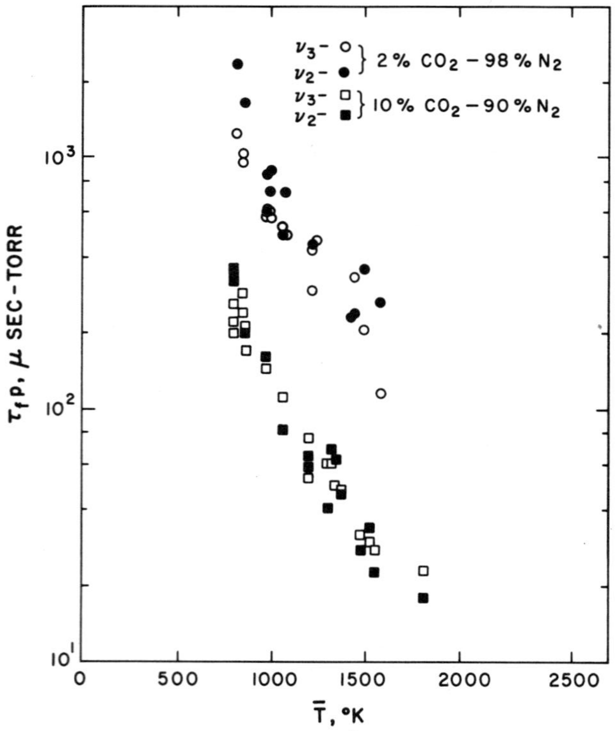

Figure 7
Data for $\tau_f p$ versus temperature, $\bar{T}$, for 2 and 10% CO_2N_2 mixtures. Shown are data obtained from both ν_2 and ν_3 emission. Within the scatter of the data it is seen that $\tau_f p$ from ν_2 or ν_3 are identical.

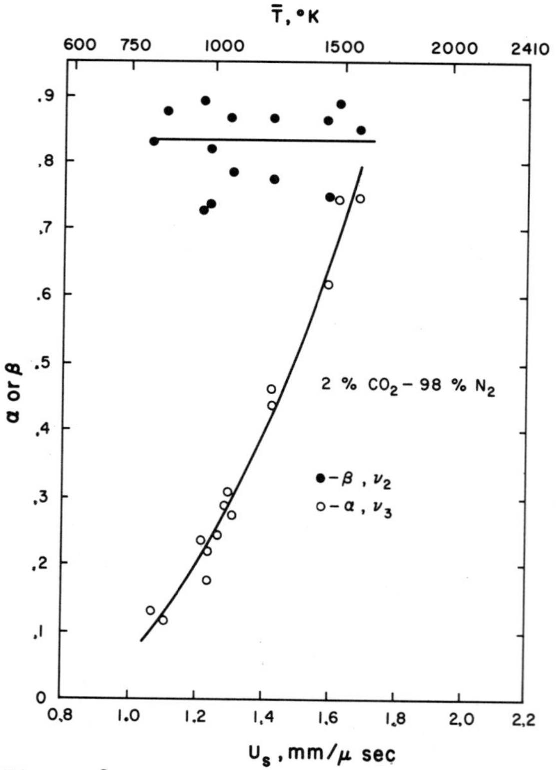

Figure 8
α and β versus $\bar{T}$ or shock velocity u_s. The data are shown for 2% CO_2 - 98% N_2 mixtures. The solid lines have been fit through the data to indicate the trends.

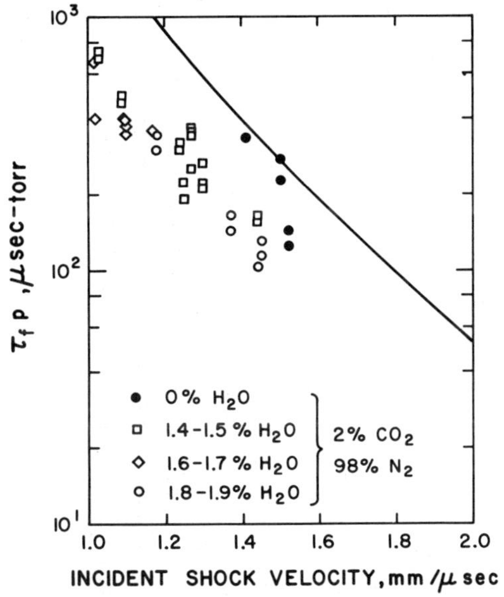

Figure 9
$\tau_f p$ versus incident shock velocity for H_2O mixtures. The solid line is a fit to $\tau_f p$ data obtained in 2% CO_2 - 98% N_2 mixtures (no H_2O). The solid symbols are also with no added H_2O, but were obtained during the course of the H_2O runs, hence some contamination with H_2O vapor may have occurred. The data are for the indicated concentrations of H_2O in a nominal mixture of 2% CO_2-98% N_2.

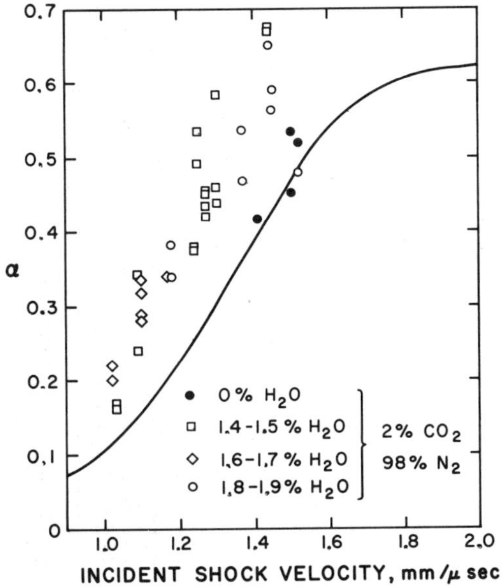

Figure 10
a data versus incident shock velocity for H_2O mixtures. The solid line is a fit to a data obtained in 2% CO_2 - 98% N_2 mixtures.

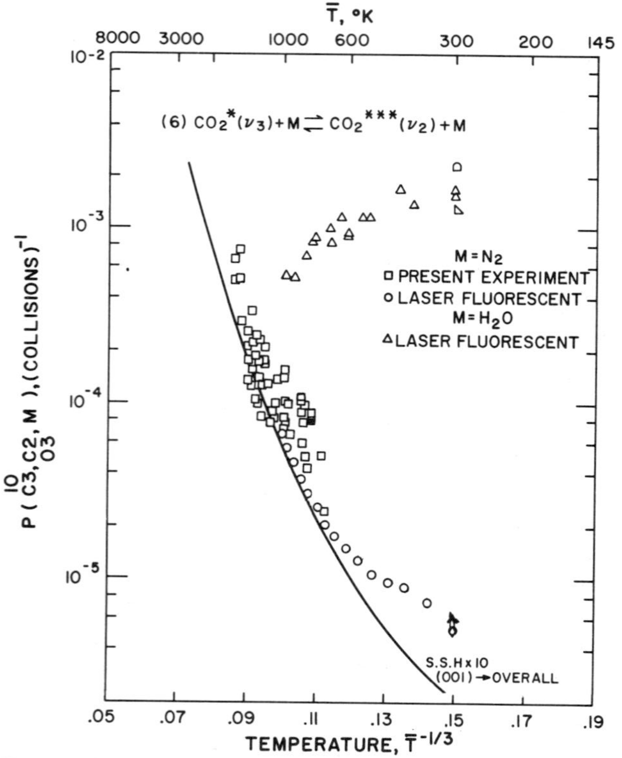

Figure 11

The probability for process (6), $P^{10}_{03}(C3, C2, M)$, versus temperature for various collision partners M. Data are shown for N_2 from the present shock tube experiment - □, as well as from the laser fluorescent experiment of Rosser et al[12] - ○. There are also a number of measurements at $T \doteq 300^{\circ}K$ by a variety of techniques. Data are also shown for $M = H_2O$ from the measurement of Rosser and Gerry[10]. The solid line is a calculation of the probability of deactivation of mode ν_3 of CO_2 by all possible processes for $M = N_2$ multiplied by 10.

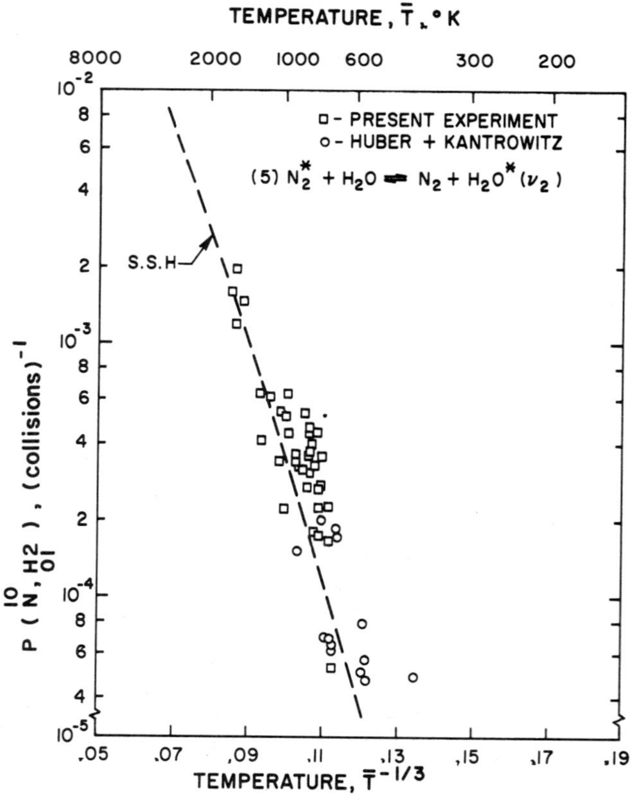

Figure 12

The probability of process (5), $P^{10}_{01}(N, H2)$, versus temperature.

Shown are the data from the present experiment - □, and the impact tube measurements of Huber and Kantrowitz[13], - ○. The dashed line is a theoretical calculation of $P^{10}_{01}(N, H2)$ by means of the S-S-H theory.

POPULATION INVERSIONS IN SHOCK-INDUCED DISSOCIATION OF ALKALI HALIDES

JAMES J. EWING, RICHARD MILSTEIN, and R. STEPHEN BERRY
Department of Chemistry and
The James Franck Institute
University of Chicago

INTRODUCTION

Alkali halide seeded shock waves have proven useful for the spectroscopic determination of halogen electron affinities,[1] for the study of radiative electron attachment to halogen atoms,[2] as a source of sodium atoms for temperature measurements by line reversal methods[3,4] and vibrational relaxation measurements[5], and for the study of the dissociation kinetics of cesium bromide in the 2000°K to 5000°K range.[6,7] The cesium bromide experiments showed that this alkali halide dissociates, in argon, to the ionic rather than the atomic dissociation limit. The ions thus produced then relax to the equilibrium mixture of ions, electrons and atoms. At temperatures exceeding 3300°K the reaction

$$Ar + CsBr \rightarrow Cs^{+} + Br^{-} + Ar$$

yields a population of Br^{-} in excess of the equilibrium value, and chemical equilibrium is attained $\sim 40\,\mu$sec after the passage of the reflected shock responsible for the dissociation. In the CsBr experiments no dissociation products are observed behind the incident shocks. This paper describes similar experiments with several other alkali halides which we are studying to understand more fully the molecular parameters that govern the over-population and relaxation processes.

Atomic overpopulations behind shock waves have been observed in the ionization of Cs[8], Cr[9], and Ti[10] in argon. These experiments have been interpreted in terms of a two step ionization model: collisions with argon on metal dominated the initial stages of relaxation; then electron collisional excitation and ionization occur when sufficient quantities of thermal electrons have been formed. Maxima in populations of excited states arise if the initial rate of ionization of a state is smaller than the rate of excitation to that state. Other shock wave experiments have shown maxima in the concentration vs time profiles of product and intermediate species, notably of OH in the lean $H_2 - O_2$ - argon system[11] and in the similar $H_2 - N_2O$-Argon system.[12] Thermal dissociation of ozone produces an overshoot of oxygen atoms, useful for the study of oxygen recombination.[13]

The possibility of experimental observation of a dissociation directly to ions in the alkali halide molecules was based on earlier theoretical studies of the curve crossing phenomena in alkali halides.[14,15] This work was primarily directed toward explaining the ultraviolet absorption spectra of these molecules. Most alkali halides exhibit continuous ultraviolet absorption, although NaI shows a banded spectrum.[16,17] This was interpreted by Berry in terms of the coupling of zero order Born-Oppenheimer states by the nuclear kinetic energy near the crossing point, R_x, for an alkali halide MX. The crossing point is the internuclear separation at which the coulombic (ionic) potential generated by $M^+ + X^-$ crosses the flat atomic (covalent) potential curve of $M^o + X^o$. R_x is given by

$$1) \qquad e^2 R_x^{-1} = \text{I. P. (M)} - \text{E.A. (X)}$$

where I. P. (M) is the metal ionization potential and E. A (X) is the halogen electron affinity. It is at the crossing point that the Born-Oppenheimer adiabatic curves will drastically change in nature. For $R < R_x$, Ψ_1, (the wave function of the lower state) is best described by an ionic function, while at large distances the electron must jump from X^- to M^+ and Ψ_1 must become an atomic function if the states behave adiabatically. The upper state Ψ_2 is covalent for $R < R_x$ and becomes ionic as it proceeds to the dissociation limit of $M^+ + X^-$. Figure 1 shows potential energy curves for NaI. The splitting Δv at the crossing point, R_x, will be determined by the matrix elements of the electrostatic attractions and repulsion operators between the diabatic states. Table 1 shows the crossing points for the alkali halides, distances in atomic units. Table 2 shows the splitting Δv in cm^{-1} between the adiabatic curves.[18] The nuclear kinetic energy operator will mix the Born-Oppenheimer states.

If the non-crossing rule were applicable to thermal dissociation of an alkali halide in an argon heat bath, the primary dissociation products would be neutral atoms, M + X. CsBr dissociates initially to the ionic limit,[6] which suggests that if the crossing distance is as large as that of CsBr, $R_x = 51.5$ a u, the potential curves are best thought of as crossing rather than obeying the adiabatic non-crossing rule. Physically this means that the electron cannot make a jump from Br^- to Cs^+ during the time that the molecule is near the crossing point. The overlap of the halide p σ orbital and s-orbital of the alkali is very small in CsBr because R_x is so large. Other alkali halides, for which R_x is considerably smaller than R_x (CsBr), have larger orbital overlap and hence larger probability for electron transfer to the metal.

The magnitude of the splitting of the adiabatic states, Δv, is a measure of the size of electron transfer probability near R_X, and also of the range through which transfer can take place. The smaller R_X is, the larger is the overlap of halide valence p σ orbital and alkali valence s orbital and the larger the probability of adiabatic electron transfer. Herschbach and Grice[18] have calculated these splittings for the alkali halides using various metal and halogen atomic wave functions. Average values for Δv for various wave functions are given in Table II, and though these numbers may be wrong by as much as a factor of 3, they are instructive in their trends. Rb Br, for example, has a $\Delta v \sim 10^{-7}$ a.u. = .02 cm^{-1} which is very much smaller than the vibrational spacings of this molecule, $\omega_e = 169\ cm^{-1}$. LiBr on the other hand has a $\Delta v = 480\ cm^{-1}$, and $\omega_e = 563\ cm^{-1}$, and clearly represents a different extreme, one where dissociation may proceed to the atoms because Δv is large.

It is useful to make a gross classification of the alkali halides into a group expected to dissociate to ions, a group expected to dissociate to atoms, and an intermediate group. We find that a very useful quantity upon which to make this correlation is the ratio of Δv_v, the vibrational level splitting in the (crossing) ionic well at the energy of the atomic limit, to the splitting, Δv, of the adiabatic electronic levels at the crossing distance. To compute Δv_v, we used the Rydberg expression

$$\Delta v_v = \frac{2C\mu R}{n^3}$$

where μ is the reduced mass of the molecule, C is a conversion constant, Ry the Rydberg constant and n the quantum number of the state in the coulomb well closest to the atomic limit.

$$n = \left(\frac{C\mu Ry}{I.P. - E.A.}\right)^{1/2}$$

This model of course will not be valid for the bottom of the potential well. Table III gives Δv_v calculated on this coulomb model. Δv_v is the splitting of vibrational states of the coulomb well at the crossing point. If Δv is considerably larger than Δv_v we expect the non-crossing rule to be followed. If Δv_v is larger than Δv, as in CsBr, we expect the crossing rule to be violated. The three molecular parameters given in the tables should effect the crossing and overpopulations. We also tried using other models such as the Morse oscillator.

Primary dissociation to the ionic limit is not a sufficient condition for the generation of maxima in the concentration profiles of the ions. To have such an overshoot (in contrast to a simple nonthermal population imbalance between ions and atoms) one must have a quenching rate of the species that is slower, initially, than rate of production of that species. The quenching rate may build up with time, or the production rate slow down, and a maximum can be produced whenever the quenching rate and production rate are balanced. CsBr at low temperatures dissociates to ions more slowly than the ions relax to atoms, and no maxima are observed. At higher temperatures the Br^- concentration goes to a maximum, often 5 x the equilibrium value.

This paper discusses general features of the dissociation kinetics of several alkali halides in the temperature range 2000°K to 7000°K. The generation of population inversions and imbalances will be discussed in terms of the crossing or avoided crossing of potential curves. The nature of the dissociation products first reached will be described in terms of the molecular parameters given in tables I, II, and III. Generation of maxima in concentration profiles will be correlated with imbalances of specific rates. The derivation of rate constants for the competitive reactions involved in shaping the species' concentration vs time profiles will be discussed along with possible mechanisms as derived from computer fitting of the experimental curves.

Table I
R_X in atomic units

	F	Cl	Br	I
Li	14.0	15.3	13.4	11.7
Na	16.1	17.8	15.3	13.1
K	30.6	37.3	27.8	21.3
Rb	37.3	47.7	33.2	24.3
Cs	61.8	97.1	51.3	32.8

Table II
Δv in cm^{-1}

	F	Cl	Br	I
Li	192	144	480	1720
Na	48	36	200	1200
K	.072	.024	.48	48
Rb	v.s.	v.s.	.024	9.6
Cs	v.s.	v.s.	v.s.	.24

Note: v.s. means less than 10^{-3} cm^{-1}.

Table III
Δv_v in cm^{-1}

	F	Cl	Br	I
Li	122	101	117	141
Na	69.6	51.8	57.3	69.2
K	24.0	14.7	19.3	27.0
Rb	16.1	8.8	11.8	16.9
Cs	7.3	2.9	5.6	9.6

Vibrational splittings of a coulomb potential at the crossing energy

EXPERIMENTAL

The shock tube and experimental procedures have been described in previous papers and will be briefly sketched here. A square tube 3-1/2" x 3-1/2" I. D. by 15' length is the driven section, filled typically with 50 torr argon. Mylar or scribed aluminium discs separated this section from the drive R section. He or H_2 at pressures ranging from 75 to 400 psi burst the diaphragms. A small amount of salt was introduced near the end wall by heating salt crystals on nichrome. Convection dispersed the resulting smoke through the final 2 feet of the driven section. The salt concentration was roughly 0.1 mole percent of the argon or less. The incident shock wave evaporated the alkali halide, the reflected shock induced dissociation. Final temperatures, calculated from Mach no. data, varied from 2000°K to 7000°K. The X^- and M^o concentrations were monitored with a rotating drum spectrograph. Light from a capillary flash lamp was used as the continuous background for the absorption measurements. Power from an inductively loaded 74 μf bank of capacitors charged to 6.3 kv gave a relatively square light pulse of 300 μsec duration. The analysis flash passed through the shock tube four times in a vertical White cell and was focused on the slit of the spectrograph. Under typical conditions of alignment, slit width and height, and drum speeds, the time resolution was 2 μsec. The spectra were recorded on Kodak 33 mm 103-0 film, developed in D-19. The data on the film was reduced by microdensitometry of the atomic lines or detachment continua of interest.

The following alkali halides have been studied: CsBr, RbBr, K Br, NaBr, LiBr, KI, NaI, CsI, RbCl and LiCl. Of these CsBr has been the most comprehensively studied, and trends in the other salts have been mapped out. Some molecular absorption experiments have been done, but the molecular spectra of most interest have not yet been successfully measured.

RESULTS AND DISCUSSION

Figure 2 shows the result of a typical CsBr experiment and a RbCl experiment for comparison. In both the halide ion concentration, represented here by logarithmic intensity differences, reaches a maximum from 6 to 14 μ sec after the passage of the reflected shock. This overshoot depends on temperature, with 3000°K the threshold for observation of a halide maximum. The ratio $I_{max}(Br^-)/I_{eq}(Br^-)$ increases with temperature, largely because the Br^- concentration at equilibrium is smaller at higher temperatures. The dissociation rates and relaxation rates are not extremely sensitive to temperature variations in the 3000°K to 5000°K range. At all temperatures the CsBr, and RbCl, dissociate initially to the ionic limit as noted by the appearance of X^- 2 to 6 μsec before the appearance of M^o. A plot of the dimensionless quantity

$$\theta = \frac{X^-(t)}{M^o(t)} \cdot \frac{M^o(eq)}{X^-(eq)}, \quad \text{or} \quad \theta^{-1},$$

shows this trend for CsBr quite well, as θ starts out very large,infinite if complete population inversion of the ionic limit is attained, and drops to 1 ~ 30 μsec after the passage of the reflected shock. θ^{-1} is zero for several μ sec after passage of the reflected shock and then rises linearly to 1. Such θ plots are attained in CsBr whether or not the Br^- concentration profile shows a maximum, indicating that one can have an inversion of the concentrations of ionic and atomic limits without having an overshoot in the Br^- concentration. Zeros in θ^{-1} plots indicate a curve crossing and inversion of ions with respect to atoms. Maxima in concentration profiles indicate a temporary imbalance of pertinent rate constants for that species. The Br^- concentrations at the maxima correspond to roughly 2×10^{14} particles/cm^3, the concentration of Cs^o at equilibrium is of order 10^{15} particles/cm^3, and the initial concentration of molecules is of order 2×10^{15} particles/cm^3. Conversion of intensities to amounts is based on published cross sections and oscillator strengths. (See ref. 2 and footnotes 4 and 12 of ref. 6.) The results for CsBr (R_X = 51.5 a u) will serve as a baseline for discussing other salts.

RbBr has a crossing point distance of 33.2 a u. This molecule dissociates into ions in the primary step. Overshoot of Br^- is observed, and in fact is strong enough to be observed with a single pass of the analysis flash through the hot gas. Attempts at observing the decay of optical absorption by RbBr molecules were unsuccessful. Since Δv_v is some 200 times Δv, the potential curves are best thought of as crossing, as in CsBr. KBr shows definite Br^- overshoot and the decay of the Br^- curve is quite similar to ones obtained for Br^- generated in CsBr shocks at similar temperatures. Figure 3 shows first order decay plots for 3 such Br^- relaxations. Such a relaxation is consistent with detachment of electrons from X^- by argon, electrons, or halide neutrals. Curve crossing upon dissociation also rules out direct charge exchange as a mechanism of neutralization. The K atoms equilibrate quickly in the reflected shock at 3900°K, and are present in to a small extent in the incident shock. The θ^{-1} plots for KBr are similar to the CsBr θ^{-1}, indicating dissociation occurs to ions. Δv_v = 19.3, and Δv = .48, indicating that ionic dissociation is favored.

The concentration profiles of LiBr are quite different from those of CsBr. Figure 4 shows that the Br^- and Li concentrations rise at the same time. No Br^- overshoot is observed, even though the CsBr at these temperatures (4400°K) produces a very large superabundance of Br^-. A plot of θ^{-1} for LiBr shows an entirely different form than one for CsBr. θ^{-1} starts above 1 for LiBr and rapidly reaches 1, while θ^{-1} for CsBr grows from zero to 1 as shown in Figure 5. These results indicate that LiBr dissociates principally to the atomic limit, and that relaxation of the Li atoms to ions occurs quickly. This is consistent with a small R_x, 13.4 a u, and Δv_v smaller than the adiabatic term splitting, 112 cm^{-1} and 480 cm^{-1} respectively.

Finally NaBr (R_x = 15.3 a u) shows no Br^- absorption at equilibrium (because of the high temperature) or in any overshoot. The Na atoms reach a maximum in 6 μsec and show a noticeable overshoot. Overshoot of metal, rapid metal rise, and lack of Br^- overshoot at temperatures where CsBr shows a huge overshoot all indicate dissociation to the atomic limit. There is a Na overshoot because its rate of production is faster than its initial rate of ionization. The small crossing distance, the large Δv (200 cm^{-1}), and the small vibrational energy gap Δv_v are all consistent with the experimentally observed dissociation to the atomic limit. The metal overshoot relaxes in roughly 15 μ sec, corresponding to the excitation of the Na ^2S state formed upon dissociation to higher terms.

Amongst the iodides, it was anticipated that the compounds with large R_x might show dissociation to the ionic limit, and strong I^- overpopulations. In fact it was extremely difficult to see any I^- absorption. Figure 6 shows data for CsI. The I^- is present before Cs atoms, and the concentration profile of I^- indeed shows a maximum but the small absorption makes quantitative chemical kinetics impossible without considerably longer optical paths. The I^- concentration reaches a maximum about 12 μ sec after the relfected shock passes. Cs^0 has an induction time of ~12 μsec at 3000°K, and equilibrates 20 μ sec after the reflected shock. The overshoot and the rate of metal equilibration increase with temperature. Thus CsI (R_x = 32.8 au) dissociates to the ionic limit, consistent with Δv(0.24cm^{-1}) being less than Δv_v (9.6 cm^{-1}). Strong maxima are not as readily seen with I^- because the rate of detachment of electrons from I^- is faster than the detachment from Br^-. This is consistent with the smaller electron affinity of I, 3.06 ev vs. 3.36 for Br, and the greater polarizability of I^- (See next section).

RbI shocks showed faint I^- absorption and only a small overpopulation. Rb was detected in the incident shock, and apparently represents an intermediate case in which dissociation is to both the atomic and ionic limit. The crossing distance of RbI is 24.3 au, less than the 27.8 a u of KBr which is known to dissociate to ions in the primary step, and more than the 15.3 a u crossing distance of NaBr. Based on magnitude of crossing distance, RbI is close to KBr, and one might be tempted to class it with those molecules which dissociate to the ionic limit. However Δv of RbI is 9.6 cm^{-1}, comparable in size to the vibrational splitting calculated in the coulomb model, 16.9 cm^{-1}. RbI is best thought of as an intermediate case in terms of these splittings. The splitting Δv is 20 x larger for RbI than Δv of K Br because Rb^+ and I^- are both considerably larger in size than K^+ and Br^- so that the orbital overlap is much larger at the crossing point.

K I also is an intermediate case, showing very faint I^- absorption, visible only as a small overshoot, and fast formation of K neutrals. Figure 7 shows data for K I at 4800°K. Absorption by K atoms reached a maximum 10 μ sec after arrival of the reflected shock and relaxed in 15 μ sec at 4800°K. Lower temperatures produce smaller overpopulations of K atoms in the ground atomic $4\,^2S$ state. This state is rapidly overpopulated by dissociation to the atomic channel, and is later excited to the equilibrium distribution of its excited and ionized states, probably by electron impact processes. The excitation and ionization processes apparently have a smaller temperature dependence than the argon-initiated dissociation process. Since an I^- overpopulation is also observed we must class K I as an intermediate system also, one in which dissociation can proceed to both the atomic and ionic limits. This is consistent with the vibrational gap being the same order of magnitude as the electronic state gap, as shown in Tables II and III.

NaI (R_X = 13.1 a u) never produced detectable amounts of I^-. Na atoms were present in incident shocks as cold as 1800°K. Na atoms rose to a maximum 6 μ sec after the passage of the reflected shock, exhibiting a small overshoot. NaI, like NaBr, clearly dissociates to the atomic limit, consistent with the very large adiabatic term splitting 1200 cm^{-1}.

RbCl, as shown in Figure 2, is very similar in its dissociation properties to CsBr. Cl^- is formed faster, consistent with the energy transfer probability from argon to RbCl being greater than from argon to CsBr. This can be seen by comparing mass factors in a classical hard sphere $T \rightarrow V$ energy transfer calculation. Rb neutrals take longer to equilibrate than Cs neutrals at a similar temperature, a fact we have not yet explained. A plot of Θ^{-1} shows clearly that the dissociation proceeds to the ionic limit. These observations are consistent with the large RbCl crossing distance, 47.7 a u, and the minuscule term splitting at the crossing point, $\Delta v \sim 10^{-3}$ cm^{-1}. Our data on other chlorides is incomplete, but we have observed a Li neutral emission overshoot at 5900°K. The 3232 Å (3p $\rightarrow$ 2s) line rises to a maximum 10 μ sec after passage of the reflected shock. This indicates an overpopulation of Li atoms, consistent with a large splitting of the adiabatic curves, 144 cm^{-1}.

Among the alkali halides experimentally examined we see the following trends:

a) If $R_X > 25$ au the ionic limit is initially populated, often producing an amount of X^- greater than the equilibrium amount.

b) If $R_x < 20$ au, the atomic limit is preferentially populated, often producing maxima in the concentration of M^o.

c) If 25 au $> R_x > 20$ au, a mixed dissociation process is involved, often leading to maxima in both the M^o and X^- profiles.

This classification by crossing point is only an approximation as one should really consider the splitting of the Born-Oppenheimer potentials and the vibrational energy spacings at the top of the well. Since the splittings and spacings can be approximated, we can classify the 3 trends accordingly

a) adiabatic dissociation to atoms for $\Delta v > \Delta v_v$

b) curves best thought of as crossing for $\Delta v_v \gg \Delta v$

c) intermediate, mixed modes of dissociation for Δv_v comparable to Δv.

A maximum in a halogen ion or metal atom concentration is caused by a dissociation to that species that is faster than the further relaxation of that species. For maxima in Br^- and Cl^-, temperatures above 3300°K are required, as well as dissociation to the ionic limit caused by the curve crossing. Dissociation to the ionic limit generates only weak I^- overpopulations implying the relaxation processes are faster for I^-. For metal maxima one requires a fast dissociation to the atomic limit, and temperatures in excess of 3500°K. The relaxation of these overpopulations we believe to be governed by electron impact excitations and ionizations. This mechanism has the favorable aspect of allowing delays in the relaxation of superabundant species because electron concentrations build up slowly at first by argon collisional detachment processes. This is consistent with the observations of electron collisional dominated ionizations and excitations of metal atoms behind shock waves.[8,9,10] The mechanism of the electron ionization and detachment are discussed in the following section.

COMPUTER SIMULATION OF CsBr DATA

In an effort to understand more fully the detailed mechanistic processes involved in the shape and size of the experimental concentration profile, a program of computer simulation of experimental data has been undertaken. The alkali halides are especially convenient systems for such kinetic analysis because of the limited number of species involved. For CsBr at reasonable temperatures the only species that can be present are CsBr, Cs^o, Cs^+, Br^-, and electrons. Cs^o, of course, will be distributed amongst its various excited states. Our experiments, to date, have concentrated on monitoring two of the species, the neutral metal M^o and the halogen ion X^-; both species have absorption features in the near ultraviolet. Monitoring these two species is enough to demonstrate disequilibria and dissociation to the atomic or ionic limits in the primary step. However, they are not enough to pin down the concentration profiles of all the other species at the same time or enable a unique solution to the coupled rate equations. Our curve fitting program requires the model system to satisfy charge balance and mass balance. An accurate measurement of a third species should allow us to determine all the pertinent rate constants. Thus far, we have attempted to fit the experimentally measured profiles (Cs^o and Br^-) to see what mechanisms are consistent with the data we have, and how the calculated profiles respond to changes in the rate constants. Without measurement of a third species, we cannot expect such a calculational program to determine all of the pertinent rate constants, but general trends and orders of magnitude can be derived. We have fit the experimental curves for several CsBr experiments, using a modified Euler method[20] to solve the six non-linear first order coupled differential equations.

In all the CsBr experiments 3 trends are obvious: 1) Cs neutrals lag Br^- in appearance b) the Br^- curves show a linear rise at early times. c) Br^- in general, reaches a maximum 10 μsec after the reflected shock. It maintains the maximum for a few microseconds, and decays to equilibrium. At sufficiently high temperatures the relaxation appears exponential, usually deviating from simple first order kinetics before equilibrating.

The mechanism we are presently using to fit the curves is:

1) $CsBr + Ar = Cs^+ + Br^- + Ar$	k_d,	k'_d	- primes for reverse rates
2) $Ar + Br^- = Br + e^- + Ar$	k_r,	k'_r	
3) $e^- + Br^- = Br + 2e^-$	k_{r2},	k'_{r2}	
4) $CsBr + e^- = Cs + Br + e^-$	k_{d2},	k'_{d2}	
5) $Cs + e^- = Cs^+ + 2e^-$	k_m,	k'_m	

The reverse rate constants for these reactions are calculated from thermodynamic equilibrium relations and the forward rate constants supplied to the program. Process 1 is the argon induced dissociation of CsBr into ions for molecules whose curves are best treated as crossing. For future salt studies we have modified the program to take into account a finite branching ratio into the atomic channel, as represented by

$$Ar + CsBr = Cs + Br + Ar.$$

The rate constant k_d is derived from the initial slope of the Br^- curves, the density of salt present in the shock tube, and the pressure of argon gas.

Process 2 is the argon detachment of electrons from bromine ions. Process 1 and 2 together cannot by themselves simulate the Br^- curve. k_r is chosen to fit the Br^- curve in combination with Process 3. Process 3 is the electron induced electron detachment that has been described for $e + H^-$.[21] The rate constant k_{r2} is adjusted and is seen to control the fit of the decay of the Br^- overpopulation. k_r controls the Br^- maximum concentration. Electron capture by Cs^+ controls the formation of Cs neutrals from the ions populated in the first step. However, even if its efficiency is unity, this process cannot make Cs^o neutrals appear in detectable amounts in a time of a few microseconds. Hence Process 4, an electron impact excitation of CsBr to the dissociative upper state was postulated to account for the rise of Cs neutrals and to provide a dissociation time of CsBr molecules faster than the very slow dissociation via argon in step 1. Using cross sections of a similar magnitude to other allowed impact excitations,[22] dissociations[23] and ionizations[8, 9, 10] we were able to fit the Cs^o neutral profiles reasonably well with processes 4 and 5. The formation of electrons first by argon-Br^- collisions, then by e^- - Br^- collisions is the profile determining feature both for the Br^- ions and the Cs neutrals. Figure 8 shows the experimental points and the calculated fit for a CsBr shock at 3960°K. The rate constants used to fit these curves are

1) $k_d = 3.3 \times 10^{-16}$ cm^3/particle sec

2) $k_r = 3.1 \times 10^{-15}$ cm^3/particle sec

3) $k_{r2} = 4 \times 10^{-11}$ cm^3/particle sec

4) $k_{d2} = 4 \times 10^{-10}$ cm^3/particle sec

5) $k_m = 2.6 \times 10^{-9}$ cm^3/particle sec

The curve fitting procedure has not yet been applied to shocks at other temperatures so temperature dependences of the rate constants, have not yet been determined. The fit of the Cs atom curve is not as good as the bromine ion curve leading us to believe that this mechanism though adequate is not a complete story. Associative detachment

$$Br + Br^- \rightarrow 2Br + e^-$$

may compete with the electron impact dissociation of step 3, though our simulations show that for reasonable cross sections[24, 25] this cannot be the dominant relaxation mechanism for Br^-. We also have not included in our mechanism dissociative attachment as a means of producing atoms and ions. Reactions of this from

$$AB + e^- \rightarrow A + B^-$$

are known to have large cross sections[26, 27] at low energies. Because of the ionic character of the alkali halides, this process may be extremely important.

CONCLUSION

We have studied the dissociation of alkali halides in reflected shock waves and found that several of these salts dissociate to the ionic limit violating the non-crossing rule. The dissociation often leads to maxima in the concentration profiles of the species produced, atoms if the atomic limit is first reached, ions if the non-crossing rule is violated, and both for some intermediate cases. The crossing distance is a reasonable measure of the propensity of a curve crossing dissociation. Large R_x lead to ions, $R_x < 20$ a u leads to atoms, and several systems are found to lie in between. The ratio of the adiabatic term splitting, Δv, to the vibrational spacing near the top of the well, Δv_v, was found to be a more reliable indication of the limit attained upon dissociation. For $\Delta v > \Delta v_v$, atoms are the initial dissociation product. For $\Delta v \sim \Delta v_v$. both ions and atoms are formed in the argon dissociative step. For $\Delta v << \Delta v_v$, usually for $R_x > 30$ a u, a population inversion of the ionic limit is found. Kinetic analysis of the CsBr experiments indicate that electrons formed 8 μsec after passage of the reflected shock are very important in determining the relaxation of ions and the dissociation of molecules.

REFERENCES

1. a) R. S. Berry and C. W. Reimann, J. Chem. Phys. 33, 917 (1960). b) R. S. Berry, C. W. Reimann and G. N. Spokes, J. Chem. Phys. 43, 3067 (1965).
2. D. E. Rothe, Phys. Rev. 177, 93 (1969).
3. J. G. Clouston, A. G. Gaydon and I. I. Glass, Proc. Roy. Soc. A, 248, 429 (1958).
4. J. G. Clouston, A. G. Gaydon, and I. R. Hurle, Proc. Roy. Soc. A, 252, 143 (1959).
5. I. R. Hurle and A. G. Gaydon, Nature 184, 1858 (1959).
6. R. S. Berry, T. Cernoch, M. Coplan and J. J. Ewing, J. Chem. Phys. 44, 127 (1968)
7. 6th International Shock Tube Symposium.
8. A. F. Haught, Phys. Fluids 5, 1337 (1962).
9. W. L. Shackleford and S. S. Penner, J. Chem. Phys. 45, 1816 (1966).
10. A. A. Boni, Jr., J. Chem. Phys. 49, 3885 (1968).
11. W. C. Gardiner, Jr., K. Morinaga, D. L. Ripley and T. Takeyama, J. Chem. Phys. 48, 1665 (1968).
12. N. Henrici and S. H. Bauer, J. Chem. Phys. 50, 1333 (1969).
13. K. L. Wray, J. Chem. Phys. 38, 1518 (1963).
14. J. L. Magee, J. Chem. Phys. 8, 667 (1940).
15. R. S. Berry, J. Chem. Phys. 27, 1288 (1957).
16. a) H. D. Schmitt-ott, Z. Physik, 69, 724 (1931). b) H. Levi, Doctoral Dissertation, Berlin, 1934.
17. R. S. Berry and W. Klemperer, J. Chem. Phys. 26, 724 (1957).
18. D. Herschbach and R. Grice, Private communication.
19. Y. P. Varshni and R. C. Shukla, J. Mol. Spec. 16, 63 (1965).
20. T. R. McCalla, Introduction to Numerical Methods and Fortran Programming, John Wiley and Sons, Inc. 1967, p. 314.
21. D. F. Dance, M. F. A. Harrison, and R. D. Rundel, Proc. Roy. Soc. (London) A299, 525 (1967).
22. F. R. Gilmore, E. Bauer and J. W. McGowan, J. Quant. Spectr. Rad. Trans. 9, 157 (1969).

23. K. K. Corvin and S. J. B. Corrigan, J. Chem. Phys. 50, 2570 (1969).
24. D. R. Bates and H. S. W. Massey, Phil. Mag. 45, 111 (1954).
25. F. C. Fehsenfeld, E. E. Ferguson and A. L. Schmeltekopf, J. Chem. Phys. 45, 1844 (1966).
26. L. G. Christophorou, R. N. Compton and H. W. Dickson, J. Chem. Phys. 48, 1949 (1968).
27. P. J. Chantry, Phys. Rev. 172, 125 (1968).

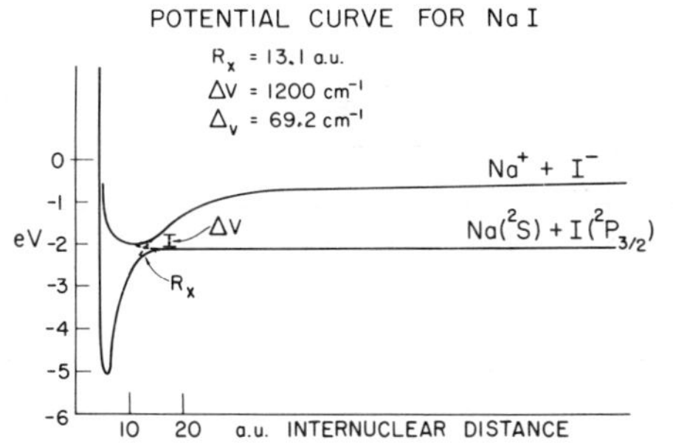

Figure 1
Potential energy diagram for NaI showing crossing point R_x and adiabatic term splitting Δv

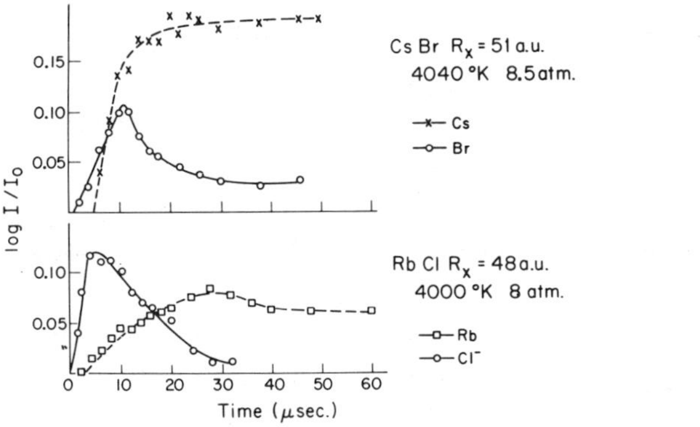

Figure 2
Intensity vs. time profiles for RbCl and CsBr under similar conditions

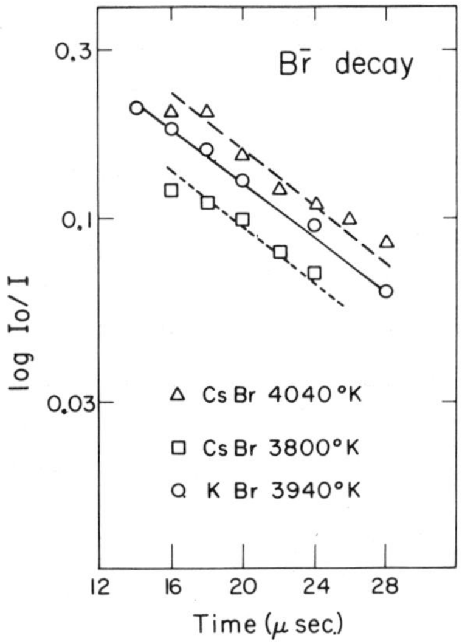

Figure 3
First-order relaxation of Br^- superabundance

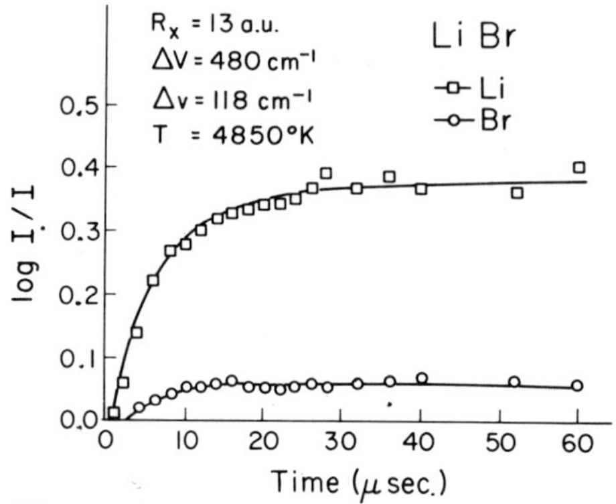

Figure 4
Intensity vs. time profiles for LiBr

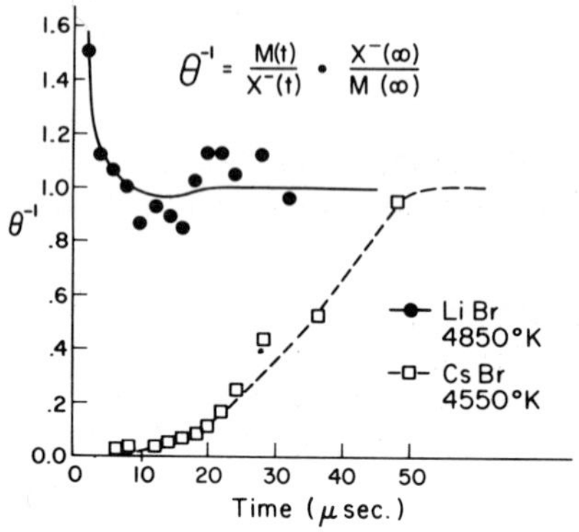

Figure 5
Plot of θ^{-1} for LiBr and CsBr

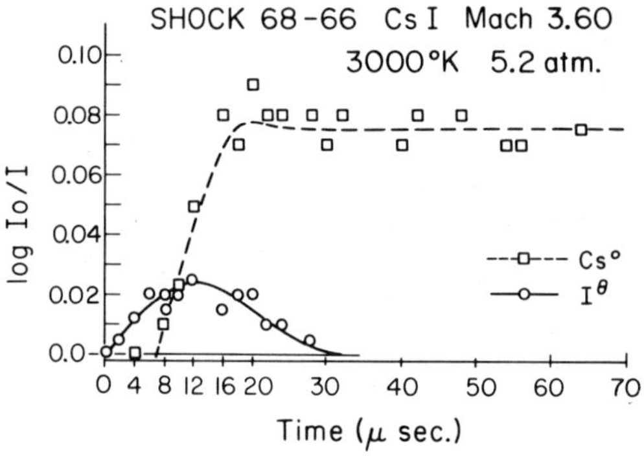

Figure 6
Intensity vs. time profile for Cs1

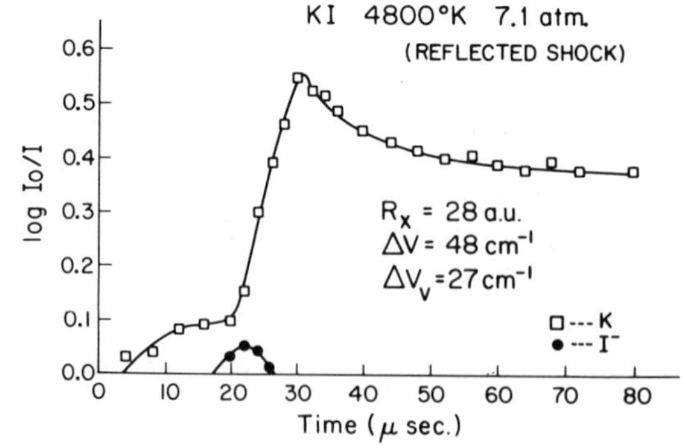

Figure 7
Intensity vs. time profile for Kl

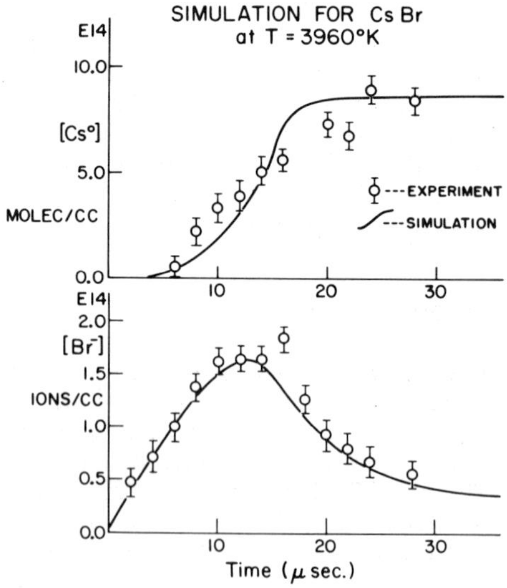

Figure 8
A computer simulation of CsBr data using the rates given in the paper

INFRARED MEASUREMENT OF CHAIN BRANCHING RATES IN HYDROGEN-OXYGEN MIXTURES IGNITED BY REFLECTED SHOCK WAVES

R. W. GETZINGER, L. S. BLAIR, and D. B. OLSON
University of California
Los Alamos Scientific Laboratory
Los Alamos, New Mexico 87544

SUMMARY

Chain branching rates in the hydrogen-oxygen reaction have been determined by measuring the exponential increase in infrared emission at 2.7 μ from H_2O and OH produced during the initial combustion of highly diluted hydrogen-oxygen mixtures ignited by reflected shock waves. Observations were made through a transparent end plate window. Data were obtained at 1266-2324°K. The results in hydrogen-lean mixtures differ significantly from recent measurements of the exponential growth of visible emission in similar mixtures containing small amounts of CO. Excellent agreement between the two sets of measurements was found in hydrogen-rich mixtures. Implications of these observations for the rate coefficients of the elementary reactions responsible for chain branching, and for the CO-O radiative recombination process in the presence of relatively large amounts of O and O_2 are considered. In particular, anomalies in the temperature dependence of the rate coefficients present in the visible emission results have been removed by the infrared experiments.

INTRODUCTION

During the induction period of the shock-initiated hydrogen-oxygen reaction the concentration of the products of the branching chain, H, O, OH, and H_2O, increase exponentially under conditions of essentially constant temperature, pressure, and reactant concentrations. The coefficient describing the exponential growth of all four species is a function of the unperturbed reaction conditions, the rate coefficients of the bimolecular chain steps I-III

$$H + O_2 \xrightarrow{k_1} OH + O, \qquad (I)$$

$$O + H_2 \xrightarrow{k_2} OH + H, \qquad (II)$$

and

$$OH + H_2 \xrightarrow{k_3} H_2O + H, \qquad (III)$$

and, at high pressures and low temperatures, the rate coefficient of the termolecular recombination reaction IV,

$$H + O_2 + M \xrightarrow{k_4} HO_2 + M. \qquad (IV)$$

In a recent series of papers[1-4] Gutman, Schott, and co-workers have reported measurements of hydrogen-oxygen exponential growth coefficients. The increased sensitivity required to detect these very small extents of reaction was achieved by utilizing a somewhat novel experimental technique. Time-resolved, spatially integrated emission from reacting gas mixtures behind reflected shock waves was monitored through a window mounted in the end wall of the shock tube. Inclusion of small amounts of carbon monoxide in the experimental hydrogen-oxygen-argon mixtures produced blue continuum emission from the radiative recombination reaction $CO + O \rightarrow CO_2 + h\nu$, which, for a substantially constant [CO], increased exponentially with time in proportion to the oxygen atom concentration. Spatial integration of the radiation from all luminous gas elements between the reflected shock front and the end wall produced a total intensity that increased exponentially with the same growth coefficient as that for oxygen atoms. In this way exponential growth coefficients were measured over a wide range of experimental variables.

The fundamental value of these coefficients lies in the direct information they yield about the rate coefficients k_1-k_4. This relationship is free of the principal complicating factors which plague efforts to extract rate coefficients from shock tube measurements of induction period lengths: the influence of initiation reactions and uncertainties in the magnitude of the concentration, or other measured property, corresponding to detectable reaction. Unfortunately, straightforward interpretation of the end-on CO-O visible emission data has led to conflicting and questionable conclusions regarding the important coefficients k_1-k_3. Early results[1-3] have now been supplemented by more recent experiments[4] in hydrogen-lean mixtures, where there is increased sensitivity to k_2 and k_3, and an effort has been made[4] to deduce a consistent set of Arrhenius parameters for k_1-k_3 that will account for measurements made over the full experimental ranges $0.10 \leq H_2/O_2 \leq 10.0$ and $1000 \leq T \leq 2500°K$. Activation energies obtained in this manner[4] are in marked disagreement with literature data at lower temperatures and, in the case of E_1, with parameters derived from the earlier end-on measurements in hydrogen-rich mixtures, where sensitivity to k_1 is greatest. Also disturbing is the failure of a conclusive test carried out in the lean experiments to confirm the assumed first order dependence of emission intensity on [CO]. These results have raised questions concerning the reliability of the visible measurements, particularly those in lean mixtures.

We report here the results of an end-on experimental investigation of the hydrogen-oxygen induction period kinetics carried out by monitoring the exponential increase in thermal emission in the infrared accompanying the progress of the branching chain. Experimental evidence indicates that the observed emission came principally from H_2O, with undetermined, but at most small, contributions from OH. As will be demonstrated later, however, the possibility that both species contributed significantly to the total intensity does not complicate the data interpretation. A common exponential growth coefficient describes the increasing concentrations of both H_2O and OH; hence for optically thin gases combined emission from both species increases exponentially with this same growth coefficient. In using infrared emission from a chain product to measure exponential growth coefficients we are following closely the method of Homer and Kistiakowsky, who recently monitored infrared emission from CO and CO_2 produced during the oxidation of acetylene[5] and ethylene[6] behind incident shock waves for the same purpose.

The present experiments represent a considerable simplification of the original end-on technique of Gutman and Schott.[1] The growth of one of the chain products is now being followed directly. Consequently, the accuracy of the measured growth coefficients is no longer dependent upon the validity of assumptions which pertain only to the chemiluminescent process, i.e., assumptions regarding the chemical step responsible for the visible emission, and the relationship between the intensity of this emission and the [CO][O] product. In addition, the requirement that CO be included in the reactant gas mixtures has been removed, thus making it possible to test for participation by CO in the branching chain chemistry. Infrared experiments carried out in comparable H_2-O_2-Ar and H_2-O_2-Ar-CO mixtures have confirmed that reactions involving CO have an insignificant influence on the observed growth coefficients. As described below, the infrared method has yielded results in lean mixtures that differ significantly from previous end-on measurements.

EXPERIMENTAL

Apparatus and Procedure

Experiments were carried out in the 10 cm dia shock tube described previously.[7-9] The expansion section of the tube was divided into two parts by the inclusion of a matching-bore rotating plug valve some 76 cm from the downstream end of the tube. Only the portion of the tube nearest the end plate was filled with the combustible mixtures, while pure argon at the same initial pressure was used as a buffer gas in the remainder of the expansion section. This procedure was adopted in an effort to minimize the amount of infrared radiation generated by reaction behind the incident shock wave before completion of the reflected shock experiment.[4] Incident shock arrival times at five piezoelectric gauge positions located 6.1, 16.1, 26.1, 46.1, and 66.1 cm from the end plate were electronically recorded[7] for later use in determining the incident shock velocity. The driver gas in all experiments was hydrogen.

Spatially integrated infrared radiation was monitored through a 3.8 cm dia (thickness: 1.9 cm) sapphire end plate window with either a room-temperature indium arsenide or a liquid-nitrogen-cooled indium antimonide photovoltaic detector. In most experiments, including all those with reactant mixtures containing CO, a Kodak 2.72 μ interference filter (half-peak width 0.7 μ) was located between the window and the sensitive surface of the detector. In order to maximize the solid angle being viewed,[4] the detector was located as near the outside surface of the window (and filter) as was practical. The intervening space was continuously flushed with dry nitrogen gas to eliminate atmospheric absorption by carbon dioxide and water vapor. For the vertically-oriented InSb detector it was necessary to construct a special holder, containing a plane mirror to reflect radiation at 90° to the shock tube axis, that could be mounted to the outside surface of the end plate. In each experiment, the output signal from the detector was amplified[10] and combined with the pulse generated on shock reflection by a piezoelectric gauge mounted flush with the surface of the end wall for display on a Tektronix 545 oscilloscope. Time response of both detector-amplifier circuits was less than 1 μsec.

Infrared emission growth measurements were made in the mixtures listed in Table 1. Nominal initial hydrogen/oxygen ratios varied between 10:1 and 1:10. Gas mixtures at initial pressures between 1.50 and 10.00 cm Hg, were shocked to reflected shock temperatures of 1266-2324°K.

Theory

For radiation proportional to the exponentially increasing concentration of a reaction product, P, end-on viewing yields the following time-dependent signal

$$I(t) = \int_0^{u_r t} \int_0^{r_o} 2\pi r\, D(x,r)\, f_p B_p \exp\{\alpha_p(t - \frac{x}{u_r})\} dr\, dx, \qquad (1)$$

where x and r denote the location of the radiating gas element with respect to the end wall (x=0) and the shock tube axis (r=0), and t is the time following shock reflection. For a constant reflected shock velocity u_r, $u_r t$ is the instantaneous position of the reflected shock front and $B_p \exp\{\alpha_p(t - \frac{x}{u_r})\}$ is the concentration of the emitting product P at t and $0 \leq x \leq u_r t$. B_p is a constant and α_p is the exponential growth coefficient. f_p in Eq. (1) is the emission intensity factor per unit concentration of P and D(x,r) is the geometric and instrumental detectivity function for emitted light. This formulation presupposes both optical thinness and the absence of any detectable luminosity in the region ahead of the reflected shock wave, i.e., in gas elements exposed to only the incident shock. Equation (1) can also be written

$$I(t) = \exp(\alpha_p t) \int_0^{u_r t} \int_0^{r_o} 2\pi r\, D(x,r)\, f_p\, B_p \exp(-\alpha_p \frac{x}{u_r}) dr\, dx, \quad (2)$$

since the integration is not carried out over time. At large t the double integral becomes a constant, and

$$I(t) = I_0 \exp(\alpha_p t). \quad (3)$$

Equations (1)-(3) have been applied[1-4] to the interpretation of end-on measurements of the exponential growth of visible emission in the hydrogen-oxygen reaction, and the quantities corresponding to f_p, B_p, and α_p (f_O, B_O, and α) have been identified. If it is supposed that the emission source is the CO-O radiative recombination reaction and f_O is proportional to [CO], the observed exponential growth coefficient, α, is the positive root of the equation

$$\left\{\frac{\alpha}{2[O_2]}\right\}^3 + 1/2\ \{k_1+\underline{n}(k_2+k_3) + k_4\ [M]\} \left\{\frac{\alpha}{2[O_2]}\right\}^2$$

$$+ 1/4\ \{k_2 k_3 \underline{n}^2 + (k_2+k_3)\ k_4\ [M]\ \underline{n}\}\left\{\frac{\alpha}{2[O_2]}\right\}$$

$$- 1/8\ k_2 k_3 (2k_1 - k_4\ [M])\ \underline{n}^2 = 0, \quad (4)$$

where $\underline{n}$ is the hydrogen/oxygen ratio. Equation (4) has been used to derive conclusions about k_1-k_3[1,2,4] and the termolecular coefficient[2] k_4 from extensive visible emission measurements of α as a function of temperature, density, and $\underline{n}$.

Similar considerations apply in the present infrared ignition experiments, in which radiation from H_2O, and possibly OH, is observed. Water vapor is produced only in reaction III. $[H_2O]_{x,t}$ is therefore given by

$$[H_2O]_{x,t} = \int_0^t k_3[H_2][OH]_{x,t}\ dt. \quad (5)$$

Since following a short initiation transient $[OH]_{x,t}$ rapidly approaches $B_{OH} \exp\{\alpha(t-\frac{x}{u_r})\}$, $[H_2O]_{x,t}$ eventually becomes

$$[H_2O]_{x,t} = \int_0^t k_3[H_2]\ B_{OH} \exp\{\alpha(t-\frac{x}{u_r})\}dt. \quad (6)$$

For constant reaction conditions behind the reflected shock, Eq. (6) can be integrated to yield

$$[H_2O]_{x,t} = \frac{k_3[H_2]B_{OH}}{\alpha} \exp\{\alpha(t-\frac{x}{u_r})\}, \quad (7)$$

where the contribution of the lower limit has been ignored since it is insignificant at the reaction times of interest. Equation (3) is therefore also applicable in the infrared experiments. For significant emission from both H_2O and OH, the pre-exponential factor I_0 is given by

$$I_0 = \int_0^{u_r t} \int_0^{r_o} 2\pi r\, D(x,r) \left\{ f_{OH} + \frac{k_3[H_2]}{\alpha} f_{H_2O} \right\} B_{OH} \times \exp\left(-\frac{\alpha x}{u_r}\right) dr\, dx, \quad (8)$$

which approaches a constant at long time. As with visible emission, a semilogarithmic plot of observed intensity against time following shock reflection should yield a straight line whose slope is α, the desired root of Eq. (4). The infrared ignition experiments make possible a measurement of α by an independent method, thus providing a means of checking earlier conclusions about the elementary reaction coefficients.

ANALYSIS AND RESULTS

An oscillogram obtained in one infrared experiment in Mixture F is reproduced in Fig. 1. The exponentially increasing deflection from the baseline is readily apparent. Less obvious, but also present, is a small component of deflection that appears to increase linearly with t from a value of essentially zero at the instant of shock reflection. Tests described below convinced us that this linear component was extraneous. Accordingly, experimental values of I(t) were obtained from the oscillogram by reducing the total deflection from the baseline by an amount determined by extrapolating this linear signal into the region where the exponentially increasing signal is dominant. A semilogarithmic graph of I(t) against t for this experiment appears in Fig. 2.

Linear corrections of this type were required in most of the infrared experiments. The effect of these corrections on the measured α's was generally small (2.4% in Fig. 2). However, the presence of this additional component of intensity provided a limit on the sensitivity of the infrared method that was considerably more restrictive than that of the noise level of either detector-amplifier circuit alone.

As a result of either this additional limitation on sensitivity or that inherent in the detector-amplifier circuitry itself, it was never possible under any experimental conditions to detect the non-exponential initiation transient seen in recent CO-O experiments.[4] Instead, in nearly all experiments infrared signals were seen to increase in a purely exponential manner over the nearly two decades of deflection observed on the oscillograms. In the worst cases, the highest temperature experiments in Mixture 4, it was possible to follow only a decade or so of clearly exponential growth before

I(t) began to fall significantly below the extrapolated line determined by the earlier points. This falloff most likely indicates the passage of the gas elements nearest the end plate from the purely exponentially branching portion of the reaction.

Another reproducible feature of the oscillograms was the appearance at the highest temperatures in all mixtures of a significant deflection from the baseline before shock reflection. As a result, at $t>0$ both the exponential and linear components described above appeared to be superimposed upon a constant deflection whose magnitude could be evaluated at $t\leq 0$. Although it was not difficult to take account of this additional signal, its presence further complicated the data analysis.

A considerable effort was made to identify and eliminate both the extraneous "prelight" and linear intensity signals. Neither appeared to be diminished in shocks in pure argon where exponential growth was absent. The presence of the 2.72 μ filter did little to enhance the exponential signal relative to these additional components. Neither trapping the gas mixtures with liquid nitrogen, using high purity, spectral grade argon, or carrying out mass spectrometric analyses yielded any clues regarding possible impurities. Use of very hot shocks in pure argon and longer pumping times between experiments proved unsuccessful in confirming the influence of adsorbed water vapor or other materials on the shock tube walls. In the end we were forced to accept and compensate for the presence of the undesired signals, which proved to be an annoying, but not serious complicating factor in the data reduction.

Conditions prevailing behind the reflected shock waves were computed using standard Hugoniot equations and the assumption of no reaction behind either the incident or reflected shock. Incident shock velocities immediately preceding reflection required in these computations were obtained by quadratically fitting the position-time data for the incident shock and evaluating the first derivative at the location of the end plate.

Experimental results from nominal 10:1, 1:3, and 1:10 mixtures (A and B, F and G, and H, respectively) appear in Fig. 3, which is a graph of $\alpha/2[O_2]$ against the reciprocal reflected shock temperature, T_R, following Gutman and Schott.[1] Since all of the infrared experiments were carried out at temperatures and densities where the influence of reaction IV on α, as seen in Eq. (4), is at most quite small, the data in Fig. 3 scale only with $\underline{n}$. The 3:1 and 1:1 results, which have been omitted from Fig. 3, fall between the 10:1 and 1:3 data as anticipated. Within the experimental scatter results obtained in Mixtures D and E with different detector-amplifier circuits are identical.

Also shown in Fig. 3 are lines representing the magnitude and temperature range of comparable CO-O visible emission data obtained in 10:1, 1:3, and 1:10 mixtures under conditions where the influence of reaction IV is negligible.[1-4] There is good agreement between the two sets of results in 10:1 (and, incidentally, 3:1) experiments. However, it is evident that significant differences, which are far too large to be accounted for by small variations in actual H_2/O_2 ratios, exist in lean mixtures. These differences first appear at temperatures above about 1600°K in mixtures when n is 1.00, and become progressively greater for smaller n. That these differences are temperature-dependent is clearly seen in the results from 1:10 mixtures in Fig. 3. That they are not a result of the participation of CO in the chemistry of the hydrogen-oxygen chain is seen in the agreement between infrared measurements in Mixtures F and G. Of these two 1:3 mixtures only G contains CO. The disagreement between infrared and visible emission measurements in lean mixtures will receive further attention later in this paper.

Figure 3 also shows the significant extension to higher temperatures made by the infrared measurements of α in 10:1 mixtures. These results are particularly important in determining the Arrhenius activation energy for reaction I, E_1, since for large n, $\alpha/2[O_2] \rightarrow k_1$. Branching coefficients from Mixtures A and B are in excellent agreement. Use of two 10:1 mixtures with significantly different reactant mole fractions made it possible to independently vary $[O_2]$ and the total concentration at any reflected shock temperature. For example, the six highest temperature 10:1 points include variations of a factor of 5 in $[O_2]$ and 2.5 in total concentration, and thus provide a means of testing for experimental inconsistencies at this extreme condition.

Schott[4] has discussed the difficulties encountered in applying Eq. (4) to the reduction of measured α's to values of k_1-k_3. Complicating factors are: (a) k_2 and k_3 enter the equation interchangeably as the sum (k_2+k_3) and product k_2k_3 and are therefore separable only by the use of independent information about their relative magnitudes, and (b) insufficient sensitivity of α to (k_2+k_3) at any n or temperature. Adequate sensitivity of α to k_2k_3, on the other hand, is assured by the fact that in very lean mixtures $\alpha \rightarrow [H_2]$ $\sqrt{2k_2k_3}$. Finally it should be noted that for real values of k_2 and and k_3, (k_2+k_3) must be $\geq 2\sqrt{k_2k_3}$. The equality of these two quantities occurs when $k_2 = k_3$.

These considerations, combined with the scatter in our experimentally determined α's, make it impossible to deduce directly the six Arrhenius parameters for k_1-k_3. Accordingly, we have adopted the following procedure for obtaining rate coefficients from the measured α's. Absolute values of k_1, $(k_2 + k_3)$, and k_2k_3 were determined at 1400, 1600, and 1800°K by least-squares solving Eq. (4) with normalized growth coefficients, $\alpha/2\ [O_2]$, interpolated from graphically smoothed representations of the data for all five n. The required quantities, $[O_2]$ and total concentration, [M], were taken from actual experiments carried out at nearly these same temperatures, and, in all cases, k_4 was taken[11] as the coefficient for M = Ar, 2.2×10^{15} cc^2/mole^2 sec. At 1600°K the derived coefficients are: $k_1 = 8.7 \pm 0.9 \times 10^{11}$ cc/mole sec, $(k_2+k_3) = 2.9 \pm 1.8 \times 10^{12}$ cc/mole sec, and $k_2k_3 = 4.0 \pm 0.9 \times 10^{24}$ cc^2/mole^2 sec^2. Within the quoted standard deviations, the sum is equal to its minimum value $(2\sqrt{k_2k_3})$. At 1400°K, (k_2+k_3) was found to be approximately twice $2\sqrt{k_2k_3}$, and at 1800°K it was 25% of the corresponding "minimum" value.

In view of the large uncertainty in (k_2+k_3) caused by the lack of sensitivity referred to earlier, the most meaningful conclusion which can be drawn from this preliminary analysis is that at these temperatures the sum is "approximately" equal to $2\sqrt{k_2k_3}$. For lack of any better information about this relationship we have assumed that, for the purpose of further analysis, $(k_2+k_3) = 2\sqrt{k_2k_3}$ throughout the entire temperature range of the infrared experiments. Although this assumption is not the only way[4] of treating the difficulty introduced by the manner in which k_2 and k_3 enter Eq. (4), it has the advantage of reducing the number of parameters in the least-squares fit from six to four (A_1, E_1, A_2A_3, and $E_2 + E_3$), with, it is believed, no significant loss in accuracy.

Incorporation of this assumption into a least-squares solution of Eq. (4) using all of the experimentally-determined infrared α's from Mixtures A-F and H, the corresponding computed T_R, $[O_2]$, and $[M]$, and the value of k_4 given earlier, yielded the following results:

$$k_1 = 9.5 \pm 0.5 \times 10^{13} \exp(-15{,}000 \pm 900/RT) \text{ cc/mole sec}$$

and $k_2k_3 = 1.4 \pm 0.2 \times 10^{27} \exp(-17{,}400 \pm 2900/RT)\ cc^2/mole^2sec^2$.

The quoted standard deviations are believed to be fair indications of the uncertainties in these parameters. The fit provided by these rate coefficients (and the assumption that $(k_2+k_3) = 2\sqrt{k_2k_3}$) for four values of n is shown in Fig. 4. The lines represent the predicted variation of $\alpha/2[O_2]$ with reciprocal reflected shock temperature. In order to avoid crowding the figure, the comparison for Mixture C (3:1) is not shown; however, the agreement in this mixture is as good as that for the 10:1 data. It is felt that these results constitute the maximum amount of accurate rate coefficient information available from the present experiments.

DISCUSSION AND CONCLUSIONS

As mentioned earlier, we believe that the infrared emission observed in the present experiments came primarily, if not exclusively, from H_2O rather than OH. This conclusion is based on observations made in recent incident shock hydrogen-oxygen recombination experiments in which 2.7 μ infrared radiation was monitored throughout the course of the entire reaction under many different conditions. With the same 2.72 μ filter used in the present investigation, the emission profiles in all recombination experiments closely resembled the anticipated H_2O profiles. At no point in time in any of the recombination experiments was there an indication of emission related to the OH profile, whose characteristics differ significantly from those of H_2O.

The present investigation has yielded the following significant results: (a) the most direct and, it is believed, the most accurate determination of E_1 for the temperature range 1300-2400°K, and (b) a considerable difference between infrared and visible emission measurements of α in lean mixtures. (a) and (b) are discussed in the following paragraphs.

E_1 is determined primarily by the temperature dependence of $\{\alpha/2[O_2]\}_{10:1}$, since as noted earlier $\{\alpha/2[O_2]\}_n \rightarrow k_1$ for large $\underline{n}$. Consequently, the accuracy of the value obtained for E_1 in the present investigation is assured by the relatively wide range of temperatures covered in the 10:1 experiments. An indication of the dominant influence of $-R\ d(\ln\{\alpha/2[O_2]\}_{10:1})/d(1/T)$ in determining E_1 is provided by the fact that the difference between these two quantities is less than the standard deviation for E_1. The influence of measurements in leaner mixtures on E_1 is negligible.

The influence of results in mixtures where $\underline{n} < 10$ on the magnitude of k_1 is somewhat greater. However, the small differences between values of $\{\alpha/2[O_2]\}_{10:1}$ and $\{\alpha/2[O_2]\}_{3:1}$ at all experimental temperatures is a strong indication that k_1 cannot be much larger than $\{\alpha/2[O_2]\}_{10:1}$. This is indeed the case: at 1600°K k_1 is only some 13% larger than $\{\alpha/2[O_2]\}_{10:1}$. The agreement between infrared and visible emission measurements in 10:1 and 3:1 mixtures provides additional confidence in the accuracy of the Arrhenius parameters deduced for k_1.

As anticipated, our values of A_1 and E_1 are in excellent agreement with those determined from CO-O experimental measurements at 1290-1667°K in mixtures where $\underline{n} \geqslant 1.0$. However, they differ considerably from values of A_1 and E_1 proposed recently[4] to reconcile visible emission measurements over the range $10.0 \geqslant \underline{n} \geqslant 0.10$. The significant influence of the lean α's on the deduced visible emission parameters for reaction I, which resulted in a surprisingly low E_1 of 12.3 kcal/mole, can be traced to the relatively poorly defined temperature dependence of $\{\alpha/2[O_2]\}_{10:1}$ in the CO-O experiments. The net effect of the differences in A_1 and E_1 on the absolute value of k_1 at these temperatures is small: $(k_1)_{IR}$ and $(k_1)_{VIS}$ agree within 20% between 1300 and 2000°K. In fact, when evaluated at 1600°K these two expressions for k_1 and two others recently obtained by less direct methods[2,12] all yield coefficients that differ by only some 16%.

An absolute reaction rate theory computation by Kurzius and Boudart[13] has predicted a temperature-dependent Arrhenius activation energy for k_1 of 16.7 kcal/mole at 300°K, 15.4 kcal/mole at 1500°K, and 14.0 kcal/mole at 2000°K. When examined from this point of view, our experimental value, 15.0 kcal/mole, is a reasonable average for the range 1300-2400°K.

The difference between exponential growth coefficients measured in lean mixtures by the two emission methods manifests itself most visibly in the corresponding differences between Arrhenius parameters derived for k_2k_3. Both the deduced magnitudes (at 1600°K: $(k_2k_3)_{IR} = 6.1 \times 10^{24}$ cc^2/mole^2 sec^2 and $(k_2k_3)_{VIS} = 1.5 \times 10^{25}$ cc^2/mole^2 sec^2) and activation energies (17.4 versus 20 kcal/mole) of k_2k_3 are in significant disagreement. Unfortunately, these coefficients have not been sufficiently well determined at high temperatures by other independent methods to justify choosing on this basis alone which, if either, experimental method is correct. Extrapolation of independent low temperature measurements[14,15] of

k_2 and k_3 yields results which tend to favor the infrared measurements (at 1600°K : $k_3k_3 = 6.0 \times 10^{24}$ cc^2 $mole^2$ sec^2; $E_2+E_3 = 15.4$ kcal/mole), but are not conclusive.

Because of its directness and greater simplicity we believe the infrared method of measuring α to be more accurate. However, it is our purpose in this paper only to point out that significant disagreement exists between measurements made in lean mixtures. These differences may indicate that the strict proportionality between what is supposed to be emission from the CO-O radiative recombination reaction and the product [CO][O] is not valid in an environment containing significant amounts of oxygen atoms and molecules. The fact that the ratio α_{VIS}/α_{IR} is of the order of two in 1:10 mixtures may reflect the dominance in quite lean mixtures of a chemiluminescent process involving two chain products.

The infrared results are not free of anomalies, however. There appears to be a certain inconsistency in the α's measured at high temperatures which makes it difficult to reconcile the data at different $\underline{n}$ within the framework of Eq. (4) and its six adjustable Arrhenius parameters for k_1-k_3. This is seen in the failure of the requirement that $(k_2+k_3) \geqslant 2\sqrt{k_2k_3}$ for values of the sum and product deduced at temperatures above 1500°K. It is also evident in Fig. 4 from the relatively poor fit for lean mixture measurements, which is a result of both the preponderance of data for $\underline{n} \geqslant 1.0$ and the assumption that $(k_2+k_3) = 2\sqrt{k_2k_3}$. This inconsistency, which we do not understand, has produced a situation in which it is impossible to find meaningful values of k_2 and k_3 that will satisfactorily reproduce all of the infrared data.

It should be emphasized, however, that this inconsistency is not believed to be serious. Only a small, systematic experimental error in a limited number of α's would be sufficient to reconcile the low and high temperature results. The effect of such an error on the principal conclusions of this work would be negligible.

ACKNOWLEDGMENTS

This work was performed under the auspices of the United States Atomic Energy Commission. The authors particularly want to acknowledge the continued interest and helpful advice provided throughout the course of this work by Dr. Garry L. Schott. The assistance of J. G. Williamson and J. L. Young in the performance of the experiments is also appreciated.

REFERENCES

1. D. Gutman and G. L. Schott, J. Chem. Phys. 46, 4576 (1967).
2. D. Gutman, E. A. Hardwidge, F. A. Dougherty, and R. W. Lutz, J. Chem. Phys. 47, 4400 (1967).
3. D. Gutman, R. W. Lutz, N. F. Jacobs, E. A. Hardwidge, and G. L. Schott, J. Chem. Phys. 48, 5689 (1968).

4. G. L. Schott, "Chain Branching and Initiation Rates Measured by Spatially Integrated Light Emission During Reflected Shock Wave Ignition", paper presented at the Twelfth (International) Symposium on Combustion, Poitiers, France, July, 1968.
5. J. B. Homer and G. B. Kistiakowsky, J. Chem. Phys. 46, 4213 (1967).
6. J. B. Homer and G. B. Kistiakowsky, J. Chem. Phys. 47, 5290 (1967).
7. L. S. Blair and R. W. Getzinger, "A Shock Tube Study of Recombination in the Hydrogen-Oxygen Reaction Using Infrared Emission from Water Vapor", Combust. Flame (to be published).
8. G. L. Schott and P. F. Bird, J. Chem. Phys. 41, 2869 (1964).
9. R. W. Getzinger and G. L. Schott, J. Chem. Phys. 43, 3237 (1965).
10. The authors wish to thank Mr. R. Hiebert of this laboratory for the design and construction of amplifiers compatible with each of the infrared detectors.
11. R. W. Getzinger and L. S. Blair, "Recombination in the Hydrogen-Oxygen Reaction: A Shock Tube Study with Nitrogen and Water Vapor as Third Bodies", Combust. Flame (In press).
12. A. L. Myerson and W. S. Watt, J. Chem. Phys. 49, 425 (1968).
13. S. C. Kurzius and M. Boudart, Combust. Flame 12, 477 (1968).
14. A. A. Westenberg and N. deHaas, J. Chem. Phys. 50, 2512 (1969).
15. G. Dixon-Lewis, W. E. Wilson, and A. A. Westenberg, J. Chem. Phys. 44, 2877 (1966).

Table 1
Experimental mixtures

Mixture	$\%H_2$	$\%O_2$	%CO	%Ar	H_2/O_2	No. Expts	Detector
A	19.94	1.87	--	78.19	10.66	10	InSb
B	5.00	0.47	--	94.53	10.64	11	InSb
C	2.25	0.75	--	97.00	3.00	22	InAs
D	1.50	1.50	--	97.00	1.00	33	InAs
E	1.44	1.44	--	97.12	1.00	15	InSb
F	0.97	2.92	--	96.11	0.333	12	InSb
G	0.95	2.86	2.02	94.17	0.333	6	InSb
H	1.38	15.15	--	83.47	0.091	12	InSb

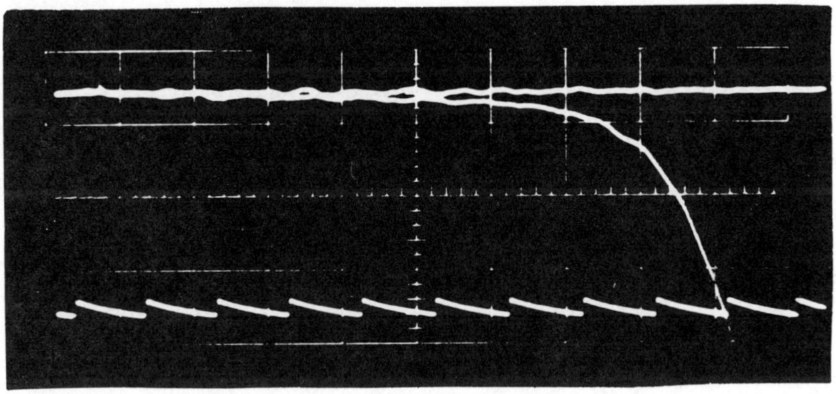

Figure 1. Oscillogram showing infrared emission growth for experiment in Mixture F. Time increases from left to right. Transducer pulse marks shock arrival at end plate. p_o = 5.00 cm Hg. T_R = 1409°K. Time marks: 10 μsec.

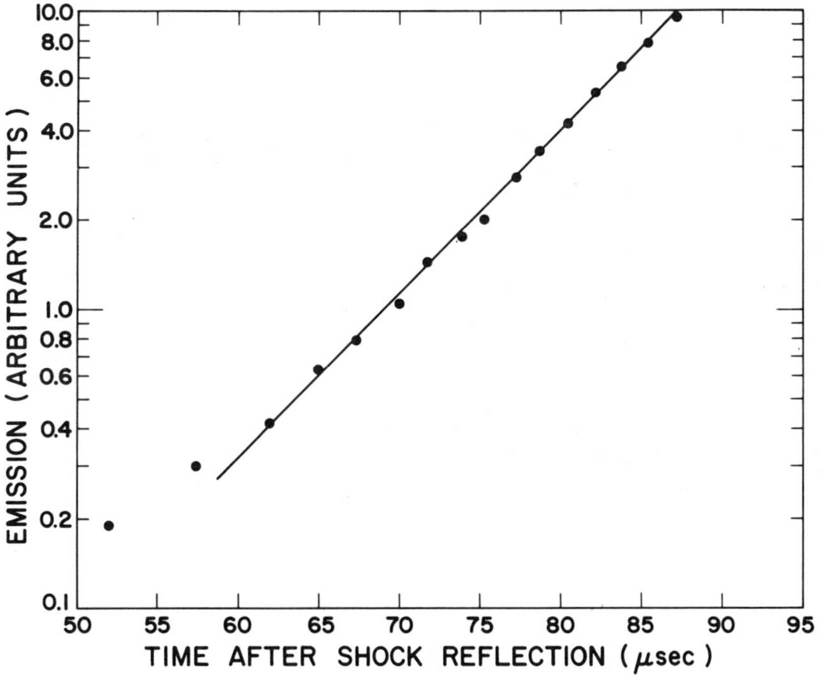

Figure 2. Exponential growth plot for experiment shown in Fig. 1. α = 1.25 x 10^5 sec^{-1}.

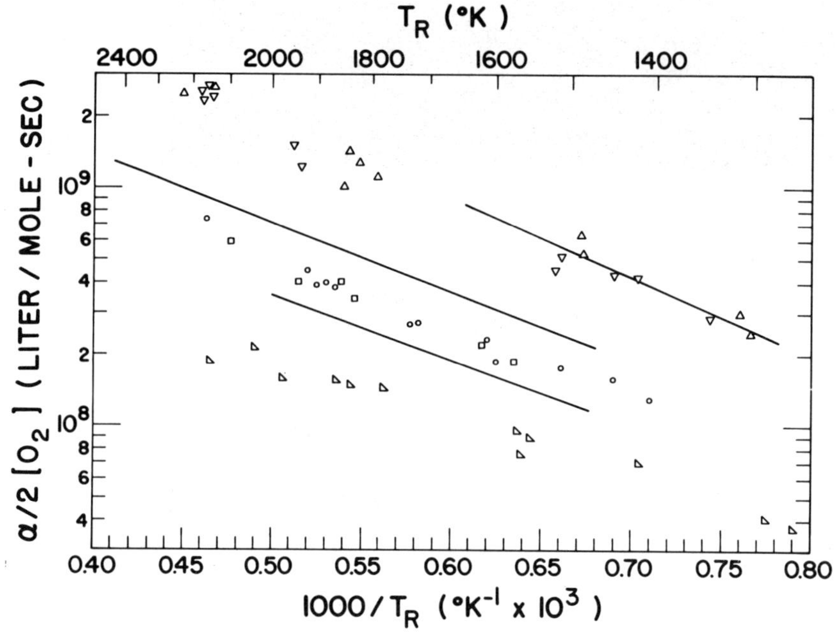

Figure 3. Normalized experimental exponential growth coefficients versus reciprocal reflected shock temperature. Lines represent visible emission measurements[1-4] made in H_2/O_2 = 10:1, 1:3, and 1:10 H_2-O_2-CO-Ar mixtures. Symbols are infrared results in mixtures: Δ-A: ∇-B; O-F, □-G; and ◺ -H.

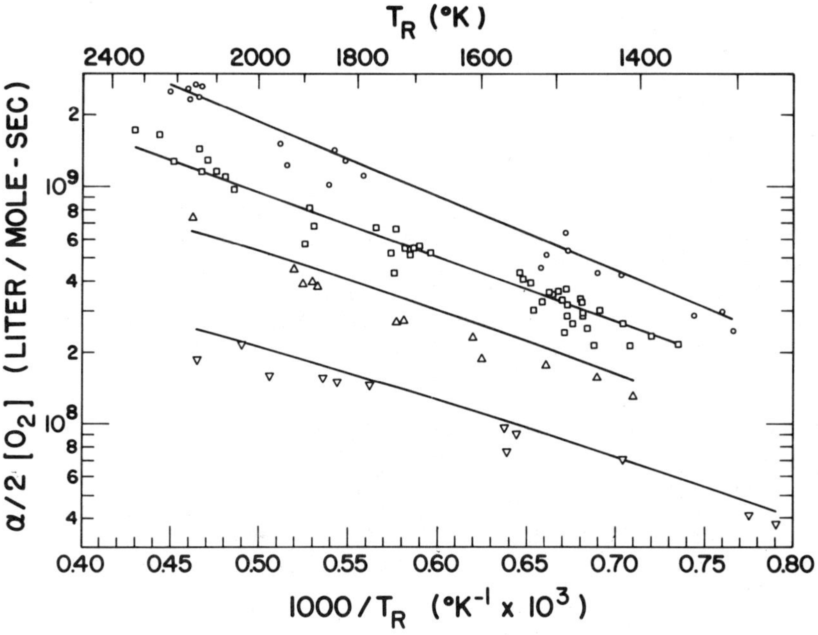

Figure 4. The dependence of measured and predicted normalized exponential growth coefficients from infrared experiments on the reciprocal reflected shock temperature. Symbols are experimental results for: O-n = 10.0 (Mixtures A and B); □-n = 1.0 (D and E); Δ-n = 0.33 (F and G); and ▽-n = 0.10 (H). Lines are coefficients predicted for these n when the least-squares-determined k_1 and k_2k_3 (where $(k_2+k_3) = 2\sqrt{k_2k_3}$) given in the text are substituted into Eq. (4).

SPECTROSCOPIC STUDY OF RADIATIVE SHOCK STRUCTURE IN HYDROGEN

Y. Nakagawa and DAVID C. WISLER
High Altitude Observatory
National Center for
Atmospheric Research*
Boulder, Colorado

The structure of a self-ionizing radiative shock wave in hydrogen is examined by means of time-resolved spectroscopic measurements of the line emission of H_α, H_β, H_γ, H_δ, and H_ϵ and the continuum emission across the Balmer series limit. The hot "dense plasma focus" formed by a coaxial accelerator is collected into a small diameter tube and used to produce a planar, practically current free shock. From the ratio and absolute intensities of the continuum emission, the temperature and number density of electrons are determined. Then the population densities of the various levels are determined from the absolute line intensities after applying the correction for the Stark and Doppler broadenings. These results are compared with the theoretical predictions based on the rate equations. It is shown that with the assumption of complete opacity of Lyman lines and continuum emission, agreements in the manner of relaxation can be obtained.

INTRODUCTION

The efficiency by which the kinetic energy of a shock wave is converted into the thermal energy of the gas is of basic importance in astrophysics in connection with the heating of stellar atmospheres.(1,2,3) Similar problems arise also in re-entry of space vehicles.(4) In order to examine such problems, it is necessary to study the shock structure and the relaxation processes behind the shock in rarified gases, particularly under non-LTE (non-local thermodynamic equilibrium) conditions. Under laboratory conditions, such shock waves can be produced by electromagnetic shock tubes. However, until recently, the shock waves thus produced were usually unstable due to the electric currents associated with the driving gas. In this study, a stable, practically current free planar shock wave is produced with a coaxial plasma accelerator and the structure

*The National Center for Atmospheric Research is sponsored by the National Science Foundation.

of this self-ionizing radiative shock wave in hydrogen is examined in detail by comparing the results of the timed-resolved spectroscopic measurements with those obtained theoretically by the use of the rate equations.

EXPERIMENTAL ARRANGEMENT

A schematic diagram of the experimental arrangement is shown in Figure 1. A copper coaxial plasma accelerator of length 27 cm with a 4.7 cm O.D. center electrode, and a 8.2 cm I.D. outer electrode, was placed in a Pyrex tube. The accelerator was connected through spark gaps to a bank of six 14-μf capacitors. The circuit was optimized so that maximum current was obtained at the time of formation of the "dense plasma focus" in front of the center electrode. A plane shock wave free from electric currents was produced by channeling this hot dense focus into a small diameter tube of 3.65 cm as shown in Figure 2. An image converter camera was used to record the shock speed and shape. In Figure 3, typical framing and streak photographs show the planar shock front and the X-T diagram.

A 1.5 M dual grating spectrograph with illuminator was placed at 45 cm from the end of the small tube. By using photomultipliers, time-resolved spectroscopic measurements of the line emission of H_α, H_β, H_γ, H_δ, and H_ε and of the continuum emission across the Balmer series limit (3635-3640 A and 5540-5550 A) were simultaneously obtained. The response time of the electronics was 10^{-8} sec. A rapid scanning piezo-electric Fabry-Perot interferometer was also used to obtain the evolution of line shapes of H_α emission.

Absolute and relative calibrations were performed by using a ribbon filament tungsten standard lamp and a standard sodium lamp. The experiments were performed with ultra-pure hydrogen gas at an initial pressure of 0.5 Torr and with a capacitor voltage of 15 kV. The nominal Mach number was 27.6.

RESULTS AND DISCUSSION

The electron temperature T_e is determined from the ratio of absolute intensities of the continuum emission, as this ratio is independent of LTE restrictions and is a function only of T_e as long as the plasma is optically thin in the continuum considered and the electron velocity distribution is Maxwellian.(5) In the present experiment, the self-collision time of the electrons was of the order of 10^{-11} sec,(6) thus clearly the requirement of a Maxwellian electron velocity distribution was satisfied. The electron number density is then determined from the absolute intensity of the continuum emission(5) with T_e determined above. The continuum emission was observed at wavelengths 5540-5550 A and 3635-3640 A where no line emission was present. Examples of the time evolution of electron temperature and density across the shock are shown in Figures 4 and

5. The data in Figure 4 indicate that the electron temperature T_e rises to a peak then decreases, reaching equilibrium after about 2.5 μsec. The electron density N_e, however, continues to decrease after reaching a maximum. This continuous decrease of N_e, together with the continuum emission lasting for at least several μsec after T_e reaches equilibrium, indicates that radiative recombination to the upper levels followed by the cascading process continues to populate the various levels. Figure 5 shows the initial rise of electron temperature and density through the shock. The faster increase of electron density may be attributed to precursor ionization,(7) as Lyman lines and continuum are found to be optically thick (Lα: $\tau \approx 4.4 \times 10^3$). On the basis of such preliminary considerations, the observed population number densities for the levels (from which Hα, Hβ, Hγ, Hδ and Hε are emitted) are compared with those given by the Saha-Boltzman relation and by the rate equation of the following form:

$$\frac{dn(p)}{dt} = \frac{\partial n(p)}{\partial t} + \frac{\partial}{\partial x}\left[n(p)v\right]$$

$$= \Big\{ \overbrace{n(c)\Big[-n(p)\sum_{q\neq p} K(p,q) - n(p)K(p,c)}^{\text{Collisional Loss}} + \overbrace{\sum_{q\neq p} n(q)K(q,p) + n(c)n(N^+) \times K(c,p)\Big]}^{\text{Collisional Gain}}\Big\}$$

$$+ \Big\{ \overbrace{\sum_{q>p} n(q)A(q,p) + \sum_{q\neq p} n(q)J_\nu B(q,p) - n(p)\sum_{q<p} A(p,\hat{q}) - n(p)\sum_{q\neq p} J_\nu B(p,q)}^{\text{Line Emission and Absorption}} \Big\}$$

$$+ \Big\{ \overbrace{n(N^+)n(c)\beta(p) - n(p)J_\nu B(p,c)}^{\text{Continuum Emission and Absorption}} \Big\}$$

where $n(i)$ is the population number density for the ith level (i=p and q)
$n(c)$ is the electron number density
v is the flow velocity
and the other symbols have their usual meaning.
This set of equations has been solved for the various $n(p)$ by assuming (1) statistical steady state $\partial/\partial t = 0$ (2) the transport term is negligible and (3) the Lyman lines and continuum are optically thick for a 20 level atom.

The population number densities for various levels are determined experimentally from the absolute line intensities,(5) after correcting the line profiles for the Doppler and Stark broadenings with the assumption that the medium is optically thin. The line center optical thickness has been found without the above broadenings to be: $\tau(H\alpha) \approx 5$, $\tau(H\beta) \approx 0.8$, and (Hγ, Hδ, Hε) all less than 0.25. With the line broadenings observed, these values should be reduced by factor 10-100. Thus the assumption that the medium is optically thin for the Balmer continuum and line emission could be justified.

The examples of the comparison between experimentally determined population number densities and those given by the theoretical relations are shown in Figure 6 for 2.5 and 5.0 μsec and in Figure 7 for 0.2 and 0.4 μsec behind the shock, respectively. It can be seen from Figure 6 that the population number densities for various levels are closely approximated by the solution of the statistically steady state rate equations with complete opacity for Lyman continuum and line radiations. The Saha-Boltzman distribution gives good approximation for upper levels, $p \geq 5$, after 2.5 μsec which could be implied from the T_e behavior (see Figure 4). The detailed examinations of the evolution of population number densities and the initial increases of T_e and N_e near the shock front obviously require time dependent rate equation calculations (see Figure 7). Presently this line of study is in progress. Also the study of time evolutions of the atomic temperature and flow velocity which can be determined in terms of the Doppler broadening and shift of the hydrogen line emission is currently being undertaken.

REFERENCES

1. Ulmschneider, P. "The Structure of the Outer Atmosphere of the Sun and of Cool Stars," Zeit. für Astrophys. 67, 193 (1967).
2. Bird, G. A. "The Equilibrium State of a Shock Heated Atmosphere," Ap. J., 141, 1455 (1965).
3. Skalafuris, A. J. "Radiative Shock Structure-Theory of Observations," J.Q.S.R.T., 8, 515 (1968).
4. Young, L. A. "Infrared Radiation as a Diagnostic Tool in Reentry Physics," J.Q.S.R.T., 8, 105 (1968).
5. Griem, H. R. "Plasma Spectroscopy," Chapt. 4-6, McGraw-Hill: New York (1964).
6. Spitzer, L., Jr. "Physics of Fully Ionized Gases," 2nd Ed. John Wiley: New York (1962).
7. Murty, S. S. R. "Effect of Line Radiation in Precursor Ionization," J.Q.S.R.T., 8, 531 (1968).
8. Bates, D. R., Kingston, A. E., and McWhirter, R. W. P. "Recombination Between Electrons and Atomic Ions I. Optically Thin Plasmas." Proc. Roy. Soc. A, 267, 296 (1962). "II Optically Thick Plasmas," Proc. Roy. Soc. A, 270, 155 (1962).
9. McWhirter, R. W. P. and Hearn, A. G. "A Calculation of the Instantaneous Population Densities of the Excited Levels of Hydrogen-Like Ions in a Plasma," Proc. Phys. Soc., 82, 647 (1963).

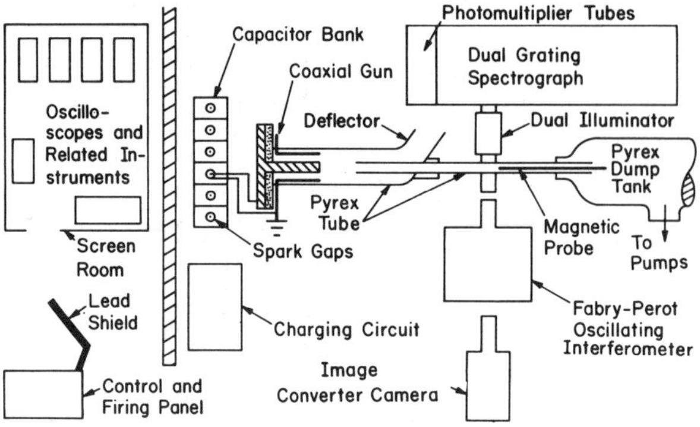

Figure 1
Experimental arrangement.

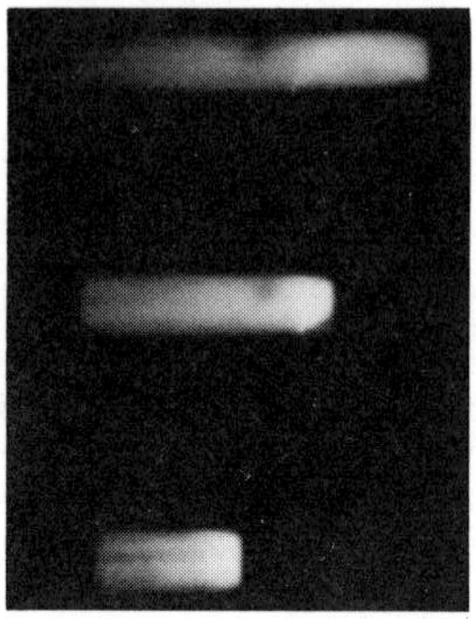

FRAMING PHOTOGRAPH
2 μsec Between Exposures
$u_s \simeq 3.6$ cm/μsec

(Note Magnetic Probe Interaction in 2nd and 3rd Photographs)

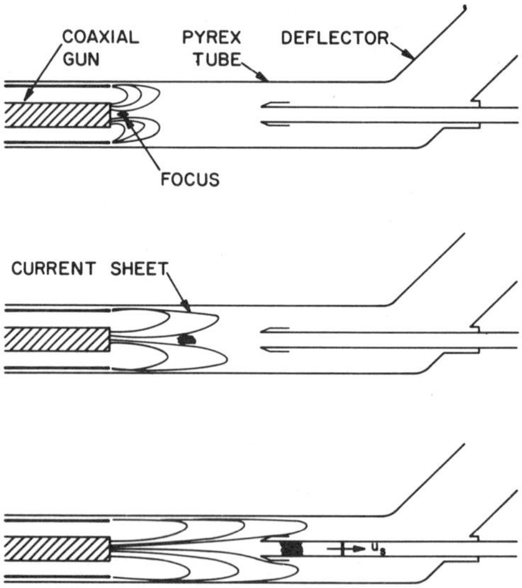

Figure 2
Formation of planar shock.

time

distance

STREAK PHOTOGRAPH

Figure 3
Framing and streak photographs of shock wave.

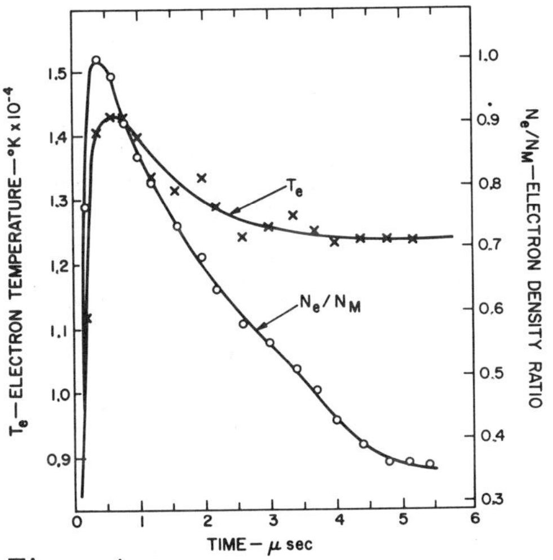

Figure 4
Time variation of electron temperature, T_e, and normalized electron number density, N_e/N_m; N_m denotes the maximum number density.

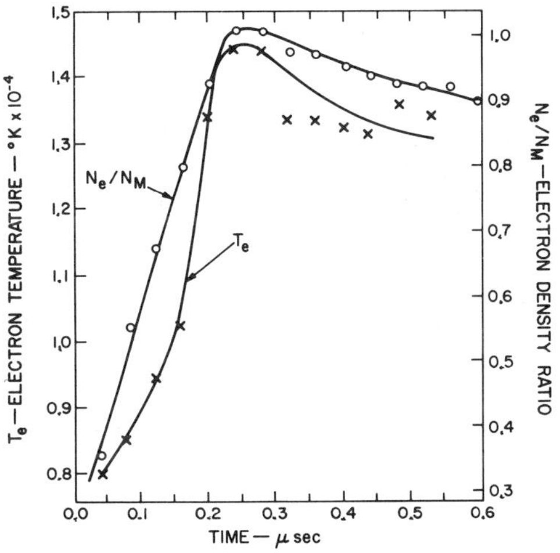

Figure 5
Time variation near the shock front of electron temperature, T_e, and normalized electron number density, N_e/N_m; N_m denotes the maximum number density.

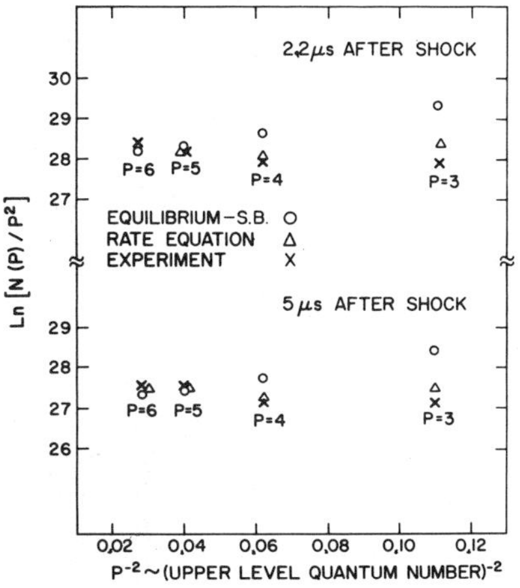

Figure 6
Comparison of experimental and theoretical results for levels p = 3, 4, 5 and 6. The theoretical results are Saha-Boltzman equilibrium (S. B.) and the statistical steady state rate equation.

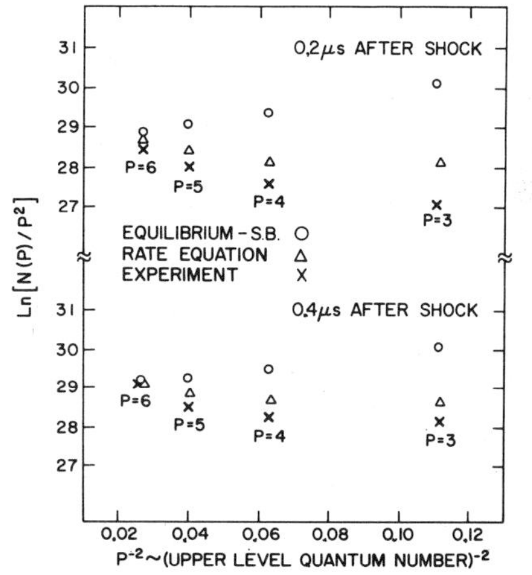

Figure 7
Comparison of experimental and theoretical results near the shock front for levels p = 3, 4, 5 and 6. The theoretical results are Saha-Boltzman equilibrium (S. B.) and the statistical steady state rate equation.

PRECISE MEASUREMENT OF RATE CONSTANT FOR THE REACTION $H + CO_2 \rightarrow OH + CO$ IN THE TEMPERATURE RANGE BETWEEN 2000°K AND 2700°K BEHIND REFLECTED SHOCKS USING AN INFRARED TECHNIQUE

TH. JUST and S. STEPANEK
Deutsche Forschungs- und Versuchsanstalt für Luft- und Raumfahrt E.V., Institut für Luftstrahlantriebe, Porz-Wahn, Germany

SUMMARY

For many elementary reactions at elevated temperatures, no precise rate constants are available. The reason for this is the large number of difficulties that one encounters when conducting kinetic measurements above 2000°K.

At present shock tube techniques together with appropriate analytical methods seem to be the most promising experimental procedures to obtain reliable results. In our work we studied the applicability of emission spectroscopy in the near infrared region to kinetic measurements. Since the development of highly sensitive photoconductors this technique can be applied successfully, and as shown in earlier works, e.g: Olschewsky, Troe, Wagner[1], the precision of the measured data obtainable with this method is very satisfactory. Wagner et al[1] applied the I.R. emission method to the unimolecular decay of small molecules, we, however, concentrated our interest on the measurement of elementary chain reactions. As a first example we selected the reaction $H + CO_2 \rightarrow CO + OH$. This reaction generally being the determining chain of the water gas reaction is of some importance in all combustion reactions with hydrocarbons. As it is well known, the mechanism of the water gas reaction is now fully understood and we have in this respect no difficulties to reckon with, but quantitative data of the rate constant of $CO_2 + H \rightarrow CO + OH$ differ relatively widely and the available data cover only a narrow temperature range. Fenimore and Jones[2], Dixon-Lewis, Sutton and Williams[3], Schecker[4]. The most accurate data up to now were obtained by Schecker[4] in the temperature range between 1400°K and 1800°K. The infrared emission technique allowed us to extend the temperature range up to 2700°K.

EXPERIMENTAL

Our measurements were carried out behind reflected shock waves. We observed the variation in time of the light emission at 4.3 μ of the reacting CO_2 molecule. For small amounts of initial CO_2 the emission originates from an optically thin layer and as a result we have strict proportionality between emission and concentration, provided that during the course of reaction the temperature remains constant. This important optical property of the test gas mixture was obtained in our case with an initial amount of 0.3 % CO_2 (upper limit: ≈ 0.55 % CO2). The other components were 10 % H_2 and 89.7 % Ar. After reflection of the initial shock at the end of the shock tube the H_2 starts to dissociate and the produced H atoms commence to react with CO_2:

$$1)\ H + CO_2 \rightarrow CO + OH$$

$$2)\ OH + H_2 \rightarrow H_2O + H$$

$$3)\ H_2 + M \rightarrow 2\,H + M$$

Because we had chosen $H_2 \gg CO_2$ reaction $\vec{2}$ is at least at the first stages of the reaction much faster than $\vec{3}$ or $\vec{1}$. Hence in a good approximation the rate determining steps are 1 and 3. If we make use of the relatively well known rate constant $\vec{3}$, then it is possible from the measured time dependent emission of CO_2 to evaluate the rate constant of reaction $\vec{1}$.

The Shock Tube and Acessories

The shock tube was made from aluminium with an inner diameter of 75 mm. Supply lines consisted of stainless steel. The test gas was let into the shock tube by means of an ultra-vacuum precision valve. The inevitable leakage of the test gas section and the supply lines never was more than 10^{-5} Torr/min. Aluminium was used for the diaphragms and CaF_2 was selected as material for the flush-mounted windows.

Test Gases

As will be discussed on page 4 , it is of extreme importance to use highly purified gases. The premixed gases containing 89.7 % Ar, 10 % H_2 and 0.3 % CO_2 were prepared for us by Deutsche L'air liquide. The guaranteed maximum contamination by O_2 and H_2O was $\leq$ 3 ppm.

Optical Devices, Recording System

The emitted light was sent through a small monochromator

(Perkin and Elmer). We found reasonable signal to noise ratios with slit widths of 0.5 mm. The latter correspond to a spectral window of ≈ 0.1 μ around 4.3 μ. The light was received by an In-Sb-cell, Valvo ORP 13. The cell was cooled by liquid nitrogen. The signals were recorded by an oscilloscope. Infrared detector and preamplifier showed a rise time of ≈ 3 μsec. The band width of the system was restricted to 100 kc with lower frequencies cut off in order to reduce the hum.

Time Resolution, Disturbances by Stray Light

The time resolution was governed completely by the optics. With suitable apertures and slit widths we measured directly an average "rise time" of 8 μsec. This is the time the reflected shock needs to cross that space in the shock tube from which the detector can receive light. In the ideal case this space is determined by geometric optics. In reality scattered light might enter this zone from outside by multireflection. Consequently one has to be very careful to avoid scattered light. In the infrared region the reflectivity of even a nonpolished aluminium surface is appreciable. The time resolution will be decreased considerably and the signal can be disturbed strongly if scattered light coming from a diffuse zone near the windows enters the monochromator. We suppressed this disturbing effect by inserting in the last part of the shock tube a steel cylinder with oxidized surface. Also the end plate of the tube was made from steel with a black surface. The reflectivity around a 4.3 μ layer of Fe_3O_4 seems to be low enough. At least we could not detect any effect after mounting these black surfaces in the shock tube.

RESULTS

Several important checks had to be carried out before we could interpret the results reasonably.

Proportionality between Emission and Concentration

We prepared several mixtures with varying amounts of CO_2 and measured the emission at various conditions before the onset of the reaction. In accordance with the results of Wagner and coworkers[13] we found in our shock tube with 75 mm inner diameter, an upper permissible limit of 1.4×10^{-7} mole/cm^3 of CO_2 (fig.1). Under our conditions this corresponds to a maximum of ≈ 0.55 % CO_2. The mixtures used for kinetic measurements contained ≈ 0.30 % CO_2.

Influence of Small Amounts of O_2

The purity of the gas mixture and the tightness of the shock tube had a marked effect on the results. On page 7 we will show a more complete reaction scheme, including the reactions with O_2 if this molecule is present as a contamination. The branching reaction, i.e. $O + H_2 \rightarrow OH + H$, will accelerate the rate of H_2-dissociation and consequently greatly affect the total rate of CO_2 disappearance during the course of reaction. The lower the initial temperature behind the reflected shock, the greater is the effect of small amounts of O_2. Figure 2 shows a computer solution of the reaction of CO_2 as a function of time with small contamination by O_2. We obtained a strong effect at 2100°K, in the neighbourhood of our lower temperature limit. With a 50% decrease of the initial CO_2 concentration and the addition of 10 ppm O_2 errors in the order of 15% arise in the time evaluation. This in turn will affect the rate constant evaluated for $H + CO_2 \rightarrow CO + OH$ markedly. The test gas has to be purified almost completely from contaminations by O_2. The best gases we could obtain commercially contained less than 3 ppm O_2 and H_2O. The inevitable further contamination due to the finite leak rate of the tube and the supply lines was not more than 0.6 ppm O_2 averaged over one filling of the shock tube. Any leakage near the observation windows would have a disastrous effect. We checked the tightness in this region several times simply by waiting a few minutes (5 - 20) before generating the shock. If the results did not differ from the common results obtained immediately after filling the tube, the shock tube was considered to be tight.

Relaxation

Kiefer and Lutz[5] recently published measurements of the relaxation of H_2-Ar mixtures. On the basis of their results we calculated the time $t_{0.98}$ needed by the gas to attain 98 % of the equilibrium distribution of the vibrational energy behind the reflected shock. With the conditions $T = 2000°K$ and $p = 4$ at, we estimated $t_{0.98} \ll 1$ μsec. Hence we may conclude that relaxation does not affect our measurements. The rate-determining step is the H_2-relaxation. CO_2 follows the equilibration of H_2 immediately.

Temperature

The initial temperature behind the reflected shock before the onset of the reaction was calculated as usual from the speed of the incident shock and the appropriate thermodynamic data. The enthalpies were taken from the Janaf tables. The velocity of the incident shock was measured with the aid of 8 thin-film-resistance gauges and electronic counters (5 Mc). The shock wave was decelerated by approximately 1.9 % per meter by boundary layer increase. The velocity of the shock near the observation windows was accurate within $\pm$ 0.2 %, which corresponds to an accuracy of the calculated initial temperature of $\pm$ 0.4 %. In connection with other research projects we also measured at the same time the temperature directly by using a band-reversal method in the infrared region. Between the direct measurement of temperature and the calculated value based on the shock velocity, we found average deviations in the order of $\pm$ 0.6 %. This result supports strongly the method of indirect temperature evaluation applied throughout this work.

We must now draw attention to another important fact. During the reaction the initial temperature may vary not only because of the enthalpy changes caused by reaction but also because of hydrodynamic effects. The latter have their source in the interaction of the reflected shock with the boundary layer that was generated by the incident shock. As is well known, this interaction can disturb the state of the gases behind the reflected shock near the observation windows considerably, and in particular cases can even prevent any reliable measurements. We decided to measure directly the constancy of the temperature behind the reflected shock in a similar but nonreacting gas mixture. For that purpose we prepared mixtures with 1% CO instead of 0.3% CO_2. This mixture again showed the emission characteristics of an optically thin layer. We can easily deduce in a good approximation for the harmonic oscillator model that the emitted intensity J satisfies

$$J \approx \frac{n_{CO}}{e^{\theta/T}-1} \qquad (1)$$

when

n_{CO} is the concentration of CO molecules in mole/cm^3 and θ = 3080oK is the characteristic temperature of the CO vibration. The formula holds for the case of the total emission of the 4.6 μ band of CO. If we observe only a part of the band with the monochromator, the rotational structure has to be taken into account. With a spectral window of 0.15 μ we found experimentally and theoretically that the simple formula 1 can be applied approximately. If we further assume nearly isentropic changes of the gas state in consequence of hydrodynamic effects, we can express the changes of n_{CO} by changes of temperature. From formula 1 we deduce:

$$\frac{dJ}{J} = \frac{dn}{n} + \frac{\theta/T}{1-e^{-\theta/T}} \cdot \frac{dT}{T}$$

$$\frac{dJ}{J} = \left(\frac{1}{\kappa-1} + \frac{\theta/T}{1-e^{-\theta/T}}\right)\frac{dT}{T}$$

with $\frac{dn}{n} = \frac{1}{\kappa-1} \cdot \frac{dT}{T}$ and κ = ratio of specific heats.

Inserting numbers, $\theta = 3080^{\circ}K$, $\kappa = 1.6$, we obtain finally:

$$\frac{dJ}{J} \approx 3.5 \frac{dT}{T}$$

A refined analysis taking into account the small effect of the rotational structure of the I.R. band leads in our case to the improved relation:

$$\frac{dJ}{J} \approx 3.0 \frac{dT}{T}$$

This formula was applied in the temperature range above $2000^{\circ}K$ to CO-mixtures and in the range below $1900^{\circ}K$ also to the original CO_2 mixtures. The latter showed no reaction during our measuring time of about 600 μsec and at low temperatures. Fig. 3 represents an experiment at $1800^{\circ}K$ with CO_2. The calibration line was obtained electronically before starting the experiment. Within $dJ/J \approx 1.5$ % the emission signal does not deviate from the calibration line. We may conclude that the temperature behind the reflected shock was at least constant within $\pm$ 0.5 % during 700 μsec. Thus boundary layer effects are very small in this case. Fig. 4 shows an experiment at $2920^{\circ}K$ with CO. Here the temperature decrease generated by the H_2 dissociation can be seen clearly. After completion of dissociation the temperature again remains almost constant. Finally fig. 5 represents an experiment at $1800^{\circ}K$ behind the incident shock! In that case the effect of the boundary layer is much more pronounced ($dT/T \approx 1.5$ %). Thus all the kinetic measurements were conducted behind reflected shocks.

Kinetic Measurements

Fig. 6 represents an oscillogram of CO_2 emission versus time during the reaction and k_1 can be calculated as a function of the intensity ratio $J/J_o = x_{CO_2}/x_{CO_{2o}}$ ($J \sim x_{CO_2}$). The initial value J_o at time $t=0$ has to be corrected for the finite time and space resolution of the recording system. The complete kinetic scheme of all important elementary reactions which have to be considered is:

1	H	$+ CO_2$	$\rightleftarrows$	CO	+ OH
2	OH	$+ H_2$	$\rightleftarrows$	H_2O	+ H
3	H_2	+ M	$\rightleftarrows$	2 H	+ M
4	H	$+ O_2$	$\rightleftarrows$	OH	+ O
5	O	$+ H_2$	$\rightleftarrows$	OH	+ H
6	CO_2	+ M	$\rightleftarrows$	CO + O + M	
7	H_2O	+ M	$\rightleftarrows$	H + OH + M	

If $O_2 \leq 5$ ppm reactions 4 and 5 can be neglected. At higher concentrations we have an increasing influence on the mechanism (Fig.2). The chain branching caused by reactions 4 and 5 will accelerate the H_2 dissociation markedly during the first stages of the reaction. A simplified data evaluation based on reactions 1 and 3 will be made impossible if there is contamination by more than 5 ppm O_2. As was described on page 4, the maximum concentration of O_2 in the test gas never exceeded 4 ppm. Reactions 6 and 7 are of minor importance at temperatures below 2700°K. This was checked by a computer program, in which we integrated the complete set of kinetic differential equations. Thus in the temperature range between 2000 and 2700°K we can describe the total reaction satisfactorily by reactions 1 to 3 only.

During the performance of the experimental work we found it much more advantageous to evaluate the rate constant k_1 by means of an approximative but closed solution including reactions 1 to 3. The alternative possibility, integrating the complete system of differential equations by a Runge-Kutta procedure, was found to be too complicated and time consuming. We used this program only for control purposes.

Reactions 1 to 3 yield the following system of kinetic differential equations:

$$\frac{dCO_2}{dt} = - k_1 \cdot H \cdot CO_2 + \bar{k}_1 \cdot CO \cdot OH$$

$$\frac{dH}{dt} = 2 \cdot k_3 \cdot M \cdot H_2 - 2 \cdot \bar{k}_3 \cdot H^2 \cdot M - k_1 . H \cdot CO_2 + \bar{k}_1 \cdot CO \cdot OH + k_2 \cdot H_2 \cdot OH - \bar{k}_2 \cdot H_2O \cdot H$$

$$\frac{dOH}{dt} = - k_2 \cdot H_2 . OH + \bar{k}_2 \cdot H_2O \cdot H + k_1 \cdot H \cdot CO_2 - \bar{k}_1 \cdot CO \cdot OH$$

The k_j with the bar correspond to the reverse reactions from "right to left", page 7. At the first stages of the reaction we may apply the steady state assumption for OH in a very good approximation. As a result of the high H_2 concentration and the small activation energy of reaction (2), OH is very quickly consumed. With $dOH/dt \simeq 0$, $OH \simeq H \cdot (k_1/k_2) \cdot (CO_2/H_2)$. H_2 can be considered practically constant throughout the reaction because of the low temperatures ($T \leq 2700°K$). It follows then that $k_1/k_2 < 1$ and $CO_2/H_2 < 1$. Therefore $OH/H << 1$, at least during the first stages of the reaction. It can now be shown that $k_1 \cdot CO \cdot OH$ can be neglected at least up to a 50% decrease in the initial CO_2 concentration. If we neglect the reverse of reaction 1 we finally have the simplified system:

$$\frac{dCO_2}{dt} = - k_1 \cdot CO_2 \cdot H$$

$$\frac{dH}{dt} = 2k_3 \cdot M \cdot H_2 - 2\bar{k}_3 \cdot M \cdot H^2$$

The solution for constant H_2 concentration and temperature is:

$$\ln \frac{x_{CO_2}}{x_{CO_{20}}} = - \frac{k_1}{2k_3 \cdot M} \cdot \ln(\cosh\sqrt{4k_3\bar{k}_3 \cdot M^3 x_{H_2} \cdot t}) \quad (2)$$

$H_2 \simeq$ const.
$T \simeq$ const.
x_{CO_2}, x_{H_2} = mole fractions
XCO_{20} = initial mole fractions
M = concentration in mole/cm^3 of the gas
t = time

With this formula values of k_1 were first calculated. The values of k_3 required were measured by several authors (Patch,[6] Rink,[7] Sutton,[8] Myerson[9]). The recently published values of Myerson, which seem to be very accurate, agree excellently with those of Patch. We selected the recommended expression of Patch:

$$k_3 = (2.9 \cdot x_{Ar} + 29 \cdot x_{H_2}) \cdot 10^{18} \cdot T^{-1} \cdot$$
$$\cdot \exp(-104200/RT)\ cm^3/mole \cdot sec$$

The efficiency of the collision partner M (reaction 3) was regarded in particular. H_2 was assumed 10 times as effective as Ar . The influence of all other components can be neglected here. The value of $\overline{k_3}$ was calculated as usual with the aid of the equilibrium constant.

At the upper limit of the temperature range the assumption $T \approx$ const. might be questionable as a result of the increasing enthalpy consumption during the H_2 dissociation. The overall reaction $CO_2 + H_2 \rightarrow CO + H_2O$ is nearly thermoneutral. The small temperature decrease is mainly due to the H_2 dissociation. A good approximation was obtained with

$$T(t) \approx T_o - \frac{H_H}{\overline{C}_p} \cdot x_H \qquad (3)$$

where, T_o = initial temperature behind the reflected shock, $\overline{C}p$ = average specific heat at constant pressure and around T_o, H_H = reaction enthalpy of reaction 3, $1/2\ H_2 \rightarrow H$, and x_H = mole fraction of H atoms.

If for the evaluation of k_1 we consider only that part of the oscillogram (Figure 6) where $CO_2/CO_{20} \geq 0.5$, we can immediately deduce from the simplified solution (2) that throughout the temperature range 2000 - 2700°K, x_H never exceeds about 0.15%. This is an important result.

In order to minimize the effects due to the decreasing temperature, for the exact evaluation of k_1 we never used intensity ratios $J/J_o = x_{CO_2} \cdot x_{CO_{20}}$ less than 0.5. For the temperature decrease in the region $1 \geq J/J_o \geq 0.5$ we get

$$\left|\frac{\Delta T}{T_o}\right| \approx \left|- \frac{H_H}{\overline{C}_p \cdot T_o} \cdot x_H\right| \approx \leq 10^{-2}$$

This is a small effect and we can find approximate but closed solutions of the scheme 1 - 3 including the temperature decrease. We approximate the expression for the rate constants:

$$k_j = k_{j_o} \exp(-E_j/RT(t))$$

by

$$k_j \approx k_j(T_o)\ (1 - x_H \cdot \frac{E_j}{RT_o} \cdot \beta + x_H^2 \cdot \frac{E_j}{RT_o} \cdot \beta^2 \cdot (\frac{E_j}{2RT_o} - 1))$$

where $$\beta = \frac{H_H}{\overline{C}_p \cdot T_o}$$

With this expression for k_1 and k_3 we obtain

$$\ln \frac{x_{CO_2}}{x_{CO_{20}}} = k_1(T_o) \cdot \frac{M\ (h + g)}{2 \cdot b \cdot h} \cdot$$

$$\left[G \cdot \ln \frac{e^{h \cdot t} + r}{1 + r} + D \cdot h \cdot t + F\ (\ 1 - \frac{1 + r}{e^{ht} + r}) \right] \qquad (4)$$

$$H_2 \approx \text{const.}$$

where,
$k_1(T_o)$ is the rate constant based on the initial temperature behind the reflected shock, M is the concentration in mole/cm^3, t the time, D, F, G, b, g, h, r are functions of E_1, E_3, M, T_o, x_{CO_2}, β and $k_3(T_o)$. (See also the Appendix.) Solution (4) was compared with an exact computer solution. The agreement was excellent if $CO_2/CO_{20} \geq 0.2$ was chosen.

Although formula (4) looks rather complicated, we found it relatively easy to calculate values of $K_1(T_o)$ directly from it. The exact Runge-Kutta solution would require a much more tedious and time-consuming trial-and-error procedure for evaluation of the correct $k_1(T_o)$.

The final results are shown in Figure 7. The solid line represents the result of a least-squares fitting procedure, based on the k_3 value of Patch.[6] The simple expression

$$k_1 = k_{10} \cdot \exp\ (-E_1/RT)$$

yields $k_{10} = 1.1 \times 10^{14} \pm 12\ \% \ \frac{cm^3}{mole \cdot sec}$

$E_1 = 24900 \ \frac{cal}{mole} \pm 1300 \ \frac{cal}{mole}$

The average error in k_{10} obtained with the least-squares fitting procedure does not include errors in k_3. The broken line in Figure 7 gives an impression of the order of magnitude of this error. This line is the result of applying a least-squares fitting procedure to the k_3 data of Rink.[7] The data of Myerson[9] fall in the same range.

Including uncertainties in k_3, the average error in k_{10} is greater than ± 12%. We write finally for k_{10}:

$$k_{10} = 1.1 \times 10^{14} \ \frac{cm^3}{mole \cdot sec} \pm 25\ \%$$

The error for a single measured point can also be estimated using the simple formula (2), p.8. Disregarding errors of k_3, we deduce that

$$-\frac{dk_1}{k_1} = \frac{1}{\ln \frac{x_{CO_2}}{x_{CO_{20}}}} \cdot \frac{dx_{CO_2}}{x_{CO_2}} + 2\,\frac{dt}{t} + 2\,\frac{dM}{M} + \frac{E_3}{RT}\,\frac{dT}{T}\,.$$

At $\frac{x_{CO_2}}{x_{CO_{20}}} \approx 0.5$ we have: $\frac{dx_{CO_2}}{x_{CO_2}} \approx \pm\, 0.04$;

$\frac{dt}{t} \approx \pm\, 0.03$; $\frac{dM}{M} \approx \pm\, 0.02$; $\frac{dT}{T} \approx 0.009$
(here the errors in shock velocity and of the temperature constancy are summed up);

$\frac{E_3}{RT} \approx 24$; $T = 2100\,^{\circ}K$

$$\frac{dk_1}{k_1} \approx \pm\, 0.36 \qquad (5)$$

This is reasonable in view of the average error ± 12% obtained from the least-squares fitting procedure. The estimation (5) seems to be a bit too pessimistic as shown by an inspection of the scatter of the measured points in Figure 7.

Comparison with other Measured Values

In Figure 8 we have plotted results of other authors together with ours. The results of Schecker[4] seem to be the most accurate values yet obtained at lower temperatures. His results were combined with ours by means of an interpolation formula. The broken line represents the best fit. Interpolation is possible, e.g. with an empirical formula of the type $k_1 = k_{10} \cdot T^m \cdot \exp(-E_1/RT)$. The appropriate numbers are

$k_{10} = 1.1 \times 10^{19}\ \frac{cm^3}{mole \cdot sec} \cdot \left[{}^{O}K\right]^{1.33} \pm 25\ \%$

$m = -\ 1.33$

$E_1 = 31300\ \frac{cal}{mole}$; Temperature range from $1400\,^{O}K$ to $2700\,^{O}K$.

The exponent m indicates that the internal degrees of freedom seem to play a role during the collision of CO_2 with H, at least at elevated temperatures.

The reported accuracy of Schecker's results is comparable to ours. It seems not too unreasonable to interpret the different measured activation energies of 24900 and 27200 as due to real effects and not random errors. The systematic deviation of both lines (Figure 8) can be well explained by the remaining uncertainties in the absolute k_3 values.

COMPARISON WITH THEORETICAL MODELS

Although the accuracy of the data is fairly high, the temperature range covered is still somewhat small for theoretical comparisons. Nevertheless, some discussion might be fruitful. The model given by Hinshelwood[10] was constructed in order to take into account inner degrees of freedom during the collision of complex molecules. Two cases are of interest here. We assume, merely by intuition, that the reaction path with the smallest activation energy should be chosen along the CO_2 axis, i.e. that the fraction of the relative energy during collision which is used to cross the potential barrier acts in that direction. Then in case I we take into account the linear (symmetric and antisymmetric) vibrations of CO_2 only. Case II includes also the bending normal vibrations. Consideration of case I leads to the approximate expression

$$k_1^{(I)} \approx k_{10}^{(I)} \cdot T^{1/2} \cdot (1-e^{-\frac{3380}{T}}) \cdot (1-e^{-\frac{1924}{T}}) \cdot \left[1+2\frac{RT}{E_1}+2\left(\frac{RT}{E_1}\right)^2\right] \cdot$$

$$\cdot \exp(-E_1^{(I)}/RT)$$

In this formula 3380°K and 1924°K are the characteristic temperatures θ_j for the two linear normal vibrations of CO_2, and the approximation is valid for $E/R > \theta_j$. Rotation was not considered; orientational effects are roughly considered as independent of temperature. Except for the inclusion of the two bending vibrations, the same statements hold also for case II:

$$k_1^{(II)} \approx k_{10}^{(II)} \cdot T^{1/2} \cdot \prod_{j=1}^{4} (1-e^{\frac{-\theta_j}{T}}) \cdot \left[1+4\left(\frac{RT}{E_1}\right)+12\left(\frac{RT}{E_1}\right)^2\right.$$

$$\left. + 24\left(\frac{RT}{E_1}\right)^3 + 24\left(\frac{RT}{E_1}\right)^4\right] \cdot \exp(-E_1^{(II)}/RT).$$

$\theta_{bending}$ (doubly degenerated) = 961°K.

Case I leads to the following numbers:

$$k_{10}^{(I)} = 4.7 \times 10^{12} \frac{cm^3}{mole \cdot sec} \left[{}^{o}K\right]^{-\frac{1}{2}}, \quad E_1^{(I)} = 26700 \frac{cal}{mole};$$

and case II yields

$$k_{10}^{(II)} = 1.0 \times 10^{14} \frac{cm^3}{mole \cdot sec} \left[{}^{o}K\right]^{-1/2}$$

$$E_1^{(II)} = 32200 \frac{cal}{mole}$$

With both expressions an excellent representation of the experimental results is possible.

That part of $k_{10}^{(I)}$ **or** $k_{10}^{(II)}$ which is proportional to the average collision number in the gas has the very reasonable value of about 2×10^{13} cm^3/mole · sec. Hence a decision whether case I or case II has some connection to reality is not possible with these measurements; however, it might be attractive to extend the measurements to lower temperatures (1000 - 800°K) and apply there the very precise ESR method (see, e.g., Fristrom and Westenberg[11]). Around 1000°K the different activation energies of cases I and II theoretically will give rise to significantly distinct absolute values of $k_1^{(I)}$ and $k_1^{(II)}$. We believe these experiments are worth undertaking, because the reaction under consideration is one of the simplest in which the contribution of the inner degrees of freedom is not negligible and has the great advantage that the molecular properties of reactants and products are well known. This reaction provides a real opportunity to improve our detailed knowledge about the mechanism of participation of inner degrees of freedom during a more complex reaction.

The transition state theory (Eyring[12]) was also applied to the experimental results. The simplest assumption of a linear collision complex yields the wrong temperature dependence of the pre-exponential factor. If we describe the results of these calculations by the expression $k_1 \approx k_{10} \exp(-E_1(T)/RT)$, than E_1 increases with increasing temperature in contradiction to the experimental observation. The source of this discrepancy is the strong temperature dependence of the partition function belonging to the collision complex $CO_2 \cdot H$. The original assumption of statistical equilibration of all inner degrees of freedom (except the degree with the imaginary frequency) during the lifetime of a great number of collision complexes might be questionable. If the "relaxation" times of the normal vibrations are equal to or longer than the lifetime of the complex, a correct description would be possible only with partly or specifically excited normal vibrations of the $CO_2 \cdot H$ complex. This would fit the experimental data better.

ACKNOWLEDGEMENTS

We would like to place on record that this work was sponsored generously by the Deutsche Forschungsgemeinschaft.

APPENDIX

The solution of

$$\frac{dCO_2}{dt} = - k_1(T) \cdot CO_2 \cdot H \qquad (6)$$

$$\frac{dH}{dt} = 2\, k_3(T)H_2 \cdot M - 2\, \bar{k}_3(T) \cdot H^2 \cdot M \qquad (7)$$

is obtained easily, because the differential equation (7) does not depend on CO_2. (7) can be integrated and the solution H(t) inserted in (6). Then

$$x_H(t) = \frac{h + g}{-2b} \cdot \left[\frac{e^{ht} - 1}{e^{ht} + \frac{h+g}{h-g}} \right]$$

with

$$h = + \sqrt{g^2 - 4\, ab}$$

$$g = - \frac{aE_3 \cdot \beta}{RT_o}$$

$$a = 2\, k_3(T_o) \cdot M \cdot x_{H_2}$$

$$b = g \cdot \beta \left(\frac{E_3}{2RT_o} + 1\right) - 2\bar{k}_3(T_o) \cdot M^2$$

$$\beta = \frac{H_H}{C_p T_o}$$

Integration of (6) with H(t) yields

$$\ln \frac{x_{CO_2}}{x_{CO_{20}}} = k_1(T_o) \cdot \frac{M(h+g)}{2\, bh} \left[G \cdot \ln \frac{e^{ht}+r}{1+r} + D \cdot h \cdot t + F \left(1 - \frac{1+r}{e^{ht}+r}\right) \right]$$

with

$$D = \left(\frac{E_1 \cdot \beta}{RT_o} \cdot \frac{h-g}{2b} - 1\right) \frac{h-g}{h+g}; \qquad F = -\frac{E_1 \cdot \beta}{RT_o} \cdot \frac{2h^2+g^2}{2b \cdot h}$$

$$G = 1 - D + \frac{E_1 \cdot \beta}{RT_o} \cdot \frac{g+h}{2b}\, ; \qquad r = \frac{g+h}{h-g}.$$

REFERENCES

1. Olschewski, H.A.; Troe, J.; Wagner, H.Gg. Zeit.physik. Chem.N.F., 47, (1965), 383.
2. Fenimore, C.P.; Jones, G.W. J.Phys.Chem.62, (1958), 1578.
3. Dixon-Lewis; Sutton, M.M.; Williams, A. Trans.Farad.Soc. 61, (1965), 255.
4. Schecker, H.G.; Diplomarbeit, Göttingen, 1964. See also: Jost, W.,; Schecker, H.G.,; Wagner, H.Gg. Zeit.f.Phys.Chem. N.F.45 (1965), 47
5. Kiefer, J.H.; Lutz, R.W. J.Chem.Phys.44, (1966), 668.
6. Patch, R.W. J.Chem.Phys., 36, (1962), 1919.
7. Rink,J.P. J.Chem.Phys.35, (1961), 1765.
8. Sutton,E.A. J.Chem.Phys.36 (1962), 2923.
9. Myerson,A.L.; Watt, W.S. J.Chem.Phys.49, (1968), 425.
10. Hinshelwood,C.N. The Kinetics of Chemical Change.Oxford, Clarendon Press, 1940.
11. Fristrom,R.M.; Westenberg,A.A. Tenth Symp.on Comb.,The Comb.Instit.,Pittsburgh, 1965
12. Eyring,H.; Glasstone,S; Laidler, K.J. The Theory of Rate Processes. McGraw-Hill Inc., New York, 1941.
13. Michel, K.W.; Olschewski, H.A.; Richtering,H; Wagner,H.Gg. Zeit. phys.Chem.44 (1965)160.

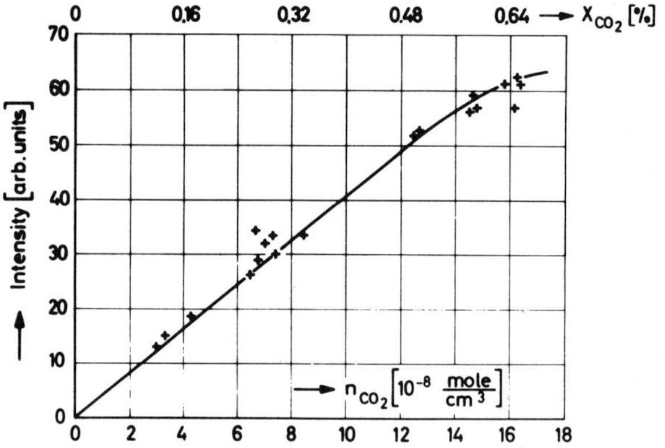

Figure 1
Intensity of CO_2 emission at 4.3 μ as a function of density

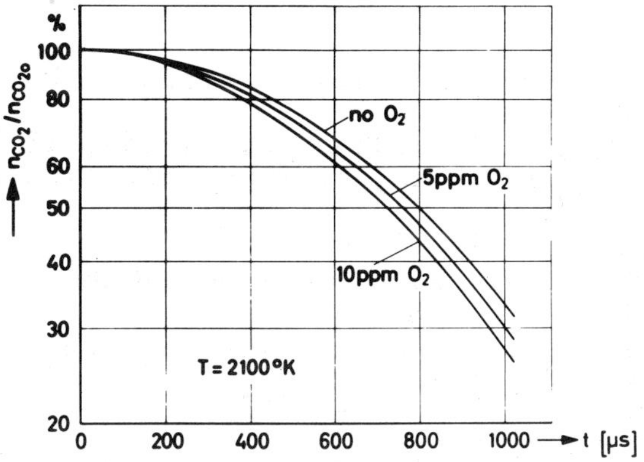

Figure 2
Influence of small amounts of O_2 on reaction rate

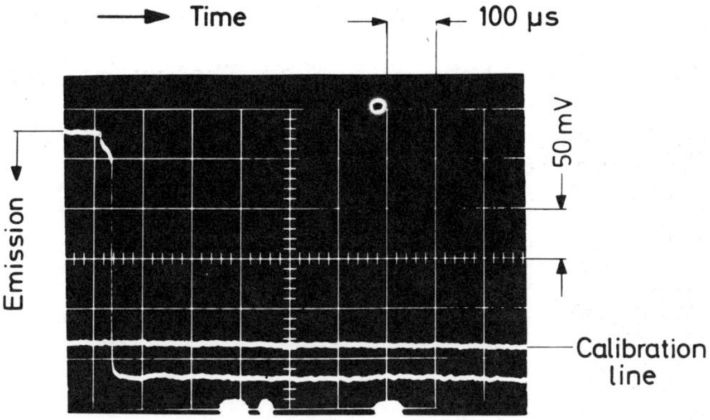

Figure 3
CO_2 emission at 1800°K. No reaction

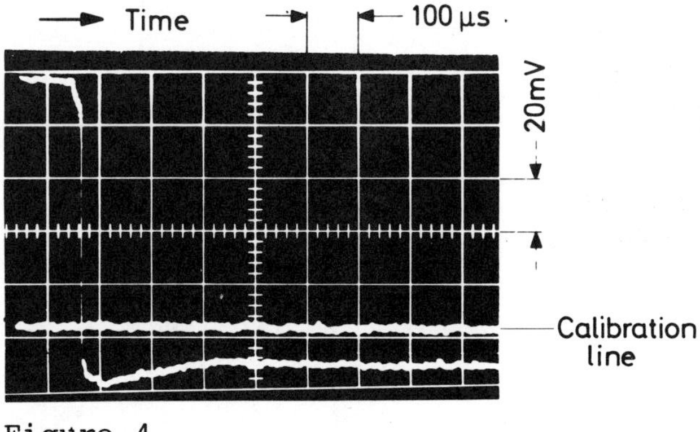

Figure 4
CO emission at 2900 °K.
(Dissociation of H_2 can be seen)

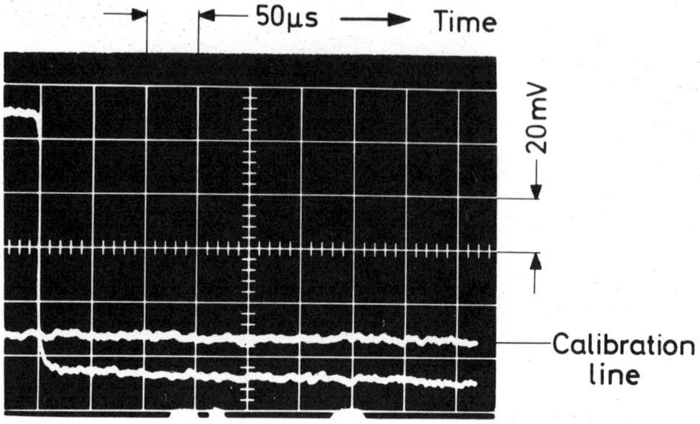

Figure 5
CO_2 emission at 1800 °K behind incident shock wave

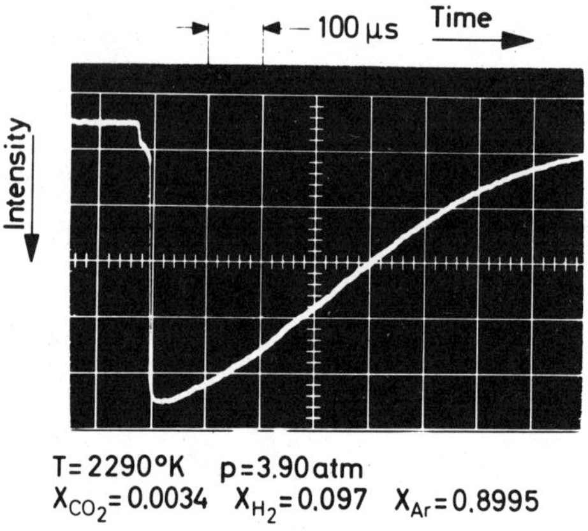

Figure 6
Decrease of CO_2 during reaction

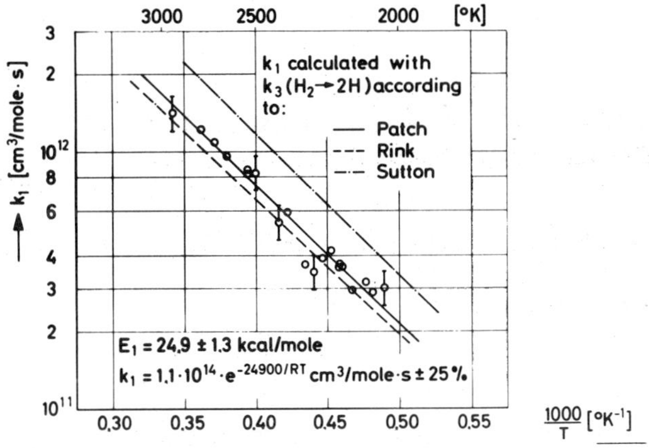

Figure 7
Measured rate constant k_1 (CO_2+H→CO+OH) as a function of temperature

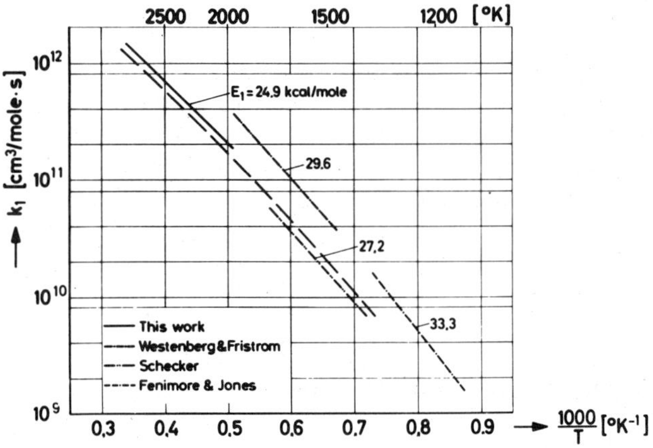

Figure 8
Comparison of measured k_1

THE EFFECT OF NH_2 RADICALS ON THE EXCHANGE REACTION BETWEEN AMMONIA AND DEUTERIUM

ASSA LIFSHITZ, PINCHAS SCHECHNER, and ALEXANDER BURCAT
Department of Inorganic Chemistry,
The Hebrew University, Jerusalem, Israel

Single pulse shock tube studies of the exchange reaction between ammonia and deuterium were performed over the temperature range 1150—1450°K, at high dilution in argon. The rate of exchange was found compatible with the following rate law:

$$d[HD]/dt = k_b[Ar]^{0.38}[NH_3]^{0.61}[D_2]^{1.01}$$

where $k_b = 10^{13.0 \pm 0.3} \exp[-(37.2 \pm 1.0) \times 10^3/RT]$ cc mole^{-1} sec^{-1}.

The effect of NH_2 radicals on the reaction rate was studied by adding 0.5% hydrazine to the reaction mixture. By a comparison of the data obtained in the presence of hydrazine and in its absence, the following rate constant for the reaction

$$NH_2 + D_2 \xrightarrow{k_7} NH_2D + D$$

was obtained:

$$k_7 = 10^{12.54} \exp[-16.4 \times 10^3/RT] \text{ cc mole}^{-1} \text{sec}^{-1}$$

It was shown that in the absence of hydrazine the reaction does not proceed via a 3-center atomic displacement mechanism owing to the very low concentration of NH_2 radicals produced by the dissociation of ammonia. A mechanism based on vibrational excitation of the deuterium molecule to a critical vibrational level as a slow step prior to the atom switching step is proposed.

INTRODUCTION

In the last few years a considerable number of isotope exchange studies have been performed in the single pulse shock tube in view of the homogeneous conditions that this technique provides. In all the exchange reactions studied, an effect of the concentration of the inert gas diluent on the reaction rate was observed. When the rate of exchange was expressed in terms of a power rate law, the reaction order with respect to the argon was as high as unity[1] in the reaction $^{14-14}N_2 + {}^{15-15}N_2 \rightarrow 2\,{}^{14-15}N_2$, and 0.4 — 0.7 in a number of exchange reactions involving H/D systems[2-5].

Prior to proposing a mechanism to account for the effect of the argon on the reaction rate, it was necessary to consider the possibility of a chain mechanism, which could be initiated by decomposition reactions of the exchange partners, the rate of which is known to be affected by the argon. In most of the exchange studies cited, it was unequivocally shown that the contribution of a chain mechanism to the observed rate of exchange was negligible. In fact, the experiments and calculations on the basis of which the chain mechanism was ruled out constituted a major part of many of these exchange studies.

This work is a reinvestigation of the exchange reaction $NH_3 + D_2 \rightarrow NH_2D + HD$,[5a] with a detailed discussion into the possibility of an exchange mechanism carried by free radicals.

EXPERIMENTAL

Apparatus and Technique

The exchange reaction was studied in a pressurized driver single pulse shock tube. The driven section was 2 in. I. D. made of "Double-tough" Pyrex tubing 13 feet long. The brass driver was about 9 feet long and its exact length was adjustable in small steps for tuning purposes. Description of the tube and details of its operation have already been given in earlier publications[1,6]. Before each experiment, the tube was pumped down to $\sim 10^{-4}$ torr, having a leak (or degassing) rate of approximately $1\,\mu$/min. The temperature range covered in this study was 1150 — 1450°K corresponding to incident shock speeds of 0.713 — 0.825 mm/μsec. The shock speed could be determined with an accuracy of $\sim\pm 0.7$ % corresponding to $\sim\pm 20$°K at the temperature range covered in this study.

Materials and Analysis

Five reaction mixtures were prepared: (a) 1% D_2 and 1% NH_3, (b) 4% D_2 and 4% NH_3, (c) 1% D_2 and 4% NH_3, (d) 0.67% D_2 and 0.67% NH_3, and (e) 4% D_2, 4% NH_3 and 0.5% NH_2-NH_2, all diluted in argon. The reaction mixtures (a)—(d) were stored at a high pressure in stainless steel cylinders which were previously pumped down to a few hundredths of a micron. The reaction mixture (e) was stored in a 12 liter pyrex bulb at atmospheric pressure.

The materials used were from the Matheson Gas Co. The ammonia was 99.99% pure, the deuterium was C. P. grade, listed as 99.5% pure. It contained 0.49% HD. The argon was prepurified grade listed as 99.998% pure. These materials were used without further purification.

The hydrazine was prepared by a reaction between a carefully dried ammonia and hydrazine hydrochloride. The anhydrous hydrazine was distilled in vacuum a few times before preparing the reaction mixture. The driver gas was pure grade helium.

Gas samples were taken from the end block of the shock tube and were analyzed by an Atlas CH4 mass spectrometer. Extents of reaction were computed from the ratio M3/M4 peak heights and corrected for the different sensitivity of the secondary electron multiplier towards HD and D_2. Attempts to compute extents of reaction from the ratios M18/M17 (NH_2D/NH_3) were unsuccessful owing to high H_2O^+ and OH^+ background, and to exchange of NH_2D with the water absorbed on the walls of the ion source.

RESULTS

The rate of production of HD can be represented in the general form of a power rate law:

$$d[HD]_t/dt = k_b[Ar]^a[NH_3]^b[D_2]^c \qquad (I)$$

where a, b and c are the reaction orders with respect to argon, ammonia and deuterium, respectively. Now define k_m, a pseudo first order rate constant, where

$$k_m = k_b[Ar]^a[NH_3]^b[D_2]^{c-1} \qquad (II)$$

The rate of exchange is then given by:

$$d[HD]_t/dt = k_m[D_2] \qquad (III)$$

To minimize the extent of the consecutive reactions which follow the first elementary step $D_2 + NH_3 \rightarrow HD + NH_2D$, the ratios HD/D_2 were kept very low in most of the runs. It is believed that under these conditions k_m represents the rate of the first elementary step of this exchange reaction. For small conversions, k_m can be fairly well evaluated from the equation:

$$k_m = \frac{([HD]_t - [HD]_0)/[D_2]_0}{\tau} \qquad \text{(IV)}$$

where τ is the time during which the reaction mixture was maintained at the high temperature and $[HD]_0$ is the concentration of HD at zero time.

From the dependence of k_m on the concentrations of the reaction partners, the reaction orders a, b, and c can be evaluated. In order to determine these values and the activation energy of the reaction, 13 groups of runs, covering a wide range of composition, pressures and temperatures, totalling close to 100 shocks were run. Table I.

TABLE I

Summary of the experimental conditions of the 13 groups of shocks

Series	No. of shocks	Composition %			$\sim p_1$ mm Hg	T_5°K
		D_2	NH_3	N_2H_4		
1A	9	1.0	1.0	–	80	1205–1487
2A	6	1.0	4.0	–	80	1150–1459
3A	7	1.0	1.0	–	185	1255–1420
4A	7	1.0	1.0	–	45	1242–1400
5A	7	4.0	4.0	–	40	1134–1402
6A	10	4.0	4.0	–	80	1111–1483
1B	10	4.0	4.0	–	160	1174–1411
2B	9	1.0	1.0	–	160	1216–1537
3B	7	4.0	4.0	–	40	1152–1437
4B	9	0.67	0.67	–	240	1223–1436
5B	7	1.0	1.0	–	160	1224–1525
1C	6	4.0	4.0	0.5	80	1060–1192
2C	3	4.0	4.0	–	80	1074–1137

summarizes the experimental conditions of the 13 groups. The shocks were run in 3 series, referred to as A, B, and C. Series B and C were run with newly prepared reaction mixtures, about one year after completion of series A. Each group of runs differed from the other by at least one parameter, such as the ammonia percentage, deuterium percentage and the total pressure. By comparing the values of the obtained first order rate constants k_m in two groups, one, or a combination of the reaction orders can be directly evaluated.

If k_m in group 1A is given by

$$k_m(1A) = k_b[Ar]^a[NH_3]^b[D_2]^{c-1} \qquad \text{(V)}$$

then

$$k_m(2A) = k_b[Ar]^a(4\times[NH_3])^b[D_2]^{c-1} \qquad \text{(VI)}$$

$$b = \frac{\log k_m(2A) - \log k_m(1A)}{\log 4} \qquad \text{(VII)}$$

In order to obtain the difference of the rate constants of two groups on a logarithm scale, the logarithms of the rate constants k_m were plotted against 1/T for all the runs in the two groups involved and $\Delta \log k_m$ was obtained from the distance between the two parallel lines drawn in each plot. In Figure 1 plots of log k_m vs 1/T for the groups 1A and 2A, and 3A and 4A are presented. Table II summarizes the results obtained by various combinations of groups.

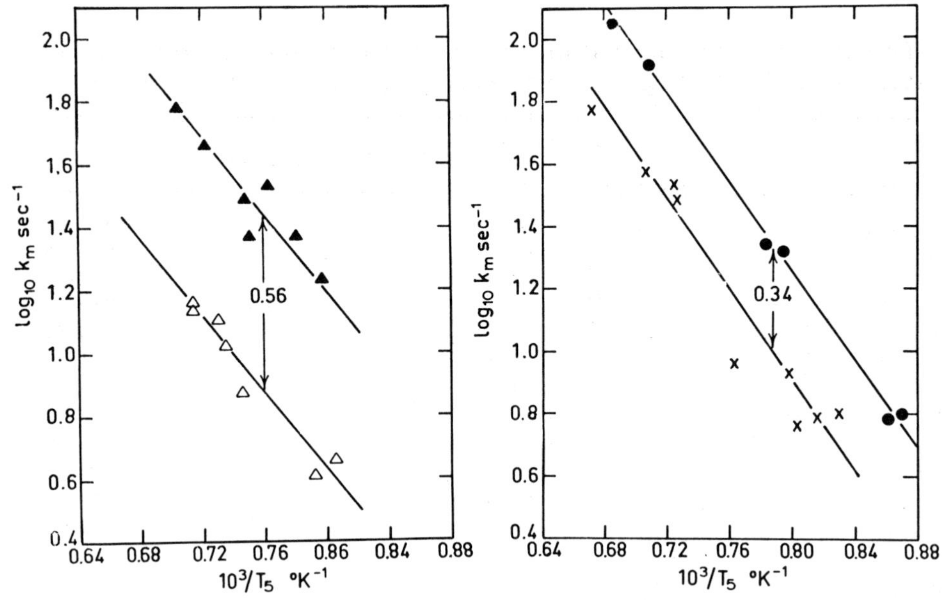

Figure 1
Plots of log k_m vs 1/T.
× – 1A; ● – 2A; ▲ – 3A; △ – 4A.

TABLE II
Reaction orders obtained from various combinations of groups of shocks

Combination	$\Delta \log k_m$	Results
1A – 2A	0.34	$b = 0.57 \pm 0.03$
3A – 4A	0.56	$a + b + (c-1) = 0.94 \pm 0.03$
3A – 5A	0.22	$a = 0.36 \pm 0.08$
2A – 6A	0.0	$(c-1) = 0.0 \pm 0.05$
1B – 2B	0.38	$b + (c-1) = 0.63 \pm 0.03$
3B – 4B	0.30	$a = 0.38 \pm 0.03$

The reaction orders obtained from the various combinations and which fit best all the data are:

$n_{Ar} = 0.37 \pm 0.03$
$n_{NH_3} = 0.60 \pm 0.03$
$n_{D_2} = 1.0 \pm 0.05$

Adjusting the individual orders to obtain an overall order of two, the following rate law is obtained.

$$d[HD]/dt = k_b[Ar]^{0.38}[NH_3]^{0.61}[D_2]^{1.01} \qquad \text{(VIII)}$$

To determine the activation energy of the reaction, log k_b was plotted vs. 1/T. In Figure 2 we present the plot for 5 groups of shocks run in series B. All the points fit close to a single line.

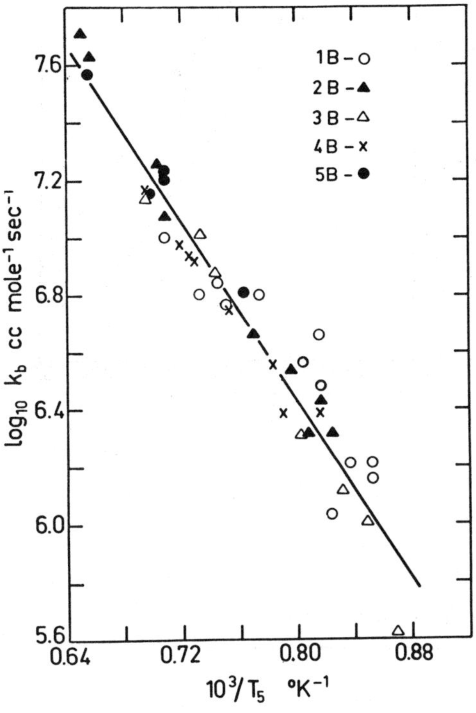

Figure 2.
A plot of log k_b vs 1/T for series B.

The value obtained for k_b is:

$$k_b = 10^{13.0 \pm 0.3}\exp[-(37.2 \pm 1.0) \times 10^3/RT] \text{ cc mole}^{-1}\text{sec}^{-1}.$$

DISCUSSION

The effect of the argon density on the reaction rate clearly indicates that the exchange between ammonia and deuterium does not represent a simple bimolecular reaction. If a hard collision had led to an exchange, an order of unity with respect to each of the exchange partners would have been observed. The first task in the interpretation of the observed rate law must be the examination of the possibility of a chain mechanism in which argon is involved as chain initiator, since the rates of decomposition of ammonia and deuterium are known to depend on the total density of the reaction mixture[7-11]. We shall discuss the initiation, propagation and termination steps of a chain reaction and compare the contribution of a chain mechanism with the observed rate of exchange.

Initiation, Propagation, and Termination Steps in a Chain Mechanism

Initiation Steps. The initiation steps that must be considered are the decomposition reactions of the reaction partners:

$$Ar + D_2 \xrightarrow{k_1} Ar + D + D \qquad (1)$$

$$Ar + NH_3 \xrightarrow{k_2} Ar + NH_2 + H \qquad (2)$$

The rate of dissociation of deuterium was measured by Sutton[7] and Rink[8] using the shock tube technique. The highest value of the two, extrapolated to 1300°K, gives 6.7×10^{-2} cc mole^{-1} sec^{-1} for k_1. The rate of decomposition of ammonia was studied by three groups of investigators[9-11] whose values, extrapolated to 1300°K, yield 2×10^2, 3.2×10^1 and 4.9×10^1 cc mole^{-1} sec^{-1} for k_2. Even if the lowest value is chosen for comparison, it is still five hundred times higher than the rate constant of the dissociation of deuterium at that temperature. Since the concentration of the ammonia and that of deuterium were always of the same order of magnitude in all our experiments, the ratio of the rate of production of the free radicals in reactions (1) and (2) is roughly equal to the ratio of the rate constants. The contribution of reaction (1) to the total initiation step is therefore negligible.

Propagation. Exchange products may be produced by the following propagation reactions:

$$H + NH_3 \xrightarrow{k_3} NH_2 + H_2 \qquad (3)$$

$$H + D_2 \xrightarrow{k_4} HD + D \qquad (4)$$

$$D + NH_3 \xrightarrow{k_5} H + NH_2D \qquad (5)$$
$$D + NH_3 \xrightarrow{k_6} NH_2 + HD \qquad (6)$$
$$NH_2 + D_2 \xrightarrow{k_7} D + NH_2D \qquad (7)$$

The free radicals that exist in the system are H, D and NH_2. D atoms are produced much faster by reaction (4) than by the initiation reaction (1). There are two parallel cycles in this set of reactions: reactions (4) and (5), in which the chain carriers are H and D atoms, and reactions (6) and (7), in which the chain carriers are D and NH_2. To what extent either of these two competitive cycles operates depends on the ratio of the rate constant k_5/k_6. Whereas reaction (6) is known to exist, [12,13] there is no indication that reaction (5) occurs at all. In a reaction between deuterium atoms and methane, the reaction

$$D + CH_4 \rightarrow H + CH_3D \qquad (8)$$

(analogous to reaction (5)) readily occurs with hot atoms (2 — 10 ev), but never with thermal ones[14]. The only reaction known to occur with thermal atoms is an abstraction reaction analogous to (6), i. e.,

$$D + CH_4 \rightarrow HD + CH_3 \qquad (9)$$

On the assumption that the behavior of hydrogen atoms towards ammonia is similar to their reaction with methane, the cycle based on reactions (4) and (5) may be eliminated from the discussion. (This assumption will be further discussed and justified at a later stage).

If there is any contribution to the exchange reaction from a free radical mechanism, the chain will be carried according to reactions (6) and (7). Neglecting isotope effects at this high temperature, the ratio k_6/k_7 is given by the equilibrium constant of reaction (3). K_3 was evaluated in the following way. Bradley et al[15] reported $\Delta H^0 = -15$ kcal/mole for the reaction:

$$NH_3 + OH \rightarrow NH_2 + H_2O \qquad (10)$$

Since the ΔH^0 of the reaction

$$OH + H_2 \rightarrow H + H_2O \qquad (11)$$

is also -15.0 kcal/mole,[16] reaction (3) is athermal. Calculating the entropy of the NH_2 radical on the rigid rotator harmonic oscillator approximation from its known moments of inertia [17] and vibrational modes,[18] we find that $\Delta S^0_{(1300°K)}$ of reaction (3) is $+4.57$ cal deg^{-1} mole^{-1}.

The equilibrium constant of reaction (3) at 1300°K is then $K_3 \sim 10$. From this value one concludes that the rate constant k_6 is about 10 times greater than the rate constant k_7. Since reactions (6) and (7) operate in a closed cycle, after reaching a stationary state, their velocities must be the same. Under these conditions $[NH_2]/[D] \approx 10$. It is true that the concentrations of the radicals increase continuously owing to the initiation reactions, but since reactions (3)—(7) are many orders of magnitude faster than the initiation reactions (1) and (2), this ratio is maintained undisturbed during the reaction. We also assume that all the back reactions are neglected owing to the very low ratio of HD to D_2 and NH_2D to NH_3.

In conclusion, the contribution of a chain mechanism to the rate of exchange can be fairly well calculated from the cycle of reactions (6) and (7), using the rate constant k_7 and assigning for the average concentration of NH_2 the value

$$[NH_2]_{av} = 2\,k_2[Ar][NH_3]\frac{\tau}{2} \qquad \text{(IX)}$$

where τ is the dwell—time of the reaction.

Termination Steps. The termination steps which lead to an exchange are:

$$Ar + NH_2 + D \rightarrow Ar + NH_2D \qquad (12)$$

and

$$Ar + H + D \rightarrow Ar + HD \qquad (13)$$

These are, however, very slow three-body processes, and owing to the very small concentration of the free radicals, their contribution to the exchange process is indeed negligible.

Determination of k_7 and the Rate of a Chain Mechanism

For the reaction $H + NH_3 \xrightarrow{k_3} NH_2 + H_2$, Oganesian and Nalbandian[12] reported a rate constant $k_3 = 4.1 \times 10^{13} \exp[-13.7 \times 10^3/RT]$ cc mole^{-1} sec^{-1}. Once again neglecting isotope effect, according to their results, $k_7 = 4.1 \times 10^{12} \exp[-13.7 \times 10^3/RT]$ cc mole^{-1} sec^{-1}. Using k_7 obtained by these authors, we can now calculate the contribution of the chain mechanism to the observed rate of exchange according to the following equations:

$$\left(\frac{d[HD]}{dt}\right)_{chain} = k_7[NH_2]_{av}[D_2] \qquad \text{(X)}$$

or

$$\left(\frac{d[HD]}{dt}\right)_{chain} = k_{m(chain)}[D_2] \qquad (XI)$$

where

$$k_{m(chain)} = k_7[NH_2]_{av} \qquad (XII)$$

Table III presents the values for $k_{m(chain)}$ calculated according to eq. (XII), and the ratios $k_{m(chain)}/k_{m(exp)}$ for comparison.

It can readily be seen from Table III that at 1300°K, in groups 1B and 4B only, the parallel chain reaction contributes more than one per cent to the observed rate of exchange, but less in all the other groups. It should be added that, to compute the average concentration

TABLE III
Contribution of a chain mechanism to the observed rate of exchange at 1300°K.
$k_7 = 2.08 \times 10^{10}$ cc mole^{-1} sec^{-1}

Group of shocks	$[NH_2]_{av} \times 10^{13}$ mole/cc	$k_{m(exp)}$ sec^{-1}	$k_{m(chain)} \times 10^2$ sec^{-1}	$\frac{k_{m(chain)}}{k_{m(exp)}} \times 10^2$
1A	11.8	13.5	2.45	0.182
2A	37.2	30.2	7.75	0.256
3A	89.5	24.6	18.6	0.755
4A	4.0	6.54	0.83	0.128
1B	266	31.6	55.4	1.75
2B	61.1	12.9	12.7	0.985
3B	19.5	6.31	4.05	0.642
4B	91.5	12.9	19.0	1.48

of the NH_2 radical, we used the rate constant k_2 reported by Michel and Wagner[9], which is higher by a factor of 5 than the average value reported by other groups[10,11]. This was done in order to calculate the upper limit of the exchange caused by the radical chain reaction Obviously, on the basis of the data presented in Table III one may rule out any contribution of a chain mechanism to the rate of exchange under our experimental conditions.

The Rate of Exchange in the Presence of Excess NH_2 Radicals

In addition to the calculations based on the reported value of k_7, it is possible to evaluate the rate of exchange on the basis of a chain mechanism by introducing a relatively high concentration of NH_2

radicals into the reacting mixture and measuring the exchange under these conditions. This can readily be performed by adding hydrazine into the reaction mixture, since its rate of decomposition is a few orders of magnitude higher than that of ammonia. In Figure 3, log k_m is plotted vs. 1/T for the two groups of shocks: (1C) to which

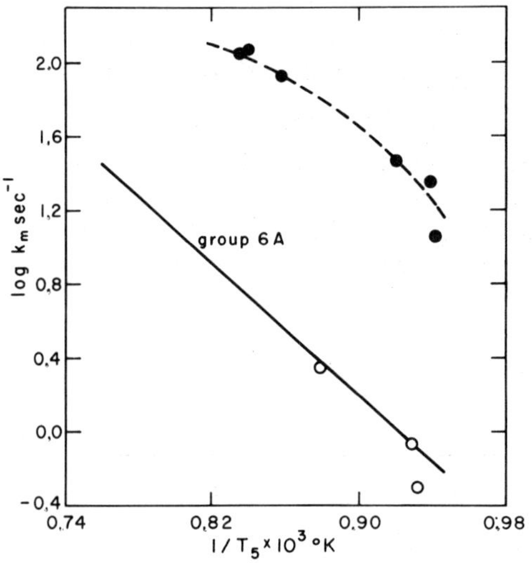

Figure 3
A plot of log k_m vs. 1/T for the two groups of shocks 1C and 2C. k_m values obtained in the presence of hydrazine (1C) are considerably higher than those obtained in its absence.
● – 1C; O – 2C

0.5 % anhydrous hydrazine was added and group 2C. Group 2C was run on the same day as 1C under the same conditions as group 6A . Groups 1C and 2C are identical in their D_2 and NH_3 concentrations and differ only by the presence of 0.5 % hydrazine in 1C. One can readily see than the k_m values obtained in the presence of hydrazine are considerably higher than those obtained in its absence. Assuming that the excess rate of exchange is due to the presence of NH_2 radicals, the value of k_7 can be evaluated with the aid of these two groups of shocks using the relation

$$k_7(T) = \frac{k_{m(N_2H_4)T} - k_{m(6A)T}}{[NH_2]_{av}(T)} \qquad \text{(XIII)}$$

We are now left with the evaluation of the average concentration of the NH_2 free radicals at each temperature. This is not equal to twice the amount of hydrazine decomposed, as NH_2 radicals disappear by a reaction with one another and with hydrazine. In order to construct a curve of the concentration of NH_2 vs. time at each temperature, a detailed Fortran IV machine program was prepared which took into account the following set of reactions:

$$Ar + N_2H_4 \xrightarrow{k_{14}} 2NH_2 + Ar \tag{14}$$

$$NH_2 + N_2H_4 \xrightarrow{k_{15}} N_2H_3 + NH_3 \tag{15}$$

$$NH_2 + N_2H_3 \xrightarrow{k_{16}} NH_3 + N_2H_2\ (H_2 + N_2) \tag{16}$$

$$NH_2 + NH_2 \xrightarrow{k_{17}} N_2 + 2H_2 \tag{17}$$

The value of k_{14} for the unimolecular region has recently been reported by Troe:[19]

$$k_{14}^{uni} = 10^{13.9}\exp[-55.0\times 10^3/RT]\ sec^{-1}$$

At high dilution in argon in our pressure range the overall reaction order of reaction 14 is 1.5.[20] k_{14}^{uni} was divided by a factor of 2 in order to fit our experimental conditions.

There are, however, no accurate values for the rate constants of reactions (15)—(17). We have taken the rate constants 15—17 suggested by Palmer et al.[20] in an article discussing the possible reactions following the dissociation of hydrazine. They proposed for k_{15} the value of the rate constant reported for the reaction:[21]

$$CH_3 + N_2H_4 \xrightarrow{k_{18}} CH_4 + N_2H_3 \tag{18}$$

where $k_{18} = 10^{11}\exp[-5\times 10^3/RT]$ cc $mole^{-1} sec^{-1}$. For the rate constant of reaction 16 they used that corresponding to the disproportionation reaction

$$C_2H_5 + C_2H_5 \xrightarrow{k_{19}} C_2H_6 + C_2H_4 \tag{19}$$

where $k_{19} = 1.7\times 10^{13}\exp[-800/RT]$ cc $mole^{-1} sec^{-1}$. The rate constant for reaction (17) was estimated by Wagner et al.[22] as

$$k_{17} = 10^{12.8}\exp[-10\times 10^3/RT]\ cc\ mole^{-1}\ sec^{-1}$$

In Figure 4 we present a plot of $[NH_2]$ vs. reaction time obtained by the computer for three shock temperatures. The dwell time of each shock was divided into two thousand steps. The program was constructed in such a way that each of the three species present in

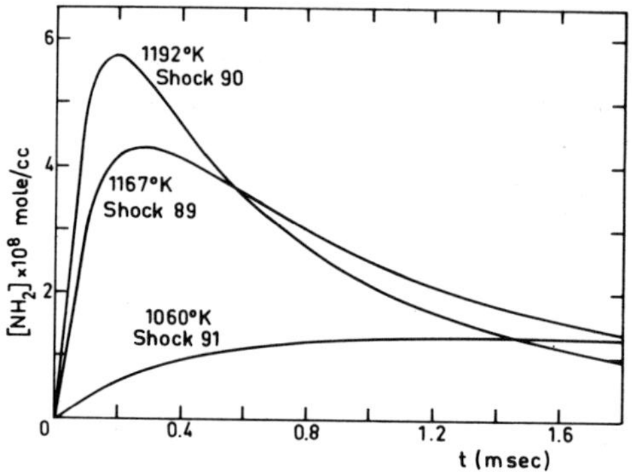

Figure 4
A plot of $[NH_2]_t$ vs. reaction time in the presence of hydrazine. $[NH_2]_{av}$ is obtained from the relation

$$[NH_2]_{av} = \frac{1}{T}\int_0^T [NH_2]_t dt$$

the reacting system (NH_2, N_2H_3 and N_2H_4) was considered as a constant value within the duration of each step (~1μsec). The values of the nth step were equal to those of the $(n-1)^{th}$ step, to which the increment (negative or positive) obtained during the $(n-1)^{th}$ step was added. This procedure is completely justified in view of the very large number of steps into which the dwell time was divided.

Before drawing any conclusion on the basis of the results of the machine calculations, we checked the sensitivity of the system towards changes in the rate constant k_{16}, because the analogy of reactions 16 and 19 is not completely obvious. It turns out that, although a rather high value was assigned for k_{16}, reaction 16 contributes very little to the depletion of NH_2 radicals from the system. Increasing or decreasing k_{16} by a factor of 10 was immaterial, as the concentration of N_2H_3 was always close to 2 orders of magnitude below the concentration of NH_2.

We have calculated the average concentration of NH_2 by numerical integration from the relation

$$[NH_2]_{av} = \frac{1}{T}\int_0^T [NH_2]_t dt \qquad \text{(XIV)}$$

for each one of the six shocks run. k_7 was calculated using eq. XIII. A plot of log k_7's vs. $1/T$ is shown in Figure 5. For comparison, the value obtained by Oganesian and Nalbandian is also presented. The value obtained by us is:

$$k_7 = 10^{12.54}\exp[-16.4\times 10^3/RT]\ \text{cc mole}^{-1}\text{sec}^{-1}$$

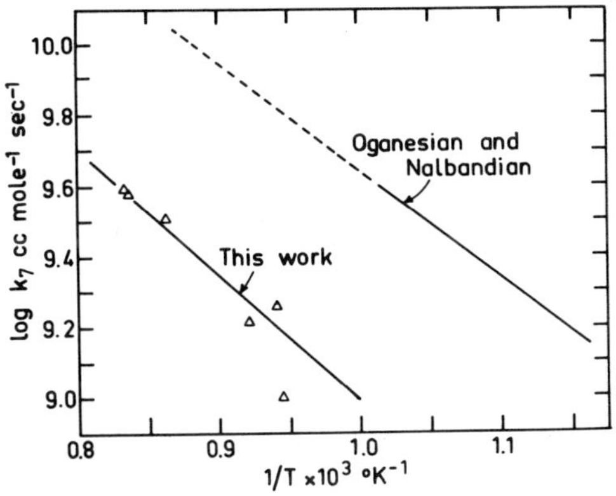

Figure 5
A plot of log k_7 (see eq. XIII) vs. 1/T for the shocks run in the presence of hydrazine. k_7 obtained by Oganesian and Nalbandian is also presented for comparison.

which is about a factor of 3 below the extrapolated value of Oganesian and Nalbandian.[13] In view of a possible slight error in our evaluation of $[NH_2]_{av}$, we do not claim that our k_7 is more correct than theirs. We do believe, however, that the data presented in Table III give the upper limit for the contribution of a chain mechanism, in the absence of hydrazine.

It should be mentioned that the data presented in Table III are based on the assumption that the chain reaction is carried mainly by NH_2 and D free radicals (reactions 6 and 7). The addition of hydrazine to the reaction mixture demonstrates that this assumption was correct.

As previously mentioned, the exchange may take place in two parallel cycles — the one based on reactions 4 and 5 and the other represented by reactions 6 and 7. If the two cycles have the same rate, it is immaterial which of the two is used for the calculations as long as the total concentration of the radicals is considered. It should be recalled that it is only the distribution of the free radicals in the system that is determined by the relative rates of the propagation reactions, but that their total concentration depends merely on the rate of reaction 2. Our assumption is incorrect only if the cycle based on reactions 4 and 5 is faster than that based on reactions 6 and 7. If our assumption is indeed incorrect, i.e., $k_5 > k_6$, then the contribution of a chain mechanism presented in Table III is underestimated. In this case, in the experiments with added hydrazine reaction 5 would have taken over and the exchange would have proceeded with a higher bimolecular rate constant. The fact that our computed k_7 is very close to that forming the basis of computation of the data in Table III, clearly indicates that $k_5 \leqslant k_6$.

In conclusion, the effect of argon on the reaction rate observed is not due to its function in initiating a chain reaction.

Discussion of an Exchange Mechanism

The only mechanism found compatible with the experimental observations, namely participation of a third body in the rate equation, assumes a molecular mechanism but imposes certain restrictions on the formation of the transition state. According to this mechanism, which will be referred to as the "vibrational excitation mechanism," there is a low probability for reaction between molecules in their ground vibrational states, even though their relative kinetic energy upon collision exceeds the activation energy. The probability for reaction increases rapidly with increasing amount of energy present in the vibrational states of one of the colliders. One must therefore separate in time two consecutive events leading to the chemical reaction: a vibrational excitation of one of the reacting molecules, and the chemical change itself. This can be represented by the following scheme:

$$D_2 + M_i \underset{k^{(D_2)}_{-(v)M_i}}{\overset{k^{(D_2)}_{(v)M_i}}{\rightleftharpoons}} D_2{}^* + M_i$$

$$D_2{}^* + NH_3 \xrightarrow{k^{(D_2)}_{(ex)}} \text{exchange products}$$

and

$$NH_3 + M_i \underset{k^{(NH_3)}_{-(v)M_i}}{\overset{k^{(NH_3)}_{(v)M_i}}{\rightleftharpoons}} NH_3{}^* + M_i$$

$$NH_3{}^* + D_2 \xrightarrow{k^{(NH_3)}_{(ex)}} \text{exchange products}$$

M_i is an exciter molecule representing each one of the three species present in the reaction mixture. Assuming a steady state for the vibrationally excited D_2 and NH_3, then for low extents of reaction the following rate law is obtained:

$$\frac{\Delta HD}{\tau} = \frac{k^{(D_2)}_{(ex)}[NH_3]\sum k^{(D_2)}_{(v)M_i}[M_i][D_2]}{k^{(D_2)}_{(ex)}[NH_3] + \sum k^{(D_2)}_{-(v)M_i}[M_i]} + \frac{k^{(NH_3)}_{(ex)}[D_2]\sum k^{(NH_3)}_{(v)M_i}[M_i][NH_3]}{k^{(NH_3)}_{(ex)}[D_2] + \sum k^{(NH_3)}_{-(v)M_i}[M_i]} \qquad \text{(XV)}$$

Using eqs. III and XV, one obtains for k_m the expression:

$$k_m = \frac{k^{(D_2)}_{(ex)} \sum k^{(D_2)}_{(v)M_i} [M_i][NH_3]}{k^{(D_2)}_{(ex)} [NH_3] + \sum k^{(D_2)}_{-(v)M_i} [M_i]} + \frac{k^{(NH_3)}_{(ex)} \sum k^{(NH_3)}_{(v)M_i} [M_i][NH_3]}{k^{(NH_3)}_{(ex)} [D_2] + \sum k^{(NH_3)}_{-(v)M_i} [M_i]} \qquad \text{(XVI)}$$

The vibrational relaxation time of ammonia is extremely short,[23] probably because of very fast v → v transitions in this polyatomic molecule. In the denominator of the second term in the r.h.s. of eq. XVI, the condition $\sum k^{(NH_3)}_{-(v)M_i} [M_i] \gg k^{(NH_3)}_{(ex)} [D_2]$ is certainly fulfilled. On the other hand, the efficiency in exciting the vibrational mode of D_2 is very low[24] owing to the high energy spacing between the levels. It may readily be assumed, as has been shown in the exchange studies between H_2 and D_2,[3] that the condition $\sum k^{(D_2)}_{-(v)M_i} [M_i] \ll k^{(D_2)}_{(ex)} [NH_3]$ is fulfilled. Imposing these two conditions on eq. XVI, the relation for k_m is:

$$k_m = k^{(D_2)}_{(v)Ar} [Ar] + k^{(D_2)}_{(v)D_2} [D_2] + k^{(D_2)}_{(v)NH_3} [NH_3] + k_{(NH_3)} [NH_3] \qquad \text{(XVII)}$$

where $k_{(NH_3)}$ is a simple bimolecular rate constant.

The dependence of k_m on the various concentrations can readily be seen from eq. XVII. The dependence on the argon comes from the term $k^{(D_2)}_{(v)Ar} [Ar]$. Since deuterium is only 5 times more effective than argon in exciting deuterium,[24] owing to the very low ratio $[D_2]/[Ar]$, the contribution of $k^{(D_2)}_{(v)D_2}$ to k_m is negligible. Indeed, k_m was found to be independent of the deuterium concentration. There are no data available on the vibrational relaxation time of deuterium in the presence of ammonia, but probably the effect on ammonia on k_m is not due to the term $k^{(D_2)}_{(v)NH_3} [NH_3]$ but to $k_{NH_3} [NH_3]$.

In our calculations leading to eq. XVII we assumed that vibrationally excited ammonia molecules maintain a Boltzman distribution, and the atom-switching step controls the rate of exchange. On the other hand, the rate of the reaction between excited deuterium molecules and ammonia is determined by the rate at which deuterium molecules are elevated to the critical vibrational level from which exchange occurs. Another assumption is, however, needed to explain the observed rate of exchange. The Boltzman concentration of the excited ammonia molecules is considerably higher than the steady state concentration of the excited deuterium. The exchange caused by an interaction between excited ammonia and deuterium may therefore be expected to provide a much faster path for the reaction than that between the excited deuterium and ammonia. In such a case, the contribution of the first term in eq. XVI to k_m would have been very small, and there should be no effect of argon on the reaction rate. It must be assumed that $k^{(NH_3)}_{(ex)} \ll k^{(D_2)}_{(ex)}$, and this is justifiable by the expected geometry of the activated complex. The N—H bond length in ammonia is greater than the D—D bond length in deuterium. The D—D bond length in a vibrationally excited deuterium becomes closer to the N—H length in an unexcited ammonia, and there is a better probability for the formation of the activated complex. On the other hand, the difference in bond lengths between vibrationally excited ammonia and unexcited deuterium is rather high, and there is a lower probability for formation of the activated complex.

Acknowledgement — The authors wish to thank M. Persky for carrying out the mass spectral analyses.

REFERENCE

1. A.Bar-Nun and A.Lifshitz, J.Chem.Phys. 47, 2878 (1967).
2. W.S.Watt, P.Borrell, D.Lewis and S.H.Bauer, J.Chem. Phys. 45, 444 (1966).
3. A.Burcat and A.Lifshitz, J.Chem.Phys. 47, 3079 (1967).
4. A.Burcat, A.Lifshitz, D.Lewis and S.H.Bauer, J.Chem. Phys. 49, 1449 (1968).
5. D.Lewis and S.H.Bauer, J.Am.Chem.Soc. 90, 5390 (1968).
5a. A.Lifshitz, C.Lifshitz and S.H.Bauer, J.Am.Chem.Soc. 87, 143 (1965).
6. A.Lifshitz, H.F.Carroll and S.H.Bauer, J.Chem.Phys. 39, 1661 (1963).
7. E.A.Sutton, J.Chem.Phys. 36, 2923 (1962).
8. J.P.Rink, J.Chem.Phys. 36, 1398 (1962).
9. K.W.Michel and H.Gg.Wagner, Technical Status Report No.24 USAF Contract A.F. 61(514) - 1142 (1963).
10. H.Henrici, Ph.D.Thesis, Gottingen (1966).
11. J.N.Bradley, R.N.Butlin and D.Lewis, Trans.Faraday Soc. 63, 2962 (1967).
12. K.T.Oganesian and A.B.Nalbandian, Dokl.Akad.Nauk SSSR 160, 162 (1965).
13. M.Schiavello and G.G.Volpi, J.Chem.Phys. 37, 1510 (1962).
14. R.Wolfgang, Ann.Rev.Phys.Chem. 16, 15 (1965).
15. J.N.Bradley, R.N.Butlin and D.Lewis, Trans.Faraday Soc. 64, 71 (1968).
16. B.J.McBride, S.Heimel, J.E.Ehlers and S.Gordon, Thermodynamic Properties to 6000°K for 210 Substances Involving the First 18 Elements. NASA Report - Sp - 3001 (1963).
17. G.Herzberg, Electronic Spectra of Polyatomic Molecules, D.Van Nostrand, New York, 1966, p.584.

18. R.T.King, Nature 181, 963 (1958).
19. J.Troe, Twelfth Symposium (International) on Cumbustion. The Combustion Institute, Pittsburgh, Pennsylvania, 1965, p.341.
21. P.Gray and J.C.J.Thynne, ibid., p.435.
22. K.W.Michel and H.Gg.Wagner, ibid., p.351.
23. W.C.Griffith, J.Appl.Phys.21, 1319 (1950).
24. J.B.Moreno, Phys.Fluids 9, 431 (1966).

SHOCK TUBE DIAGNOSTICS, INSTRUMENTATION, AND FUNDAMENTAL DATA

R. I. SOLOUKHIN
University of Novosibirsk, USSR

SUMMARY

The full exploitation of shock-tube techniques in chemical-physics research strongly depends on the development of a variety of specific procedures for the rapid measurement of the temporal shocked gas thermodynamic parameters in a nonequilibrium zone, e.g. the temperature or density. The sensitivity of the method used, the response time, the range of applications, and the extent of interference with the phenomena under study should be considered in comparing various available techniques of measurement. Each of the methods should also be characterized by an adequate and suitable choice of the gas parameters to be effectively detected in a shock-tube flow. In this paper, some of the recently developed experimental techniques for obtaining fundamental shock-tube data have been summarized in regard to the following:

1. Evaluations of the limits of variation of pressures, temperatures and densities in relaxation zones; hydrodynamic effects of chemical reactions in the simplest shock tube flows.
2. Advances in rapid photography and instrumentation; laser techniques and compensating methods; electronics and detectors with rapid-time resolution.
3. Gas temperature measurements; spectroscopic techniques, recent developments, and summary data; hydrodynamic methods; plasma-physics experiments; X-ray spectroscopy, and Doppler broadening at high temperatures; laser scattering techniques.
4. Gas density measurements I: brief review of advances in quantitative schlieren and interferometer methods; interferometer-chronometers; laser-interferometers; microwave reflection technique.
5. Gas density measurements II: laser beam scattering and spectral distribution of the scattered radiation; electron, ion and neutral particle beams; advances in ultra violet and infra-red absorption techniques.

6. Conductivity measurements in shock and detonation waves; coil impedance technique for reflected wave studies; comparison with probe and microwave diagnostics.
7. Pressure measurements in complex shock tube experiments; expansion wave technique and recombination kinetic data.
8. Shock tube experiments in exothermal systems; ignition kinetic data; nonstationary reaction zones and the role of combined measurements; illustrations.
9. Summary tables of experimental techniques (density and temperature measurements, applications and recommendations); tables of basic data obtained from shock tube experiments (molecular relaxation, reaction kinetics, recombination processes).

INTRODUCTION

Shock tube techniques, which have developed rapidly and are now widely employed, have recently gained importance in gasdynamics and kinetics experiments. Extremely short response times are required in shock tube studies, but no serious limitations have arisen in this regard in applications of laser techniques, optical spectroscopy, interferometry, scattering of light and particle beams, and other experimental methods. The sensitivity and accuracy are not worse than in stationary measurements. Gas density and temperature measurement techniques have advanced especially in recent years. The simultaneous determination of the concentration of chemical species as a function of time and of the gross variables in the experiment seems to be the most important diagnostic technique in shock tube kinetic studies.

As an introduction to Session Six, this survey will cover only a few specific topics in shock tube operations and present some illustrations of several experimental methods. More detailed reviews and particular discussions are already available,[1-7] but certain features of some novel techniques and their related plasma physics experimental methods should be considered in more detail to point out the importance of extending these methods to new ranges of gas parameters and better time resolutions.

In a shocked gas, the total change of a gas parameter in a nonequilibrium zone behind a shock front essentially depends on the amount of absorbed or released energy. Therefore, the choice of a suitable variable to be determined behind shock waves is of particular interest to the method of measurement. For example, it is easy to show that pressure measurement cannot be the method best suited to vibrational relaxation studies because the pressure deviations are very small compared with density or temperature changes.

In contrast, in detonations and in exothermal reactions, determination of the pressure appears to be quite an effective and appropriate method.

To evaluate the pressure and density variations in a relaxation zone, a strong shock approximation can be considered:

$$\frac{2Q}{V_s^2} = \left(1-\frac{\rho_1}{\rho}\right)\left(\frac{\gamma+1}{\gamma-1}\frac{\rho_1}{\rho} - 1\right)$$

Close to the shock front, the pressure and density changes, in a reaction zone can be estimated by

$$\frac{\Delta p}{p} = -\frac{\rho_1}{\rho}\frac{\Delta Q}{\varepsilon}, \qquad \frac{\Delta\rho}{\rho} = -\frac{2}{\gamma+1}\frac{\Delta Q}{\varepsilon},$$

where ΔQ is the energy released (or absorbed) and ε is the internal energy of the shocked gas. Thus, for vibrational relexation in a diatomic gas, $\Delta\rho/\rho \sim 0.3$ whereas $\Delta p/p \sim 0.05$ only. The same proportions would be valid in highly diluted exothermal systems ignited behind incident shock waves, though in detonation regimes both pressure and density variations across the reaction zone appear to be quite significant. It should be noted, however, that a thermodynamic gas state behind an incident shock wave propagating in a gas system of small ΔQ closely corresponds to isobaric conditions.

As an example, typical simultaneous recordings of the density and pressure variations within an induction zone behind a shock wave propagated in an oxygen-hydrogen mixture diluted with argon are shown in Figure 1. The density change is quite notable in the interferogram, whereas the decrease in pressure in the course of the reaction is negligible.[6]

Generally, the extent of change in the measured variable in the course of the process under study should be considered in comparing various experimental methods as well as their time resolution, sensitivity, range of application, reproducibility, accuracy, etc. A comparison with the methods used in plasma physics seems to be quite instructive bearing in mind the similar purposes of these methods and their possible applications to shock tube studies at moderate temperatures.

ADVANCES IN RAPID PHOTOGRAPHY AND INSTRUMENTATION

Different methods for rapid photography, oscillography, and fast electronics are available for shock tube studies. Significant and noteworthy improvements of these methods as well as new advances have, however, been developed in a number of recent experimental works. The use of a pulsed solid-state laser as a light source in high-speed stroboscopic photography has made it possible to obtain very clean schlieren records and interferograms[8-12] with exposures of 10^{-8} sec and with an interval between flashes of 10^{-6} sec. The monochromatic laser light illuminator is of particular importance for interferometer methods, and it may be nicely synchronized with the process investigated using a Kerr-cell in Q-spoiled laser generation regimes. A laser light pulse may be used also as a spark gas trigger with a response time of 10^{-8} sec.[12]

A simple technique which uses an ordinary flash lamp as a light source still seems to be quite advantageous in obtaining "frame" photographs of quasi-stationary, shock-tube flows.[6, 13] The optical flow visualization is recorded by a streak camera through a narrow slit placed normal to the direction of movement, providing that the film velocity is equal to the image velocity of the process. A typical record of a shock tube flow is shown in Figure 2. Quite lengthy gas flow patterns might be registered using this technique and only narrow wall windows are required in the observation chamber.

An ordinary framing camera should be recommended for use in studies of quasi-stationary, two-dimensional, shock-tube flows because it is possible to select a suitable stationary stage of the process. As an example, a series of interferograms of the shock wave expansion flow of dissociated oxygen obtained by a zero-fringe interferometer[14] is shown in Figure 3. Constant density lines in the interferogram are used for a time-resolved measurement of the density field distributions in this nonsteady supersonic flow.

Spark photography has been shown to be useful in ballistic-type, shock-tube experiments.[15] Typical shadowgraph records of a shock wave diffracted around a cylinder are shown in Figure 4.

Although shock velocity measurement techniques have been widely described, a number of new advances in this field should be mentioned. Schlieren or shadow light beam detectors have been utilized.[16, 17] A simple photomultiplier system might be adopted in luminous shock front recordings using a number of wall slits focused on the same photomultiplier cathode. Shock velocities ranging from 20 to 100 km/sec have been measured employing this technique. Microwave Doppler techniques have also been applied to shock velocity measurements.[18, 19] In addition mention should be made of a simple electronic trigger device[20] in which shock velocity data are used to provide an automatic change in the time delay.

Hall-effect methods for shocked gas velocity measurements in the presence of a magnetic field have been shown to be suitable for shock-tube studies at moderate temperatures.[21-23] No appreciable force interactions between a gas flow and magnetic field occurred in this case, and electric fields of about 10 volts/cm are induced at flow velocities of 1 km/sec in a magnetic field of 10 kilogauss. In gaseous detonation studies[23] an accuracy of 2 to 4 per cent has been obtained using this technique. Precise flow velocity measurements are essential to hydrodynamic methods of measuring gas temperatures which are discussed later.*

Various electronic scanning systems have been applied in rapid spectroscopic methods. The image converter technique has been employed in a device shown in Figure 5a, which can provide rapid scans of spectral line profiles.[6] An electron beam is transmitted through a narrow slit-diaphragm and detected directly by a photomultiplier system. The image of a spectral line on the cathode is converted from the cathode plane into the diaphragm plane while being magnified 10^4-10^5 times. Scanning regimes and suitable time resolutions are provided by an electrical deflection of the converted line image across the slit. Line recording times of 5×10^{-8} sec might be easily achieved by employing this method. Similar principles may be developed to record a spectral line profile by scanning Fabry-Perot interferometer fringes through an annular slit.

A different method has been proposed for the same purpose on the basis of a fiber optic technique[24] (see Figure 5b). An image of a spectral line is focused on a number of fiber-optic guides connected with a photomultiplier system. The light intensity distribution is then determined using a fast electronic commutator with seven independent information channels which provide the successive detection of light intensity, photomultiplier signals reproducing the line profile. A system of 350 light guides has been shown to be sufficient to obtain a record of a spectral interval of 0.5 Å with a time resolution less than one microsecond.

In conclusion, a rotating-mirror, multiple-scan spectrometer has recently been employed for recording hydrogen-line profiles with time intervals of 1.6 microseconds at a spectral scan speed of 119.2Å/microsecond in order to determine electron concentrations in an electromagnetic shock tube.[170]

*It is interesting to note the flow velocity measurements made by combining the Doppler principle with the schlieren method using laser light.[166]

GAS TEMPERATURE MEASUREMENTS

Numerous techniques have been developed for gas temperature measurements in shock waves, and quite detailed surveys are now available in this area.[3, 4, 6, 16] It appears to be reasonable to restrict our consideration here only to some of the recent results in applications of line reversal and relative line intensity methods; a brief summary will be made of hydrodynamic temperature measurements as well as of plasma physics techniques applied at high temperatures.

Spectral-line reversal measurements behind incident shock waves have been performed successfully in various gaseous systems. Pioneering attempts made with sodium line emission and absorption measurements[3, 25, 26] have yielded quite acceptable shocked gas temperatures in air and in nitrogen but the measured temperatures in argon were significantly lower than the calculated values.[27, 28] The difference observed was about 600°K at T_s = 2900°K and gas pressure at one atm [28], this result was also obtained in experiments[16] where Ba II - line reversal temperatures have been measured in argon ($\Delta T \sim 2000$°K at T_s = 6000°K). The difference vanished only at shock pressures higher than 12 atm.

This problem has stimulated a number of interesting investigations into electronic excitation and radiative equilibrium of sodium atoms and barium ions in shock-heated argon and in molecular gases.[16, 28] It has been suggested that collisions with argon atoms could not maintain the thermal equilibrium in the system because of their small cross-sections. Comparing measured and equilibrium temperatures, one could estimate collision resonance quenching rates from the ratio of the collisional transition probability to the classical probability of radiative escape (approximately 10^8 sec^{-1}).

The effective quenching cross-section of sodium atom-argon collisions evaluated[28] was lower than 10^{-19} - 10^{-20} cm^2 whereas for Ba^+ - Ar collisions σ = 4.10^{-17} cm^2 was obtained.[16] Thus, nonequilibrium radiation will be observed during periods of the order of $\tau \sim (nv_T\sigma)^{-1} \sim 10^{-4}$ sec ($n \sim 10^{18}$ cm^{-3}) if excited levels are depopulated by spontaneous radiative transitions. Vibrationally excited molecules of oxygen or nitrogen were found to be much more effective than neutral atoms in the thermal excitation of Na in shock waves.[28-30] For example, only 0.5-1 per cent of nitrogen or carbon monoxide added to argon was found to be sufficient to reach equilibrium gas temperatures within a rise time of the order of vibration relaxation periods of the impurity gas molecules.[28] The effective cross-section for the excitation of Na by energy transfer from the vibrational mode of the nitrogen molecules was found to be greated than 10^{-15} cm^2.[29] Spectral-line reversal temperatures currently measured within a vibrational relaxation zone appear to be quite a suitable source of data on instantaneous vibrational temperatures and relaxation times in high-temperature gases. In addition, nonequilibrium expansion shock tube flows have been examined using similar methods of N_2 vibrational temperature measurements in a N_2-Ar mixture.[30]

Although the spectral-line reversal technique provides an overall accuracy of 1-2 percent of a measured temperature, some specific difficulties and limitations exist owing to the influence of the added emissive species. In particular, long delays in the evaporation and excitation of the radiative material are undesirable. The utilization of chromium lines[31-33] essentially improves the method. Observation with emission bands of the shocked gas itself[34-36] made it possible to measure electronic temperatures of excited radicals (C_2, CN) with a high time resolution.*

Spectral-line reversal techniques have been applied to study the temperature distributions behind reflected shock waves.[37-39] Equilibrium temperatures have been observed in nitrogen, air, oxygen and carbon dioxide in a temperature range between 1500° and 7000°K, temperature deviations being within 10 percent, probably owing to hydrodynamic interactions between the boundary layer and reflected wave. Again, nonequilibrium temperatures have been observed in argon using the sodium line method at pressures lower than 10 atm,[38] whereas equilibrium temperatures have been measured in argon containing 0.5 percent carbon monoxide with an accuracy of ± 35° at 2500°K.[39]

Measurements of relative intensities of selected lines for which transition probabilities are known provide quite a reliable method which is applicable at temperatures above 4000°K.[1, 40-46] Hydrogen, helium, carbon, argon and other lines might be used in measurements and quite high concentrations of free electrons are required in order to provide equilibrium emission of the heated gas. The higher the temperature of the gas under test the larger should be the energy difference between the upper levels of lines selected for measurements in order to obtain an appropriate accuracy. The following rough estimate of the accuracy of the method can be made:

$$\frac{\Delta T}{T} = \frac{kT}{\varepsilon_2 - \varepsilon_1} \frac{\Delta X}{X}$$

where $X = I_2/I_1$ is the measured ratio of the line intensities. Therefore, vacuum ultraviolet lines have been utilized in electromagnetic shock tube experiments[47] where temperatures of 1.5-4 ev were measured in a hydrogen plasma using C IV ion lines (419Å and 1549Å).

*Note the use of infrared CO_2 bands for spectral-line reversal measurements.[149]

The absolute line intensity technique as well as the method based on comparison of the line intensity with the continuum intensity in the same spectral region will not be discussed here since these methods have been adequately described in the literature.[1, 40]

Spectroscopic methods of temperature measurements are based on the thermal equilibrium of radiative species with the gas under test and mainly electron temperatures are being determined using these techniques. Therefore, an independent verification of the data obtained, which might be made by "hydrodynamic" methods, seems to be an important tool in shock tube experiments.[45-49] In these methods, flow Mach numbers are determined by means of detached shock distance measurements, a blunt obstacle being placed in the supersonic flow existing behind an incident shock wave. The gas (or ion) temperature is then determined from the sound velocity of the shocked gas.

As was shown in Reference 46, the measured temperatures coincide within the limits of experimental errors when three independent techniques are employed (Figure 6): the relative line intensity method (1), hydrodynamic measurements (2), and line broadening estimations (3). Shocked gas temperatures of 1-2 ev and initial pressures of 0.05-1 torr in air and argon have been tested in the experiments. As an example, the temperature profiles behind a shock wave propagated in argon are shown in Figure 6. The separation between thermalized and discharge (driver) plasmas is clearly detected by both spectroscopic and hydrodynamic methods in this run while mixing between these plasmas occurs at lower pressures and at higher shock Mach numbers.

New advances in measurements of high temperatures have been developed in plasma physics experiments. Note first that measurements of the Doppler width of a spectral line have become a quite reliable method yielding ion temperatures with an accuracy of about 10 percent at a temperature around 10^2 ev and densities of 10^{14}-10^{15} cm^{-3}. The half-width of a line is $\Delta\lambda/\lambda = 7.16\times10^{-7}(T/\mu)^{1/2}$ where μ is the atomic weight and the temperature is in $^{\circ}$K. Therefore, $\Delta\lambda$ might be a few angstroms at temperatures of 10^2 ev. This method is restricted to moderate densities ($\leq 10^{14}$cm^{-3}) because of the Stark effect at higher densities.

At temperatures above 100 ev the technique of soft X-ray spectroscopy (2-50 Å) is particularly valuable.[1, 50, 51] Special crystal spectroscopy techniques and detectors have been developed in these experiments. Continuum soft X-ray emission records have also been used for plasma temperature measurements. Electron temperatures of laser[52] and theta-pinch[53,54] plasmas have been derived from the relative intensities of the X-ray flux transmitted through a number of metal foils of different thicknesses.

The time delay between the onset of simple kinetic processes (ionizations, nuclear reactions, etc.) which occur at high temperatures with different temperature-dependent cross-sections can be employed in plasma temperature evaluations.[55] Time delays in the appearance of impurity line emission in various stages of carbon and oxygen ionization as well as relative rates of proton and neutron production in deuterium and helium-3 plasmas have been used in these measurements, which also require knowledge of the plasma density.

Scattering of a ruby laser beam from plasmas and spectroscopic analysis of the scattered light has proved to be an extremely effective method for measurements of ion and electron temperatures and densities[56-62] (see summary in Ref. 7). The spectral distribution of the scattered light is a function of the parameter $\alpha = (\lambda_0/\lambda_D)(4\pi \sin\theta/2)^{-1}$ where λ_D is the Debye length and θ is the angle of observation. For $\alpha \ll 1$ the electron temperature may be determined from a line width of the Thomson scattering spectrum (electron Doppler broadening) whereas the ion temperature distribution is mainly reflected by the scattered radiation spectrum when $\alpha \gg 1$. A pair of line satellites displaced from the wavelength of the incident beam by the plasma frequency appears in the latter case owing to the collective effect of ions and electrons (see Figure 7). The distance of these lines from the ion mode is determined by longitudinal plasma fluctuations:

$$\Delta\lambda = \frac{\lambda_0}{2\pi c}\left[\omega_p^2 + \frac{3kT_e}{m_e}\frac{4\pi}{\lambda_0}\sin\frac{\theta}{2}{}^2\right]^{\frac{1}{2}}$$

where ω_p is the plasma frequency. The shape of the central ion line of the scattered spectrum yields information on the difference between the electron and ion temperatures.[56]

When considering laser beam scattering methods one should take into account the low cross-sections of the Thomson light scattering,

$$\sigma_T = \frac{8\pi}{3}\left(\frac{e^2}{mc^2}\right)^2 = 6.7\times10^{-25}\text{cm}^2,$$

which requires both very high light intensity and high detection sensitivity (10^{-15}-10^{-10} of the incident beam radiation). Nevertheless, Q-spoiled laser techniques have made it possible to obtain satisfactory results from this method with a response time of the order of 10^{-8} sec.

Multichannel recording systems or scanning monochromators are usually used to resolve a scattered radiation spectrum.[56, 62] The use of an image converter camera[59] with an electronic shutter

(10-20 n sec) and multistage light amplification made it possible to obtain a suitable photographic film image of the relevant spectral region. From accurate quantitative measurements of the exposed film quite precise information on the line profiles was obtained.

GAS DENSITY MEASUREMENTS I (WAVE OPTICS)

Schlieren and interferometer techniques have been widely applied to density measurements in gasdynamics and plasma physics research. Though interferometer methods seem to be preferable in quantitative investigations, a brief summary of quantitative schlieren techniques developed in shock tube studies should be given.[6, 63-66] The use of a suitable light source in a schlieren system (flash lamp, laser, etc.) and elimination of the self-luminosity of the process make possible photomultiplier measurements of the intensity of light passing a knife edge which yield information on the density gradient distribution behind a shock front. The sensitivity of the method may be increased by using a gas laser light source and by placing the knife edge a long distance from the object (in one case 7 m[65]). Detection sensitivities of about $dn_a/dx \sim 10^{17}\,cm^{-4}$ and a time resolution of 0.2 μ sec have been achieved in these experiments.

The density ratio across a shock front can be determined from measurements of the refraction of a schlieren light beam by the shock front if the angle between the light beam and the wave front is acute.[64] For example, the oscilloscope record of the schlieren light output is shown in Figure 8, where density jumps were detected for incident and reflected shock waves in air.

Detection sensitivities of quantitative schlieren methods may be evaluated as follows:

$$\varepsilon = L\frac{dn}{dx} = \begin{cases} -L\dfrac{d}{dx}\left(4.46\times10^{-14}\lambda^2 n_e\right) & \text{(plasma)} \\ +L\dfrac{d}{dx}\left(3\times10^{-4}\dfrac{n_a}{n_o}\right) & \text{(neutral gas)} \end{cases}$$

where ε is a total angle of the beam deflection and L is the extent of an object. For suitable conditions particle density gradients of 10^{18} and $10^{17}\,cm^{-4}$ for electrons or neutral atoms and molecules, respectively, may be detected.

Many applications of optical interferometry to shock tubes and plasma studies have been reviewed by R. A. Alpher and D. R. White.[1] Therefore, only some recent results should be mentioned here. The double-frequency technique[67] was improved by use of a Q-spoiled ruby laser to make simultaneous measurements at both the fundamental wavelength 6943 Å and its second harmonic 3472 Å which was generated by passing the beam through a nonlinear optical medium (KDP or ADP crystals). This technique, which was developed for laser-induced spark interferometry,[11] has been applied to shock tube studies of ionization relaxation in hydrogen.[68]

A general note might be made on the use of the zero-fringe interferometer technique. Constant density lines yield very useful information on two-dimensional density field distributions, and it should be noted that the zero-fringe record is more easily interpreted than the parallel-fringe record, though similar interpretations can be obtained using a superposition of a parallel-fringe interferogram on one of no flow. As a sample, two streak interferograms of the same detonation expansion process in a shock tube are shown in Figure 9. The zero-fringe interferogram technique has also been applied to measure the gas density distribution in quasi-stationary shock tube flows of recombining gases[14] (see Figure 3).

Complete three-dimensional records of the interference pattern with complicated geometry have been obtained by means of the laser holographic technique.[69, 70] Interferograms are reconstructed from a hologram which is made by means of a double exposure on the same photographic plate using a pulsed ruby laser (time-lapse interferometry), the first exposure being made with the model absent. The resulting interferograms look quite similar to zero-fringe records and permit post-exposure examinations of the flow patterns from various directions of observation.

The increasing availability of laser techniques has stimulated the wide use of optical interferometer-chronometer arrangements. The shock tube (or other object to be studied) is placed in the optical path of an interferometer which is illuminated with monochromatic (laser) light, and the rapid shifts of parallel fringes are detected by a photomultiplier. Quite small fringe shifts may be recorded because of the high monochromaticity of laser light which also minimizes the effects of self-luminosity of the test gas or plasma.[14, 71] This technique provides electrical signals which are quite convenient since the changes in other gas parameters may be recorded simultaneously using a multiple beam oscilloscope. A typical record of the pressure and density change in an expanding sample of a dissociated gas heated by a reflected wave is shown in Figure 10. This technique has been used in experiments on direct measurements of oxygen-atom recombination rates[14] where a partial reflection of the shock tube flow provided an expansion regime similar to that provided by single pulse shock tube techniques.

A simple evaluation of the detection sensitivity of the method can be made as follows:

$$N_f = (n-1)\frac{L}{\lambda} \sim \begin{matrix} -4.5\times10^{-14} L\lambda n_e \\ +10^{-23} n_a L/\lambda \end{matrix} \quad \text{or} \quad \begin{matrix} n_e \sim 10^{15}\,\text{cm}^{-3} \\ n_a \sim 10^{16}\,\text{cm}^{-3} \end{matrix}$$

where N_f is the number of shifted fringes, $\lambda \sim 5\text{x}10^{-5}$ cm, $L \sim 10$/cm, n_a and n_e are respectively the neutral gas particle and electron densities, and the accuracy of determination of the fringe shifts is $2\text{x}10^{-2}$. A Q-spoiled CO_2 infrared laser[72] seems to be suitable for this technique since electron densities of the order of 10^{14} cm^{-3} can be measured using a wavelength of 10 microns. A rotation of a barium fluoride plate, which provides an external spoiling of a Michelson-type, laser interferometer with a coupled cavity, was shown to be the most effective technique for the laser light modulation.

In recent years there have been a number of laser-interferometer studies of the electron density behavior in ionizing shock waves and discharge plasmas.[7, 73-76] A scheme using three mirrors was shown to be very promising (see Figure 11). A helium-neon laser is utilized in a two-wavelength regime in which the intensities of the 0.63 and 3.39 micron laser beams are coupled and the first beam is used to follow interference in the infrared. Interference occurs between the beam reflected from M_3 and the optical cavity oscillations in the M_2-M_3 region, and modulation in the 3.39 micron laser beam intensity produces a supplementary modulation in the 0.63 μ beam radiation.* Thus intensity modulation of the output beam of the laser itself is used to detect the fringe shifts arising because of time variations in the indices of refraction of a gas under investigation. The time resolution of this method was found to be limited to $3\text{x}10^6$ fringes per second and electron densities of $3\text{x}10^{16}$ cm^{-3} have been measured. The sensitivity of the interferometer has been improved by using a spherical mirror for M, rather than a plane mirror[73] and also employing a multipass mirror system which increases the sensitivity by at least a factor of 20.[73]

Faraday rotation of polarized laser radiation in a plasma which has a superimposed axial magnetic field was used to measure the line integral of the electron density along some axial path in a theta-pinch.[77-79] The angle of rotation of the plane of polarization is

---

*Infrared radiation has been monitored directly by using the three mirror scheme in shock tube experiments[76] with the variations in the red beam (0.63 μ) usually being recorded.

$$\pi \frac{\omega_p^2 \omega_H L}{\omega^3 \lambda} = 2.6\times10^{-17} HL\lambda^2 n_e$$

Therefore the infrared region of the spectrum is used to measure electron densities of 10^{16}cm^{-3}. The far infrared technique[80] seems to be promising for lower densities measurements.

Generally, electron densities in the range from 10^{14} to 10^{15} cm^{-3} are not easily accessible to suitable measuring techniques since this region lies below the range of optical spectroscopy or interferometry and above the useful range of microwave systems. Even with a 3 mm wave the maximum measurable density is of the order of 10^{14} cm^{-3}. Nevertheless, a microwave reflection method has been developed[81, 82] to measure electron densities whose plasma frequency exceeds the probing one. In this method, the electron density is determined from the phase angle of the reflection coefficient at the sharp boundary of the plasma, for example, at the end of a shock tube.[82] Electron densities up to 10^{16} cm^{-3} have been measured using this technique, though a discrepancy of one order of magnitude was observed in comparison with calculated equilibrium data on shock wave electron concentrations.

GAS DENSITY MEASUREMENTS II (RADIATION SCATTERING, EMISSION, AND ABSORPTION TECHNIQUES)

Laser beam scattering techniques have been used to measure the neutral gas densities and electron concentrations in dense plasmas by means of measurements of absolute intensities of the scattered radiation.[56-62] Rayleigh scattering from molecular gases of known cross-section was found to be a suitable standard of calibration for the method. Thomson scattering computations for various laboratory plasmas have also been presented.[60] Density measurements of shock tube or laser produced plasmas of 10^{16} -10^{17} cm^{-3} have been carried out in detail with a time resolution of 10^{-7} sec. It is possible to obtain additional information on electron concentrations from spectral characteristics of the scattered radiation as mentioned earlier.

Electron beam transmission techniques have been widely applied for neutral gas density measurements in the range of 10^{17}-10^{18} cm^{-3} in shock tube studies.[83-85] The exponential attenuation of a beam due to electron elastic scattering, excitation and ionization processes yields accurate gas density profiles behind shock waves. Electron guns with energies of 10^4-10^5 ev and appropriate scintillator and photomultiplier systems are employed to measure instantaneous beam attenuation coefficients.

A rough evaluation of the appropriate range of the beam absorption technique applications can be made using the following approximation for cross-sections of ionization and electron impact excitation processes at energies higher than 10^3 ev: $\sigma_{i,e} \cong A/\varepsilon$, where the beam energy ε is in electron volts and A is in the range of $(1-5)\times10^{-14}$ for the simplest atomic and molecular gases. Thus, the absorption coefficient of the beam attenuation at a pressure of 1 torr and beam energy of 10^4 ev is about $n\sigma \sim 0.1$ cm^{-1}:

The large-angle (Rutherford) scattering of an electron beam has also been employed to investigate shock front structure in rarefied systems at initial pressures lower than 10^{-1} torr in order to measure density profiles within the viscous or molecular relaxation transient zone.[86] The beam absorption technique is not applicable at these densities because only a small fraction of the beam is absorbed. Methods based on the absorption of soft X-rays have been widely employed for measuring densities behind shock and detonation waves[152, 153] in gases diluted with a heavy inert component like xenon. This technique is particularly valuable for experiments in dense gases and heterogeneous media.

The transmission of an ion and neutral particle beam across a shock tube can be applied also to density measurements in gases and plasmas.[1, 6, 87-89] Magnetized plasmas are particularly suitable since any changes in the electric charges of the neutral beam may be easily detected because of the effect of the magnetic field on charged particle trajectories. Three types of particle interactions appear to determine the value of the atomic beam absorption coefficients, namely, elastic scattering, ionization of a gas, and charge transfer between a positive ion and a neutral atom followed by any kind of deflection in electric and magnetic fields.

Cross-sections for elastic scattering of neutral particles have been measured;[90] for example, for H_e-H_2 the following approximation may be used at energies of 0.5-1.9 kev: $\log \sigma = -0.329 \log \varepsilon + 1.896$, where σ is in 10^{-16} cm^2 and ε is in ev. Thus the scattering cross-sections are of the order of 5×10^{-17} cm^2 at energies of 10^4 ev. Ionization cross-sections at beam energies of 10^4 ev and $T_e \sim 10^2$ ev appear to be lower than 10^{-16} cm^2. Quite large cross-sections are observed for the resonant charge exchange process in which the charge transfer between a positive ion and an atom of equal nuclear charge results in no change in the electronic state. Cross-sections for symmetric resonant charge transfer between protons and atomic hydrogen vary almost linearly from 2×10^{-16} to 36×10^{-16} cm^2 on a semilogarithmic scale of energies between 100 kev and 10 ev respectively.[91] At energies of 10 kev, the cross-sections are 10^{-15} or 7.10^{-16} cm^2, respectively, for charge exchange between the hydrogen atom and H^+ or O^+. Provided that only neutral beam particles which have passed through the sample under test are detected, an absorption coefficient of 0.1 cm^{-1} will correspond to plasma densities of 10^{14} cm^{-3} at beam energies of 10 kev. Plasma and neutral gas densities of 10^{15}-10^{16} cm^{-3} can be determined on the basis of the ionization energy losses of a particle or ion beam provided that it is not affected by deflecting fields.[89, 92]

An apparatus for neutral atomic beam density measurements is shown schematically in Figure 12. A multicomponent ion beam which may consist of light particles (H_2^+, H_e^+, H^+, Li^+) is transformed into a neutral particle beam using a gas target for the charge transfer neutralization. The neutral beam which has traversed the plasma is detected by means of separate mass-spectrometry of the different ions obtained after the partial re-charging of the neutral beam in a "strip-chamber". Electrostatic cleaning of the beam is used twice, before and after the beam passes through the plasma. A typical record of H^o- absorption when a beam passed through a discharge plasma of $10^{14}$$cm^{-3}$ density is shown in Figure 13. The He^o-beam attenuation profile is recorded simultaneously. He-beam absorption coefficients yield additional information on plasma electron temperatures owing to the temperature dependence of the electron impact ionization cross-sections of helium atoms (10^{-16} cm^2).

Near-ultraviolet and vacuum ultraviolet emission and absorption techniques are very significant diagnostic methods in shock tube studies of vibrational relaxation, dissociation and ionization kinetics in oxygen, nitrogen, air, carbon dioxide and other systems. As summarized in Figure 14, the molecular bands have large photo-absorption cross-sections in this spectral region since electronic excitations are involved, as for example, in the Schumann-Runge band of molecular oxygen.[4, 93-98] Note that the vacuum ultraviolet absorption coefficients at 1470 Å are decreased at high temperatures, while in the "quartz" region at 2245 Å only temperatures higher than 1500°K are accessible for measurements since the vibrationally excited molecules are responsible for the light absorption in this region. This technique allows the use of quite low gas number densities in the range of 10^{16} to 10^{18} cm^{-3}. It is necessary to slow down the relaxation processes to improve the time resolution of the method. Detailed studies of the vibrational relaxation times of oxygen molecules[93, 100] as well as dissociation rates of oxygen[93-96] and nitrogen[101] have been carried out using uv absorption techniques over the temperature range between 1500 and 20, 000°K.

The absorption oscillator strength for the hydroxyl electronic band system at 3050-3150 Å has been evaluated[2,102] by using measurements of the dissociation of H_2O vapors behind reflected shock waves. In the course of the wave reflection, the linear rate of increase of the absolute spectral intensity observed axially in the increasing transparent gas region was detected by monitoring the uv-emission of the shocked gas layer at the end of the shock tube. From the absolute intensity data the oscillator strength (f-number) was found to be $(3.9 \pm 0.9) \times 10^{-3}$.

Ultraviolet emission techniques have been used for equilibrium radiation measurements in air at temperatures of 8000-14000°K.[103] A tungsten photoelectric cell has been used to make photometric measurements at wavelengths shorter than 2000 Å in incident shock waves. Equilibrium air radiation measurements have been performed in a shock tube combined with a projectile gun,[104] a system in which total shock velocities of 11.3 km/sec have been achieved. This technique was applied to obtain high-speed recording of a wide spectral region. As the projectile flies by in the focal plane of the light-collecting mirror, the luminous shock layer acts as a moving entrance slit providing the swept spectrum in the exit slit plane. The spectral region of 3000-4000 Å has been measured for air.[104] The spectrum sweeping techniques have also been used to yield f-numbers for the CN-violet system at high temperatures in a ballistic device.[105]

The absolute emission-intensity in the wavelength range between 2300 Å and 4511 Å was found to be proportional to the square of the oxygen atom concentration in shocked ozone-argon mixtures at temperatures of 2500°-3800°K.[106] As was shown earlier,[107,108] the emission intensity which is directly proportional to the oxygen atom concentration was observed in the presence of O and CO at wavelengths below 4500 Å. Thus the emission-intensity records might be employed for instantaneous measurements of concentrations of the components of shock heated gases. In addition, note the possibility of the application of the Lewis-Rayleigh afterglow to atom concentration measurements in shock waves.[109]

The afterglow intensity of a shocked gas or of nitrogen molecules which were predissociated by a pulsed electric discharge was found to be proportional to the rate of recombination of atoms at temperatures in the range between 7000° and 18,600°K.

Emission and absorption measurements in the infrared region of the spectrum have become important experimental aids for shock tube kinetic studies. The instantaneous rate of change of a molecular gas concentration behind a shock front can be reliably derived from emission or absorption profiles obtained by the use of a rapid response time infrared detector. Some of the specific shock tube instrumentations for the infrared studies have been described.[2] The possibility of using liquid-nitrogen-cooled, gold-doped germanium or In Sb photoconductive detectors should be mentioned particularly.

As an example, two records of the infrared emission around 4.5 microns from a shocked N_2O-Ar mixture are shown in Figure 15. The emission is fairly constant in the equilibrium vibrational temperature region at rather low temperatures, whereas a drop is seen at higher temperatures due to a relaxation process of the N_2O thermal decomposition. Thus both vibration relaxation times (the emission rise time) and decomposition rates can be determined from the ir-emission records. Infrared light absorption records in pure N_2O (Figure 15c) show behavior similar to that of the shocked gas system. BaF_2 infrared optics have been employed in these experiments together with a gold-doped germanium ir-detector.

The infrared emission yields a direct measurement of the instantaneous vibrational energy of a shocked gas in the case of an optically thin molecular gas system.[86] Carbon monoxide and CO_2-vibration relaxation times as well as CO_2 and HF-dissociation rates[86, 110-112] have been determined by observations of infrared emission behind incident shock waves. Water vapor absorption coefficients in the vicinity of 2.7 microns and H_2O-decomposition rates have been measured in incident and reflected shock waves at high temperatures.[113, 114]

In addition, ionization rates have been measured from the infrared emission of air around 6 microns.[115] Far-infrared radiation from a high-temperature air plasma was shown[103] to be predominantly that produced by free-free transitions of electrons (bremsstrahlung); therefore, the electron density behind the shock front might be determined by using this technique.

ELECTRICAL CONDUCTIVITY MEASUREMENTS IN SHOCK WAVES

Measurements of the electrical conductivity in a partially ionized gas can yield information on electron concentrations as well as on electron-atom collision frequencies:

$$\chi = \frac{n_e e^2}{m_e \nu_{ec}} \cong \frac{n_e e^2}{\sqrt{8/\pi}\, m_e n_a \sigma_{ec} \sqrt{kT}} \cong \frac{4\times 10^{-10} n_e}{n_a \sigma_{ec} \sqrt{T}}$$

where x is the electrical conductivity, n_e and n_a the electron density and the density of other particles, respectively, and σ_{ec} is the electron-particle collisional cross-section.

In addition to the well-known method of Lin,[116] which is based upon the dynamic interaction of the shock tube flow and the magnetic field, different probe techniques developed for measurements in slightly ionized gases should be mentioned. Coil impedance and Q-factor probes have been used to measure the electric conductivity behind incident and reflected shock waves in air-argon and in argon seeded with alkali metals.[117-119] The principles of the methods are based on the reactance or resistance change of a radiofrequency tuned circuit in the presence of a plasma sample inside or around the probe coil. A miniature conductivity probe which was calibrated in a shock tube was employed[118] in plasma jet conductivity measurements in the range of 0.1-110 $(\text{ohm-cm})^{-1}$.

The simplest dc-probe resistance measurements yield reasonable values for the average gas conductivity between probe and wall or for samples in a gap probe in the case of gaseous detonation studies,[23, 120-122] whereas the Langmuir probe technique when used in shock tube measurements[123] does not yield a meaningful voltage-current relation. Meanwhile, a double-probe method with a high-frequency multistep pulse voltage has been effectively applied to conductivity measurements in detonation waves.[124] In addition, some attempts to study electric conduction mechanisms between cold electrodes in a high-velocity, shock ionized air plasma at temperatures above 6000°K and electron densitites of the order of 10^{14}cm^{-3} [125,126] should be mentioned.

The conductivity data obtained by the dc-gap probe and 3 cm microwave absorption in detonation waves[122] in $C_2H_2+O_2$ are in close agreement and are seen to be slightly higher than the calculated equilibrium values. Similar results were obtained in our unpublished experiments where electron densities of 10^{12}cm^{-3} were detected in $C_2H_2+O_2$ at initial pressures of 0.1 atm while using an 8 mm wavelength microwave interferometer.

Conductivity and microwave absorption measurements in shock-tube flows with well-known equilibrium gas parameters may be used to determine the value of the electron collisional cross-sections at high temperatures for a variety of molecular and atomic systems. This method is very suitable for electron collision studies since purely thermal heating is employed as a plasma source, strong electric fields are absent, and the electron densities are high enough to ensure equilibration by electron impacts in sufficiently short times. Effective electron-particle collisional cross-sections for the oxygen, nitrogen and other species have been determined in the temperature range between 3000° and 5000°K by employing the microwave absorption technique in shock tube experiments.[127] For example, σ_{ec} for the nitrogen molecule varies from 0.3 to $3\text{x}10^{-15}$ cm^2 in the range between 2800° and 5000°K.

PRESSURE MEASUREMENTS AND APPLICATIONS

Various methods of construction of pressure transducers have been developed for shock tube studies, and the most successful types of piezoelectric gauges now provide pressure data in complex shock tube studies with microsecond time resolutions. An accuracy of 2-5 per cent and detection sensitivities up to a few volts per atmosphere have been achieved. The use of an acoustic absorbing rod with the acoustic impedance closely matched to the impedance of the sensing element was shown to be the important detail of a device with microsecond time resolutions.[6,128-134] Barium titanate-zinc, quartz-duralumin and lead metaniobate-tin pairs seem to be the most suitable combinations.

An advantage of this design is the small size of the sensing elements; for example, piezoceramic disks 1 mm in diameter may be easily used without serious losses in sensitivity. We have found also that by placing the sensing element directly on the front face of an absorbing rod we avoid the stress wave dispersion arising through coupling of longitudinal and radial elastic waves produced in the system when a step increase in pressure is applied. The rod assembly should be potted in beeswax or in silicon rubber[132] in order to provide electrical and mechanical isolation. Direct contact of the sensing element face with the shocked gas sample is very much recommended.

The role of pressure measurements is quite substantial in complex shock-tube experiments since it is the most appropriate dynamic characteristic to be measured with high time resolutions in complicated non-steady gas flows behind shock waves. For example, pressure records obtained in electromagnetic shock tubes[48] yielded important information on the onset of mixing between discharge and thermalized plasmas behind a shock front. Pressure measurements are particularly important in single-pulse shock tubes and in expansion wave techniques developed to study non-equilibrium processes in shocked gas sample subjected to a negative temperature pulse.[14,135,137]

Cooling rates ranging from 10^5 to 10^6 $^{o}K/sec$ might be achieved in an ordinary single-pulse shock tube depending on its length. Therefore, rate constants of relatively slow processes, for example, of carbon monoxide-atomic oxygen recombination ($k_R \sim 2 \times 10^{13}$-$10^{14}$ cm^6 $mole^{-2}$ sec^{-1}) have been determined using this technique.[138] Cylindrical diaphragms at the end of the shock tube have been used to get higher cooling rates in direct measurements of chlorine-atom recombination rates.[139] Quite steep negative temperature-gradient pulses have been obtained using the reflected wave interaction at the contact surface provided that the reflected wave is an expansion wave.[137] Pressure records were used in this case in order to find an appropriate wave generation regime by removing secondary compression waves.

The highest cooling rates of $(0.5-5)10^8$ °K/sec have been obtained in direct measurements of oxygen-atom recombination rates using shock tube end expansion flows[140, 141] and a special expansion wave technique[14] mentioned also in Reference 137. Pressure traces were taken together with laser-chronometer records to follow the course of a recombination process in a cooled gas sample (Figure 10). A summary of data on recombination rate constants derived from direct shock tube measurements in various systems is shown in Figure 16. These data are of special interest in high temperature dissociation kinetic studies since the common method for determining the recombination rate constant from the dissociation rate constant and the equilibrium constant should be applied with caution in studies of fast processes or of complicated reacting systems.

SHOCK TUBE STUDIES OF EXOTHERMAL SYSTEMS

Shock tube flow patterns associated with a positive energy release might also be investigated by combining pressure measurements with schlieren or interferometer observations. Figure 17 shows typical pressure records and streak interferograms taken of the double-headed detonation waves proceeding into a $2CO + O_2$ mixture. The pressure profiles and density distribution fields closely correspond to each other, clearly indicating the secondary compression of the shocked gas in transverse waves propagating behind a primary shock in a shock tube of square cross-section. Note the nonsteady induction zones which arose behind the curved primary shock wave, the transverse wave structure, and the behavior of the attenuating compression "tail." The purpose of this experiment was to determine a sequence of three-dimensional wave processes occurring in spinning-type detonations in a relatively slowly reacting system. Since single-spin and multiheaded detonation studies have been summarized in numerous survey papers and monographs, only one-dimensional, shock-tube flows with regular exothermal reaction zones should be considered here in order to analyze recent data on ignition kinetics in a set of complex reactive systems. In addition to studies of oxy-hydrogen[142, 143] and oxy-methane[144, 145] systems, shock tube ignition experiments have been carried out recently in NH_3-O_2 and N_2O-H_2 exothermal systems. Induction period data for oxy-ammonia systems are shown in Figure 18[146, 147] for the temperature range from 1500° to 5000°K. A typical interferogram and a hydroxyl uv-emission record obtained simultaneously in a NH_3-O_2

mixture diluted with argon are shown in Figure 19 with the same time scale. The induction period and the narrow zone of heat release are clearly apparent. It is also seen that over a large range of 1/T, the activation energy of a rate-controlling reaction may be fixed at 42 kcal/mole, although different experimental techniques have been applied at low and high temperatures. It is interesting to note that the reaction rate depends essentially on the N_2 and NH_3, concentrations (Figure 18).

The ignition kinetic scheme was found to be significantly different in a nitrous oxide-hydrogen system.[148] In the temperature range below 2500°K, the overall activation energy corresponds to 22 kcal/mole, while at temperatures higher than 2500°K the apparent activation energy is about 60 kcal/mole. These data have been explained in terms of a chain branched reaction mechanism in which the initiation process of the N_2O decomposition appears to be the rate-controlling stage at higher temperatures.

Infrared emission traces of N_2O around 4.5 microns taken together with OH ultraviolet emission records (Figure 20) are quite descriptive in illustrating the reaction course behind incident shock waves. If a highly diluted mixture is taken, energy deposition effects on pressure and density are negligible; thus, a classical one-dimensional reaction zone would easily be obtained. Spectroscopic measurements of selective absorption or emission appear to be more useful in these experiments than schlieren, interferometer, and pressure measurements.

CONCLUSIONS

In order to summarize in a brief and descriptive form the variety of procedures and available measurement techniques which have been developed recently in shock tube studies, typical methods are listed in Table I together with recommendations and range of appropriate applications. Some of the most important experimental data on molecular and chemical kinetics obtained in shock tube studies are presented in Table II. These tabulations will not cover all the numerous basic results obtained in shock-tube experiments and serve mainly as illustrations of various applications of specific diagnostic techniques in nonequilibrium gasdynamics. Based on the work of the last few years, we can expect that further developments of shock-tube operations and techniques of experimental measurements will lead to rapid progress in shock-wave physics and chemistry.

ACKNOWLEDGEMENTS

The author is indebted to Dr. R. E. Duff for valuable remarks and recommendations.

REFERENCES

1. Huddlestone, R. H. and Leonard, S. L. (Eds.). 1965. Plasma Diagnostic Techniques (Academic Press, New York-London).
2. Bauer, S. H., Rol, N. C. and Kiefer, J. H. 1961. A survey of the Methods for the Chemical Analysis of Transient Species with Microsecond Time Resolution, in: Physical Chemistry in Aerodynamics and Space Flight, Vol. 3, pp. 118-131.
3. Gaydon, A. G. and Hurle, I. R. 1963. The Shock Tube in High-Temperature Chemical Physics (Chapman and Hall Ltd., London). Hurle, I. R. 1967, Reports on Prog. in Phys. V. 30, 1491 London.
4. Stupochenko, E. V., Losev, S. A. Osipov, A. I. 1965. Relaxatsionnye Processy v Udarnykh Volnakh (Nauka, Moscow).
5. Strehlow, R. A. 1967. Shock Tube Chemistry Techn. Report AAE 67-2 (University of Illinois, Urbana). Greene, E. F. and Toennis, J. P. 1964 Chemical Reactions in Shock Waves, Edward Arnold Ltd., London.
6. Soloukhin, R. I. 1966. Shock Waves and Detonations in Gases (Mono Book Corp., Baltimore, Maryland); Nesterikhin, Yu. E. and Soloukhin, R. I. 1967. Methody Scorostnykh Izmerenii v Gasodynamike i Fizike Plasmy (Nauka, Moscow).
7. Dushin, L. A. and Pavlichenko, O. S. 1968. Issledovanie Plasmy s Pomoshju Laserov (Atomizdat, Moscow).
8. Hecht, G. J., Steel, G. B. and Oppenheim, A. K. 1966. ISA Transactions, 5, 133.
9. Oppenheim, A. K., Urtiew, P. A. and Weinberg, F. J. 1966. Proc. Roy. Soc. A291, 279; A295, 13.
10. Krugliakov, E. P., Malinovsky, V. K. and Nesterikhin, Yu. E. 1965. Magnitnaja Hydrodynamika 1, 80; 2, 31.
11. Alcock, A. J. Panarella, E. and Ramsden, S. A. 1966. Proc. Seventh Intern. Confer. Phenomena in Ionized Gases: Belgrade, P. 224; Appl. Phys. Letters 8, 187.
12. Panarella, E. 1966. Optical Interferometry for Plasma Measurements in a Theta-Pinch Device, Mech. Eng. Report No. MT-57, NRC, Ottawa.
13. Soloukhin, R. I. 1967. Eleventh Symposium (International) on Combustion, p. 671 (The Combustion Institute, Pittsburgh, Pa.).
14. Soloukhin, R. I. 1967. Combustion and Flame, 11, 489.

15. Syshikova, M. P., Bereskina, M. K. and Semenov, A. N. 1967. In: Aerofizicheskie Issledovanija Sverkhzwukovykh Techenii, p. 7 (Nauka, Moscow-Leningrad).
16. Faizullov, E. S. 1962. Proc. P. N. Lebedev Phys. Institute, USSR Academy of Sciences (Moscow) 18, 105.
17. Bereskina, M. K., Dobrynin, B. M., Semenov, A. N. and Syshikova, M. P. 1967. In: Aerofizicheskie Issledovanija Sverkhzwukovykh Techenii, P. 54 (Nauka, Moscow-Leningrad).
18. Brossard, J., Manson, N., Pelhate, M. and Vautheir, R. 1966. Compt. Rend. Acad. Sci. Paris B263, 593. 1967. Le Journal de Physique, 28, 487.
19. Aro, T. O. and Walsh, D. 1966. J. Sci. Instr. 43, 572.
20. Nemkov, R. G. 1967. In: Svoistva Gasov pri Vysokikh Temperaturakh, p. 151 (Nauka, Moscow).
21. Vasilieva, R. V. 1965. Zh. Prikl. Mekh. Tekhn. Fiz. 5, 127.
22. Croce, P. A. 1965. Rev. Sci. Instr. 36, 1561.
23. Soloukhin, R. I. 1966. Teplofizika Vysokikh Temperatur 4, 176. See also: Veyssiere, M. and Brochet, C. 1968. Compt. Rend. Acad. Sci. Paris, B267, 924.
24. Gluskin, E. S., Savchenko, O. Ja. and Shun'ko E. V. 1968. Preprint No. 226, Inst. Nucl. Phys., USSR Acad. Sci., Novosibirsk.
25. Sobolev, N. N., Potapov, A. V. Kitaevy, V. F., Faizullov, F. S. Aljamovskii, V. N., Antropov, E. T. and Isaev, I. L. 1958. Izv. Akad. Nauk SSSR (Phys. Ser.) 22, 730.
26. Clouston, J. G., Gaydon, A. G. and Glass, I. I. 1958. Proc. Roy. Soc. A248, 429.
27. Clouston, J. G., Gaydon, A. G. and Hurle, I. R. 1959. Proc. Roy. Soc. A252, 143.
28. Tsuchiya, S. and Kuratani, K. 1964. Combustion and Flame, 8, 299.
29. Hurle, I. R. 1964. J. Chem. Phys. 41, 3911.
30. Hurle, I. R. and Russo, A. L. 1965. J. Chem. Phys. 43, 4434.
31. Gaydon, A. G. and Hurle, I. R. 1961. Proc. Roy. Soc. A262, 38.
32. Coates, P. B. and Gaydon, A. G. 1965. Proc. Roy. Soc. A283, 18.
34. Fairbairn, A. R. 1962. Proc. Roy. Soc. A267, 88.
35. Fairbairn, A. R. 1963 Proc. Roy. Soc. A276, 513.
36. Soloukhin, R. I. and Sharapova, T. A. 1962 Zh. Prikl. Mekh. Tekhn. Fiz. 2, 37.
37. Kudryavtsev, E. M., Sobolev, N. N., Tunitskii, L. N. and Faizullov, F. S. 1962. Proc. P. N. Lebedev Phys. Inst., USSR Acad. Sciences (Moscow), 18, 159.
38. Borisov, A. A., Zaslonko, I. S. and Kogarko, S. M. 1964. Zh. Prikl. Mekh. Tekhn. Fiz. 6, 104.

39. Tsuchiya, S. and Kuratani, K. 1965. J. Chem. Phys. 42, 2986.
40. Griem, H. R. 1964. Plasma Spectroscopy (McGraw-Hill Book Co. Inc., New York). 1969 Wilkerson, T. D. et al, Phys. Fluid (Suppl) 12, I-22; 1969, Wood, A. D. and Wilson, K. H., Seventh International Shock Tube Symposium, Toronto.
41. Barnard, A. J., Cormack, G. D. and Simpkinson, W. V. Can. J. Phys. 40, 531.
42. Soloukhin, R. I. and Toktomyshev, S. D. 1965. Zh. Prikl. Mekh. Tekhn. Fiz. 5, 124.
43. Coates, P. B. and Gaydon, A. G. 1966. Proc. Roy. Soc. A293, 452.
44. Fukuda, K., Ishii, K., Okasaka, R. and Fujimoto, T. 1965. J. Phys. Soc. Japan, 20, 2309.
45. Soloukhin, R. I. 1966 Proc. Seventh Intern. Confer. Phenomena in Ionized Gases, Belgrade, p. 800.
46. Pietrzyk, Z. A., Soloukhin, R. I. and Toktomyshev, S. D. 1967, Zh Prikl. Mekh. Tekhn. Fiz. 4, 104.
47. Dolgov-Saveljev, G. G. Krugliakov, E. P., Kurbatov, A. I., Malinovskii, V. K., Nesterikhin, Yu. E. and Sagdeev, R. Z. 1968. Issledovanie Structury Quasistationarnykh Udanykh Voln v Plasme, in: Annual Review, Inst. Nucl. Phys. USSR Academy of Sciences, Novosibirsk, P. 14.
48. Vorotnikova, M. I. and Soloukhin, R. I. 1964. Zh. Prikl. Mekh. Tekhn. Fiz. 5, 138.
49. Pietrzyk, Z. A. 1964. Arch. Mech. Stos. 2, 16; 1967. Fluid Dyn. Trans. (IBTP, Polish Acad. Sci.), 3, 343.
50. Sawyer, G. A., Bearden, A. J., Henins, I., Jahoda, F. C. and Ribe, F. L. 1963. Phys. Rev. 131, 1891.
51. Dolgov-Saveljev, G. G. and Panchenko, V. E. 1967. Tekhnika Issledovanija Mjagkogo Rentgenovskogo Izlutchenija Plasmy, Prepr. 175, Inst. Nucl. Phys., USSR Academy of Sciences, Novosibirsk.
52. Alcock, A. J. Pashinin, P. P. and Ramsden, S. A. 1966. Phys. Rev. Letters, 17, 528.
53. Bogen, P., Lie, Y. T., Rusbuldt, D. and Schluter, J. 1968. Paper CN-24/K-10. Third IAEA Fusion Conference, Novosibirsk.
54. Alinovskii, N. I., Nesterikhin, Yu. E., Papyrin, A. N. and Ponomarenko, A. G. 1968 Diagnostika Plasmy, vol. 2, p.137 (Atomizdat, Moscow).
55. Goldman, L. M., Kilb, R. W. and Pollock, A. C. 1964. Phys. Fluids, 7, 1005; Plasma Physics (J. Nucl. Energy, Part C), 6, 217. Also: Prasad, A. N. and Abd el-Karim, K. 1968. Proc. Phys. Soc. 2, 1, 77.
56. Funfer, E., Kronast, B. and Kunze, H. 1963. Phys. Letters, 5, 125; 1964. Phys. Letters, 11, 42; 1966. Phys. Rev. Letters, 16, 1082; 1967. Report IPP 1/70, Inst. Plasmaphysik, Garching.

57. Ascoli-Bartoli, U., Katzebstein, J. and Lovisetto, L. 1964. Bull. Amer. Phys. Soc. 9, 495; 1965. Nature, 207, 63.
58. Ramsden, S. A. and Davies, W. E. R. 1965. Bull. Amer. Phys. Soc. 10, 227; 1966. IEEE J. Quant. Electr. 2, 267.
59. Dolgov-Saveljev, G. G., Iskol 'dskii, A. M., Krougliakov, E. P. and Malinovskii, V. K. 1967. Laser and Unconventional Optics J. 1, 10.
60. Gerry, E. T. and Patrick, R. M. 1965. Phys. Fluids 8, 208.
61. Izawa, Ya., Yokoyama, M. and Yamanaka, C. 1966. J. Phys. Soc. Japan 21, 1610; 1967. ibid. 23, 1185; 1968. Japan. J. Appl. Phys. 7, 954.
62. Lubin, M. J., Dunn, H. S. and Friedman, W. 1968. Paper CN-24/F-8, Third IAEA Fusion Conference, Novosibirsk.
63. Resler, E. L. and Sheibe, M. 1955. J. Acoust. Soc. Amer. 27, 932; Ind. Eng. Chem. 47, 1182. Also: Witteman, W. J. 1961. Rev. Sci. Instr. 32, 292.
64. Soloukhin, R. I. 1965. Fizika Goreniya i Vzryva, 1, 112, see also Reference 6.
65. Kiefer, J. H. and Lutz, R. W. 1966. J. Chem. Phys. 44, 668; Phys. Fluids, 9, 1638; 1967. Eleventh Symp. (Intern.) on Combustion, P. 67 (The Combustion Institute, Pittsburgh, Pa.).
66. Vasil'ev, L. A. and Ershov, I. V. 1964. Doklady Akad. Nauk SSSR, 157, 317.
67. Alpher, R. A. and White, D. R. 1959. Phys. Fluids, 2, 162.
68. Belozerov, A. 1968. Study of the Initial Ionization Processes in a Strong Shock Wave, UTIAS Report No. 131, Toronto.
69. Heflinger, L. O., Wuerker, R. F. and Brooks, R. E. 1966. J. Appl. Phys. 37, 642.
70. Tanner, L. H. 1966. J. Sci. Instr. 43, 878; 1967. ibid. 44, 725; 44, 1011.
71. Jones, N. R. and McChesney, M. 1964. J. Sci. Instr. 41, 682.
72. Miyamoto, K., Kon, S. and Morimoto, S. 1968. Japan. J. Appl. Phys. 7, 1304. See also: Witteman, W. J. 1968. Appl. Phys. Letters, 10, 347; Herold, H. and Jahoda, F. C., 1969 Rev. Sci. Inst. 40, 145; Soloukhin, R. I. and Jakoby, Ju. A., Opt. Spectr. (in press).
73. Ashby, D. E. T. F. and Jephcott, D. F. 1963. Appl. Phys. Letters, 3, 13; 1965. J. Appl. Phys. 36, 29.
74. Thomas, K. S. 1968. Phys. Fluids, 11, 1125.
75. Gerardo, J. B. Verdeyen, J. T. and Gusinow, M. A. 1963. Appl. Phys. Letters, 3, 121; 1965. J. Appl. Phys, 36, 2146.
76. Fontaine, B., Inglesakis, G. and Valensi, J. 1967. Compt. Rend. Acad. Sci. Paris, 265, 1230.
77. Dougal, A. A. Craig, J. F. and Gribbe, P. F. 1964. Phys. Rev. Letters 13, 156; 1966. AIAA J. 4, 12.

78. Jahoda, F. C. Little, E. M., Quinn, W. E., Ribe, F. L. and Sawyer, G. A. 1964. J. Appl. Phys. 35, 2351.
79. Gribbe, R. F., Little, E. M. Morse, R. L. and Quinn, W. E. 1968. Phys. Fluids, 11, 1221.
80. Kon, S., Yamanaka, M., Yamamoto, J. and Yoshinaga, H. Far Infrared CN Laser, in: Annual Review, April 1966 - March 1967, Inst. Plasma Phys., Nagoya University, Japan, P. 89.
81. Tsukishima, T. and Takeda, S. 1962. J. Appl. Phys. 33, 3290; 1965. Techn. Note 256, Natl. Bureau Stand., Washington.
82. Takeda, S. and Tsukishima, T. 1963. J. Phys. Soc. Japan 18, 426; 1965. ibid. 20, 1090; 1968. ibid. 25, 297.
83. Duff, R. E. 1959. Phys. Fluids, 2, 207; Bull. Amer. Phys. Soc. 4, 283.
84. Busygin, E. P. and Tumakaev, G. K. 1967. In: Aerofizischeskie Issledovanija Sverkhzwukovykh Techenii, p. 66 (Nauka, Moscow-Leningrad).
85. Lashkov, A. I. 1965. Inzhenernyi J. 5, 1114.
86. Camac, M. 1964. Phys. Fluids, 7, 1076; Avco-Everett Research Report 194, Everett, Mass.
87. Chernetskii, A. V. Zinov'ev, O. A. and Kozlov, O. V. 1965. Apparatura i Methody Plasmennykh Issledovanii (Atomizdat, Moscow).
88. Podgornyi, I. M. 1968. Lekzii po Diagnostike Plasmy (Atomizdat, Moscow).
89. Alinovskii, N. I., Alkhimov, A. P., Kornilov, V. A., Papyrin, A. N., Ponomarenko, A. G. and Rogozin, A. I. 1968. Opredelenije Parametrov Plasmy Puchkami Neutral'nykh Atomov, Preprint 240, Inst. Nucl. Phys., USSR Acad. Sciences, Novosibirsk.
90. Admur, J. and Smith, A. L. 1968. J. Chem. Phys. 48, 565.
91. Fite, W. L., Smith, A. C. H. and Stebbings, R. F. 1962. Proc. Roy. Soc. A268, 527.
92. Alinovskii, N. I., Nesterikhin, Yu. E. and Pakhtusov, B. K. 1967. Proc. Eighth Intern. Confer. Phenomena in Ionized Gases, Vienna, P. 515.
93. Losev, S. A. 1958. Doklady Akad. Nauk SSSR, 120, 1291; Soviet Phys-Dikl. Vys. Shkoly, 3, 197. Losev, S. A. and Generalov, N. A. 1966. J. Quant. Spectr. Rad. Transfer, 6, 101. Kuksenko, B. V. and Losev, S. A. 1969. Doklady Acad. Nauk SSSR, 185, 69; 185, 293.
94. Camac, M. 1961. J. Chem. Phys. 34, 448. Camac, M. and Vaughan, A., ibid 34, 460.
95. Wray, K. L. 1962. J. Chem. Phys. 37, 2354; 1963. ibid. 38, 1518; 1965. Tenth Symp. (Intern.) Combust., p. 523. (The Combustion Institute, Pittsburgh, Pa.).

96. Wray, K. L. and Freeman, T. S. 1964. J. Chem. Phys. 40, 2785.
97. Blake, A. J., Carver, J. H. and Haddad, G. N. 1966. J. Quant. Spectr. Rad. Transfer, 6, 451.
98. Metzger, P. H. and Cook, G. R. 1964. J. Quant. Spectr. Rad. Transfer, 4, 107.
99. Jacoby, Yu. A. Komin, A. V. and Ivanija, S. P. 1968. J. Quant. Spectr. Rad. Transfer, 8, 805.
101. Appleton, J. P., Steinberg, M. and Liquornic, D. J. 1968. J. Chem. Phys. 48, 599.
102. Bauer, S. H., Schott, G. L. and Duff, R. E. 1958. J. Chem. Phys. 28, 1089. Watson, R. 1964. J. Quant. Spectr. Rad. Transfer 4, 1. Watson, R. and Ferguson, W. R. 1965. ibid. 5, 595, See also Reference 114.
103. Allen, R. A. Textoris, A. and Wilson, J. 1965. J. Quant. Spectr. Rad. Transfer, 5, 95.
105. Reis, V. H. 1946. J. Quant. Spectr. Rad. Transfer, 4, 789; 1965, ibid. 5, 585.
106. Myers, B. F. and Bartle, E. R. 1968. J. Chem. Phys. 48, 3935.
107. White, D. R. 1961. Phys. Fluids, 4, 465.
108. Myers, B. F. and Bartle, E. R. 1967. J. Chem. Phys. 47, 1783.
109. Wray, K. L. 1966. J. Chem. Phys. 44, 623.
110. Hooker, W. J. and Millikan, R. C. 1963. J. Chem. Phys. 38, 214. Millikan, R. C. 1963. ibid. 38, 2855; 1964. ibid. 40, 2594.
111. Camac, M. and Feinberg, R. M. 1967. Eleventh Symp. (Interns.) Combust., p. 137 (The Combustion Institute, Pittsburgh, Pa.).
112. Jacobs, T. A., Giedt, R. R. and Cohen, N. 1965. J. Chem. Phys. 43, 3688.
113. Olschewski, H. A., Troe, J. and Wagner, H. Gg. 1967. Eleventh Symp. (Intern.) Combust. p.155 (The Combustion Institute, Pittsburgh, Pa.). Also: Penzias, G. J. and Maclay, G. J. 1965. Tenth Symp. (Intern.) Combust. p.189 (The Combustion Institute, Pittsburgh).
114. Patch, R. W. 1965. J. Quant. Spectr. Rad. Transfer, 5, 137.
115. Wilson, J. 1966. Phys. Fluids 9, 1913.
116. Lin, S. C., Resler, E. L. and Kantrowitz, A. 1955. J. Appl. Phys. 26, 95. Lamb, L. and Lin, S. C. 1957. ibid. 28, 754.
117. Lau, J. 1964. Can. J. Phys. 43, 1548; 1965. ibid. 43, 1334.
118. Stubbe, E. J. 1966. AIAA Plasmadynamic Conference Monterey, California, Paper 66-181, See also 1968 Proc. IEEE 56, 1483.

119. Poberezhskii, L. P. 1968. Teplofizika Vysokikh Temperatur, 6, 973.
120. Basu, S. and Fay, J. A. 1959. Seventh Symp. (Intern.) Combust. p. 277 (The Combustion Institute, Butterworth, London).
121. Bellet, J. C., Brossard, J., Manson, N. and Veyssiere, M. 1964. J. Combust. Convers. Energies, IFCE, Paris, p. 95. Veyssiere, M. and Manson, N. 1967. Compt. Rend. Acad. Sci. Paris, 264, 199; 1968, ibid. B 266, 990.
122. Edwards, D. H. and Lawrence, T. R. 1965 Proc. Roy. Soc. A 286, 415.
123. Hand, C. W. and Kistiakowsky, G. B. 1962. J. Chem. Phys. 6, 1154.
124. Terao, K., Nishida, Y. and Sato, T. 1967. Japan. J. Appl. Phys. 37, 1239.
125. Hoppmann, R. F. 1968. Phys. Fluids 11, 1092.
126. Gorelova, N. A., Gorelov, V. A. and Kil'djuschova, L. A. 1969. Teplofizika Vysokikh Temperatur 7, 18.
127. Lobastov, Yu. S. 1966. In: Issledovanija po Fizicheskoy Gasodynamike, p. 119 (Nauka, Moscow).
128. Turetric, W. 1941. Problem der Detonation (Berlin).
129. Zaitsev, S. G. 1958. Pribory i Techn. Exper. 6, 97.
130. Soloukhin, R. I. 1961. Pribory i Techn. Exper. 3, 583.
131. Edwards, D. H. 1958. J. Sci. Instr. 35, 346; 1959. J. Fluid Mech. 6, 497; 1962. Paper No. 109, Coll. Intern. CNRS, Gif-sur-Yvette (France); 1964. J. Sci. Instr. 41, 609.
132. Ragland, K. W. and Cullen, R. E. 1967. Rev. Sci. Instr. 38, 740.
133. Topchian, M. E. 1962. Zh. Prikl. Mekh. Tekhn. Fiz. 4, 94.
134. Guerraud, C., Leyer, J. C. and Brochet, C. 1967 Compt. Rend. Acad. Sci. Paris B 264, 5. Brochet, C., Guerraud, C., Manson, N. and Veyssiere, M. 1969. ibid. B 268, 361.
135. Glick, H. S., Klein, J. and Square, W. 1957. J. Chem. Phys. 27, 850.
136. Wilson, J. 1963. J. Fluid Mech. 15, 497.
137. Resler, E. L. 1966. Chemical Studies Using Gasdynamic Wave Interactions, in: Dynamics of Fluids and Plasmas, p. 135 (Academic Press Inc., New York). Bauer, S. H. Hou, K. C. and Resler, E. L. 1969 Phys. Fluids (Suppl.) 12, I-125.
138. Brabbs, T. A. and Belles, F. E. 1967. Eleventh Symp. (Intern.) on Combustion, p. 125 (The Combustion Institute, Pittsburgh, Pa.).
139. Jacobs, T. A., Kartunian, R. A., Giedt, R. R. and Wilkins, R. 1963. Phys. Fluids 6, 972.

140. Glass, I. I. and Takano, A. 1965. Nonequilibrium Expansion Flows of Dissociated Oxygen and Ionized Argon around a Corner, in: Progr. Aeron. Sci., ed. by Kuchemann and Sterne, vol. 6, pp. 163-249. (Pergamon Press). Glass, I. I. 1967. Can. Aeronaut. Space J. 13, 347-426.

141. Drewry, J. E. 1967. An Experimental Investigation of Non-equilibrium Corner Expansion Flows of Dissociated Oxygen, UTIAS Report No. 124, Toronto.

142. White, D. R. and Moore, G. E. 1965. Tenth Symp. (Intern.) on Combustion, p. 785 (The Combustion Institute, Pittsburgh, Pa.); 1967. Eleventh Symp. (Intern.) on Combustion, p. 147, (The Combustion Institute, Pittsburgh, Pa.).

143. Hamilton, C. W. and Schott, G. L. 1967. Eleventh Symp. (Intern.) on Combustion, p. 635 (The Combustion Institute, Pittsburgh, Pa.).

144. Miyama, H. and Takeyama, T. 1965. Bull. Chem. Soc. Japan, 38, 37.

145. Voevodsky, V. V. and Soloukhin, R. I. 1965. Dokl. Akad. Nauk SSSR 161, 1118.

146. Takeyama, T. and Miyama, H. 1965. Bull. Chem. Soc. Japan, 38, 1670; 1966. ibid. 39, 2352; 39, 2609; 1967. Eleventh Symp. (Intern.) on Combustion, p. 845 (The Combustion Institute, Pittsburgh, Pa.). Miyama, H. 1968. Bull. Chem. Soc. Japan 41, 1761.

147. Soloukhin, R. I. 1966. Some Shock Tube Methods in Chemical Physics (Nauka (Siberian Division), Novosibirsk).

148. Soloukhin, R. I. and Van Tiggelen, P. J. 1969. Bull. Soc. Chim. Belg. 78, 179.

149. Galaktionov, I. I., Korovkina, T. D., Mikhalevsky, V. D. and Podmoshensky, I. V. 1969. Teplophysica Vysokikh Temperatur, 1, 85.

150. Suckewer, S. 1968. Plasma Phys. 10, 527.

151. Elton, R. C. and Roth, N. V. 1967. Appl. Optics. 6, 2071.

152. Kistiakowsky, G. B. and Kydd, P. H. 1956. J. Chem. Phys. 25, 824.

153. Duff, R. E. Knight, H. T. and Rink, J. P. 1958. Phys. Fluids, 1, 393.

154. Measures, R. M., Jaskolka, A., Rodrigo, A. B. and Pettitt, B. A. 1968. Laser Excitation Diagnostics, in: Annual Progress Report, UTIAS, Toronto, pp. 90-93

155. Bristow, M. 1968. An Experimental Determination of the Polarizability of Singly-Ionized Argon, in: Annual Progress Report, UTIAS, Toronto, pp. 24-25.

156. Benson, S. W. and Berend, G. C. 1966. J. Chem. Phys. 44, 471.

157. Sutton, E. 1962. J. Chem. Phys. 36, 2923.
158. Seery, D. J. and Bowman, C. T. 1968. J. Chem. Phys. 48, 4314.
159. Jones, W. M. and Davidson, N. 1962. J. Amer. Chem. Soc. 84, 2868.
160. Schott, G. L. and Bird, P. F. 1964. J. Chem. Phys. 41, 2869. Getzinger, R. W. and Schott, G. L. 1965. ibid. 43, 3237.
161. Hurle, I. R. 1967. Eleventh Symp. (Intern.) on Combustion, p. 86 (The Combustion Institute, Pittsburgh, Pa.).
162. Gutman, D. and Schott, G. L. 1967. J. Chem. Phys. 46, 4576.
163. Miyama, H. and Takeyama, T. 1965. J. Chem. Phys. 42, 2636.
164. Bull, D. C. 1968. Combustion and Flame, 12, 603.
165. Sulzmann, K. G. P., Myers, B. F. and Bartle, E. R. 1965. J. Chem. Phys. 42, 3969; 43, 1220. Brokaw, R. S. 1967. Eleventh Symp. (Intern.) on Combustion, p. 1063 (The Combustion Institute, Pittsburgh, Pa.).
166. Schwar, M. J. R. and Weinberg, F. J. 1969. Nature, 221, 357; see also Proc. Roy. Soc. A 311, 469, 1969.
167. R. A. Creswell, U. A. DiValentin, and J. E. Dove 1969 Phys. Fluids (Suppl.) 12, I-125
168. Watt, W. S. and Myerson, A. L. 1969. J. Chem. Phys. 51, 1638.
169. Lin, M. C. and Bauer, S. H. J. Chem. Phys. 50, 3377.
170. Hill, R. A. and Bannister, J. R. 1969. J. Quant. Spectr. Rad. Transfer 9, 1083.
171. Clark, T. C. Garnett, S. H. and Kistiakowsky, G. B. 1969. J. Chem. Phys. 51, 2885.

Table 1
Tabulation of experimental techniques

Gas parameter	Measurement method	Range of applications	Detection sensitivity and accuracy, per cent	Time resolution, μ sec	Comments and recommendations	References
1	2	3	4	5	6	7
Equilibrium gas temperature	Spectrum-line reversal technique	1500°–7000°K	1–2	1–5	Nonequilibrium effects in argon at densities lower than $3 \cdot 10^{19}$ cm^{-3}	3,4,6, 25–39,149
	Measurements of relative line intensities	4000°–40000°K	2	0.1–1	Ultraviolet region is used at high temperatures	40–47, 150
	Doppler line width	10–100 ev	10–30	1	X-ray spectroscopy of impurity lines	1,7, 50,53
	Laser cooperative scattering (ions Doppler width)	100–1000 ev	5–10	0.02–0.1	Plasma densities of 10^{16}–10^{18} cm^{-3}	53,56,58, 61,62
Electron temperature	Stark broadening of spectrum-lines	0.5–10	5–10	1	Are determined from equilibrium electron densities	1,40,46
	laser beam (Thomson) scattering	1–10000	10–20	0.02–01	Satellite line width is measured	56,58,61
	Neutral beam absorption	20–1000	10–20	0.5–1	Temperature is determined from measurements of ionization cross-sections	1,92

Table 1 (continued)

1	2	3	4	5	6	7
	Soft X-ray emission technique	50–10000 ev	5–50	0.1–1	Used for evaluations	52,53, 56,151
	Radiation emitted by the succesive ionization stages	5–300 ev	25	0.1	Electron density should be known	55
Gas density	Quantitative schlieren technique (density, and density gradients)	$5\cdot10^{14}$–10^{20} cm^{-3} 10^{14}–10^{18} cm^{-4}	2–5 0.2–0.5	0.5–2 0.2–1		6,63–66
	Optical interferometry	10^{16}–10^{21}	0.1–0.2	0.02–0.5		1,6,67–70
	Laser beam scattering	10^{18}–10^{21}	5–10	0.02–0.1	Rayleigh cross-sections $\sigma_R \sim 10^{-26}$ cm^2	7,58–61
	Electron beam absorption	10^{16}–$5\cdot10^{18}$	8–10	0.1		83–84
	Electron beam scattering	10^{15}–$2\cdot10^{18}$	1–5	0.5–1		85–86
	X-ray absorption	$5\cdot10^{18}$–$5\cdot10^{19}$	2–5	1–2	A noble gas is added	152,153
Mixture component concentrations	Ultraviolet and vacuum uv absorption	10^{16}–10^{19}	5–15	0.1	Absorption of O_2, N_2, CO_2, OH, NO, H_2 is employed	2,4, 93–103
	Infrared emission and absorption	$5\cdot10^{14}$–10^{20}	5–20	0.1–0.5	Systems of CO, CO_2, N_2O, NO, H_2O, NH_3, HF are recommended	2, 110–115
	Light emission in the visible; afterglow emission	10^{14}–10^{19}	5–15	1	Atomic oxygen, nitrogen, carbon monoxide	106–109

Table 1 (concluded)

1	2	3	4	5	6	7
	Mass spectrometry	10^{18}–10^{20} (to 10^{11} in a nozzle)	0.1–1	1–10		2,137, 167, 171
	Laser induced fluorescence			0.1		154
Electron concentrations	Microwave techniques	10^{11}–10^{16}	0.5–5	1	Microwave reflection method is recommended at high densities	1,6,7, 81,82
	Optical interferometry	10^{15}–10^{20}	0.1–0.5	0.02–0.1	A double frequency technique is recommended	67,68,155
	Laser-interferometers	10^{14}–10^{18}	0.1–0.5	0.5–1		72–76
	Faraday rotation	10^{16}–10^{17}	1–10	1	Magnetic field is applied	77–79
	Laser beam scattering	$5\cdot10^{15}$–10^{17}	5–15	0.1	Both the satellite line width and the distance from the center are used for measurements	58,59,61
	Spectral line broadening	5.10^{15}–10^{19}	5–10	2–5		1,40,46 151
	Neutral beam absorption	10^{13}– 10^{15}	10–15	2–5	Determined from the ionsdensity	1,6,92
	Infrared emission	10^{13}–10^{16}	10–20	0.5–1	free-free transitions in ionized air	103,115

Table 2
Molecular kinetics data obtained from shock tube experiments

Gas system	Kinetic process	Temperature range, °K	Measurement method	Rate constants and relaxation times	References
1	2	3	4	5	6
O_2, N_2, CO, Cl_2, Br_2, I_2, and gas mixtures (generalized,[100] to 50 per cent accuracy)	Vibrational relaxation	300-8000	Interferometer, uv-absorption, and other techniques	$p\tau_v = 10^{-8}\exp[1.15\times10^{-3}\mu^{\frac{1}{2}}\theta^{\frac{4}{3}}(T^{-\frac{1}{3}}-0.015\mu^{\frac{1}{4}})]$ atm-sec where p is the pressure in atm, μ is the reduced mass of the colliding partners, θ = hν/k is the characteristic temperature of the oscillator.	4,65,93 94,100, 110,156
Oxygen		1000-3700	Quantitative schlieren,	$p\tau_{O_2-O_2}=(2.92\pm0.20)\times10^{-10}\exp[(126.0\pm0.9)T^{-\frac{1}{3}}]$ atm-sec	65, 93,94,100
		1200-10500	uv-absorption, interferometer	$p\tau_{O_2-O_2}=3.43\times10^{-12}\exp(153T^{-\frac{1}{3}})$ atm-sec	
		300-850	Interferometer	$p\tau_{O_2-O_2}=4.47\times10^{-9}\exp(6.3\,T^{\frac{1}{3}})$ atm-sec	
		1000-10000	UV-absorption	$p\tau_{O_2-Ar}=4.33\times10^{12}\exp(173.13\,T^{-\frac{1}{3}})$ atm-sec	
		1600-3300	Quantitative schlieren	$p\tau_{O_2-O}=4.35\times10^{-8}-7.75\times10^{-12}T$ atm-sec	65

Table 2 (continued)

1	2	3	4	5	6
Hydrogen		1100-2700	Quantitative	$p\tau_{H_2-H_2} = (3.9\pm0.8)\times10^{-10} \exp[(100.0\pm2.6)T^{\frac{1}{3}}]$ atm-sec	65
			schlieren	$p\tau_{H_2-Ar} = 4.1\ p\tau_{H_2-H_2}$	
Deuterium		1100-3000	Quantitative schlieren	$p\tau_{D_2-D_2} = (2.7\pm0.3)\times10^{-10} \exp[(110.5\pm1.5)T^{-\frac{1}{3}}]$ atm-sec	65
Carbon dioxide		300-5000	IR emission, electron beam scattering	$p\tau_v = 2.02\times10^{-8} \exp(36.5\ T^{-\frac{1}{3}})$ atm-sec	86
Oxygen	$O_2 + M \xrightarrow{k_D} O + O + M$	3000-7000	UV absorption	$k_{O_2-O_2} = 5.6\times10^{10} T^{\frac{1}{2}} (\frac{D}{RT})^3 \exp(-\frac{D}{RT})$ cc/mole-sec	93
	$D = 118\ \frac{kcal}{mole}$	4200-16000		$k_{O_2-Ar} = 1.5\times10^{11} T^{\frac{1}{2}} (\frac{D}{RT})^{\frac{3}{2}} \exp(-\frac{D}{RT})$ cc/mole-sec	93,95
				$k_{O_2-Ar} = 1.85\times10^{11} \cdot T^{1/2} \exp(45{,}700/RT)$ cc/mole-sec	168
		3700-7000		$k_{O_2-N_2} = 2.5\times10^{11} T^{\frac{1}{2}} (\frac{D}{RT})^2 \exp(-\frac{D}{RT})$ cc/mole-sec	93
		5000-7500		$k_{O_2-O} = 1.5\times10^{15} (\frac{D}{RT}) \exp(-\frac{D}{RT})$ cc/mole-sec	94
		3000-10000		$k_{O_2-O_2} = [1+600\ \theta^{-23} \exp(-4\times10^{-9}\ \theta^{92})) k_{O_2-Ar} (\theta > 3)$	93,95
				$k_{O_2-O} = [1+108\ \theta^{-058} \exp(-1.7\times10^{-3}\ \theta^{33})) k_{O_2-Ar} (\theta > 4)$	
				where $\theta = 10^{-3}\ T\ ^oK$	

Table 2 (continued)

1	2	3	4	5	6
Nitrogen	$N_2 + M \overset{k_D}{\rightarrow} N + N + M$; $D = 225\ \frac{kcal}{mole}$	8000-15000	UV absorption	$k_{N_2-N_2} = (1.67 \pm 0.4) \times 10^{17} T^{-1.6} exp(-\frac{D}{RT})$ cc/mole-sec	101
				$k_{N_2-Ar} = (6.2 \pm 0.5) \times 10^{16} T^{-16} exp(-\frac{D}{RT})$ cc/mole-sec	
				$k_{N_2-N} = (7.25 \pm 2.7) \times 10^{16} T^{-16} exp(-\frac{D}{RT})$ cc/mole-sec	
Hydrogen	$H_2 + M \overset{k_D}{\rightarrow} H + H + M$; $D = 103.2\frac{kcal}{mole}$			$k_{H_2-Ar} = 6.17 \times 10^{14} exp(-\frac{D}{RT})$ cc/mole-sec	157
NO	$NO + M \overset{k_D}{\rightarrow} N + O + M$; $D = 150\ \frac{kcal}{mole}$	3000-8000	UV absorption	$k_D = 7.0 \times 10^{10} T^{\frac{1}{2}} (\frac{D}{RT})^2 exp(-\frac{D}{RT})$ cc/mole-sec	95
Carbon dioxide	$CO_2 + M \overset{k_D}{\rightarrow} CO + O + A$; $D = 125.7\ \frac{kcal}{mole}$	3000-5500	UV absorption	$k_{CO_2-CO_2} = (1.5 \pm 0.8) \times 10^{14} exp(-\frac{87000}{RT})$ cc/mole-sec	93
		2800-4400	IR emission	$k_{CO_2-Ar} = [Ar] \times 5.10^{14} exp(-\frac{99000}{RT})$ cc/mole-sec	113
HCl	$HCl + M \overset{k_D}{\rightarrow} H + Cl + M$; $D = 102.2\ \frac{kcal}{mole}$	2900-4000	UV absorption	$k_{HCl} = 4.2 \times 10^{13} exp(-\frac{81000}{RT})$ cc/mole-sec	158
HF	$HF + M \overset{k_D}{\rightarrow} H + F + M$; $D = 134.1\ \frac{kcal}{mole}$	3800-5300	IR emission	$k_{HF-Ar} = (1.1 \pm 0.3) \times 10^{19} exp(-\frac{134100}{RT})$ cc/mole-sec	112
O_3	$O_3 + M \overset{k_D}{\rightarrow} O + O_2 + M$	770-910		$k_{O_3-N_2} = (5.8 \pm 0.6) \times 10^{14} exp(-\frac{23150}{RT})$ cc/mole-sec	159

Table 2 (continued)

1	2	3	4	5	6
Oxygen	$O+O+M \overset{k_D}{\rightarrow} O_2+M$	3000-3700	Interferometer	$k_{O_2} = (2 \div 0.5)\times10^{15}\ cc^2/mole^2\text{-}sec$	14
		1500-3000	UV absorption	$k_{Ar} \cong 2.5\times10^{13}\ cc^2/mole^2\text{-}sec$	95
	$O+CO+M \rightarrow CO_2+M$	2800-3600	Visible light emission	$(2.6 \div 1.2)\times10^{13}\ cc^2/mole^2\text{-}sec$	138
Hydrogen	$H+H+M \overset{k_R}{\rightarrow} H_2+M$	1400-2000	UV absorption	$k_{Ar} = (4\pm2)\times10^{14}\ cc^2/mole^2\text{-}sec$	160
		2500-7000	Spectrum-line reversal	$k_{H_2} = (1.4 \div 3.9)\times10^{15}\ cc^2/mole^2\text{-}sec$	161
		2500-5500	measurements	$k_H = 9\times10^{16} \div 3.7.10^{14}\ cc^2/mole^2\text{-}sec$	161
	$H+OH+M \overset{k_R}{\rightarrow} H_2O+M$	1400-2000	UV absorption	$k_{Ar} = (4\pm2)\times10^{15}\ cc^2/mole^2\text{-}sec$	160
	$H+O_2+M \overset{k_R}{\rightarrow} HO_2+M$	1150-1850		$k_{Ar} = 1.4\times10^{15}\ cc^2/mole^2\text{-}sec$	160
Hydrogen-Oxygen	$H+O_2 \overset{k_1}{\rightarrow} OH+O$	1000-2000	UV absorption	$k_1 = 2.14\times10^{14} \exp\left(-\frac{16600}{RT}\right)\ cc/mole\text{-}sec$	143
				$k_1 = 7.76\times10^{13} \exp\left(-\frac{14400}{RT}\right)\ cc/mole\text{-}sec$	162
	$O+H_2 \overset{k_2}{\rightarrow} OH+H$	1000-2000	UV absorption	$k_2 = (0.4 \div 3.6)\times10^{13} \exp\left(-\frac{9200}{RT}\right) cc/mole\text{-}sec$	143
				$k_2 = 3.24\times10^{13} \exp\left(-\frac{10000}{RT}\right)\ cc/mole\text{-}sec$	162
	$OH+H_2 \overset{k_3}{\rightarrow} H_2O+H$	1000-2000	UV absorption	$k_3 = (2.1 \div 18.9)\times10^{13} \exp\left(-\frac{5900}{RT}\right) cc/mole\text{-}sec$	143
				$k_3 = 6.15\times10^{13} \exp\left(-\frac{5000}{RT}\right)\ cc/mole\text{-}sec$	162

Table 2 (concluded)

1	2	3	4	5	6
	Induction times	1100-2200	Interferometer	$\tau^{-1}[O_2]^{-\frac{1}{3}}[H_2]^{-\frac{2}{3}} = 2.75\times10^{13}\ \exp\left(-\frac{17200}{RT}\right)$ cc/mole-sec	142
				$\tau^{-1}[O_2]^{\frac{1}{2}}[H_2]^{-\frac{1}{2}} = 0.63\times10^{13}\ \exp\left(-\frac{14200}{RT}\right)$ cc/mole-sec	142
		1000-1700	Schlieren, interferometer	$\tau^{-1}[O_2]^{-1} = 1.12\times10^{13}\ \exp\left(-\frac{19500}{RT}\right)$ cc/mole-sec	6,13
Acetylene-Oxygen	Induction times	1400-2400	Ionization measurements	$\tau^{-1}[O_2]^{-1} = 2.51\times10^{14}\ \exp\left(-\frac{14450}{RT}\right)$ cc/mole-sec	123
				$\tau^{-1}[O_2]^{-1} = 3.02\times10^{14}\ \exp\left(-\frac{17100}{RT}\right)$ cc/mole-sec	123
		1100-2200	Interferometer	$\tau^{-1}[O_2]^{-\frac{1}{3}}[C_2H_2]^{-\frac{2}{3}} = 7.6\times10^{13}\ \exp\left(-\frac{17300}{RT}\right)$ cc/mole-sec	142
	Summary data	850-2500	Various methods	$\tau^{-1}[O_2]^{-1} = 2.75\times10^{14}\ \exp\left(-\frac{23900}{RT}\right)$ cc/mole-sec	123,147, 163
Methane-Oxygen	Induction times	1200-2400	Schlieren, interferometer	$\tau^{-1}[O_2]^{-1} = 1.66\times10^{14}\ \exp\left(-\frac{33200}{RT}\right)$ cc/mole-sec	145,147
Ammonia-Oxygen	Induction times	1500-4500	Infrared emission	$\tau^{-1}[O_2]^{-1} = 1.86\times10^{14}\ \exp\left(\frac{37700}{RT}\right)$ cc/mole-sec	164
	Summary data	1400-4000	Interferometer, light emission and absorption	$\tau^{-1}[O_2]^{-1} = 3.9\times10^{14}\ \exp\left(-\frac{41200}{RT}\right)$ cc/mole-sec	146,147
Garbon monoxide-Oxygen	Induction kinetics (initiation rate constant)	1500-3000		$k_i = 3.5\times10^{12}\ \exp\left(-\frac{51000}{RT}\right)$ cc/mole-sec	165
				$k_i = 2.5\times10^{12}\ \exp\left(-\frac{48000}{RT}\right)$ cc/mole-sec	

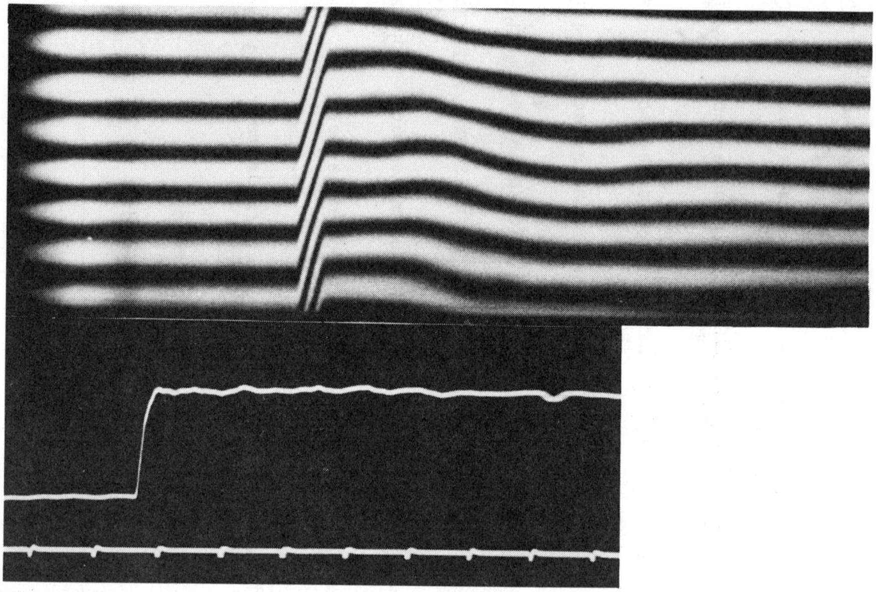

Figure 1
Interferogram and pressure record of a shock wave in 0.1 (H_2+O_2) + 0.9 Ar at $p_1 = 0.1$ atm and $T_s = 1500^oK$; time marks 10 μsec apart

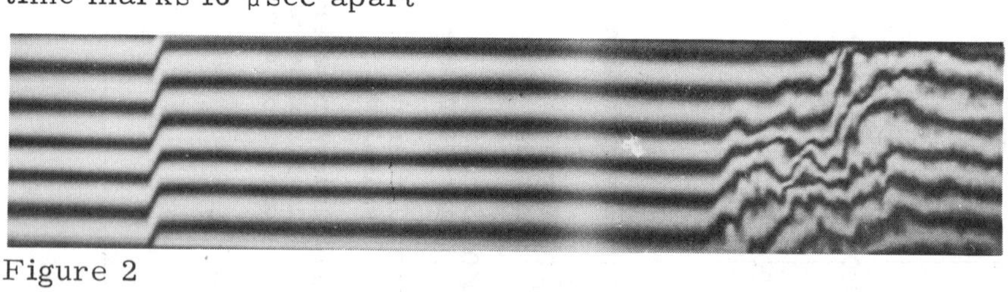

Figure 2
Interferogram of a shock-tube flow made by the compensating method

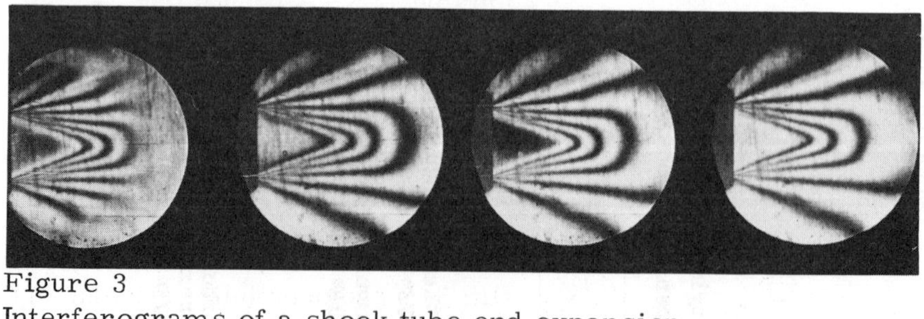

Figure 3
Interferograms of a shock tube end expansion flow of dissociated oxygen: $M_s = 10$, $p_1 = 10$ torr, 3 μsec between frames

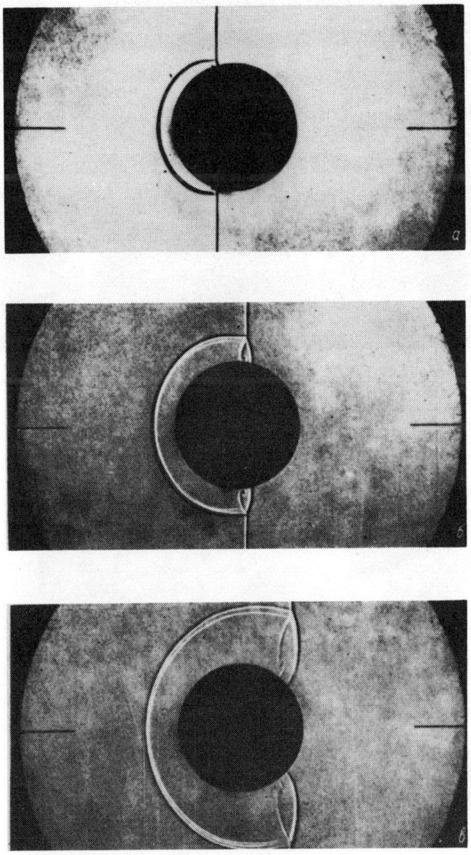

Figure 4
Schlieren frames of shock diffraction around a cylinder: $M_s = 2.42$, $p_1 = 200$ torr[15]

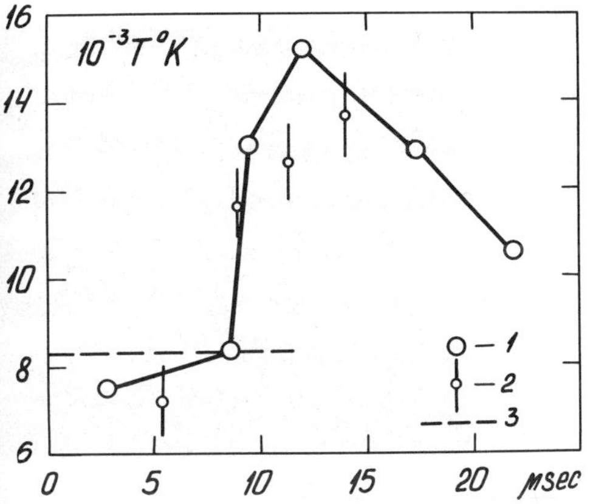

Figure 6
Gas-temperature profiles determined by different methods: 1, hydrodynamic method; 2, relative line intensity measurements, hydrogen lines; 3, equilibrium driven gas temperature; argon, $p_1 = 0.08$ torr

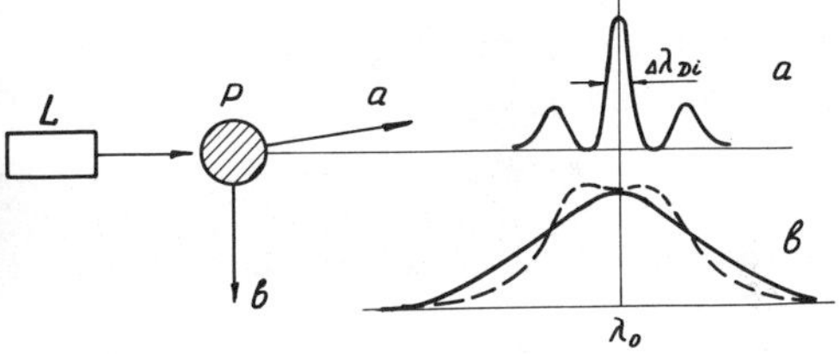

Figure 7
Laser light scattering technique:
a) scattering through small angles, ion Doppler width of central line;
b) scattering through a right angle: solid line $T_e \sim T_i$, dashed line $T_e \ll T_i$

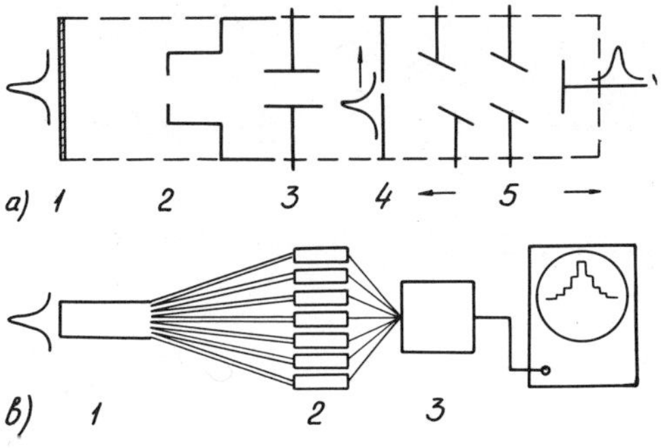

Figure 5
a) Image converter device combined with a pm - system for rapid recording of spectrum-line profiles; 1, cathode; 2, electrode; 3, deflecting plates; 4, narrow slit; 5, a photomultiplier system. b) Fiber optic technique for obtaining line profiles: 1, light pipes; 2, set of photomultipliers; 3, electonic commutator

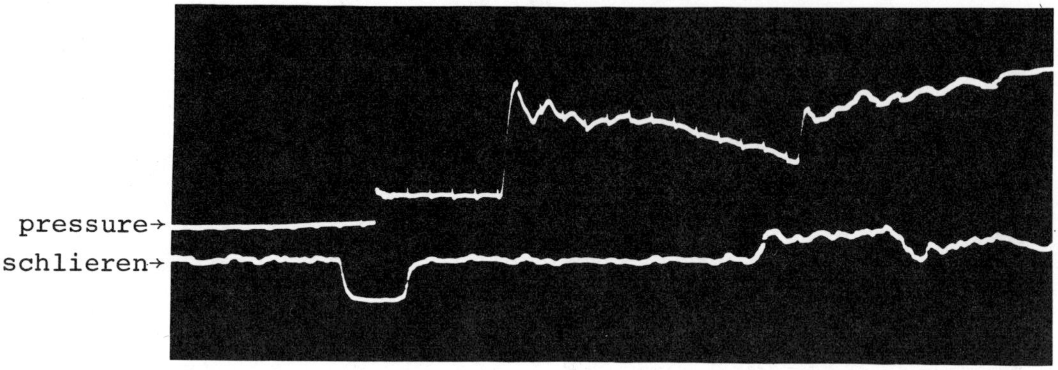

Figure 8
Quantitative schlieren oscillogram of the light refraction in incident and reflected shock waves in air. Pressures are recorded simultaneously

(a)

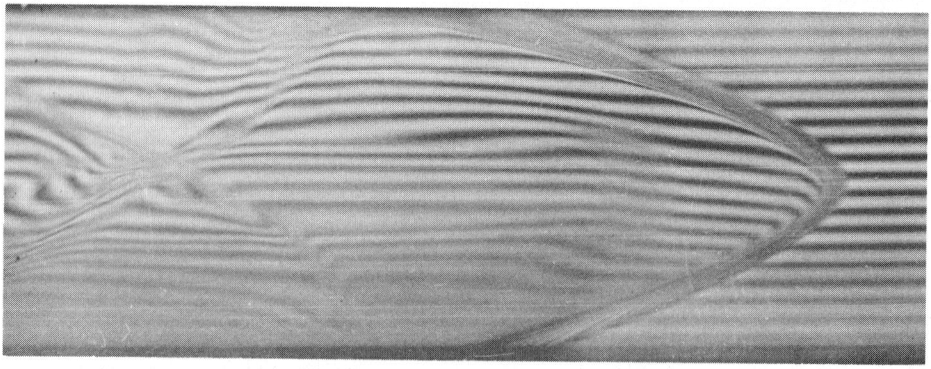

(b)

Figure 9
Streak interferograms of expanding detonations in $C_2H_2 + O_2$, $p_1 = 0.08$ atm:
a) zero-fringe regime,
b) ordinary multifringe regime

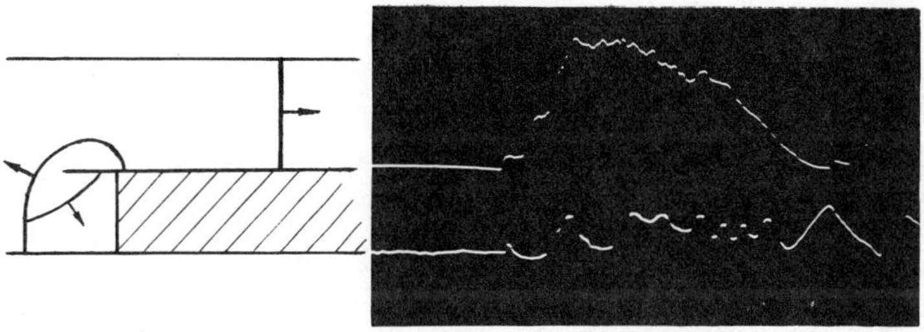

Figure 10
Pressure (upper trace) and density (lower trace) records of the behavior of the expanding gas sample in oxygen atom recombination experiments. A scheme for producing partial reflection of the shock-tube flow is also presented

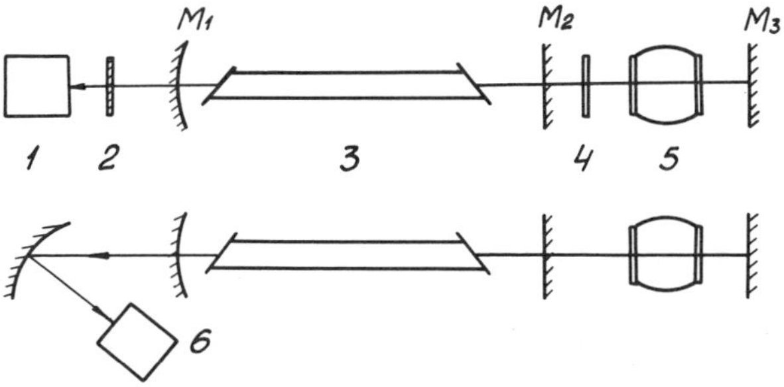

Figure 11
Schemes of the three-mirror laser-interferometers:
1, beam p.m.-detector; 2, red-light filter;
3, helium-neon laser tube; 4, infrared filter;
5, shock tube; 6, infrared beam detector

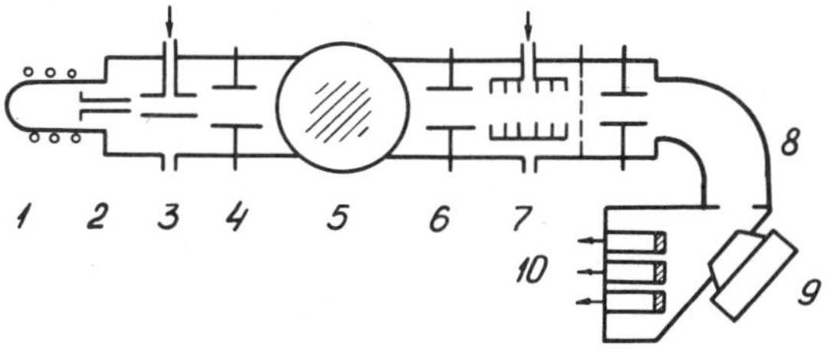

Figure 12
Neutral beam diagnostic technique:
1, ion source; 2, electrode; 3, charge transfer neutralization chamber;
4, 6, electrostatic cleaning plates;
5, gas sample; 7, "stripping-chamber";
8, 9, mass-spectrometer-type beam separation device; 10, ion detection system.

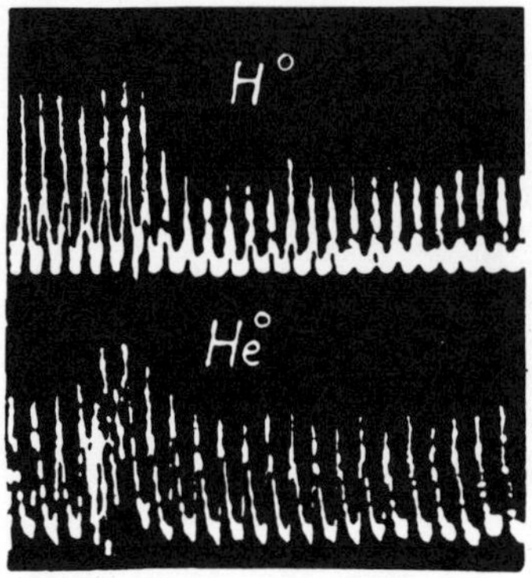

Figure 13
Oscillogram showing absorption of hydrogen and helium atomic beams in a discharge in hydrogen, $p_1 = 3\times10^{-3}$ torr, V = 30 kv, I = 8 ka; time marks 2 μ sec apart.

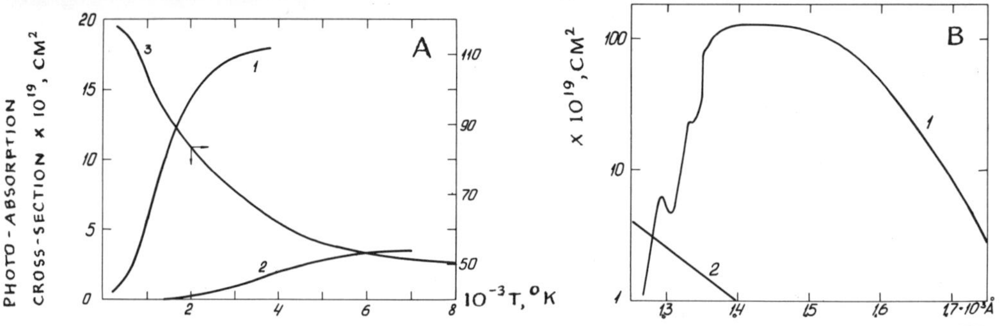

Figure 14
Photoabsorption cross-sections of molecular gases in the uv region A) Temperature dependence: 1, oxygen, λ = 2245 Å; 2, carbon dioxide, λ = 2380 Å; 3, oxygen, λ = 1470 Å.
B) Spectral distribution of absorption coefficients: 1, oxygen, atmospheric conditions; 2, nitrogen, T = 10,000°K.

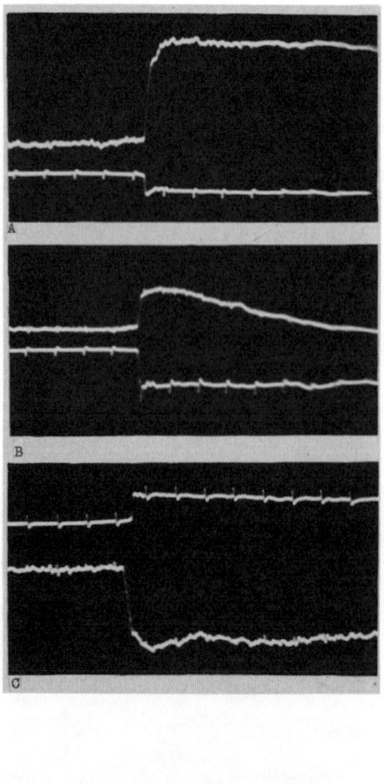

Figure 15
Infrared emission and absorption measurements, λ = 4.5
A) Emission, $0.1\ N_2O + 0.9\ Ar$, $T_s = 1150°K$.
B) Emission, $0.1\ N_2O + 0.9\ Ar$, $T_s = 1750°K$.
C) Absorption, pure N_2O, $T_s = 1100°K$.

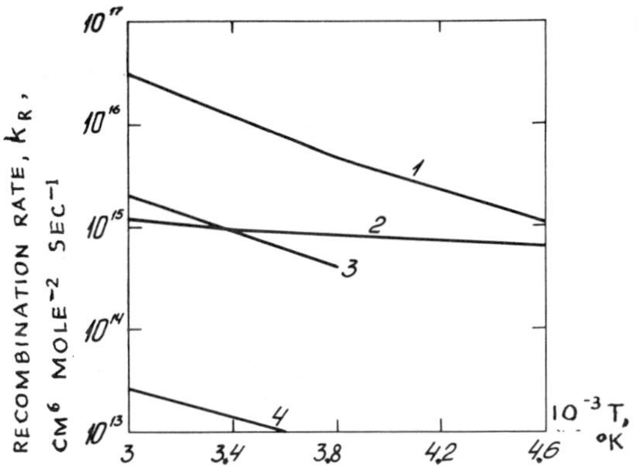

Figure 16
Summary of data for recombination rate constants in various atomic gas systems: 1, H + H + H; 2, H + H + H_2; 3, O + O + O_2; 4, CO + O + Ar

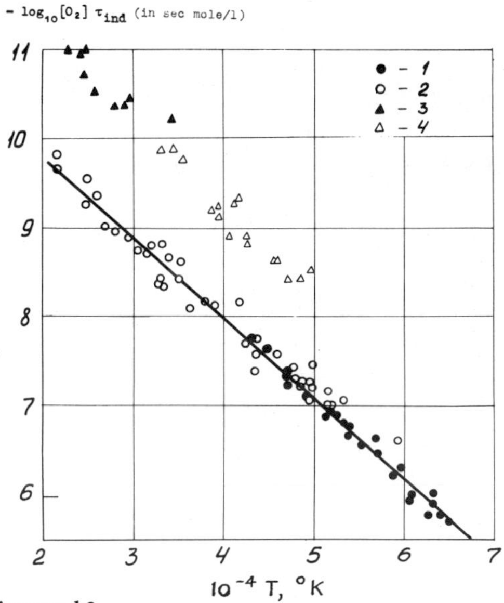

Figure 18
Summary of data on ignition delays in NH_3-O_2 systems: 1, measurements in reflected waves;[146] 2, our measurements, incident waves in a stiochiometric mixture dilutes with argon; 3, mixture $9NH_3 + O_2$; 4, mixture 4/9 NH_3 + 5/9 O_2 + 10 N_2

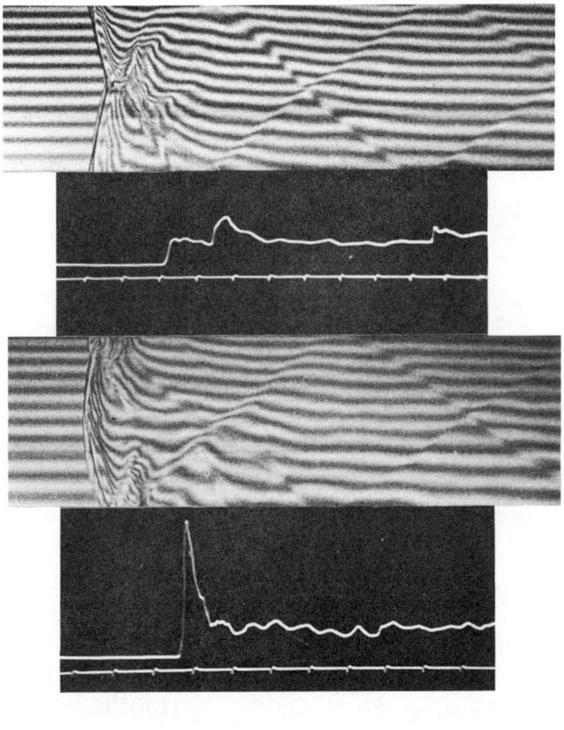

Figure 17
Interferograms and pressure profiles of shock tube detonations in 2 CO + O_2 at p_1 = 0. 38 atm

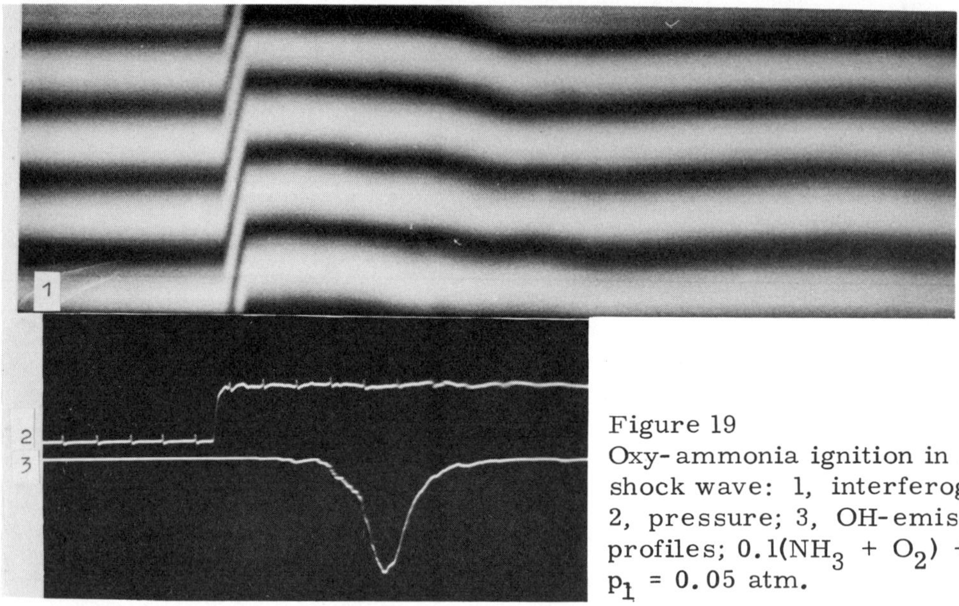

Figure 19
Oxy-ammonia ignition in a shock wave: 1, interferogram; 2, pressure; 3, OH-emission profiles; $0.1(NH_3 + O_2) + 0.9$ Ar, $p_1 = 0.05$ atm.

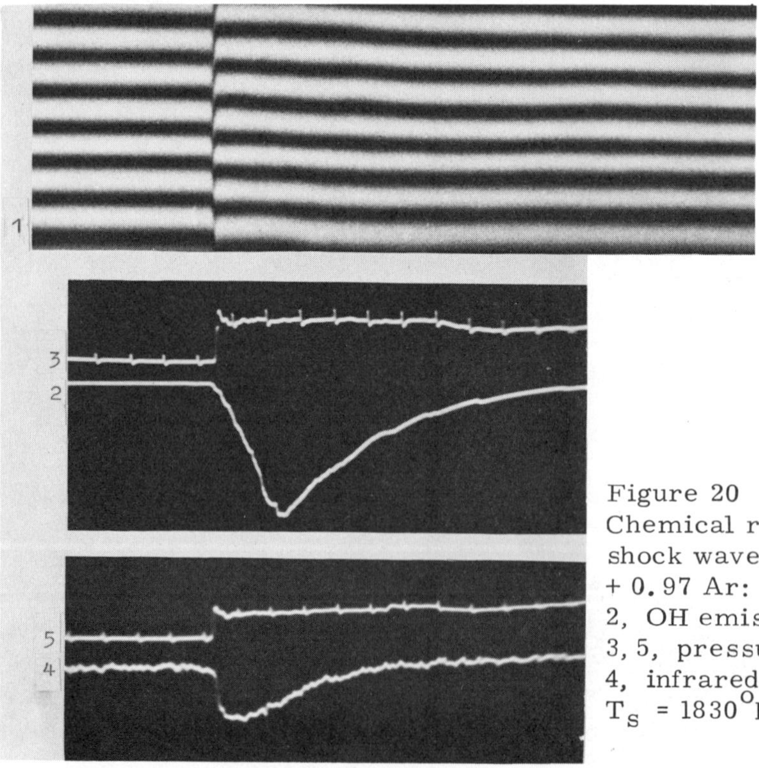

Figure 20
Chemical reaction zone behind a shock wave in $0.03 (N_2O+H_2) + 0.97$ Ar: 1, interferogram; 2, OH emission at 3075 Å; 3, 5, pressure records, 4, infrared emission at 4.5μ, $T_s = 1830\,^{\circ}K$, $p_1 = 0.05$ atm.

INSTRUMENTATION FOR MEASURING FLOW PROPERTIES IN AN EXPLOSIVELY DRIVEN SHOCK TUBE

WILLIAM J. BORUCKI, DAVID M. COOPER, and WILLIAM A. PAGE
National Aeronautics and Space Administration,
Moffett Field, California

SUMMARY

Gas flows in high-performance shock tubes, such as those with magnetic, electric, or explosive drivers, are often inhomogeneous and of very short duration (see Ref. 1). Consequently, the characterization of the flow in these facilities is difficult. This paper describes the instrumentation used to study the flow and radiative properties of the test gas produced by the Ames High-Explosive Shock Tube. This facility produces flows which simulate the stagnation regions of shock layers formed about manned entry vehicles returning from nearby planets. A detailed description of the facility and its capabilities are presented in a companion paper (Ref. 2).

Shock velocities between 14.7 and 16.9 km/sec at initial pressures of 0.2 to 0.8 torr have been produced in a glass shock tube 5 cm in diameter and 7.3 m in length. Temperatures and pressures behind the shock front range from 14,000 to 18,000° K and 0.8 to 3.5 atm, respectively, depending on the shock velocity and the initial loading pressure. All tests were conducted using dry air and 1 to 2 percent hydrogen as the test gas. The hydrogen was included in the test gas for diagnostic purposes.

INSTRUMENTATION AND DIAGNOSTIC TECHNIQUES

Figure 1 shows the arrangement of instrumentation around the shock tube. The instruments consist of a Kerr-cell-shuttered camera, a Kerr-cell-shuttered spectrograph, five narrow-band radiometers, and one broad-band radiometer. Six shock wave detectors, spaced along the shock-tube upstream of the test section, are used to measure the shock wave velocity. The spectrograph and the Kerr-cell camera view the same longitudinal region of the shock tube, along light paths separated by less than 10 degrees, while the radiometers are displaced some 30 cm downstream.

Kerr-Cell-Shuttered Camera

The Kerr-cell camera consists of a camera back with an f/1.8, 85-mm lens (stopped to f/3), and a large-aperture Nitrobenzene Kerr-cell shutter with HN22 polarizers and a 3-71 Kodak filter. A 50 μsec exposure is sufficient to provide a picture with 20,000 resolution bits of the shock-tube flow on Kodak Tri-X film. Since the Kerr-cell shutter has a small residual transmission when closed, an opaque shutter is required in order to exclude radiation from the bright piston material which comes into view some 30-50 μsec after the passage of the shock front. This opaque shutter (Fig. 2) is explosively driven, has a 2.5-cm-diameter aperture, a closing time of 35 μsec, and is located in front of the Kerr-cell shutter. Two explosive squibs are ignited by a capacitor discharge power supply. The shock produced by the detonation moves a 5-mil-thick piece of brass in hinge-like fashion across the aperture. The capping shutter and the physical dimensions of the Kerr-cell limit the field of view of the camera to a 20-cm-diameter area in the plane of the shock-tube centerline. A front surface mirror is placed beside the glass tube to allow the Kerr-cell camera to view the flow simultaneously in orthogonal planes.

The two views obtained by the Kerr-cell-shuttered camera permit the three-dimensional shape and structure of the luminous shocked gas to be determined. The shock-tube test time and the spatial distribution of radiation of the shocked gas are also deduced from these photographs.

Stigmatic Spectrograph

A specially developed Kerr-cell-shuttered f/1.5 stigmatic spectrograph which covers the spectral region from 3900 to 6700 Å on each exposure is diagrammed in Figure 3. A 100- by 125-mm plane grating and an f/1.5, 150-mm Aerojet Delft aerial reconnaissance lens give the instrument unusual light-gathering power. The grating has 600 lines/mm and is blazed for 4600 Å. AeroEktar lenses (300 mm, f/2.5) are used for the collimator and the light collector. Kovar spacers are used for temperature compensation of critical optical components in the spectrograph. The instrument plate factor is approximately 80 Å/mm while the best resolution of the instrument is 1.5 Å. The long dimension of the spectrograph slit is focused along the shock-tube centerline, where the slit image is 100-mm long and 1/4-mm wide. A 0.2 μsec exposure of the shock-tube flow produces spatially resolved spectra on Kodak 2475 film (ASA 1000). The theoretical spatial resolution varies along the shock-tube centerline from 4.2-mm to 5.7-mm. (The term spatial resolution as used in this paper is the minimum distance that two points A and B must be separated in order that light from point A does not fall on the image of point B and vice versa.) A special Benzonitrile Kerr-cell shutter with ultraviolet polarizers allows the instrument to view the near UV portion of the spectrum. To minimize the required aperture of the Kerr-cell shutter, the cell is of non-symmetrical design so that it will pass the entire optical beam from the light collector without vignetting. Again, because of the bright radiation from the piston material following the driven gas, the spectrograph is equipped with an exploding-wire capping shutter having a closing time of approximately 30 μsec. A 6-mm-wide flat surface mirror was placed on the optical axis of the spectrograph directly behind the shock tube in some of the tests to double the intensity of the light entering the spectrograph.

Radiometers

The absolute intensities of selected spectral features of the radiation as a function of time are measured by five calibrated narrow pass-band radiometers (Ref. 3) and one broad-band total radiation detector. The arrangement by which the radiometers view the shock-tube flow is depicted in Fig. 4. The radiometers consist of Bausch and Lomb 1/4-m monochromators which have been adjusted to 30 to 50 Å half-intensity bandpasses and use photomultiplier tube detectors. The broad-band total radiation detector is an S-20 phototube.

The glass shock tube terminates in a small stainless-steel test section containing splitter plates (Fig. 4). These plates are used to exclude from view the relatively thick, variable-temperature, boundary layer of the glass shock-tube wall. One-mm-wide slits in the splitter plates allow the radiometers to view only the center portion of the flow passing between the plates and the thin boundary layer formed on the splitter plates. Quartz windows are flush mounted on the surface of the splitter plates directly in front of the slits. The spatial resolution of the radiometer system is approximately 1-mm, whereas the optical rise time is 0.1 μsec. The path length between the splitter plates is 2 cm.

The narrow-band radiometers complement the other instruments in that they provide absolute intensity measurements for comparison with theoretical predictions. Their high-spatial resolution also allows a precise determination of the onset of impurity radiation in the center of the flow. The radiometers are always used in pairs with one measuring the spectral feature of interest and one measuring the background radiation nearby. In the present experimental configuration, one radiometer always measures the intensity of the continuum radiation at 3357 Å. Two others monitor spectral features expected to be of special interest in a particular test, such as sodium or copper lines.

Shock-Velocity Detectors

The velocity of the shock wave is measured by a series of six S-20 photomultiplier tubes receiving light beams from the luminous shock front through highly collimated apertures 1-mm in diameter. The output from each photomultiplier tube is amplified, twice differentiated, and then used to trigger a 100 mHz electronic chronograph. The detectors are spaced along a 2.4 m section of the shock tube just ahead of the test-section region. Since each records the local time variation of the radiation intensity, the attainment of steady-state-flow conditions can be verified for each test. A more detailed description of the shock velocity detectors may be found in Ref. 2.

RESULTS

Shock-Flow Structure

A sample Kerr-cell photograph, displaying orthogonal views of the flow, is shown in Fig. 5. The lower image is a direct view, whereas the upper is obtained with a mirror set at 45° adjacent to the shock tube. The heavily exposed rectangular strip in the direct (i.e., lower) view is the result of placing a narrow mirror directly beneath the shock tube.

The photograph shows a cloud-shaped dark area about 2.5 cm behind the planar shock front, which, in Fig. 5, appears slightly curved because it had passed the optical axis of the camera when the photograph was taken. The cloud appears turbulent, is highly structured, and is nonsymmetrical about the shock-tube centerline. The use of orthogonal views of the gas flow has shown that the location of this region varies both radially and axially from test to test.

Spectroscopic data (similar to that on Fig. 6) show that the material which appears as the dark cloud in the photograph is composed primarily of the initial test gas and a small amount of sodium. It will be shown later that the dark clouds are less luminous than the surrounding gas because the clouds are a portion of the downstream test gas that has been cooled by radiation and convection and was then swept forward toward the shock front. The sodium doublet at 5893 Å first appears in the spatially resolved spectra at the same location as the dark cloud in the Kerr-cell photograph. Since the sodium may: (1) come from the walls of the glass tube, (2) be present in the unshocked air, or (3) come from the vaporized piston, produced by the explosive driver (Ref. 2), a test was conducted to determine the origin of the sodium. (It should be noted that the absence of neutral sodium lines in the first part of the spectrum is not proof that sodium is not present. The shocked gas directly behind the shock front is so hot that it tends to suppress the neutral sodium population and hence to suppress the line radiation from neutral sodium.) A strip of copper foil was placed inside the glass shock tube directly within the explosive charge. Fig. 6 is a combination of the Kerr-cell photograph and the spectrum obtained in this test. Both are to the same longitudinal scale. The spectral lines radiated by the vaporized copper (in particular, the bright pair at 5153 and 5218 Å) are observed to appear at the same spatial location as that of the dark cloud and of the sodium lines. It may be further noted that at 2.7 cm behind the shock front, there is an abrupt discontinuity in the intensity of the continuum radiation and a narrowing of the hydrogen lines (meaning an abrupt reduction in the electron density and temperature) at the same location as the dark cloud in the Kerr-cell photograph. All of this evidence supports the conclusion that the cloudy area is a mixing region of air and vaporized piston material and that the length of the test gas, free of contamination and of mixing, is the region shown in the Kerr-cell photograph between the shock front and the foremost portion of the turbulent region. The test time is determined by dividing this length by the shock wave velocity.

The position of the mixing region is found to depend upon the initial loading pressure. Figure 7 is a composite of three Kerr-cell photographs of the flow in which the velocity was approximately 16 km/sec and the initial loading pressure varied from 0.2 to 0.8 torr. At 0.2 torr, the mixing region is so far forward that it has interacted with the shock front to produce a nonplanar shock surface. Consequently, the test time is zero at this loading pressure. Photographs of tests performed at initial loading pressures of 0.4 and 0.8 torr reveal undisturbed shock fronts and non-zero test times. The test times so derived are plotted against initial loading pressure in Fig. 8. The time corresponding to the first appearance of Na in the spectrographic and radiometric (which will be discussed later) data are also plotted on Fig. 8 for comparison. (The symbols for the three types of observations in Fig. 8 are separated in pressure for clarity. They were actually taken as parts of the same tests — i.e., at identical pressures.) Mirels' laminar flow theory (Ref. 4) is also shown on the figure. It is seen that actual test times are substantially shorter than predicted by that theory. The difference between theory and experiment may be due to turbulent mixing, leakage of the test gas through the piston particles, and radiative cooling. None of these effects is accounted for in Mirels' theory.

Spectral Features

The spectrum presented in Fig. 6 is typical of the spectra obtained with the f/1.5 spectrograph. (However, the abrupt longitudinal discontinuity in intensity observed is this spectrum is not always present.) The main features are strong atomic line radiation from nitrogen and oxygen, hydrogen alpha and beta line radiation, and the $3p^2P$ sodium doublet radiation. The width of the spectral features is due primarily to Stark broadening resulting from high-electron concentrations. The bright band of light throughout the spectrum is produced by continuum radiation from free-free and free-bound transitions of electrons in the fields of ions and atoms.

The presence of line multiplets from different energy levels and a strong continuum allow the evaluation of line intensity and line-to-continuum ratios which may be used to determine the spatial variation of the electron density and temperature. These thermodynamic properties may also be obtained from the line shapes of neutral nitrogen and oxygen. Since the theoretical shape of the hydrogen alpha line is well known and the experimental shape may be accurately measured, the hydrogen alpha line was chosen to measure the electron density.

Electron Density

The electron density of the shocked gas as a function of distance behind the shock front is determined as follows: The theoretical H_α profile reported by Kepple (Ref. 5) is first broadened by the instrument function and then fitted to the experimental shape by iterating on the electron density until a best fit between the experimental and theoretical curves is obtained. Since the static pressure is nearly constant behind the shock front, a change in electron density signifies a change in the thermodynamic state of the gas, including temperature, number densities of neutral hydrogen, nitrogen, and oxygen, and hence a change in the continuum radiation background. These changes plus changes in the optical depth of the hydrogen alpha line are carefully taken into account for each iteration. The computer program used for this task was developed by Wood and Wilson and is described in Ref. 6. With the pressure behind the shock front computed from the initial loading pressure and the shock velocity, two thermodynamic properties (electron density and pressure) are available to compute the equilibrium state of the gas as a function of distance behind the shock front.

Figure 9 is a plot of the electron density as a function of distance behind the shock for one experimental test. The electron density obtained from the shape of the hydrogen alpha line is represented by the solid curve. The dashed curve will be explained later. The electron density corresponding to Rankine Hugoniot equilibrium conditions is also plotted on Fig. 9. This value was obtained from the graphs of Refs. 7 and 8 by using the initial test section pressure and shock velocity. It is evident from this figure that there is good agreement between experiment and theory immediately behind the shock. Thereafter, the electron density drops rapidly to lower values. With the measured variation of the electron density and the computed pressure behind the shock, the temperature variation may be obtained as shown in Fig. 10. The rapid decay of the temperature from the Rankine-Hugoniot value is attributed to radiative cooling of the test gas. The turbulent region also appears to be cooled by cold downstream gas that has been brought forward by the mixing process. (See Fig. 7.) The relative importance of the two cooling mechanisms in this region is not quantitatively understood at this time. The increase in temperature at a distance of 5 cm behind the shock appears to correlate with a bright area observed at the same spatial location in the Kerr-cell photograph.

An error analysis has been made which indicates the electron density has been measured to within 10 percent. At a temperature of 15,000° K and a pressure of 3 atm, a 10 percent error in the electron density is associated with only a 3 percent error in temperature.

Radiometric Data

The absolute intensity of the spectral radiation as a function of time after shock wave passage obtained from two of the radiometers (with 40 Å half-intensity band passes) is shown in Fig. 11. The continuum radiation is measured by a radiometer centered at 5820 Å. This radiation, represented by the dashed curve, rises abruptly to a peak as the shock passes and then decreases. The second radiometer was centered at 5890 Å and measures the $3p^2P$ sodium doublet as well as the underlying continuum. The radiation measured by both radiometers is very similar in detail for the first 1.6 μsec at which time the radiation from the radiometer observing both sodium and continuum rises rapidly, indicating the presence of strong sodium line radiation. The spectrographic data for this test also recorded the sodium doublet as first observable 1.6 μsec behind the shock front.

Also included on the figure is a theoretical prediction of the equilibrium continuum radiation for an adiabatic shock layer by Biberman and Norman (Ref. 9), which agrees well with the measured peak intensity.

By using Kramer's theory for continuum radiation, as modified by Biberman and Norman, (i.e., the intensity of the radiation is proportional to the square of the electron density and inversely proportional to the square root of the temperature), electron densities were deduced from the continuum radiation measurements. Inspection of Fig. 9 shows that these electron densities are in good agreement with those obtained spectrographically.

The rate of cooling of the plasma behind the shock front has been measured both in the center of the flow using splitter plates and across the entire diameter of the shock tube using the shock velocity detector system. Although the results of the two systems are not always in agreement, both systems indicate that the cooling rate varies from test to test even though the initial loading pressures and shock velocities are nearly identical. The data on Figs. 8 and 9 are taken from a test where there was good agreement between the cooling rates from both systems.

Flow Between the Splitter Plates

Since the shock detectors along the tube invariably produced identical records for a single test, it was decided to investigate the flow between the splitter plates. Fig. 12 is a Kerr-cell photograph of the flow between the splitter plates and a diagram of the splitter-plate assembly. The splitter plates were attached to steel struts and moved forward in the shock tube to allow them to be photographed. The illumination of the glass tube to the left of the shock front is a result of light leakage through the Kerr-cell shutter before the capping shutter closed. This illumination is a result of the enhanced radiation intensity from the gas which has passed over the splitter plates. The irregular dark lines in the photograph are due to cracking of the glass shock tube after the passage of the shock front.

With careful examination of the Kerr-cell photograph, faint lines originating at the splitter plate edges can be seen. The lines between the splitter plates may be Mach lines. It is apparent that no flow irregularities such as choking or flow over the sides of the splitter plates are observed. The inconsistency of the cooling rates is presently undergoing further study.

CONCLUSIONS

Four types of instrumentation have been used to obtain detailed information on radiatively cooled flows produced in a high performance shock tube. In particular, spectrograph data have yielded the variation of purity and electron temperature along the flow direction; radiometric data have given the absolute levels of radiation; the orthogonal Kerr-cell photographs have shown the presence of turbulent structure in the flow close behind the shock front; and the shock velocity detector system has verified that steady-state flow has been obtained in each test.

REFERENCES

1. Griem, H. R.: Plasma Spectroscopy, McGraw-Hill, San Francisco, 1964.
2. Berggren, Robert E.; Compton, Dale L.; Canning, Thomas N.; and Page, William A.: Ames' High-Explosive Shock-Tube Facility. Seventh International Shock Tube Symposium, Toronto, Canada, June 23-25, 1969.
3. Arnold, J. O.: A Shock Tube Determination of the Electronic Transition Moment of the C_2 (Swan) Bands. J. of Quant. Spectrosc. and Rad. Transfer, vol. 8, 1968, pp. 1781-1794.
4. Mirels, Harold: Test Time in Low Pressure Shock Tubes. Physics of Fluids, vol. 6, No. 9, Sept. 1963, pp. 1201-1214.
5. Kepple, Paul C.: Improved Stark Profile Calculations for the First Four Members of the Hydrogen Lyman and Balmer Series. Univ. of Maryland Tech. Rep. 831, 1968.
6. Wood, Allen D.; Wilson, Kennetch H.: Radiant Energy Transfer Measurements in Air. Final Rep. NASA/Lockheed Contract NAS7-579, Feb. 1969.
7. Zeimer, Richard W.: Extended Hypervelocity Gas Dynamic Charts for Equilibrium Air. Space Tech. Labs. Inc., Tech. Rep. 60-0000-09093, April 1960.
8. Staff of Chance Vought Research Center: The Thermodynamic Properties of High Temperature Air. Rep. No. RE-IR-14, June 1961.
9. Biberman, L. M.; and Norman, G. E.: Emission of Recombination Radiation and Bremsstrahlung from a Plasma. J. of Quant. Spectrosc. and Rad. Transfer, vol. 3, 1963, pp. 221-245.

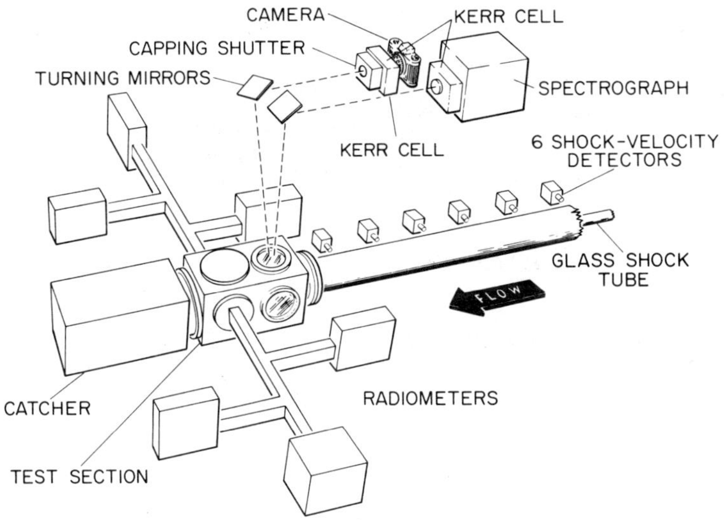

Figure 1
Arrangement of instrumentation about shock tube.

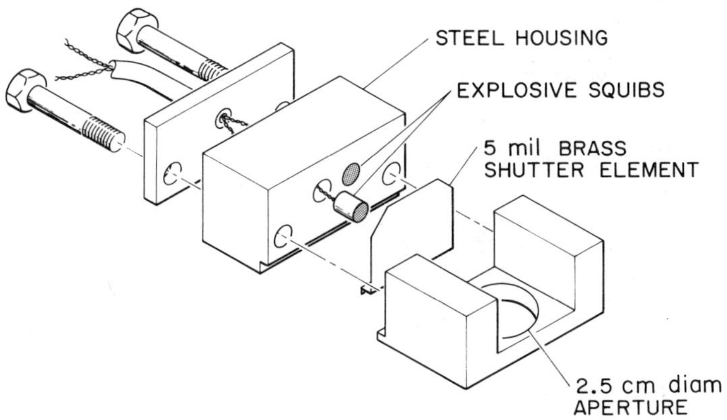

Figure 2
Sketch of explosively driven capping shutter.

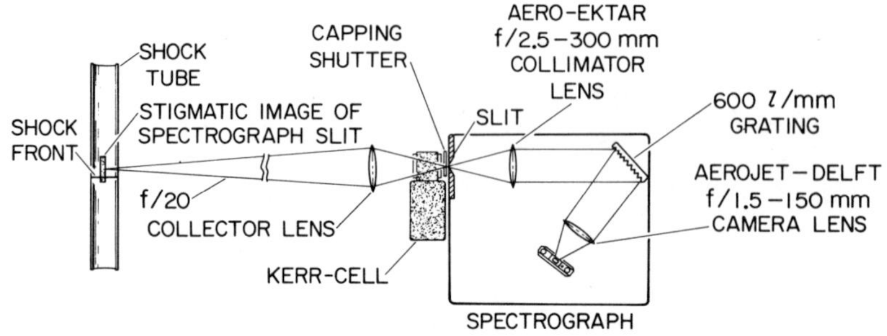

Figure 3
Layout of Kerr-cell-shuttered f/1.5 stigmatic spectrograph.

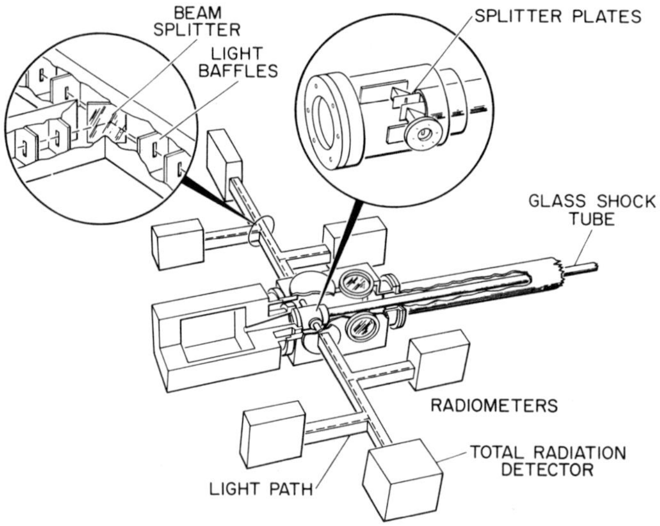

Figure 4
Narrow-band radiometer arrangement for viewing shock wave.

Figure 5
Kerr-cell-shuttered photograph displaying orthogonal views of shock structure; $V_\infty = 15$ km/sec, Exposure time = 0.05 μsec, Test No. 21, $P_1 = 0.8$ torr.

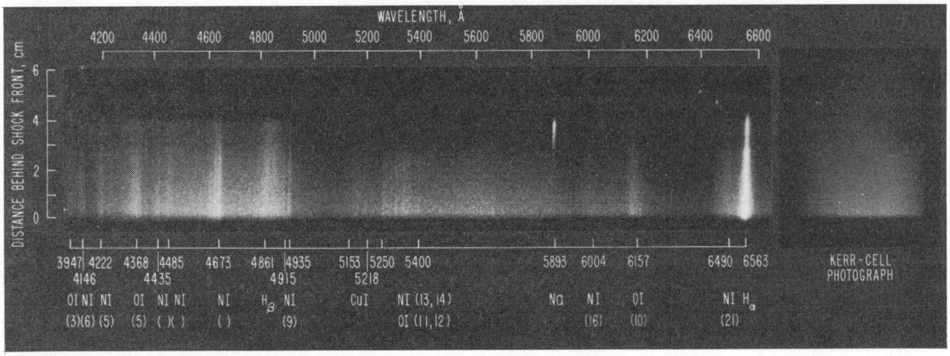

Figure 6
Kerr-cell photograph and spatially resolved spectrum of the shock-tube flow; $V_\infty = 15$ km/sec, $P_1 = 0.8$ torr, Test No. 21.

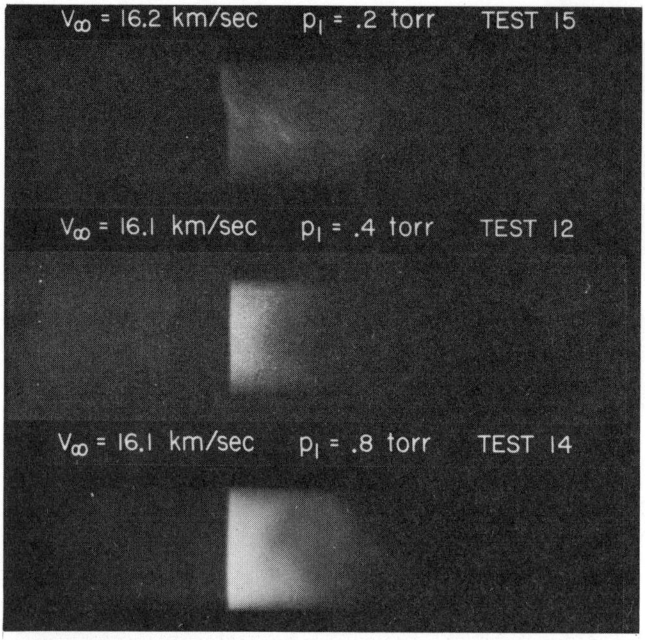

Figure 7
Kerr-cell-shuttered photographs.

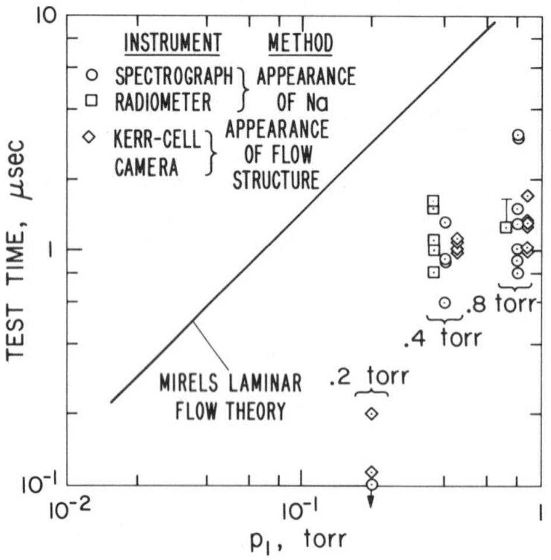

Figure 8
Test time results; V_∞ = 14.5 to 16.4 km/sec.

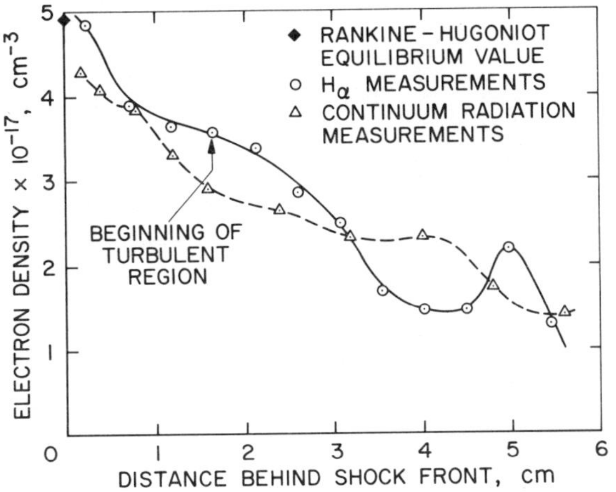

Figure 9
Variation of electron density behind shock front;
V_∞ = 16 km/sec, P_1 = 0.8 torr, Test No. 14.

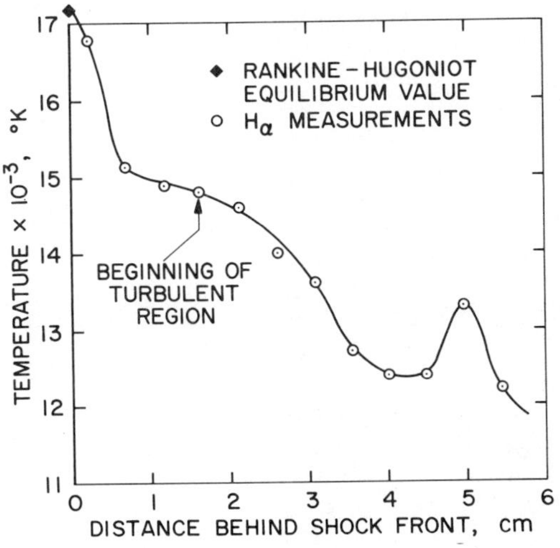

Figure 10.
Variation of temperature behind shock front.
Test No. 14.

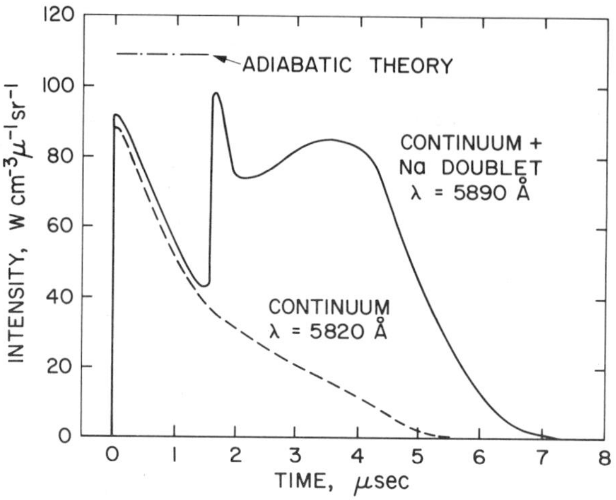

Figure 11
Radiometer measurements behind shock front;
V_∞ = 16 km/sec, P_1 = 0.4 torr, Test No. 12.

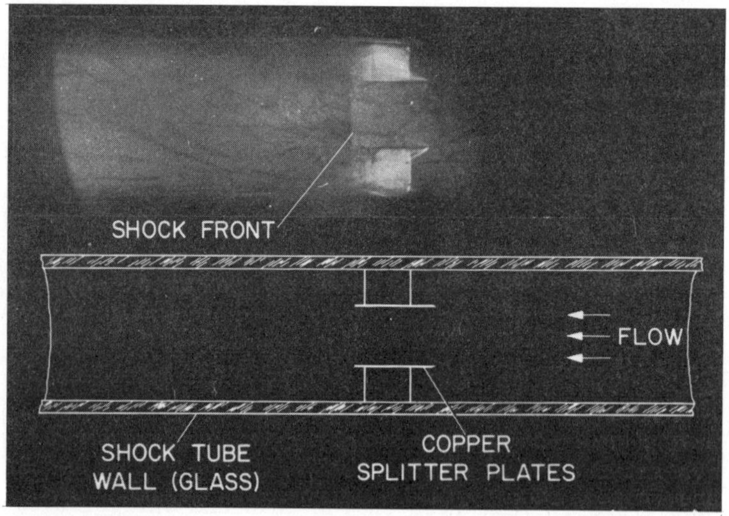

Figure 12
Diagram of splitter plate assembly and Kerr-cell photograph of flow between splitter plates,
Test No. 23.

INTERFEROMETRIC STUDY OF THE END-WALL THERMAL LAYER IN IONIZING ARGON

R. A. KUIPER*
Stanford University
Stanford, California

SUMMARY

Experimental electron concentration and mass density profiles through an ionized argon thermal layer are presented. The thermal layer is formed on the end wall of a combustion driven shock tube following the reflection of a strong shock wave. The profiles were obtained 100 μsec after reflection from two-wavelength snapshot interferograms with closely spaced fringes parallel to the end wall, and reference thermodynamic properties at the edge of the boundary layer were determined from time-resolved, two-wavelength interferograms. The reflected-shock temperatures, pressures, and electron concentrations were respectively on the order of 10^4 °K, 1 atm, and $10^{17}cm^{-3}$. Calibration experiments for nonionized thermal layers indicated that the neutral thermal conductivity was proportional to the 3/4 power of the local temperature. Measured electron concentrations in the thermal layer were consistently higher than predicted by present theories. This discrepancy is primarily attributed to failure of the theories to include nonequilibrium effects.

I. INTRODUCTION

Theoretical calculations of the transport properties of ionized gases have long been in need of experimental verification. The present experiments were intended to provide such a verification for the thermal conductivity of ionized argon by comparing theoretical and experimental boundary-layer profiles. These spatially resolved distributions of the thermodynamic properties were obtained by application of two-wavelength interferometry. The thermal layer considered is the thin transition region that grows in thickness on the cold end wall of a shock tube following the reflection of a strong, ionizing shock wave. Because the high temperature plasma between the end wall and reflected shock is essentially stagnant, there are no fluid shear stresses effecting the

---

*Presently a Member of the Technical Staff, The Aerospace Corporation, El Segundo, California

boundary-layer development. The experimental profiles could thus be compared with theoretical profiles from a less complex theory than that for a flat plate parallel to an ionized stream.

Fay and Kemp[1] and Knöös[2] provide the boundary-layer theory with which the present experiments are compared, although Fay and Kemp were concerned primarily with the heat transfer rates to the end wall and Knöös with the profiles for ionized side-wall boundary layers. Knöös' analysis is briefly outlined below with an emphasis on the theory's primary assumptions. The experimental results indicate that these assumptions in the theory are oversimplifying, and that a more complex analysis is required.

Previous experiments[3,4,5] have indicated that the reflection process in vibrationally and chemically reacting gases, as well as ionizing gases is complicated by nonequilibrium wave interactions. This fact, and the requirement to know the properties at the edge of the thermal layer, prompted an initial study of the shock reflection process. The results of this interferometric study, which are briefly described below, include time histories of the reflected shock properties that are used in the boundary layer data reduction.

Earlier end-wall measurements, such as those of Camac and Feinberg,[3] were concerned with heat transfer rates at the wall surface for both ionized and nonionized gases. The results from such experiments for nonionized argon are in agreement with the generally accepted results of Amdur and Mason.[6] From molecular beam experiments and kinetic theory, Amdur and Mason conclude that neutral argon thermal conductivity is proportional to the local temperature raised to the 3/4 power. Prior to performing the present experiments for ionized thermal-layers, the same experimental arrangement was used in a regime where ionization could be considered negligible. The results of these "calibration" experiments were in excellent agreement with the above temperature dependence. This agreement tends to support the experimental profiles obtained for ionized argon.

II. END-WALL BOUNDARY-LAYER THEORY

Several theoretical calculations have been developed for the heat transfer in a monatomic gas including ionization effects.[1,2,7,8] The theoretical calculations outlined here and compared below with the experimental results are those of Knöös,[2] although they are essentially similar to the equilibrium calculations of Fay and Kemp.[1] Relative to the present considerations, the most important assumptions in the analysis are:

1. Immediately following shock reflection ($t = 0$), a uniform equilibrium argon plasma is placed in contact with the room-temperature end wall.
2. Thermodynamic and chemical equilibrium is assumed in the boundary layer.
3. Radiation effects in both the free stream and boundary layer are neglected.

4. The wall is at a floating potential, i.e., no net current to the wall.

5. It is assumed that the calculated transport properties for partially ionized argon are correct. As noted above, the original intent of this study was to examine this last assumption.

With C_i, h_i, and $\bar{V}_i$ equaling the mass concentration, enthalpy, and diffusion velocities of species i, respectively, the conservation equation are

$$\frac{\partial \rho}{\partial t} + \frac{\partial (\rho v)}{\partial y} = 0$$

$$\frac{\partial h}{\partial y} + v \frac{\partial h}{\partial y} = \frac{1}{\rho} \frac{\partial}{\partial y} \left[K \frac{\partial T}{\partial y} + \sum_i \rho C_i h_i \bar{V}_i \right]$$

where ρ, v, y, t, and T are the density, velocity, distance normal to the wall, time after reflection, and local temperature. The right-hand side of the energy equation can be written in terms of a total thermal conductivity so that the heat flux is given by

$$q = -K_t \frac{\partial T}{\partial y}$$

where

$$K_t = K + \frac{\alpha(1 - \alpha)}{2} D_{amb} \frac{p}{T} \left[\frac{5}{2} + \frac{I}{kT} \right]$$

and α, D_{amb}, p, and I are the degree of ionization, ambipolar diffusion coefficient, pressure, and ionization potential. The unsubscripted K is the translational thermal conductivity of the mixture of electrons, ions, and neutrals and is obtained from the individual components through a mixture rule.(1) The above equations can be written in terms of the similarity variable defined by

$$\eta = \frac{1}{2} \left(\frac{\rho_\infty \quad C_{p\infty}}{t \; K_{t_\infty}} \right)^{\frac{1}{2}} \int \frac{\rho}{\rho_\infty} \, dy$$

where C_p is the ratio of specific heats and the subscript ∞ refers to the free stream conditions. The transformed equations become

$$2\eta \frac{dH}{d\eta} + \frac{d}{d\eta}\left(\frac{\rho K_t C_{p\infty}}{\rho_\infty K_{t_\infty} C_p} \frac{dH}{d\eta}\right) = 0$$

where

$$H = \frac{h - h_w}{h_\infty - h_w}$$

and the boundary conditions are

$$H(0,t) = 0, \quad H(\infty,t) = 1; \quad t > 0$$

It should be noted that the above transformed equation is also applicable to the unionized boundary layer with $C_p = C_{p\infty}$. The additional assumption of charge neutrality and ambipolar diffusion in the boundary layer appears to be well justified by the small Debye length in the boundary layer, i.e., 10^{-5}mm. Deviations from the assumptions already listed above, however, are believed to be primarily responsible for the discrepancies between the theoretical and experimental profiles presented below.

III. EXPERIMENTAL TECHNIQUES

Two-wavelength, optical interferometry applied with a Mach-Zehnder interferometer is a well established diagnostic technique for studying ionized monatomic gases.(9) When no disturbance is present in the test section beam of the adjusted interferometer, a uniformly spaced fringe pattern results. A refractivity disturbance, $\Delta(\mu - 1)$, from some reference condition in the test section will result in a change in the order of interference at the point x,y of the image plane. Neglecting optical imperfections and refractivity gradients allows one to express this change in the order of interference, or "fringe shift" as

$$S(x,y) = L\Delta(\mu - 1)/\lambda$$

where L is the test section width and λ is the wavelength passed by the optical interference filters. The application of this equation to plasma diagnostics arises from the highly dispersive nature of the electron refractivity, which to a good approximation, depends on the wavelength squared, as expressed in the following equation:

$$(\mu - 1) = A(\lambda)[\rho - B(\lambda^2)n_e]$$

where ρ is the local mass density, n_e is the electron concentration, and $A(\lambda)$ and $B(\lambda^2)$ are constant for a given wavelength. If one can measure the fringe shifts, $S(x,y)$, for each of two wavelengths relative to uniform reference conditions, then there result two algebraic equations for the unknowns ρ and n_e.

The effect of refractivity gradients and the presence of a side-wall boundary layer on the end wall studies were considered in an error analysis and found to be negligible beyond 0.4 mm from the end wall. At 0.3 mm from the wall, the fringe shift error rises rapidly to 0.03 of a fringe. These errors are included in the estimated uncertainties shown in the boundary layer profiles at 0.3 mm from the wall.

Examples of two types of snapshot interferograms for the end-wall thermal layer are shown in Fig. 1. The field of view is split in half by optical filters to obtain simultaneous fringe shifts at two wavelengths. While the interferogram on the left best illustrates the presence of the thermal layer, the one on the right is best suited for quantitative spatial resolution of the thermodynamic properties. It may be noted that the boundary layer thickness is on the order of 1 to 2 mm and that the fringe shift upward indicates an increase in refractivity. The fringe shift at each of the two wavelengths (4530 and 5720 A) are obtained from magnified snapshot interferograms with narrow fringes, approximately five per mm, adjacent and parallel to the end wall. The test section light beam is oriented parallel to the end wall, and the interferometer is adjusted so that the fringes move away from the wall for an increase in refractivity. The fringe shifts in the boundary layer are determined by the fringe number displacement from a least squares straight line fit through the points outside the layer on a fringe number versus distance plane. In practice, this is accomplished by measuring with a traveling microscope the distances of the numbered maxima and minima of fringe intensities from cross-wires mounted in the object plane at a known distance from the wall face. Mounting these crosswires in the end wall and knowing the magnification removes the uncertainty in end-wall location resulting from single edge diffraction patterns. Uncertainty estimates indicated that the end wall could be located relative to the fringe profile to within 0.016 mm (unmagnified distance). Figure 2 shows a plot of the measured fringe shifts obtained from the snapshot interferogram on the right in Fig. 1.

In the unionized boundary layer experiments, 2 μsec snapshots were taken 300 μsec after reflection, while for the ionized experiments they were taken 100 μsec after reflection. The time resolution is obtained by shuttering the exploding wire light source with a Kerr cell. The reasons for selecting these times will be discussed in the next section along with the method used to determine the free stream properties at the edge of the boundary layer. As implied above, these properties are necessary for reference conditions in order to derive the electron concentration and mass density profiles from fringe shift profiles such as those of Fig. 2.

IV. SHOCK REFLECTION PROCESS

A preliminary study of the shock reflection process was made in order to measure the reflected shock properties and to determine the effect of possible wave interactions on the boundary-layer studies. In this study, time resolved interferograms were obtained at two wavelengths with a rotating mirror camera and the Mach-Zehnder interferometer. With the camera slit normal to the end wall and the fringes, a distance versus time (x-t) diagram is produced by the slit image sweeping across the film. The fringes and test section are focused on the camera slit so that, when there is no disturbance in the test section, the swept image produces a series of parallel fringes with the end wall forming the time axis. Fringe shifts across the incident and reflected shock and the interacting wave phenomena make their trajectories visible in the interferograms.

Figure 3 shows two streak interferograms of the shock reflection process in ionizing argon. The interferogram on the right is a so called infinite fringe or zero order interferogram wherein a single fringe fills the field of view, so that subsequently the areas of constant exposure represent contours of constant refractivity. As can be seen fromthe regular interferogram on the left, the fringe shift across the incident shock, propagating to the right in the lower portion of the photo, is approximately one half fringe. The exposure in the infinite fringe photo thus shifts from light to dark. The various shock and nonequilibrium wave interactions that are visible in the streak interferograms can best be discussed with the aid of the x-t diagram in Fig. 4.

Following passage of the incident shock, traveling at U_1, the shocked gas attains a "frozen" state 2F and moves at a velocity U_{2F} toward the end wall, as indicated by the sloping particle paths. Following an incubation period τ_{2L}, (L = laboratory time), the gas attains an ionized equilibrium state 2E. The sudden onset of ionization is very evident in the infinite-fringe streak interferogram in Fig. 3 where the exposure changes from dark to light. This suddenness is characteristic of the monatomic gases such as argon where approximately 80 percent of the electron production takes place in the last 20 percent of the time τ_{2L}. This ionization "front" propagates at the same speed as the incident shock and the conditions immediately following its passage can be calculated assuming Saha equilibrium behind a normal shock propagating into a quiescent gas.(10)

The reflected shock initially travels away from the wall at the frozen speed U_{5F}. This speed and the state of the gas adjacent to the wall immediately following reflection can be calculated assuming a perfect gas, $\gamma = 5/3$. Because the pressure and temperature are considerably higher behind the reflected shock, the time required to attain equilibrium τ_5, is much shorter than τ_{2L}. The reflected ionization front is also quite sharp in character and propagates away from the wall at a speed I_5. The effect of the negative contribution to the refractivity from the electrons is quite striking in the streak interferograms where it can be seen that the fringes shift toward the wall, as one would expect for a sharp decrease in the refractivity. Because of the density increase across the reflected ionization front, an expansion wave front propagates away from the wall at the acoustic speed into the quiescent gas in region 5F. The reflected shock can be seen to decelerate upon being overtaken by the expansion wave front.

The observed interaction between the reflected shock and the incident ionization front is similar to the head-on collision of two normal shocks. The ionization front is transmitted as a compression wave, which reflects from the end wall and eventually overtakes and coalesces with the transmitted reflected shock. A contact surface forms in the interaction region because the gas on either side of it is processed by different entropy producing shock waves. The bifurcation or formation of a "lambda foot" on the reflected shock usually occurs in more complex gases than argon. In the present case, it can be shown to result from the onset of ionization behind the incident shock and is directly caused by the inability of the boundary layer fluid to negotiate the pressure jump across the reflected normal shock.

Experimental limitations precluded determiningthe effect of the interaction wave on the end-wall boundary-layer development. To avoid this complication, the boundary-layer experiments were performed for incident-shock strengths and times after reflection such that either the interaction wave had not arrived at the wall before the boundary-layer interferograms were taken or its influence was very weak. The time after reflection was arbitrarily selected as 100 μsec and the shock strengths were selected such that τ_{2L} was approximately equal to 100 μsec. This provided thermal layers that were thick enough to be spatially resolved and still have enough electrons present in the free stream to be of interest. An example of a streak interferogram at these weaker conditions is shown in Fig. 5.

The properties at the edge of the boundary layer were determined by plotting fringe shifts as a function of time and incident-shock Mach number from interferograms at two different wavelengths. Figure 6 compares the measured electron concentrations in the region 5E with the usually accepted theory for frozen conditions 2F behind the reflect shock. The theoretical values are too low because the analysis assumes that τ_5 is zero and that the reflected shock speed is initially equal to U_{F5E} (see Figs. 4 and 5). An example of the resulting time histories of the measured densities and electron concentrations are shown in Fig. 7 along with the pressure and temperature histories obtained by assuming Saha equilibrium. Radiation cooling appears to be a significant factor. The streak interferograms also showed that τ_5 was much less than 100 μsec for the thermal-layer test conditions. The time histories thus determined provided the reference conditions that allowed the interpretation of the measured boundary-layer fringe shifts.

V. EXPERIMENTAL RESULTS AND DISCUSSION

As noted above, experiments were first performed for conditions in which negligible ionization took place behind the reflected shock. In these cases, the boundary-layer interferograms were taken at 300 μsec after reflection. Streak interferograms for these conditions showed that the reflected-shock conditions were given correctly by perfect gas calculations and that these conditions prevailed for at least 300 μsec after reflection. These statements appear to be applicable for $M_s < 7$ and $P_1 < 10$ torr.

The neutral argon thermal conductivity was assumed to be given by $K_A = K_\omega (T/T_\omega)^\nu$, where ω refers to the wall conditions and $K_\omega = 4.25 \times 10^{-5}$ cal/cm-sec-°K, $T_\omega = 297$°K. For a perfect gas, then C_p = constant and the boundary layer equation could be written for the nondimensional temperature profiles with the exponent ν as a parameter. The rms and mean deviation of the theoretical from the experimental profiles were then determined for different values of the exponent. Typical results for such procedures are shown in Fig. 8 along with the experimental and theoretical profiles for the indicated conditions. Because there is no ionization present, only one wavelength is necessary to reduce the fringe data, and a series of four runs gave eight values of ν. The mean value was 0.75 with a 0.01 rms deviation from the mean. As noted above, this is in agreement with the most widely accepted value for argon and provided substantial justification for extending the experimental technique to ionized boundary layers.

The ionized boundary-layer profiles are obtained from simultaneous fringe shifts at two wavelengths. It may be noted in Fig. 2 that the fringe shifts at 4530 and 5720 A are close to being equal, unlike what would be expected for an unionized layer. This simply illustrates the dispersive influence of the electrons on the refractivity. The density and electron distributions in this case are obtained from

$$\frac{\rho}{\rho_\infty} = 1 + \frac{\lambda_1 A_2 B_2 S_1 - \lambda_2 A_1 B_1 S_2}{L(B_2 - B_1)\ A_1 A_2 \rho_\infty}$$

$$\frac{n_e}{n_{e\infty}} = 1 - \frac{\lambda_2 A_1 S_2 - \lambda_1 A_2 S_1}{L(B_2 - B_1)\ A_1 A_2 n_{e\infty}}$$

where the subscripts 1 and 2 correspond to evaluation at the two different wavelengths.

The results from three of the ionized boundary-layer experiments are shown in Figs. 9-a, b, and c along with the theoretical curves obtained from the analysis of Ref. 2 outlined above. The degree of ionization plotted in the figures is obtained from $\alpha = n_e m_A/\rho$ where m_A is the mass of an argon atom. Experimental uncertainties in the results were calculated at 0.3 mm from the wall and are indicated by the horizontal and vertical bars. These uncertainties are based on: (1) an overall fringe shift uncertainty at each wavelength of 0.1; (2) nominal uncertainties in the free stream values of 7, 11, and 30 percent for initial pressures of 10, 5, and 2 torr respectively; (3) a wall position uncertainty of 0.015 mm; (4) and a time uncertainty of 4 μsec. It is readily apparent that the experimental uncertainty does not make up the discrepancy between theory and experiment, and one must look to the theoretical treatment for the source of the disagreement.

The theoretical curves are based on the assumption that the free stream properties prevailed from time t = 0. A rough estimate of the effect of varying free stream conditions in the theory was obtained by performing the calculations with the conditions at the onset of ionization behind the reflected shock as inputs to the analysis, i.e., the conditions at 5E (Fig. 5) instead of those at t = 100 μsec. Although this is an artificial approach, the resulting negligible difference was interpreted as implying that varying freestream conditions are not a primary source of the observed discrepancy.

The undetermined influence of the initial boundary layer development under the frozen conditions 5F is a similar problem. The facts that τ_5 is much less than 100 μsec and that Camac and Feinberg(3) detected no change in the heat transfer rate at the onset of equilibrium in region 5 indicate that the initial portion of the boundary-layer development has an insignificant effect on the experimental or theoretical profiles at t = 100 μsec.

The assumption of local thermodynamic equilibrium and a one temperature boundary layer theory may be subject to question. The first indication of this occurs, if the Saha equation is used to calculate a temperature from the measured ρ and n_e and then a pressure from $p = (1 + \alpha)\ \rho RT$. It is then found that this pressure increases monotonically towards the wall by factors as high as 2.0 and 1.2, respectively, for the initial pressures of 2 and 10 torr. This is contrary to the constant pressure assumption that is arrived at by an order of magnitude comparison of the various terms in the normal momentum equation. In view of the fact that the gas is highly ionized, it may no longer be true that the body forces acting on the fluid are negligible. Therefore, even though the induced velocity in the boundary layer is small, adequate body forces (arising from induced electric fields) may be acting to support the negative pressure gradient dp/dy. However, the full equations including the normal momentum equation would have to be solved to verify this conjecture.

Whether or not the rates of production or depletion of the various species through collisions can keep up with the rates due to temperature gradients was applied as a criterion to determine if equilibrium prevailed in the boundary layer. The value of $(\partial n_e/\partial t)/n_e$ was computed with the electron continuity equation and was then compared with the number of available collisions, Q_a, for the three body recombination reaction

$$e + e + A^+ \rightarrow e + A^* + h\nu$$

Equilibrium is assumed if the following criterion is satisfied:

$$\frac{Q_a}{(\partial n_e/\partial t)/n_e} >> 1$$

The techniques and expressions for calculating the terms of this criterion are discussed in Refs. 2 and 11. Calculations have shown that the criterion is not satisfied during the entire 100 μsec, even at distances of up to 80 percent of the boundary-layer thickness from the wall. It should be pointed out that the criterion is more easily satisfied at longer times after reflection, but its use is questionable since the denominator is calculated based on the assumption of equilibrium during all stages of development.

With respect to the question of temperature equilibrium between electrons and heavy particles, it is generally argued that because of the disparity in mass, $m_A/m_e \sim 10^5$ collisions are required for equilibration to a single temperature. Temperature equilibrium may be assumed if the atom temperature does not change much over a distance L given in terms of the collision mean free path by $(L/\ell)^2 \sim 10^5$. The criterion may be expressed as

$$(\delta T_A/T_A)/L \sim \text{grad } (\ell n T_A) << 1/10^{5/2}\ell$$

Using a conservative estimate for the mean free path gives 1 mm^{-1} for the right-hand side, which says that the atom temperature should not change much over distances on the order of one millimeter. It would therefore appear that the criterion is not satisfied close to the wall where the gradients are steep. In the one solution presented by Camac and Kemp,[7] for a two-temperature end-wall analysis, the electron temperature at the sheath edge was still 50 percent of the free stream value. Unfortunately, their analysis was not performed for conditions applicable for direct comparison with the present study.

Little can be said regarding the nature and extent of the plasma sheath at the wall in the present problem because it is not well understood. It is readily apparent that the two usual assumptions of no heat transfer, and that the problem is "steady" with no time derivatives, are not applicable in the present problem. Therefore, one cannot reasonably expect that the net current to the wall will be zero. The resulting effect on the boundary-layer profile is not clear, except that now there may not be charge neutrality at the sheath edge.

The complications discussed above severely cloud the sole issue of thermal transport. Therefore, analytic variation of the transport properties to obtain agreement with experiment, as was done with the unionized case, would be of dubious value. It may be said, however, that the relative agreement observed for the density profiles -- primarily in the outer portion of the layer -- indicates that the theoretical thermal conductivity is at least reasonable. The theoretical total thermal conductivity of Knöös(2) was found to be a maximum of 30 percent lower than the theoretical values of Devoto.[11] Devoto found that his results were on the order of 10 to 20 percent lower than the arc measurements of Knopp and Cambel[12].

The experimental results of this study have shown that the present theories do not adequately describe the end-wall boundary layer for ionized argon. The reasons for the observed discrepancies have been discussed in terms of the inapplicability of the basic assumptions inherent in the theories. The present experimental profiles appear to be sufficiently accurate to provide a basis for a more comprehensive analysis which should, in turn, offer some insight into nonsteady, nonequilibrium boundary layers in ionized gases.

REFERENCES

1. Fay, J. A. and Kemp, N. H., "Theory of Heat Transfer to a Shock-Tube End-Wall from an Ionized Monatomic Gas," J. Fluid Mech., Vol. 21, Part 4, 659-672 (1965).
2. Knöös, S., "Boundary Layer Structure in a Shock-Generated Plasma Flow, Part 1. Analysis for Equilibrium Ionization, "J. Plasma Physics, Vol. 2, Part 2, 243-255 (1968).
3. Camac, M. and Feinberg, R. M., Thermal Conductivity of Argon at High Temperatures, Avco-Everett Research Report 168 (1963).
4. Baganoff, D., "Experiments on the Wall Pressure History in Shock Reflection Processes," J. Fluid Mechanics, Vol. 23, 209 (1965).
5. Smith, J. A. , "Experimentally Determined Structure of the Shock Reflection Process in Ionizing Xenon," Physics of Fluids, Vol. 11, No. (10) 2150 (1968).
6. Amdur, I. and Mason, E. A., "Properties of Gases at Very High Temperatures," Physics of Fluids, Vol. 2, 370 (1958).

7. Camac, M. and Kemp, N. H., A Multitemperature Boundary Layer, Avco-Everett, Research Report 184 (1964).
8. Back, L. H., "Laminar Boundary-Layer Heat Transfer from a Partially Ionized Monatomic Gas," Physics of Fluids, Vol. 10, No. (4), 807 (1967).
9. Alpher, R. A. and White, D. R., "Optical Refractivity of High Temperature Gases. II Effects Resulting from Ionization of Monatomic Gases," Physics of Fluids, Vol. 2, 162 (1959).
10. Wong, H. and Bershader, D., "Thermal Equilibration Behind an Ionizing Shock," J. of Fluid Mechanics, Vol. 26, 459 (1966).
11. Devoto, R. S. "Transport Properties of Partially Ionized Argon," Physics of Fluids, Vol. 10, 354 (1967).
12. Knopp, C. F. and Cambel, A. B., "Experimental and Theoretical Determination of Atmospheric Argon Plasma," Physics of Fluids, Vol. 9, 989 (1966).

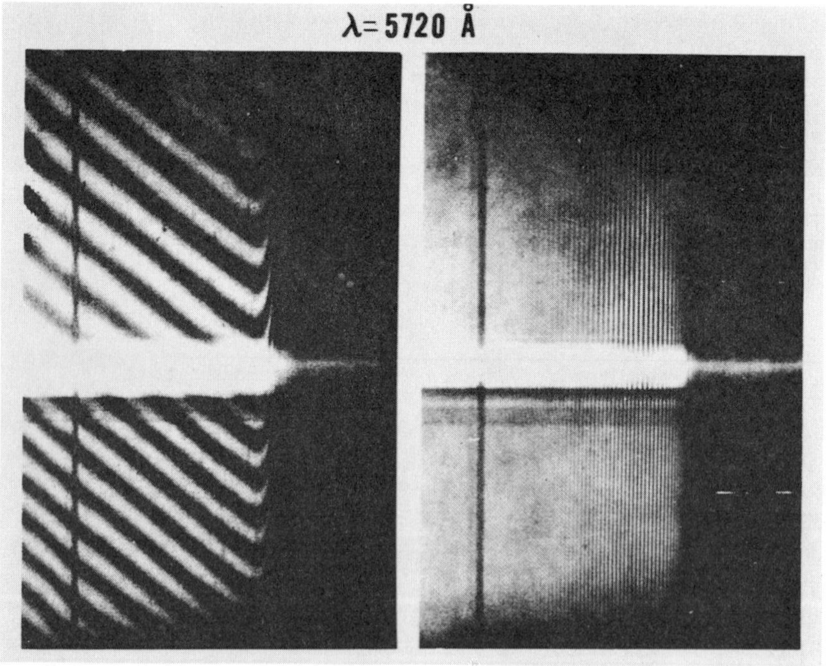

Figure 1
Snapshot interferograms of the ionized end-wall boundary layer in argon, $p_1 = 5$ torr. Free stream properties $T = 12{,}400°K$, $n_e = 2.5 \times 10^{17} cm^{-3}$, $p = 7.2$ atm, $\alpha = 0.067$, $t = 100$ μsec.

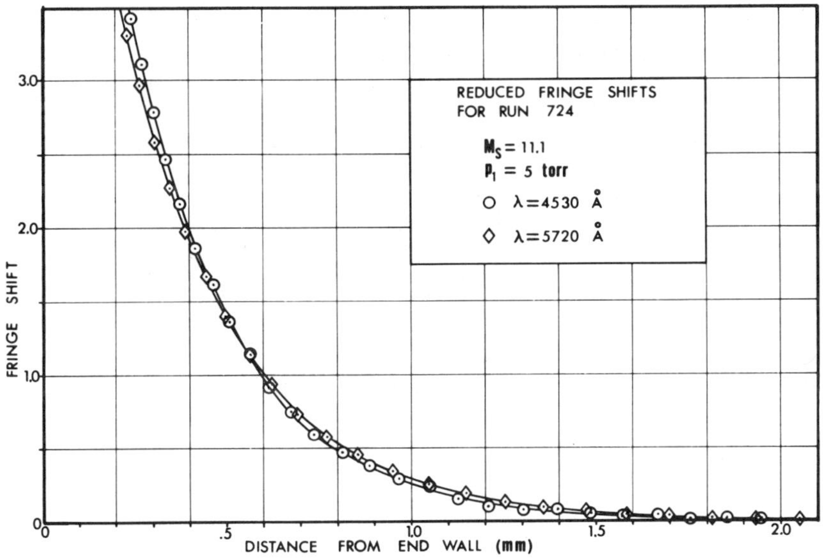

Figure 2
Measured fringe shifts at two wavelengths in an ionized argon boundary layer.

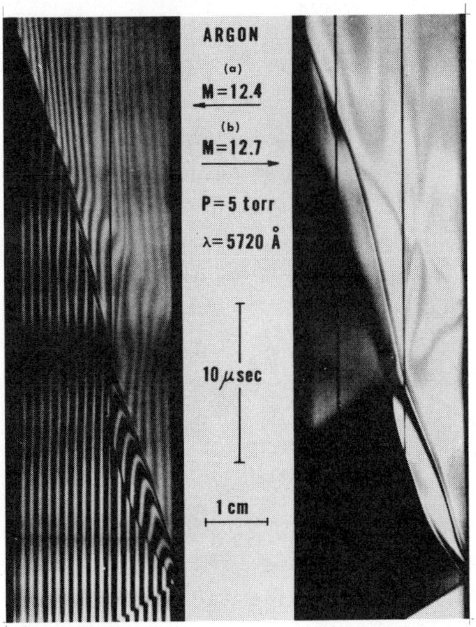

Figure 3
Regular and infinite-fringe streak interferograms of the shock-reflection process in ionizing argon.

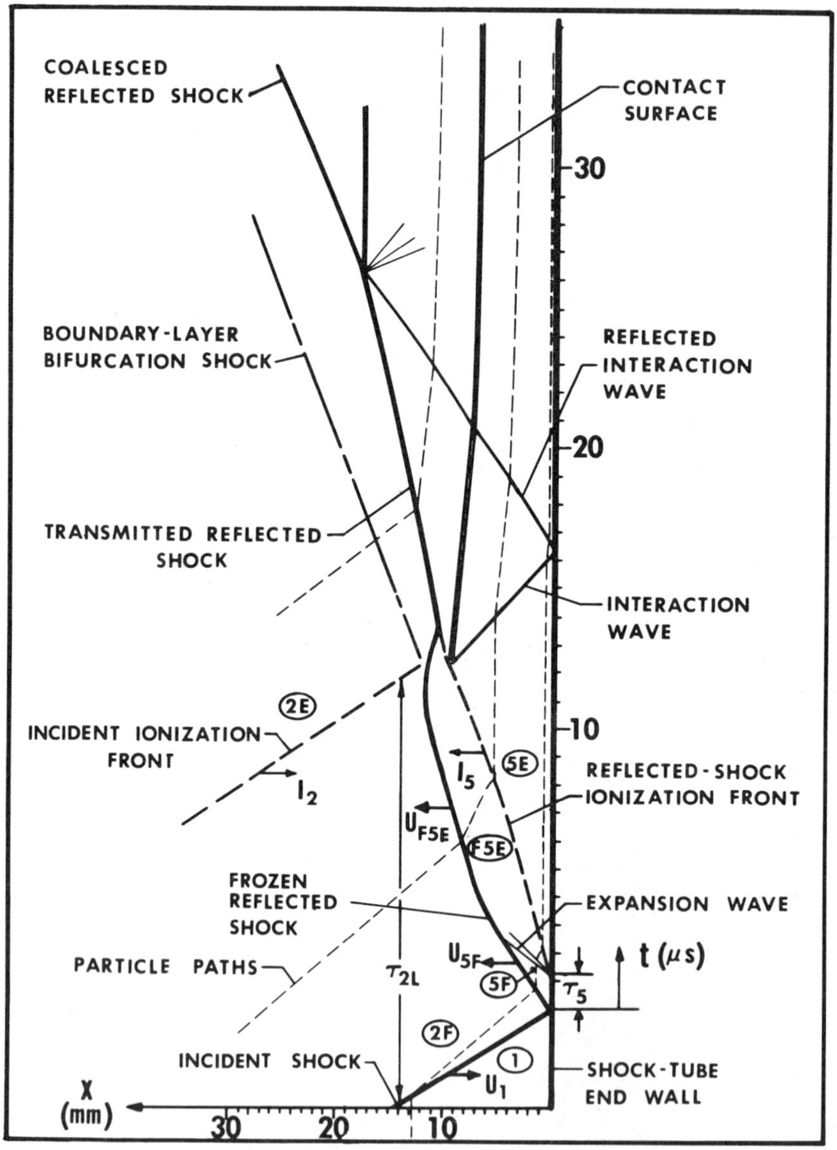

Figure 4
Distance versus time (x-t) diagram of the shock-reflection process in ionizing argon ($M = 12.7$, $p_1 = 5$ torr).

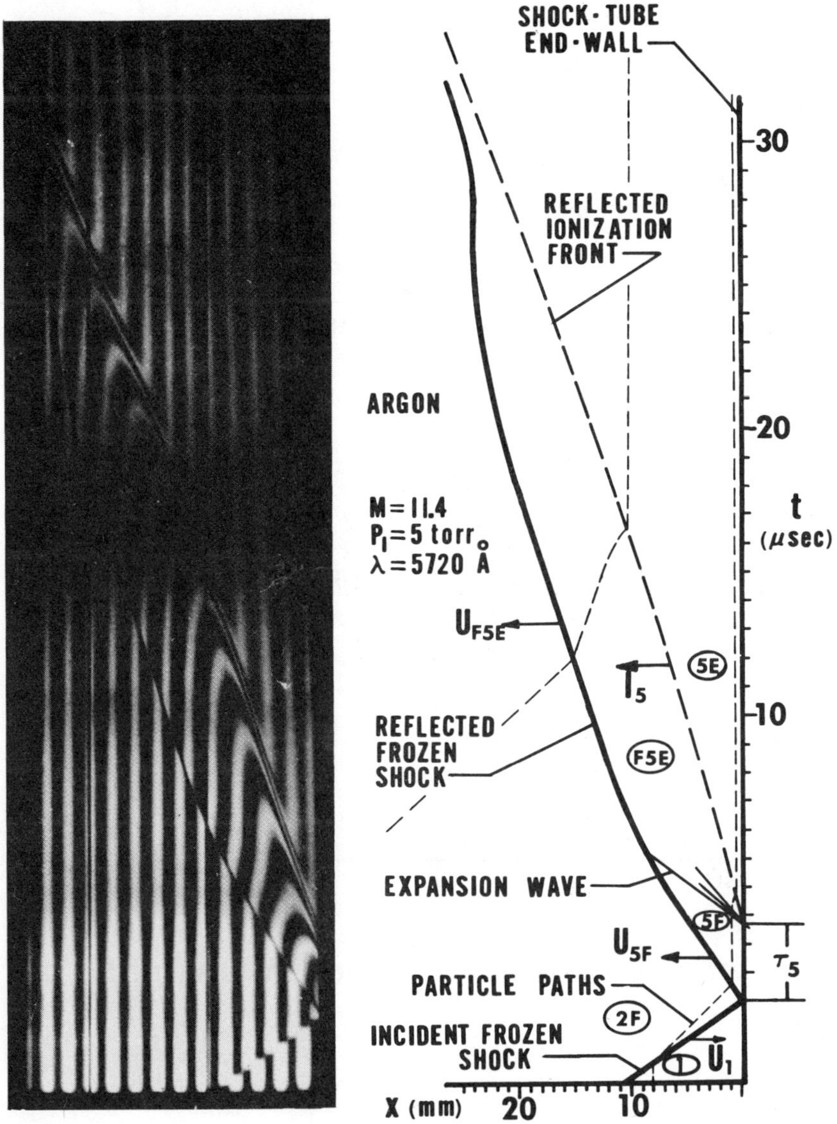

Figure 5
Streak interferogram and x-t diagram of the shock-reflection process in ionizing argon.

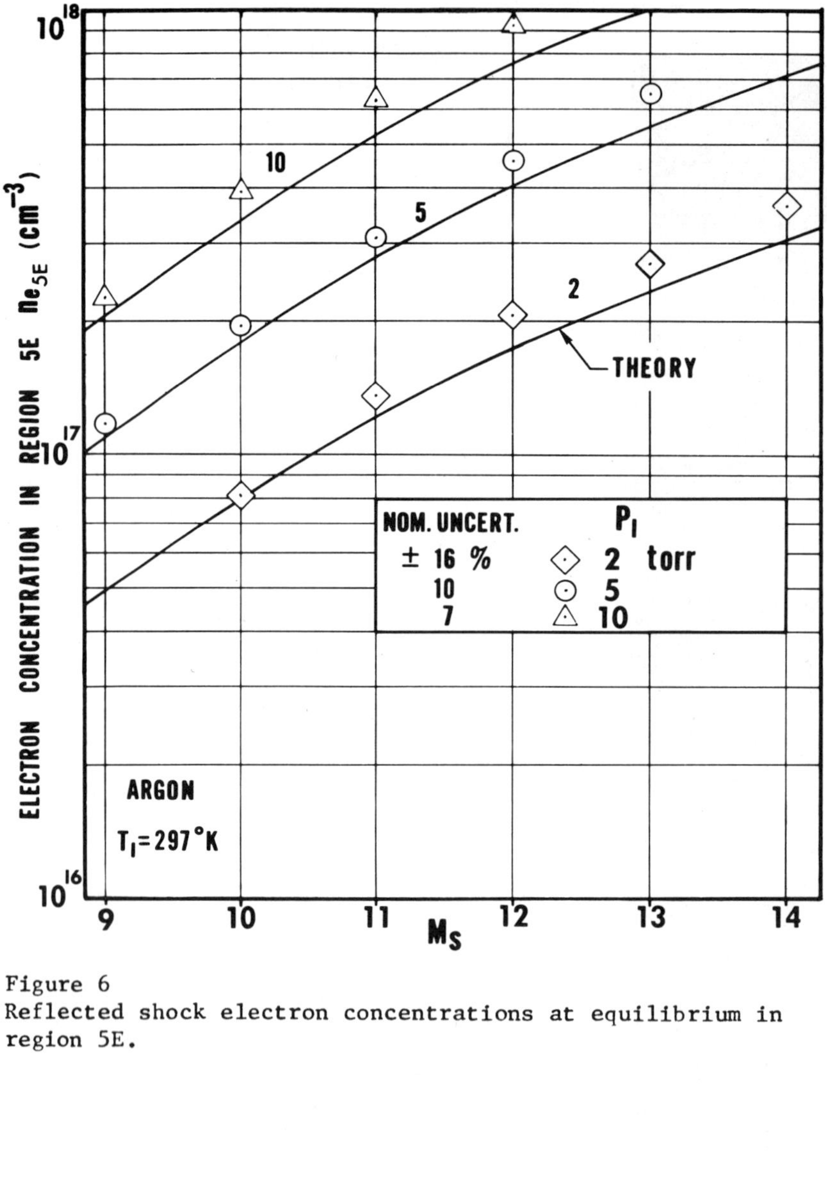

Figure 6
Reflected shock electron concentrations at equilibrium in region 5E.

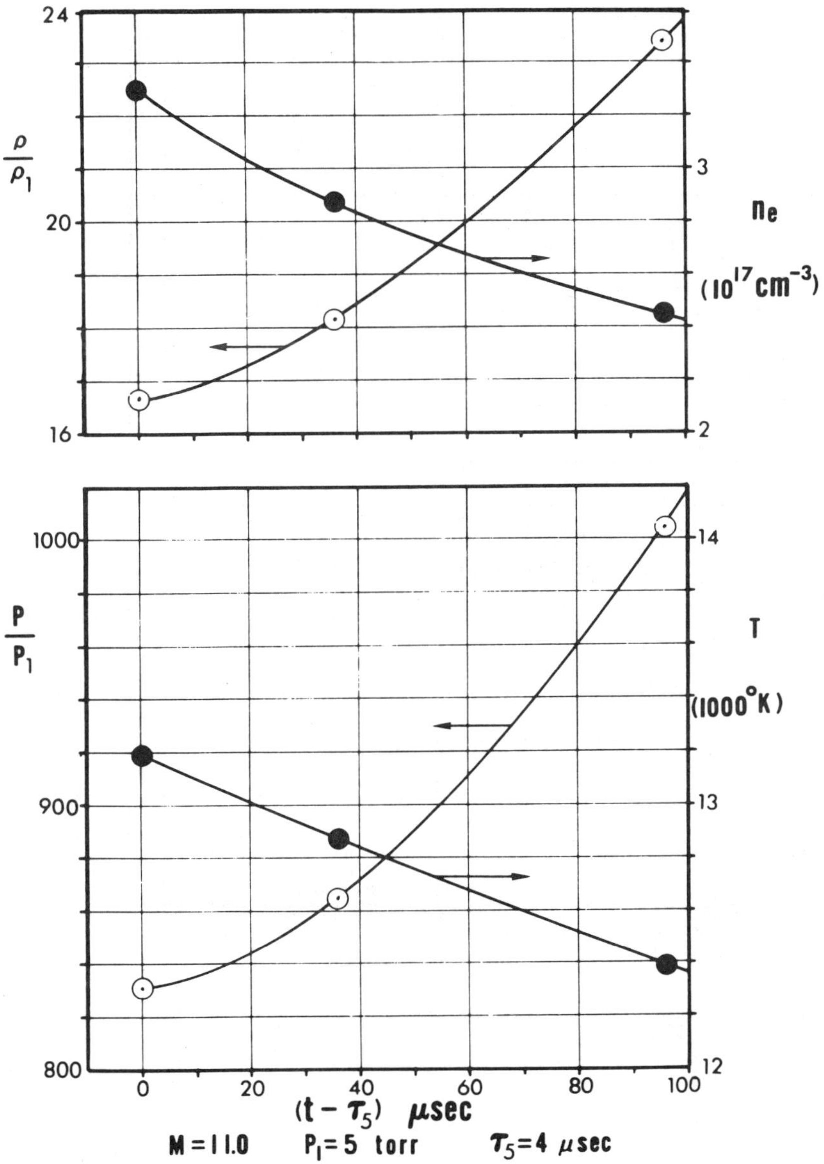

Figure 7
Time history of argon thermodynamic properties in the reflected-shock region with ionization.

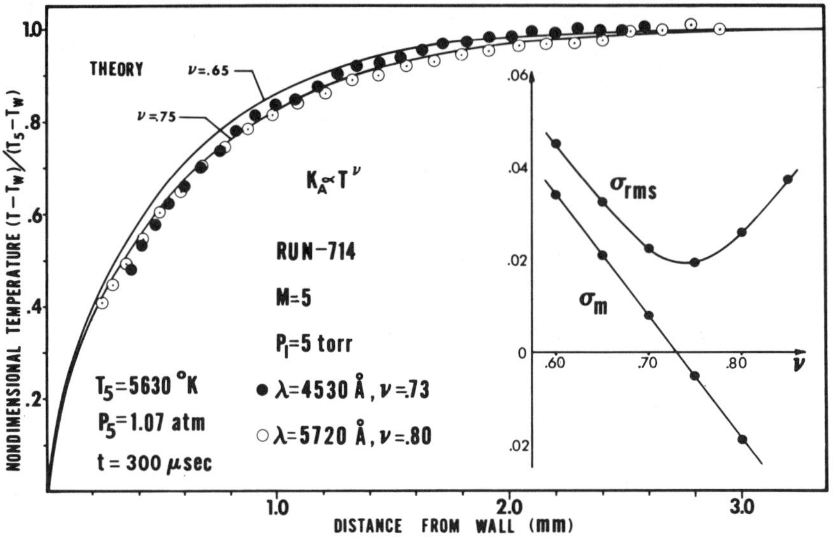

Figure 8
Experimental and theoretical temperature profiles for the end-wall thermal layer in unionized argon.

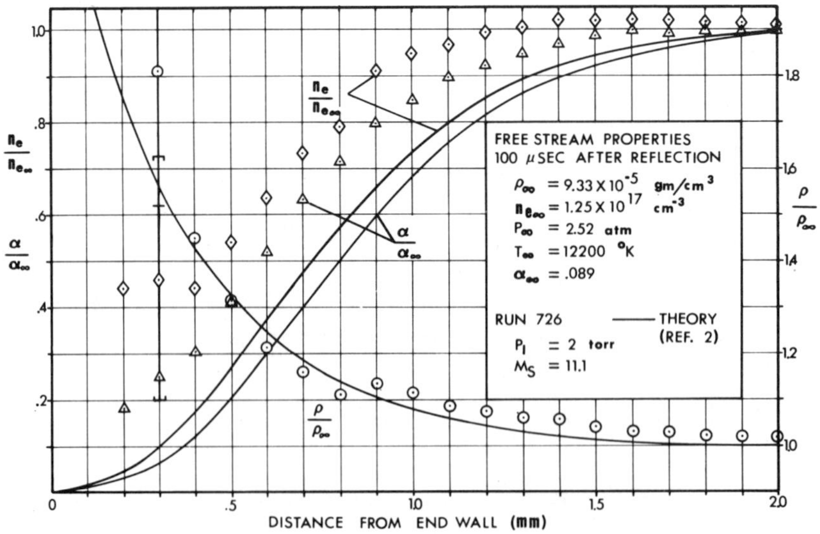

Figure 9a
Experimental and theoretical temperature profiles for the end-wall electron concentrations in ionized argon.

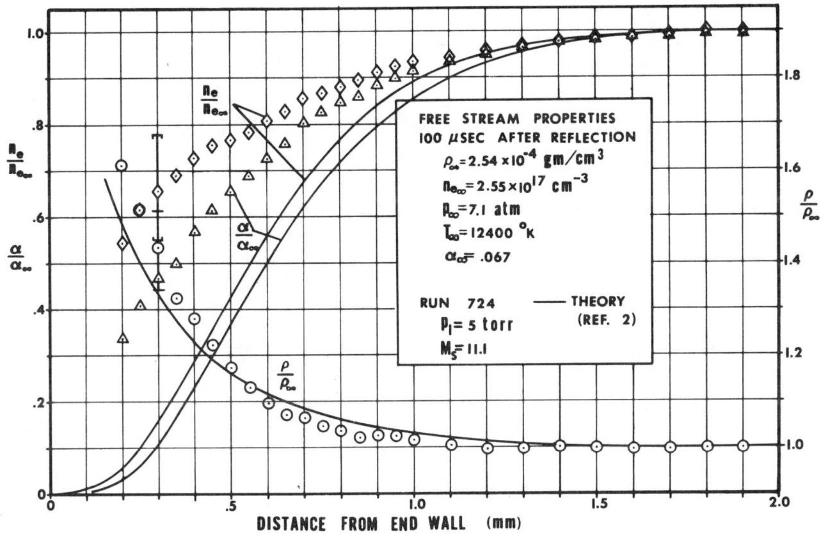

Figure 9b
Experimental and theoretical end-wall profiles of density and electron concentrations in ionized argon.

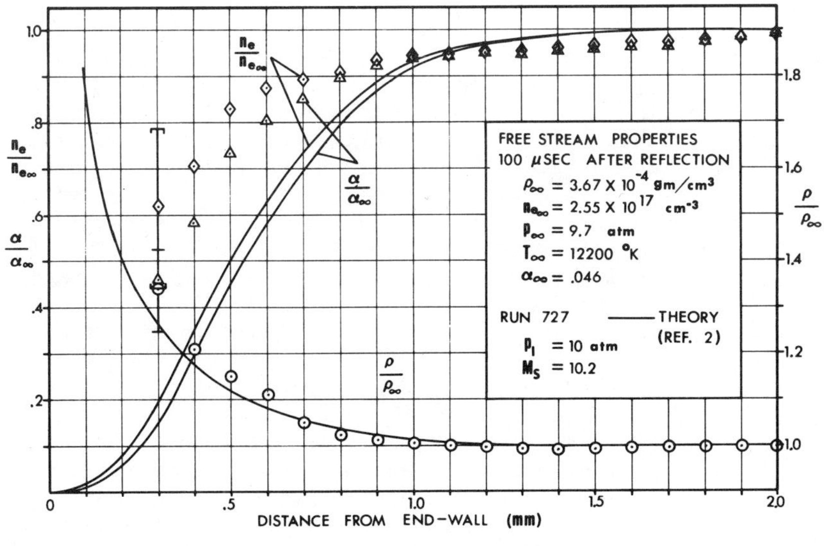

Figure 9c
Experimental and theoretical end-wall profiles of density and electron concentrations in ionized argon.

PHOTOGRAPHIC STUDY OF EARLY STAGES OF VORTEX FORMATION BEHIND AN EDGE

RAYMOND J. EMRICH
Department of Physics
Lehigh University
Bethlehem, Penna. 18015, USA

H. REICHENBACH
Ernst-Mach Institut.
78 Freiburg im Breisgau, Germany

SUMMARY

Flow separation behind an edge combined with formation of a concentrated vortex is frequently seen when a fluid starts to move from rest. The interaction of the vortex with the remainder of the flow field has been of interest for a long time -- behind the canoe paddle and behind the spoon in the teacup, for example. Detailed calculations have been carried out, starting with a contribution by Prandtl[1] in 1922, and summarized in a 1965 paper by Küchemann and Weber[2]. These calculations involve assuming, for the outer part of the vortex, an inviscid spiral form with slip layers forming the jelly of the jelly-roll and, for the inner part, an axially symmetric continuous viscous flow. Parameters are included in the theoretical formulations to fit the two regions together. Rather precise experimental data obtained from shock tube interferograms have been used to choose the parameters.

The characteristics of the first separation and early formation of the spiral vortex have not been described. The calculations have assumed the separated flow and the vortex to have existed from the beginning of the motion. In their article, Küchemann and Weber[2] state, "The behavior of the flow at and in the neighborhood of the separation line where the hypothetical vortex sheet originates poses a problem which remains largely unsolved." In an earlier article, Rott[3] states, ". . . . . Qualitatively, however, by considering the motion of newly produced vorticity under the influence of the vorticity generated previously, it is easily understood that the discontinuity sheet will be wound into a narrow spiral, outside of which the flow may be taken as irrotational."

Experimentally, in the shock tube, the vortex grows in size at a linear rate with time[4] after a more rapid initial growth. It seemed possible to us that some previously unobserved processes might precede the formation of the spiral. Our search for such processes has not revealed any, however. At times as short as 2 microseconds after the shock starts the flow, a region appears behind the edge which is soon seen to have a circular shape in high resolution spark photographs, and the center of the circle is observed to move with nearly constant speed in a straight line away from the point. Analysis of the concentrated vortex reveals no basis for doubting that a spiral vortex exists at times as short as a continuum picture of the events is valid.

FLOW STARTED BY SHOCK DIFFRACTING OVER AN EDGE

Whenever a shock strikes an object and starts a subsonic flow over or around a sharp corner, a spiral vortex appears on the downstream side of the corner. A surface across which the gas velocity is discontinuous (slip stream) originates at the edge and curls into the downstream shadow region. A typical shadow photograph appears in Figure 1. This vortex and the surrounding flow have been observed and studied by many authors[3-8].

An essential result of these studies was the observation that the flow is nearly self-similar (pseudostationary). This property is that the flow field at any time is a spatially magnified replica of earlier flow fields, which means that all flow variables and wavefronts are functions of x/t and y/t alone and not of x, y and t separately. Time is measured from the instant the shock front strikes the knife edge. Nearly self-similar means, however, that in contrast to the behavior of the wave fronts, the dimensions of the vortex do not strictly conform to the requirement of self-similarity; rather, the vortex grows at a constant rate as though from an earlier origin in time after a characteristic formation time. The distance-time graph of Figure 2 illustrates this. In the example shown, the formation period lasts about 20 microsec. With decreasing initial gas density but constant shock strength, this formation period lengthens.

From numerous spark interferograms, of which Figure 3 is a typical example, the density distribution throughout the field has been evaluated. The measurements obtained this way may be interpreted as revealing that the vortex has an inner viscous core which constitutes a constant fraction of the entire vortex as it grows. This fraction increases with increasing shock strength[8].

An independent experimental determination of the velocity distribution within a spiral vortex using photography of submicron-sized oil drop tracers was attempted[9]. This showed that, even in the rather weak shock ($M_S = 1.15$) flow used, particles were thrown out of the center of the vortex in the initial stages. Since the oil drops follow the flow in which they find themselves with a response time of only a few microseconds, it is clear that the rotational speeds within the vortex are many thousands of revolutions per second.

The flow field initiated by the diffracting shock has features which can be understood in more detail by reference to Figure 4(a). There in region b the gas has a constant horizontally directed velocity. The diffracted shock wave behind the knife edge imparts a downward motion to the gas previously at rest in region a . From the pointed edge, an outward moving rarefaction wave is generated; its front is denoted as a dashed line behind the incident shock front. A second compression of the gas in region c has been performed by the reflected shock which is moving back into the gas in region b already compressed by the incident shock. The gas in region c is at rest. Near the pointed edge, in region bc , the gas expands and generates an upward motion. Above the edge, the gas flow is bent around into the horizontal direction by a strong local pressure gradient which can be seen on the interference photograph of Figure 3. The direction of the gas velocity, induced by the shock in the neighborhood of the pointed edge, is sketched in Figure 4(b).

One might guess that a part of the horizontally directed gas put into motion above the pointed edge by the incident shock wave will be drawn down into the lower pressure region ab , while the remainder of the horizontally moving gas will separate from the downward directed part. Such a bifurcation of the flow would occur at a place somewhat behind but very near to the edge. At this place, which one can call a bubble, the pressure and temperature possess minimum values. The gas just above the separation bubble is strongly deflected and can wrap around the center of the bubble.

The time and distance scales involved in the bubble formation and the first revolution of gas around it should depend on the contour of the pointed edge[10], on the shock strength and on the kinematic viscosity of the gas. There seems to be no dynamical restriction on the size of a self-similar vortex which can exist at even the shortest times. These topics will be discussed in turn.

Effect of Edge Contour on Time and Distance Scales

At large scale, the propagation of a shock over a rounded contour, such as a sphere or cylinder, is associated with a sequence of shock reflections followed by a separation of the flow behind the edge[11,12]. Since, with sufficient magnification, any corner takes on a rounded contour, the separation and vortex formation behind an edge is, in part, affected by the sharpness of the contour. By employing a very sharp edge, this effect can be made smaller than the resolution of the instruments used for study.

Effect of Shock Strength on Time and Distance Scales

At sufficiently high shock strengths, when the flow behind the primary shock is supersonic, no vortex can be seen behind a corner.[5] For weak shocks ($M_s = 1.1$) the vortex appears to be perfectly circular. In the intermediate shock strength region, the shape of the vortex departs progressively from circular with increasing shock strength. For study of the early stages of the vortex formation, a small shock strength (and small gas velocity) is desirable.

Effect of Viscosity on Time and Distance Scales

A gradual accretion of gas occurs in the early stages at the separation bubble with strong velocity gradients caused by the successive arrival and stopping of gas samples previously moving horizontally. The strong velocity gradients are, in turn, associated with strong viscous forces. Viscosity also plays a role in affecting the gas flowing up along the surface in region bc of Figure 4(a). When this gas arrives at the separation bubble, it will have acquired vorticity from the strong shear in the boundary layer.

The scale at which either or both of these processes are of importance in a small bubble or in a small spiral vortex can be estimated. Figure 5 is a schlieren photograph of the spiral vortex after 60 μsec of development. The slip layers winding in on each other can be seen. It is clear that when the diffusion of vorticity starting from these layers has progressed from one layer to the next, the effect of viscosity on the vortex structure and on its further growth will be important. J. T. Stuart[13] proposes that vorticity diffuses after a time t over the characteristic distance $\delta = \sqrt{\nu t}$ where ν is kinematic viscosity. For the conditions chosen for the study reported here, we have observed that the vortex center moves with a constant speed $v_c = (0.16)\, v_s$ where v_s is the shock speed. With r representing the vortex radius, the time available for diffusion of vorticity within the vortex is $t = r/v_c$. Therefore,

$$\delta = \sqrt{\frac{\nu}{v_c}}\sqrt{r} \quad ,$$

and a measure of the size of region in the vortex influenced by viscosity, relative to the dimension of the entire vortex is

$$\frac{\delta}{r} = \sqrt{\frac{\nu}{v_c}}\,\frac{1}{\sqrt{r}} = \sqrt{\frac{\nu}{v_c^{\,2}}}\,\frac{1}{\sqrt{t}} \,.$$

For the conditions of the experiments reported here, v_s is 500 m/s and for air, ν is 1.5×10^{-5} m^2/s. The characteristic thickness of region affected by viscosity, after 10 microseconds, is $\delta = 12 \times 10^{-6}$ m which is a fraction of the vortex size $\delta/r = 0.015$. This thickness of 12 microns and a fraction of only 1-1/2 percent indicates that viscosity plays an extremely small role.

The studies should, therefore, be directed to observing the smallest possible dimensions. From the dependence of the diffusion distance δ on the vortex center speed v_c, we can see that values as small as possible for v_c should be sought. This means flows started by weak shocks should be studied.

Dynamical Relations in Self-Similar Vortex

The structure proposed by those calculating details of an incompressible vortex[2] bears some resemblance to a self-similar flow with the tangential component of velocity remaining constant and independent of r throughout the vortex and the radial component of velocity remaining near zero. Equating the centripetal force on an element of fluid to the pressure change across it times its area, we have $(v^2/r)dr = dp/\rho$ where v is the tangential velocity component at distance r from the axis and p and ρ are gas pressure and density respectively. Since all the gas comes from a uniform reservoir in region c of Figure 4(a) and since the effects of viscosity are restricted to small regions, the gas entropy may be considered constant. Then $dp/\rho = dh$ where h is enthalpy. For an ideal gas, $dh = C_p dT$ where C_p is specific heat and T is temperature.

Integrating the equation $(v^2/r)dr = C_p dT$ with v constant, we obtain

$$- v^2 \ln \frac{r}{r_2} = C_p (T_2 - T)$$

where T_2 and r_2 are values at the outside edge of the vortex. This formula implies that at a sufficiently small value of r, T can be below absolute zero. While this is impossible, we may estimate the value of r for which $T_2 - T = 300K$, a place where there is no longer available adequate pressure to provide the needed centripetal force. This value of r is many times smaller than the gas mean free path. From this, we gather that, no matter how small the bubble or vortex may be, there is still enough pressure gradient available to provide the needed centripetal force.

EXPERIMENTAL STUDY OF EARLY FLOW

The aim of the experiments was to photograph with as high magnification as possible the pattern in the neighborhood of the pointed edge just after a shock strikes. These experiments were carried out in a 4 cm by 11 cm shock tube at the Ernst-Mach-Institut.

With high magnification, two optical design problems arise. First, non-axial rays associated with an extended light source follow slanted paths through the window section of the shock tube and their deviations (schlieren) or optical path differences (interferometric) do not carry a one-to-one correspondence between points in the object plane and on the camera image plane. This results in lack of definition, and correction for this fault requires reducing the source size to more nearly a point source, or reducing the active path length in the window section. In careful optical design, the source size is always reduced as much as possible in the presence of the available source luminance and film sensitivity, to give optimum image quality. Usually the active path length in the window section cannot be made small because detection sensitivity is directly proportional to it. The second optical design problem arising with high magnification is the need for shorter exposure times to "freeze" the motion. A shorter exposure time must be accompanied by a higher source luminance.

Advantage has been taken of the quite large density gradients to be studied in the flow about the knife edge, and the detection sensitivity sharply curtailed. The window section thickness used was only 5 mm. An 8 degree steel blade of this thickness was held between two interferometer quality windows of 40 mm diameter. One window formed a part of the regular shock tube wall, and the other was set into a partition inside the shock tube which had a "cookie cutter" leading edge to select a 5 mm wide slice out of the main shock front. The ray paths for schlieren and interference optical viewing are shown in Figure 6 and Figure 7. The image size on the film was approximately half the object size (magnification = 0.5). The spark light source, using a battery of barium titanate capacitors, had a duration of about 0.3 microsec. Quite detailed photographs are obtained in air at densities as low as 0.1 atm at $M_S = 1.4$. The overall quality of the photographic image permitted resolution of 0.2 mm in object space in the shock tube.

A special color filter and color film were used with interference optic observations. The filter has two transmission maxima at different wavelengths. Two interference patterns with different fringe spacings are thus superposed. When used with initial fringe spacing adjusted to infinity, very small density changes are displayed as vivid color shifts.

RESULTS

Schlieren optical photographs of flows during the first 20 microsec, with schlieren blades parallel and perpendicular to the object blade, at initial densities ranging from 0.1 atm to 1 atm, mostly at low Mach numbers, indicated consistently a perfectly formed circular bubble or vortex appearing closely behind the pointed edge. The slip stream rolling into spiral form is visible in the schlieren optical photographs. The earliest circular bubble is seen at about 2 microsec, the earliest slip stream at about 6 microsec, and the earliest spiral form of slip stream at about 12 microsec.

Interference optical photographs also reveal the circular bubble or vortex 2 microsec after the shock strikes the knife edge. An interferogram, with initial fringe spacing infinite, appears in Figure 8. The changing density in the rarefactions, as well as the jumps in density across the shocks, can be seen. The perfectly circular vortex with a different density at its center is also seen.

From a series of interferograms made at $M_S = 1.4$, initial density 0.6 atm, the position of the center of the vortex is recorded at measured times. The center moves across and downward in a straight line at 68° to the vertical at a constant speed equal to $0.16\ v_S$ where v_S is the shock speed. The vortex center motion seems to satisfy the condition of self-similarity (pseudo-stationarity.)

CONCLUSION

Study of early stages (2 microseconds and later) of the vortex formation behind an edge has provided no evidence of processes different from those previously observed and described for later times (20 to 150 microsec.)

The studies support previous assumptions that there is flow separation and vortex formation at arbitrarily early times. The formation of a spiral in less than 12 microseconds after the shock hits the knife edge may be recognized as one of the most rapid and intricate flow processes in nature. Speculation concerning the physical processes involved suggests that the process may occur in even shorter times.

REFERENCES

1. Prandtl, L. Über die Enstehung von Wirbeln in der idealen Flüssigkeit mit Anwendung auf die Tragflugeltheorie und andere Aufgaben. Gesammelte Abhandlungen Ludwig Prandtl. Edited by Tollmien, Schlichting, Görtler and Riegels. Springer, 1961. Pages 697-713.
2. Küchemann, D. and Weber, J. Vortex Motions. Zeits. für angew. Math. und Mech. 45, 457 (1965).
3. Rott, N. Diffraction of a Weak Shock with Vortex Generation. J. Fluid Mech. 1, 111 (1956).
4. Reichenbach, H. and Merzkirch, W. Untersuchungen über das Ähnlichkeitsverhalten einer instationären Wirbelspirale. Zeits. Flugwiss. 12, 219 (1964).
5. Waldron, H. E. An Experimental Study of a Spiral Vortex Formed by Shock Wave Diffraction. UTIA Technical Note No. 2, University of Toronto Institute for Aerospace Studies, Toronto, Canada (September 1954).
6. Howard, L. N. and Matthews, D. L. On the Vortices in Shock Diffraction, J. Appl. Phys. 27, 223 (1956). Earlier version is Princeton University Department of Physics Technical Report II-21 (July 1955).
7. Schardin, H. Ein Beispiel zur Verwendung des Stosswellenrohres für Probleme der instationären Gasdynamik. Z. angew. Math. Physik 9, 606 (1958).
8. Merzkirch, W. Theoretische und experimentelle Untersuchungen an einer instationären Wirbelströmung. Zeits. Flugwiss. 12, 395 (1964).
9. Breslin, John A. and Emrich, Raymond J. Tracer Photography Used for the Study of Vortex Formation at the Open End of a Shock Tube. Bulletin of APS 13, 821 (1968).
10. Griffith, Wayland and Brickl, David E. The Diffraction of Strong Shock Waves. Phys. Rev. 89, 451 (1953).
11. Heilig, W. Diffraction of a Shock Wave by a Cylinder. Phys. of Fluids 12, I-154 (1969).
12. Amann, H. O. Experimental Study of the Starting Process in a Reflection Nozzle. Phys. of Fluids 12, I-150 (1969).
13. Rosenhead, L., Ed. Laminar Boundary Layers. Oxford University Press, 1963. Page 349.

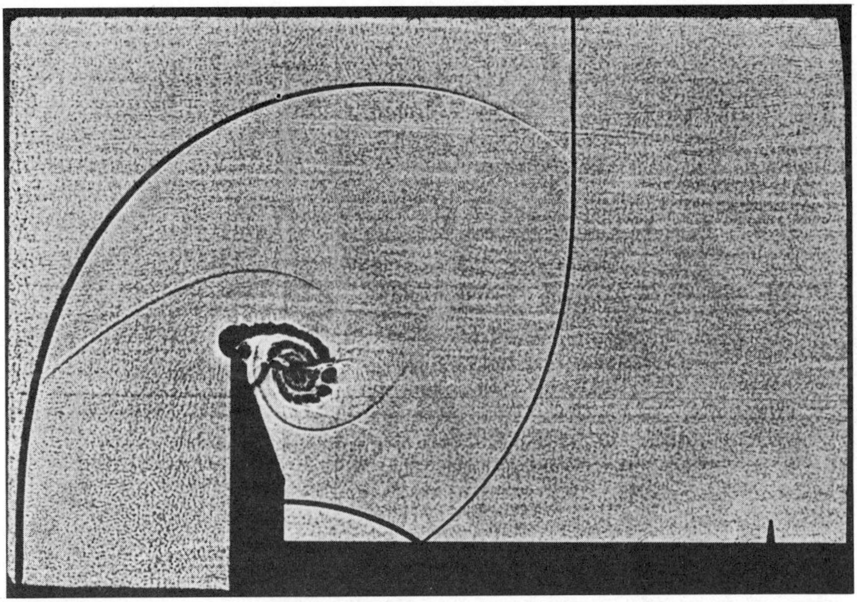

Figure 1
Shock diffracting over a knife edge. Shadow photograph, $M_S = 1.4$ (from Ref. 7.)

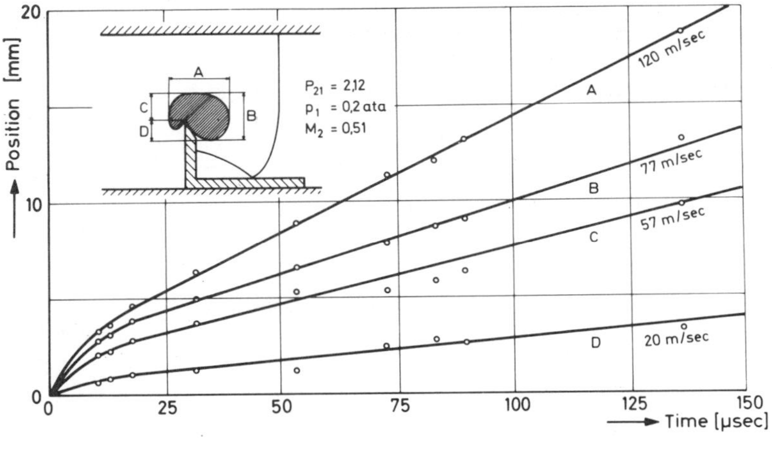

Figure 2
Growth of various dimensions of vortex behind knife edge measured from series of shadowgrams made of a single flow by the Cranz-Schardin multiple spark camera. Initial pressure p_1 = 0.2 atm, $M_S = 1.4$ (from Ref. 4.)

Figure 3
Interferogram of flow field produced by shock diffracting over a knife edge. The interferometer adjustment is for infinite fringe breadth so that lines of constant blackness in the picture display lines of constant density. Initial pressure $p_1 = 0.6$ atm, $M_S = 1.4$. Depth of flow section 5 mm. Note the strong gradient at the upstream edge of the corner where the flow up over the edge turns sharply.

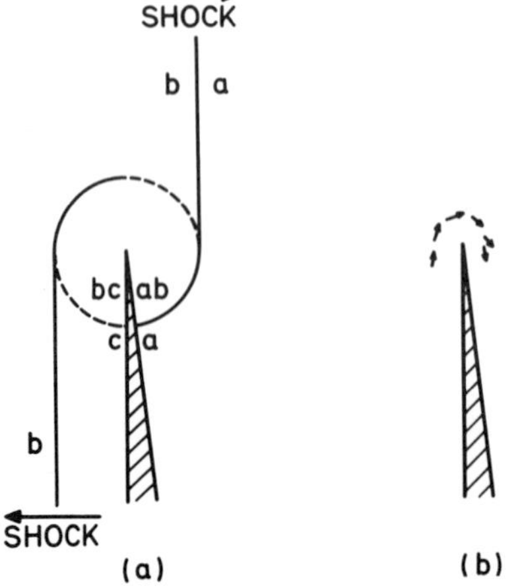

Figure 4
Shock wave initiation of flow field over a rigid wall. (a) wave fronts. (b) gas motion induced by wave.

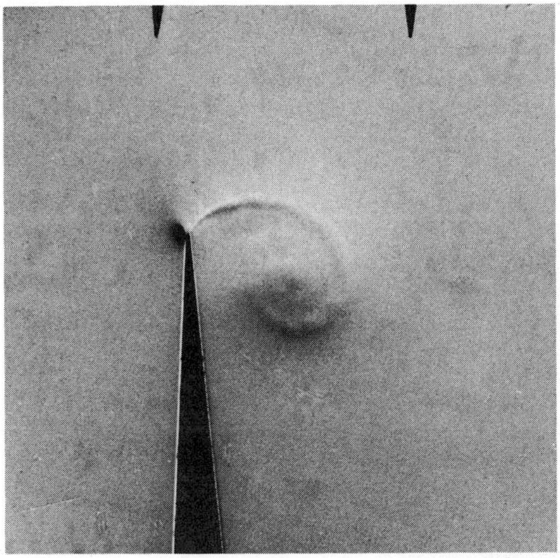

Figure 5
Spark photograph with schlieren optics of the spiral vortex 60 microseconds after the flow is started. Fiducial points in picture are 10 mm apart. The shock has passed out of the field of view. $M_S = 1.4$, $p_1 = 0.6$ atm air.

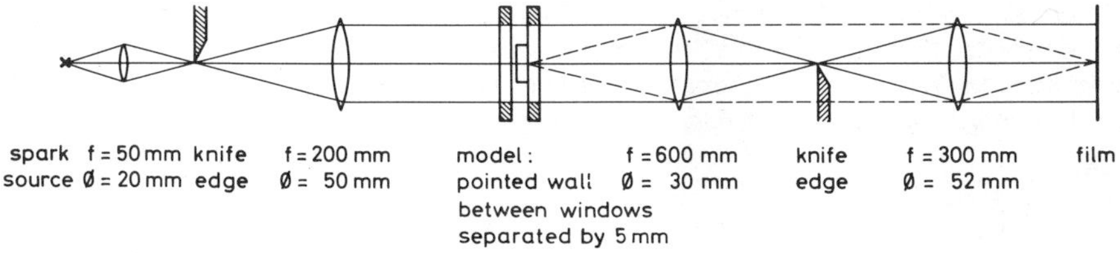

Figure 6
Schlieren optic arrangement.

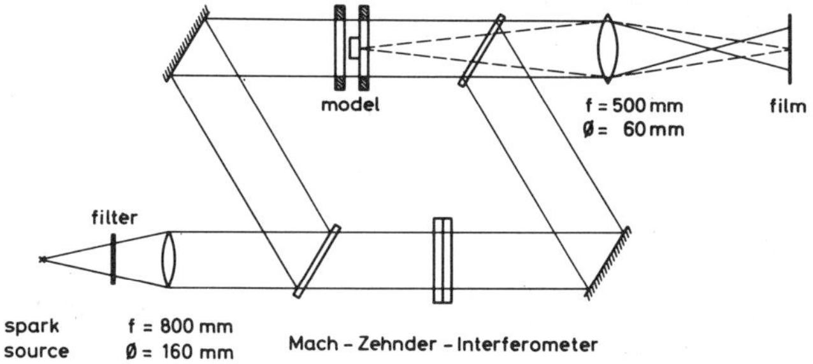

Figure 7
Interference optic arrangement.

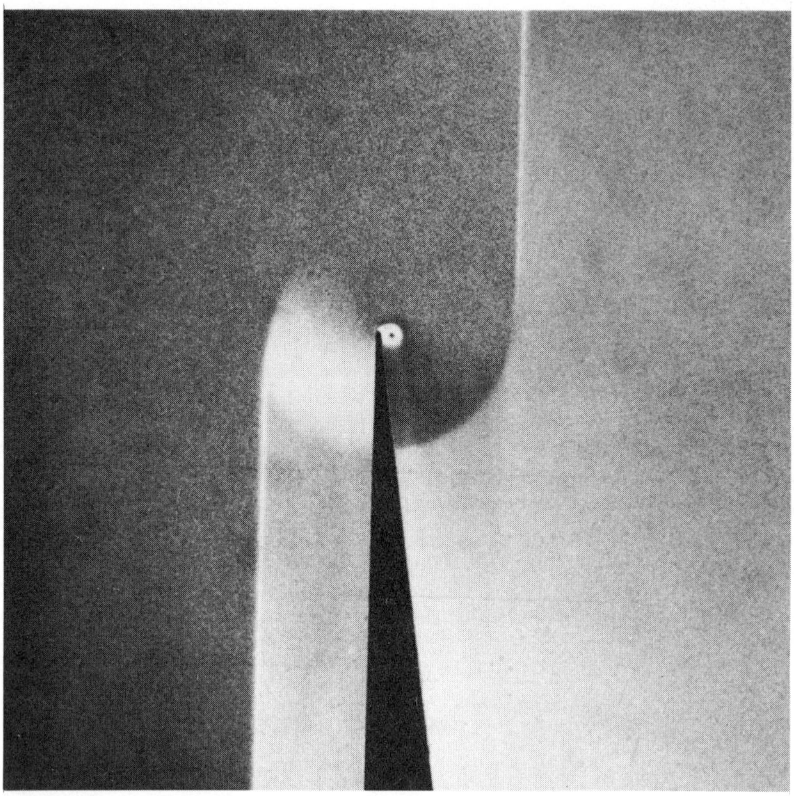

Figure 8
Interferogram of flow 7 μsec after shock. $M_S = 1.16$, initial pressure $p_1 = 0.625$ atm. The shock has traveled 3 mm in physical object space after meeting the knife edge.

EXPERIMENTAL STUDIES OF THE SHOCK REFLECTION AND INTERACTION IN A SHOCK TUBE

L. Z. DUMITRESCU and
C. POPESCU
Institute of Fluid Mechanics
Bucharest, Rumania

R. BRUN
Institut de Mécanique des Fluides
Marseille, France

SUMMARY

Some new experimental results are presented on the interaction of the reflected shock wave with the boundary layer and the contact surface in a shock tube. The pressure recordings show remarkable flow uniformity. However, heat-transfer signals indicate strong differences between points on the end-plate in the center and near the shock tube wall. In some cases, the hot region produced by the shock is less disturbed near the lateral wall.

INTRODUCTION

The discussion of the complex phenomena taking place after the reflection of a shock at the end of a shock tube is far from settled, although numerous papers have been issued on the subject (see, e.g. Reference 1 and the cited literature). Here, some new evidence is put forward in the hope of throwing light on the processes of interaction between the reflected shock wave, the boundary layer and the contact surface. Particular attention is paid to the shock bifurcation and boundary-layer detachment, on the one hand, and to the contact surface instability, on the other.

The effects of these phenomena on the duration and uniformity of the hot region in the shock tube should be clearly assessed, bearing in mind applications to hypersonic aerodynamics and chemical kinetics studies.

The tests have been carried out in the facilities of both institutes. Pressure and heat-transfer gauges, placed at various stations along the tubes and on the end-wall, and hot-wire probe traverses of the flow were used.

Various combinations of driver/driven gases were used: air/air, air/N_2, air/A, air/CO_2, N_2/A, He/A, H_2/A, H_2/N_2, at initial pressures p_1 of 0.5 to 760 mm Hg, covering a range of shock Mach numbers $2 < M_s < 8.5$.

The positions of the various gauges are identified in Figure 1. Pressure and heat transfer were recorded in the centre and near the edge of the shock tube end-plate (stations A, B), as well as on the lateral wall, near the end-plate (station C) and further upstream (D). Hot-wire traverses were made at various distances x from the diaphragm (stations E). In addition, the shock wave velocity was monitored. The heat-transfer gauges were of the usual platinum (or gold) film type. Pressure measurements used sensors and techniques described in Ref. 2, and wire traverses were based on the theory of Ref. 3.

SHOCK WAVE REFLECTION PROCESS

In this section the process immediately after the reflection of the primary shock is discussed. The effects of boundary-layer detachment and shock bifurcation are separated from those of the shock-contact surface interaction, which will be considered later on. The tests have been made in a cylindrical tube (100 mm diameter x 6 m long) having a honed and chromed inner surface.

Pressure histories

End-plate pressure measurements. In Figs. 2,3, and 4 (upper traces) several oscilloscope records are given as examples. It can be seen from the smooth signals that a regular behavior of the pressure exists at the end-plate. No differences occur between records at stations A and B. The gradual rise of pressure attributable to primary shock-wave attenuation is visible. A slight change in slope, owing to the increased attenuation after the transition from a laminar to a turbulent boundary layer[4], may be discerned. Finally, all records have failed to show any "dip" in the pressure, as reported by other authors,[4,5] although the utmost care was exercised in carrying out the experiments so as to avoid spurious signals. All the above findings are valid for all gas combinations, throughout the full range of M_s (from 1.1 to 8.5) and p_1 (0.5 to 760 mm Hg).

Lateral-wall pressure records. By contrast with the above, the pressure signals at station C depend on the experimental conditions. In argon at all initial pressures and shock strengths and in air at lower M_s and higher p_1, the traces show two clear pressure steps, corresponding to the incident and reflected shocks. At lower initial pressures (under 50 mm Hg) in air, however, the second step is followed by a gradual rise, which becomes longer as p_1 is lowered and the shock strength is increased (Fig.5). In the extreme, the picture is completely distorted, the pressure plateau behind the reflected shock not being reached until the arrival of the contact surface. However, the phenomenon is reproducible and records taken in successive tests are similar even in their details (Figs. 6,7).

It may be concluded that these features are due to the bifurcation of the reflected shock, which replaces the step rise in pressure with a gradual one. The effect is very marked in CO_2 even at $p_1 = 1$ atm.

Heat-transfer gauge signals

End-wall records. These show rather complex behavior, depending on the nature of the gas, the initial pressure and shock strength. In argon, where the reflected shock does not bifurcate, and in air and nitrogen at $p_1 > 100$ mm Hg, the traces at stations A and B are regular and similar, showing a gradual rise, owing to primary shock attenuation, with a slight change in slope marking the transition from a laminar to a turbulent boundary layer. The trace at B rises slightly less than at A (Fig.8). The influence on the attenuation is, however, much greater than on pressures. At high Mach numbers especially, the wall temperature may increase by more than 200 percent due to attenuation of the incident shock wave.

By contrast, in CO_2 at all pressures and Mach numbers, and in air and N_2 below 25 mm Hg, the shape of trace B changes dramatically. The slight rise due to attenuation remains; but, after a certain time T_τ, which depends on M_s, the wall temperature begins to increase markedly, almost linearly (see Figs. 3 and 4, lower traces). Later, it levels off for a short time before the arrival of the re-reflected wave from the interaction between the reflected shock and the contact surface (this is more evident at the lower Mach numbers).

For comparison, in Fig. 9, the two traces A and B from Figs. 3 and 4 have been superposed, adjusting the scales to fit the sensitivities of the two gauges. The points corresponding to breaks in slope are identified as P, Q and R. It should be pointed out that, before the arrival of the disturbance at Q, the trace at station B reveals a better uniformity of the hot region than in the centre of the end-plate; later, this situation is reversed.

In Fig. 10, the duration T_τ of the temperature plateau has been plotted against the Mach number, in air and nitrogen. Although the spread of the experimental points is rather important, a clear tendency can be discerned, and an empirical curve can be drawn through the results.

The dependence of T_τ on the initial pressure is less obvious, and attempts to correlate the results in the form $p_1^{\frac{1}{2}}T_\tau = f(M_s)$ have failed. The influence of p_1 shows up rather as a threshold (about 25 mm Hg for air) above which the break in the slope of the signals disappears, the trace B showing instead a gradual temperature increase, while above 100 mm Hg the two temperature records at A and B become quite similar.

These effects are no doubt due to the bifurcation of the reflected shock, but their interpretation is not clear. The rise in temperature after T_τ may be due to a residual flow of hot gas, after passing through the foot of the bifurcated shock. The time delay T_τ may correspond to an induction time, necessary for the vortex behind the bifurcated foot[1] to break down and arrive at the end-wall. It is, however, surprising that no disturbances reach the centre of the end-plate. A detailed visualization of the flow in this region would be very useful.

The third break in the slope (point R) and the subsequent levelling off in temperature might be explained by the penetration of cold driver gas through the bifurcated foot, shortly before the arrival of the reflected shock.

It should be emphasized that the above phenomena are quite sensitive to the experimental conditions. A number of tests have been carried out in which the boundary layer has been artificially rendered turbulent, by putting inside the tube several layers of Scotch tape (0.27 mm thick) about 100 mm upstream of the end-plate. The effect of this disturbance was to destroy the sharpness of the slope break Q. However, the trace at B still had a more pronounced increase (at low p_1) than at A.

Boundary-layer transition. A limited number of tests were made at station D (Fig. 1) with a heat-transfer gauge very carefully adjusted to fit smoothly to the shock tube inner surface. A typical record (Fig. 11) shows the transition from a laminar to a turbulent boundary layer. The duration of the laminar flow, correlated as $T_\tau p_1^{\frac{1}{2}}=f\ (M_s)$, agrees, within a spread of about 25 percent, with the curve given by Hartunian (cited in Ref. 4). This shows that, in our shock tube, the boundary layer behaves regularly.

SHAPE OF THE CONTACT SURFACE

In order to ascertain the influence of the interaction between the reflected shock and the contact surface on the properties of the hot region, a study of the shape of the contact surface and of the influence of the boundary layer upon it has been carried out. Tests have been performed in a cylindrical shock tube (50 mm dia. x 9 m long), with N_2/A, at p_1 = 0.5 to 2 mm Hg, giving shock Mach numbers between 3 and 4. In these conditions, the boundary layer remains laminar and is sufficiently thick to have a non-negligible effect on the flow.

The arrival of the contact surface was detected by a small (2 mm long) hot-wire probe placed at various distances from the diaphragm station, and set in various depths across the diameter. It can be shown[3] that, when the wire is submitted to a constant-velocity flow, the voltage drop varies exponentially; if, moreover, the time-constant of the wire is much longer than the flow duration, the trace is practically straight, being interrupted by the arrival of the contact surface (Fig. 12, upper trace). When the wire is close to the tube wall, within the boundary layer, where the flow is nonsteady, its response is no longer linear. Thus, the boundary-layer thickness can be approximately determined, as well as the arrival of the cold driver gas (Fig. 12, lower trace).

In Fig. 13 the results are presented, for various distances from the diaphragm x. One can see the attenuation of the primary shock, as well as the increasing convexity of the contact surface, owing to the development of the boundary-layer. The response of a heat-transfer gauge placed on the tube wall (y = o) is also shown; it gives an apparently shorter time of arrival of the contact surface, but this may be due to boundary-layer transition being produced by the gauge itself.

Our results substantiate the theoretical prediction of the hot flow duration, based on the assumption of a leaky piston.[6] It follows from this theory that the flow duration tends towards a limit, T_m, which has actually been reached in our tests. Figure 14 is a comparison of measured test durations (in the centre of the tube) with theoretical curves[6, 7] and other experimental results.[6,8,9]

INTERACTION BETWEEN THE REFLECTED SHOCK WAVE AND THE CONTACT SURFACE

The difference between the durations of the constant-pressure and constant-temperature regimes at the shock-tube end has been noticed by many authors[1], However, conflicting opinions are expressed as to the reasons. Some point to the instability of the contact surface, while others discuss the effects of the reflected shock bifurcation. The results of our series of tests are summarized below.

Experiments with helium/argon

These have been carried out in a cylindrical tube (100 mm dia. x 4 m long) at rather low initial pressures (p_1 = 3 to 50 mm Hg) in order to overemphasize the boundary-layer effects, if they exist at $M_s = 2.5$ to 4.5. At these conditions, the reflected shock does not bifurcate in argon or in the helium driver gas below $M_s = 4.3$. Thus, any arrival of cold gas at the end plate through the foot of the bifurcated shock can be excluded, while the contact surface instability begins, theoretically[1], for $M_s \geq 3.85$.

The pressure measurements, made at the centre of the end plate, show the arrival of the re-reflected wave from the contact surface, which below $M_s = 3.65$ (experimentally determined tailored shock Mach number) is a rarefaction wave (Fig. 15 a, upper trace), and above that value a shock wave (Fig. 15 c). At the tailored condition (Fig. 15 b), the pressure remains practically constant for about 3 msec, until the arrival of the head of the reflected rarefaction wave from the driver chamber end.

The heat-transfer records at station A also show the arrival of a disturbance, almost coincident with that appearing on the pressure traces. However, two important features must be emphasized: first, that this disturbance always corresponds to an increase in wall temperature, either below or above tailoring (Fig. 15 a and c, lower trace); secondly, that it appears even at $M_s = 3.65$ (Fig. 15 b) when no perturbation is present on the pressure records.

A possible explanation of these facts could be that the disturbance is produced by the arrival of cold driver gas at the end plate, due to the instability of the contact surface. This gas, being denser and having a residual velocity, might produce the observed rise in the wall temperature. The "bubbles" of cold gas, detached from the contact surface, and bearing no pressure differentials, are somehow channeled by the boundary layer towards the center of the end plate. The convexity of the contact surface may also contribute to the observed behavior.

On the other hand, heat-transfer records taken at station B are markedly different from those at the centre of the end plate (Fig. 16). At $M_s = 3.7$ and below, the duration of the undisturbed signal is much longer (about 3 msec), showing that the first disturbance reaches the centre of the end-plate long before the edges. Above tailoring, however (Fig. 16 b), the duration of the undisturbed signals coincides for the two heat-transfer gauges and the pressure pick-up.

Figure 17 summarizes the results. Finally, in Fig. 18, the duration of the undisturbed regime at the shock tube end is compared with Mirels's theory.[7]

Tests with hydrogen/argon

These experiments have been carried out in a cylindrical shock tube (100 mm in dia. x 6 m long). The behavior is very regular. On all traces of pressures and temperatures at stations A and B, the appearance of disturbances are coicident, corresponding to a rarefaction wave below tailoring, and to a shock wave above. At the tailoring Mach number (about 6.5), the duration of the undisturbed regime is limited only by the head of the reflected rarefaction wave, to about 1 msec (Fig. 19). It should be pointed out, however, that the pressures, and especially the temperatures, are affected by the attenuation of the primary shock, the constant plateau being replaced by a gradual increase of the signals. We conclude that no instability of the contact surface is evident in this case.

Tests with hydrogen/nitrogen

Here the pressure records again are regular, in the under-tailored (Fig. 20 a), tailored (Fig. 20 b) and over-tailored (Fig. 20 c) cases. The tailored shock Mach number is at $M_s = 5.75$.

The interpretation of the heat-transfer traces is made more difficult by the effects discussed in Section 2.2.1 End-wall records. We tried to avoid them by increasing the intial pressure, but safety reasons limited p_1 to about 50 mm Hg, to obtain the necessary shock strengths. In these conditions, the centre trace always shows a disturbance, appearing approximately at the same time as on the pressure records, but this corresponds in all cases to a fall in temperature (even above tailoring, at Mach numbers as high as 7.2), and is present also at the tailored Mach number. The lateral trace, although somewhat disturbed, exhibits a continuing rise in temperature (Fig. 21).

These results suggest that, in the case of H_2/N_2 also, the contact surface is unstable and, even at the tailored condition, the arrival of cold gas disturbs the hot regime at the shock tube end. The delay in the appearance of disturbances at station B seems to show that the bifurcation of the reflected shock is not responsible for the phenomenon. It would be interesting to extend the experiments with N_2/H_2 at higher initial pressure, when the lateral temperature trace is less disturbed by the bifurcation, and where the effects of the contact surface instability should be more evident.

In Fig. 22, the duration T_τ of the undisturbed pressure plateau is plotted against the shock Mach number, for H_2/A and H_2/N_2. No attempt has been made to correlate these points with durations measured on the heat-transfer traces, since the perturbations appearing on these would make the scatter of the points unacceptably wide.

DISCUSSION

The collected data permits us to make some interpretations and draw some conclusions, as follows:

1. The very regular behavior of the pressure traces at the end plate is a striking feature. This can be explained by the fact that any pressure non-uniformity gives rise to waves, which quickly re-establish equilibrium. By constrast, temperature gradients are much more slowly attenuated, by slow gas motions and diffusion.

The results also show the inconsistency of drawing any conclusions concerning the pressure history at the end of the shock tube from records taken from gauges mounted in the side wall (Sec. 2.1.2. Lateral-wall pressure records) due to the effects of bifurcations. The "dip" in pressure reported by other authors[4,5] may be explained by a faulty mounting of the pick-ups, leaving a small cavity in front of the sensing surface.

2. By contrast, the strong temperature non-uniformities which are evident across the end-plate deserve detailed consideration. Our experiments have led us to the tentative conclusion that the bifurcation of the reflected shock, and the subsequent boundary-layer detachment, are responsible for the temperature disturbances appearing shortly after the reflection of the primary shock.

On the other hand, the fact that in certain cases, and in contrast to the pressures, the temperatures fail to exhibit tailoring seems to be due to the instability of the contact surface, as evidenced especially by the results with He/A (where the reflected shock is not bifurcated) and by the fact that the first disturbance appears at the centre of the end plate. This should also be connected with the convexity of the contact surface (Sec. 3). In all cases, any study of shock-tube phenomena should not neglect this latter effect.

Summarizing, it appears that the contact surface is stable, allowing operation in the tailored mode, for H_2/A. However, it is unstable for He/A and H_2/N_2.

3. As pointed out, the uniformity of the hot region parameters is, in certain cases, better near the edge of the end-plate than at its centre, at least for short durations (see Figures 9 and 16). This might have some practical applications, e.g., for choosing the optimum

position of sampling orifices for mass spectrometric studies with shock tubes.

4. Finally, the importance of the gradual rise in pressure, and especially in temperature, due to primary shock attenuation, must be stressed. The effects on the quality of hypersonic flows generated in shock tubes should be carefully assessed. Perhaps these non-uniformities could be minimized by using a tapered shock tube. However, this requires further study.

ACKNOWLEDGEMENTS

The authors gratefully acknowledge the continuous support of Professor J. Valensi, Director of the Institut de Mécanique des Fluides de Marseille, and Professor E. Carafoli, Director of the Institute of Fluid Mechanics, Bucharest, towards the completion of this work.

REFERENCES

1. Davies, L.and Wilson, J.L. 6th International Shock Tube Symposium, Freiburg, 1967, pp. 213-235.
2. Dumitrescu, L.Z. 6th International Shock Tube Symposium, Freiburg, 1967, pp. 574-588.
3. Christiansen, W.H. Galcit Hypersonic Research Memo 62 (1961).
4. Rudinger, G. Phys. Fluids 4,(1961), 1463-1474.
5. Center, R. E. Phys. Fluids 6,(1963), 307-308.
6. Roshko, A. Phys. Fluids 3, (1960), 835-842.
7. Mirels, H. Phys. Fluids, 6, (1963), 1201-1214.
8. Hooker, W.J. Phys. Fluids, 4, (1961) 1451-1463.
9. Fox, J.N., McLaren, T.I., and Hobson, R.M. Phys. Fluids 9, (1966).
10. Dumitrescu, L.Z. Cercetâri în tuburile de soc (Studies in Shock Tubes), Editura Academiei, Bucharest, 1969.

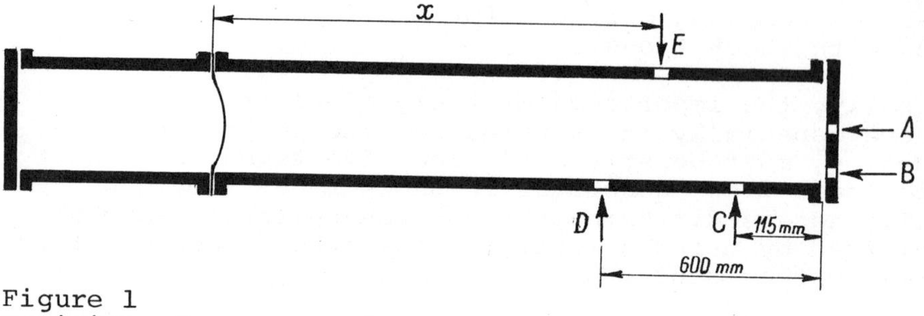

Figure 1
Position of sensors along the shock tube

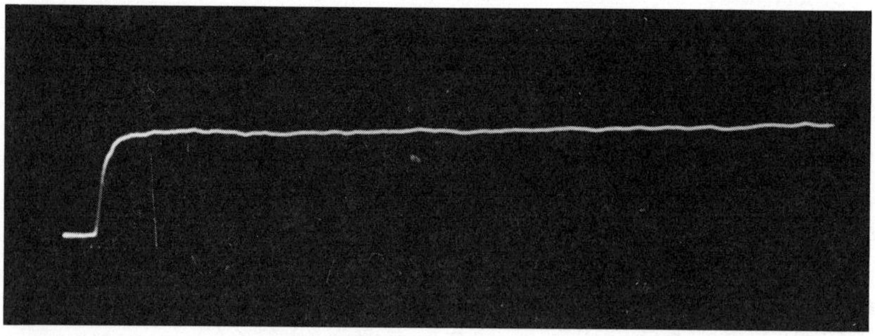

Figure 2
Pressure records at station A, in air p_1=4.9 mm Hg, M_s=3.86
Time base 500 μsec

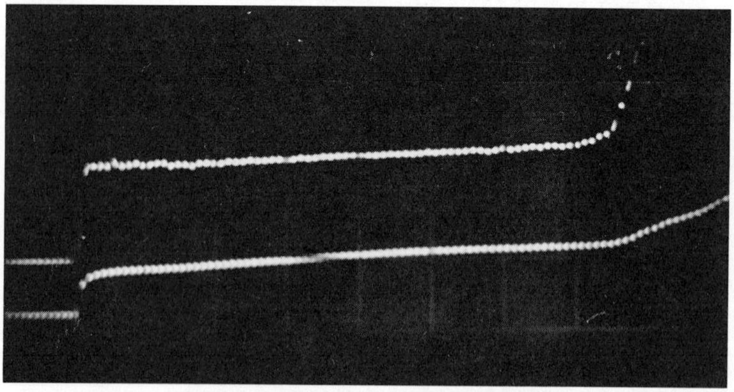

Figure 3
Pressure and temperature records at station A, in air
p_1=8.5 mm Hg, M_s=3.80
Sweep rate, 100 μsec/sq

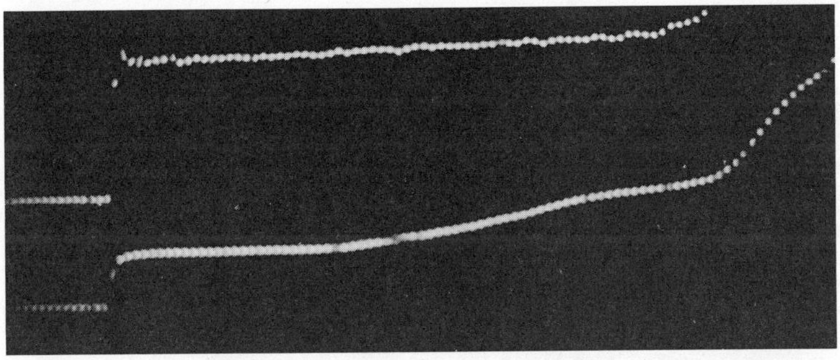

Figure 4
Pressure and temperature records at station B, in air p_1=9.3 mm Hg, M_s=3.75 Sweep rate, 100 µsec/sq

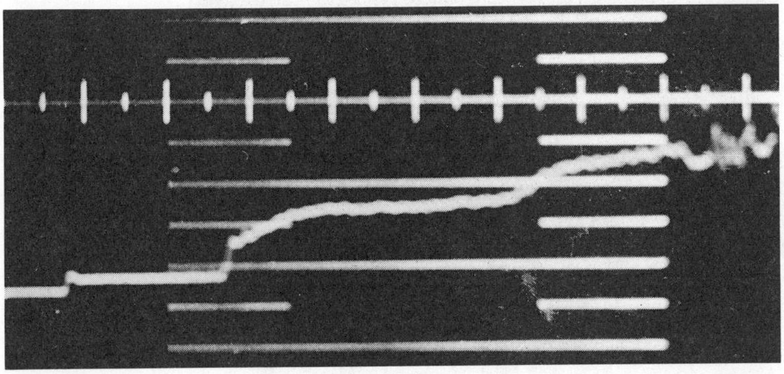

Figure 5
Lateral Wall pressure variation station C, in air p_1=10.2 mm Hg, M_s=2.96 Sweep rate, 200 µsec/sq

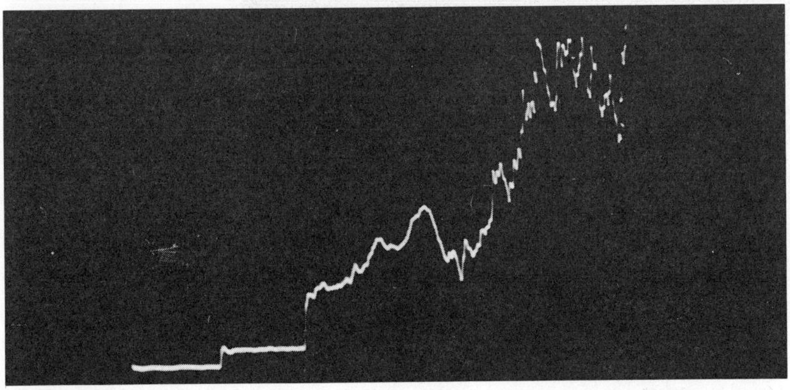

Figure 6
Lateral Wall pressure variationstation C, in air at low pressures p_1=9.9 mm Hg, M_s=3.66 Time base, 2 msec

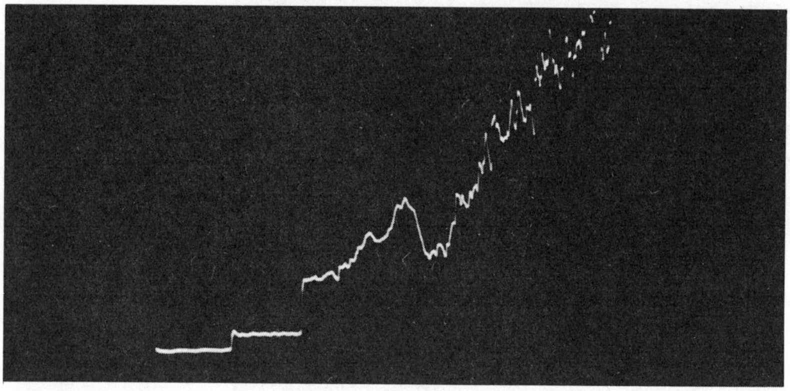

Figure 7
Repeated result (Figure 6)
p_1=10.2 mm Hg, M_s=3.67

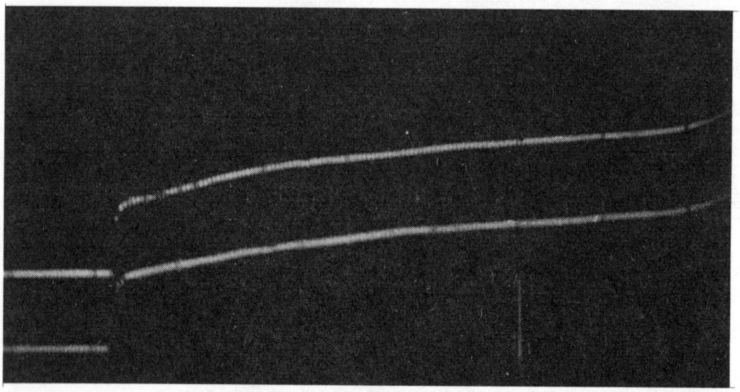

Figure 8
End-plate temperature
records in air.
Upper trace, station A
Lower trace, station B
p_1=104 mm Hg, M_s=2.85
Sweep rate, 200 μsec/sq

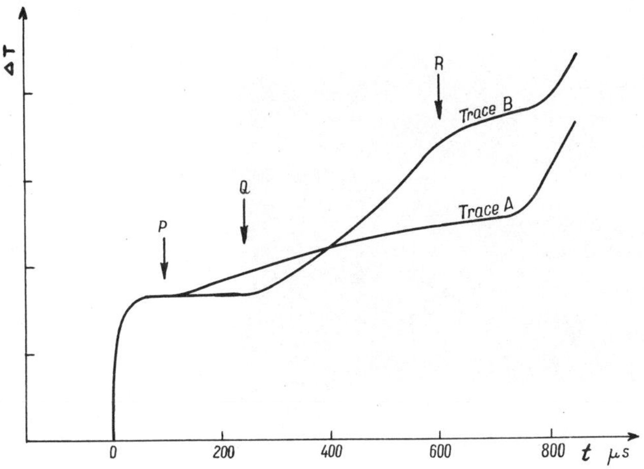

Figure 9
Comparison of temperature traces
at stations A and B,
from Figures 3 and 4

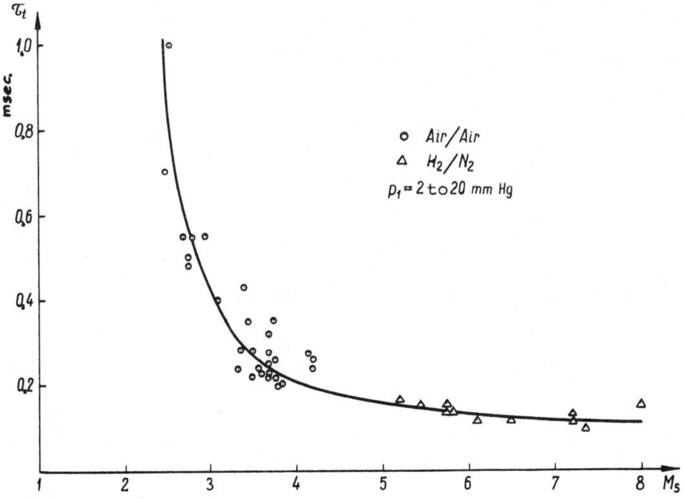

Figure 10
Duration of undisturbed
temperature regime at station B,
in air, for varying Mach number

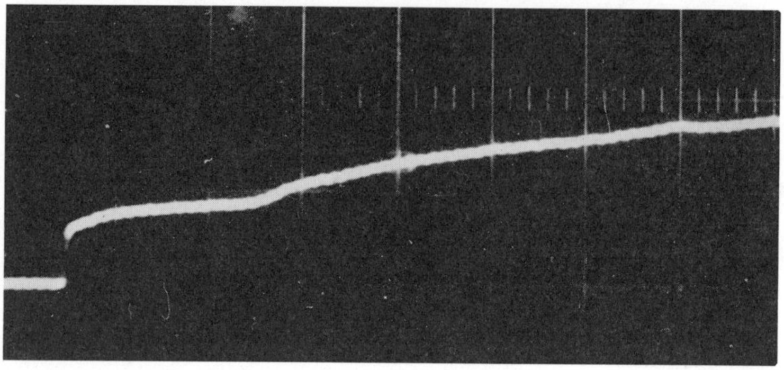

Figure 11
Laminar-turbulent boundary-layer transition in air, evidenced by heat-transfer record at station D.
p_1=15 mm Hg, M_s=3.22
Sweep rate, 100 μsec/sq

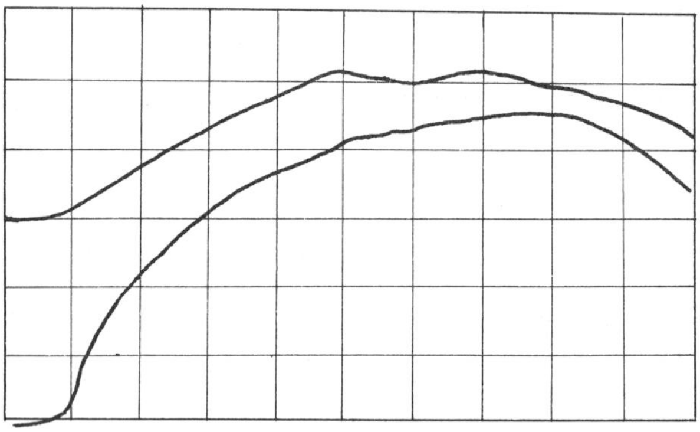

Figure 12
Example of hot-wire probe response, in argon, at x = 5.68 m (station E),
Upper trace: probe mounted on the centreline (y= 24.7 mm).
Lower trace: probe near the wall (y=3.1 mm).
p_1=2mm Hg, M_s=3.30
Sweep rate: 200 μsec/sq

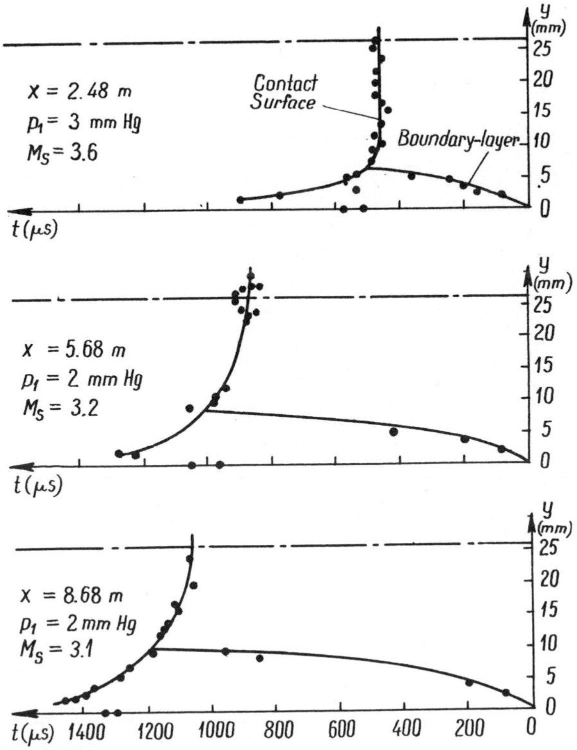

Figure 13
Evolution of the boundary-layer and contact surface along the shock tube, in argon.

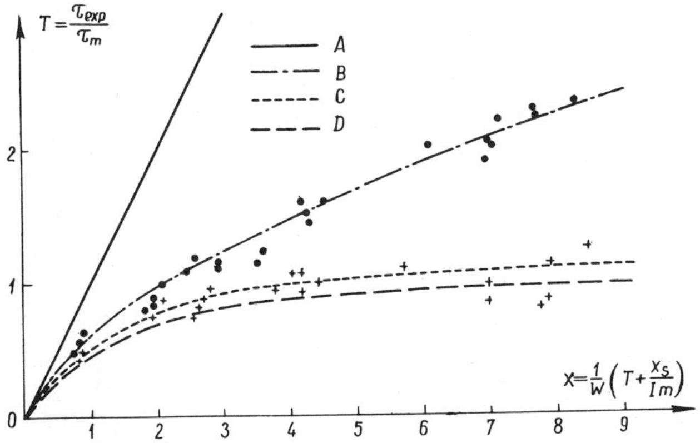

Figure 14
Flow duration in argon
A: Theory (ideal)
B: Flow duration measured at the wall
C: Flow duration at the shock-tube centre
D: $\frac{1}{2}X = \log(1 - \sqrt{T}) + \sqrt{T}$

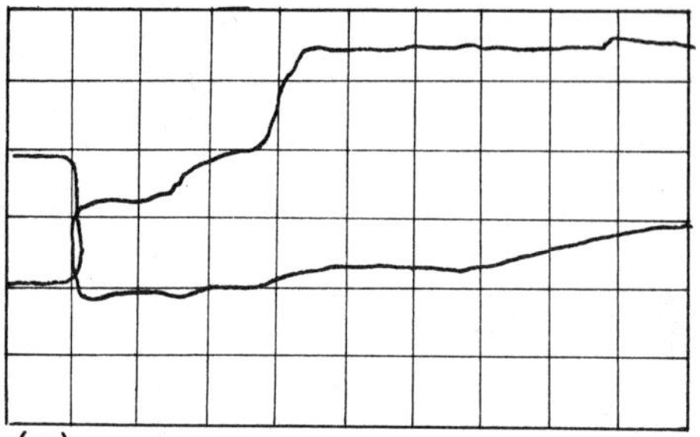

(a)

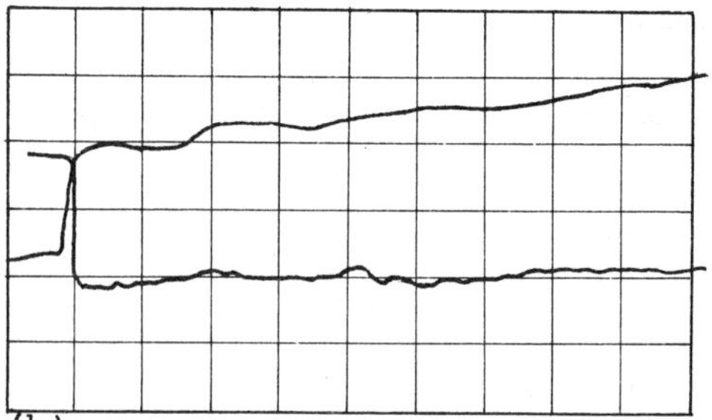

(b)

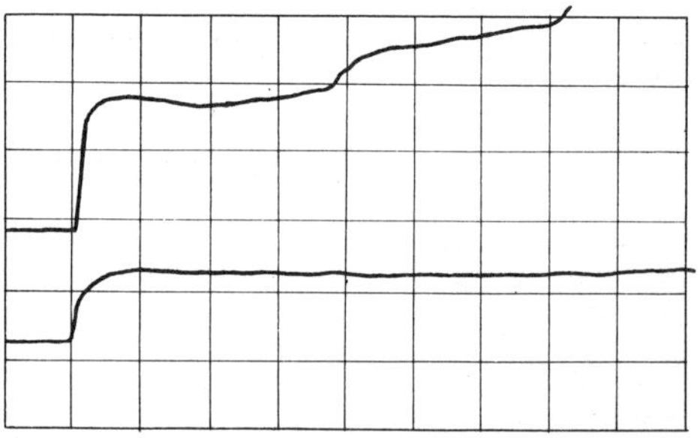

(a)

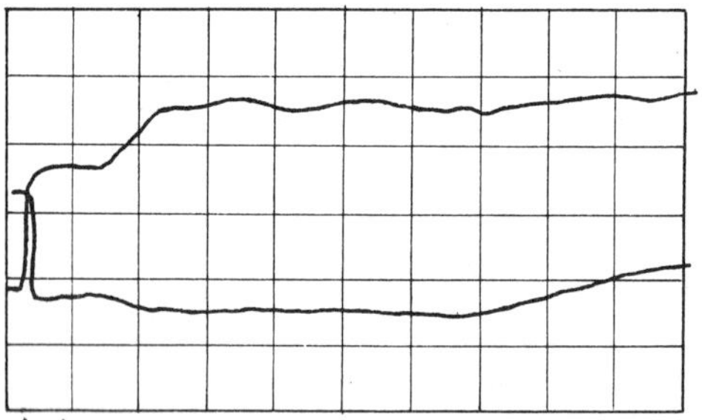

(c)

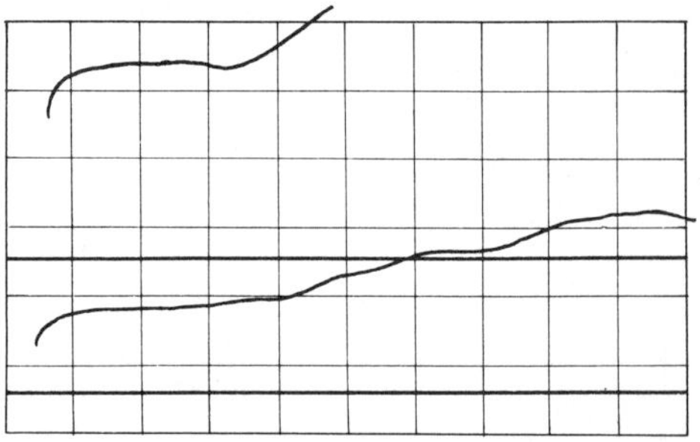

(b)

Figure 15
Examples of pressure and temperature records at the centre of the shock tube end-plate, with He/A
Upper trace: pressure (increasing downwards)
Lower trace: temperature (station A)
Sweep rate: 500 μsec/sq.
a: M_S=3.47, undertailored
b) M_S=3.66, tailored
c) M_S=4.35, overtailored

Figure 16
Examples of temperature records He/A
Upper trace: Station A
Lower trace: station B
Sweep rate: 200 μsec/sq.
a) M_S=3.47
b) M_S= 4.35

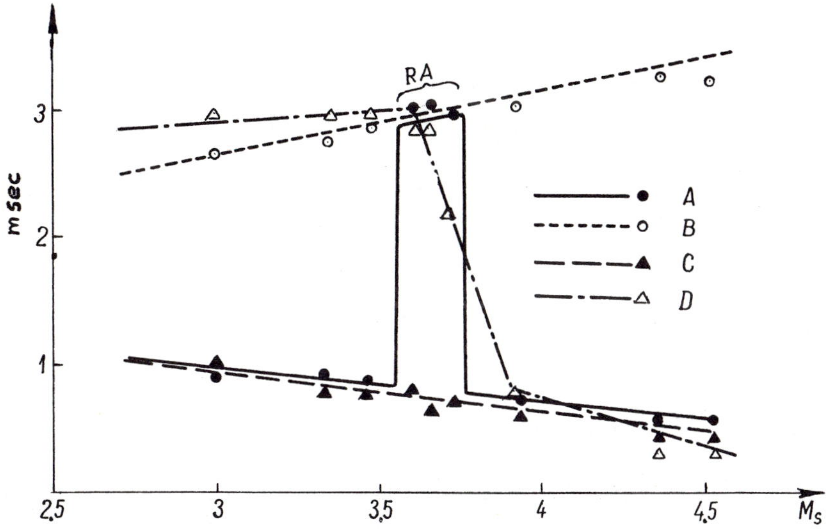

Figure 17
Duration of undisturbed temperature and pressure signals at the end of the shock tube, He/A:

A: Duration of the steady pressure regime between arrival of the primary shock and that of the reflected wave, resulting from the interaction between the reflected shock and the contact surface; R.A: tailoring regime.

B: Duration of constant pressure within 15 percent (Ref.1)

C: Duration of constant temperature at station A (end-plate centre)

D: Duration of constant temperature at station B (end-plate edge)

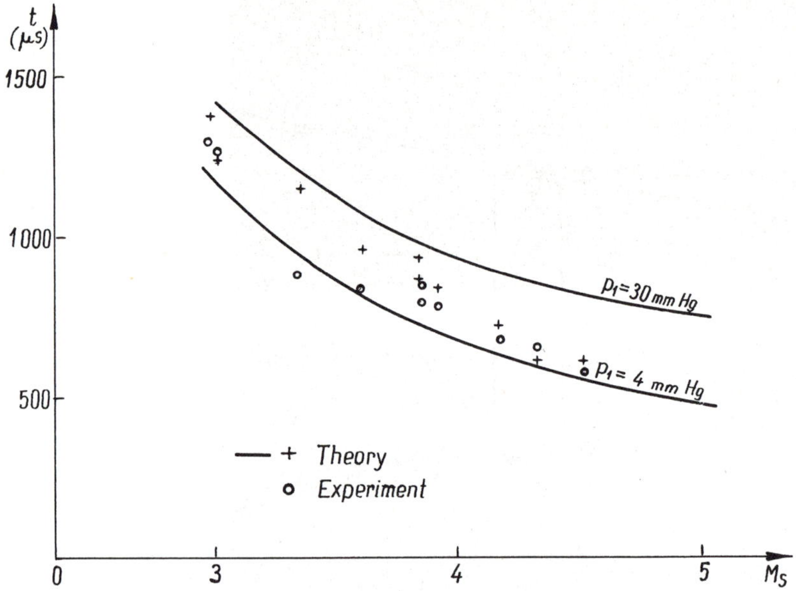

Figure 18
Duration of undisturbed pressure signal
at the end of the shock tube, He/A.

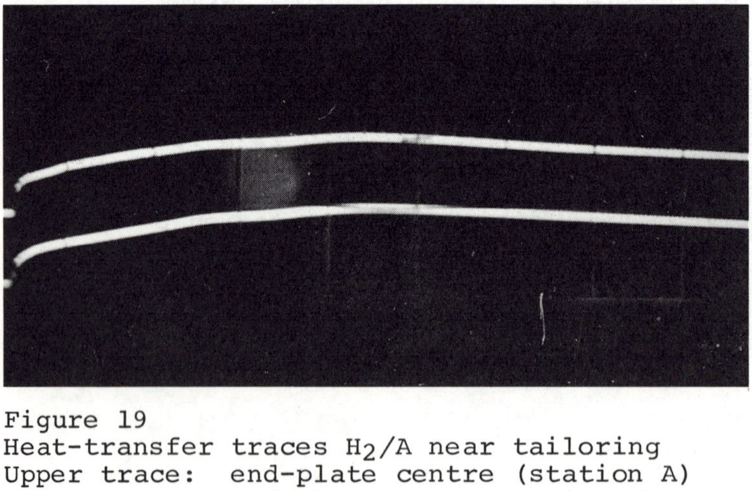

Figure 19
Heat-transfer traces H_2/A near tailoring
Upper trace: end-plate centre (station A)
Lower trace: end-plate edge (station B)
p_1=15.8 mm Hg, M_s=6.8
Sweep rate: 200 μsec/sq

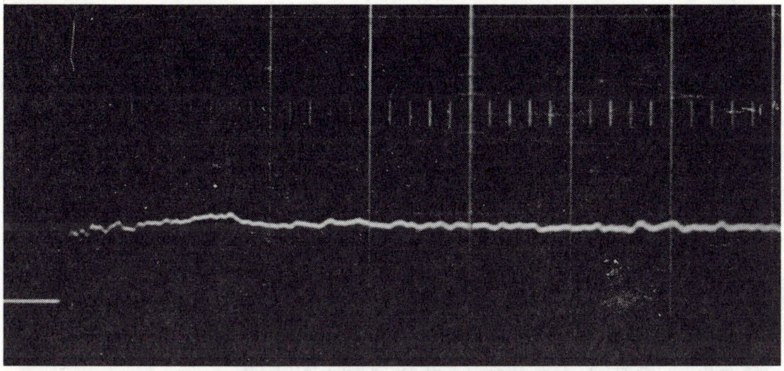

(a)

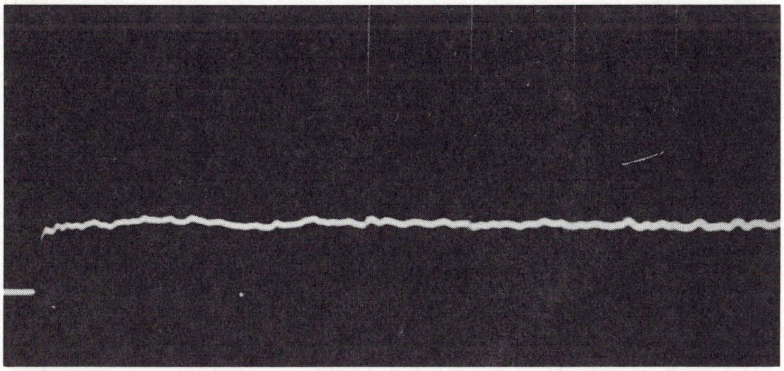

(b)

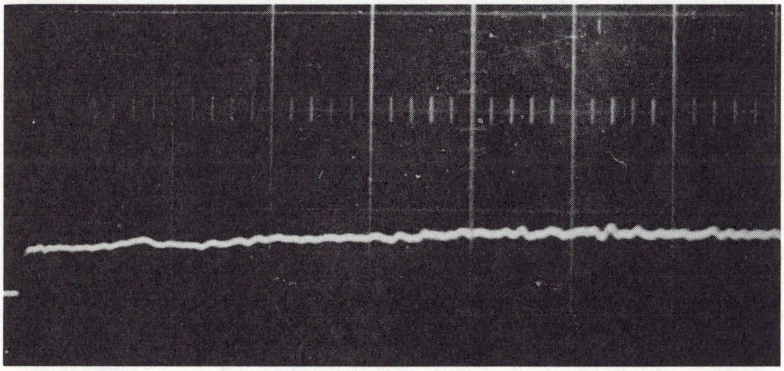

(c)

Figure 20
Examples of pressure records
H_2/N_2 at the end-plate
Sweep rate: 200 μsec/sq.

a) Undertailored: p_1=25 mm Hg, M_s=5.2
b) Tailored: p_1=19.5 mm Hg, M_s=5.47
c) Overtailored: p_1=10.4 mm Hg M_s=6.1

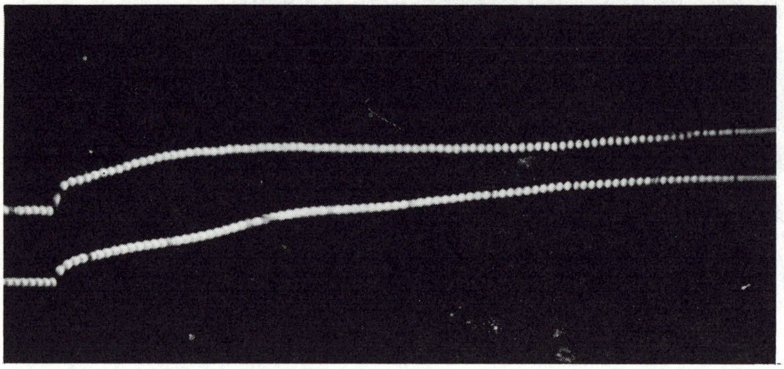

Figure 21
Heat-transfer records H_2/N_2 near tailoring, at the shock tube end
Upper trace: centre (station A)
Lower trace: edge (station B)
p_1=51 mm Hg, M_s=5.66. Sweep rate: 100 μsec/sq

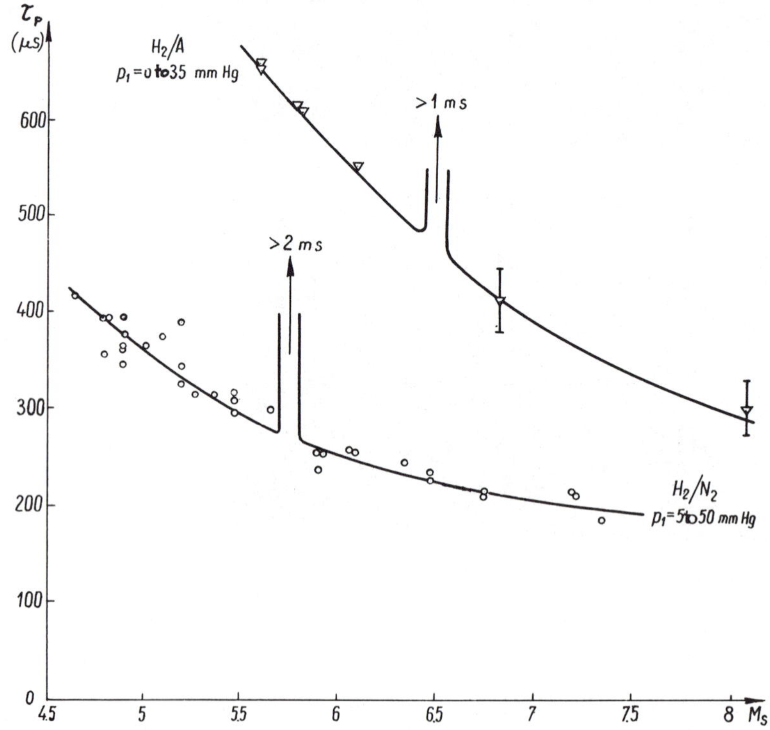

Figure 22
Duration of undisturbed pressure signal as a function of M_s, H_2/A and H_2/N_2.

TEMPERATURE MEASUREMENTS OF AIR PLASMAS IN THE REFLECTED SHOCK REGION*

ALLEN D. WOOD and K. H. WILSON
Lockheed Palo Alto Research Laboratory
Palo Alto, California

SUMMARY

The temperatures of air plasmas behind a reflected shock wave were measured over the interval 11 – 13,000 K. These measurements had a twofold purpose. First, very few diagnostic investigations have been carried out in the region above 10,000 K, and radiative cooling effects may become important here.[1] Second, the plasma temperature is vital for a meaningful comparison of experimental measurements and theoretical predictions for the total radiant intensity of high-temperature air.[2] The latter purpose demands accurate measurements because the total intensity varies as about the tenth power of temperature at the conditions of interest.

To eliminate systematic errors, three different spectroscopic measurement techniques were employed. These were measurements of the integrated intensities from three NI and OI lines, the spectral continuum intensity at 4935 A, and the shapes of an NI line and H_α and H_β of the Balmer series. The requisite data were taken simultaneously with a spectrograph utilizing time-integrated recording on photographic film and with a multichannel monochromator (i.e., a polychromator) utilizing time-resolved photoelectric recording.

DESCRIPTION OF APPARATUS

The experiments were conducted in a 12-in.-diameter shock tube with electric-arc heated helium as the driver gas. The driven gas was air at an initial pressure of 0.2 Torr. The reflected shock region encompassed the full diameter (30.5 cm), and the spectroscopic measurements were made through flush-mounted quartz windows along a slab of gas 1.5 mm wide and located 7 mm out from the end wall. The incident shock velocities, which were 7 – 8 mm/μsec, were measured over the last 3 ft using ionization pickups. The incident shock test times were determined photoelectrically[2] and were typically 25 μsec. These times were comparable to the observation times at the measurement location in the reflected shock region.

---

* This research was supported by the National Aeronautics and Space Administration under Contract NAS 7-579 issued by the Fluid Physics Branch, Research Division, Office of Advanced Research and Technology.

The spectroscopic measurements utilized two Model 78-000 Jarrell-Ash 1.5-meter Wadsworth spectrographs equipped with 590 groove/mm gratings. Each had an effective aperture of f/29 and a first-order dispersion of about 11.2 A/mm. One, the spectrograph, was equipped with a quartz optical system and a rapid-closing (~ 5 μsec) capping shutter which yielded exposure times of about 25 μsec. By using Kodak Type 2475 Recording Film and various order-blocking filters along the focal plane, the entire 2500 − 7000 A range was covered at one time in first order. A photoelectric monitor of the zero order (white light) gave information about the temporal behavior of the recorded radiation and monitored the action of the shutter. A six-step optical filter was placed across the entrance slit so that relative spectral intensity profiles could be generated by conventional photometric techniques.

The other instrument was converted into a polychromator by placing five RCA 1P28 photomultiplier tubes and color filters behind fixed exit slits located along the focal plane. The entire system was calibrated absolutely by using a standard lamp with a tungsten ribbon filament. Both instruments were used with 50 μ entrance slits, while the spectral widths of the polychromator exit slits were 5 − 50 A. They viewed the same plasma but from opposite ends of the 30.5-cm path length.

PRELIMINARY SPECTRAL SURVEYS

Two prerequisite steps before undertaking the thermometric measurements were an identification of the radiative features and a determination of the steadiness of the emergent radiation.

Spectrograms were obtained at temperatures of 9 − 14,000 K which encompassed a wide range of electron densities ($1\times10^{16} - 3\times10^{17}\,cm^{-3}$). The nature of the radiation changed markedly. At the low temperature end there was a considerable amount of molecular radiation, mostly from the N_2^+ (1st Neg) band system, very little continuum, and a few narrow atomic lines. At intermediate temperatures the molecular radiation had largely disappeared and was supplanted by a continuum and many prominent atomic lines and multiplets. At high temperatures the continuum became very strong and the lines became quite broad and overlapping and were difficult to see above the continuum.

Nearly all the NI and OI lines and multiplets that were expected based on the tabulations of Wilson and Nicolet[3] were identified. Four untabulated multiplets, lying in the 4350 − 4670 A region, were identified as belonging to atomic nitrogen by comparison with the spectral scans of Morris et al.[4] The only impurity radiation ever seen in emission was that of hydrogen (H_α and H_β) and the intensities indicated an initial H_2 concentration of a small fraction of 1%. On the several occasions when the shutter failed to operate, lines from Fe, Ni, Cr, and Al were identified in absorption.

The polychromator was used initially to monitor a number of spectral features of the plasma, including various NI and OI lines, H_α, the continuum at 5469 A, and the N_2^+ (1st Neg) band head at 3914 A, as well as the nonspectral zero-order light. These monitors were all remarkably similar and reasonably constant over the first 25 μsec following the 2 μsec transient caused by the passage of the reflected shock. This similarity was taken as an indication of the expected presence of chemical equilibrium. The zero-order light was included to show that white-light photoelectric monitors are valid indicators of temporal variations in plasmas of this type.

To illustrate the temporal behavior, various photoelectric traces from 12 shots are arranged in Figure 1 in order of increasing incident shock velocity. The traces are principally from either the 5469 A continuum polychromator channel or a white light photoelectric monitor (labeled the model profile), discussed elsewhere.[2] The action of the rapid closing shutter is shown directly on the two spectrograph zero-order traces and by the dashed lines on the others. Although shot-to-shot variations are apparent, the broad trend is that the traces are generally flat for $U_s < 7.5$ mm/μsec. Those at higher velocities show a gradual decrease with time which indicates a cooling phenomenon. However, because the radiation is such a sensitive function of temperature, the corresponding temperature decreases are quite small. For example, the visible continuum at 14,000 K (U_s=8.3) varies as the eighth power of temperature and hence the 15% decrease in the model profile trace over 25 μsec implies a temperature drop of less than 2%. Since a result obtained from a time-integrated spectrogram would vary by about half this amount, the effects of radiative cooling were small when compared with the ± 4% uncertainty of the thermometric measurements.

PLASMA THERMOMETRY

The spectral surveys showed that both chemical equilibrium and steady state conditions would prevail in the reflected shock region. This meant that a time-integrated spectrogram was a valid representation of the plasma radiation. Therefore, the thermometric techniques discussed below depended to some extent on data obtained from both the spectrograph and the polychromator.

Line-Shape Thermometry

The shape of an atomic line at the conditions of interest is governed by Stark broadening. Thus, a shape measurement yields the electron density and hence temperature through the Saha equation. The 4935 A NI line and H_α (6563 A) were selected for study because they are prominent, well-known, single lines located in fairly clear spectral regions. In addition, H_β (4861 A) was used on the one occasion when an excessive 1.3% initial H_2 concentration made it measurable.

The experimental line shapes were obtained from the spectrogram by the techniques of homochromatic photometry.[5] Then with the aid of a digital computer, these were compared against theoretical line profiles which were broadened by the instrument function and which included the effects of opacity and continuum radiation. For the Lorentzian shape of the NI line, the half-width data of Wilson and Nicolet[3] were used. For H_β, the shapes of Griem[6] and for H_α, both Griem and the somewhat wider half-widths recommended by Morris et al.[4] were used. (These Balmer line shapes have recently been supplanted by the work of Kepple.[7])

Examples of this shape-fitting process are shown in Figures 2, 3, and 4. Figure 2 shows the measured profile of the NI line at one of the lowest temperatures, the instrument function, and the points from the computer-generated shape that best fit the measured profile. The ordinate of relative spectral exposure is similar to a relative spectral intensity except for the broadening of the instrument function which, in this case, was quite important. The two profiles agreed to an rms value of 0.011 units of this ordinate.

This same line at the highest measured temperature is shown by Figure 3. Here the line is considerably broader, the continuum level is much higher, and the fit is even better. Some overlapping from the 4915 A NI line is apparent at this higher electron density. Figure 4 shows the result of a similar fitting process applied to H_β. Here the shapes corresponding to ±20% perturbations of the best-fit electron density are partially shown to indicate the sensitivity of the fit.

An error analysis revealed that the temperatures obtained from the NI line and H_α were uncertain to ± 4.5% at 12,000 K. This rose to ± 7% at 15,000 K because the electron density becomes a less-sensitive function of temperature. Because the shape of H_β is better known, it was a ± 2% thermometer at 11,000 K.

Continuum-Intensity Thermometry

In general, the continuum radiation of high-temperature air is considerably less certain than the strengths of many lines and multiplets and hence a continuum measurement might be considered to be an inferior thermometer. However, the 4955 A continuum of pure nitrogen has been extensively studied by Morris et al.,[8] who recommend its use in plasma thermometry. Therefore, the spectral intensity of the continuum underlying the 4935 A NI line was used to yield a temperature based largely on their measured recombination and attachment cross sections. This intensity was obtained from the continuum level determined by the shape-fitting process together with the absolute intensity obtained from the polychromator as discussed below.

Line-Intensity Thermometry

This technique involved an absolute measurement of the integrated line intensity from a particular atomic line or multiplet. This directly yields a temperature based on a Boltzmann distribution of the upper state population. The 4935 A NI line, the 6157 A OI multiplet, and the 6490 A NI multiplet were selected because they are prominent, well-known, and fairly clear spectrally. The requisite atomic data were taken from Wiese[9] if possible or from the tabulations of Wilson and Nicolet.[3]

The integrated line intensities were obtained essentially by calibrating the spectrogram via the absolute intensities from the polychromator. This technique permitted the continuum level to be ascertained and provided a means of accounting for the line wings. The polychromator channels were arranged as follows:

Spectral feature	Center wavelength of exit slit, A	Full width of exit slit, A
NI line	4935	4.6
Continuum ?	5469	16.6
OI multiplet	6163	45.8
NI multiplet	6488	47.1
Continuum ?	7032	16.7

For the lines, the full widths of the exit slits were about six times the line or multiplet half-widths at nominal conditions. The two questionable continuum channels were employed to provide a consistency check, as will be demonstrated below. The spectral surveys revealed very few places in the visible region where one could unequivocally state that an exit slit measured only the continuum. The 5469 A region comes close, while the 7032 A region represents a trial and error minimum point since it could not be photographed. It will suffice to say that these two channels provided an upper bound on the continuum level.

An example of this consistency check is shown by Figure 5. The five solid symbols represent average spectral intensities over the width of the polychromator exit slits. The open symbols represent peak and continuum spectral intensities obtained with the aid of the spectrogram. A smeared-model continuum prediction at the temperature indicated by the 4935 A continuum intensity is shown, as is the Planck function at this same temperature. The other four continuum intensities consistently lay either quite close to or slightly above the continuum curve.

Also shown in Figure 5 is the continuum prediction evaluated at the gasdynamic temperature. This is the temperature obtained in the conventional manner from the measured incident shock velocity and initial pressure together with the Rankine-Hugoniot equations and a suitable thermochemical equilibrium calculation. It is clear from this figure that the gasdynamic temperature is too high.

The computations necessary to obtain temperatures from integrated line intensities were performed on a digital computer by an iterative procedure which correctly accounted for opacity effects. An error analysis revealed that the temperatures obtained in this manner were uncertain to ± 4% at 12,000 K and ± 9% at 15,000 K under optimum conditions. This increase occurs because the rate of loss of neutral particles by ionization is becoming appreciable and diminishes the domination of the Boltzmann factor. In practice, other difficulties arise. The continuum increases much more rapidly with temperature than does the spectral intensity of a line center because the lines must also broaden because of the increased electron density. The result is that the relative height of the continuum increases as about the sixth power of temperature. (Compare Figures 2 and 3.) Hence, the subtraction of this ever-increasing continuum results in a value for the integrated line intensity which becomes the difference of two large numbers. In addition, as the attendant line broadening increases the overlapping of adjacent lines and multiplets, the separation of their contributions becomes more difficult.

All these difficulties, when considered together, limited the practical usefulness of the line-intensity technique to temperatures below about 13,000 K. This is typified by Figure 6, which shows relative exposure profiles for the 6410-6570 A region for three shots representing an intermediate and two extremes in incident shock velocity. The superposition of these three profiles clearly demonstrates the rising continuum and the effects of line broadening, shifting, and overlapping. No attempt was made to separate the contributions from the various multiplets or to fit the H_α shape on the upper trace.

RESULTS

Using the above thermometric techniques, plasma temperatures were obtained for six shots covering the region 11,000−13,000 K. The results are shown in Figure 7, where the measured temperatures are plotted against the gasdynamic temperature. This figure reveals that the measured values lie generally within the heights of the error bars but that these values do not cluster about the gasdynamic temperature. Also some erratic behavior is indicated. (For example, the measured temperatures on shots 601 and 605 are nearly the same, while the gasdynamic temperatures differ by almost 1000 K.)

To locate possible systematic deviations, the measured temperatures on any one shot were averaged and the deviations of each value from this mean are shown in Figure 8. The results of some techniques do lie entirely on one side of the mean, but the deviations are within the uncertainties stemming from the respective f-numbers and half-widths. The use of the Griem[6] shapes for H_α would result in H_α temperatures higher by 3–4%.

More important, however, is the last column of Figure 8 which shows the deviations of the gasdynamic temperatures. These are definitely one-sided and average about 4% above the mean. The erratic nature of this temperature is exemplified by the lack of correspondence between its deviation and the incident shock velocity.

CONCLUSIONS AND RECOMMENDATIONS

The plasma temperatures measured during this experiment are in reasonable agreement with each other considering the expected errors. However, the mean of these measured values lies, on the average, about 4% below the gasdynamic temperature. Further, although the number of shots is not as high nor the scatter as low as might be desired, there are strong indications that the actual temperatures are randomly scattered in a manner not predictable from the incident shock velocity.

This average 4% difference observed between the measured and gasdynamic temperatures is certainly in the right direction and of about the correct magnitude to explain the 50% deviation between the theoretical prediction and the experimental measurements of the total radiant intensity of air at these conditions.[2] However, it is difficult to justify any stronger conclusion on the basis of six ± 4% measurements when a 4% effect is all-important.

The measurements reported herein were not significantly affected by radiative cooling phenomena in the reflected shock region. The analysis presented earlier showed this to be less than a 1% effect at temperatures higher than measured herein. The presence of a trace amount of hydrogen could not lower the gasdynamic temperature sufficiently. A calculation showed that 1% hydrogen in the driven gas would lower the gasdynamic temperature by only 50 K.

Currently, these thermometric measurements are being extended to higher plasma temperatures. Although the line-intensity technique is unsuitable because of the reasons discussed earlier, other techniques are being employed. The driven gas is seeded with 10% hydrogen and the techniques utilize the shape of H_β, the 4955 A continuum intensity, and the spectral intensity at the center of H_α which equals the Planck function with this high concentration of hydrogen.

Whether the average 4% difference between the measured and gasdynamic temperatures is significant depends on the type of investigation being performed in the shock tube. For some, the 4% difference might be regarded as a substantiation of the gasdynamic temperature. However, for others, and especially for those which involve measurements of radiative properties, this difference is quite significant. For these, the gasdynamic temperature may not adequately define the state of the reflected shock plasma and separate thermometric measurements may be required. Further, because of the erratic behavior, these measurements should be performed simultaneously.

REFERENCES

1. W. C. Griffith and A. D. Wood, Radiation Gas Dynamics in the Shock Tube. 6th Intl. Shock Tube Symposium, 1967.
2. A. D. Wood et al., Measurements of the Total Radiant Intensity of Air. AIAA J. 7, 130–139, 1969.
3. K. H. Wilson and W. E. Nicolet, Spectral Absorption Coefficients of Carbon, Nitrogen, and Oxygen Atoms. JQSRT 7, 891–941, 1967.
4. J. C. Morris, R. U. Krey, and R. L. Garrison, Radiation Studies of Arc Heated Nitrogen, Oxygen, and Argon Plasmas. U.S. Air Force Report ARL 68-0103, 1968.
5. R. A. Sawyer, Experimental Spectroscopy. Prentice-Hall, New York, chap. 10, 1951.
6. H. R. Griem, Plasma Spectroscopy. McGraw-Hill, San Francisco, 1964.
7. Paul C. Kepple, Improved Stark Profile Calculations for the First Four Members of the Hydrogen Lyman and Balmer Series. Univ. of Maryland Tech. Rept. 831, 1968.
8. J. C. Morris, R. U. Krey, and G. R. Bach, The Continuum Radiation of Oxygen and Nitrogen for Use in Plasma Temperature Determination. JQSRT 6, 727–740, 1966.
9. W. L. Wiese, M. W. Smith, and B. M. Glennon, Atomic Transition Probabilities, Vol. I, Hydrogen Through Neon. NBS Rept. No. NSRDS-NBS4, May 1966.

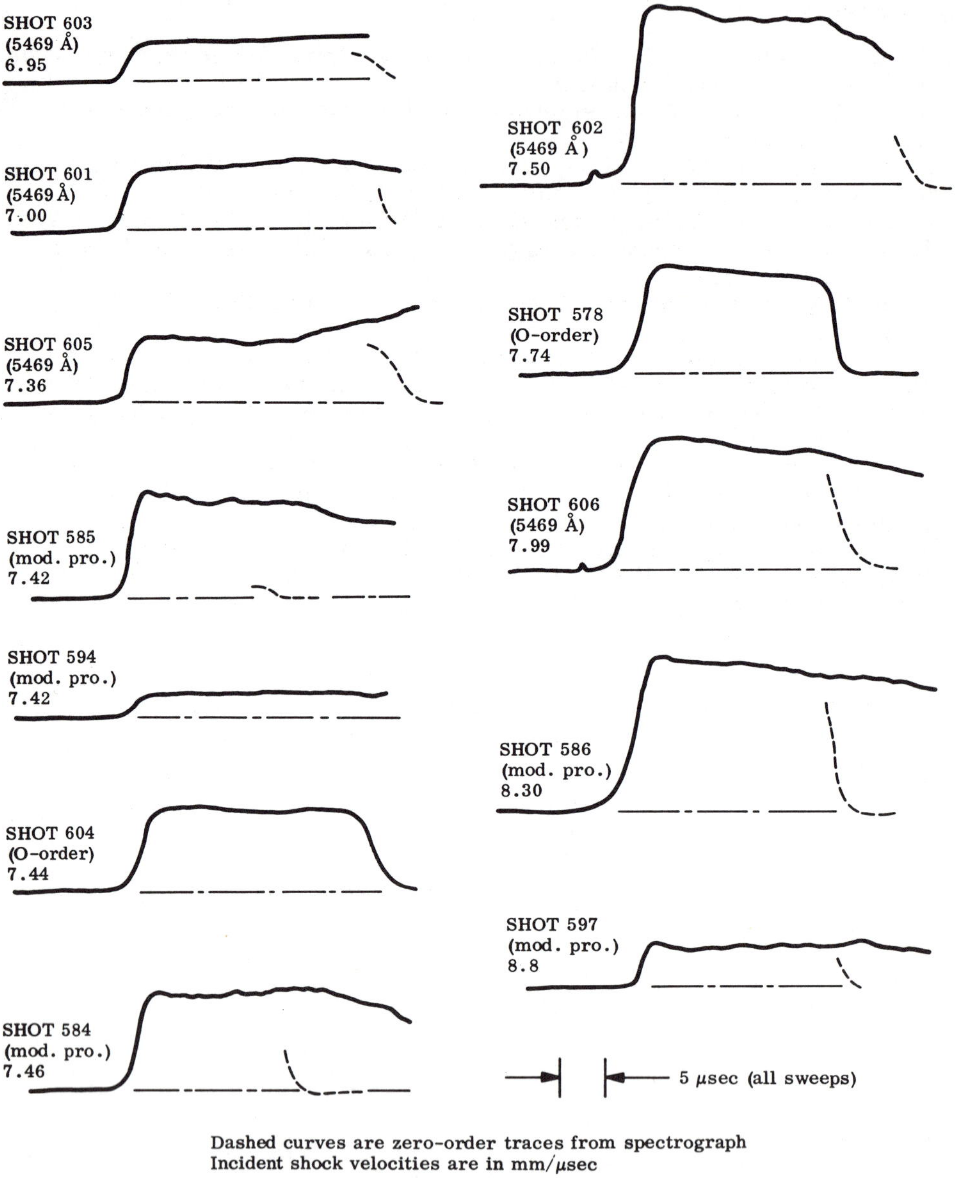

Figure 1
Various photoelectric traces from twelve shots to illustrate the temporal behavior of the test gas radiation

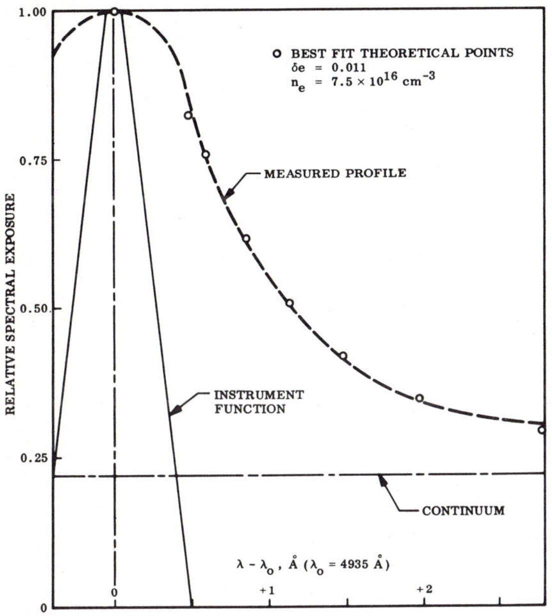

Figure 2
Example of the shape-fitting process for the NI line on shot #601 (U_S = 7.00 mm/μsec)

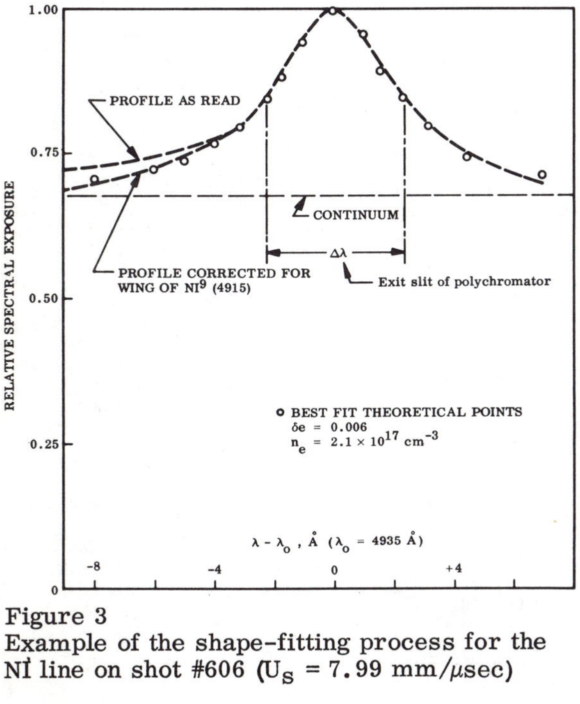

Figure 3
Example of the shape-fitting process for the NI line on shot #606 (U_S = 7.99 mm/μsec)

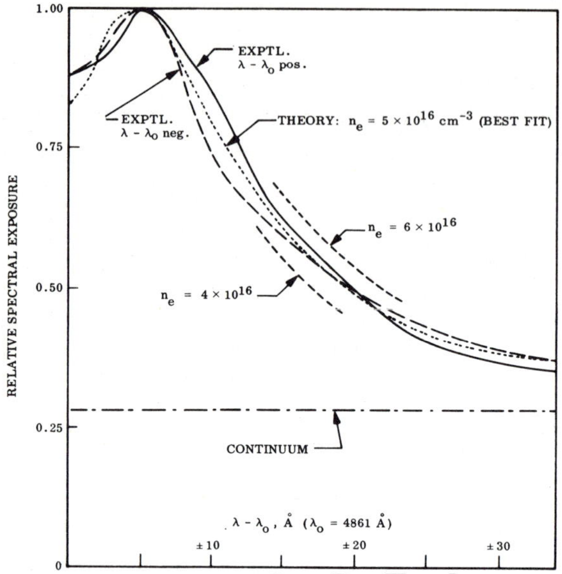

Figure 4
The best-fit shape for H_β on shot #603
(U_S = 6.95 mm/μsec)

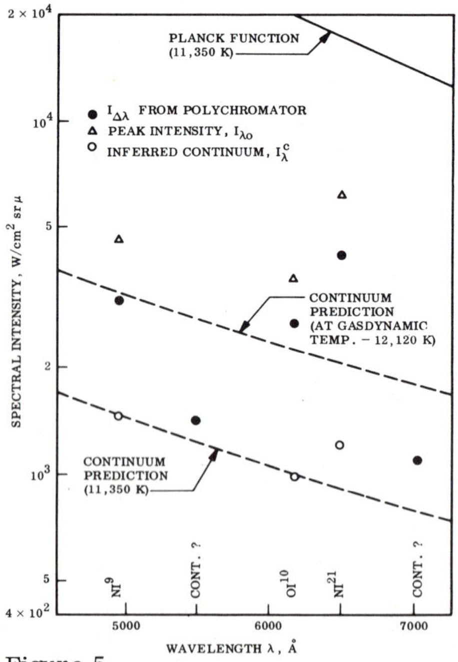

Figure 5
The five polychromatic intensities and the spectrographic peak and continuum spectral intensities on shot #605 (U_S = 7.36 mm/μsec)

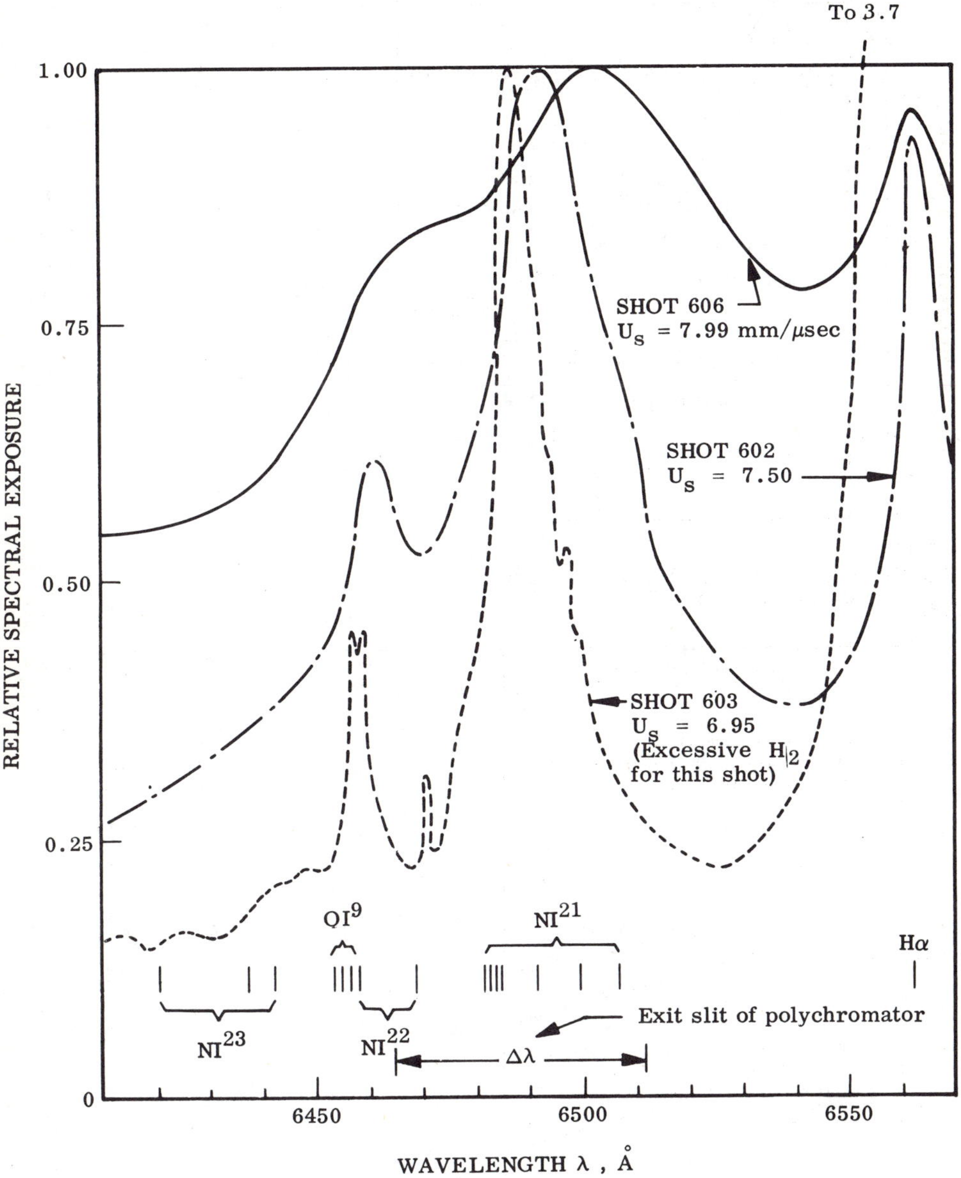

Figure 6
Relative exposure profiles for the 6410-6570 A region for three different incident shock velocities

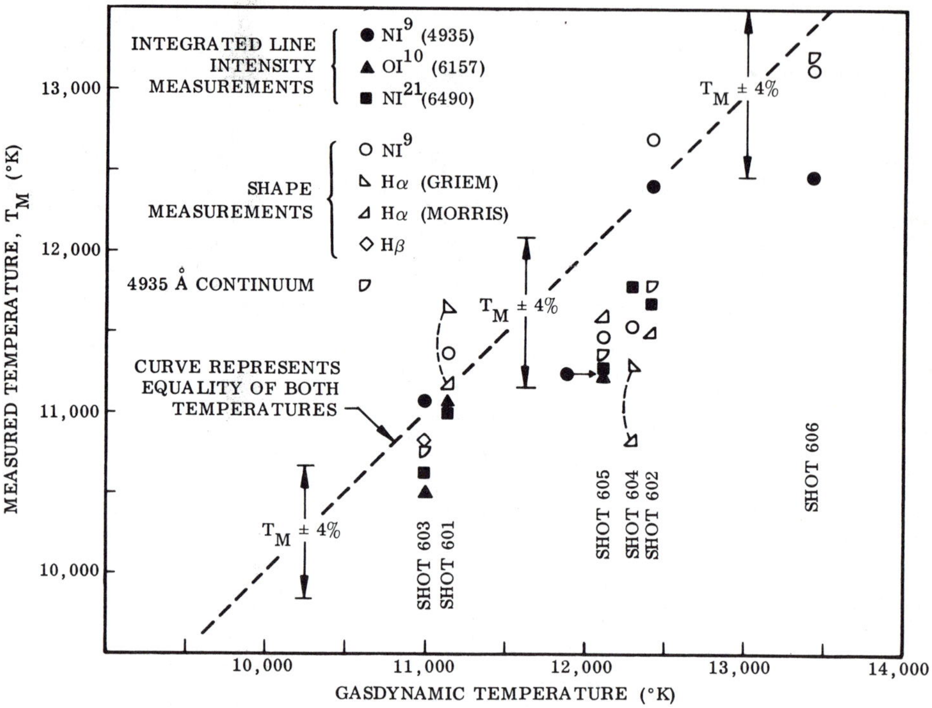

Figure 7
Results of the plasma temperature measurements

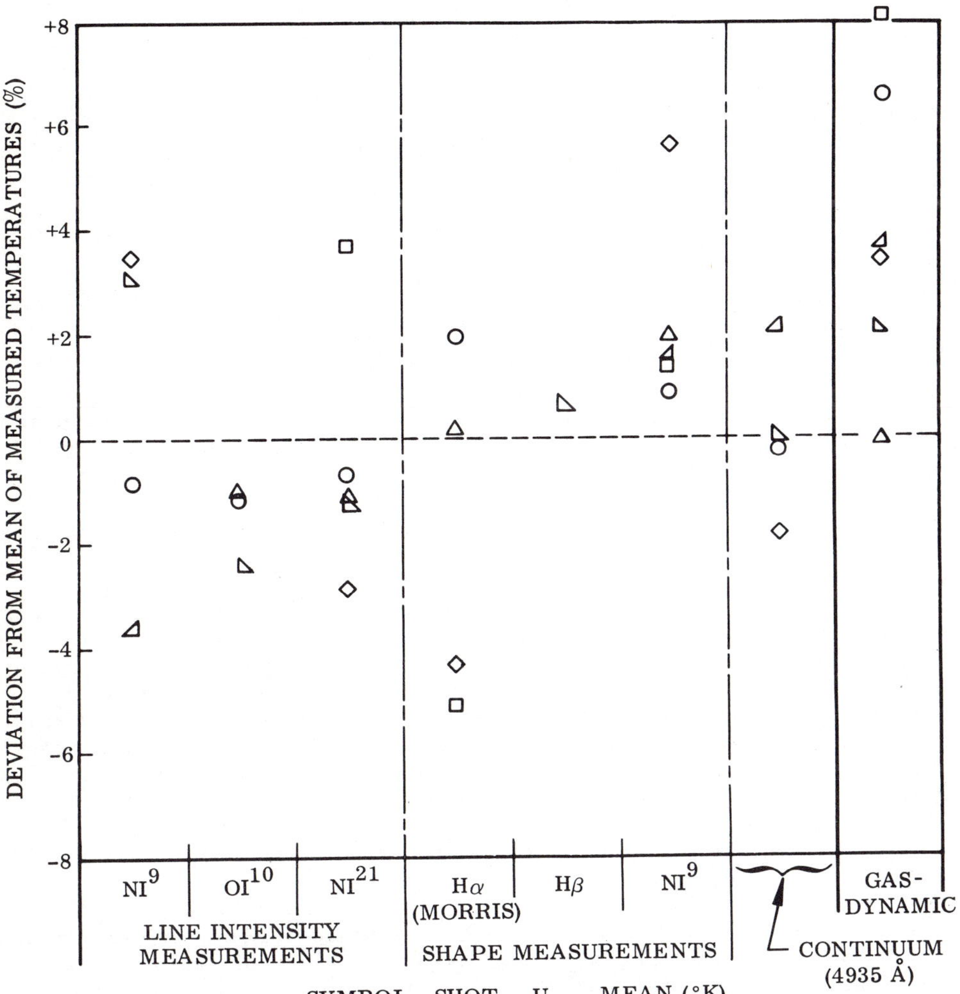

SYMBOL	SHOT	U_s	MEAN (°K)
◺	603	6.95	10,760
△	601	7.00	11,160
○	605	7.36	11,380
□	604	7.44	11,385
◇	602	7.50	12,025
◿	606	7.99	12,950

Figure 8
Comparison of the results obtained by the various thermometric techniques

MEASUREMENT OF ELECTRON DENSITY BEHIND SHOCK WAVES BY FREE-MOLECULAR LANGMUIR PROBES

R. E. CENTER
Avco Everett Research Laboratory
Everett, Massachusetts

SUMMARY

There exists a lack of experimental confirmation of the collisionless electrostatic probe theory for ratios of the probe radius to Debye length of order unit or less which corresponds to orbital-motion-limited current to the probe. These conditions are needed to fulfill the desirability of free molecular flow for many applications, e. g., ballistic ranges. This paper describes the development and the experiments on a free molecular cylindrical probe in the region behind the incident shock wave to compare the measurements with collisionless probe theory. The measurements were made in an air/argon mixture with an initial pressure of 0. 2 torr and a shock speed of 2. 8 mm/μsec resulting in temperatures of 7000°K and electron densities up to 10^{12} cm^{-3} far downstream of the shock wave. The probe radii were of the same order as the Debye length. The experimental results confirm the theoretical predictions for electron collection in the orbital-motion-limited regime.

INTRODUCTION

The use of electrostatic probes in high enthalpy shock tube and many other unsteady flows has been limited because of the difficulty of quantitative interpretation. However, the interpretation is simplified if the probe diameter is sufficiently small so as to be in essentially free molecular flow. This paper describes the development and experiments on a free-molecular cylindrical probe, behind the incident shock wave to measure the local electron density and to compare the measurements with collisionless probe theory.

Although the electron density can theoretically be inferred from either electron or ion current measurements, there are many advantages in operating the probes in the electron current region. First, by operating the probes under orbit limiting conditions the electron current to the probe can be made insensitive to the electron temperature[1] for large applied potentials. Second, because of the high electron thermal speed the electron current to the probe is independent of the probe orientation to the flow whereas ion collection is dependent on the orientation for directed ion velocities comparable to the thermal velocity.[2] This is of considerable importance because the use of slender cylindrical cantilevered probes, whose diameters are small enough to be in the free molecular regime, can result in excessive bending of the probe when placed in a high speed flow. This problem of bending can be avoided in the electron collection regime by supporting the probe at either end and placing it transverse to the flow as was done in the present experiments. Finally, the relevant collision cross sections for electrons are generally smaller than the corresponding cross sections for ions which can limit the probe dimensions for collisionless flow.

The theoretical electron current density to an infinitely long Langmuir probe was described by Langmuir and Mott-Smith.[3] More recently collisionless probe theory has been studied by Laframboise[4] who has obtained numerical results for probe characteristics over a wide range of ratios of the probe radius to Debye length and ratios of electron to ion temperature. However, there exists a lack of confirmation of the collisionless probe theory for ratios of the probe radius to Debye length which are of the order of unity or smaller resulting in orbital-motion-limited operation. This corresponds to the conditions needed to fulfill the desirability of free molecular flow for many applications, e. g. , ballistic ranges. The present experimental measurements extended into the region of orbit-limiting operation of the probe and the results confirm the theoretical predictions[3, 4] for electron collection.

EXPERIMENTAL APPARATUS AND TECHNIQUES

A 24-inch diameter pressure-driven shock tube, described by Lin and Fyfe,[5] was used to produce a plasma behind the incident shock wave. This facility was large enough to allow four different size probes to be mounted at the same axial location without mutual interference as was verified experimentally. For an axisymmetric shock wave each probe was subject to the same temporal variation of the gas conditions. Each probe was mounted on a sting so as to locate the probe six inches from the wall, as shown in Fig. 1a, and well outside the wall boundary layer.

The probes were constructed by suspending a platinum-rhodium alloy wire between two needles mounted in a thin insulating wedge as shown schematically in Fig. 1b. This suspension was essential for the small diameter probes in order to achieve reasonable values of the slenderness ratio. The probe diameters ranged from 0. 0005 inch to 0. 005 inch for which the pertinent dimensions are listed in Table I. The choice of probe length was a compromise between large slenderness ratio to avoid end effects and physical limitations of strength and current depletion. The maximum current drain of the largest probe was less than 1% of the net current flow past the test station. Although the platinum alloy will not oxidize readily as compared with tungsten, it was still necessary to clean the probes carefully between runs because of surface contamination. Before each run the probes were cleaned in hot concentrated caustic soda solution and then placed in the shock tube which was pumped down to less than 10^{-4} torr. Immediately before each run the probes were heated by passage of a current through the probe. Reproducible results were obtained with this procedure. Because of the rapidly varying conditions behind the shock wave, it was not feasible to sweep the probe voltage without incurring transient electronic effects. All probes were biased to the same voltage relative to the shock tube wall in a given test and the bias varied over the range of +3 to +16 volts in a series of tests at the same conditions. Data from different runs could be compared because of the good reproducibility of the shock tube operating conditions.

Table I
Cylindrical probe dimensions

r_p (Inches)	$\ell_p/2r_p$	$2r_p/\lambda_{n-n}$	r_p/λ_D (maximum)
0.00025	220	0.1	0.91
0.00050	193	0.2	1.8
0.0010	98	0.4	3.6
0.0025	80	1.0	9.1

Measurements were made in a 5% air/95% argon mixture with an initial pressure of 0.2 torr. This mixture was chosen in order to produce a controllable electron density at temperatures of approximately 7000°K at which the ionization is predominantly due to the $N + O \rightleftarrows NO^+ + e$ reaction. In this way it was possible to obtain a small but calculable electron concentration and thereby obtain a relatively large Debye length. The argon (Linde) used in these experiments was obtained from cylinders without further purification. Room air was purified by an ascarite packed column to remove water vapor and carbon dioxide. The gases were mixed in a stainless steel mixing tank which was designed to ensure adequate mixing overnight. The maximum mole fraction of impurities was 5×10^{-4} as a result of the ultimate pumping pressure and outgassing rate of the shock tube. Use of a cold hydrogen driver produced shock velocities of approximately 2.8 mm/μsec and a test slug of 0.9 msec duration, the test slug being limited by the effects of the boundary layer growth behind the incident shock wave. The shock speed was measured with thin film heat transfer gauges mounted in the side wall and was reproducible within ± 1%. Under these conditions the maximum boundary layer displacement thickness[6] upstream of the contact surface was 4 cm. The neutral-neutral and electron-neutral mean free paths were approximately 0.013 cm and 0.5 cm respectively.

PROBE-DATA REDUCTION

A probe characteristic as shown schematically in Fig. 2 is defined by the probe current response to varying the probe potential, the latter usually being referred to the plasma potential, i.e., that potential at which there are no electric fields at the probe. In the present experiment the applied potential is referred to the shock tube wall and it is necessary to relate the plasma potential to the wall potential. With reference to Fig. 2 it is evident that the plasma-wall potential is the sum of the floating potential V_f and the potential V_w due to aerodynamic flow effects.

The floating potential adjusts itself to maintain equal electron and ion currents to the probe so that

$$\frac{1}{4} N_i \bar{c}_i f(M_i) = \frac{1}{4} N_e \bar{c}_e e^{-eV_f/kT_e} \tag{1}$$

where the subscripts i and e refer to the ions and electrons respectively and $N\bar{c}/4$ is the random thermal flux. The term $f(M_i)$ is included to allow for the convective ion flux to the probe, M_i being the ion Mach number. For equal electron and ion temperatures the above result can be expressed in the form

$$\frac{eV_f}{kT_e} = \ln \sqrt{\frac{m_i}{m_e}} \frac{1}{f(M_i)} \tag{2}$$

It can be readily shown[7] that $f(M_i) \sim 0(1)$ for $M_i \sim 0(1)$ and the effects of convection are negligible in the evaluation of V_f because of the large ion to electron mass ratio, m_i/m_e.

The potential V_w between a floating probe and the shock tube wall can arise due to several aerodynamic effects. The temperature of the heavy ions reaching the wall is lower than the free stream temperature due to collisions in the wall boundary layer. However, the electrons arrive at the wall at essentially the free stream temperature because of the large electron-neutral mean free path (of the order of 1 cm). Thus in order to maintain zero net current to the wall the wall potential must decrease relative to the plasma potential resulting in a positive value of V_w (c. f. Eq. (1)). At the same time the potential V_w is affected by the varying charged particle density behind the shock wave and it is necessary to measure this quantity experimentally. This was done for all the probes by measuring V_w across a large resistance between the probe and the wall, the resistance being made large enough to prevent significant current flow. A sample trace is shown in Fig. 3. The transient response immediately following the arrival of the shock wave may be due to varying plasma effects over the insulated probe support but the long term behavior appears consistent with the hypothesis of a low ion temperature at the wall.

In the comparison of experimental and theoretical probe response data shown in the following section, the probe potentials have been corrected by the sum of the floating potential derived from Eq. (2) and the experimentally determined wall potential V_w.

THEORETICAL FLOW FIELD CALCULATIONS

The gas properties in the relaxation region behind the normal shock wave were computed for the known initial gas composition and pressure as well as the experimentally measured shock speed. In the conventional normal shock chemistry for constant area flow it is usual to integrate the various chemical rate equations subject to the constraints imposed by the shock conservation conditions.[8] However for low initial pressures the boundary layer growth behind the incident shock wave can result in appreciable acceleration of the flow relative to the shock.[9] In the present experiments the estimated boundary layer growth results in an effective area increase of a factor of 2 between the shock wave and the contact surface due to the negative displacement thickness. The resultant increase in the enthalpy of the test gas has a marked effect on the chemistry because of the temperature dependence of the reaction rates. A modified stream-tube program[10] was used to permit integration of the chemical rate equations subject to the constraints imposed by the variation of effective free stream area with distance behind the shock wave. This program uses the air chemistry reaction data of Lin and Teare[11] with the exception of the rate for the dominant electron-producing reaction $N + O \rightleftarrows NO^+ + e$ which has been increased by a factor of 3 in agreement with recent experimental data.[12,13]

In the relaxation region behind the shock wave the translational temperature of the argon and the nitrogen vibrational temperature are effectively coupled in 70 μsec after the passage of the shock wave. The electron temperature was assumed to be given by the nitrogen vibrational temperature as a result of the large inelastic collision cross section as indicated by Schultz.[14] For the experimental conditions of gas mixture, initial pressure and shock speed the maximum theoretical electron density in the test gas was 8×10^{11} electrons/cm^3 at a temperature of 6700°K. This corresponded to a minimum Debye length, λ_D, of 7×10^{-4} cm where

$$\lambda_D = 6.9\,(T_e/N_e)^{1/2},$$

the electron temperature and density being in units of °K and cm^{-3} respectively. This was comparable with the smallest probe diameter and the maximum ratios of probe radius to Debye length are listed in Table I for each probe. The theoretical electron density and temperature calculations have been used in the analysis of the experimental results.

DISCUSSION OF RESULTS

Typical probe current signals are shown in Fig. 4 for the two smaller probes with an applied bias of 10 volts relative to the shock tube wall. The small transient evident in the smaller probe signal shortly after the passage of the shock wave is a result of the reversal in the floating probe-wall potential as shown in Fig. 3. This transient was used to locate the time of arrival of the shock wave which is evident in the floating probe signal data of Fig. 3. The abrupt fall in probe current (electron density) some 900 μsec after the arrival of the shock wave corresponds to the passage of the contact surface. This is in good agreement with the theoretical test time based on boundary layer calculations.[9]

A comparison between the probe current variation and the theoretical probe response is shown in Fig. 5 for the 0.00025 inch radius probe signal of Fig. 4. The applied probe potential has been corrected for the variation in the floating probe-wall potential as described earlier in order to obtain the probe-plasma potential, V. The theoretical electron current density to the probe is given by the expression[15]

$$\frac{I_p}{2\pi r_p \ell_p} = \frac{e N_e \bar{c}_e}{4} \left(1 + \frac{4}{\pi} \frac{eV}{kT_e}\right)^{1/2} \quad (3)$$

which is a close approximation to the Langmuir-Mott-Smith result[3] and the numerical calculations of Laframboise[4] for $r_p/\lambda_D \lesssim 0(1)$. The theoretical curves for the probe current shown in Fig. 5 are based on the numerical calculations of electron density, N_e, behind the shock wave. The presence of the laminar boundary layer results in an increase of a factor of 2 in the effective free stream area and causes a significant change in the slow chemistry. There is excellent agreement between theory and experiment for 600 μsec after the passage of the shock wave. The departure between theory and experiment at longer times near the contact surface may be due to diffusion across the contact surface in the presence of a flame front due to the hydrogen, air, argon mixture. It should be emphasized that the transient response in the floating probe-wall potential had a negligible effect on this data since the probe response is dominated by the rate of change of electron density and this effect was much larger than that due to the rate of change of V (equal to dV_w/dt).

Some experimental current-voltage characteristics are shown in Figs. 6 (a, b) for the probe responses at 200 and 400 μsec after the passage of the shock wave. The probe voltage, V, is referred to the plasma potential. The probe currents have been normalized by the computed random electron flux to the probe surfaces $A_p N_e \bar{c}_e/4$. The computed electron densities were 3.3×10^{11} and 5.1×10^{11} electrons/cm^3 corresponding to the two positions relative to the shock wave and the electron (neutral) temperature was approximately 6700°K. Also shown in the figure are the theoretical probe characteristics for several values of r_p/λ_D as calculated by Laframboise[4] in the framework of the collisionless probe theory. The small probe results are in excellent agreement with the theoretical characteristics while the larger probe currents are slightly less than the theoretical predictions. This may have been due in part to end effects resulting from the smaller slenderness ratios of the larger probes as indicated in Table I. Furthermore the current drain to the larger probes may have resulted in a significant depletion of electrons and so disturbed the electron density in a length scale comparable with the electron-neutral mean free path.

For large probe potentials the smaller probe results agree with the predicted square root dependence on V for the orbital-motion-limited current density, Eq. (3). It is evident under these conditions that the electron current to the probe is independent of electron temperature as a result of the electron thermal velocity being proportional to $T_e^{1/2}$. Demetriades[1] has indicated the advantages of probe operation under these conditions for experiments in turbulent flows.

CONCLUSIONS

The results of this investigation demonstrate the usefulness of cylindrical Langmuir probes for the measurement of electron density in non-steady flows. The experimental results confirm the theoretical predictions for electron collection in the orbital-motion-limited region of operation where the probe radius is of the order of the Debye length. In this regime the probe current is directly proportional to the electron density and independent of electron temperature for large probe potentials. This method of probe operation has an evident advantage over ion collection probes for which the current depends on both the ion density and the electron temperature thereby necessitating a separate measurement of electron temperature in order to deduce the ion density. The only obvious limitation to the use of the electron collection probe is the necessity to restrict the probe surface area in order to prevent a significant loss of electrons from the plasma. This condition can be readily satisfied in most non-steady flows but may be a limitation in quasi-steady plasmas.

ACKNOWLEDGMENT

The author gratefully acknowledges the advice and assistance of J. D. Teare in obtaining the theoretical flow field results. Some initial experiments were performed by E. G. Plett.

REFERENCES

1. Demetriades, A. and Doughman, E. L., AIAA 4, 451 (1966).
2. Sonin, A. A., AIAA 4, 1588 (1966).
3. Langmuir, I. and Mott-Smith, H. M., General Electric Review 27, 454 (1924).
4. Laframboise, J., Univ. of Toronto, Institute for Aerospace Studies, Rept. 100 (1966).
5. Lin, S. C. and Fyfe, W. I., Phys. Fluids 4, 238 (1961).
6. Mirels, H., Phys. Fluids 9, 1265 (1966).
7. Smetana, F. O., Proc. Third Symposium on Rarified Gas Dynamics, Vol. 11, pp. 65-91, Ed. by J. A. Laurmann, Academic Press, New York (1963).
8. Wray, K. L. and Teare, J. D., J. Chem. Phys. 36, 2582 (1962).
9. Mirels, H., Phys. Fluids 6, 1201 (1963).
10. Camac, M., Feinberg, R. M. and Teare, J. D., Avco Everett Research Laboratory Research Report 245 (Dec. 1966).
11. Lin, S. C. and Teare, J. D., Phys. Fluids 6, 355 (1963).
12. Frohn, A. and deBoer, P. C. T., AIAA 5, 261 (1967).
13. Thompson, W. P., Bull. Am. Phys. Soc. 10, 727 (1965).
14. Schultz, G. J., Phys. Rev. 135A, 988 (1964).
15. Sutton, G. W., AIAA I, 193 (1969).

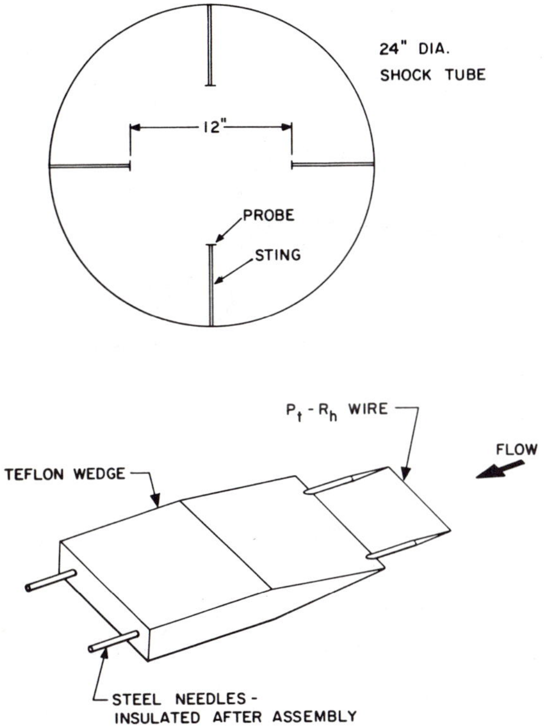

Figure 1
Schematic of the probe construction and assembly in the shock tube

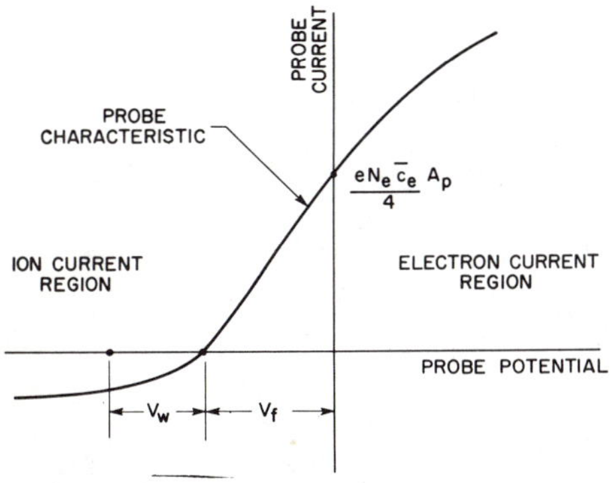

Figure 2
Probe characteristic with the probe potential being referred to the plasma potential. The experimental probe potential has to be corrected for the potential difference between the wall and the plasma, $V_w + V_f$

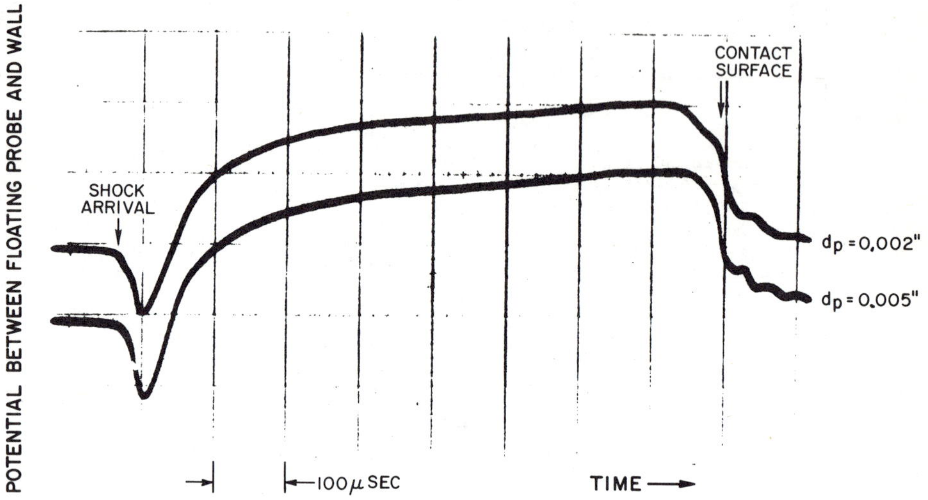

Figure 3
Sample trace of the potential difference between the floating probes and the shock tube wall, V_W

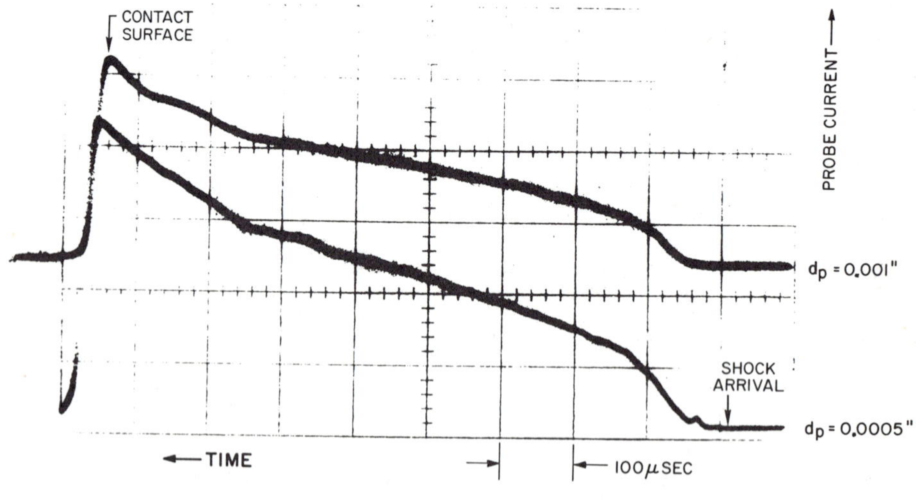

Figure 4
Typical probe current histories behind the incident shock wave for the two smaller probes for an applied potential of +10 volts relative to the shock tube wall

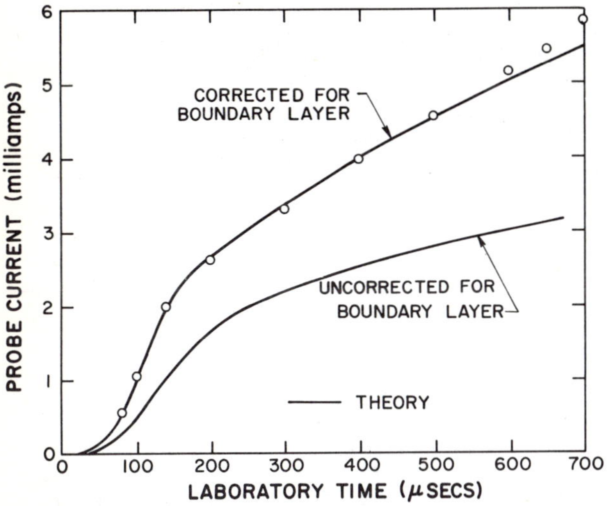

Figure 5
Probe current history (open circles) compared with the orbital-motion-limited theory for the smaller probe history of Fig. 4. The theoretical curves are based on the calculated electron density profile with and without a correction for the laminár boundary layer.

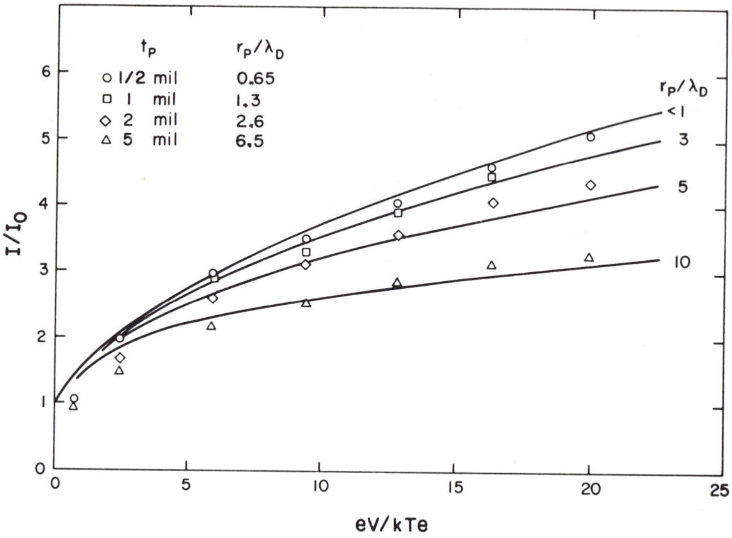

Figure 6a
Comparison between experimental and theoretical probe characteristics 200 μsec after the passage of the incident shock wave. The probe currents have been normalized by the computed random thermal current, $I_o = eN_e\bar{c}_e A_p/4$

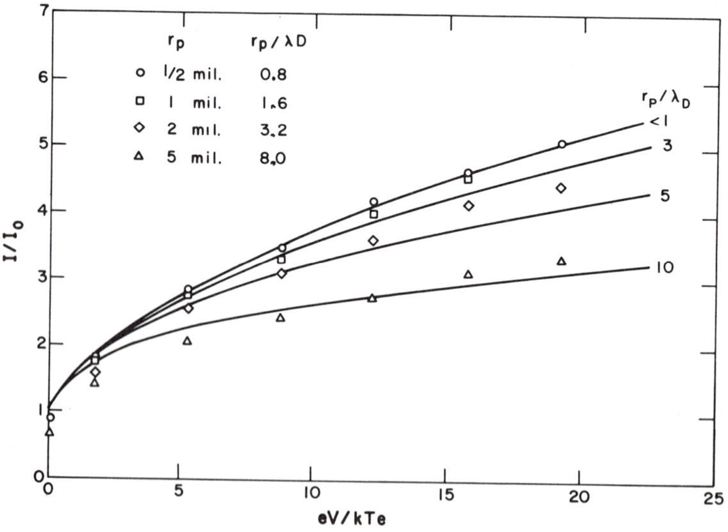

Figure 6b
Comparison between experimental and theoretical probe characteristics 400 μsec after the passage of the incident shock wave

ELECTRIC ION-COLLECTING PROBES GOVERNED BY CONVECTION AND PRODUCTION

P. C. T. de BOER, R. A. JOHNSON and P. R. GRIMWOOD
Cornell University

SUMMARY

A theory is presented for the functioning of a flat-plate ion-density probe for the case that ion production inside the electric sheath cannot be neglected. The production arises because the number density of electrons in the sheath is far below its equilibrium value. It is assumed that the production rate is constant throughout the sheath, which is a good approximation for many cases of interest. Results have been obtained in non-dimensionalized form for the current density at the electrode as a function of distance, for various values of the production parameter. The theory is compared with the results of experimental measurements behind shock waves in air. The comparison allowed a crude evaluation of the mobility of NO^+ ions in high temperature air, and of the reaction constant for the reaction $N + O \rightarrow NO^+ + e$.

Another probe configuration considered is that of a needle aligned with the flow. Here, ion production inside the sheath was assumed negligibly small. Both the continuum as well as the free molecular case are considered. Conditions are given under which the ion flux reaching the needle is determined mostly by convective rather than thermal flux. Results for the current density again are presented in dimensionless form, and are compared with available experimental data.

INTRODUCTION

The use of electric probes for the measurement of plasma properties offers some important advantages over other possible methods. The measurements are local ones and are relatively easy to make. Furthermore, the sensitivity can be made very high, and is easily variable. The main difficulty with electric probes lies in the interpretation of the data obtained. Much theoretical and experimental work has been done for the case that the plasma is at rest, and is free-molecular with the respect to the probe. For a review of this work, we refer to an article by Chen[1]. The use of electric probes in the continuum regime is not very common, and meets with a number of additional difficulties[1]. Important among these difficulties are cooling of the plasma near the probe, a large possible influence of diffusive effects on the currents collected, and a dependence of the results on ion and electron production near the probe. In the case of a flowing plasma, these effects may be further complicated by the presence of a viscous boundary layer and of a bow shock, which can cause large changes in the plasma properties.

The present work is a continuation of some previous researches[2-4] undertaken in order to achieve functioning of electric probes in high speed, slightly ionized continuum flows. At sufficiently low densities of the charged particles ($< 10^{-9} cm^{-3}$, say), it is possible to use a hollow probe, which collects all entering ions.[2] Essentially, the achievement of total collection depends on the probe being completely filled by the electric sheath which is formed around the ion collecting electrode. In order that the main flow can pass freely through the probe, the viscous boundary layer must be thin compared with the probe diameter. The hollow, total collector probe functions, therefore, under conditions where the electric sheath is much thicker than the viscous boundary layer. As shown in references 3 and 4, the sheath may be much thicker than a Debye length, provided the potential applied to the collecting electrode is large compared with the ion temperature.

For higher densities of the charged particles, the sheath thickness is smaller. It can still be made substantially larger than the Debye length by using a large potential on the electrode, but there is an upper limit to this potential arising from current amplification and breakdown phenomena. This upper limit is about 100 - 200 volt. There results a situation where the sheath no longer fills the probe, but where it still is thick compared with the viscous boundary layer. To lowest order, the effect of the latter layer on the collection current can be neglected. This greatly simplifies the electric boundary layer problem to be solved. For the case of an electrode imbedded on a flat plate, we have obtained solutions by using a quasi-one-dimensional method[3], an integral method[4], and very recently by a method of characteristics. Differences between the solutions obtained from these various methods were found to be minor. The two-dimensional ion flow pattern near the leading edge of the flat plate was studied by Dukowicz[5] who used numerical techniques. Another "thick sheath" problem was studied by Kulgein[6] who obtained an approximate result for a wire probe perpendicular to the flow. Most other investigators of electric continuum probes have assumed the electric sheath to be thin compared with the viscous boundary layer (see, for example, references 7-11).

In the present paper, we consider the influence of ion production inside the electric sheath on the flat plate probe. The production arises because the number density of electrons in the sheath is far below its equilibrium value. While the production mechanism must be considered separately for each case, we assume that the production rate is constant throughout the sheath. This is a reasonable assumption for many gases, and specifically for air under conditions of our experiments. An approximate algebraic relation is obtained between sheath thickness and current density. The current density at the electrode is determined as a function of distance by using the integral method of reference 4. The results are compared with experimental data obtained behind shock waves of mach number 9.7 in air at an initial pressure of 3 torr. At an electrode potential of 190 volt, the calculated sheath thickness at the end of the first electrode (which has a length of 10 mm) was about 0.6 mm, while the boundary layer displacement thickness at that point was about 0.2 mm. The free stream Debye length was about 6×10^{-3} mm, the free stream mean free path 3×10^{-3} mm, and the non-dimensionalized electric potential $-eV/kT_i = 640$.

Further calculations in the present paper concern the case of a "needle" probe, i.e., a thin cylindrical probe aligned with the flow. Production effects are neglected in these calculations. Results are presented for both the free-molecular and the continuum case.

FLAT PLATE ION SHEATH WITH PRODUCTION

Theory

The situation to be analyzed is sketched in Fig. 1. The potential on the electrode is supposed to be large and negative with respect to the plasma potential. The degree of ionization of the plasma is supposed to be small. The viscous boundary layer which forms on the plate is neglected, and the treatment is restricted to the ion sheath region, across which the major drop in voltage takes place. The steady state ion flow is governed by the combined ion continuity and ion momentum equation

$$\nabla\cdot(n_i\vec{u}_i) = \nabla\cdot(n_i\vec{U} - D_i\nabla n_i - n_i K\nabla\varphi) = (\partial n_i/\partial t)_{prod}, \qquad (1)$$

together with the Poisson equation

$$\nabla^2\varphi = -e(n_i - n_e)/\varepsilon_0. \qquad (2)$$

Here, n_i is the ion density, $\vec{u}_i$ the ion velocity, $\vec{U}$ the free stream velocity, D_i the ion diffusion coefficient, K the ion mobility, φ the electric potential, n_e the electron density, e the electron charge, and ε_0 the dielectric constant for vacuum. In principle, these equations must be supplemented with an equation for n_e. However, to very good accuracy we can neglect n_e with respect to n_i in equation (2). The reason for this is that the drift velocity of the electrons is much higher than that of the ions, while the currents carried by ions and electrons are comparable. The current collected at the electrode is carried by ions, **only**. The presence of the electron current allows the net current to be divergence-free. In other respects, the presence of the electrons does not enter into the calculations.

We restrict our treatment to the main part of the sheath, where the electric field is sufficiently large to ensure that mobility effects dominate over diffusion effects, so that

$$|D_i\nabla n_i| \ll |n_i K\nabla\varphi|.$$

If desired, the extent to which this inequality is satisfied can be found later. Furthermore, we assume that the ion production term $(\partial n_i/\partial t)_{prod}$ is constant throughout the sheath. This is in accordance with our assumption $n_e \ll n_i$, together with the assumption that the production mechanism involves reacting species having a constant number density inside the sheath. This situation may be approximated by air at not too high temperatures, where the dominant production mechanism is

$$N + O \rightarrow NO^+ + e.$$

The equilibrium concentrations of N and O may be several factors of 10 larger than that of NO^+. Under suitable conditions of pressure and temperature, the N and O concentrations will not be noticeably depleted by the occurrence of the production reaction. The reverse of the production reaction is neglected in view of the assumption that the number density of the electrons is far below its equilibrium value.

With these assumptions, equations (1) and (2) become

$$R\frac{\partial \nu}{\partial \xi} + \frac{\partial}{\partial \xi}\left(\nu \frac{\partial \omega}{\partial \xi}\right) + \frac{\partial}{\partial \eta}\left(\nu \frac{\partial \omega}{\partial \eta}\right) = P, \tag{3}$$

$$\frac{\partial^2 \omega}{\partial \eta^2} + \frac{\partial^2 \omega}{\partial \xi^2} = \nu, \tag{4}$$

respectively. Here, we have introduced dimensionless quantities as follows

$$\xi = x/\lambda_D,$$

$$\eta = y/\lambda_D,$$

Debye length $\lambda_D = \left(\frac{\varepsilon_0 k T_i}{n_{io} e^2}\right)^{1/2}$,

$\nu = n_i/n_{io}$, $\quad n_{io}$ = free stream ion density,

$$\omega = -e\varphi/kT_i,$$

Electric Reynolds number $R = U\lambda_D / D_i$,
Production parameter $P = \lambda_D^2 / \tau D_i$, where

$\tau = n_{io} / (\partial n_i / \partial t)_{prod}$.

We also have used the Einstein relation $K = e D_i / k T_i$.

We now make the further approximation that the sheath is thin, so that derivatives with respect to ξ are much smaller than corresponding derivatives with respect to η. Equations (3) and (4) then become

$$R \frac{\partial \nu}{\partial \xi} + \frac{\partial}{\partial \eta}\left(\nu \frac{\partial \omega}{\partial \eta}\right) = P, \tag{5}$$

$$\frac{\partial^2 \omega}{\partial \eta^2} = \nu . \tag{6}$$

These equations must be solved under the boundary conditions

$$\eta = 0: \qquad \omega = \omega_w = -eV/kT_i,$$

$$\eta = \eta_s: \qquad \nu = 1, \quad \omega = 0.$$

We first integrate (5) across the sheath:

$$R \frac{d}{d\xi}\left[\left(\frac{\partial \omega}{\partial \eta}\right)_s - \left(\frac{\partial \omega}{\partial \eta}\right)_w\right] - R \frac{d\eta_s}{d\xi} + \left(\nu \frac{\partial \omega}{\partial \eta}\right)_s - \left(\nu \frac{\partial \omega}{\partial \eta}\right)_w = P\eta_s . \tag{7}$$

Here, subscript s indicates quantities evaluated at the sheath edge $(\eta = \eta_s)$, and subscript w quantities evaluated at the wall $(\eta = 0)$. The terms with a factor R represent convective effects. The term on the right hand side represents production inside the sheath. The last two terms on the left hand side represent motion of ions arising from the electric field. The electric field is much larger at the wall than at the sheath edge, and we neglect $\nu(\partial \omega / \partial \eta)_s$ with respect to $(\nu \, \partial \omega / \partial \eta)_w$.

For the same reason, we might neglect $(\partial w/\partial\eta)_s$ with respect to $(\partial w/\partial\eta)_w$ in the first term. However, keeping the term $(\partial w/\partial\eta)_s$ causes no problems (whereas keeping $\nu(\partial w/\partial\eta)_s$ does), and, in fact, improves the agreement of the results for $P=0$ near the leading edge with the numerical results of Dukowicz[5]. The quantity

$$K = R^{-1}(\nu\,\partial w/\partial\eta)_w = j_w/(e\,n_{io}U), \qquad (<0),$$

is the non-dimensionalized current density at the wall. Equation (7) may be regarded as a relation for η_s as a function of ξ. In order to solve this relation, we use an approximate result, from which η_s, $(\partial w/\partial\eta)_s$ and $(\partial w/\partial\eta)_w$ follow as a function of K. This approximate result is obtained by considering an equivalent one-dimensional situation, i.e., a situation in which derivatives with respect to ξ in (5) and (6) are zero. Integrating twice gives

$$\partial w/\partial\eta = -(P\eta^2 + C_1\eta + C_2)^{1/2}, \tag{8}$$

where C_1 and C_2 are constants of integration. It is easily shown that

$$C_1 = 2KR, \qquad (<0),$$

$$C_2 = (P\eta_s + KR)^2 - P\eta_s^2 - 2KR\eta_s, \quad (>0).$$

The latter result follows from setting $\nu = \partial^2 w/\partial\eta^2 = 1$ at $\eta = \eta_s$. Equation (8) now may be written

$$\beta\frac{\partial(w/w_w)}{\partial\zeta} = -\left[\zeta^2 + 2\alpha\zeta - 1 - 2\alpha + P(1+\alpha)^2\right]^{1/2}, \tag{9}$$

where $\zeta = \eta/\eta_s$, $\alpha = KR/(P\eta_s)$, $\beta = w_w/(P^{1/2}\eta_s^2)$. Equations (8) and (9) give $(\partial w/\partial\eta)_s$ and $(\partial w/\partial\eta)_w$ as a function of α, and hence of K. Furthermore, integrating (9) across the sheath, and using the boundary conditions $w/w_w=0$ at $\zeta=1$, $w/w_w=1$ at $\zeta=0$ yields α as a function of β and P (and hence K as a function of η_s and P):

$$2\beta + P^{1/2}(1+\alpha)^2 + \alpha\left[P(1+\alpha)^2 - 1 - 2\alpha\right]^{1/2} + \tag{10}$$

$$+ (P-1)(1+\alpha)^2 \ln\frac{\left[P(1+\alpha)^2 - 1 - 2\alpha\right]^{1/2} + \alpha}{(1+\alpha)(1-P^{1/2})} = 0.$$

Equation (7) with the term $(v\, \partial\omega/\partial\eta)_S$ neglected becomes in the present notation

$$\frac{d}{dz}\left\{\frac{P(1+\alpha)-1}{P\beta^{1/2}} + \left[\frac{P(1+\alpha)^2-1-2\alpha}{P\beta}\right]^{1/2}\right\} = \frac{\alpha+1}{\beta^{1/2}} . \tag{11}$$

Near the leading edge $z \ll 1$ and $-\alpha \gg 1$, which gives

$$\alpha = -\left(\frac{3}{8z}\right)^{1/2} \frac{(1-P)^{1/2}}{P} , \quad \beta = \left(\frac{3}{8z}\right)^{1/2}\left(\frac{1-P}{P}\right)^{1/2} , \tag{12}$$

and hence

$$y \equiv \alpha\beta^{-1/2}P^{3/4} = \kappa R\, \omega_w^{-1/2} = -\left(\frac{3}{8}\right)^{1/4} \frac{1-P^{1/4}}{z^{1/4}} , \tag{13}$$

where it is assumed that $P<1$. The asymptotic result for large z can be obtained by setting d/dz in (11) equal to zero, which gives $\alpha=-1$. Using (10) then gives $\beta=\frac{1}{2}$ and

$$y \equiv \alpha\beta^{-\frac{1}{2}}P^{3/4} = -2^{1/2}P^{3/4} . \tag{14}$$

Physically the reason for the constancy of the current density at large z is that, once the sheath is sufficiently thick, all of the necessary electrostatic shielding is done by ions produced inside the sheath, so that no supply of ions by convection is needed. The asymptotic behaviour would be different if the production rate inside the sheath does not remain constant as assumed. The asymptotic value for large z of the sheath thickness can also be found by noting that the current density at the sheath edge must be zero. This leads to

$$j = e n_{io} \delta_s / \tau ,$$

$$E = (\delta_s - y)\left(\frac{e n_{io}}{\tau \varepsilon_o K}\right)^{1/2} ,$$

$$\delta_s = (-2V)^{1/2}\left(\frac{\tau \varepsilon_o K}{e n_{io}}\right)^{1/4}$$

For given P, (11) can be solved for intermediate values of z by numerical techniques. The solution is started at small z by using (12). Equation (10) then is solved for all β by using a Newton - Raphson iteration scheme. This allows solving (11) by numerical integration, which yields α and β as functions of z, with P as a parameter. A useful result is obtained by plotting the non-dimensionalized current density $\mathcal{J} \equiv \alpha \beta^{-1/2} P^{3/4} = \kappa R \omega_w^{-1/2}$ as a function of z, with P as a parameter. This plot is shown in Fig. 2. It is seen that the current density approaches a constant value for large z. Fig. 3 shows the dimensionless current density integrated along the electrode, i.e. the dimensionless current per unit width of the electrode, as a function of z. The curves were obtained by numerical integration of the result shown in Fig. 2.

The no-production results shown in Figures 2 and 3 were obtained in reference 4. Briefly, they follow by setting $P=0$ in (8), which then assumes the form

$$\partial \omega/\partial \eta = -\left[2\kappa R(\eta-\eta_s)+\kappa^2 R^2\right]^{1/2}. \tag{15}$$

Integrating this equation, and applying the boundary conditions $\omega(0)=\omega_w$ and $\omega(\eta_s)=0$ yields

$$\eta_s/\omega_w^{1/2} = \tfrac{1}{2}\mathcal{J}\left[1-(1+3/\mathcal{J}^2)^{2/3}\right], \tag{16}$$

where $\mathcal{J}=\kappa R/\omega_w^{1/2}$, as before. Using (15) and (16), and setting $P=0$, equation (7) may be written

$$\frac{d}{dz}\left\{\mathcal{J}\left[\frac{1}{2}-\left(1+\frac{3}{\mathcal{J}^2}\right)^{1/3}+\frac{1}{2}\left(1+\frac{3}{\mathcal{J}^2}\right)^{2/3}\right]\right\}=\mathcal{J}. \tag{17}$$

The approximate solution for small z now is

$$\mathcal{J} = -\left(\frac{3}{8z}\right)^{1/4},$$

while the large z solution is

$$\mathcal{J} = -\left(\frac{9}{2z^3}\right)^{1/4}$$

The result for intermediate z again is found by numerical integration.

A simplified version of the theory given for $P\neq 0$ may be constructed by using the quasi-one-dimensional method of reference 3. As shown in reference 4, the results for $P=0$ obtained by this method are quite close to the results of the integral method, provided $z \gtrsim 1$
The method consists of replacing (1) by a "global"ion continuity equation

$$-j = e\,\mathcal{U}\,n_{io}\frac{d\delta_s}{dx} + \frac{e n_{io}\delta_s}{\tau} \tag{18}$$

Non-dimensionalizing as before, this becomes

$$-\alpha = \frac{\beta^{1/2}}{P}\frac{d\beta^{-1/2}}{dz} + 1 \tag{19}$$

This equation must be supplemented with a relation between α and β. Again we use (8), but we make a further simplification by setting the electric field at the sheath edge $(\partial\omega/\partial\eta)_s$ equal to zero, rather than the number density at the sheath edge ν_s equal to one. This leads to $C_2 = -2KR\eta_s - P\eta_s^{\,2}$. Integration of (8) gives

$$2\beta + \alpha(-1-2\alpha)^{1/2} + (1+\alpha)^2 \ln\frac{1+\alpha}{\alpha + (-1-2\alpha)^{1/2}} = 0. \tag{20}$$

It follows from (19) and (20) that α and β now are universal functions of $Pz = x/(\tau\mathcal{U})$. The most useful plot is that of $\alpha\beta^{-1/2} = KR/(\omega_w^{\,1/2}P^{3/4})$ as a universal function of Pz (see fig. 4). The asymptotic result for large z is found by setting $d/dz = 0$ in (19), which leads to the previous result (14). In general, the results agree quite well with the previous results, provided z is not too small. The universality of the curve shown in Fig. 4 is an important advantage in reducing experimental data.

In order to gain some insight into the influence of the viscous boundary layer on the ion flow pattern, we have made some calculations of the current density at the electrode in which the mobility K was allowed to be a function of x and y. Production effects were neglected in these calculations, and use was made of the quasi-one-dimensional method of reference 3. The mobility was found from the relation $K = K_1\rho_1/\rho$, where K_1 and ρ_1 were the free stream values, and where $\rho(x,y)$ was found from an approximate solution for a compressible boundary layer on a flat plate, under conditions corresponding to the experiment described below. In the absence of production, the sheath thickness grows as $x^{1/4}$, and the boundary layer as $x^{1/2}$. As a consequence, the boundary layer always becomes thicker than the sheath for sufficiently large x. The calculation was carried out from $x=0$ to the value of x where the boundary layer displacement thickness was about one-tenth of the sheath thickness. The current collected up to this point was found to be decreased by only 1%, compared to the constant mobility-case. The smallness of this change finds its origin in the small dependence of current collected on mobility K: as can be easily verified, the current is proportional to $K^{1/4}$. Work on the corrections arising from the viscous boundary layer continues. It may be expected that these corrections are relatively small so long as the boundary layer remains smaller than the sheath.

Experimental Details

In order to test the theory presented, a number of experiments were carried out in a shock tube using air as the test gas. In many aspects these experiments were similar to those of reference 12, where a detailed description may be found of the shock tube, the techniques for maintaining gas purity, the measurement of shock velocity, and the composition and control of the test gas. An important difference between the present work and that of reference 12 is the use of a different electric probe. The probe of reference 12 consisted of glass and was known to have a rather blunt leading edge. Therefore, a new probe was designed which was fabricated out of stainless steel. This allowed making the leading edge very sharp, the estimated radius of curvature being about 0.001". The probe had the form and size shown in Fig. 5. In the design of the probe, the analysis of reference 2 was followed to ensure a satisfactory flow pattern about the probe. The two ring electrodes were mounted flush with the inner wall of the probe, and were separated from the probe body and from each other by nylon insulators. Ring electrodes were chosen in order to have a maximum collection current for a given electrode length, and hence for a given spatial resolution. Silver-soldered leads connected the ring electrodes to the external circuitry. The leads were imbedded in the nylon insulators and were brought out of the shock tube through channels in the rear of the wedge-shaped fins supporting the probe. These fins, as well as the position of the probe were identical to those described in reference 12.

The electrical circuit is shown in Figure 6. The value of the load resistance R_L was chosen as a compromise between two requirements. On the one hand, the resistance was to be large compared to the internal resistance of the voltage source. This source consisted of dry batteries. Their internal resistance was of the order of 160 Ω, but depended on the number of batteries used in a given experiment, and on the time history of the current drawn. On the other hand, the load resistance was to be small enough so that no appreciable voltage drop would occur at the collecting electrode during the measurement. A value of 10 kΩ was selected for the load resistance. This made the influence of the internal resistance of the dry cells on the measurements negligibly small and lead to a non-negligible, but easily calculated, potential drop of the electrodes during the measurements. The ring electrodes were biased negatively with respect to the grounded body of the probe.

Experimental Results

All data were obtained for the following standard conditions: shock mach number 9.7 ± 0.2, initial pressure 3 torr. The nominal collector potential ranged from -20 to -190 V. For larger potentials, breakdown phenomena were observed. A representative oscillogram of the currents collected by the two ring electrodes is shown in Fig. 7. The up-stream and down-stream electrodes have been labelled 1 and 2, respectively.

After passage of the shock wave, the currents rise rather steeply, and overshoot their final values, which are approximately constant. The data obtained for the electrode currents are plotted against electrode potential in Fig. 8. For each run, the maximum values achieved during the overshoot are shown as the top of a vertical bar, while the approximate final values are shown as the bottom of the bar. The horizontal bars indicate the estimated uncertainty in the determination of electrode potential. Since the current J_1 collected by electrode 1 is different from the current J_2 collected by electrode 2, the voltage drops at the two electrodes also are different. As a consequence, corresponding data points for J_1 and J_2 are offset horizontally.

In order to reduce the experimental data, use was made of the results given in Fig. 4. The procedure used was as follows. Assuming a value of τU, the dimensionless current density $K R \omega_w^{-1/2} P^{-3/4}$ can be integrated over the length L_1 of the first electrode, and also over the length L_2 of the second electrode. Division of the results provides the ratio J_1/J_2 of the currents collected, for the value of τU assumed. By repeating the calculation, J_1/J_2 is obtained as a function of τU (see Fig. 9). The measurement of J_1/J_2 thus directly furnishes τU, and hence $L_1/\tau U$. The dimensionless current

$$F\left(\frac{L_1}{\tau U}\right) = \int_0^{L_1/(\tau U)} \frac{K R}{\omega_w^{1/2} P^{3/4}} \, d(Pz)$$

follows by integrating the result shown in Fig. 4. The actual current J_1 is related to the dimensionless current by

$$J_1 = \pi d (e n_{i0})^{3/4} U (-V)^{1/2} (\varepsilon_0 \tau K)^{1/4} F(L_1/\tau U), \qquad (21)$$

where d is the electrode diameter. The value of U can be found from the measured shock velocity. For K we took the low temperature result [4,12]

$$K = 2.5\, \rho_0/\rho \quad \text{cm}^2/(\text{volt sec}), \qquad (\rho_0 = 1.3\ \text{kg}/\text{m}^3). \qquad (22)$$

With these values, (21) allows evaluation of n_{i0}. Alternatively, if n_{i0} is known, either from equilibrium calculations or by other means, use of (21) allows a determination of K.

The values obtained for n_{i0} at different electrode potentials are shown in Fig. 10. The horizontal dashed line represents the result calculated from the shock velocity with the aid of curves[13] and tables[14]. Also indicated are the values of the wall potential for which the sheath thickness at the end of the first electrode is approximately twice and three times the boundary layer displacement thickness at that point. For sufficiently large voltages, the agreement between expected and measured value of n_{i0} is quite good. This leads to the conclusion that, contrary to a suggestion in reference 12, the low temperature result (22) is a reasonably good approximation also at the present temperatures ($\simeq 3450°K$). It is believed that the results for equilibrium ion density n_{i0} obtained in reference 12 were to some extent influenced by additional ionization caused by the bow shock on the relatively blunt leading edge of the probe.

Fig. 11 shows the values obtained for n_{io}/τ for various electrode potentials. For sufficiently large potentials, n_{io}/τ has an approximately constant value of $5 \times 10^{16}\ cm^{-3} sec^{-1}$. Using the values for equilibrium number density of the oxygen and nitrogen atoms found from reference 14, this would correspond to a forward rate constant of $14 \times 10^{-16}\ cm^3/sec$, at $T = 3450°K$. This value must be compared with the value $2 \times 10^{-16}\ cm^3/sec$ suggested in reference 15. The discrepancy between the two values can be explained partly by the circumstance that during each run the potential on the second electrode was somewhat higher than on the first one. Furthermore, it is likely that some current amplification occurred within the sheath. This would cause a larger relative increase in J_2 than in J_1, because J_1 is determined mostly by convective effects. Further experiments are under way in which the potential on the two electrodes is more nearly equal, and in which current amplification is significantly reduced. Hopefully, these will resolve the point.

AXISYMMETRIC PROBES

The theory presented in the previous section can be extended to axisymmetric configurations. In doing so, we limit ourselves to the case of no production $(P=0)$. We give the main results, without restating all of the assumptions made in deriving them. The equations governing the steady state ion flow are

$$\nabla(n_i \vec{u}_i) = 0 , \tag{23}$$

$$\nabla^2 \varphi = -e n_i/\varepsilon_o . \tag{24}$$

The velocity $\vec{u}_i$ is given by different expressions in the continuum and in the free molecular case. For the continuum case $\vec{u}_i = \vec{U} + K\vec{E} = \vec{\imath} U + \vec{\jmath} K E_y$, while for the free-molecular case $\vec{u} = \vec{\imath} U + \vec{\jmath} v_i$, $\frac{1}{2} m v_i^2 = -e\varphi$, where m is the ion mass, and where we have neglected the y-component of ion velocity outside the sheath.

Continuum Case

The main equations are

$$U \frac{\partial n_i}{\partial x} - \frac{K}{r}\frac{\partial}{\partial r}\left(r n_i \frac{\partial \varphi}{\partial r}\right) = 0 , \tag{25}$$

$$\frac{1}{r}\frac{\partial}{\partial r}\left(r \frac{\partial \varphi}{\partial r}\right) = -\frac{e n_i}{\varepsilon_o} . \tag{26}$$

Non-dimensionalizing as before, and setting in addition $\rho = r/\lambda_D$, these equations assume the form

$$\frac{\partial \nu}{\partial z} = -\frac{1}{\rho}\frac{\partial}{\partial \rho}\left(\rho \nu \frac{\partial \omega}{\partial \rho}\right), \tag{27}$$

$$\frac{1}{\rho}\frac{\partial}{\partial \rho}\left(\rho \frac{\partial \omega}{\partial \rho}\right) = \nu . \tag{28}$$

Operating on (27) with $\int_{\rho_s}^{\rho_p} \rho \, d\rho$, and using (28) gives

$$\frac{d}{dz}\left[\left(\rho\frac{\partial \omega}{\partial \rho}\right)_s - \left(\rho\frac{\partial \omega}{\partial \rho}\right)_p\right] - \rho_s \frac{\partial \rho_s}{\partial z} = -\left(\rho \nu \frac{\partial \omega}{\partial \rho}\right)_s + \left(\rho \nu \frac{\partial \omega}{\partial \rho}\right)_p . \tag{29}$$

(p = probe)

As before, we neglect the first term on the right hand side with respect to the second one, and we note that the second term equals $\rho_p R \kappa = \rho_p R j_w/(e n_{io} U)$. We find an approximate profile $\nu(\rho)$ by using the axisymmetric solution which is independent of ξ. This follows from (27) and (28) with $\partial \nu/\partial z = 0$ as $\partial \omega/\partial \rho = -(C_1 + C_2^2/\rho^2)^{1/2}$, with $C_1 = \rho R \kappa$, $\kappa = j_w/(e n_{io} U)$ $(C_1 < 0, \ \kappa < 0)$. From the boundary condition $\nu = 1$ at $\rho = \rho_s$ it follows that $C_2^2 = C_1 \rho_s - C_1^2$. Integration gives

$$\omega = -(C_1\rho^2 + C_2^2)^{1/2} + C_3 + C_2 \ln\left\{\left[C_2 + (C_2^2 + C_1\rho^2)^{1/2}\right]\rho^{-1}(-C_1)^{-1/2}\right\}.$$

Applying the boundary conditions $\omega = \omega_w$ at $\rho = \rho_p$, $\omega = 0$ at $\rho = \rho_s$, we find the desired approximate relation between current density and sheath thickness:

$$\frac{-\mathcal{V}}{(-C_1\sigma_s^2 + C^2)^{1/2}} = \ln\left\{\frac{(-C)^{1/2} + (R^2 - C)^{1/2}}{\sigma_s\left[(\sigma_s^2 C - 1)^{1/2} + (\sigma_s^2 - C)^{1/2}\right]}\right\} + \frac{-(-C)^{1/2} + (-C + R^2 - 1)^{1/2}}{(\sigma_s^2 - C)^{1/2}}, \tag{30}$$

where $\sigma_s = \rho_s/\rho_p$, $C = C_1/\rho_p^2 = \kappa R/\rho_p$, and $\mathcal{V} = \omega_w/\rho_p^2 = \omega_w \lambda_D^2/r_p^2$. In this notation, (29) becomes

$$\frac{dC}{dz} + \frac{d}{dz}\left(C^2 + C - CR^2\right)^{1/2} - R\frac{dR}{dz} = C . \tag{31}$$

Near the leading edge $\sigma_s \to 1$ and $-C \ll 1$, which gives

$$C = -\mathcal{V}/\ln \sigma_s \tag{32}$$

$$\sigma_s - 1 = \left(\tfrac{8}{3}\mathcal{V}^{-2} z\right)^{1/4}. \tag{33}$$

It can be shown that these results agree with the flat plate result following equation (17). This is to be expected because the curvature effect vanishes near the leading edge where $Y_s = r_s - r_p \to 0$. The downstream asymptotic solution may be found by assuming $C \ll 1$ and $\sigma \gg 1$, which leads to

$$-C = \mathcal{V}^2 \sigma^{-2} \left(-1 + \ln 2\sigma_s\right)^2, \tag{34}$$

$$\sigma^4 \left[(\ln 2\sigma)^2 - \tfrac{5}{2} \ln 2\sigma + \tfrac{13}{8} \right] = 4\mathcal{V}^2 z. \tag{35}$$

Equations (30) and (31) have been solved numerically, using (32) and (33) to start the solution at small z. The results for the non-dimensionalized current density and sheath thickness as a function of z are shown in figures 12 and 13, respectively. It is interesting to note that for given flow conditions and electrode potential, the flat plate sheath is thicker than the axi-symmetric sheath. This can be seen by comparing the quantity

$$Y_s(\text{flat plate})/r_p = \mathcal{V}^{1/2}(9z/2)^{1/4}$$

with $\sigma_s - 1$, where σ_s is the result shown in Fig. 13. The explanation of the difference is that there is better shielding in the axisymmetric case resulting from the convergence of ion streamlines towards the probe.

Free Molecular Case

When the ion mean free path is large compared with the sheath thickness, the equations to be solved can be written

$$\frac{\partial}{\partial \xi}(\sigma \nu_i) + \frac{\partial}{\partial \sigma}(\sigma \nu w_i) = 0 \tag{36}$$

$$\frac{1}{\sigma}\frac{\partial}{\partial \sigma}\sigma\frac{\partial}{\partial \sigma} w_i^2 = \nu_i \Omega^2, \tag{37}$$

where $\xi = x/r_p$, $\sigma = r/r_p$, $\nu = n_i/n_{io}$, $w_i = -v_i/U$ while Ω is the non-dimensionalized plasma frequency,

$$\Omega = \left(\frac{n_{io} e^2}{m \varepsilon_o}\right)^{1/2} \frac{r_p}{U} . \qquad (38)$$

Operating on (36) with $\int_1^{\sigma_s} d\sigma$, and using (37) yields

$$\frac{1}{2\Omega^2} \frac{d}{d\xi}\left[\left(\sigma \frac{\partial}{\partial \sigma} w_i^2\right)_s - \left(\sigma \frac{\partial}{\partial \sigma} w_i^2\right)_p\right] - \sigma_s \frac{d\sigma_s}{d\xi} + \sigma_s j_t - \nu_p w_{ip} = 0 . \qquad (39)$$

Here, $w_{ip} = (-2eV/m)^{1/2} U^{-1} = -v_{ip}/U$,
while j_t is the non-dimensionalized thermal flux entering the sheath,
$j_t = \frac{1}{4}(8kT_i/\pi m)^{1/2} U^{-1}$.
It is useful to keep this flux as a parameter, because it plays a crucial part in the downstream solution. This downstream solution is the conventional Langmuir result. We assume that the electron temperature is not larger than the ion temperature, so that there are no complications arising from the Bohm sheath criterion. Except for allowing j_t to be non-zero, the present solution is based on neglecting the ion temperature compared with the electric potential inside the sheath. In order to solve (39), we again make use of an approximate result for current density as a function of sheath thickness. This approximate result can be obtained by using (37) and (36) with $d/d\xi = 0$
As shown in references 16 and 17, this leads to

$$j = \frac{4\varepsilon_o}{9 r^2 \beta(\gamma)} \left[-\varphi(r)\right]^{3/2} \left(\frac{2e}{m}\right)^{1/2} . \qquad (40)$$

Here, β is a function of $\gamma = \ln r/r_s = \ln\sigma - \ln\sigma_s$, which function is plotted in Figures 43 and 44 of reference 16. Defining the sheath edge as the value of r for which the electric field is zero, (40) yields

$$\left(\sigma \frac{\partial}{\partial \sigma} w_i^2\right)_s = 0$$

$$\left(\sigma \frac{\partial}{\partial \sigma} w_i^2\right)_p = w_{ip}^2 \left[\frac{2}{3} + \frac{4}{3\beta(\gamma_p)} \left.\frac{\partial \beta(\gamma)}{\partial \sigma}\right|_p\right]$$
$$= w_{ip}^2 f(\sigma_s) , \qquad (41)$$

where f is a known function of σ_s . Substituting (41) into (39), and rearranging, we obtain

$$\mathcal{V}\frac{df(\sigma_s)}{d\sigma_s}\frac{d\sigma_s}{d\hat{x}}+\sigma_s\frac{d\sigma_s}{d\hat{x}}+\sigma_s Q-\frac{1}{[\beta(-\ln\sigma_s)]^2}=0, \tag{42}$$

where $\hat{x}=(4/9)\mathcal{V}w_{ip}\xi$, $Q=9j_t/(4\mathcal{V}w_{ip})$, $(Q>0)$, and $\mathcal{V}=\frac{1}{2}w_{ip}^2/\Omega^2=\omega_w/\rho_p^2$, as before. The asymptotic solution for large $\hat{x}$ is

$$\sigma_s[\beta(-\ln\sigma_s)]^2=Q^{-1}, \tag{43}$$

$$-\kappa/(\sigma_s j_t)=1. \tag{44}$$

The latter equation implies that all current collected by the probe enters the sheath as thermal flux, rather than as convective flux. For small $\hat{x}$

$$\sigma=\exp(3\hat{x}/4\mathcal{V}), \tag{45}$$

$$-\kappa/(\sigma_s j_t)=16\,\mathcal{V}^2/(9\,Q\,\hat{x}^2). \tag{46}$$

Equation (42) was solved by numerical methods for a number of values of $\mathcal{V}$ and Q. Some results for σ_s as a function of $\hat{x}$ are shown in Fig. 14. As follows follows from (43), the large $\hat{x}$ value of σ_s depends on Q, only. Furthermore, as follows from (45), the small $\hat{x}$ value of σ_s depends on $\mathcal{V}$, only.

In applying the results just described to calculating the current collected by the probe, there arises a serious difficulty: the integral over the current density diverges at small $\hat{x}$ (cf. equation (46)). The reason for this divergence lies in the way the sheath edge has been defined. The definition based on zero electric field causes a divergence in n_i as σ_s approaches zero. The difficulty can be remedied by defining the sheath edge as the point where $\nu_i=1$. This implies that there is a finite electric field and vertical component of ion velocity at the sheath edge and requires a new computation of the current density vs. sheath thickness relation. Alternatively, the quasi-one-dimensional method can be used. An equation analogous to (42) is obtained, the only difference being the absence of the term with $\mathcal{V}$. This means that $\mathcal{V}$ has disappeared as a parameter. The small $\hat{x}$ behaviour now is

$$\sigma_s=1+(3\hat{x})^{1/3}, \tag{47}$$

$$\frac{-K}{\sigma_s j_t} = \frac{1}{Q(3\hat{x})^{2/3}} . \qquad (48)$$

Using these results, meaningful curves can be obtained for the current density as a function of $\hat{x}$ (see Fig. 15)

It is interesting to find a criterion which determines when convective effects must be taken into account, and when the downstream asymptotic solution applies. The latter is, of course, simply the conventional Langmuir result. Such a criterion can be found by comparing (44) with (48). Equating the two expressions yields

$$\hat{x}_c = \frac{1}{3} Q^{-3/2} \qquad (49)$$

Crudely speaking, convective effects are important when $\hat{x} < \hat{x}_c$, while they can be neglected when $\hat{x} > \hat{x}_c$. When the mean free path becomes of the order of the sheath thickness or smaller, the diffusive flux is decreased, while the convective flux remains essentially unchanged. As a consequence, the relative importance of convective effects increases when the mean free path decreases.

The results of the present section are applicable to some of the conditions in experiments reported by Dunn and Lordi.[18] A preliminary analysis indicates that the present method results in a significant improvement in the measurement of n_{io} over an analysis neglecting convective effects. A detailed description of this comparison will be reported elsewhere.

CONCLUDING REMARKS

The assumptions under which the probe theory presented in this paper is valid are satisfied over quite a wide range of experimental conditions. Depending on circumstances, either a hollow, total collector probe,[19] a flat plate probe, or a needle probe may be most suitable. Each of these geometries makes it possible to turn the motion of the plasma from a nuisance into an advantage. In view of various other advantages such as simplicity and high spatial resolution, we believe that probes of the types described are very useful instruments for measuring the ion density in high speed flows of ionized gases and plasmas.

REFERENCES

1. F. F. Chen, in "Plasma Diagnostic Techniques", edited by R. H. Huddlestone and S. L. Leonard, Academic Press, N.Y. (1965), Ch. 4.
2. P. C. T. de Boer, Rev. Sci. Instr. 37,775 (1966).
3. P. C. T. de Boer and R. A. Johnson, Phys. Fluids 11, 909 (1968).
4. P. C. T. de Boer, W. P. Thompson and R. A. Johnson, AIAA preprint No. 68 - 165 (Jan. 1968).
5. J. Dukowicz, Cornell Aeronautical Laboratory Technical Report RA-2641-Y-1, Jan. 1969.

6. N. Kulgein, AIAA J. 6, 151 (1968).
7. S. H. Lam, AIAA J. 2, 256 (1968).
8. P. M. Chung, Phys. Fluids 7, 110 (1969).
9. P. M. Chung, AIAA J. 3, 817 (1965).
10. P. M. Chung and V. D. Blankenship, AIAA J. 4, 442 (1966) and AIAA J. 4, 1312 (1966).
11. C. H. Su, AIAA J. 3,842 (1965).
12. A. Frohn and P. C. T. de Boer, AIAA J. 5, 261 (1967).
13. C. L. Lewis and E. G. Burgess III, AEDC-TR-65-196 (1965).
14. J. Hilsenrath and M. Klein, AEDC-TR-65-58 (1965).
15. S. C. Lin and J. D. Teare, Phys. Fluids 6, 355 (1963).
16. I. Langmuir and K. T. Compton, Rev. Mod. Phys. 3, 191 (1931).
17. I. Langmuir and K. B. Blodgett, Phys. Rev. 22, 347 (1923).
18. M. G. Dunn and J. A. Lordi, Cornell Aeronautical Laboratory, Technical Report AI-2187-A-10 (1968).
19. J. N. Bradley and P. A. Robinson, Proc. Roy. Phys. Soc. A 301, 285 (1967).

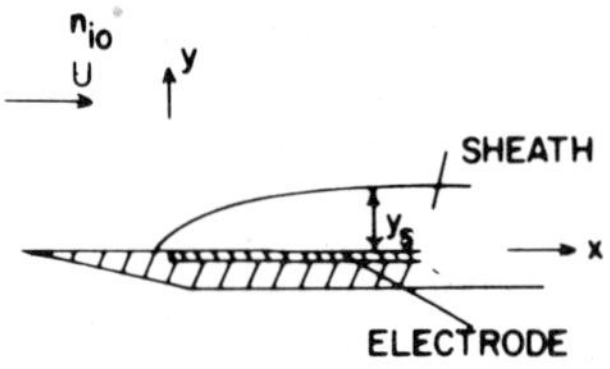

Figure 1
Sketch of flat plate probe geometry.

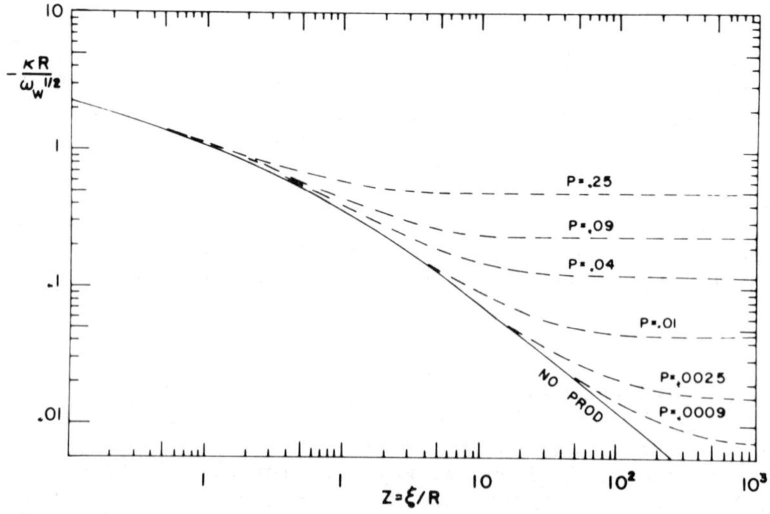

Figure 2
Dimensionless current density as a function of dimensionless distance, for various values of production parameter P.

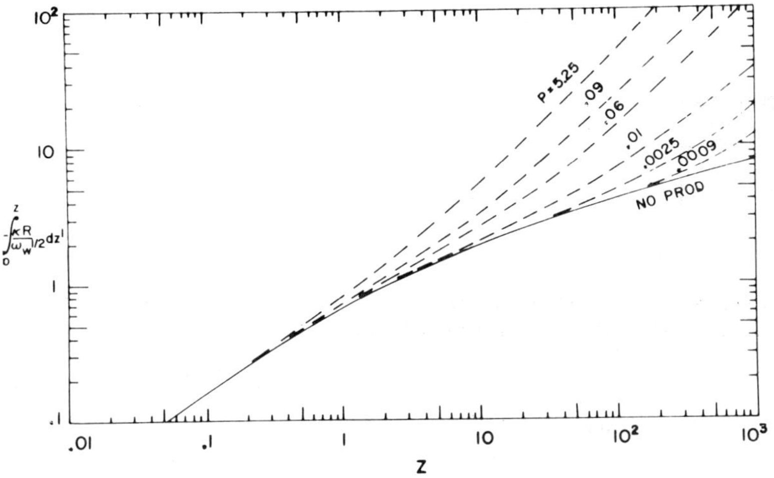

Figure 3
Dimensionless current per unit width of the electrode, as a function of dimensionless distance, for various values of production parameter P.

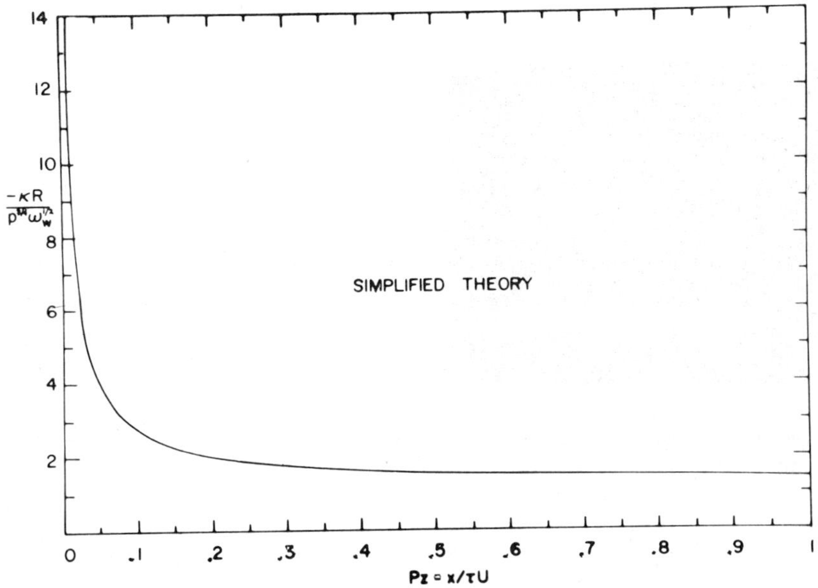

Figure 4
Universal curve of dimensionless current density as a function of dimensionless distance, based on simplified theory.

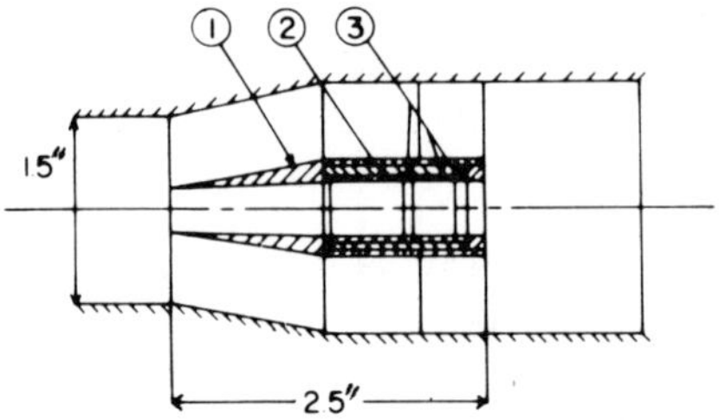

Figure 5
Drawing of probe used: 1. stainless steel probe body, 2. electrode, 3. nylon insulators.

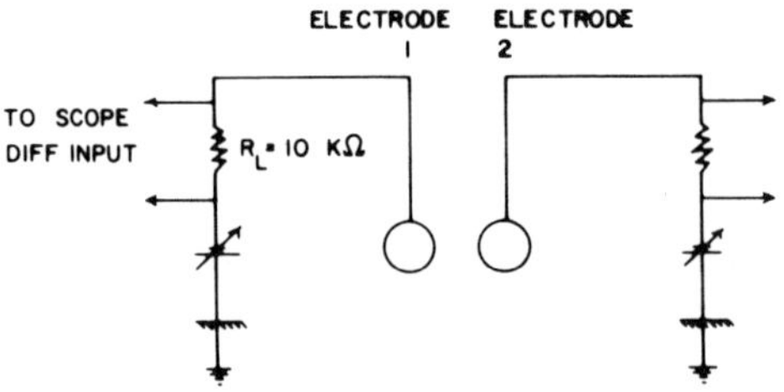

Figure 6
Electrical circuit.

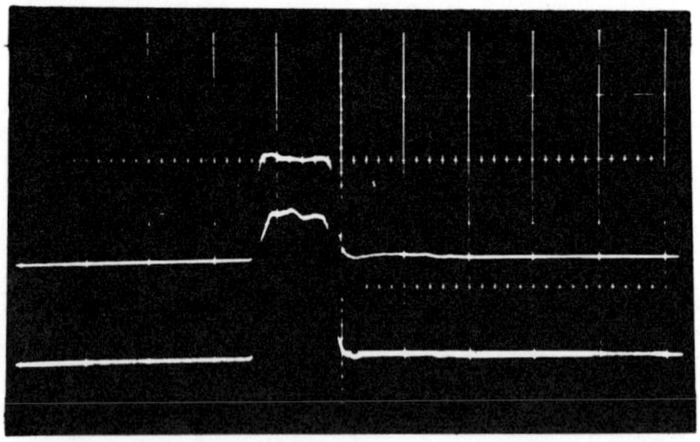

Figure 7
Representative oscillogram. Upper and lower traces show currents J_1 and J_2 collected on first and second electrode, respectively. Scales: upper 20 V/div., lower 5 V/div., horizontal 50 μsec/div, from left to right. Shock mach number 9.7, initial pressure 3 torr, nominal electrode potential 92 V.

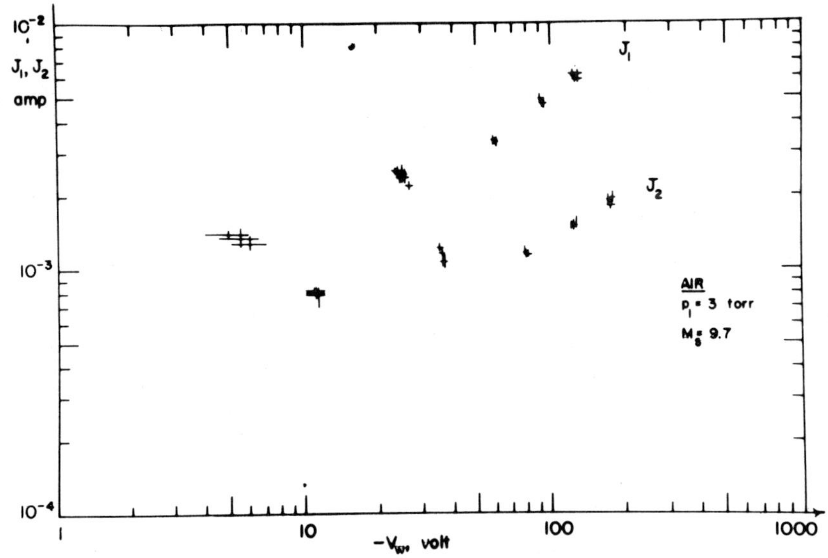

Figure 8
Data obtained for J_1 and J_2, at various electrode potentials V_w.

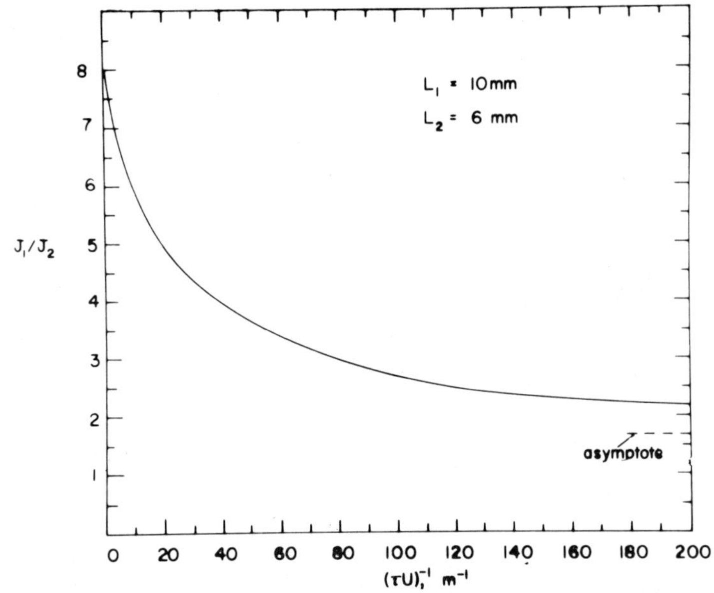

Figure 9
Current ratio J_1/J_2 as function of $(\tau U)^{-1}$.

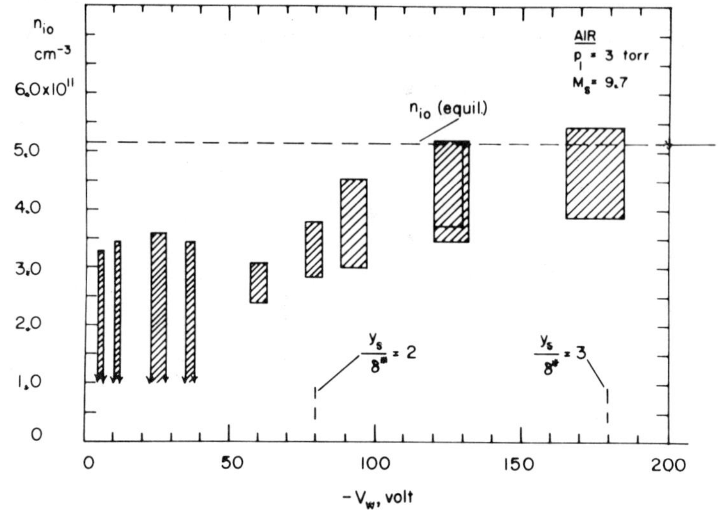

Figure 10
Results obtained for n_{io}, at different corrected electrode potentials V_w.

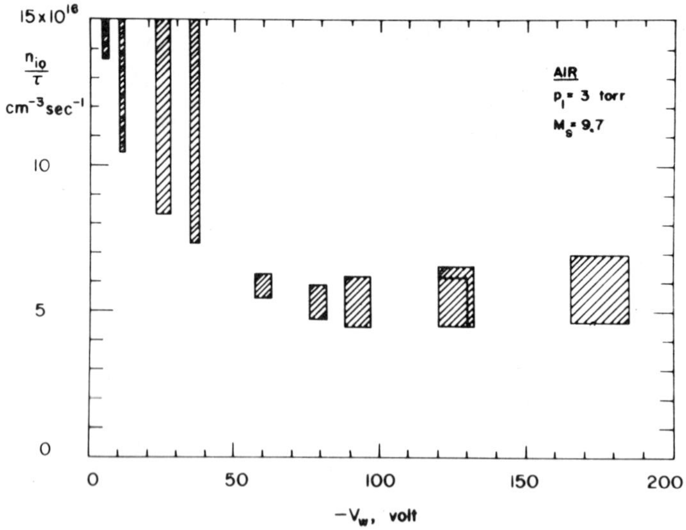

Figure 11
Results obtained for the production rate n_{io}/τ, for various corrected electrode potentials V_w.

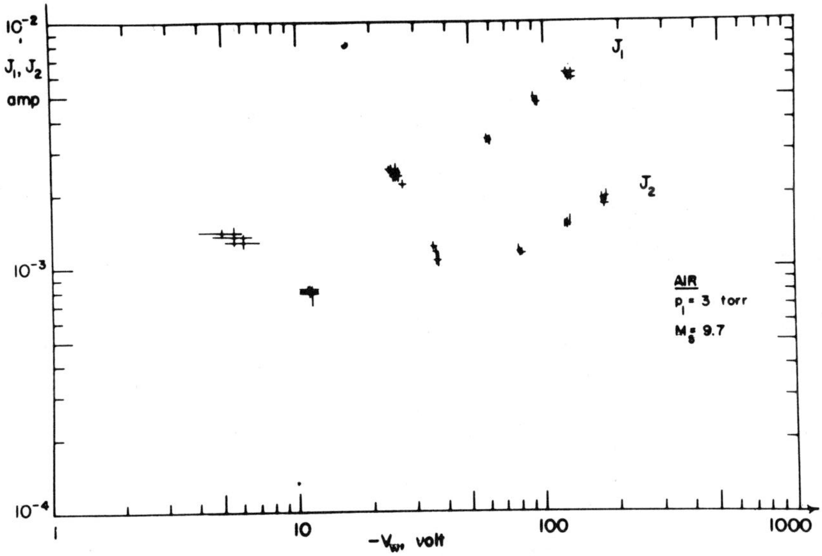

Figure 8
Data obtained for J_1 and J_2, at various electrode potentials V_{w}.

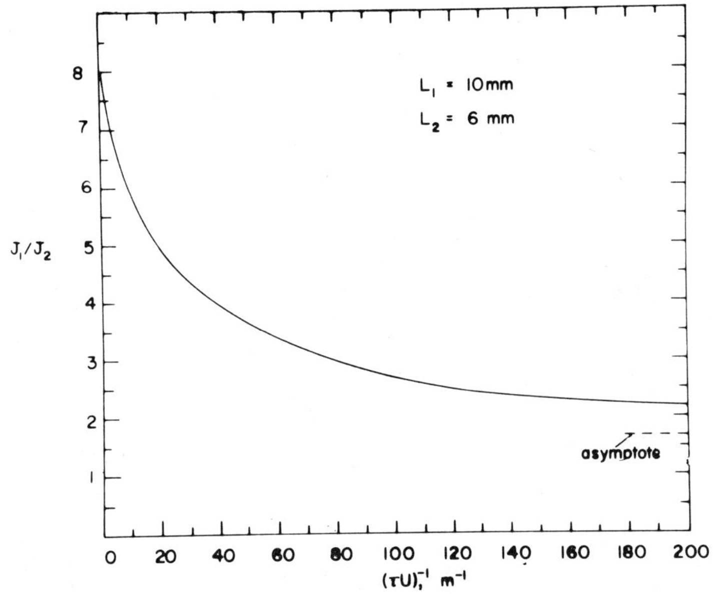

Figure 9
Current ratio J_1/J_2 as function of $(\tau U)^{-1}$.

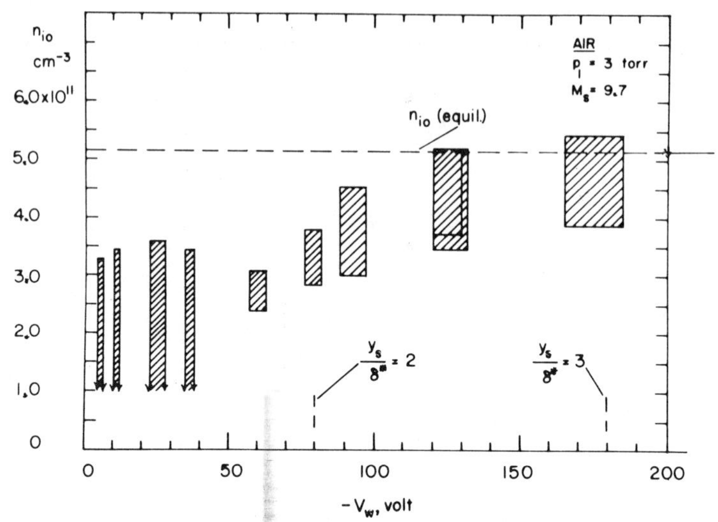

Figure 10
Results obtained for n_{io}, at different corrected electrode potentials V_w.

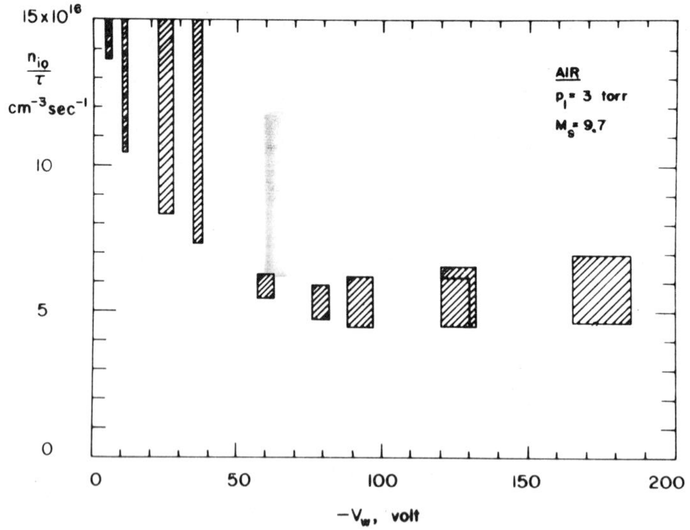

Figure 11
Results obtained for the production rate n_{io}/τ, for various corrected electrode potentials V_w.

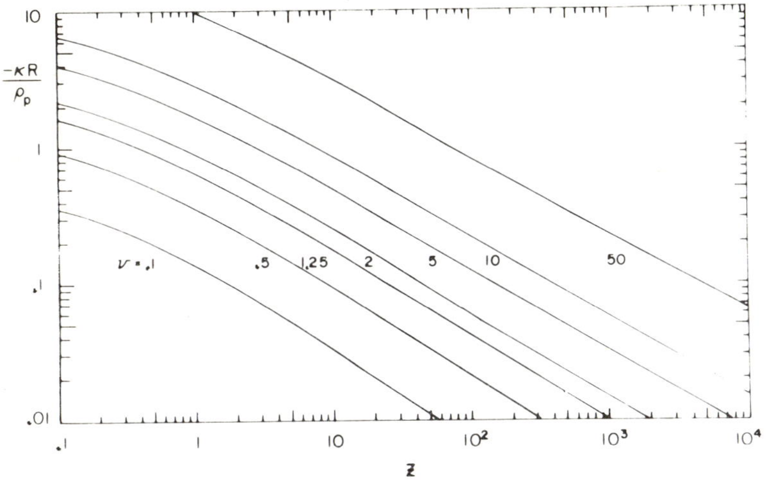

Figure 12
Axisymmetric probe in continuum case: non-dimensionalized current density $-\kappa R/\rho_p$ as a function of dimensionless distance z, at various values of the parameter ν.

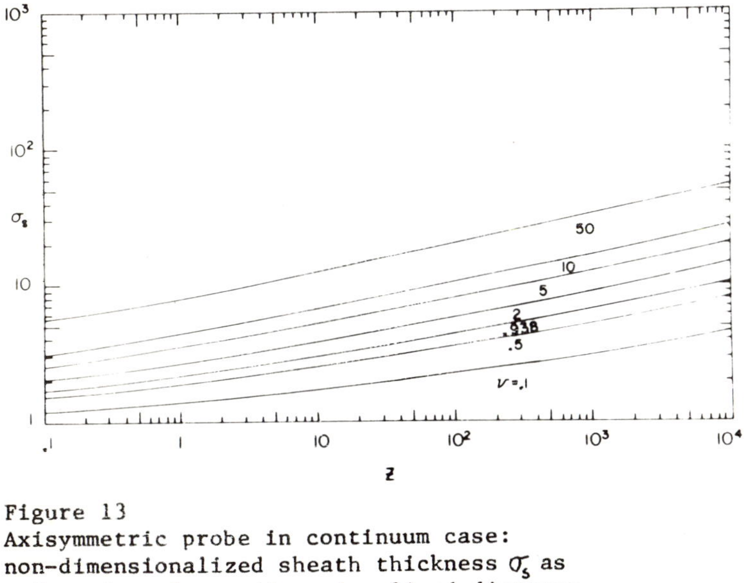

Figure 13
Axisymmetric probe in continuum case: non-dimensionalized sheath thickness σ_s as a function of non-dimensionalized distance z, for various values of the parameter ν.

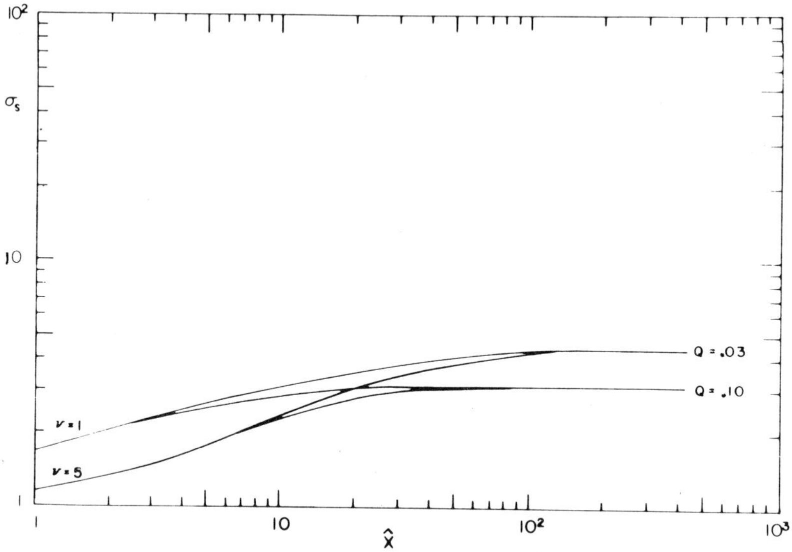

Figure 14
Axisymmetric probe in free-molecular case: non-dimensionalized sheath thickness σ_s as a function of non-dimensionalized distance $\hat{x}$, for various values of the parameters ν and Q. This result is obtained using integral method.

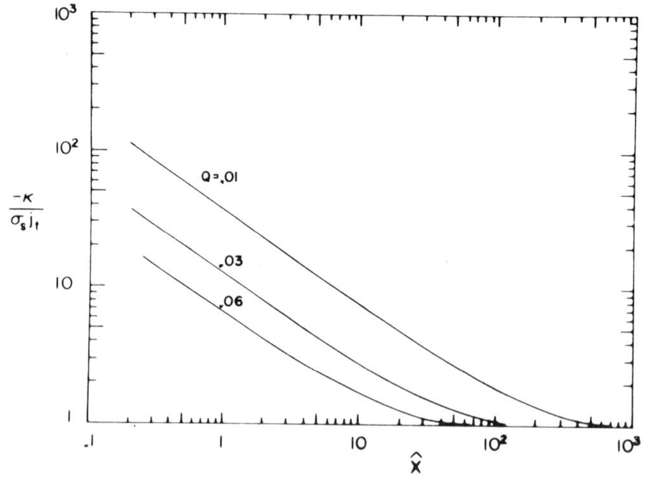

Figure 15
Axi-symmetric probe in free-molecular case: non-dimensionalized current density $-\kappa/\sigma_s j_t$ as a function of non-dimensionalized distance $\hat{x}$, for various values of the parameter Q.

not exist. The council recognized that these requirements must be met first even though it would face criticism for slow progress. In 1968 the Science Council issued Report No. 4 entitled "Towards a National Science Policy for Canada". Despite its modest title this report outlined Canada's national goals, and recommended the establishment of some comprehensive, mission-oriented major programs to ensure a "multi-disciplinary approach to the solution of important national problems". Perhaps the most significant fact that emerged from this study was that policy is needed to enable science to compete effectively with other alternatives requiring allocation of natural resources - a startling reminder of how far we have gone from the easy days of funding science for its own sake.

CANADA'S NATIONAL GOALS

Science policy can be established only after the national goals have been agreed upon. These goals must contain the main aspirations of the Canadian people. Then the factors leading to the attainment of the goal must be identified and the role played by science and technology ascertained.

Is it so difficult to formulate national goals? The following goals could apply to many countries:

(1) National Prosperity: This means a high rate of economic growth, reasonable price stability, an equitable distribution of rising income, a viable balance of payments, full employment and a reduction of regional economic disparities. It is not difficult to see that a major requirement from science and technology to achieve this goal is increased industrial productivity, innovation, improved management practices, more efficient service industries, conservation and replenishment of resources, application of new technology to improve communications, transportation and other basic elements of the economy.

(2) Health: This requires the provision of medical services of steadily increasing quality, the improvement of the environment in which we live and the development and improvement of practices conductive to public health. The contribution of science and technology to this goal would encompass continued medical research and to ensure the highest possible standards of training and practice, the application of systems science to health services, the study of individual and group behaviour in relation to physical and mental well-being and the control of health hazards already produced by the misuse of science and technology.

(3) Education: Opportunities are needed for high quality education at all levels up to and including adults themselves. Science and technology is especially important in this area. Pure and applied research is a vehicle for graduate training. The scientific method can be used to improve educational systems. The use of advanced, computer-based educational aids will increase the quality of education. Systems science can increase its productivity.

(4) Freedom, unity and security: In Canada it is essential to promote a better understanding and co-operation between the different parts of the country. The rights and safety of the individual must be preserved. In particular crime prevention must receive increased attention. The contributions of science and technology include military science to support Canada's peace-keeping role, improving man's capacity to travel, facilitating communications of all kinds and the development of new techniques to combat crime.

(5) Leisure and Personal Development: This goal involves the reduction of hours of work, the development of mechanical devices to reduce menial labor, and the provision of opportunities for culture, recreation and challenging personal endeavour. Here the application of science and technology will satisfy man's continual effort to understand himself and the universe, create more leisure time through automation, provide efficient transportation systems, develop advanced communication systems and invent accessories needed for modern leisure.

Other goals can be established and in each case the role of science and technology specified. Thus the need to fund an area of science is tied to accepted national goals and only in this way can science be properly supported by governments that are continually under pressure by other competing demands for support.

MAJOR NATIONAL PROGRAMS

It is evident that national goals will not be achieved without the organization of the scientific effort on a national basis. Research and development will receive considerable attention, but it is the implementation of the solutions to problems leading to new products and services that may have to be emphasized. In Report No. 4 the Science Council says: "The greatest concern of the Science Council is to see that the growth of science and technology in Canada is channelled in appropriate degrees towards specific broad objectives as they are defined. To permit this channelling, it is proposed that most new undertakings in Canadian science be organized as large, multidisciplinary, mission-oriented projects having as a goal the solution of some important economic or social problem and in which all sectors of the scientific community must participate on an equal footing. This report refers to these initiatives as major programs."

The major program approach provides a coordinated, multidisciplinary, total attack on the solution of large-scale practical problems. The increasing degree of organization, which society continually experiences as part of modern progress, indicates that a piecemeal method will not be satisfactory. Nowadays people gather in large urban concentrations with an ever increasing degree of interdependence and an efficient overview of energy supply, transportation and communications must be evolved.

Essentially the major program must be socially and economically oriented, able to advance effectively our current knowledge and designed to exploit efficiently the available resources of manpower and money. But most important of all, the major program must be well timed. The real problem is the determination of priorities which must often take into account conflicting interests.

To assist in the process of deciding priorities, certain criteria can be established: (1) The problem should be special to the country such as that associated with characteristic climatic conditions, availability of resources, etc. (2) There should be no duplication of work already done in other developed nations. (3) Direct social or economic benefit must be established. (4) The program must present technological challenges and stimulate innovation. (5) The program must be funded at a sufficiently high level to ensure that participating groups will be viable and above-critical in size.

Let us compare two eminently successful major programs: the Apollo space program and the "rice revolution in Asia". While one hears the question asked, "Why put a man on the moon"?, it is becoming generally recognized that the proper question is "What are the economic benefits"? There is no doubt that the space program provided enormous stimulus to technological development. In the United States it was a mechanism by which government could finance indudstry on a massive scale such that new and successful industries were created with a resulting favourable export balance due largely to sophisticated electronics, computers, aircraft etc. The great challenge of space was the demand to satisfy standards of workmanship, reliability, materials integrity and management far beyond those of normal needs.

While much less spectacular, the "rice revolution of Asia" is just as interesting. Experiments on new strains of rice began under an international cooperative program in the Philippines. A new three-crops-a-year strain of rice was developed. Other experiments produced a variety of dwarf wheat with extraordinarily high yield. These new species, the wide use of irrigation, and a more systematic application of fertilizers have initiated a "green revolution" in India and Pakistan. In West Pakistan sufficiency in rice, wheat and other food grains has been achieved and these products are being sent to East Pakistan. In both India and Pakistan, despite problems of local administration, the green revolution is inching ahead. The long range forecast is that India will be self-sufficient in rice and wheat by 1977 even though the population explosion is such that by the turn of the century there will be a million new mouths to feed every month. We see that this major program was undertaken for very important social reasons.

The advancement of a country through science is a matter for serious and detailed study. This fact has been recognized in many ways, one of which was the establishment of the Organization for Economic Cooperation and Development in 1960 under an agreement signed in Paris by member countries of the Organization for European Economic Cooperation and by Canada and the United States. The OECD has been preparing reviews of national science policy for various countries after on-site investigations by a team of examiners. These studies are made for two purposes: first, to enable the countries concerned to appraise the role of scientific research in achieving national goals, and second, to establish a general source of information on science policies and the part they play as an instrument of government.

A study of Canada has recently been completed by OECD. The advice to Canada is simple and direct: the scientific, economic and political sectors of the country must combine to produce an atmosphere conductive to effective innovation. This is also true of many other industrial countries. Canada has not yet committed itself in depth to traditional national programs, and a considerable scientific and technical potential exists for a well-planned development of the country through science.

INNOVATION

One hears this word so much nowadays! Technological innovation denotes the technical, industrial and commercial steps leading to the manufacturing and marketing of new products or the introduction of new processes. Simple investment in new manufacturing equipment or techniques to improve production lies at one end of the innovative spectrum while the coordinated sequence of scientific research, market investigation, invention, development, design, tooling, prototypes and marketing of a new product is at the other end. Basically the essential purpose of technological innovation is the commercial exploitation of technical knowledge to open up new markets or hold existing ones against competition.

Any systematic procedure to improve innovation must begin with the recognition that the total resources of the country will not permit an involvement in every kind of scientific enterprise. The choice of a restricted number of major programs will certainly be made differently in different countries. Thus Britain has decided against following the two major powers into the field of space research in favour of support of critical industries such as computers, machine tools and atomic energy to promote innovation in these fields. But for bicultural, economic and political reasons Canada decided to pursue space as one major program. A new Communications Department has been established and a company has been approved by parliament to open the way for the first domestic communications satellite in a stationary orbit by 1971. Attention will also be given to an earth resources survey satellite. On the other hand we must admire the deliberate policy of the Japanese industry in assessing the prospective overseas market demand and making these the focal point of their export policy.

What are the conditions for successful innovation? I think the following are important:

(1) Management must be technically competent, market-oriented and astute commercially.

(2) Research and development must be closely coordinated with manufacturing, financial and marketing activities.

(3) Programs leading to innovation must be based on a reliable "market awareness" through "task force studies" of market potential.

(4) The cost of launching a new project must bear a proper relationship to the scale of production and size of the market.

(5) Most practical of all is the necessity for a short lead-time from the initiation of the project to the marketing of the product.

ECONOMIC GROWTH

It cannot be stressed too much that research and development, production and marketing must form a single innovative unit. But in many countries research and development are divided between university, government and industrial laboratories. It is generally agreed that basic research is the prerogative of universities, while the remaining two sectors of the economy will engage in it according to the requirements of their missions. But is basic research always the best vehicle to carry a program of advanced education? In Canada it is apparent that the Ph. D. trained in basic science shows little interest in the practical short-range problems of industry while the latter has come to regard the Ph. D. graduate as over-specialized whose services are difficult to retain. If universities are to play a a useful role in the innovative sequence, they must become more conscious of the needs of the community that supports them, in particular more aware of the problems of industry "that lays the golden egg". More generally the present emphasis on original research as the major requirement for an advanced degree tends to over-supply the research and development field and leave undermanned such important areas as industrial management.

Canada is like some European nations in that the relationship of its universities with industry is virtually non-existent. This is strange in view of the fact that the reverse is true in the United States. This problem finds its expression in the present fear of overproduction of Ph. D. s now assumed to be occurring in Canada. Two questions arise: (1) Has it happened because Canadian industry does not accept an advanced degree as a desirable educational qualification because the Ph. D. is indicative of a person of limited specialist interest? (2) Is the traditional education in Canadian universities the best preparation for future employment? It is a matter of good educational practice as well as public service for universities to direct more of their basic research toward the innovative needs of the country - to indulge in more socalled "mission oriented basic research". No real loss of academic freedom or intellectual standards should result if the program is properly managed.

In the wider sphere of cooperation between industry, government and universities, a prime objective must necessarily be economic growth. Dr. O. M. Solandt, Chairman of the Science Council of Canada has said "First of all, we must support economic growth because it is the growth of the economy that pays for everything else. If our economy is not healthy and expanding, there is no use talking about even fundamental research". It is quite clear that successful economies have been developed through a proper relationship between government and industry. One can say much on this subject but in the end the basic Canadian problem is the lack of an effective entrepreneurial capability in industry needed to parallel an active innovating spirit. The pioneering enterprises on Route 128 in the Boston area of U. S. A. find no counterpart in Canada. On the other hand the research community such as that at Sheridan Park near Toronto, recently organized to directly involve industry, government and universities in the innovative sequence, indicates that Canada may be finding its own solution to this problem, with the help of some government initiative. Another method of approaching this problem is a direct incentive program introduced by the Canadian government, designed to promote research and development in industry under the control of industrial management.

In what I have said above I have tried to indicate some of the elements of modern science policy particularly as it pertains in Canada.

But the same problems occur in other countries and it may have been useful to you to hear something of our Canadian experience. One thing we all know: the world is moving into a long period of increasing challenge from many directions. Perhaps the great value of the space program lies in the integrated view of our planet Earth as seen from the moon. It is worthwhile to recognize that we are, in fact, Earthmen and Earthwomen, that we are moving through a great void on a planet that clearly has a large but limited potential for the support of a prosperous and happy humanity. Political, social and economic problems come from many parts of the globe to challenge man's intelligence, inspiration, adaptability and goodwill. Surely we need world-wide goals and an associated international science policy. There is concern that the development of weapons of mass destruction continues unchecked, populations increased out of control, pollution threatens to create a hostile environment, the gulf between the developed and under developed nations is ever widening, racial tensions continue unabated, standards of public and private morality have dropped, crime is on the increase and the tensions of modern living continue to rise. Solutions must be and will be found, but I am convinced that the basis of these solutions is the international brotherhood of man and that a major contribution will be made by international science policy established on this foundation.

SOME CANDID SHOTS AT THE SYMPOSIUM
23 - 26 June 1969

Concert Hall, Edward Johnson Building

Prof. R. J. Emrich and Academician R. I. Soloukhin

Concert Hall, Edward Johnson Building

Dr. W. C. Griffith and Mrs. Barbara K. Crowley

Cocktail Hour, Hart House Quadrangle

Mr. I. Hiraki, Dr. J. Pain, Dr. G. N. Patterson, Mr. & Mrs. E. Ryan, Mrs. G. N. Patterson, Mr. M. Daschuk

Presidential Reception, University College Quadrangle

Dr. S. Knoos, Prof. D. S. Dosanjh, Prof. D. Bershader, Mr. M. Rogers

Banquet Head Table, Great Hall, Hart House

Prof. H. G. Hopkins, Prof. R.J. Emrich, Mr. M. Rogers, Prof. G. N. Patterson (Banquet Speaker) Prof. I. I. Glass (Chairman) Prof. G. de B. Robinson, Prof. G. B. Kistiakowsky, Dr. P. Savic, Mr. R. R. Birukoff, Prof. A. Hertzberg (Paul Vieille Lecturer), Prof. O. Laporte, Dr. R. E. Duff, Prof. R. I. Soloukhin and Dr. A. C. Kolb

Banquet Head Table, Great Hall, Hart House

Mr. C. H. Luttman, Prof. H. O. Pritchard, Prof. J. G. Hall, Dr. W. R. Warren, Mr. J.L. Stollery, Prof. R.W. Nicholls, Prof. H.G. Hopkins, Prof. R.J. Emrich, Mr. M. Rogers, Dr. G.N. Patterson, Dr. I. I. Glass

Presidential Reception, U. C. Quadrangle

Dr. and Mrs. A. J. Howsmon, Mr. M. Rogers

UTIAS Tour, Hypervelocity Launcher and Shock Tube Laboratory

Mr. R. R. Birukoff, Mr. I Hiraki, Dr. I. Wada

Presidential Reception, U.C. Quadrangle

Mr. & Mrs. W. H. Kubbinga, Prof. J. A. Steketee, Mrs. I. Wada, Mrs. G.N. Patterson, Prof. R. J. Emrich

Presidential Reception, U.C. Quadrangle

Mr. T. Kashiwagi, Dr. O. Laporte, Dr. I. I. Glass, Dr. I. Wada

INDEX OF AUTHORS